Regenerative Medicine, Artificial Cells and Nanomedicine – Vol. 4

HEMOPERFUSION, PLASMAPERFUSION AND OTHER CLINICAL USES OF GENERAL, BIOSPECIFIC, IMMUNO AND LEUCOCYTE ADSORBENTS

Regenerative Medicine, Artificial Cells and Nanomedicine
ISSN: 2010-2836

Series Editor: Thomas Ming Swi Chang

Published

Vol. 1: Artificial Cells: Biotechnology, Nanomedicine, Regenerative Medicine, Blood Substitutes, Bioencapsulation, and Cell/Stem Cell Therapy
by T. M. S. Chang

Vol. 2: Present and Future Therapies for End-Stage Renal Disease
edited by E. A. Friedman and M. C. Mallappallil

Vol. 3: Selected Topics in Nanomedicine
edited by T. M. S. Chang

Vol. 4: Hemoperfusion, Plasmaperfusion and Other Clinical Uses of General, Biospecific, Immuno and Leucocyte Adsorbents
edited by T. M. S. Chang, Y. Endo, V. G. Nikolaev, T. Tani, Y. Yu and W.-H. Zheng

Regenerative Medicine, Artificial Cells and Nanomedicine – Vol. 4

HEMOPERFUSION, PLASMAPERFUSION AND OTHER CLINICAL USES OF GENERAL, BIOSPECIFIC, IMMUNO AND LEUCOCYTE ADSORBENTS

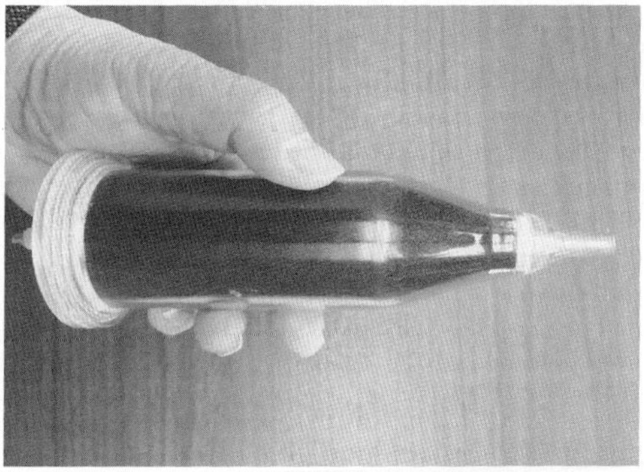

Edited by

Thomas Ming Swi Chang
McGill University, Canada

Yoshihiro Endo
Shiga University of Medical Science, Japan

Volodymyr G Nikolaev
Ukrainian Academy of Sciences, Ukraine

Tohru Tani
Shiga University of Medical Science, Japan

Yaoting Yu
Nankai University, China

Wen-Hui Zheng

World Scientific

NEW JERSEY • LONDON • SINGAPORE • BEIJING • SHANGHAI • HONG KONG • TAIPEI • CHENNAI

Published by

World Scientific Publishing Co. Pte. Ltd.
5 Toh Tuck Link, Singapore 596224
USA office: 27 Warren Street, Suite 401-402, Hackensack, NJ 07601
UK office: 57 Shelton Street, Covent Garden, London WC2H 9HE

Library of Congress Cataloging-in-Publication Data
Names: Chang, Thomas Ming Swi, editor.
Title: Hemoperfusion, plasmaperfusion, and other clinical uses of general, biospecific, immuno, and leucocyte adsorbents / editors Thomas Ming Swi Chang, Yoshihiro Endo, Volodymyr G. Nikolaev, Tohru Tani, Yaoting Yu, Wen-Hui Zheng.
Other titles: Regenerative medicine, artificial cells and nanomedicine ; v. 4.
Description: New Jersey : World Scientific, 2017. | Series: Regenerative medicine, artificial cells and nanomedicine ; vol. 4 | Includes bibliographical references and index.
Identifiers: LCCN 2016033860| ISBN 9789814749077 (hardcover : alk. paper) | ISBN 9814749079 (hardcover : alk. paper)
Subjects: | MESH: Hemoperfusion--methods | Adsorption
Classification: LCC RM174 | NLM WG 168.5 | DDC 615.39--dc23
LC record available at https://lccn.loc.gov/2016033860

British Library Cataloguing-in-Publication Data
A catalogue record for this book is available from the British Library.

Printed in Singapore

Contents

I

Introduction

Chapter 1

First Design and Clinical Use in Patients of Surface Modified Sorbent Hemoperfusion Based on Artificial Cells for Poisoning, Kidney Failure, Liver Failure and Immunology

Thomas M.S. Chang*

1. Introduction

The introduction section starts with a brief historical overview and brief summary of the basic principles and the different areas of clinical use. The sections to follow will be detailed descriptions of the different areas. This chapter will be followed by 30 chapters that go over the present status in detail.

1.1. *A Brief Historical Overview*

Adverse effects of particle embolism and damage to blood cells had prevented the use of hemoperfusion using adsorbents like activated charcoal (Yatzidis, 1964; Dunea and Kolff, 1965) or resin (Winchester,

*Artificial Cells & Organs Research Centre, Department of Physiology, Medicine & Biomedical Engineering, Faculty of Medicine, McGill University, Montreal, QC, Canada, H3G1Y6.

1996). While carrying out research on artificial cells (Chang, 1964), my theoretical analysis showed artificial cells can be the basis of a very simple and inexpensive detoxifier that can be more effective than the standard hemodialyser (Chang, 1966). While continuing with basic research on the other aspects of artificial cells, I spent a major part of my early 15-year research career developing this idea followed by successful clinical trials and routine use in patients (Chang, 1966, 1969a, 1971d, 1973a, 1974c, 1974h, 1975i, 1976g, 1977b, 1980e).

In order to allow for more widespread development and use for patients, the following was carried out:

(1) The details of methods, procedures and results were immediately published in journals and in a monograph (Chang, 1972) that contained even more details. Thus when I was a Chinese Academy of Science visiting professor in 1978, I found that Professor W.H. Zheng in China was able to successfully prepare the device and use it in patients in China based only on my published details.

(2) Welcomed trainees and visiting scientists and clinicians from around the world.

(3) Freely gave advice to different groups and discussed the advantages and disadvantages of different approaches. The Medical Research Council of Canada supported research and clinical trials, which were carried out at McGill University. Thus instead of accepting the many foreign offers, I encouraged them to start a company in Montreal. Finally, Mr. Messeur sold his dialysis company in France and started a company in Montreal. I agreed to help if I did not have to sign an exclusive agreement. In return, I did not receive any consulting fee or stocks, and the company would give McGill University a research grant. The company also agreed to sell the device for not more than $35 each. Mr. Messeur was an excellent developer and collaborator, and together we scaled this up for production and clinical trials. My clinical trials at the McGill University teaching hospitals, especially the Royal Victoria Hospital, resulted in FDA approval for clinical use. It was so successful at $35 per device that a large US dialysis company bought it over despite my objection. Since I did not hold

any stock I had no vote on this. In any case, the US dialysis company increased the price for the device to about $300 almost overnight. At this price, it is no longer affordable or competitive with standard dialysis therapy. As a result, it is not being used much in Canada and the US. Fortunately, other countries have developed and produced this device with modifications and extensions. For instance, the Chinese products are effective and not costly, and as a result it is being used extensively there.

(4) Organized and coordinated an international symposium series on hemoperfusion rotating around the world starting at McGill (Chang) followed by Italy (Bonamini, Bologna U): Israel (Sideman, Technion); Turkey (Piskin, Ankara U); China (Huang, president of the Chinese Academy of Medical Sciences); Mexico (Trevino Becerra, Mexico U); USSR (Nikolaev, USSR Academy of Science); Germany (Klinkmann & Falkenhagen, Rostock); Japan (Odaka, Chiba U); Italy (Casciani & Splendiani, Rome U) and the US (Langer, MIT).

(5) When other centers and companies around the world became fully established to develop and extend this area, I discontinued the above series and moved my attention to other areas of artificial cells that need to be developed. This way, other centers can develop and extend this area without my interference in any way. This is important because the pioneer and originator can be very controlling and protective about his original approach.

As a result of 1 to 5 above, there is extensive development, extension and clinical use of this approach around the world as shown by the many chapters in this book.

1.2. *Basic Principle of Surface Modifications Based on the Principle of Artificial Cells*

The principle will be described in detail later. Briefly it is based on thousands of 90 micron diameter adsorbent artificial cells in the form of microscopic surface modified adsorbent granules. These are retained inside a small container by screens at either end (Fig. 1.1).

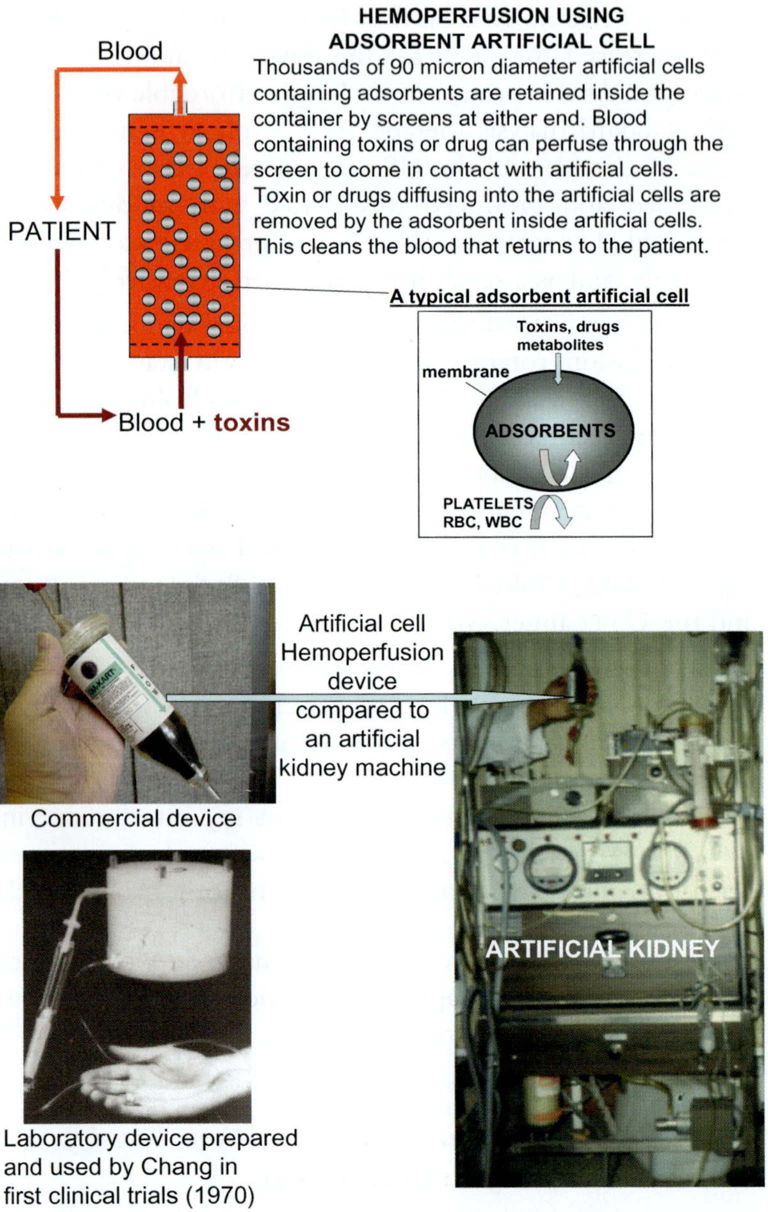

Fig. 1.1. *Upper*: Basic principle of adsorbent artificial cells in hemoperfusion. Lower left: Laboratory and commercial hemoperfusion devices. *Lower right*: Much smaller and simpler hemoperfusion device compared to an artificial kidney machine.

Blood from patients containing toxins or drugs perfuses through the screen of the container to come in contact with the artificial cells. Toxins or drugs diffusing into the artificial cells are removed by the adsorbent inside the artificial cells. This cleans the blood that returns to the patients. The membrane of the artificial cells prevents particles from the adsorbent from being released into the body and also prevents the adsorbent from damaging the blood cells (Chang, 1966) (Fig. 1.1). This solves the earlier problem of particle embolism and severe damage to blood cells when using unmodified adsorbents in hemoperfusion (Yatzidis, 1964; Dunea and Kolff, 1965).

1.3. *Clinical Trials*

There was extensive international and local support for this approach. Thus, Professor Kolff, the inventor of the artificial kidney, wrote the Canadian Medical Research Council that this approach should be extensively supported. Locally, McGill's human ethics committee having reviewed in detail all our preclinical safety and efficacy animal studies (Chang, 1969a, 1972a; Chang and Malave, 1970; Chang *et al.*, 1971a) approved the use of this device for clinical trials at McGill's teaching hospitals. At that time this was a very new idea, and although physicians were very supportive, they were concerned about liability and thus not ready to carry out this procedure in clinical trials in patients. Thus, the arrangement was for me to prepare the hemoperfusion devices in my laboratory then use them to personally carry out clinical trials on patients myself under the following protocol. The nephrologists at the dialysis units would refer patients to me. I would then make the final decision whether to carry out the procedure and to do the treatment on the patients myself. The Phase I safety study showed the safety of this in patients (Chang and Malave, 1970; Chang *et al.*, 1971a; Chang, 1972a), and as a result clinical trials on efficacy followed. The subsequent clinical trials meant being on call 24 hours a day over a number of years, and this was possible only because of the understanding of my wife, Lancy.

1.4. *Treatment of Patients with Severe Accidental or Suicidal Poisoning*

Clinical trial on its use in suicidal or accidental poisoning is more straightforward. The efficacy can be shown after one or two treatments, and the results can be assessed both quantitatively and clinically. Thus this was first successfully carried out in three patients (Chang *et al.*, 1973a, 1973b), then continued to a total of 11 adult patients (Chang 1975g, 1976c, 1980e) plus one pediatric patient (Chang *et al.*, 1980a) with conclusive clinical results. These will be discussed in more detail in a later section of this chapter. The results of these clinical trials have led to the routine use of this approach around the world for the treatment of drug poisoning (Better *et al.*, 1979; Biberstein *et al.*, 1983; Cohan *et al.*, 1982; Verpooten *et al.*, 1984; Gelfand *et al.*, 1977; Gibson *et al.*, 1978; Lorch and Garella, 1979; Winchester, 1996; Diaz-Buxo *et al.*, 1978; Sideman and Chang, 1980b; Piskin and Chang, 1982; Chang and Zheng, 1983; Chang and Ho, 1985; Chang and Nicolaev, 1987; Chang and Odaka, 1991; Klinmann *et al.*, 1990; Kawasaki, 2000; Lin *et al.*, 2002; Lopez-Lago *et al.*, 2002; Peng *et al.*, 2004; Chinese Symposium on Hemoperfusion, 2005, also chapters in this book).

The exact method of preparing artificial cells containing adsorbent has been published in reproducible detail for all to use (Chang, 1972a, 1976g). As a result, a number of countries, for instance, Canada, China, the US, the UK, Japan, Sweden, Italy, Spain and the former USSR were able to use this principle to produce their own variations of commercial hemoperfusion devices with the usual variations in efficacy and biocompatibility. While the inferior ones are no longer in use, those that are safe and effective are being used in routine clinical treatment of accidental and suicidal poisoning. For some unknown reasons, in some areas of the world, these devices are extremely expensive and as a result are too costly for widespread use. In other areas of the world, inexpensive but safe and efficient devices are being produced and used extensively, saving thousands of patients with potentially fatal accidental or suicidal poisoning.

1.5. *Hemoperfusion for the Removal of Unwanted or Toxic Substances from Blood in Other Conditions Including Terminal Renal Failure, Liver Failure and Immunology*

Hemoperfusion has been an established routine clinical method for the treatment of patients with severe suicidal and accidental poisoning for many years. Its ability to remove unwanted or toxic substances from the blood is also applicable in other clinical conditions. This includes its use in kidney failure, liver failure and use as an immunosorbent.

1.6. *Treatment of Patients with Terminal Kidney Failure*

Hemoperfusion in terminal kidney failure patients results in improvement in well-being and uremic symptoms (Chang *et al.*, 1971a, 1972g, 1974). It efficiently removes uremic wastes and toxins, especially the larger middle molecules (Chang, 1972e; Chang and Migchelsen, 1973; Chang and Lister, 1980, 1981). However, hemoperfusion does not remove electrolytes, water or urea. Thus, it has been used in series with hemodialysers with reported clinical improvements in "well-being," nerve conduction velocity, pruritis, pericarditis, peripheral neuropathy and, what is also important, reduction in treatment time (Chang *et al.*, 1975, 1982b; Martin *et al.*, 1979; Inou *et al.*, 1979; Odaka *et al.*, 1980; Agishi *et al.*, 1980; Stefoni *et al.*, 1980, also chapters in this book). It is also used for uremic patients with aluminium or iron overload based on the use of deferoxamine to bind these heavy metals and for hemoperfusion to remove the complex (Chang and Barre, 1983; Chang, 1986d; Chang *et al.*, 1984a, 1989a; Hakim *et al.*, 1985; Winchester, 1996).

In an attempt to eliminate the need or cost for the bulky and expensive hemodialysis machine, hemoperfusion has been used in series with a small ultrafiltrator with oral adsorbents to control potassium and phosphates (Chang *et al.*, 1975, 1977b, 1979a; Chang, 1976c). As will be described later, an oral urea removal system is being developed to complete the hemoperfusion-ultrafiltrator approach.

1.7. *Treatment of Patients with Hepatic Coma*

Its effectiveness in removing toxins from the circulating blood has also been shown in hepatic coma patients. Thus, the first use of this in a grade IV hepatic coma patient resulted in the recovery of consciousness within an hour (Chang, 1972b). As will be described later, this was followed by extensive clinical trials in hepatic coma patients around the world, showing that this is effective in removing hepatic toxins resulting in the recovery of consciousness in a significant number of patients (Chang, 1972b, 1975e, 1975j, 1976c, 1982a; Gazzard *et al.*, 1974; Blume *et al.*, 1976; Chang *et al.*, 1977a; Bartels *et al.*, 1977, 1981; Silk and Williams, 1978; Gelfand *et al.*, 1978; Gimson *et al.*, 1978; Odaka *et al.*, 1978; Amano *et al.*, 1978; Maeda *et al.*, 1980, Cordoatri *et al.*, 1982; William, 1983; O'Grady *et al.*, 1988, also chapters in this book). However, the liver is a complex organ with many other functions besides detoxification. Thus hemoperfusion has an important role in removing hepatic toxins to allow the use of dialysis or plasmapheresis to look after the other functions. Artificial cells containing hepatocytes and stem cells may also contribute to liver regeneration. These will be discussed later in this chapter and in even more detail in other chapters in this book.

1.8. *Immunoadsorbent and Specific Absorbents*

Albumin can bind tightly to the ultrathin collodion membrane of adsorbent artificial cells to increase the blood compatibility for hemoperfusion (Chang, 1969a). This albumin-coated collodion-activated charcoal (ACAC) was found to effectively remove antibodies to albumin in animal studies (Terman *et al.*, 1977). This has become a basis of one line of research in which other types of antigens or antibodies are applied to the collodion coating of the activated charcoal to form immunosorbents. Other immunosorbents based on this principle have also been developed for the treatment of human systemic lupus erythematosus, hyperacute renal xenograft rejection, and removal of antiHLA antibodies in transplant candidate treatment of familial hypercholesterolemia with monoclonal antibodies to low-density lipoproteins (Terman, 1980; Terman *et al.*, 1979a, 1979b; Hakim *et al.*, 1990; Wingard *et al.*, 1991; Yang *et al.*, 2004, also chapters in this book). This surface modification was applied to synthetic immunosorbents resulting in blood compatible synthetic blood group immunosorbents (Chang, 1980d) and used clinically for removing blood group antibodies from plasma for bone marrow transplantation (Bensinger *et al.*, 1981). This has led to the use of different synthetic immunosorbents with modified blood compatible surfaces for the removal of antigen, antibody and even leucocytes. Synthetic adsorbents for the removal of endotoxin are also being used clinically. These will all be discussed in detail later in this chapter and in even more detail by other groups in a number of chapters in this book.

2. Detailed Analysis of the Principle of Artificial Cells in Surface Modification for Adsorbent Hemoperfusion

2.1. *Permeability and Transport Characteristics*

The artificial cell has an ultrathin membrane of less than 0.05 micron when compared to the 2.0 micron thickness of the dialysis membrane. Furthermore, the small size of artificial cells means that 30 mL of artificial cells can have a total surface membrane area of 2 m^2 compared to the 1 to 2 m^2 of a whole dialysis machine (Table 1.1, Fig. 1.2). This means that 30 mL of artificial cells can have a theoretical mass transfer that is 100 to 200 times that of a whole artificial kidney machine (Chang, 1966, 1972a) (Table 1.1).

Table 1.1. Analysis of Mass Transfer

	30 mL Artificial Cells	1Artificial Kidney Machine
Total surface area:	2 m^2	2 m^2
Membrane thickness:	0.02 μ ultrathin	2.00 μ
Mass transfer	100 ×	1

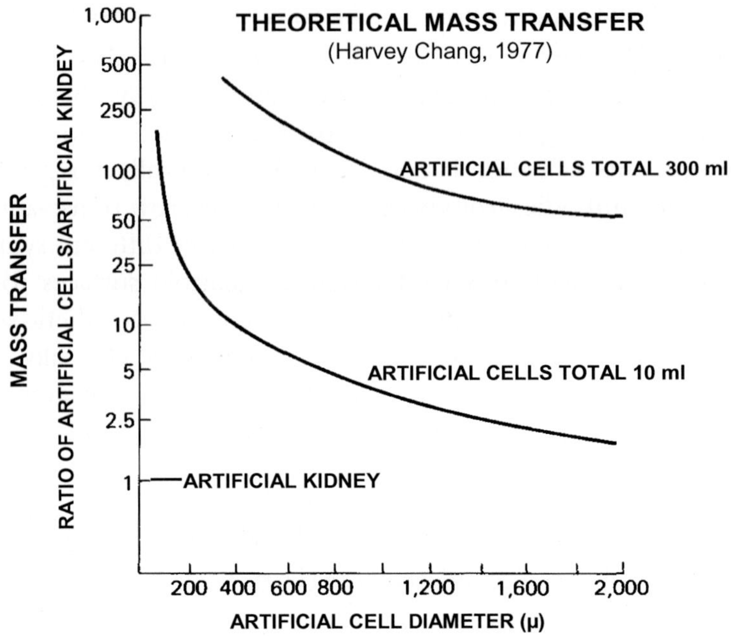

Fig. 1.2. Mass transfer of artificial kidney represented as 1. Ratio of mass transfer of 10 mL and 300 mL of artificial cells of different diameters. Hemoperfusion uses artificial cells of about 100 micron diameter (computer simulation by H. Chang).

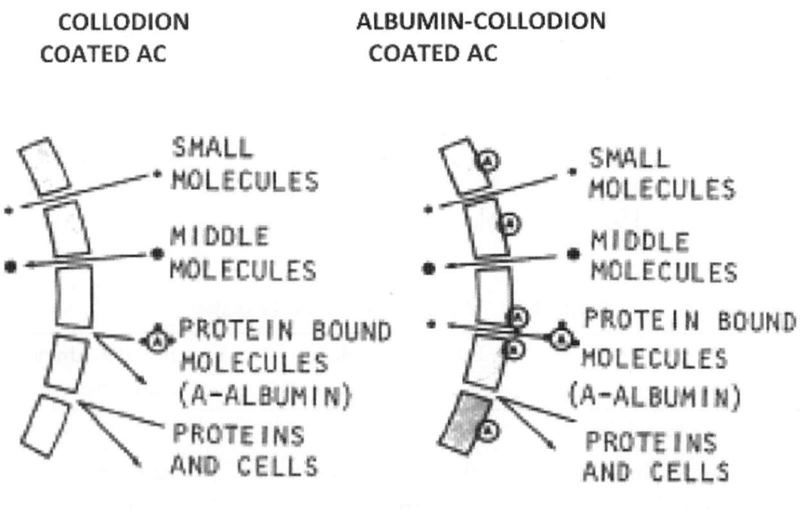

Fig. 1.3. *Left*: Ultrathin polymeric (collodion) membrane coating. *Right*: Ultrathin albumin-collodion membrane coating.

We also analyzed the effects of varying the diameter, and the total volume of artificial cells can be varied over a wide range. Figure 1.2 compares the mass transfer of artificial cells with that of a standard hemodialysis machine. Here, the mass transfer of a standard hemodialysis machine is taken as unity and compared to that of 10 ml and 300 ml of artificial cells of different mean diameters. Even with 10 ml of artificial cells, the mass transfer is many times that of the standard hemodialysis machine. With 300 ml of artificial cells, the mass transfer is even higher.

2.2. *Experimental Analysis*

Experimental study of typical artificial cells shows that the equivalent pore radius is 18 Å (Chang, 1965, 1972a). This means that the membrane would be permeable and also permeable to metabolites and peptides normally present in the body fluid. Detailed analysis of the rate of movement of different molecules has also been carried out (Chang and Poznansky, 1968a) (Table 1.2). These results show that metabolites normally present in the body can equilibrate very rapidly across the artificial cell membrane.

Table 1.2. Experimental Permeability Data for Artificial Cells

Solutes	Half-time for Equilibration (sec)	Permeability Constant (*P*) (cm/sec)	Solute Permeability Coefficient (*W*) (moles/dyne-sec.)
Urea	4.3	2.01×10^{-4}	8.23×10^{-15}
Creatinine	17.5	0.61×10^{-4}	2.52×10^{-15}
Uric acid	42.5	0.19×10^{-4}	0.77×10^{-15}
Creatine	16.6	0.75×10^{-4}	3.08×10^{-15}
Glucose	26.2	0.54×10^{-4}	2.17×10^{-15}
Sucrose	35.5	0.37×10^{-4}	1.62×10^{-15}
Actylsalicylate	39.0	0.32×10^{-4}	1.31×10^{-15}
Trtiated water	<1.0	N/A	N/A

This being the case, 30 ml of artificial cells retained in a shunt perfused by circulating blood will have the mass transfer of a whole artificial kidney machine (Chang, 1966) (Figs. 1.1, 1.2). However, in the case of the artificial kidney machine, once solutes cross the membrane they are "washed" away by 100–200 L of dialysis fluid. For artificial cells with a total internal volume of 30 ml, within a very short time, there will not be a concentration gradient for further solute diffusion. However, artificial cells are not meant to function as dialysers. They are made to function as microscopic bioreactors. This way, solute entering the artificial cells can be rapidly converted or removed, thus maintaining a concentration gradient. Thus our earlier study shows that artificial cells containing urease in a hemoperfusion chamber can rapidly lower blood urea, converting it to ammonium (Chang 1966). Adsorbents, like activated charcoal, placed inside artificial cells can also remove solutes that equilibrate rapidly inside (Chang, 1966). Why do we choose to use activated charcoal?

2.3. *Artificial Cells Containing Activated Charcoal in Hemoperfusion*

Activated charcoal has the ability to adsorb and remove a large spectrum of molecules. Since ancient times, the Chinese and the Greeks have used this for removing toxic substances. Activated charcoal has also been placed in a hemoperfusion device for direct perfusion by blood (Yatzidas, 1964). However, Dunea and Kolff (1965) showed in animal study that this resulted in (1) embolism due to the release of charcoal powder into the body and (2) this also damaged and removed blood cells, especially platelets.

By putting activated charcoal inside artificial cells, one can prevent particle release and also prevent the direct contact of blood cells (Chang, 1966) (Fig. 1.1). At the same time, unwanted molecules can move rapidly into the artificial cell and be removed by the activated charcoal.

In order to put this into practice, artificial cells were first prepared to contain activated charcoal powder (Chang, 1966). However, in a hemoperfusion device, the ultrathin membrane of the artificial cells cannot withstand the larger pressure gradient during perfusion.

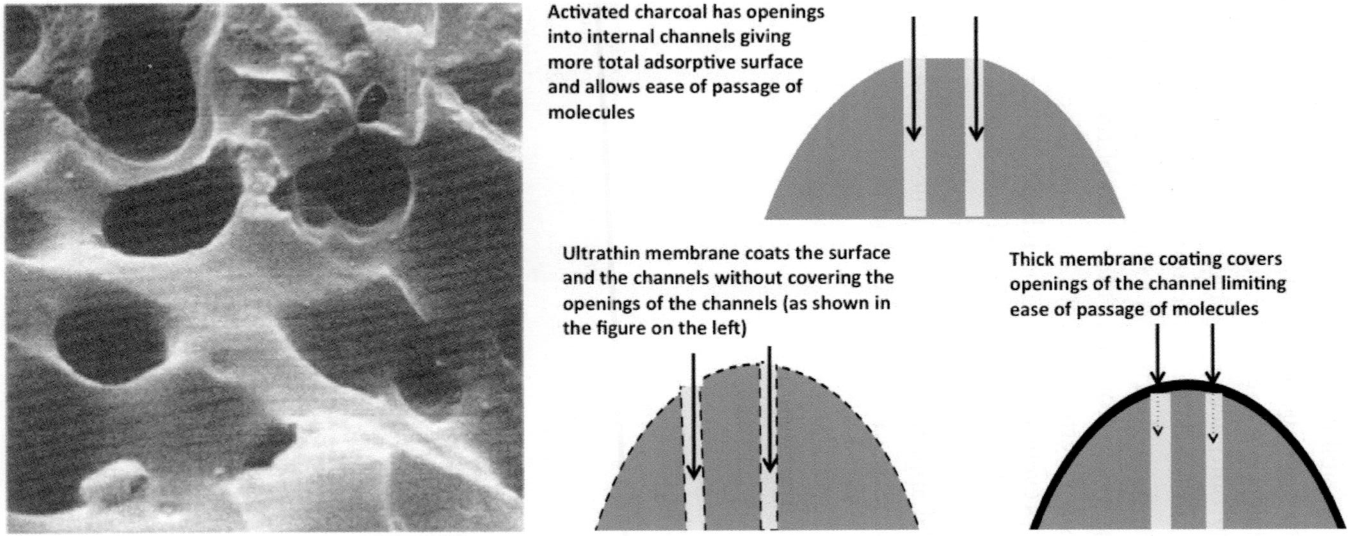

Fig. 1.4. A scanning electron-microscopic picture of ultrathin albumin-collodion membrane coated activated charcoal. The activated charcoal pores are not covered by the ultrathin membrane allowing for more efficient passage of molecules into the adsorbent.

Furthermore, the flexibility of the ultrathin membrane increases flow resistance and also packing and obstruction at the exit port of the hemoperfusion device. A more successful method is to coat the ultrathin membrane directly onto activated charcoal granules (Chang, 1969a). This way, there is no breakage of the ultrathin membrane, no increase in flow resistance and no obstruction to flow. Furthermore, an ultrathin membrane coating would not cover the pores of the activated charcoal surface and allows for better contact (Fig. 1.4). The ultrathin coating also results in the prevention of particle release. At the same time, this surface modification also results in a blood compatible surface with no damage to circulating blood cells. This forms the basis of all subsequent successful industrial approaches. However, those using a thick coating would result in decrease in permeability and mass transfer.

2.4. *Device Configuration*

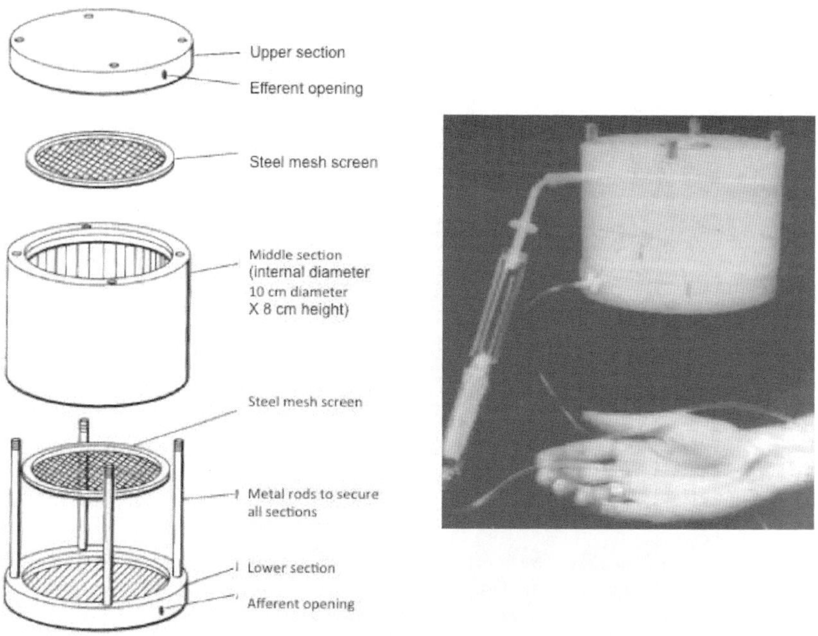

Fig. 1.5. Laboratory hemoperfusion device configuration. Screen on both sides retained the adsorbent artificial cells inside the column.

2.5. *Effects on Embolism*

In vitro studies show that there is no particle release since there is no significant difference in particle counts in the fluid entering the hemoperfusion device and that from the fluid leaving the device (Fig. 1.6) (Chang and Malave, 1970). In animal studies blood smear from blood leaving the device did not show any embolic particles. Histological studies in a total of more than 20 dogs did not show any evidence of embolism, even in artificial cell hemoperfusion devices stored for up to one month (Chang and Malave, 1970).

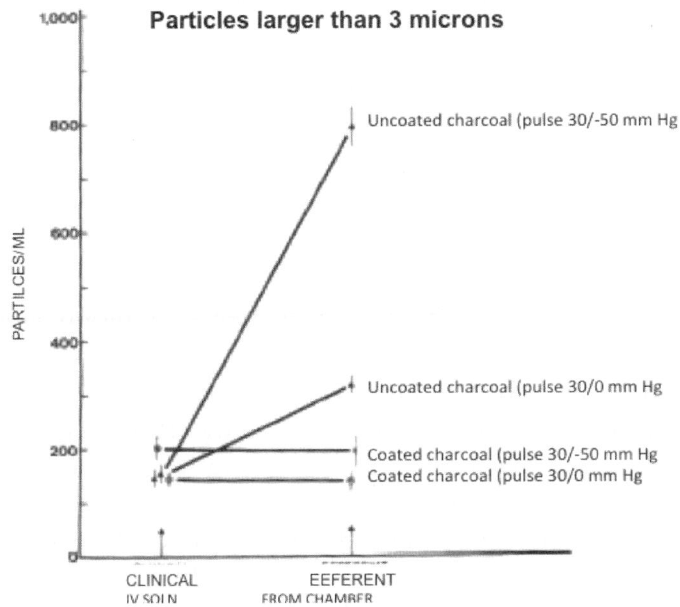

Fig. 1.6. Ultrathin membrane coating modified surface prevents release of particles.

2.6. *Effects of Hemoperfusion on Platelets*

Figure 1.7 shows that uncoated activated charcoal after contact with heparinized blood, when looked at under a scanning electron microscope, shows fibrin entrapment of red blood cells, leukocytes and platelets. On the other hand, the artificial cell in the form of

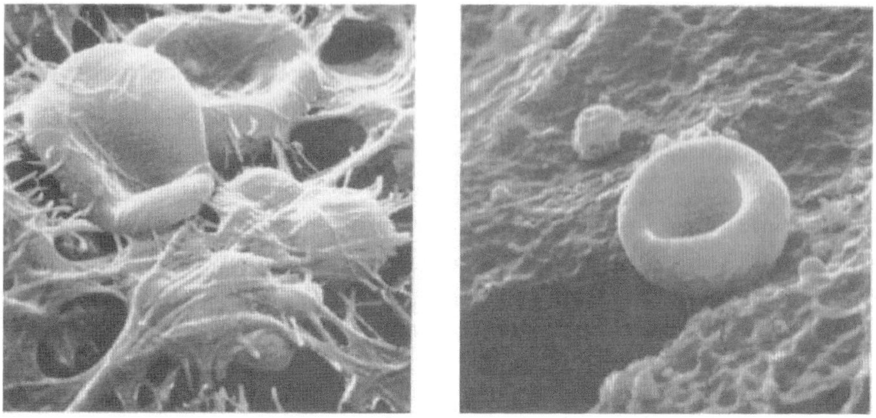

Fig. 1.7. *Left*: Uncoated activated charcoal after contact with heparinized blood looked at under a scanning electron microscope, shows fibrin entrapment of red blood cells, leukocytes and platelets. *Right*: On the other hand, artificial cell in the form of albumin-cellulose nitrate-coated activated charcoal granules did not show fibrin entrapment of formed elements of blood.

albumin-cellulose nitrate-coated activated charcoal granules did not show fibrin entrapment of formed elements of blood (Fig. 1.7) (Chang, 1974b). This scanning electron microscopic study supported the finding that albumin-cellulose nitrate-coated activated charcoal, used both in animal studies (Chang, 1969, 1974b) and in patients (Chang and Malave, 1970; Chang 1974b), prevents the depletion of platelets from the blood. Unlike the 50% decrease in platelet level when using unmodified activated charcoal, platelet levels in patients before and after two hours of hemoperfusion using albumin-cellulose nitrated coated activated charcoal artificial cells remained at $91.8 \pm 11.8\%$ (Fig. 1.8). Depending on the blood compatibility of the polymers used for surface modification used by other groups, there can be great variations of down to 50%.

There is no hemolysis or removal of red blood cells by the albumin-cellulose nitrate-coated activated charcoal. Unlike other polymers, cellulose nitrate can bind albumin tightly by adsorption even in the presence of plasma. There were no significant changes in post hemoperfusion plasma hemoglobin levels or leukocyte levels (Chang and Malave, 1974b).

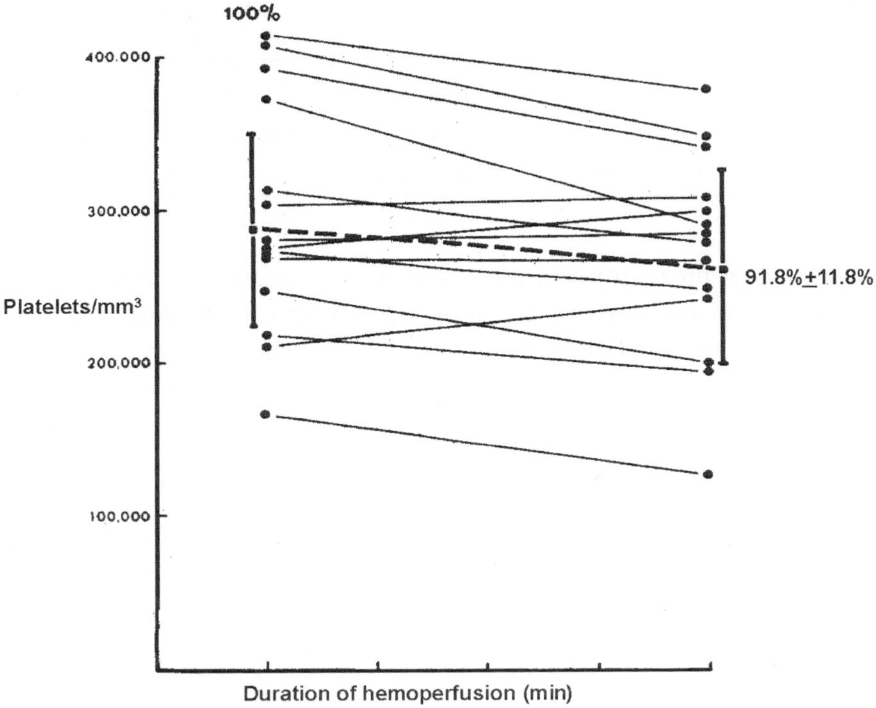

Fig. 1.8. Platelet levels in patients before and after two hours of hemoperfusion using albumin-cellulose nitrated coated activated charcoal artificial cells. There is no significant decrease in platelet level. Earlier report elsewhere using nature activated charcoal without membrane coating resulted in marked decrease in platelet levels.

2.7. *Clearance of the ACAC Artificial Cell Artificial Kidney*

On the basis of the mass transfer analysis (Table 1.1 and Fig. 1.2) one should not be too surprised at the very high clearance value of artificial cells containing activated charcoal (Table 1.3). Activated charcoal adsorbs substances that rapidly cross the membrane. Furthermore, the albumin coating of the artificial cell membrane also takes part in the removal of protein-bound molecules in the circulation (Fig. 1.3). As a result, the clearance for the different metabolites or drugs is many times higher than that for the standard hemodialysis machines.

For instance, as shown in Table 1.3, creatinine clearance is 230 mL/min as compared to 120 mL/min for standard hemodialysis machines.

Table 1.3. Clearance ml/min at Blood Flow 300ml/min

	Artificial Kidney HD	Artificial Cell Hemoperfusion
Creatinine	120 mL/min	230 mL/min
Middle molecules	30 mL/min	140 mL/min

What is even more important is the clearance for middle molecules of 140 mL/min as compared to 30 mL/min with the standard hemodialysis machines available up to 1980. The clearance for drugs like phenobarbiturate, methaprylon, methaqualone and glutethimide of between 210 to 230 mL/min is even more striking when compared to the much lower clearance by the standard hemodialysis machines. Clearance for protein-bound molecules, such as glutethimide, is many times higher using artificial cell hemoperfusion. On the other hand, it should be noted that the clearance depends a great deal on the ability of the charcoal to adsorb the material that crosses the artificial cell membrane. Thus, urea and electrolytes are not adsorbed to any extent by the charcoal, and very small clearance is obtained.

2.8. *Improvements in Hydrodynamic*

The laboratory ACAC hemoperfusion device as shown in Fig. 1.5 has configurations and inflow characteristics that allow for good hydrodynamics. However, this configuration is not convenient for industrial scale up. For industrial scale up, the configuration is shown in Fig. 1.9. In this configuration, it is necessary to analyze the optimal hydrodynamics (Victor Chang *et al.*, 1987a). Figure 1.9 shows hydrodynamic measurements of flow with the configuration of the industrial scaled up ACAC hemoperfusion device. When inflow enters the larger end of the device (left figure), there is channeling and stagnation of flow near the wall. This decreases the efficiency of extraction of unwanted molecules from the fluid. Furthermore, this can contribute to deposition of platelets and white cells. When inflow enters the narrow end of the device there is improved flow hydrodynamic with less channeling and less stagnation of flow close to the wall of the device.

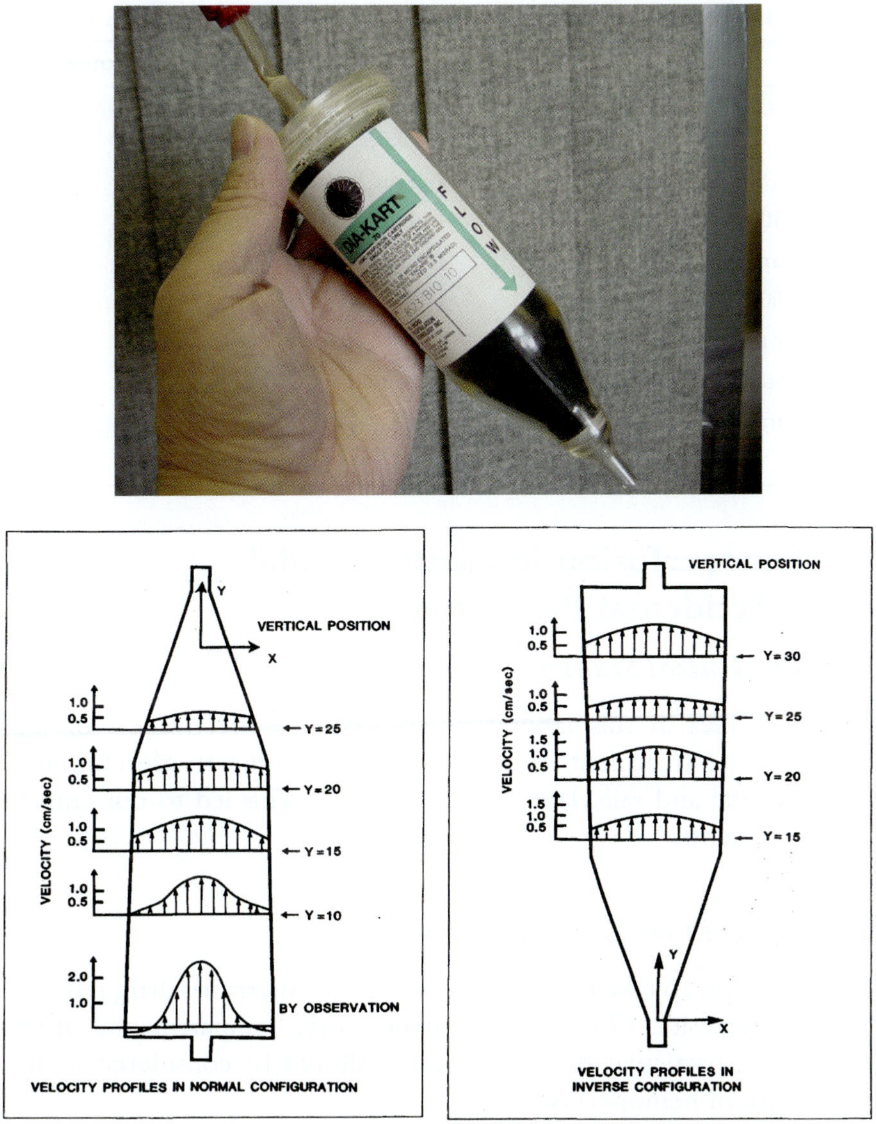

(From Victor Chang *et al.*, 1987)

Fig. 1.9. Industrial scaled up ACAC hemoperfusion device. *Left*: Flow pattern when inflow is into the larger end of the device. *Right*: Inflow into the narrow end of the device resulted in improved flow hyrodynamic with less channeling and less stagnation of flow closer to the wall of the device (From V. Chang *et al.*, 1987a).

Table 1.4. Earlier Clinically Used Coated Charcoal Hemoperfusion Devices

Systems	Membrane Material (thickness)
ACAC & CAC (Chang research)	Albumin-collodion or collodion (0.05 μm)
Hemacol (Smith and Nephew)	Acrylic hydrogel (3.0 μm)
Adsorba 300C (Gambro)	Cellulose (3.0 μm)
Nipro Hemocarbo (Nissho)	Collodion (thin)
CAC Dialaid (Montreal, Canada)	Collodion (0.05 μm)
Detoxifier type 1 (Shanghai, China)	Cross-linked gelatin
HNAC (Stefoni, Italy)	Modified acrylic (1.5 μm)
DPH (Kuraray)	Hema-based (0.1 μm?)
Others	

3. Hemoperfusion in Acute Suicidal or Accidental Poisoning

3.1. *Preclinical Studies*

Animal studies in this laboratory showed the effectiveness of this system in the treatment of acute intoxication due to glutethimide, pentobarbital and salicylate (Chang, 1972a). This led to our clinical trials in patients.

3.2. *Criteria for Drug Removal*

It is important to take into consideration the criteria for drug removal by hemoperfusion (Table 1.5). Furthermore, compartmental distribution of a particular drug (Fig. 1.10) should be considered in any clinical use of hemoperfusion.

3.3. *Protocol for Clinical Trials*

Other physicians at that time were not willing to carry out the procedure because of the potential for liability. As a result, I had to prepare the hemoperfusion devices in my laboratory at McGill, bring each of

Table 1.5. Criteria for Drug Removal by Hemoperfusion

1. Permeability of membrane to drug
   (E.g. drugs tightly bound to plasma albumin)
2. Removal after permeation by the adsorbent
   (e.g. highly lipid soluble or protein bound)
4. Removal compared to kidney, liver, GI
5. Time-dose cytotoxic relationship:
   (e.g. If Px delayed theophylline brain damage & paraquat lung damage)
6. Other factors: salicylate (aspirin) removed very efficiently by hemoperfusion but it also causes acidosis that dialysis to adjust acid base balance.

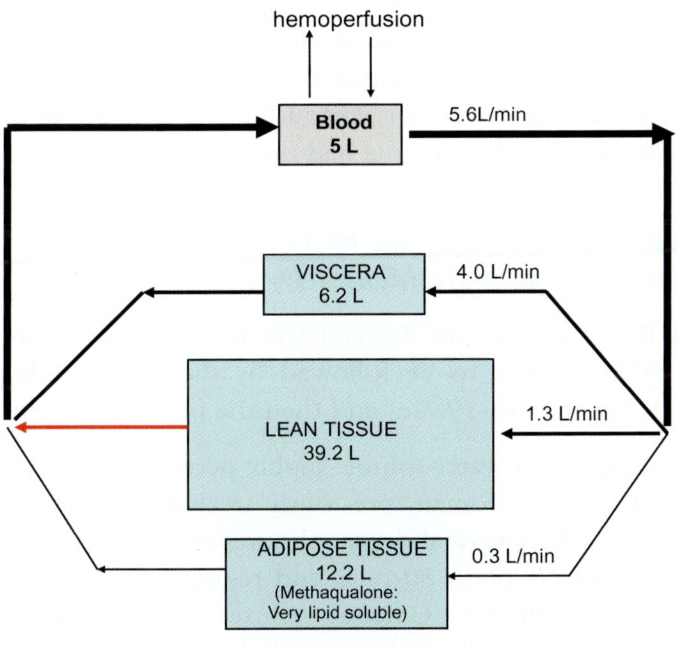

Fig. 1.10. Compartmental distribution of drugs in the body.

them to the hospital and carry out the treatment of the patients myself. The arrangement was for the nephrologists to refer patients to me. I would then make the final decision based on the following criteria (Table 1.5).

For the first three adult patients, I would agree to carry out the procedures only in those patients who were seriously intoxicated with systemic drug level well above the lethal levels or who had serious complications requiring rapid recovery from the effects of the drugs. For example, in one case, before treatment the patient had cardiac arrests on two occasions with possible pulmonary embolism requiring heparinization. Her blood pressure was not obtainable by auscultation and her methyprylon level was twice the lethal level. In this particular case, the desirability of rapid elimination of the drug and its effects is obvious. In another case, a combined intoxication existed — the methyprylon level was three times, and the methaqualone level at least twice the lethal level, furthermore, it is well known that glutethimide is cleared poorly by the standard artificial kidney available at that time. In the case of a pediatric patient with accidental theophylline overdose, irreversible brain damage would have been the result if treatment was not started immediately. In all cases, the potential benefits to the patients had to be high.

3.4. *Result of Clinical Trial on 11 Adult Patients with Suicidal or Accidental Drug Poisoning*

The first three patients are described in more detail (Chang *et al.*, 1973a, 1973b) below to be followed by the other adult patients (Chang 1975g, 1976c, 1980e) and then the pediatric patient.

Case 1: Example of a water-soluble highly permeable and adsorbable drug: A 50-year-old woman, previously treated for depression, was admitted after taking several drugs. She was comatose and her blood pressure was *90/60,* pulse *80/*min. and regular, respiration *12/min.* and temperature 98.4°F. Corneal reflexes were absent. She was treated by intubation followed by gastric lavage containing activated charcoal. Heparin, 5,000 units IV q4h, was started since pulmonary embolism was suspected. Four hr later two cardiac arrests occurred and were reversed with cardiac massage and direct current electroshock. A subsequent chest X-ray showed fracture of four ribs. Ventilatory assistance was required. Standard hemodialysis treatment did not result in any improvement. Ten hr after admission her blood

Table 1.6. First 11 Adult Patients with Suicidal Overdose (Chang *et al.*, 1973a, b; Chang 1975g)

	Clinical Drugs (Coma Grade)		Clearance (ml/min)	Hemoperfusion (Number of 2hr Px)	Outcome
1	IV	Methyprylon	230	2	recovery
2	IV	Glutethimide	150	1	recovery
3	IV	Methyprylon+	230	2	recovery
		Methaqualone	230		
4	IV	Glutethimide	230	4	recovery
5	IV	Phenobarbital	228	1	recovery
6	IV	Phenobarbital	162	1	recovery
7	IV	Glutethimide	—	1	recovery
8	IV	salicylate	150	1	recovery
9	IV	methyprylon	—	1	recovery
10	IV	phencyclidine	—	1	recovery
11	IV	secobarbital	200	1	recovery

pressure could not be determined by auscultation. At this time her blood methyprylon level was 9.4 mg/dL and six hr later it was 9.6 mg/dL. Direct measurement of arterial blood pressure from the A V shunt gave a systolic pressure of 80 mm. Hg (Fig. 1.11).

The patient was referred to me and hemoperfusion was carried out. For the first 20 min the systolic blood pressure was maintained at 90–100 mm Hg with levarterenol. Then the levarterenol was discontinued and the pressure remained at 130–150 mm Hg throughout the three hr of hemoperfusion. The initial arterial blood methyprylon level was 9.6/dL. Hemoperfusion for two hours lowered this to 4.1 mg/dL, which is below the dangerous level of 6 mg/dL (Fig. 1.11). An initial clearance of 250 mL/min was obtained. As a result, the recovery was dramatic and straightforward. For example, voluntary respiration became more regular and increased to *40/min.* with tidal volume increasing to 325 mL during hemoperfusion. Corneal reflexes and voluntary eye opening and blinking returned and response to painful stimuli appeared. Limb movements were detected 27 hr after admission and 4 hr later the patient responded to verbal

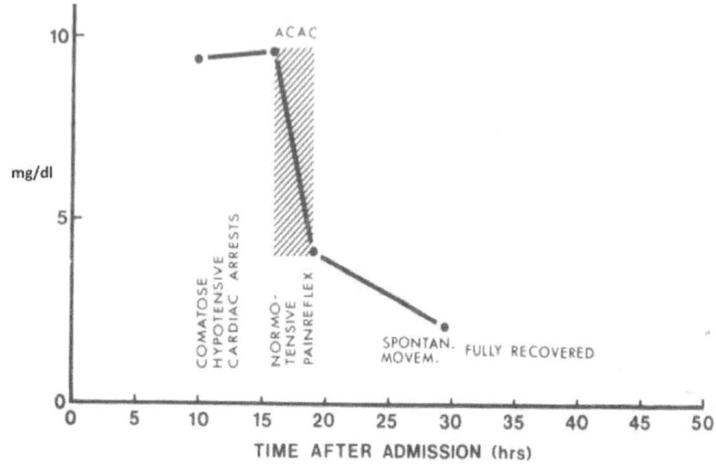

Fig. 1.11. Clinical and laboratory results of hemoperfusion in a patient with severe suicidal methryprylon overdose.

commands. She no longer required ventilatory assistance and started talking four hr later.

Case 2: Example of compartmental distribution of highly lipid-soluble drug: A 26-year-old woman was admitted after ingesting a large amount of glutethimide. She was comatose and areflexic. Her blood pressure was 90/60 and she required ventilatory support. Six hr after admission her systolic blood pressure was less than 70 mm. Hg and isoproterenol was infused. Twenty-four hr after admission her condition was unchanged. Her rectal temperature was 87°F, blood pressure 90/50, and she still required isoproterenol and ventilatory support. At this time she had clinical and X-ray signs of bilateral bronchopneumonia. Her blood glutethimide level was 7 mg/dL. Standard hemodialysis did not result in any improvement.

At 27 hr after admission she was referred to me and treated by hemoperfusion through 300 g of the ACAC microcapsules for two hours. In the course of the hemoperfusion her blood pressure rose from 90/50 to 120/80 in 30 min without isoproterenol infusion. Her blood glutethimide level fell to 3 mg/dL, but 10 hours after completion of the hemoperfusion it rose to 5 mg/dL (Fig. 1.12). Except for

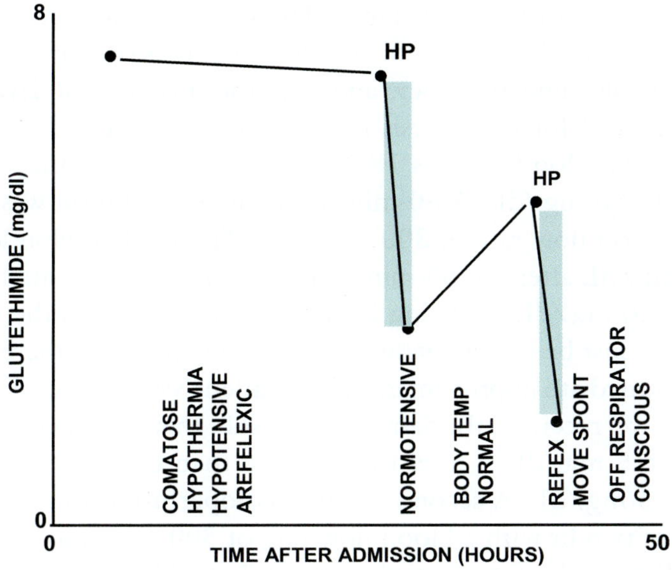

Fig. 1.12. Clinical and laboratory results of hemoperfusion in a patient with severe suicidal glutethimide overdose.

a very slight pupillary reflex and gag reflex, she remained comatose. Her blood pressure was 110/55. Since glutethimide has a high lipid coefficient, a large part of the drug is accumulated in lipoid tissue. The rebound in its level after hemoperfusion likely reflects a redistribution of glutethimide from the lipoid tissue into the bloodstream. A second hemoperfusion with another 300 g of ACAC microcapsules was carried out. Within one hour a 2+ patellar reflex and a slight —plantar reflex could be elicited. After one and one-half hours of hemoperfusion, her tidal volume increased from 250 mL (pretreatment) to 475 mL. In addition, she exhibited some spontaneous limb movements. Her blood pressure increased to 130/75. The posthemoperfusion blood glutethimide level was 1.5 mg/dL (Fig. 1.12). Two hr after completion of the second hemoperfusion she no longer required ventilatory support. She was up and about the following day.

Case 3: A 27-year-old man was admitted after ingesting a mixture of medications. He was comatose and unresponsive to painful stimuli and required ventilatory assistance. He also showed signs of methaqualone intoxication: tonic convulsions, hypotension, cardiac failure, myoclonia, bleeding tendency, and hypothermia. Gastric lavage was carried out and forced diuresis started. Eight hr after admission his blood methyprylon level was 18.3 mg/dL, three times the "dangerous" level of 6 mg/dL. A 60-min ACAC hemoperfusion was carried out at a blood flow rate of 300 mL/min. The methyprylon level fell to 12.9 mg/dL after the 60-min treatment and by 32 hr after admission was 9.0 mg/dL, lowering its level from three times the dangerous level to just below the dangerous level. Despite this, there was no marked immediate improvement in the patient's clinical condition. At this time we received the laboratory result that the methaqualone level was 6.5 mg/dL, a level that is more than twice the dangerous level of 2.5 mg/dL. A second ACAC hemoperfusion was given over a period of two hr with a blood flow rate of 300 mL/min. The levels fell to 1.8 mg/dL of methyprylon and 2.7 mg/dL of methaqualone. By 48 hr his pupils reacted sluggishly to light, and the gag reflex, deep pain response and bowel sounds were present. At 96 hr his breathing was independent; all reflexes were normal. By 130 hr after admission he was oriented, rational and up and about.

Case 4: This patient was admitted comatose, areflexic, with fixed dilated pupils, and requiring respiratory assist and pressor drug to maintain her blood pressure. She had taken a large mixture of drugs, including glutethimide, thioridazine, salicylate, phenobarbital, and tetracycline, in addition to a large intake of alcohol. She was hemodialyzed three times in the first 48 hr with no change in her condition. She was hemoperfused for 1-1/2 hr with 300 g of ACAC, followed immediately by another 300 g of ACAC hemoperfusion column for 1 hr. With this hemoperfusion her corneal reflex returned and her blood pressure could be maintained without pressor drug. In the next two days she was treated by two hemodialyses but without any changes in her clinical condition. A third hemoperfusion was carried out 10 hr after the last hemodialysis. With this hemoperfusion gag reflex and cough reflex returned and there was occasional spontaneous breathing. This was followed by a fourth hemoperfusion after which spontaneous breathing returned. One day after, she was responsive and following this she recovered and was up and about. Her phenobarbital level was 20 mg% on admission and clearance obtained was 228.6 mL/min. Her glutethimide clearance obtained was 251 mL/min.

Case 5: A 79-year-old patient was admitted with a history of ingestion of phenobarbital with a blood level of 30 mg%. The patient was comatose, areflexic, and required ventilatory assist when hemoperfusion was carried out. With 2 hr of hemoperfusion, pupil reflex, corneal reflex, and patellar reflex returned. Later the patient started to trigger the respirator, have spontaneous movement and recovered shortly after this. Clearance obtained for hemoperfusion at a blood flow rate of 200 mL/min was 188 mL/min clearance.

Case 6: This patient was admitted with Grade IV coma due to phenobarbital intoxication. The patient received ACAC hemoperfusion at 180–200 mL/min blood flow rate. The clearance was 162 mL/min. Hemoperfusion resulted in recovery of the patient.

Case 7: This 21-year-old patient was admitted with a history of ingestion of glutethimide and phenobarbital. She was admitted in Grade IV coma. There was X-ray evidence of right upper lobe atelectasis and right lower lobe consolidation. One 2 hr hemoperfusion resulted in the improvement of the patient and complete recovery and discharge subsequently.

Case 8: This 22-year-old man was admitted with a history of overdose of salicylate and sodium bromide. The salicylate level was 52.85 mg% when hemoperfusion was started. The clearance obtained at the flow rate of 300 mL/min was 165.9 mL/min. Hemoperfusion of 2 hr lowered the systemic salicylate level to 37.76 mg%. At initiation of hemoperfusion the patient was in Grade II coma responding to painful stimuli. After 1 hr of hemoperfusion the patient recovered consciousness and complained of thirst and hunger and wanted to get up and also carried on conversations. He recovered after this.

Case 9: This was a 51-year-old patient who had taken an overdose of methyprylon. The patient was comatose, areflexic, hypotensive, and on ventilatory assist. The patient received 2 hr of hemoperfusion with ACAC hemoperfusion at a blood flow rate of 160 mL/min. Two hours after starting hemoperfusion there was bilateral patellar reflex 2+, ankle reflex 2+, light reflex, and some voluntary movement. Shortly afterwards the ventilatory assist was not required and he recovered shortly after this.

Case 10: This was a 30-year-old patient admitted with a history of having taken phencyclidine. Phencyclidine was qualitatively demonstrated in the blood and urine of this patient. The patient was comatose, areflexic, hypotensive, and a 2 hr ACAC hemoperfusion was carried out in this patient. Ater the hemoperfusion the patient still remained comatose. Three days later the patient was off the respiratory and started to have voluntary movement and then recovered after this. In this patient it is not possible to measure phencyclidine level quantitatively to calculate clearance or changes in level because the method available at that time could only give a qualitative assessment of the presence of phencyclidine in blood and urine.

Case 11: This 55-year-old patient was admitted having taken a bottle of Seconal (secobarbital). Shortly after admission he developed respiratory arrest. He was resuscitated, put on a respirator, and blood pressure was maintained with pressor drug. He developed ventricular tachycardia which responded to lidocaine. His secobarbital level was 6 mg%. Two hr of hemoperfusion was carried out. At the beginning the patient was comatose with no reflexes. After 1 hr of hemoperfusion, there was 2+ achilles tendon reflex. And 2 hr after hemoperfusion, both patellar and Achilles reflexes were present. Two days later he recovered completely and was transferred out of the Intensive Care Unit. This hemoperfusion was carried out with the Hemosorba system and the clearance obtained was 200 mL/min.

3.5. *Time–dose Cytotoxicity Relationship: Pediatric Patient with Accidental Theophylline Overdose*

A three-year-10-month female child with a history of asthma developed tachynea, respiratory difficulty and a fever. Instead of pediatric dose, she was accidentally given the adult dose of 750 mg of theophylline by her family at home, an amount 10 times in excess of that prescribed. She became restless, developed tachycardia, and vomited coffee ground material positive for occult blood. Her serum theophylline level was 67 μg/mL one hour and 45 min after this error. She was admitted to the intensive care unit. The hospital immediately referred this patient to me for possible treatment with ACAC hemoperfusion. It was decided to carry out this procedure immediately because: (1) The theophylline level varied from 67 to 74 μg/mL, with 40 μg/mL being considered a potentially lethal dose. (2) It is known that if the level is not lowered quickly, the child would suffer irreversible brain damage — as in an earlier case when the child was not treated by hemoperfusion. (3) Available dialysis systems at that time were not effective in removing theophylline.

A three-hour hemoperfusion procedure was carried out (Chang *et al.*, 1980a). The maximal blood flow possible in this 15.4 kg child was 60–75mL/min in the first hr. After the first hour, the flow rate could be increased to 90 mL/min, and this rate was maintained for the remainder of the treatment. Before hemoperfusion the child had a tachycardia of 187/min and a blood pressure of 120/70 mm Hg. As hemoperfusion lowered the systemic theophylline level, the heart rate decreased to 145/min at 1 hr, 135/min at two hr and was 127/min at the end of the procedure (Fig. 1.13). Blood pressure was unchanged throughout the hemoperfusion.

Figure 1.13 shows that the blood level of theophylline before HP was 74 μg/mL. Hemoperfusion lowered the level within 1.5 hours to 40 μg/mL and within three hours to 15 μg/mL. Hemoperfusion was discontinued. The theophylline level in the efferent from the hemoperfusion device was zero showing a 100% extract of the drug on one circulation through the device. The removal of the drug from the tissue compartment must also have been high since before hemoperfuion the blood level remained high and stable but after hemoperfusion was completed, the serum level continued to remain at the low level. The child recovered uneventfully.

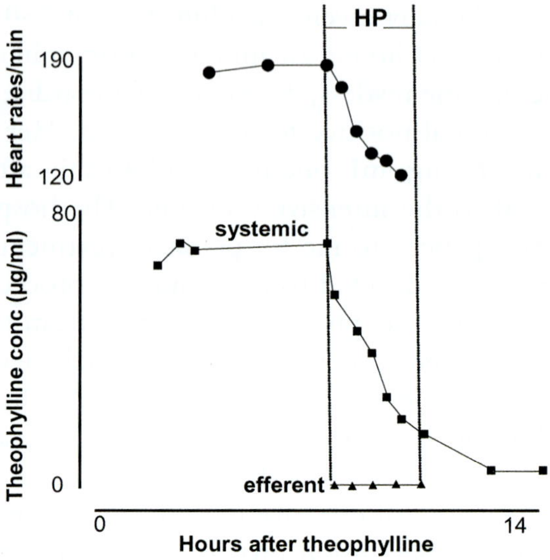

Fig. 1.13. Clinical and laboratory results of hemoperfusion in a three-year-old pediatric patient with severe accidental theophylline overdose. Her level was well above the dangerous level of 40 μg/mL that could have resulted in irreversible brain damage. Hemoperfusion rapidly lowered this level (Chang *et al.*, 1980a).

3.6. *Aluminium–desferal: Solution to Permeation and Adsorbent Problem*

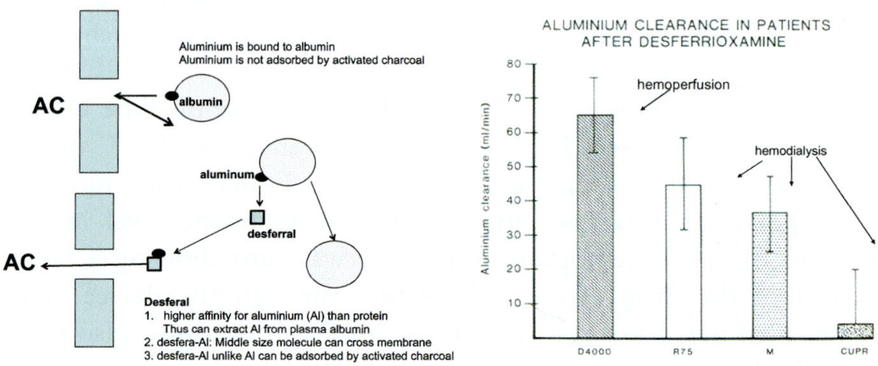

Fig. 1.14. Left: Aluminium is bound to albumin and cannot cross the membrane to be removed by the AC absorbent. With the addition of desferral, aluminium bound to desferral, which is a small molecule that can be adsorbed by AC. As a result the aluminium–desferral complex can cross the membrane to be removed by the AC. Right: Our clinical trial shows that by doing this, aluminium can be more efficiently removed when compared to dialysis (Chang and Barre, 1983, *Lancet*).

3.7. *Routine Clinical Uses in Patients Around the World*

Activated charcoal is a well-known adsorbent for a large number of drugs and toxins. Artificial cells containing activated charcoal, as demonstrated here, take advantage of this adsorbing property and at the same time prevent the two major problems of charcoal embolism and reduction in platelets. The results of these first clinical trials have led to the use of this approach around the world for the treatment of patients with accidental or suicidal poisoning (Better *et al.*, 1979; Biberstein *et al.*, 1983; Cohan *et al.*, 1982; Verpooten *et al.*, 1984; Gelfand *et al.*, 1977; Gibson *et al.*, 1978; Lorch and Garella, 1979; Winchester, 1996; Diaz-Buxo *et al.*, 1978; Sideman and Chang, 1980b; Piskin and Chang, 1982; Chang and Zheng, 1983; Chang and Ho, 1985; Chang and Nicolaev, 1987; Chang and Odaka, 1991; Klinmann *et al.*, 1990; Chang and Odaka, 1991; Kawasaki, 2000; Lin *et al.*, 2002; Lopez-Lago *et al.*, 2002; Peng *et al.*, 2004, Chinese symposium in hemoperfusion, 2005). More recent results are available in chapters in the present book.

The exact methods of preparation of artificial cells containing activated charcoal have been published in reproducible detail for all to use (Chang, 1972a). As a result, a number of countries started to produce their own industrial hemoperfusion devices. As would be expected, there are variations and extension of Chang's original method resulting in variations in efficacy and biocompatibility. While the inferior ones are no longer in use, those that are safe and effective are being used in routine clinical treatment of accidental and suicidal poisoning. In some areas of the world, for some unknown reasons, these devices are extremely expensive and as a result are too costly for wide spread use. In other areas of the world, especially where hemodialysis machines are not easily accessible, inexpensive but safe and efficient devices are being produced and used extensively. All the very extensive experience in the use of adsorbent artificial cells in hemoperfusion for acute poisoning has led the nephrologists to establish the following criteria for using this in acute poisoning (Table 1.7).

Table 1.7. Clinical Criteria for Hemoperfusion (modified from Winchester, 1996)

1. Progressive deterioration despite intensive care.
2. Severe intoxication with mid-brain dysfunction.
3. Development of complications of coma.
4. Impairment of normal drug excretory function.
5. Intoxication with agents producing metabolic and/or delayed effects.
6. Intoxication with an extractable drug that can be removed at a greater rate than endogenous elimination

Because of its efficient adsorbing properties for numerous drugs the system may be even more useful in cases of acute intoxication involving a number of drugs, especially when the exact drugs are not known at the time that treatment is required. Being portable and having a constant extracorporeal volume, it is particularly useful in areas where hemodialysis machines are not easily accessible or in pediatric practice.

4. Hemoperfusion in Terminal Renal Failure Patients

4.1. *Rationales*

Hemodialysis machines are effective in the treatment of chronic renal failure patients. However, at that time there were not enough machines; furthermore it was extremely expensive. Only a small number of patients can receive treatment in a few countries. In most other countries, the high cost and unavailability of machines prevented the use of this in terminal renal failure patients. Since hemoperfusion is efficient in removing toxin or unwanted waste from the blood, we carried out study into its possible use in uremic patients.

This author has carried out 55 hemoperfusion procedures in 14 renal failure patients. Initially, only one hemoperfusion procedure was carried out on each of four terminal renal failure patients. Having shown the safety of this procedure, two hemoperfusions were carried out on the fifth patient and three procedures were then carried out on each of the next three patients. After this, a 72-year-old woman, who could not be managed by peritoneal dialysis or hemodialysis, was placed on an eight-month regime of hemoperfusions supplemented with hemodialysis as required, the latter for the removal of urea, water, and electrolytes. Long-term hemoperfusion in this patient was safe and effective, and her condition was stabilized sufficiently for her to be maintained on standard dialysis for the next 10 years. This led to the inclusion of other patients for long-term hemoperfusion.

Since commercial devices were not available until later, the total number of patients was limited by my having to prepare the devices myself and also to carry out the procedure at that time. To allow for objective clinical assessment, I was responsible for the preparation of the hemoperfusion device and the carrying out of the procedures on patients. A group of nephrologists would refer the patients and independently assess the clinical and laboratory results. The patients were otherwise treated as any other hemodialysis patient in the McGill teaching hospital dialysis units, where medications, frequencies of transfusions, and number of supplemental hemodialyses were determined independently by the nephrologists. The standard chronic hemodialysis program at that time was for each patient to receive an average of two 6 h hemodialyses each week, usually on the EX 01 hemodialyzer.

4.2. *Hemoperfusion Alone Supplemented by Hemodialysis*

General clinical results

No adverse effects from hemoperfusions were observed in the patients. There was little or no change in blood pressure throughout the hemoperfusion procedure, even in patients who normally experienced severe hypotensive episodes when on the EX 01 hemodialyzers.

Immediately after the 2 hr hemoperfusion, the patients feel much less fatigued than when they undergo the 6 hr hemodialysis with the EX 01 artificial kidney. This is especially so in a patient who normally stays home in the evening to rest after the 6 hr EX 01 hemodialysis treatment. In contrast, 1 hr after the 2 hr ACAC hemoperfusion, she goes out to have dinner, make social visits or go to the theater. At the initiation of the clinical trial, each patient receiving hemoperfusion is required to stay in the hospital overnight for observation before being allowed to return home the next day. When the procedure is found to be safe, patients treated with the hemoperfusion procedure are allowed to go home 1 hr after hemoperfusion. The patients' general well being is at least as good on the weekly 2 hr ACAC hemoperfusion and 6 hr hemodialysis (EX 01) as on the twice-weekly 6 hr EX 01 hemodialysis. The following are examples of patients on longer term treatment with hemoperfusion.

First patient on long-term hemoperfusion

This is a 72-year-old woman with chronic renal failure and congestive heart failure. She was admitted with complaints of nausea, vomiting, low back pain, diarrhea and hiccups. She was bedridden in the hospital, with a BUN of 186 mg/dL, creatinine of 24 mg/dL, and a 24 hr creatinine excretion of 80 mg. She was placed on peritoneal dialysis, but this resulted in massive intra-abdominal bleeding and severe shock. Hemodialysis treatment resulted in hypotension. She was referred to this author since there was no other way to treat her condition. At that time, because of lack of facilities, patients of her age were not generally accepted into a long-term hemodialysis program. Thus, I first had the dialysis unit agree that if she improved on the hemoperfusion program, the dialysis unit would accept her on the standard dialysis program. For the next 50 days, she was treated with nine hemoperfusion procedures each lasting for 2 hr. During these 50 days she received only one hemodialysis for the removal of water and electrolytes. After this she continued for a total of eight months on hemoperfusion combined with hemodialysis. The schedule of treatment and response for the first 50 days of her eight-month treatment on hemoperfusion are summarized in Fig. 1.15.

On this regime, her uremic symptoms of nausea, vomiting, diarrhea and hiccups cleared. Surprisingly, after two weeks on this regime, she felt well enough to be allowed by the nephrologists to leave the hospital and return home to her husband. During this time, she no longer had uremic symptoms and was able to do her shopping and housework. At the beginning she was asked to stay in the hospital for 24 hr after each procedure. Later, she only came to the hospital 1 hr before her hemoperfusion and then returned home 1 hr after the completion of the hemoperfusion. There were no side effects throughout all her hemoperfusions. There was little or no change in her blood pressure, nor other clinical side effects. This is contrary to her standard hemodialysis treatment with the EX 01, where she had

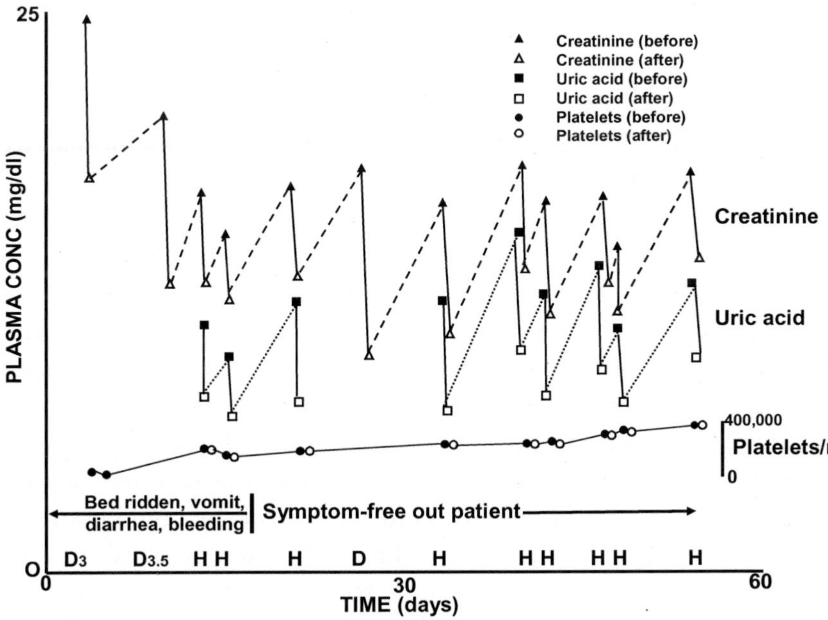

Fig. 1.15. First terminal renal failure patient on long-term hemoperfusion. *Graph shows the results of the first 50 days of the eight-month treatment.*

hypotensive episodes with hemodialysis especially during fluid removal.

After the 50-day period, the dialysis unit admitted her on a half time basis with one 6 hr hemodialysis per week, with a 2 hr hemoperfusion per week. During the eight-month period, she received a total of 24 hemoperfusion procedures and all were uneventful with no side effects. The only problem was fluid retention since hemoperfusion does not remove fluid. After a total of eight months on the hemoperfusion program, she was placed on a standard two-weekly 6 hr hemodialysis program for the next 10 years until she passed away at over 80 years of age. Independent inquiry by the nephrologists showed that her feeling of well-being was the same when she was on an average weekly regime of 2 hr hemoperfusion plus 6 hr hemodialysis as compared to her twice-a-week 6 hr hemodialysis. Thus 2 hr on hemoperfusion seems to be as effective as 6 hr of hemodialysis.

A 51-year-old woman was admitted with a long history of hypertension and chronic renal failure caused by chronic pyelonephritis. Her urine output remains at about 500 mL/day with creatinine excretion of 0.4 g/24 hr. She was started on occasional peritoneal dialysis and then on 6 hr twice-weekly EX 01 hemodialysis while waiting for renal transplantation. She developed severe pruritis which disturbed her sleep at night and was a nuisance in the daytime. Because of this, she was started on hemoperfusion to treat her pruritis. She continued on the 6 hr twice-weekly hemodialysis program with the addition of a 2 hr weekly hemoperfusion. This treatment resulted in a marked decrease of her pruritis. In addition, her acceptance of the hemoperfusion procedure was such that she was placed on the long-term hemoperfusion regime of one 2 hr hemoperfusion plus one 6 hr EX 01 hemodialysis weekly. In addition to her diminished pruritis, she found that with hemoperfusion she could come in 1 hr before hemoperfusion, have 2 hr ACAC hemoperfusion, and after the treatment, feel well enough to go directly to social evenings, dancing, or playing cards with her friends. On the other hand, with EX 01 hemodialysis she felt tired and had to rest at home in the evening. Furthermore, she experienced cramps in her legs during or soon after each 6 hr EX 01 hemodialysis. She remained symptom free and with no pruritis on this hemoperfusion and hemodialysis program for four months. After this, she returned to the standard hemodialysis program because one hemodialysis each week was not enough to control her body fluid.

A 53-year-old male with chronic renal failure was started on the 6 hr twice-weekly EX 01 hemodialysis. His daily urine output was 1L and creatinine excretion was 0.5 g. He was started on a weekly 2 hr hemoperfusion and 6 hr EX 01 hemodialysis program. Being a very active business man, he accepted this program because it meant that on the day of treatment with hemoperfusion, he could work in the morning, have his lunch and come in for treatment then return in the same afternoon, sometimes to work further in the evening. At the end of three weeks he received a renal transplant.

A 21-year-old man was admitted to the hospital with symptoms of cramps, headaches, nausea, vomiting and diarrhea. His creatinine clearance was 1.6 mL/min and urinary output was 700 mL/day. This patient, with spina bifida at birth, was paraplegic with a neurogenic bladder, recurring urinary tract infection and acute pyelonephritis. At that time, there was no opening for him in the chronic hemodialysis program, and he was not considered suitable for renal transplantation. The patient was treated with 2 hr of hemoperfusion weekly, supplemented by very occasional hemodialysis as needed for control of fluid and electrolytes. This treatment resulted in marked improvement of the patient's symptoms — except for occasional diarrhea. He was well enough to return to work. Three months later, an opening became available in the chronic hemodialysis unit and the patient was treated with an average of once a week 2 hr hemoperfusion plus once a week 6 hr hemodialysis. After a total of six months on hemoperfusion, the patient continued to be symptom free. There was improvement in nerve conduction velocity. Each 2 hr hemoperfusion lowered the serum middle molecule (300–1500 mol wt) level by 55%. He functioned well and was able to work and swim. The only problem was that one hemodialysis per week was not enough to prevent fluid retention. Thus, after six months on this program he was placed on the standard twice a week hemodialysis program.

4.3. *Hemoperfusion and Removal of Uremic Metabolites*

Since then other artificial-cell-based hemoperfusion devices have been produced industrially. The following table (Table 1.8) shows that the better hemoperfusion devices are much more effective than the dialysers available at that time for the removal of uremic metabolites as shown by the clearance for creatinine, uric acid and middle molecules. Also, with the better hemoperfusion systems like ACAC and Adsorba 300C, there was no significant effect on the platelet levels.

It has been strongly proposed that middle molecular weight substances (300–5,000 m.w.) including ß-2-microglobulin (11,800 m.w.), may be responsible for uremic symptoms. However, the hemodialysis membrane available at that time had a very low clearance for middle molecules and required up to eight hours for effective removal. On the other hand, the better hemoperfusion devices are much more efficient in removing middle molecules (Table 1.8; Fig. 1.6) (Chang, 1972e; Chang and Migchelsen, 1973; Chang and Lister, 1980, 1981).

4.4. *Conjoint Hemoperfusion-Hemodialysis*

The above result shows that hemoperfusion is able to carry out its function in removing toxin or unwanted waste metabolite to keep patients symptom free (Table 1.9). On the other hand, the above clinical results also showed that hemoperfusion, even with occasional hemodialysis, cannot look after electrolytes or fluid retention. For the these reasons, clinical investigators around the world started to carry out clinical trials using hemoperfusion in series with hemodialysis for patients with dialysis-resistant uremic symptoms and also to reduce the time needed for treatment (Chang *et al.*, 1975, 1982b; Martin *et al.*, 1979; Inou *et al.*, 1979; Odaka *et al.*, 1980; Agishi *et al.*, 1980; Stefoni *et al.*, 1980, others in this book). The use of the better commercially prepared hemoperfusion devices in series with hemodialysers resulted in clinical improvements in "well-being," nerve conduction velocity, pruritis, pericarditis, peripheral neuropathy and, what is also important, reduction in treatment time (Table 1.9) (Chang *et al.*, 1975, 1982b; Martin *et al.*, 1979; Inou *et al.*, 1979; Odaka *et al.*, 1980;

Table 1.8. Characteristic of Artificial-Cell-Based Hemoperfusion Devices

	ACAC	Hemacol	Adsorba 300C
Hemoperfusion system	Albumin collodion coated (300 g) (0.05–0.5 micron)	Acrylic hydrogel coated (300 g) (3–5 micron)	Cellulose acetate coated (300 g) (3–5 micron)
Max. clearance (ml/min)			
creatinine	235 (QB 300)	181 (QB 300)	185 (QB 200)
uric acid	230 (QB 300)	116 (QB 300)	186 (QB 200)
middle molecules			
300–1500	144 (QB 300)	as in dialyzer	no data
1000–2000	no data	no data	120 (QB 300)
(peak 7c)			
2000–5000	93 (QB 300)	no data	no data
PTH (free)			
% of pre-perfusion			
platelet level	92–100%	70–80%	90–100%
References	Chang *et al.*	Winchester *et al.*	Martin *et al.*
	Odaka *et al.*		Oulès *et al.*

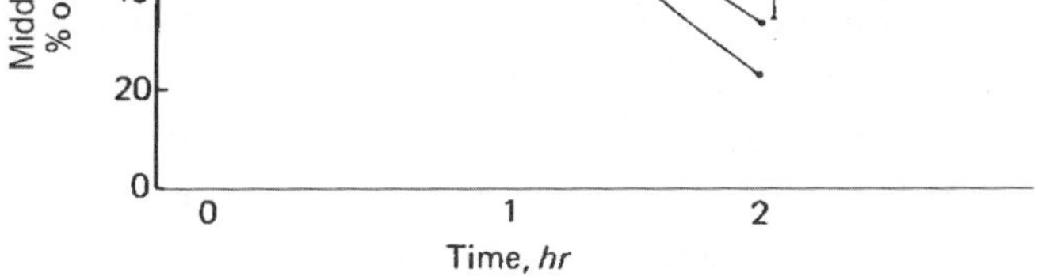

Fig. 1.16. Clinical studies in uremic patients show that middle molecules can be lowered rapidly after two hours of ACAC hemoperfusion. On the other hand, hemodialysis machines available at that time were not effective.

Table 1.9. Effects of Hemoperfusion in Uremic Patients

Authors	Clinical Improvements in the Following Symptoms
Chang *et al.*, 1971–1974	Pruritis, well-being, nausea, vomiting
Chang *et al.*, 1975	Peripheral neuropathy
Odaka *et al.*, 1978	Pericarditis, symptom free
Chang *et al.*, 1979	Hematocrit, less fluid retention
Martin *et al.*, 1979	Pericarditis
Inou *et al.*, 1979	Peripheral neuropathy
Odaka *et al.*, 1980	Pericarditis
Agishi *et al.*, 1980	Peripheral neuropathy
Stefoni *et al.*, 1980	Peripheral neuropathy, pruritis, pericarditis, well-being

Agishi *et al.*, 1980; Stefoni *et al.*, 1980, others in this book). These clinical trial results led many centers to use the combined hemoperfusion-hemodialysis, especially in those patients with resistant uremic symptoms under the standard hemodialysis program.

We designed a device, Dialaid D4000, combining hemoperfusion in series with a capillary membrane system (Fig. 1.17, Table 1.10) to carry out both functions (Chang *et al.*, 1982a and 1982b).

Since then, hemodialysis membranes have improved so that the present high flux membranes are better than the standard dialysis membrane in removing the larger middle molecules. Even then, the clearance is still much less than for the better hemoperfusion devices. Hemoperfusion devices are usually manufactured by manufacturers of hemodialysis machines and dialysis membranes. In those countries with strong hemodialysis companies, the hemoperfusion devices are extremely expensive. On the other hand, in those countries with no large dialysis industries, hemoperfusion devices are not expensive and therefore continue to be used for renal failure patients. More recent results are available in this book. In all countries, hemoperfusion continues to be commonly used for uremic patients with aluminium or iron overload. This is based on the earlier clinical demonstration of

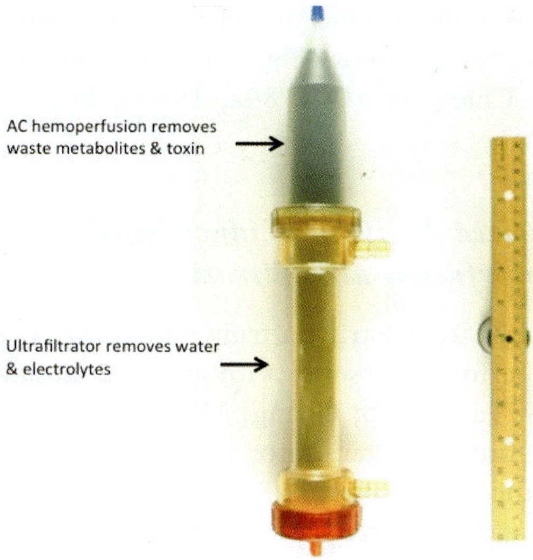

Fig. 1.17. Dialaid D4000 combining hemoperfusion (upper section) with a capillary membrane system for dialysis (lower section).

Table 1.10. Clinical Result

	Clearance (mL/min)	Pre CAK	2 1/2 HR CAK
Middle Molecules (peak 7)	165.2 ± 1735	40.8 ± 8.04	16.16 ± 7.18
Creatinine (μM/L)	194.0 ± 26.5	1519 ± 231	792 ± 192
Uric Acid (μM/L)	167.9 ± 22.2	589 ± 97	276 ± 66
Urea (μM/L)	142.5 ± 20.4	31.0 ± 11.9	16.9 ± 6.6
Phosphate (μM/L)	112.3 ± 34.6	1.7 ± 0.4	1.0 ± 0.3
Platelet ($\times 10^3$)		312 ± 54	265 ± 57 (n.s.)
Leucocytes ($\times 10^3$)		7.4 ± 1.0	7.9 ± 1.2 (n.s.)
Neutrophil (z)		82.3 ± 6.7	82.6 ± 7.9 (n.s.)
Lymphocytes (z)		13.9 ± 4.3	13.7 ± 6.6 (n.s.)
RBC ($\times 10^6$)		3.5 ± 0.7	4.0 ± 1.0 (n.s.)

the use of deferoxamine to bind these heavy metals and for hemoperfusion to remove the complex (Fig. 1.14) (Chang and Barre, 1983; Chang, 1986d; Chang *et al.*, 1984a, 1989a; Hakim *et al.*, 1985; Winchester 1996).

4.5. *Miniaturized Artificial Kidney-based on Hemoperfusion-ultrafiltration*

A third approach is to construct a truly miniaturized artificial kidney based on hemoperfusion in series with an small ultrafiltrator (Chang *et al.*, 1975, 1979a, 1977b; Chang, 1976c) (Fig. 1.18). In clinical studies, a small Amicon ultratiltrator was used (Fig. 18). Hydrostatic pressure from the blood pump alone gave effective ultrafiltration. Dialysis fluid is not required; the ultrafiltrate flows directly into a measuring cylinder.

This ultrafiltrator in series with ACAC hemoperfusion was studied in two patients, one for three and the other for six months (Chang *et al.*, 1979). The patient treated for six months has a creatinine clearance of 0.2 mL/min and a urine volume of 50 mL/day. He was followed for a six-month control period of thrice weekly hemodialysis treatments. This was followed by a six-month test period when a 2.5 hr hemoperfusion-ultrafiltration replaced one of the three weekly 6 hr hemodialysis treatments (Table 1.11, Fig. 1.19).

The patient felt well throughout. Hematocrit increased slightly from 25.15% ± 3.61 to 26.6% ± 1.73. Fluid retention diminished as body weight fell from 82.6 ± 1.2 to the optimal dry weight of 78.0 ± 0.8 kg and hypertension improved. More fluid could be removed without side effects in 2.5 hr of hemoperfusion-ultrafiltration than that by 6.0 hr of hemodialysis. Middle molecule clearance in this patient was 113 mL/min at a blood flow rate (QB) of 200 and 160 mL/min at QB of 300. This was significantly higher than for hemodialysis. Nerve conduction velocity did not change. Hemoperfusion-ultrafiltration is more effective than hemodialysis in removing middle molecules, creatinine, uric acid, sodium chloride and water (Table 1.11). On the other hand, it does not remove

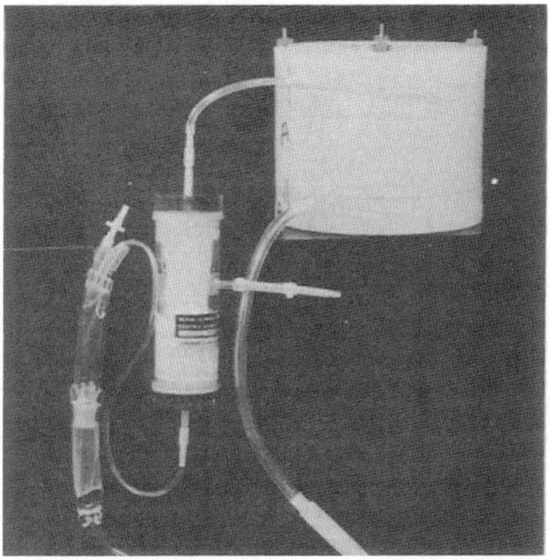

Fig. 1.18. ACAC hemoperfusion in series with a small ultrafiltrator.
Promising clinical results led us to modify the D4000 so that it has a ultrafiltrator in series with the hemoperfusion.

Table 1.11. Patient on Hemoperfusion-Ultrafiltraion (Creatinine Clearance: 0.2 mL/min; Urine Output: 50 mL/24 hr)

Clearance (mL/min) (at QB 300 mL/min)	Hemoperfusion	Hemoperfusion-Ultrafiltration
Creatinine	230	235
Uric acid	235	235
300–1500 MW	120	134
PTH free	73	81
Fluid removal/2hr	0	2,500–2,700 mL
NaCl removal/2hr	0	17.0–18.4 g

sufficient potassium and urea since after six months on this regime, there was an increase in mean blood urea nitrogen from 86 ± 14 to 105 ± 18 mg/dL and potassium from 5.2 ± 0.57 to 5.8 ± 0.08 mEq/L. Oral adsorbents can remove potassium and phosphate.

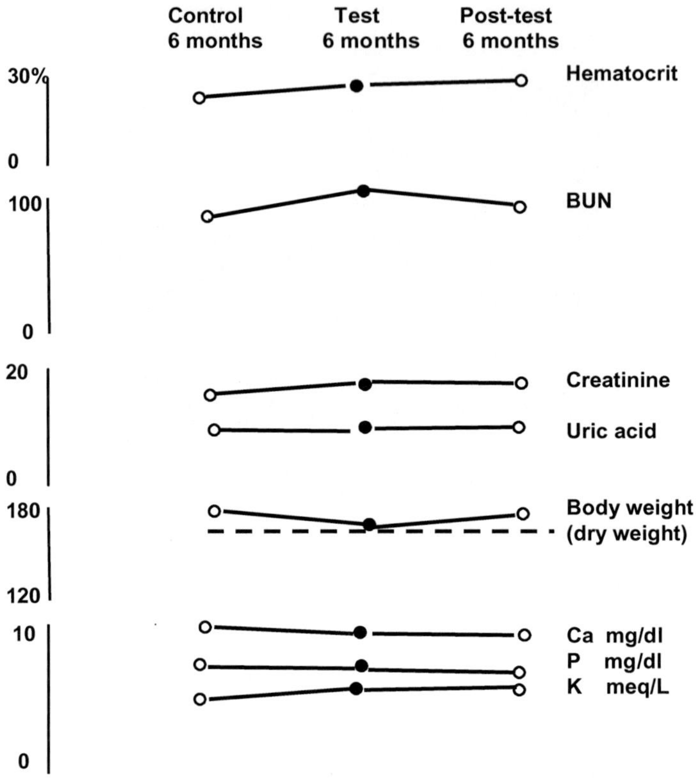

Fig. 1.19. Laboratory data of patient. A six-month control period followed by a six-month test period. During the test period, one of the weekly six-hour hemodialysis treatments was replaced by 2.5 hours of treatment using the miniaturized hemoperfusion-ultrafiltration device.

4.6. *Urea Removal Systems*

A urea removal system is being developed to complete this miniaturized artificial kidney system. This is based on the finding that artificial cells containing the enzyme urease can convert urea into ammonia that is then removed by ammonia adsorbent (Chang, 1966). This was developed and tested in patients (Kjellstrand *et al.*, 1981). However, too much ammonium adsorbent was needed. In order to solve this problem, artificial cells containing multi-enzyme systems were studied (Gu and Chang, 1988a, 1988b, 1990a, 1990b, 1990c, 1991).

Another approach is the use of artificial cells containing urea-removing microorganisms (Prakash and Chang, 1996, 1998, 1999, 2000a, 2000b) including our use of artificial cells containing modified *lactobacillus* (Chow, Liu, Prakash, Chang, 2003). *Lactobacillus* is commonly used in yogurt for human consumption. Professor Friedman's group has extended this and tested clinically in patients with promising results (Friedman, 2011).

5. Hemoperfusion in Liver Failure

5.1. *First Observation of Recovery of Consciousness in Hepatic Coma*

As shown in acute poisoning patients in the above section, adsorbent artificial cell hemoperfusion is effective in removing suicidal and accidental overdoses of medications in patients. It is also effective in removing other toxic substances from the circulating blood. Its effectiveness in removing hepatic toxins from the circulating blood was first shown in a grade IV hepatic coma patient with recovery of consciousness within an hour after the initiation of hemoperfusion (Chang, 1972b).

A 50-year-old woman was admitted with a history of alcoholic abuse, onset of jaundice, fatigue, nausea, vomiting, and dark urine. There was no history of contact with hepatitis or of intravenous or intramuscular medication. On admission, she had spider nevi and ascites. The diagnosis was acute alcoholic hepatitis. Her condition deteriorated after admission and she became comatose and unresponsive. After she remained comatose for two days, her condition was considered as terminal, and with the insistence of her relatives she was referred by her physician to me for possible hemoperfusion since nothing else could be done. One hour after hemoperfusion her consciousness started to improve, and she began to recognize her relative and answer questions in sentences. Hemoperfusion was carried out for a total of 80 minutes. She remained conscious for about an hour after the end of the hemoperfusion, but lapsed into coma again.

Three days later she was still comatose, and a second hemoperfusion was initiated. At the start of perfusion she was comatose, an hour later she looked at people when spoken to and there were increases in voluntary movements and response to pain. She did not, however, recover full consciousness as in the first hemoperfusion. Shortly after this a liver biopsy specimen showed cirrhosis, acute hepatitis, with small foci of regeneration. Thereafter consciousness fluctuated between stupor and coma. A third hemoperfusion was carried out. Before this hemoperfusion, E.E.G. background activity was irregular and diffuse, with continuous theta and delta activity and occasional high-amplitude single slow-wave and rare complex biphasic or triphasic configurations. At this time the patient was semi-comatose and did not respond to questions. An hour and a half after hemoperfusion the patient's consciousness improved and she started to respond and answer questions; she also complained of thirst and heaviness in the leg. E.E.G. recording after hemoperfusion showed a minimum improvement in the background activity.

Biochemical evaluation of the treatment of hepatic coma is difficult, because the precise pathological mechanism is not known. On the other band, the three hemoperfusions each produced a clinical response in consciousness. In one case when the E.E.G. was recorded there was a slight improvement in background activity. Since there was no change in the blood-ammonia level after hemoperfusion, we cannot attribute the improvement in consciousness to the removal of blood-ammonia. However, whereas standard hemodialysers at that time were not efficient in removing large molecules from blood, hemoperfusion is very efficient in removing molecules up to 5,000 molecular weight (Chang, 1972e, Chang and Migchelsen, 1973; Chang and Lister, 1980, 1981) and also protein-bound molecules. Hemoperfusion is also effective in removing other chemicals related to hepatic coma.

5.2. *Results around the World on Effect of Hemoperfusion on Hepatic Coma*

This led to further studies around the world in a large number of patients (Gazzard *et al.*, 1974; Chang, 1976c, 1982a; Blume *et al.*,

1976; Bartels *et al.*, 1977, 1981; Silk and Williams, 1978; Gelfand *et al.*, 1978, Gimson *et al.*, 1978; Odaka *et al.*, 1978; Amano *et al.*, 1978; Maeda *et al.*, 1980; Cordoatri *et al.*, 1982; William, 1983; O'Grady *et al.*, 1988, others in this book). These results support the finding that a significant number of hepatic coma patients recovered consciousness. However, this did not translated into improved survival rate. Furthermore, these clinical results are not conclusive since the pathogenesis of hepatic coma and survival rate is complicated. In addition to hepatic toxins other factors like brain edema, standard of care also play important roles (William, 1983; O'Grady *et al.*, 1988). Furthermore, survival rate is related to age, etiology, grades of coma, and other complications, especially brain edema. In addition, hemoperfusion can only remove toxins from the body, but the liver is a very complex organ with many other metabolic and synthetic functions. We should not expect hemoperfusion alone to be a complete artificial liver. Hemoperfusion will more likely be an important component of a more complex system (Chang, 1983a, 1983e, 1986c, 1992b).

5.3. *Control Studies in Galactosamine-induced Hepatic Failure Rats*

We therefore carried out detailed control studies in galactosamine-induced hepatic failure rats to allow for a more valid statistical analysis (Chang *et al.*, 1978; Chang, 1971a, 1978b; Chirito *et al.*, 1978, 1979; Mohsini *et al.*, 1980; Tabata and Chang, 1980). With this model we can avoid variations in age, etiology and grades of coma. However, this model may not be the counterpart of fulminant hepatic failure in human. In addition the survival rate in this model of grade III coma is more like grade IV coma in human. This research included detailed studies in regard to the time of initiation of treatment; the need for supplements of essential factors; comparison of hemoperfusion with exchange transfusion, liver perfusion, and hormones; and other approaches. Male Wistar rats weighing 275–309 g and 47–67 days old were used. Forty-eight hr after galactosamine injection (1.1 g/kg), those rats in Grade III coma were used in this study. Grade III coma rats were those which "sleep most of the time, but arousable." Of the 356 rats, 122 reached Grade III coma 48 h after injection with galactosamine. To avoid variations between batches, all the animals in Grade III coma in each batch were randomly divided into equal numbers of test and control animals. Similar studies were also carried out in rats in grade II coma. The results of all the studies (Chang *et al.*, 1978; Mohsini *et al.*, 1980; Tabata and Chang, 1980) are very briefly summarized in Table 1.12.

In grade II coma rats, either one hemoperfusion or one liver perfusion resulted in significant increase in survival time and survival rates when compared to the control group (Table 1.12). However, cross circulation did not have any significant effects.

In grade III coma rats, one of hemoperfusion, liver perfusion or blood exchange transfusion did not result in any significant increase in survival time or survival rates. On the other hand, combined use of one hemoperfusion with one blood exchange transfusion resulted in significant increase in survival time but not survival rate. Two hemoperfusions also resulted in significant increase in survival time but not survival rate (Table 1.12). Another approach was carried out with two hemoperfusions after one blood exchange, with the second hemoperfusion 4 h after the first. When this regime was used, both the survival time and the survival rate of Grade III coma rats were significantly increased ($p < 0.01$ and $p < 0.05$, respectively) (Table 1.12).

Table 1.12. Survival Time and Rates and Grades of Coma in Rats Treated by Hemoperfusion (ACAC) Alone or in Combination

Coma	Treatment	Survival Time	Survival Rate
II	ACAC hemoperfusion	↑	↑
II	Liver perfusion	↑	↑
II	Cross-circulation	NS	NS
III	ACAC hemoperfusion		NS
III	Liver perfusion	NS	NS
III	Blood exchange	NS	NS
III	Blood exchange + ACAC	↑	NS
III	ACAC + glucagon + insulin	NS	NS
III	2 × ACAC + Blood exchange	↑	↑

Our research in fulminant hepatic rat models shows that hemoperfusion started in the earlier grades of coma can result in significant increase in both survival time and survival rates (Chang *et al.*, 1978; Chirito *et al.*, 1979). This has been supported by the result of another group that treated patients in Grade III coma rather than in Grade IV coma, as shown in Table 1.12 above (Gimson *et al.*, 1978). Another group also used galactosamine rats for hemoperfusion (Niu *et al.*, 1978). Their results corroborated our findings.

Our laboratory research also shows that hemoperfusion alone increases the survival time but not the survival rates in Grade III hepatic coma rats. This may correspond to the transient recovery of consciousness in hepatic coma patients. Fresh blood exchange transfusion alone did not improve the survival time or survival rate of FHF Grade III coma rats. Two ACAC hemoperfusion combined with fresh blood exchange transfusion significantly improved the survival time of FHF Grade III coma rats, but not the survival rate. An additional hemoperfusion after the first hemoperfusion combined with blood exchange transfusion improved both the survival time and survival rate in Grade III coma rats. This seems to suggest that hemoperfusion could be one part of a more complete liver support system. This book contains chapters on more recent clinical results of its use in combination with dialysis or hemofiltration. Studies are also ongoing in using hemoperfusion to remove hepatic toxins to allow the liver to regenerate. One line of research is to combine this with the use of hepatocytes perfusion system (Rozga *et al.*, 1993; Sussman *et al.*, 1994). Another approach is to combine this with artificial cells containing hepatocytes or stem cells (Chang, 2007a, 2007b, 2007c, 2007d, 2010, 2014; Liu and Chang, 2005, 2006a, 2006b, 2009, 2012).

6. Selective Adsorbents and Immunosorbents

6.1. *Surface Modification in Immunoadsorption*

Albumin can bind tightly to the ultrathin collodion membrane of adsorbent artificial cells (Fig. 1.20) (Chang 1969a). This is initially used to increase the blood compatibility of the adsorbent artificial cells for hemoperfusion (Chang, 1969a). This albumin coating has

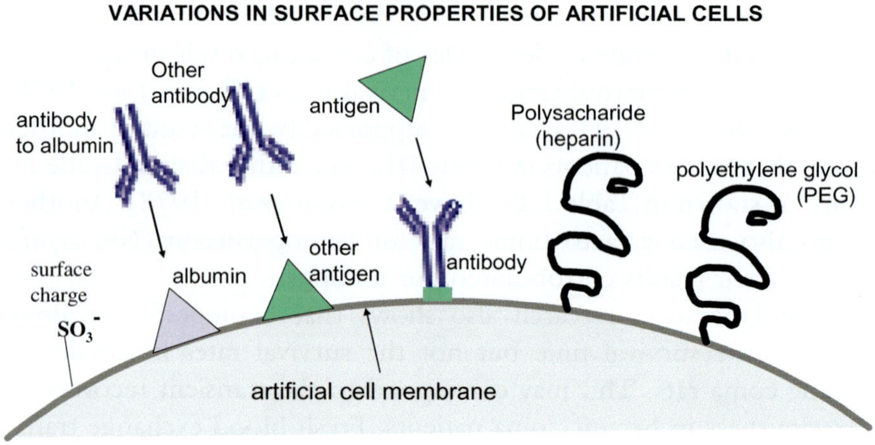

Fig. 1.20. Surface properties of artificial cell membrane can be varied by (1) incorporation of negative or positive charge; (2) incorporation of albumin to increase blood compatibility; (3) incorporation of antigens to bind antibodies or antibodies to bind antigen; (4) incorporation of polysaccharide like heparin or polyethylene glycol (PEG) to increase compatibility or retention time in circulation.

also been applied to synthetic immunosorbents resulting in blood compatible synthetic blood group immunosorbents (Chang, 1980d). In addition, Terman *et al.* (1977) show in animal studies that albumin-coated collodion-activated charcoal (ACAC) can remove antibodies to albumin. This has become a basis of one line of his research in which other types of antigens or antibodies are applied to the collodion coating of the artificial cells to form immunosorbents (Fig. 1.20). Other immunosorbents based on this principle have also been developed for the treatment of human systemic lupus erythematosus, removal of antiHLA antibodies in transplant candidates, and treatment of familial hypercholesterolemia with monoclonal antibodies to low-density lipoproteins (Terman, 1980; Terman *et al.*, 1979a, 1979b; Hakim *et al.*, 1990; Wingard *et al.*, 1991; Yang *et al.*, 2004). Antibodies have also been incorporated to the surface of lipid liposomes to allow for drug targeting to cells bearing the corresponding antigen (Torchilin, 2005). Much more recent details are available in this book by Professor Yu's group and many other centers.

Mucopolysaccharides are important components of the surface of cell membranes. Incorporating a polyscharide, heparin, to the artificial cell membrane increases the biocompatibility and circulation time of artificial cells (Chang *et al.*, 1967, 1968). Incorporation of a synthetic polymer, polyethylene glycol (PEG), into the membrane of artificial cells, nanocapsules and lipid vesicles is even more effective for increasing biocompatibility and circulation time (Chang *et al.*, 2003; LaVan *et al.*, 2002; Torchilin, 2005). More recent extensions of this are described in detail in this book.

Albumin can bind tightly to the ultrathin collodion membrane of adsorbent artificial cells (Chang, 1969a). This was initially used to increase the blood compatibility of the adsorbent artificial cells for hemoperfusion (Chang, 1969a). In addition, albumin-coated collodion-activated charcoal (ACAC) was found to effectively remove antibodies to albumin in animal studies (Terman *et al.*, 1977). This has become the basis of one line of research in which other types of antigens or antibodies are applied to the collodion coating of the activated charcoal to form immunosorbents. Other immonosorbent based on this principle have also been developed for the treatment of human systemic lupus erythematosus, removal of antiHLA antibodies in transplant candidates, treatment of familial hypercholesterolemia with monoclonal antibodies to low-density lipoproteins and other conditions (Terman, 1980; Terman *et al.*, 1979a, 1979b; Hakim *et al.*, 1990; Wingard *et al.*, 1991; Yang *et al.*, 2004). Immunosorbents are now a very important and extensive area, and a number of chapters are devoted to them in this book.

6.2. *Synthetic Immunosorbents for Blood Group*

We have carried out a study on surface modification of synthetic immunosorbents to prevent particulate release (Table 1.13) and blood incompatibility. We used the albumin-collodion coating to synthetic immunosorbents resulting in blood compatible synthetic blood

Table 1.13. Particles Released from Uncoated and Coated Immunosorbent

Test System	Particles Released (mean + SD)
(1) Control (saline)	33 ± 7.6 (n=5)
(2) Synthetic B-immunosorbent	307 ± 12 (n=5)
(3) Albumin-Collodion coated synthetic B-immunosorbent	43 ± 8.5 (n=5)

Table 1.14. Synthetic Immunosorbent: Remove Blood Group Antibody from Plasma

	Group A	Group B
B antibodies (control)	None	Yes
B antibodies (synthetic B-immunosorbent)	None	None
B antibodies (albumin-collodion coated Synthetic B-immunosorbent)	None	None
A antibodies (control)	Yes	None
A antibodies (synthetic B-immunosorbent)	Yes	None
A antibodies (albumin-collodion coated Synthetic B-immunosorbent)	Yes	None

group immunosorbents (Chang, 1980d). The surface modified synthetic immunorsobent continues to effectively remove plasma antibody to blood groups (Table 1.14, Fig. 1.21) (Chang, 1980d). This albumin-coated synthetic adsorbent has been applied clinically for removing blood group antibodies from plasma for bone marrow transplantation (Bensinger *et al.*, 1981).

6.3. *Selective Synthetic Adsorbent for Endotoxin*

Synthetic adsorbent for the selective removal of endotoxin is also possible (Tani *et al.*, 1992). This will be discussed in detail by Professor Tani and other centers in this book.

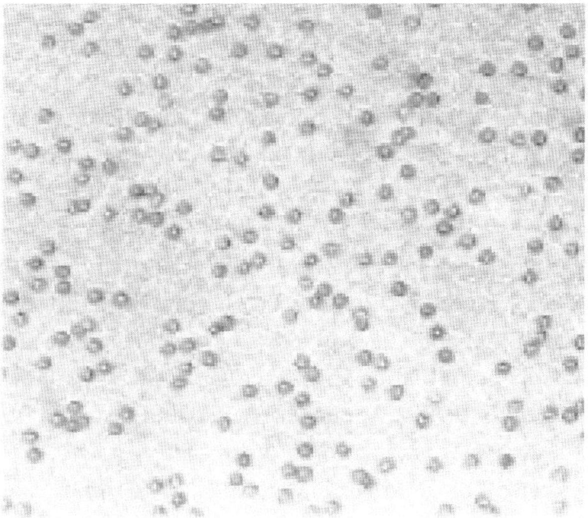

Fig. 1.21. Plasma from group A blood treated with albumin-collodion synthetic B-immunosorbent added to Group B erythrocytes. No longer caused agglutination of group B erythrocytes.

7. APPENDIX

7.1. *Detailed Procedures for the Laboratory Preparation of the ACAC Hemoperfusion Device*

(Chang, 2007a)

7.2. *Procedure for Hemoperfusion*

(Chang 2007a)

8. Useful Websites

http://www.medicine.mcgill.ca/artcell
http://www.artificialcell.info
http://www.artificialcell.org

9. References

Agishi T, Yamashita N, Ota K. (1980) Clinical results of direct charcoal hemoperfusion for endogenous and exogenous intoxication. In: Sideman and Chang (eds.), *Hemoperfusion: Kidney and Liver Support and Detoxification*, Part I. Hemisphere, Washington, D. C., pp. 255–263.

Amano I, Kano H, Takahira H, *et al.* (1978) Hepatic assist system using bead-type charcoal. In: Chang TMS (ed.) *Artificial Kidney, Artificial Liver and Artificial Cells.* Plenum Press, New York, pp. 89–98.

Barre P, Gonda A, Chang TMS. (1986) Routine clinical applications of hemoperfusion-hemodialysis in uremia. *Int J Artif Organs* **9**: 305–308.

Barre P, Chang TMS. (1984) Application of coated charcoal hemoperfusion to clinical practice. *Renal Family J* **6**: 20–25.

Bartels O, Neidhardt B, Neidhardt M, *et al.* (1977) Untersuchungen und Entahrungen mit der Kohlehaemoperfusion bei Leberkoma (Investigation and experience with carbon hemoperfusion for liver coma). In: Demling L, Bartels O (eds.), *Entgiftung mit Hamoperfusion. (Detoxification by Hemoperfusion)*, Freiburg, Bundemagel-Verlag, p. 110.

Better OS, Brunner G, Chang TMS, *et al.* (1979) Controlled trials of hemoperfusion for intoxication. *Ann Int Med* **91(6):** 925.

Biberstein MP, Ward DM, Ziegler MG. (1983) Use of BetaBlockade and Hemoperfusion for Acute Theophylline poisoning. *Western J Med* **141:** 485–490.

Blume V, Helmsteadt D, Sybrecht G, *et al.* (1976a) Haperfusions therpie des akuten Leberversagens. *Dtsch Med Wochenschr* **14:** 559.

Bonomini V, Chang TMS (eds.). (1982) *Hemoperfusion.* Contributions to Nephrology Series. S. Karger AG, Basel, Switzerland, 149 pp.

Chang TMS. (1957) Hemoglobin Corpuscles. Report of a research project for Honours Physiology, Medical Library, McGill University. Also reprinted 1988 as part of "30th anniversary in Artificial Red Blood Cells Research." *J Biomaterials, Artificial Cells & Artificial Organs* **16:** 1–9.

Chang TMS. (1964) Semipermeable microcapsules. *Science* **146(3643):** 524–525.

Chang TMS. (1965) Semipermeable aqueous microcapsules. *PhD Thesis*, McGill University.

Chang TMS, Macintosh FC, Mason SG. (1966) Semipermeable aqueous microcapsules: I. Preparation and properties. *Can J Physiol Pharmacol* **44:** 115–128.

Chang TMS. (1966) Semipermeable aqueous microcapsules ("artificial cells"): with emphasis on experiments in an extracorporeal shunt system. *Trans Am Soc Artif Intern Organs* **12:** 13–19.

Chang TMS, Johnson LJ, Ransome OJ. (1967) Semipermeable aqueous microcapsules: IV. Nonthrombogenic microcapsules with heparin-complexed membranes. *Can J Physiol Pharmacol* **45:** 705–715.

Chang TMS, Poznansky MJ. (1968a) Semipermeable aqueous microcapsules (artificial cells): V. Permeability characteristics. *Biomed Mater Res* **2:** 187–199.

Chang TMS, Poznansky MJ. (1968b) Semipermeable microcapsules containing catalase for enzyme replacement in acatalsaemic mice. *Nature* **218(5138):** 242–45.

Chang TMS, Pont A, Johnson LJ, Malave N. (1968) Response to intermittent extracorporeal perfusion through shunts containing semipermeable microcapsules. *Trans Am Soc Artif Intern Organs* **14:** 163–168.

Chang TMS. (1969a) Removal of endogenous and exogenous toxins by a microencapsulated absorbent. *Can J Physiol Pharmacol* **47(12):** 1043–5.

Chang TMS, Malave N. (1970) The development and first clinical use of semipermeable microcapsules (artificial cells) as a compact artificial kidney. *Trans Am Soc Artif Intern Organs* **16:** 141–148.

Chang TMS. (1970) "Nonthrombogenic Microcapsules." *U.S. Patent,* 3, 522, 346.

Chang TMS. (1971c) Nonthrombogenic Microcapsules." *Canadian Patent,* 876, 100.

Chang TMS. (1971d.) A new concept in artificial kidneys using semipermeable microcapsules. *Chemeca '70*, pp. 48–54, Butterworths, Australia.

Chang TMS, Gonda A, Dirks JH, Malave N. (1971a) Clinical evaluation of chronic, intermittent, and short term hemoperfusions in patients with chronic renal failure using semipermeable microcapsules (artificial cells) formed from membrane-coated activated charcoal. *Trans Am Soc Artif Intern Organs* **17:** 246–252.

Chang TMS, MacIntosh FC, Mason SG. (1971b) "Encapsulated hydrophilic compositions and Methods of Making them." *Canadian Patent*, 873, 815.

Chang TMS. (1972a) Artificial Cells, monograph, 212 pp., 77 illustrations, five tables, Charles C. Thomas, Publisher, Springfield, Illinois, USA. (Out of print but available for free online viewing at www.medicine.mcgill.ca/artcell) Book reviewed by: *Nature* **242:** 211, 1973; *Scientific American*, Vol 227, November 1972; *Mayo Clinic Proceedings*, Vol 47, November 1972; *Chemical Engineering Journal*, May 14, 1973; *Biological Abstracts* **55(6)** March 15, 1973; *Biomedical Engineering Journal*, April 1973; *Quarterly Journal of Experimental Physiology*; Nephron, Vol 10, No 23, 1973; *Annuals, Royal College of Surgeons of England*, July 1974; *J Biomat Med Dev Artif Organs* **1:** 427, 1973; *New Zealand J Med Lab Tech, 26*, November 1972. Book translated into: Japanese (1976) by Prof. Kondo T, Japan; Russian (1979) by Prof. Nicolaev V, USSR.

Chang TMS. (1972b) Haemoperfusions over microencapsulated adsorbent in a patient with hepatic coma**.** *Lancet* **ii:** 1371–1372.

Chang TMS. (1972c) A new approach to separation using semipermeable microcapsules (artificial cells): Combined dialysis, catalysis, and absorption. In: Li N (ed.), *Recent Developments in Separation Science.* Chemical Rubber Co. Press, Cleveland, OH, USA, pp. 203–216.

Chang TMS. (1972e) Microcapsule artificial kidney and medium molecular weight clearance. *Proc Eur Dia Transplant Assoc* **9:** 568–574.

Chang TMS. Gonda A, Dirks JH, *et al.* (1972g) ACAC microcapsule artificial kidney for the long term and short term management of eleven patients with chronic renal failure. *Trans Am Soc Artif Intern Organs* **18:** 465–472.

Chang TMS, Coffey JF, Barre P, *et al.* (1973a) Microcapsule artificial kidney: treatment of patients with acute drug intoxication. *Can Med Assoc J* **108:** 429–433.

Chang TMS, Coffey JF, Lister C, *et al.* (1973b) Methaqualone, methyprylon, and glutethimide clearance by the ACAC microcapsule artificial kidney: *In vitro* and in patients with acute intoxication. *Trans Am Soc Artif Intern Organs* **19:** 87–91.

Chang TMS. (1973a) "Blood compatible microcapsules containing detoxicants." U.S. Patent, 3, 725, 113.

Chang TMS. (1973b) The use of semipermeable microcapsules (artificial cells) as microsystems for separation. *Am Chem Soc* **33:** 581–583.

Chang TMS. (1973e) Recent trends in biocompatible materials and microcapsular absorbents for kidney function replacement. In: Kenedi RM (ed.), *Perspectives in Biomedical Engineering.* University Park Press (MacMillan), Baltimore, MD, pp. 39–44.

Chang TMS, Migchelsen M. (1973) Characterization of possible "toxic" metabolites in uremia and hepatic coma based on the clearance spectrum for larger molecules by the ACAC microcapsule artificial kidney. *Trans Am Soc Artif Intern Organs* **19:** 314–319.

Chang TMS, Migchelsen M, Coffey JF, Stark A. (1974) Serum middle molecule levels in uremia during long term intermittent hemoperfusions with the ACAC (coated charcoal) microcapsule artificial kidney. *Trans Am Soc Artif Intern Organs* **20:** 364–371.

Chang TMS. (1974b) Platelet-surface interaction: Effect of albumin coating or heparin complexing on thrombogenic surfaces. *Can J Physiol Pharmacol* **52(2):** 275–285.

Chang TMS. (1974c) A comparison of semipermeable microcapsules and standard dialysers for use in separation. *J Separation & Purification Methods* **3(2):** 245–262.

Chang TMS. (1974g) Future prospectives: pediatric hemodialysis based on artificial cells. *Clin Proc Children's Hospital National Medical Center, Washington, DC* **30(10):** 304–311.

Chang TMS. (1974h) Performance characteristics of the microcapsule artificial kidney. In: Whelpton D (ed.) *Renal Dialysis.* Sector Publisher, London, pp. 135–147.

Chang TMS. (1975a) Rationale for the use of the ACAC microcapsule artificial kidney for the treatment of patients with chronic renal failure. *L'Ateneo Parmense, Acta Bio-Medica* **46:** 35368.

Chang TMS. (1975d) Artificial kidney and artificial liver. *Lancet* **iv:** 1451–1452.

Chang TMS. (1975e) Experience with treatment of acute liver failure patients by hemoperfusion over biocompatible microencapsulated (coated) charcoal. In: Williams R, Murray-Lyon M (eds.), *Artificial Liver Support.* London, Pitman Medical Publishers, p. 229.

Chang TMS. (1975g) Microencapsulated adsorbent hemoperfusion for uremia, intoxication and hepatic failure. *Kidney Int* **7:** S387–S392.

Chang TMS. (1975i) A microcapsule artificial kidney: a step toward miniaturization. *J Dial Transplant* **4**: 23–25.

Chang TMS. (1975j) Experience with the treatment of acute liver failure patients by haemoperfusion over biocompatible microencapsulated (coated) charcoal. In: Williams R (ed.), *Artificial Support Systems for Acute Hepatic Failure*. Whitefriars Press, Ltd., London, UK, pp. 229–233.

Chang TMS, Chirito E, Barre P, *et al.* (1975) Clinical performance characteristics of a new combined system for simultaneous hemoperfusion hemodialysis-ultrafiltration in series. *Trans Am Soc Artif Intern Organs* **21:** 502–508.

Chang TMS. (1976c) Hemoperfusion alone and in series with ultrafiltration or dialysis for uremia, poisoning and liver failure. *Kidney Int* **10:** S305–S311.

Chang TMS. (1976g) Microcapsule artificial kidney: including updated preparative procedures and properties. *Kidney Int* **10:** S218–S224.

Chang TMS. (1976h) "Blood compatible microcapsules containing detoxicants." Canadian Patent, 982, 941.

Chang TMS. (1976) Hemoperfusion alone and in series with ultrafiltration or dialysis for uremia, poisoning and liver failure. *Kidney Int* **10(Suppl 7):** S305.

Chang TMS. (1977b) Protective effects of microencapsulation (coating) on platelet depletion and particulate embolism in the clinical applications of charcoal haemoperfusion. In: Kenedi RM, Courtney JM, Gaylor JDS, Gilchrist T (eds.), *Artificial Organs*. MacMillan Press Ltd., London, UK, pp. 164–177.

Chang TMS, Chirito E, Barre P, *et al.* (1977b) Clinical evaluation of the clearance profiles of a portable, compact, dialysate-free system incorporating microencapsulated charcoal hemoperfusion for blood purification with ultrafiltration for fluid removal. *J Dialysis* **1(3):** 239–259.

Chang TMS. (1977f) Criteria, evaluation, and perspectives of various microencapsulated charcoal hemoperfusion systems. *J Dial Transplant* **6:** 50–53.

Chang TMS. (1977j) Novel approaches in artificial kidney, artificial liver, and detoxification. *J Dialysis & Transplant* **8:** 20–26.

Chang TMS. (1978a) A 1978 perspective of hemoperfusion. *Artif Organs* **2(4):** 359–362.

Chang TMS. (1978b) Artificial cells and charcoal hemoperfusion. *Int J Artif Organs* **1:** 111–112.

Chang TMS. (1978c) Artificial cells for artificial kidney, artificial liver and detoxification. In: Chang TMS (ed.), *Artificial Kidney, Artificial Liver and Artificial Cells.* Plenum Press, New York, pp. 57–77.

Chang TMS. (1978e) Future of haemodialysis. In: Davidson A (ed.), *Dialysis '77.* Pitman Medical Publisher, London, UK, pp. 269–277.

Chang TMS. (1978f) Hemoperfusion in chronic schizophrenia. *Int J Artif Organs* **1(6):** 253–255.

Chang TMS. (1978g) Introductory chapter. In: Chang TMS (ed.), *Artificial Kidney, Artificial Liver and Artificial Cells.* Plenum Press, New York, pp. 3–8.

Chang TMS. (1978h) Microcapsule artificial kidney in replacement of renal function: with emphasis on adsorbent hemoperfusion. In: Drukker W, Parsons FM, Maher JF (eds.), *Replacement of Renal Function by Dialysis.* Martinus Nijhoff Publishers, The Hague, pp. 217–231.

Chang TMS. (1978i) Microcapsule artificial kidney. *Proc Canadian Clinical Engineering Conference,* National Research Council, Ottawa, Canada, pp. 134–137.

Chang TMS (ed.). (1978l) *Artificial Kidney, Artificial Liver and Artificial Cells.* Plenum Publishing Corporation, New York, 315 p.

Chang TMS. (1978m) Adsorbent hemoperfusion — general discussion. S (ed.), *Artificial Kidney, Artificial Liver and Artificial Cells.* Plenum Press, New York, pp. 301–304.

Chang TMS, Lister C, Chirito E, *et al.* (1978) Effects of hemoperfusion rate and time of initiation of ACAC charcoal hemoperfusion on the survival of fulminant hepatic failure rats. *Trans Am Soc Artif Intern Organs* **24:** 243–245.

Chang TMS, Malouf C. (1978) Artificial cells microencapsulated multienzyme system for converting urea and ammonia to amino acid using alpha-ketoglutarate and glucose as substrate. *Trans Am Soc Artif Intern Organs* **24:** 18–20.

Chang TMS, Chirito E, Barre P, *et al.* (1979a) Long-term clinical assessment of combined ACAC hemoperfusion-ultrafiltration in uremia. *Artif Organs* **3(2):** 127–131.

Chang TMS, Malouf C, Resurreccion E. (1979b) Artificial cells containing multienzyme systems for the sequential conversion of urea into ammonia, glutamate, then alanine. *Artif Organs* **3(supp):** 284–287.

Chang TMS, Malouf C. (1979) Effects of glucose dehydrogenase in converting urea and ammonia into amino acid using artificial cells. *Artif Organs* **3(1):** 38–41.

Chang TMS. (1979c) Assessments of clinical trials of charcoal hemoperfusion in uremic patients. *Clin Nephrol* **11(2):** 111–119.

Chang TMS. (1979d) Hemoperfusion: clearance, blood compatibility and safety. *Int J Artif Organs* **2(6):** 276–277.

Chang TMS. (1980d) Blood compatible coating of synthetic immunoadsorbents. *Trans Am Soc Artif Intern Organs* **26:** 546–549.

Chang TMS. (1980e) Clinical experience with ACAC coated charcoal hemoperfusion in acute intoxication. *Clin Toxicol* **17:** 529–542.

Chang TMS. (1980h) Present status and future prospectives of artificial cells in hemoperfusion. In: Sideman S, Chang TMS (eds.), *Hemoperfusion: Artificial Kidney and Liver Support and Detoxification*. Hemisphere Publishing Corp., Washington, D.C., pp. 93–104.

Chang TMS. (1980k) Microencapsulation of specific sorbents for medical applications. In: Giordano C (ed.), *Sorbents and Their Clinical Applications*. Academic Press, Inc., New York, pp. 195–217.

Chang TMS. (1980m) Hemoperfusion: Closing remarks. In: Sideman S, Chang TMS (eds.), *Hemoperfusion: Artificial Kidney and Liver Support and Detoxification*. Hemisphere Publishing Corp., Washington, D.C., pp. 403–405.

Chang TMS. (1978n) Adsorbent hemoperfusion — general discussion. In: Chang TMS (ed.), *Artificial Kidney, Artificial Liver and Artificial Cells*. Plenum Press, New York, pp. 301–304.

Chang TMS, Lister C, Chirito E, *et al.* (1978) Effects of hemoperfusion rate and time of initiation of ACAC charcoal hemoperfusion on the survival of fulminant hepatic failure rats. *Trans Am Soc Artif Intern Organs* **24:** 243–245.

Chang TMS, Espinosa-Melendez E, Francoeur TE, Eade NR. (1980a) Albumin-collodion activated coated charcoal hemoperfusion in the treatment of severe theophylline intoxication in a 3-year-old patient. *Pediatrics* **65(4):** 811–814.

Chang TMS, Lister C. (1980) Analysis of possible toxins in hepatic coma including the removal of mercaptan by albumin-collodion charcoal. *Int J Artif Organs* **3(2):** 108–112.

Chang TMS, Winchester J, Rosenbaum J, *et al.* (1980b) Adsorbent hemoperfusion for blood purification. *Trans Am Soc Artif Intern Organs* **26:** 593–597.

Chang TMS. (1981a) Lecture series on artificial cells. Monograph in Chinese of a 12 hour lecture series by Chang TMS at the Chinese Academy of Sciences, Beijing, China. Monograph prepared and translated into Chinese by Wang HZ, Zheng WH, Miao EH, *et al.*, Chinese Medical Association, Hopei, Shijiazhuang.

Chang TMS. (1981b) Hemoperfusion, exchange transfusion, cross circulation, liver perfusion, hormones and immobilized enzymes. In: Brunner G, Schmidt FW (eds.), *Artificial Liver Support.* Springer-Verlag, Berlin. Germany, pp. 126–133.

Chang TMS. (1981e) Controversies and issues in hemoperfusion. *Artif Organs* **5(supp):** 10–15.

Chang TMS. (1981f) Current status of sorbent microencapsulation. *Advances in Basic and Clinical Nephrology* Proc. 8th Int. Congress of Nephrology. S. Karger Publisher, Basel, Switzerland, pp. 400–406.

Chang TMS, Shu CD. (1981) Artificial cells for hemoperfusion. *Chinese J. Organs Transplant, Zhonghua Qiguan Yizhi Zazhi* **2:** 50–55.

Chang TMS, Lister C. (1981) Middle molecules in hepatic coma and uremia. *Artif Organs* **4(supp):** 169–172.

Chang TMS, Lacaille Y, Picart X, *et al.* (1981) Composite artificial kidney: a single unit combining hemodialysis and hemoperfusion. *Artif Organs* **5(supp):** 200–203.

Chang TMS. (1982a) Earlier hemoperfusion in fulminant failure. *Lancet* **2(8306):** 1039.

Chang TMS. (1982c) Blood compatible adsorbent hemoperfusion for extracorporeal blood treatment. In: Criban TCJ, Vissar J, Nivard RJF (eds.), *Theoretical Aspects, Industrial and Biomedical Applications of Affinity Chromatography and Related Techniques.* Elsevier Scientific Publishing Co., Amsterdam, Netherlands, pp. 357–364.

Chang TMS. (1982d) Hemoperfusion and artificial cells. *Contr Nephrol* **29:** 3–6.

Chang TMS. (1982e) Hemoperfusion in 1981. *Contr Nephrol* **29:** 11–22.

Chang TMS. (1982g) Plasma perfused over immobilized protein A for breast cancer. *New Engl J Med* **306:** 936.

Chang TMS, Barre P, Kuruvilla S, *et al.* (1982a) Hemoperfusion-hemodialysis in a single unit: composite artificial kidney. *J Artificial Support Systems* **1:** 63–67.

Chang TMS, Barre P, Kuruvilla S, *et al.* (1982b) Phase one clinical trial of a new composite artificial kidney: A single unit combining dialysis with hemoperfusion. *Trans Am Soc Artif Intern Organs* **28:** 43–48.

Chang TMS. (1983a) A multifaceted approach to artificial liver support. In: "Liver support/transplants and artificial organs," Panel Conference (Starzl TE, Chang TMS, Williams R, Nose Y). *Trans Am Soc Artif Intern Organs* **29:** 795–799.

Chang TMS. (1983d) Past, present and future of hemoperfusion. In: Piskin E, Chang TMS (eds.), *Past, Present and Future of Artificial Organs.* Meteksan Publ. Co., Ankara, Turkey, pp. 109–127.

Chang TMS. (1983e) Artificial liver support: requirements and approaches. *Int J Artif Organs* **6(4):** 178–182.

Chang TMS. (1983f) Past, present and future of artificial cells. In: Piskin E, Chang TMS (eds.), *Past, Present and Future of Artificial Organs.* Meteksan Publ. Co., Ankara, Turkey, pp. 10–24.

Chang TMS, Barre P. (1983) Effect of desferrioxamine on removal of aluminium and iron by coated charcoal haemoperfusion and haemodialysis. *Lancet* Nov: 1051–1053.

Chang TMS, Barre P, Kuruvilla S, *et al.* (1983) Laboratory and clinical assessments of hemoperfusion and hemodialysis in a single unit: Composite artificial kidney. In: Piskin E, Chang TMS (eds.), *Past, Present and Future of Artificial Organs.* Meteksan Publ. Co., Ankara, Turkey, pp. 196–200.

Chang TMS, Zheng WH. (1983) Hemoperfusion with collodion coated Chinese petroleum based spherical charcoal. *Chinese J Biomed Eng* **2:** 170–177.

Chang TMS, Espinosa-Melendez E, Lal S, *et al.* (1983) B endorphin and other middle molecules in chronic schizophrenia and removal by adsorbent hemoperfusion. In: Ernst K, Seidel K (eds.), *Detoxifications in Psychiatric Disorders.* Veb Verlag Volk und Gesundheit, Berlin, pp. 69–77.

Chang TMS. (1984c) Artificial cells in hemoperfusion and other applications. *Nephron* **36:** 161–168. (special festschrift issue for Professor Kolff).

Chang TMS. (1984g) Coated charcoal hemoperfusion. *Life Support Systems* **2:** 99–106.

Chang TMS. (1984h) Liver support systems. In: Bartlett (ed.), pp. 461–485. *Life Support Systems in Intensive Care.* Year Book Publisher.

Chang TMS, Lister C, Varma R. (1984a) Kinetics of coated charcoal hemoperfusion in the removal of Desferal for the elimination of aluminum. In: Atsumi K, Maekawa M, Ota K (eds.), *Progress in Artificial Organs.* ISAO Press, Cleveland, USA, pp. 602–605.

Chang TMS, Odaka M, Schreiner GE, *et al.* (1984b) Adsorbents in hemoperfusion. In: Atsumi K, Maekawa M, Ota K (eds.), *Progress in Artificial Organs.* ISAO Press, Cleveland, USA, pp. 624–627.

Chang TMS, Barre P, Kuruvilla S. (1985) Long-term reduced time hemoperfusion-hemodialysis compared to standard dialysis: A preliminary crossover analysis. *Trans Am Soc Artif Intern Organs* **31:** 572–576.

Chang TMS, Ho BL (eds.). (1985) *Hemoperfusion and Artificial Organs.* China Academic Publishers, Beijing, China, 464 p.

Chang TMS. (1986a) Application of artificial cells in hemoperfusion, blood substitutes, immunosorption, artificial pancreas, and enzyme replacement. In: Nose Y, Kjellstrand C, Ivanovich P (eds.), *Progress in Artificial Organs.* ISAO Press, Cleveland, USA, pp. 232–238.

Chang TMS. (1986b) Experimental artificial liver support with emphasis on fulminant hepatic failure: Concepts and reviews. *Seminars in Liver Diseases Series on Fulminant Hepatic Failure* **6:** 148–159.

Chang TMS. (1986c) Experimental evaluation of artificial liver support. In: Williams R (ed.), *Liver Failure.* Clinics in Critical Medicine Series, Churchill Livingstone Inc., UK, pp. 191–206.

Chang TMS. (1986d) Hemoperfusion in poisoning, fulminant hepatic failure, aluminium and iron removal, uremia and immunosorption. In: Nose Y, Kjellstrand C, Ivanovich P (eds.), *Progress in Artificial Organs.* ISAO Press, Cleveland, USA, pp. 595–600.

Chang TMS, Trevino Becerra A (eds.). (1986) Hemoperfusion Special Issue. *Int J Artif Organs* **9:** 279–368.

Chang TMS, Nicolaev N (eds.). (1987) Hemoperfusion, sorbent and immobilized bioreactants. Special Issue, *Int J Biomaterials, Artificial Cells and Artificial Organs* **15:** 1321.

Chang TMS, Barre P, Lister C, Kuruvilla C. (1989a) Artificial cells in medical applications with emphasis on hemoperfusion for aluminium removal and crossover control clinical trial on hemoperfusion-hemodialysis. *Contrib Nephrol* **70:** 237–249.

Chang TMS, Daka J. (1990) "A novel method for bilirubin removal," Canadian Patent granted.

Chang TMS, Odaka M (eds.). (1991) Hemoperfusion, sorbent and immobilized bioreactants. Special issue, *Biomaterials, Artificial Cells and Immobilization Biotechnology* **19:** 1–298.

Chang TMS. (1991a) Artificial Cells. In: Dulbecco R (ed.), *Encyclopedia of Human Biology,* Vol 1. Academic Press, Inc., San Diego, CA, 377–383.

Chang TMS. (1992b) Artificial cells for artificial liver support. In: Brunner G (ed.), *Artificial Liver Support.* Springer-Verlag, Berlin, 261–270.

Chang TMS. (1992e) Artificial Cells — 35 years. *J Artificial Organs* **16:** 8–12.

Chang TMS. (1992f) Artificial liver support based on artificial cells with emphasis on encapsulated hepatocytes. *J Artificial Organs* **16:** 71–74.

Chang TMS. (1997b) Artificial Cells. In: Dulbecco R (ed.), *Encyclopedia of Human Biology,* 2nd ed. Academic Press, Inc., San Diego, CA, USA, pp. 457–463.

Chang TMS. (1997l) Live E. coli cells to treatment uremia: Replies to letters to the editor. *Nature Med* **3:** 2–3.

Chang TMS, Prakash S. (1998b) Microencapsulated genetically engineered cells: comparison with other strategies and recent progress. *Molecular Med Today* **4:** 221–227.

Chang TMS, Prakash S. (1999) Chapter on removal of urea in uremia and ammonia in liver failure with emphasis on the use of artificial cells containing genetically engineered cells. In: Kuhtreiber WM, Lanza RP, Chick WL (eds.), *Handbook of Cell Encapsulation.* Birkhauser & Springer-Verlag, pp. 379–416.

Chang TMS, Malave N. (2000) The development and first clinical use of semipermeable microcapsules (artificial cells) as a compact artificial kidney (special milliniium republication of previous key papers). *Therapeutic Apheresis* **4:** 108–116.

Chang TMS. (2001) Bioencapsulated hepatocytes for experimental liver support. *J Hepatol* **34:** 148–149.

Chang TMS. (2005i) Therapeutic applications of polymeric artificial cells. *Nature Review: Drug Discovery* **4:** 221–235.

Chang TMS. (2007a) Monograph on "Artificial Cells: Biotechnology, nanotechnology, blood substitutes, regenerative medicine, bioencapsulation, cell/stem cell therapy." World Scientific Publisher/Imperial College Press, 435 p. [http://www.medicine.mcgill.ca/artcell/2007%20ebook%20artcell%20web.pdf]

Chang TMS. (2007b) Artificial cells containing hepatocytes and/or stem cells in regenerative medicine. In: "Artificial Cells" *Regenerative Medicine, Artificial Cells & Nanomedicine* **1:** 225–251. [http://www.medicine.mcgill.ca/artcell/511.pdf]

Chang, TMS. (2007c) Hemoperfusion in poisoning, kidney failure, liver failure and immunology. In: "Artificial Cells" *Regenerative Medicine, Artificial Cells & Nanomedicine* **1:** 252–299. [http://www.medicine.mcgill.ca/artcell/512.pdf]

Chang TMS. (2007d) Perspectives on the future of artificial cells as suggested by past research. In: "Artificial Cells" *Regenerative Medicine, Artificial Cells & Nanomedicine* **1:** 299–334. [http://www.medicine.mcgill.ca/artcell/513.pdf]

Chang TMS. (2010) Artificial cells containing stem cells for liver failure. Feature article in Special Issue on stem cells, *Asia Pacific Biotechnology (APBN)* **14(3):** 17–19.

Chang TMS. (2012) The history, development and promise of artificial cells: microdialyser and nanodialysers with bioreactive properties. In: Ing TS, Rahman MA, Kjellstrand CM (eds.), *Dialysis: History, Development and Promise.* World Scientific Publisher/Imperial College Press, Singapore/London, pp. 873–888.

Chang TMS. (2013) Artificial cells that started nanomedicine. In: Chang TMS (ed.), *Selected Topics in Nanomedicine.* World Science Publisher/Imperial College Press, pp. 1–21.

Chang V, Battistin R, Rodrigue A, Chang TMS. (1987a) Flow analysis for optimal configuration of hemoperfusion device. *Int J Artif Organs* **10:** 115–120.

Chawla AS, Chang TMS. (1973) Nonthrombogenic polymeric membrane prepared by gamma-radiation grafting of heparin. *Am Chem Soc* **33:** 379–385.

Chawla AS, Chang TMS. (1974) Nonthrombogenic surface by radiation grafting of heparin: Preparation, *in-vitro* and *in-vivo* studies. *Int J Biomaterial Medical Devices & Artif Organs* **2(2):** 157–169.

Chawla AS, Chang TMS. (1975a) A new method for the preparation of nonthrombogenic surface by radiation grafting of heparin: Preparation and *in-vitro* studies. In: Gregor HP (ed.), *Biomedical Applications of Polymers.* Plenum Press, New York, pp. 147–157.

Chawla AS, Chang TMS. (1975b) Use of solubility parameters for the preparation of hemodialysis membranes. *J Applied Polymer Sci* **19:** 1723–1730.

Chirito E, Reiter B, Lister C, Chang TMS. (1977a) Artificial liver: The effect of ACAC microencapsulated charcoal hemoperfusion on fulminant hepatic failure. *Artif Organs* **1(1):** 76–83.

Chirito E, Reiter B, Lister C, Chang TMS. (1978) In: Chang TMS (ed.), Assessment of two rat models of acute hepatic failure for testing artificial liver devices. *Artificial Kidney, Artificial Liver and Artificial Cells.* Plenum Press, New York, pp. 239–243.

Chirito E, Lister C, Chang TMS. (1979) Biochemical, hematological and histological changes in a fulminant hepatic failure rat model for artificial liver assessment. *Artif Organs* **3(1):** 42–46.

Chow KM, Liu ZC, Prakash S, Chang TMS. (2003) Free and microencapsulated Lactobacillus and effects of metabolic induction on Urea Removal. *Artificial Cells, Blood Substitutes & Biotechnology.* **4:** 425–434.

Cohan SL, Winchester JF, Gelfand MC. (1982) Treatment of intoxications by charcoal hemadsorption. *Drug Metab B Rev* **13:** 681.

Cohan SL, Winchester JF, Gelfand MC. (1978) Hemoperfusion. *Annu Rev Pharmacol Toxicol* **27:** 169.

Cordopatri F, Boncinelli S, Marsili M, *et al.* (1982) Effects of charcoal haemoperfusion with prostacydin on the coagulation-fibrinolysis system and platelets. *Int J Artif Organs* **5:** 243.

Daka JN, Chang TMS. (1989) Bilirubin removal by the pseudoperoxidase activity of free and immobilized hemoglobin and hemoglobin co-immobilized with glucose oxidase. *Biomat Artif Cells and Artif Org* **17:** 553–562.

Daka NJ, Sipehia R, Chang TMS. (1989) Enhanced oxidation of bilirubin by an immobilized tri-enzyme system of glucose oxidase, bilirubin oxidase and horseradish peroxidase. *Biochim Biophys Acta* **991:** 487–489.

Diaz-Buxo JA, Farmer CD, Chandler TY. (1978) Hemopertusion in the treatment of amitriptyline poisoning. *Trans Am Soc Artif Intern Organs* **24:** 699.

Dixit V, Chang TMS. (1981) Soluble liver extracts as supplement to detoxification by hemoperfusion. *Int J Artif Organs* **4:** 291–294.

Dixit V, Chang TMS. (1982) Effects of prostaglandine on the survival time of fulminant hepatic failure rats. *Int J Artif Organs* **5:** 388.

Dixit V, Chang TMS. (1985) Preliminary report on effects of prostaglandine on brain edema in fulminant hepatic failure rats. *Int J Artificial Organs* **8(1):** 55–56.

Dixit V, Chang TMS. (1987) Effects of prostaglandine on brain edema and liver histopathology in a galactosamine- induced fulminant hepatic failure rat model. *J Biomat Art Cells Art Org* **15:** 559–574.

Dixit V, Chang TMS. (1990) Brain edema and blood brain barrier in galactosamine induced fulminant hepatic failure rats: animal models for evaluation of liver support systems. *J Amer Soc Artif Int Organs* **36:** 21–27.

Dunea G, Kolff WJ. (1965) Clinical experience with the Yatzidis charcoal artificial kidney. *Trans Am Soc Artif Intern Organs* **11:** 178.

Espinosa-Melendez E, Zelman M, Barre P, *et al.* (1982) Oxystarch modified by boiling: Effects of oral administration in stable hemodialysis patients. Proceedings, International Symposium on "Hemoperfusion and Artificial Organs," 83–85.

Espinosa-Melendez E, Chang TMS. (1983) State-of-the-art in hemoperfusion. *Nephrologica Mexicana* **4:** 51–55.

Espinosa-Melendez E, Bourgouin PA, Chang TMS. (1986) Combined hemoperfusion/hemodialysis in the treatment of uremic patients with vascular access problems. *In J Artif Organs* **9:** 309–312.

Espinosa-Melendez E, Chang TMS. (1980) ACAC (albumin-collodion coated activated charcoal) hemoperfusion: *In-vitro* removal and clearance of theophylline. *J Dialysis* **4(1):** 21–29.

Espinosa-Melendez E, Lal S, Nair NPV, Chang TMS. (1979) Plasma beta endorphin levels in chronic schizophrenic patients and normal controls. *Artif Organs* **3(4):** 375–376.

Friedman E (ed.). (2011) *Novel Therapies in Terminal Renal Failure*, 2nd in the book series on "Regenerative Medicine, Artificial Cells and Nanomedicine." World Scientific Publisher/Imperial College Press.

Gazzard BG, Portmann BA, Weston MJ, *et al.* (1974) Charcoal haemoperfusion in the treatment of fulminant hepatic failure. *Lancet* **1:** 1301–1307.

Gelfand MC, Winchester JF, Knepshield JH, *et al.* (1977) Treatment of severe drug overdose with charcoal hemoperfusion. *Trans Am Soc Artif Intern Organs* **23:** 599.

Gibson TP, Lucas SV, Nelson HA, *et al.* (1978) Hemoperfusion removal of digoxin from dogs. *J Lab CUn Med* **91:** 673.

Gimson AES, Braud S, Mellon PJ, *et al.* (1982) Earlier charcoal haemoperfusion in fulminant hepatic failure. *Lancet* **2:** 681.

Gu KF, Chang TMS. (1988a) Conversion of ammonia or urea into L-leucine, L-valine and L- isoleucine using artificial cell immobilising multienzyme system and dextran-NADH+. I. Glucose dehydrogenase for cofactor recycling. *J Am Soc Artif Int Organs* **11:** 24–28.

Gu KF, Chang TMS. (1988b) Conversion of ketoglutarate into L-glutamic acid with urea as ammonium source using multienzyme system and dextran-NAD+ immobilised by microencapsulation with artificial cells in a bioreactor. *J Bioeng Biotech* **32:** 363–368.

Gu KF, Chang TMS. (1990a) Conversion of ammonia or urea into essential amino acids, L- leucine, L-valine and L-isoleucine using artificial cells containing an immobilized multienzyme system and dextran-NAD+. 2. Yeast alcohol dehydrogenase for coenzyme recycling. *J Biotechnology and Applied Biochemistry* **12:** 227–236.

Gu KF, Chang TMS. (1990b) Conversion of ammonia or urea into essential amino acids. L- leucine, L- valine and L-isoleucine using artificial cells containing an immobilized multienzyme system and dextran-NAD+. IV. Malate dehydrogenase for co-enzyme recycling. *J Catalysis* **62:** 331–339.

Gu KF, Chang TMS. (1990c) Production of essential L-branched-chained amino acids, in bioreactors containing containing artificial cells immobilized multienzyme systems and dextran-NAD+. *Appl Biochem Biotechnol* **26:** 263–269.

Gu KF, Chang TMS. (1991) Conversion of urea into essential amino acids, L-leucine, L- valine and L- isoleucine using artificial cells containing an immobilized multienzyme system and dextran-NAD+. III. L-lactic dehydrogenase for co-enzyme recycling. *J Appl Biochem Biotechn* **27:** 75–91.

Hakim RM, Schulman IM, Lazarus JM. (1985) *abstracts Am Soc Nephrol* **18:** 65A.

Hakim RM, Milford E, Himmelfarb J, *et al.* (1990) Extracorporeal removal of antiHLA antibodies in transplant candidates. *Am J Kidney Dis* **16:** 423.

Kashani SA, Chang TMS. (1988) Release of hepatic stimulatory substance from cultures of free and microencapsulated hepatocytes: Preliminary report. *J Biomaterials, Artificial Cells and Artificial Organs* **16:** 741–746.

Kashani S, Chang TMS. (1991a) Physical chemical characteristics of hepatic stimulatory factor prepared from cell free supernatant of hepatocyte cultures. *Biomaterials, Artificial Cells and Immobilization Biotechnology* **19:** 565–578.

Kashani S, Chang TMS. (1991b) Effects of hepatic stimulatory factor released from free or microencapsulated hepatocytes on galactosamine induced fulminant hepatic failure animal model. *Biomaterials, Artificial Cells and Immobilization Biotechnology* **19:** 579–598.

Kawasaki C, Nishi R, Uekihara S, *et al.* (2000). Charcoal hemoperfusion in the treatment of phenytoin overdose. *Am J Kidney Dis* **35:** 323–326.

Kaziuka EN, Chang TMS. (1979) *In-vitro* assessment of the removal of phenols by ACAC hemoperfusion. *Int J Artif Organs* **2(4):** 215–221.

Kjellstrand C, Borges H, Pru C, *et al.* (1981) On the clinical use of micro-encapsulated zirconium phosphate-urease for the treatment of chronic uremia. *Trans Am Soc Artif Intern Org* **27:** 24–30.

Klinkmann H, Falkenhagen D, Chang TMS (eds.). (1990) Hemoperfusion, sorbent and immobilized bioreactants. Special issue, *Int J Biomaterials, Artificial Cells and Artificial Organs* **18:** 455–568.

Kolff WJ, Berk HTJ. (1944) The artificial kidney: A dialyser with a great area. *Acta Med Scand* **117:** 121.

Legallais C, David B, Doré E. (2001) Bioartificial livers (BAL): Current technological aspects and future developments. *J Membrane Sci* **181:** 81–95.

Lin CC, Chou HL, Lin JL. (2002) Acute aconitine poisoned patients with ventricular arrhythmias successfully reversed by charcoal hemoperfusion. *Am J Emerg Med* **20:** 66–67.

Liu JP, Gluud LL, Als-Nielsen B, Gluud C. (2004) Artificial and bioartificial support systems for liver failure. *Cochrane Database Syst Rev* (**1**): CD003628.

Liu ZC, Chang TMS. (2005) Transplantation of bioencapsulated bone marrow stem cells improves hepatic regeneration and survival of 90% hepatectomized rats: a preliminary report. *Artif Cell Blood Sub* **33(4):** 405–410.

Liu ZC, Chang TMS. (2006a) Transdifferentiation of bioencapsulated bone marrow cells into hepatocyte-like cells in the 90% hepatectomized rat model. *J Liver Transplant* **12:** 566–572.

Liu ZC, Chang TMS. (2006b) Artificial cell microencapsulated stem cells in regenerative medicine, tissue engineering and cell therapy. In: Pedraz (ed.), *Cell Encapsulation* (in press).

Liu ZC, Chang TMS. (2009) Preliminary study on intrasplenic implantation of artificial cell bioencapsulated stem cells to increase the survival of 90% hepatectomized rats. *Artif Cell Blood Sub* **37:** 53–55. [http://capmc.ca/CAPMC/pmc/articlerender.cgi?artid=2517]

Liu ZC, Chang TMS. (2012) Intrasplenic transplantation of bioencapsulated mesenchymal stem cells enhances the regeneration of remnant livers and improves the recovery rates of 90% partial hepatectomized rats. *Stem Cells International,* Volume 2012, Article ID 697094, 6 p. doi: 10.1155/2012/697094 PMC3515999 http://www.hindawi.com/journals/sci/2012/697094/

Lopez Lago AM, *et al.* (2002) Paraquat poisoning and hemoperfusion with activated charcoal. *An Med Intern* **19**: 310–312.

Lorch JA, Garella S. (1979) Hemoperfusion to treat intoxications. *Ann Intern Med* **91:** 301.

Martin AM, Gibbins JK, Oduro A, Herbert R. (1978) Clinical experience with cellulose-coated carbon hemoperfusion. In: Chang TMS (ed.), *Artificial Kidney, Artificial Liver, and Artificial Cells.* Plenum, New York, pp. 143–152.

Morley D, Chang TMS. (1987) Effect of Membrane Thickness of Collodion Coated Activated Charcoal on Adsorption of Small and Middle Molecules. *J Biomat Art Cells Art Org* **15:** 617–632.

Nasielski P, Chang TMS. (1978) Coated charcoal hemoperfusion for the removal of galactose in rats. *Int J Artif Organs* **1(3):** 143–147.

Nasielski P, Chang TMS. (1978) Microencapsulated charcoal hemoperfusion forgalactosemia. In: Chang TMS (ed.), ***Artificial Kidney, Artificial Liver and Artificial Cells.*** Plenum Press, New York, pp. 255–257.

Niu Z, Jia SR, Zhang DY, ***et al.*** (1980) The effects of hemoperfusion in galactosamine induced FHF rats. ***Chongqing Medical College Bull***, November 1980, pp. 1–6.

Odaka M, Tabata Y, Kobayashi H, ***et al.*** (1978) Cinical experience of bead-shaped charcoal hemoperfusion in chronic renal failure and fulminant hepatic failure. In: Chang TMS (ed.), ***Artificial Kidney, Artificial Liver and Artificial Cells.*** Plenum Press, New York, pp. 79–88.

Odaka M, Hirasawa H, Kobayashi H, ***et al.*** (1980) Clinical and fundamental studies of cellulose coated bead-shaped charcoal hemoperfusion in chronic renal failure. In: Sideman, Chang (eds.), ***Hemoperfusion: Kidney and liver support and detoxification***, Part I. Hemisphere, Washington, D. C., pp. 45–55.

O'Grady JG, Gimson AE, O'Brien CJ, ***et al.*** (1988) Controlled trials of charcoal hemoperfusion and prognostic factors in fulminant hepatic failure. *Gastroenterology* **94:** 1186.

Peng A, Meng FQ, Sun LF, ***et al.*** (2004) Therapeutic efficacy of charcoal hemoperfusion in patients with acute severe dichlorvos poisoning. ***Acta Pharmacol Sin*** **25:** 15–21.

Piskin E, Chang TMS (eds.). (1982) ***Hemoperfusion and Artificial Organs.*** Artificial Organs Society, 187.

Piskin E, Chang TMS (eds.). (1983) Past, Present and Future of Artificial Organs. Meteksan Publisher Co., Ankara, Turkey, 262.

Prakash S, Chang TMS. (1996) Microencapsulated genetically engineered live E coli DH5 cells administered orally to maintain normal plasma urea level in uremic rats. ***Nature Med*** **2:** 883–887.

Prakash S, Chang TMS. (1998) Growth and survival of renal failure rats that received oral microencapsulated genetically engineered E. coli DH5 cells for urea removal. ***Artif Cell Blood Sub*** **26:** 35–51.

Prakash S, Chang TMS. (1999) Artificial cell microcapsules containing genetically engineered ***e.coli dh5*** cells for ***in-vitro*** lowering of plasma potassium, phosphate, magnesium, sodium, chloride, uric acid, cholesterol, and creatinine: A preliminary report. ***Artif Cell Blood Sub*** **28(5–6):** 475–482.

Prakash S, Chang TMS. (2000a) Artificial cells microencapsulated genetically engineered E. coli DH5 cells for the lowering of plasma creatinine *in-vitro* and *in-vivo*. *Artif Cell Blood Sub* **28(5):** 397–408.

Prakash S, Chang TMS. (2000b) Lowering of uric acid *in-vitro* and *in-vivo* by artificial cell microcapsulated genetically engineered *E. coli DH 5* cells. *Int J Artificial Organs* **23:** 429–435.

Prichard S, Chirito E, Chang TMS, Sniderman AD. (1977) Microencapsulated charcoal hemoperfusion: A possible therapeutic adjunct in digoxin toxicity. *J Dialysis* **1(4):** 367–377.

Prichard S, Chirito E, Sniderman AD, Chang TMS. (1978) Use of albumin cellulose nitrate microencapsulated charcoal hemoperfusion: In acute digoxin toxicity in dogs. In: Chang TMS (ed.), *Artificial Kidney, Artificial Liver and Artificial Cells.* Plenum Press, New York, pp. 245–247.

Shi ZQ, Chang TMS. (1983) *In vitro* adsorption spectrum of plasma amino acids by coated charcoal hemoperfusion. *Int J Artif Organs* **6(5):** 267–270.

Shi ZQ, Chang TMS. (1984b) Effect of charcoal hemoperfusion on amino acid disturbance in experimental hepatic coma. *Int J Artif Organs* **7:** 203–208.

Shi ZQ, Chang TMS. (1985) Amino acid disturbance in experimental hepatic coma: The effects of charcoal hemoperfusion. In: Chang TMS, Ho BL (eds.), *Hemoperfusion and Artificial Organs.* China Academic Publisher, China, pp. 301–310.

Shu CD, Chang TMS. (1980) Tyrosinase immobilized within artificial cells for detoxification in liver failure. I. Preparation and *in vitro* studies. *Int J Artif Organs* **3(5):** 287–291.

Shu CD, Chang TMS. (1981) Tyrosinase immobilized within artificial cells for detoxification in liver failure: II. *In vivo* studies in fulminant hepatic failure rats. *Int J Artif Organs* **4:** 82–84.

Sideman S, Chang TMS. (1979) Hemoperfusion: Kidney and liver supports and detoxification. *Artif Organs* **3(4):** 393–402.

Sideman S, Chang TMS. (1980a) Hemoperfusion: State of the art and future requirements. *Artif Organs* **4:** 70–75.

Sideman S, Chang TMS (eds.). (1980b) *Hemoperfusion: I. Artificial Kidney and Liver Support and Detoxification.* Hemisphere Publishing Corp., Washington, D.C., USA, 473 p.

Sideman S, Chang TMS (eds.). (1981) *Hemoperfusion: II. Devices and Clinical Applications.* Samuel Neaman Institute for Advanced Studies in Science and Technology, Technion, Haifa, Israel, 167 p.

Singh SM, McCormick BB, Mustata S, *et al.* (2004) Extracorporeal management of valproic acid overdose: A large regional experience. *J Nephrol* **17:** 43–49.

Sipehia R, Bannard RAB, Chang TMS. (1986a) Adsorption of large lipophilic molecules with exclusion of small hydrophilic molecules by microencapsulated activated charcoal formed by coating with polyethylene membrane. *Journal of Membrane Science* **29:** 277–286.

Sipehia R, Chawla AS, Chang TMS. (1986b) Enhanced albumin binding to polypropylene beads via anhydrous ammonia gaseous phase. *Biomaterials* **1:** 471–473.

Sipehia R, Bannard RAB, Chang TMS. (1989a) Poly(vinylidene fluoride) orpoly(vinylidene chloride/vinyl chloride)-coated activated charcoal for the adsorption of large lipophilic molecules with exclusion of small hydrophilic molecules. *J Membrane Science* **47**: 293–301.

Sussman NL, Gislason GT, Kelly JH. (1994) Extracorporelal liver support. Application to fulminant hepatic failure. *J Clin Gastroenterol* **18:** 320.

Tabata Y, Chang TMS. (1980) Comparisons of six artificial liver support regimes in fulminant hepatic coma rats. *Trans Am Soc Artif Intern Organs* **26:** 394–399.

Tabata Y, Chang TMS. (1982b) Intermittent vascular access for extracorporeal circulation in conscious rats: A new technique. *Artif Organs* **6:** 213–214.

Tani T, Chang TMS, Kodama M, Tsuchiya M. (1992) Endotoxin removed from hemoglobin solution using polymyxin-B immobilized fibre (PMX-F) followed by a new turbidometric endotoxin assay. *Biomaterials, Artificial Cells and Immobilization Biotechnology* **20:** 394–399.

Terman DS, Tavel T, Petty D, *et al.* (1977) Specific removal of antibody by extracorporeal circulation over antigen immobilized in colodion charcoal. *Clin Exp Immunol* **28:** 180.

Terman DS, Buffaloe G, Mattioli C, *et al.* (1979) Extracorporeal immunoabsorption: initial experience in human systemic lupus erythematosus. *Lancet* **2:** 824.

Terman DS, Garcia-Rinaldi R, McCalma R, *et al.* (1979) Modification of hyperacute renal xenograft rejection after extracorporeal immunoadsorption of heterospecific antibody. *Int J Artif Organs* **2:** 35.

Terman DS. (1980) Extracorporeal immunoadsorbents for extraction of circulating immune reactants. In: Giordano E (ed.), *Sorbents* and *Their Clinical Applications.* New York, Academic Press Inc, p. 470.

Tominaga M, *et al.* (1997) Pharmacological evaluation of portal venous isolation and charcoal haemoperfusion for high-dose intra–arterial chemotherapy of the pancreas. *Br J Surg* **84:** 1072–1076.

Vasilakakis DM, D'Haese PC, Lamberts LV, *et al.* (1992) Removal of aluminoxamine and ferrioxamine by charcoal hemoperfusion and hemodialysis. *Kidney Int* **41:** 1400.

Verpooten GA, De Brae ME. (1984) Combined hemoperfusion-hemodialysis in severe poisoning: Kinetics of drug extraction. *Resuscitation* **11:** 275.

Williams R. (1983) Fulminant hepatic failure. *Postgraduate* **59(Suppl4):** 33.

Winchester JF, Apiliga MT, MacKay JM, Kennedy AC. (1976) Combined hemodialysis-charcoal hemoperfusion in the dialysis patient. *Kidney Int* **10(Suppl 7):** S315–S319.

Winchester JF. (1986) Management of iron overload. *Semin Nephrol* **4(Suppl 1):** 22.

Winchester JF (editor-in-chief). (1996) *Replacement of Renal Function by Dialysis,* 4th edition. Kluwer Academic Publishers, Boston.

Winchester JF. (1990) Active methods for detoxification: Oral sorbents, forced diuresis, hemoperfusion and hemodialysis. In: Haddad LM, Winchester JF (eds.), *Clinical Management of Poisoning and Drug Overdose,* 2nd edition. WB Saunders Co, Philadelphia, PA, p. 148.

Wolfe EA, Chang TMS. (1987) Orally ingested microencapsulated urease and an adsorbent, zirconium phosphate, to remove urea in kidney failure. *Int J Artif Organs* **10:** 269–275.

Wong H, Chang TMS. (1986) Bioartificial liver: Implanted artificial cells microencapsulated living hepatocytes increases survival of liver failure rats. *Int J Artif Organs* **9:** 335–336.

Wong H, Chang TMS. (1988) The viability and regeneration of artificial cell microencapsulated rat hepatocyte xenograft transplants in mice. *J Biomat Artif Cells Artif Organs* **16:** 731–740.

Yang L, Cheng Y, Yan WR, Yu YY. (2004) Extracorporeal whole blood immunoadsorption of autoimmune myasthenia gravis by cellulose tryptopan adsorbent. *Artif Cell Blood Sub* **32:** 519–518.

Yatzidis H. (1964) A convenient hemoperfusion micro-apparatus over charcoal for the treatment of endogenous and exogenous intoxications. *Proc Eur Dialysis Transplant Assoc* **1:** 83.

Yu YT, Chang TMS. (1980) The effects of polymer-solvent compositions on the formation of collodion membrane artificial cells. *Int J of Biomaterial, Medical Devices & Artif Organs* **8:** 273–281

Zhou MX, Chang TMS. (1987) Effects of Polylactic Acid Microcapsules Containing Prostaglandin E2 on the Survival Rates of Grade II Coma Galactosamine-induced Fulminant Hepatic Failure Rats. *J Biomat Art Cells Art Org* **15:** 549–558.

II

General Adsorbents

Chapter 2

Hemoperfusion

Yingming Gu* and Ying Huang*

Hemoperfusion (HP) is an extracorporeal technique that removes exotoxin and/or endotoxin and purifies blood by means of an adsorptive mechanism during the passage of the blood through a perfusion apparatus with adsorptive agent. It is a new way of treatment for drug poisoning in clinical practice. The combination of hemoperfusion and hemodialysis method has adsorption function and can adjust water, electrolytes, and the acid–base balance of blood, significantly increasing the quality of clinical treatment. Additionally, it also increases the wake-up rate of hepatic encephalopathy patients.

1. The Progress of Hemoperfusion

In 1948, Muirhead and Reid conducted an adsorption experiment with resin for the first time. They mixed the anion and cation exchange resin and directly applied this mixture to conduct the hemoperfusion experiment on dogs. The hemoperfusion removed urea nitrogen in blood of nephrectomized dogs. This had serious complications, such as muscle twitching, restlessness and respiratory failure. In 1958, Schreiner used hemoperfusion therapy with lactic acid cation exchange column to treat a barbiturate-poisoned patient. After 15 min of hemoperfusion, it

*The Second Affiliated Hospital of Guangzhou Medical University, Guangzhou, China.

was proved that the hemoperfusion therapy was effective in clearing drugs. However, in the early resin experiment, there was fever, electrolyte disorder, hemolysis and other serious side effects, thus the experiment was not continued.

In 1964, the Greek researcher Yatzidis used hemoperfusion therapy with activated charcoal to treat acute poisoned and renal failure patients. The clinical trials provided the evidence that activated charcoal has strong absorption ability on creatinine, uric acid, phenols, indole, barbitone guanidine, salicylic acid and barbiturate. At the same time, they found the hemoperfusion therapy significantly reduced the symptoms of pericarditis, gastrointestinal and neuron suppression disease. However, the activated charcoal released fine particles and induced thrombosis and serious damage of platelets in the patients. Thus this method was not used in clinical practice either.

In 1966, Chinese Canadian researcher Chang (1966–1969) used ultrathin semipermeable membrane of collodion or albumin-collodion to coat activated charcoal granules to conduct hemoperfusion for the first time. This method prevented the release of charcoal particles and had better blood compatibility. People thus resumed research on hemoperfusion therapy for the treatment of acute poisoning and uremia (Chang *et al*, 1971, 1972, 1973; Chang, 1975; He and Zhang 1999; Martin *et al.*, 1978; Zi *et al.*, 1995). It was found that this way, hemoperfusion could clear not only uremic toxins but also medium molecular substances and polymer amino acid. In the late 1970s, absorption and coating materials underwent enormous improvement and development. Because the coated activated charcoal device had high efficiency and it could avoid adverse events, the method was applied not only in drug poisoning and uremia but also in acute hepatic encephalopathy, schizophrenia, psoriasis, and a variety of autoimmune diseases.

In 1979, the first central hospital in Tianjin applied hemoperfusion therapy for treating drug poisoning for the first time in China. Afterwards, Beijing, Shanghai, Guangzhou, Chongqing and other places also carried out the research of this work. To date, many kinds of hemoperfusion devices have been manufactured in China, and have been widely used in clinical practice.

2. The Principle of Hemoperfusion

The principle of hemoperfusion is to perfuse the blood of patients in contact with nonspecific solid adsorptive agents which can remove some metabolites, exogenous drugs or toxic material from the human body by adsorptive mechanism, then send the purified blood back to the patient to achieve the purpose of treatment.

3. The Hemoperfusion Device

3.1. *Adsorptive Agent and Coating Technique*

The adsorptive agent of hemoperfusion must comply with the following standards:

1) Non-toxic and no allergic reaction induced when in contact with human blood.
2) No physical change and chemical change during hemoperfusion process.
3) Possess good mechanical strength, no particle release, no shape change.
4) Good blood compatibility.

3.1.1. *Activated charcoal*

Activated charcoal is a kind of granular or powdery adsorbent which is prepared from animal or plant material using high temperature carbonization and activation. Activated charcoal has the characteristics of large area (more than 1,000 m^2/g), high porosity and broad pore diameter distribution. The pore diameter can be classified into micropore area (< 20A), transition-pore area (20~500A) and large pore area (diameter > 500A). The greater the molecular weight, the more adsorption capacity it has. The basis of absorption ability are more pores and large internal surface. The activated charcoal is a kind of broad spectrum absorbent — it can absorb many kinds of chemical compounds, especially compounds which are difficult to dissolve in water.

It also has good adsorptive performance (fast absorption and high absorption capability) on creatinine, uric acid and barbiturate drugs. However, the low selectivity of activated charcoal, poor mechanical strength and the ease of release of particles which can result in capillary embolism limited its application in clinical practice. In 1969, TMS Chang solved this problem by using the artificial cell. Thereafter, hemoperfusion therapy with activated charcoal was widely used in clinical practice.

3.1.2. *Resin*

Resin is a high-molecular polymer with three-dimensional mesh structure. There are many different types of resin based on differences in synthetic unit and cross-linking agent. The synthetic resin is made of styrene or cyclic-polymer which is made through suspension polymerization of acrylic ester and divinylbenzene. The styrene frame with exchange group was called absorptive resin, which was used as the blood purification absorbent. Adsorptive resins include polar absorptive resin and nonpolar absorptive resin. The polar absorptive resin has strong adsorbability on materials which are easy to dissolve in water and have great polarity. The nonpolar absorptive resin has strong adsorbability for liposoluble substances. According to the demand, absorptive resin with different physical structures can be made by improving the synthesis technique, which can induce different pore size and different superficial area. The pore diameter and superficial area are two important factors for the absorptive function of absorptive resin.

The resin is a multi-aperture organic polymer, which often induces the reduction of components of blood, especially platelets. According to Rosebaum's report, one hemoperfusion could reduce platelets up to 50%, but after hemoperfusion, they could recover to the original level. When treating drug or toxin poisoned patients, this is acceptable because the patients usually have no major organ failure or clotting mechanism failure. If resin hemoperfusion is used in other diseases, such as hepatic failure or renal failure, it can easily result in serious hemorrhage and other complications. Therefore, many researchers used biocompatible materials to microencapsulate resin or used the methods of blood plasma separation to avoid contact between blood components and the adsorbent. Dekoning used cellulose acetate and polymer electrolyte with anticlotting activity to cover XAD-4; Brumer and Falkehage used agarose and albumin to coat XAD-4 and Y56 non ion exchange resin, respectively. Hughes and Williams used albumin to coat XAD-7. These techniques can improve the blood compatibility of resin.

3.1.3. *Microcapsule technique*

In 1969, TMS Chang used the microcapsule technique (ACAC) to solve the problem of carbon particle release and improve blood compatibility. The technique preserves the absorptive efficiency of activated charcoal, reduces the release of particles and increases the blood compatibility significantly by coating the semipermeable membrane on the surface of the carbon particle. The characteristics of the microcapsule material are:

(1) The thickness of the membranes is generally below 5 um; it must have sufficient strength in order to prevent the carbon particles falling off.
(2) The micropore diameter of membrane is 5~45A, to allow the toxic substance to penetrate the microcapsule membrane.
(3) Coating materials do not adhere or condense with blood components, so as to guarantee the blood flow and low perfusion pressure.
(4) Good blood compatibility. The coating materials of microcapsule include collodion, albumin and albumin-collodion, cellulose acetate membrane, cross-linked polyvinyl alcohol, methyl acrylic hydroxyl ethyl ester, cross-linked gelatin, etc.

3.1.4. *Hemoditoxifier*

Hemoditoxifier have various different types, including cylindrical, drum-shaped and spindle type; those hemoperfusion devices comply

Table 2.1. Commonly Used Hemoditoxifiers

Manufacturer	Device	Adsorbent Type	Absorption Dose	Material of Microcapsule
Gambro	Adsorba 300c	Activated charcoal	300g	Cellulose
	Adsorba 150c	Activated charcoal	150g	Cellulose
Fresenius	Hemochol	Activated charcoal	300g	Acrylic acid aquogel
Asahi	Hemochol	Activated charcoal	200g	Collodion
Bioencapsulation	Diakart	Activated charcoal	70g	Collodion
Organot-Teknika	Hemopar260	Activated charcoal	250g	Cellulose acetate membrane
Clark	Biocomapatible System	Activated charcoal	50, 100, 250g	Heparinized polymer
Exreraeorporeal	Hemoresin	XAD-4 resin	350g	Collodion
Aier-Longfang	YTS-200	Activated charcoal	200g	Polyvinyl alcohol
Lizhu-Zhuhai	HA type	Pore absorptive resin	230mL	Collodion

with hydrodynamic characteristics, which can minimize dead space and lower the resistance. The capacities of device are generally 100–300g charcoal volume. The material is plastic and disposable. Generally marketed hemoditoxifiers are summarized in Table 2.1.

3.2. *Hemoperfusion Device*

Hemoperfusion devices have been manufactured with arterial and venous monitoring, liquid level monitoring, bubble monitoring, and blood thermostat. When using the hemodialysis device to conduct hemoperfusion, the hemodialysis device should be on the preparation mode which makes the air leakage monitoring function available through air leakage alarm system of hemodialysis device. The warming equipment in venous circuit can be added or the device immersed in thermostat water bath if warming is needed. Another way is to combine the hemodialysis device with the hemoperfusion device (connecting hemodialysis device and hemoperfusion device in series). The hemoperfusion should be put in front of the hemodialysis device, using the warming system of the hemodialysis device without additional warming system.

4. The Clinical Application of Hemoperfusion

Hemoperfusion is commonly used in drug or toxin poisoned patients. It is also used in treatment of uremia, hepatic coma, immunological and infection disease.

4.1. *Drug or Toxin Poisoned Patients*

According to their different solubility, drugs or toxins can be divided into water-soluble, liposoluble and protein-binding. The drugs and toxins which can be removed by hemoperfusion are described in Table 2.2.

Table 2.2. Toxins and Drugs Removed by Hemoperfusion

Classification	Drugs
Hyponotics	Barbiturates, sleep can lead, methaqualone, stability, chlorpromazine, chlordiazepoxide
Analgesic-antipyretic drugs	Aspirin, acetaminophen
Cardiovascular drugs	Digoxin, quinidine
Antibacterial drugs	Gentamicin, ampicillin, clindamycin
Antineoplastic drugs	
Alcohol	
Poisonous weeds paraquat potassium dichromate	
Organophosphorus organochlorine	

4.1.1. *Characteristics of hemoperfusion*

(1) Drug concentration of plasma already reached lethal concentration in patients.
(2) The drugs and toxins are absorbed continuously.
(3) Seriously poisoned patients who cannot be treated by medical methods.
(4) Pneumonia patients in long-time coma.
(5) Patients who have liver, heart, kidney function insufficiency, which induces the reduction of ability to excrete drugs.
(6) Poisoned patients who have metabolic delay effect, such as ethylene glycol and paraquat poisoning.

4.1.2. *Factors which impact the curative effect of hemoperfusion*

(1) Drugs which affect humans quickly, such as cyanide.
(2) Rate of drug metabolism clearance greater than rate of hemoperfusion metabolism clearance.
(3) Drugs which induce severe acidosis.
(4) Drugs with irreversible effects.
(5) Drugs which have large volume distribution, such as tricyclic antidepressant.
(6) Drugs which have no serious toxicity.

4.1.3. *Curative effect evaluation in clinical*

(1) During or after the period of hemoperfusion therapy, patients have different levels of recovery, such as recovering from coma, reflect blood pressure becoming balanced, heart sounds enhancement, deep and fast breathing, limb activation.
(2) Patients waking up does not mean that they are out of danger, because the toxins can travel through gut, tissue space, internal organs and muscles after a few hours; especially adipose tissue with less blood supply can diffuse toxins into the bloodstream.

(3) Hemoperfusion therapy can only clear the toxins, it cannot reverse the effect of toxins, such as change of pathology and physiology. It is different from the mechanism of pharmaceutic detoxification process. For example, when patient have organic phosphorus pesticide poisoning, hemoperfusion therapy can recover the cholinesterase activity, but cannot replace the reactivators of cholinesterase agent and atropine treatment.
(4) The inhibition of breathing, cardiovascular function insufficiency and imbalance of water, electrolytes and acid-base induced by toxins needs to be treated with corresponding methods.
(5) Hemoperfusion treatment can save drug or toxin poisoned patients. But long-term poisoned patients and those who have complications with lung edema and brain edema, should be treated with hemoperfusion/hemodialysis combined methods early to prevent patient death from circulatory and breathing failure.

4.2. *Uremia*

The Greek researcher Yatzides first used activated charcoal in hemoperfusion therapy to treat uremia. He also proved that activated charcoal can absorb creatinine, uric acid, indican, phenol, indole, guanidine and middle molecule material. It could improve somnolence and gastrointestinal symptoms in patients and decreased the serum creatinine and uric acid concentration significantly in patients. The method could not be used for a long time because of disadvantages such as bad blood compatibility and charcoal thrombosis. In 1969, albumin-collodion packaging activated charcoal (ACAC) overcame the aforementioned disadvantages and research of activated charcoal hemoperfusion therapy on uremia developed progressively (Chang, 1969). When the blood flow rate was 300mL/min, for Chang's ACAC (300g) system (Chang *et al.*, 1971, 1972, 1973; Chang, 1972, 1975), the clearance rate of middle molecule was 144mL/min, creatinine was 230mL/min, and uric acid was 235mL/min. After two hours' hemoperfusion the creatinine concentration of blood decreased from 1184.6 to 822.12 mL/min, the uric acid decreased from 440.15 to 255.70 μmol/L. The middle molecule material concentration in artery decreased to 55%. When the blood flow rate was 200 mL/min with Adsorba 300C, the clear rate of middle molecule material was 120mL/min, creatinine was 180mL/min, and uric acid was 180mL/min. The hemoperfusion conducted for two hours was equal to the effect of hemodialysis for 12 hr.

Martin *et al.* (1978) compared the curative effect between traditional hemodialysis and hemoperfusion/hemodialysis combined therapy. The time needed for recovery of uremic symptoms for the hemodialysis group was 16–28 days and 14–35 times hemodialysis process. For hemoperfusion/hemodialysis combined therapy, the symptom recovery time was 6–20 days, and the hemodialysis process decreased to 4–14 times. This indicated that the hemoperfusion process shortened the recovery time from uremia pericarditis significantly, and reduced the necessary hemodialysis times. Stefoni *et al.*, researched if the hemoperfusion/hemodialysis combined therapy can reduce the hemodialysis times per week in long-term patients. They treated five patients of 350mL urine volume per day with two times/week hemoperfusion/hemodialysis combined therapy to replace the standard three times/week hemodialysis. They found the concentration of plasma creatinine, uric acid and urine had no significant change. The health condition of the patients was good. It demonstrated that the combined therapy could shorten the total time of treatment to one-third.

Hemoperfusion is good at clearing uric acid, creatinine and middle molecule material, but it is bad at clearing urea. Moreover, it has no effect on water, electrolyte and acid-base imbalance. Therefore, it cannot be used by itself to treat uremic patients. Activated charcoal hemoperfusion and hemoperfusion/hemodialysis combined therapy when used together could effectively clear the uric acid, creatinine and middle molecule substance. It can also fix the water, electrolyte and acid-base imbalance. But this kind of therapy will increase the extracorporeal blood volume, so the doctors need to pay attention to monitoring of blood pressure during the treatment process.

4.3. *Hepatic Encephalopathy*

The mechanism of hepatic encephalopathy is still unknown; it is generally acknowledged that the mechanism may be related to a number of factors, such as increased blood ammonia, false neurochemical transmitters and higher aromatic amino acid concentration and the imbalance in the ratio of branched chain amino acids and aromatic amino acids in the blood. Hemoperfusion can clear ammonia, false neurochemical transmitters, such as hydroxyl ethanolamine, free fatty acid, phenol, mercaptan, and aromatic amino acids. It can also increase the ratio of branched chain amino acids and aromatic amino acids, and increase the cAMP concentration in cerebrospinal fluid, so it can be used to treat the hepatic coma.

In 1972, Chang (Chang, 1972) first successfully treated a patient with grade IV hepatic encephalopathy with ACAC. This resulted in the recovery of consciousness within two hours. Gelfand *et al.* reported that doctors used the 300g acrylic acid hydrogel to package the activated charcoal then conducted the hemoperfusion process for 10 grade IV hepatic encephalopathy patients. In the result, nine patients recovered consciousness, four cases survived, blood ammonia decreased 23%, blood aromatic amino acids, such as styrene acrylic amino acids, tyrosine and histidine decreased 29–44%. The ratio of branched chain amino acids and aromatic amino acids in the blood increased from 1.08 ± 0.23 to 1.3 ± 0.4, The ratio of branched chain amino acids and aromatic amino acids in the cerebrospinal fluid increased from 0.26 ± 0.2 to 0.32 ± 0.03. At the same time, the cAMP concentration increased significantly in the CSF, from 2.3 ± 0.7 to 3.4 ± 1.2ng/mL. Some researchers consider that it may improve the energy metabolism of brain by increasing the cAMP concentration in CSF resulting in the recovery of symptom of hepatic encephalopathy. In China, it was proved that in NK-110 resin *in vitro* experiment, it has good absorptive ability for bile acid. The researcher from Nankai University synthesized the macroporous resin which has excellent absorptive ability — the max absorptive value of bilirubin reached 24 mg per gram. The clinical investigation proved that the bilirubin concentration of jaundice patients decreased from 256.6 to 126.5 μmol/L. Agishi used hemoperfusion to treat 18 patients with acute hepatic failure and 23 patients with chronic hepatic failure exacerbation. Thirteen of the 18 acute hepatic failure patients recovered consciousness and four survived. Twenty-two of the 23 chronic hepatic failure patients recovered consciousness and seven survived. Gazzard used hemoperfusion therapy to treat hepatic encephalopathy patients; the survival rate was 38%, In the group without hemoperfusion, the survival rate was only 10–15%. According to the afore-mentioned data analysis, the effect of hemoperfusion treatment on the recovery of consciousness in hepatic encephalopathy is certain. But in the treatment process, the doctor should take note of the blood compatibility. Because the hepatic disease patients already have the defect of deceasing platelets and blood coagulation factors, some researchers thought the patient should have fresh plasma or platelets imported before hemoperfusion.

4.4. *Immune Diseases*

The principle of this treatment is the use of an adsorbent to absorb the antibodies in the blood, antigen composite system, the decomposition product and other components, to remove the immune substance in blood. For example, we can use the adsorbents that contain protein A to remove the antibody IgG. We also can use the IgE compositive absorptive substance to treat asthma. Some researchers designed an adsorbent to absorb the some component of disassembled products to block the complement system. The activated charcoal adsorbent which contains DNA could remove the DNA antibody, and use the vector material of blood antigen (A or B) to remove the anti A or B antibody in the blood so it can be used for the treatment of connective tissue diseases. Some people conduct the immune absorption before renal transplant to remove the anti-A and anti-B antibody in the blood (the adsorbent contains the composite A or B blood type antigen).

Most patients' antibody concentration is very low after transplant. Moriconi *et al.* used protein A immune absorption perfusion to absorb anti-neutrophil cytoplasmic antibodies (ANCA). This treatment may work with diseases with companying ANCA antibodies. The experiment proved that in chronic renal failure, immune absorption could absorb the elevated β_2 microglobulin in the plasma. Using chromatographic theory *in vitro* perfusion to conduct the LDL immune adsorption could decrease the total blood cholesterol in blood to 40%; it could improve the metabolism of lipoprotein.

4.5. *The Application of Hemoperfusion in Combined Tumor Chemotherapy*

In 1985, Nakana *et al.* infused anti-cancer drugs in the artery of dogs (cis-platinum, Mitomycin, MMC+ADM), with hemoperfusion in the venous outflow to remove the drug: regional hemoperfusion therapy. The anti-cancer drugs' concentration in the blood and urine of the canines decreased significantly. This result had been considered to indicate combined therapy was beneficial to reducing renal toxicity. The five-year survival rate increased from 48.6 to 69.8%; it had been thought valuable for late phase tumors. There was another report which was about brain tumor and hepatic tumor using regional hemoperfusion therapy. Some researchers thought that therapy was a hopeful assisting treatment for late-phase brain tumor and hepatic tumor.

4.6. *Infection-related Conditions*

In the present, most of researchers consider that the hemoperfusion process has certain absorptive effect on toxins in plasma. Someone put PMX-B and charcoal or other vectors together. The *in vitro* experiment proved that there was a selective endotoxin combination effect. When using the cation group to remove the bacterial endotoxin in the plasma, the activated charcoal and non-ion resin (XAD-7) can absorb various cell factors and endotoxin in plasma. The activated charcoal and resin can absorb the endotoxins TNF, IL-1 and IL-6, which were marked with iodine. In China, researchers have already conducted an investigation on the treatment of burns in large and middle molecular peptides disease. They thought it could improve the symptoms of poisoning, and middle molecular substances and protein-dissolved active components were reduced in blood.

4.7. *Other Diseases*

Maeda *et al.* reported that six cases of psoriasis patients had been treated with hemoperfusion, After treatment, four cases recovered and two cases were improved; the clinical outcome was satisfactory. Hermann *et al.* used hemoperfusion to rescue three thyroid crisis patients; after four hours' hemoperfusion, the total concentration of T4 and T3 in blood of patients decreased significantly. Two cases had improved significantly. After 12 hr, the patients recovered consciousness. The application of hemoperfusion to treat psychiatric patients has already been used for a long time. After the treatment, the psychiatric symptoms were improved. The mechanism of the effect is still unclear. It could be related to the removal of β endophin within the blood. Wu *et al.* reported that after hemoperfusion therapy, drug overdose patients and toxin poisoning psychiatric patients showed significant improvement in their psychiatric symptoms. Recently, researchers in China used hemoperfusion to treat bronchial asthma, improving the symptoms and decreasing the frequency of asthmatic attacks (Yingming Gu and Ying Huang, 2015).

References

Chang TMS. (1966) Semipermeable aqueous microcapsules ("artificial cells"): with emphasis on experiments in an extracorporeal shunt system. *Trans Am Soc Artif Intern Organs* **12:** 13–19.

Chang TMS. (1969) Removal of endogenous and exogenous toxins by a microencapsulated absorbent. *Can J Physiol Pharmacol* **47(12):** 1043–5.

Chang TMS, Gonda A, Dirks JH, Malave N. (1971) Clinical evaluation of chronic, intermittent, and short term hemoperfusions in patients with chronic renal failure using semipermeable microcapsules (artificial cells) formed from membrane-coated activated charcoal. *Trans Am Soc Artif Intern Organs* **17:** 246–252.

Chang TMS. (1972) Haemoperfusions over microencapsulated adsorbent in a patient with hepatic coma. *Lancet* **2:** 1371–1372.

Chang TMS, Gonda A, Dirks JH, *et al.* (1972) ACAC microcapsule artificial kidney for the long term and short term management of eleven patients with chronic renal failure. *Trans Am Soc Artif Intern Organs* **18:** 465–472.

Chang TMS, Coffey JF, Barre P, *et al.* (1973) Microcapsule artificial kidney: treatment of patients with acute drug intoxication. *Can Med Assoc J* **108:** 429–433.

Chang TMS. (1975) Microencapsulated adsorbent hemoperfusion for uremia, intoxication and hepatic failure. *Kidney Int* **7:** S387–S392.

Gazzard BG, Portmann BA, Weston MJ, *et al.* (1974) Charcoal haemoperfusion in the treatment of fulminant hepatic failure. *Lancet* **1:** 1301–1307.

Gu Y, Huang Y. (2015) Personal communication.

Gurland H, Samtleben W, Lysaght MU, *et al.* (1996) Extracorporeal blood purification techniques. In: Jacobs C, Kjellstrand CM, Koch KM (eds), *Replacement of Renal Function by Dialysis*, 4th ed. Dordrech: Klurwer Academic Publishers, p. 472.

He, Changmi and Xun Zhang (1999). Renal replacement therapy. Shanghai: science and technology literature, pp. 193–198.

Martin AM, Gibbins JK, Oduro A, Herbert R. (1978) Clinical experience with cellulose-coated carbon hemoperfusion. In: Chang TMS (ed), *Artificial Kidney, Artificial Liver, and Artificial Cells.* Plenum, New York, pp. 143–152.

Wang, Zhi Gang. Blood purification science. The first edition. Beijing: Beijing science and technology **92:** 219–230.

Yatzidis H. (1964) A convenient hemoperfusion micro-apparatus over charcoal for the treatment of endogenous and exogenous intoxications. *Proc Eur Dialysis Transplant Assoc* **1:** 83.

Zi DX, *et al.* (1995) Blood purification technology in the treatment of acute drug or poison intoxication. *Journal of Nanjing University* **31:** 8.

Chapter 3

Pathophysiology and Types of Poisoning

Han Li* and Shixiang Wang*,†

1. Introduction

Hemoperfusion is a blood purification method based on adsorption, and is an extracorporeal form of treatment because the blood is pumped through a device outside the patient's body. Hemoperfusion was first introduced in the 1940s, and was refined between 1950 to 1970, and then introduced clinically for the treatment of poisoning in the 1970s and 1980s (Rahman and others, 2006). Despite its availability, this technique is only infrequently utilized as a medical process used to remove toxic substances from a patient's blood (Tyagi *et al.*, 2008). During hemoperfusion, the blood passes through a column with absorptive properties aiming at removing specific toxic substances from the patient's blood (Shalkham and others, 2006). It especially targets small- to medium-sized molecules that tend to be more difficult to remove via conventional hemodialysis. The adsorbent substances most commonly used in hemoperfusion are resins and activated carbon (Ghannoum and Gosselin, 2013).

Its major uses include removing drugs or poisons from the blood in emergency situations, removing waste products from the blood in

*Department of Blood Purification, Beijing Chao-Yang Hospital, Capital Medical University, Beijing, China.

*,†Corresponding author: Shixiang Wang, E-mail: wxy1988@263.net.

patients with renal failure, and so on. This chapter mainly introduces toxins and detoxification via hemoperfusion.

2. Poisoning Profile

Poisoning induced by all kinds of toxins is not uncommon in clinic. In 2010, poisoning resulted in about 180,000 deaths, down from 200,000 in 1990 (Lozano, 2012). There were approximately 727,500 emergency department visits in the United States involving poisonings — 3.3% of all injury-related encounters (Villaveces *et al.*, 2013).

In 2004, poisoning was second only to motor vehicle crashes as a cause of death from unintentional injury in the US. Nearly all poisoning deaths in the US are attributed to drugs, and most drug poisonings result from the abuse of prescription and illegal drugs (CDC, 2007). The CDC reported that poisoning mortality rates in the US increased each year from 1999 to 2004, rising 62.5% during the five-year period. The largest increases were among females (103.0%), whites (75.8%), persons living in the southern United States (113.6%), and persons aged 15–24 years (113.3%). Larger rate increases occurred in states with mostly rural populations. Rates for drug poisoning deaths increased 68.3%, and mortality rates for poisonings by other substances increased 1.3% (CDC, 2007).

The toxins leading to poisoning include bio-toxins and chemical toxins. Some toxins may be poisons that occur naturally in plants and animals, chemical contaminants, or toxic products of microorganisms. Poisoning causes include drug overdose, chemical intoxication and contact poisoning caused by mistake (Malhi and others, 2014).

Acute poisoning is exposure to a poison on one occasion or during a short period of time. Symptoms develop in close relation to the exposure. Absorption of a poison is necessary for systemic poisoning. In contrast, substances that destroy tissue but are not absorbed, such as lye, are classified as corrosives rather than poisons. Furthermore, many common household medications are not labeled with skull and crossbones, although they can cause severe illness or even death.

Chronic poisoning is long-term repeated or continuous exposure to a poison where symptoms do not occur immediately or after each exposure. The patient gradually becomes ill, or becomes ill after a long latent period. Chronic poisoning most commonly occurs following exposure to poisons that are bio-accumulate or bio-magnified, such as mercury and lead.

Contact with or absorption of poisons can cause rapid death or impairment. Agents that act on the nervous system can cause paralysis in seconds or less, and include biologically derived neurotoxins.

Most biocides, including pesticides, are created to act as poisons to target organisms, although acute or less observable chronic poisoning can also occur in non-target organisms, including the humans who apply the biocides and other beneficial organisms. Many substances regarded as poisons are toxic only indirectly. An example is "wood alcohol", or methanol, which is not poisonous itself, but is chemically converted to toxic formaldehyde and formic acid in the liver. Many drug molecules are made toxic in the liver, and the genetic variability of certain liver enzymes makes the toxicity of many compounds differ between individuals.

Poisoning resulting in toxic effects have mainly three aspects: one is direct cause cellular damage, leading to organ dysfunction such as hepatotoxicity and nephrotoxicity; the second is blocking the body's normal physiological reflex, inhibiting the body's normal functions, such as how antipsychotic drugs inhibit nerve function and cause respiratory and cardiac function inhibition; the third is breaking the stability of the internal environment, inducing water and electrolyte disorder, such as serious organic acid poisoning, and so on.

The treatment principle of acute poisoning is immediately suspending contact with the poison, clearing the toxins which have entered the body but have not yet been absorbed, using the specific antidote, symptomatic and supportive treatment, and promoting excretion of the absorbed poison using blood purification, such as hemoperfusion (Gil *et al.*, 2010).

Many acute poisoning have no special detoxification therapy. Support therapy is very important — it can help critically ill patients survive, protect important organs and restore their function. Acute poisoning patients should stay in bed and keep warm. Medical staff should closely observe the patient's mind, breathing, circulation, and give the corresponding processing.

Blood purification can correct the internal environment of disturbances caused by poisoning. Because of toxic liver and kidney dysfunction in critical patients limiting the ability to remove toxins, removing poison by the method of blood purification is the best choice. The therapeutic effects of blood purification in patients with poisoning include maintaining and replacing organ function, renal function in particular, and keeping the balance of the internal environment. The most important is removing toxins, thus reducing the toxicity (Ghannoum *et al.*, 2014).

3. Influencing Factors of Blood Purification in Poison Removal

3.1. ***Volume of distribution.*** The volume of distribution (V_D) is toxic doses divided by toxicant concentration under steady state distribution volume; it is a mathematical value, rather than the actual volume of distribution, but it represents the proportion of poison, both inside and outside the blood vessels distribution. Heparin, for example, a drug confined to the blood compartment, has a V_D of approximately 0.6 L/kg. Drugs distributed primarily in the extracellular water (e.g. cephalothin) will have a V_D of approximately 0.2 L/kg. Some poisons will have V_D values exceeding the volume of total body water (0.6 L/kg) because they are extensively bound to, or stored in the tissue.

Substances with high rate of organization have a large volume distribution, mainly distributing in the extravascular blood vessels. Substances with high protein binding in the blood have a small volume distribution, mainly distributing in the intravascular blood vessels. Generally the smaller the volume of distribution, the faster the excretion of toxic substances, and the shorter the residence time in the body. Greater volume of distribution causes slower excretion and longer retention in the body. With poisons that have a high V_D (e.g. digoxin, glutethinmide, tricyclis), the amount of poison present in the blood represents only a small fraction of the total body load. Thus, even if a hemoperfusion treatment extracts most of the poison present in the blood flowing through the extracorporeal circuit, the amount of drug removed during a single treatment session will represent only a small percentage of the total body load. Subsequently, additional amounts of the drug will enter the blood from tissue stores, sometimes causing a recurrence of the toxic manifestations. On the other hand, even transiently lowering the blood concentration of many drugs may mitigate certain important toxic effects of these agents. Hence, hemoperfusion can sometimes effectively reduce drug toxicity even when the V_D is large.

3.2. *Protein binding rate.* Protein binding rate refers to the poisons in the blood with different plasma protein (mainly albumin) combining ability. Poison in the body, whether in the blood or tissue in all free form and not up to two kind of forms, the combination is loose, most toxic is reversible, therefore make the free type and combination type of the poison keep a dynamic balance. Combination type is the storage form of the poison — only free poison can play an active role and can be cleared by convection or diffusion methods. Type and junction station poison does not play an active role, cannot easily be cleared by the body itself, but can be cleared by convection and diffusion methods.

3.3. *Half-life.* Half-life refers to toxicant concentration in plasma decreased half the time required. Half-life = 0.693 x distribution volume/clearance, usually more than five half-life, 3% toxic residue remaining in the body. Half-life depends on the volume distribution and clearance.

3.4. *Clearance.* Clearance refers to how many ml of the poison in the blood plasma is cleared in the unit of time, including renal clearance, hepatic clearance and blood purification clearance. Blood purification clearance does not reflect the actual situation of the drug or poison removal, because many materials of blood drug concentration in blood purification fall sharply, resulting in a decline in the amount of the actual clearance. Therefore, generally using per unit time to remove the amount of drug or toxicant evaluation of the effect of blood purification.

4. Technical Points of Hemoperfusion

4.1. ***Vascular access.*** In patients without permanent vascular access in place, oercutaneous cannulation of a large central vein using a dialysis catheter is required.

4.2. ***Anticoagulation method during hemoperfusion.*** The adsorbent surface in hemoperfusion is rougher than dialyzer membrane in hemodialysis, and the specific surface area is large, thus increasing the contact area of the blood significantly, so the anticoagulant dosage is more than hemodialysis. Generally, heparin dosage is 0.5–1.0 mg/kg for the first time, with an additional 8–10 mg per hr after. Because the patients have individual differences, it is better to adjust the dosage of the anticoagulant according to APTT or blood coagulation time.

4.3. ***Choice of hemoperfusion cartridge.*** Typical sorbents are activated carbons (charcoals), ion exchange resins, or nonionic exchange macroporous resins. Sorbent particles have been rendered biocompatible by coating the surface with a polymer membrane. The cartridges contain various amounts of a sorbent, the smaller ones being designed for pediatric use. No controlled comparative evaluation *in vivo* performance of the various brands of cartridges has been published.

4.4. ***The hemoperfusion circuit.*** The hemoperfusion circuit is similar to the blood side of a hemodialysis circuit and includes an air detector and a venous air trap. Standard hemodialysis blood pumps and machines (without use of dialysis solution) are often used to drive the blood through the tubing and cartridge.

4.5. ***Priming the hemoperfusion circuit.*** Setup and priming procedures differ depending on the brand of cartridge used and the manufacturer's literature should be consulted in all instances. The hemoperfusion cartridge must be primed in a vertical position with the arterial (blood inlet) side facing down. One manufacturer (Gambro) recommends that its cartridges be rinsed initially with 500 mL of 5% dextrose in water to load the charcoal with glucose.

This maneuver is alleged to result in a lesser drop in the plasma glucose level during the hemoperfusion treatment. Not all manufacturers recommend a glucose rinse.

4.6. *Duration of hemoperfusion.* Hemoperfusion duration depends on the saturated adsorption capacity and speed of adsorption materials used. The adsorption using activated carbon as adsorbent will be close to saturation after two to three hr perfusion for most of the solute, so you need to switch to another perfusion after two hours, but a perfusion time not more than six hr should be conducted. For high levels of lipotropy drugs or toxic poisonings, after a blood perfusion the drug or poison will be released from fat tissue into the blood, causing poisoning symptoms, therefore there is a need after a period of time between each blood perfusion to clear the drug or poison two to three times. A single three-hour treatment will substantially lower the blood levels of most poisons for which hemoperfusion is effective. Prolonged use of one hemoperfusion cartridge is inefficient because the charcoal tends to become saturated, especially when cartridges containing less than 150 g charcoal are used. Replacement of saturated devices with fresh ones is not usually required, and any rebound in blood drug concentrations consequent to tissue release can be treated with a second hemoperfusion session. On the other hand, a continuous hemoperfusion treatment may need to be prolonged for several days until clinical improvement or nontoxic blood level is achieved. Hemoperfusion devices may need to be changed every four hours in the course of continuous treatment.

5. Detoxification by Hemoperfusion

5.1. *Organophosphate Poisoning*

Organophosphorus pesticides are fat-soluble macromolecular substances, which can easily bind to proteins. Once ingested, they can rapidly distribute across organs and tissues. By inhibiting cholinesterase, these pesticides cause massive accumulation of acetylcholine. The resulting cholinergic hyperactivity causes nicotinic and muscarinic signs, as well as central nervous system signs and symptoms. After absorption, organophosphorus pesticides can be stored in adipose tissues, brain, liver, kidney, lung, and gastrointestinal wall, forming gut-blood-gut circulation, which allows organophosphorus pesticides to be stored in the fat and other tissues, and to be released back into the blood. Recent research compared organophosphate poisoning patients who received one hemoperfusion with organophosphate poisoning patients with repeated hemoperfusion, who had significantly less atropine use, shorter time of awakening from coma, higher cure rates, shorter time until normalization of cholinesterase levels, lower appearance of intermediate myasthenia syndrome and higher survival rates. The author concluded that early repeated hemoperfusion is more efficient than single hemoperfusion in treating organophosphate poisoning (Bo, 2014).

5.2. *Caffeine Poisoning*

Caffeine, 1, 3, 7-trimethylxanthine, is widely sold as a stimulant, anorectic, and diuretic, and it can be conveniently obtained in drug stores. Caffeine is rapidly absorbed with a peak level occurring 0.5 to 2 hours after ingestion. It is metabolized primarily in the liver by the cytochrome P4501A2 system, and the elimination half-life is typically three to six hours in adults. However, this is not the case with caffeine overdoses, and the half-life may be several-fold longer. A number of cases of severe caffeine intoxication have been reported, and the estimated lethal dose is 10 to 20 g; a possible lethal caffeine concentration is estimated to be 80 mg/L.

Ishigaki S *et al.* (Ishigaki and others, 2014) recently reported a case of a 32-year-old man who ingested approximately 15.6 g of caffeine in a suicide attempt, who suffered from sustained ventricular tachycardia despite conservative treatment. Therefore, the patient was given a combined therapy of hemoperfusion and hemodialysis which resulted in rapid improvement of the patient's ventricular tachycardia as well as a reduction of the patient's plasma caffeine level.

5.3. *Herbicide Poisoning*

Herbicides like paraquat and glyphosate are the most common poisons responsible for high morbidity and mortality (Roberts and others, 2010). Paraquat (1, 1'-dimethyl-4, 4'-bipyridium dichloride) is a non-selective contact herbicide that is relatively widely used (Liu and others, 2014). It is a very dangerous pollutant of the environment, as it readily binds to both clay and organic matter in soil and is very slowly biodegraded. Delayed toxicity with pulmonary fibriosis, renal and multi-organ failure can occur following ingestion of more than 10 ml of paraquat (Khodayar *et al.*, 2014). Plasma levels of above 3 mg/L, regardless of when they are measured, are usually fatal. Initial management includes gastric lavage with administration of activated charcoal. Hemoperfusion is effective in drug removal and should be considered when the plasma paraquat level is 0.1 mg/L or above. Repeated or continuous hemoperfusion may be needed for several days to maintain plasma levels below 0.1 mg/L as paraquat has a large V_D and a slow intercompartmental transfer rate. Recently, Cherukuri H *et al.* reported that hemoperfusion may be useful in reduction of high mortality rate of herbicide poisoning (Cherukuri *et al.*, 2014).

5.4. *Acetaminophen Poisoning*

Activated charcoal perfusion should be given to acetaminophen poisoned patients within four hours of ingestion. Serum levels should be measured and plotted against the Rumack–Matthew nomogram to establish the risk for hepatotoxicity and requirement for N-acetyl cysteine (NAC) therapy. If serum acetaminophen levels are above 150 mg/L at 4 hr, the likelihood of toxicity is high and NAC should be given. Liver toxicity may be found at lower serum levels when acetaminophen ingestion is combined with even moderate amounts of ethanol. NAC, by increasing reduced glutathione stores, prevents the accumulation of the toxic byproducts. The efficacy to prevent liver failure declines if perfusion started more than 10 hours after ingestion. It may still be worth it if the treatment is conducted after 24 hours. Acetaminophen is moderately water soluble and is minimally protein bound and thus can be removed by hemoperfusion (Bryczkowski and Geib, 2012).

5.5. *Digoxin Poisoning*

Digoxin is widely used but has a narrow therapeutic index (Sanaei-Zadeh *et al.*, 2011). Serum digoxin levels below 0.8 ng/ml are considered sub-therapeutic whereas levels more than 2.0 ng/ml may be toxic. Toxic manifestations include cardiac and non-cardiac manifestations. Patient may either develop brady or tachy-arrhythmias. Electrocardiographic changes may include premature ventricular contractions, atrial fibrillation or flutter, non-paroxysmal atrioventricular junctional tachycardia or atrioventricular blocks. But none of these are pathognomonic of digoxin toxicity. Non-cardiac symptoms may include nausea, vomiting, lethargy, anorexia, headache or confusion. Rarely, patients may develop visual disturbances including diplopia, blurred or yellow vision, as was in the present case. Co-existent electrolyte imbalance, especially hypercalcemia, hypokalemia and hypomagnesemia, can potentiate digoxin toxicity. The V_D of digoxin is large (8 L/kg in normal patients, 4.2 L/kg in dialysis patients), and the drug is 25% protein bound. For these reasons, only 5% of the body load will be removed by a four-hour hemodialysis treatment. Hemoperfusion is more effectively and has been shown to improve symptoms of digoxin poisoning.

Juneja D *et al.* (Juneja and others, 2012) recently presented a case of a young female who presented to our casualty with alleged history of consumption of 17.5 mg of digoxin tablets. After admission to ICU, the patient developed atrioventricular blocks with hemodynamic instability which had to be managed with temporary pacemaker. Her serum digoxin levels were high (12.63 ng/mL) and in the absence of Fab fragments, resin hemoperfusion was done which drastically reduced the serum digoxin levels and reverted the symptoms.

6. Hemoperfusion-related Side Effects

Hemoperfusion-related side effects are hypocalcaemia, thrombocytopenia, coagulopathies, and hypothermia. Mild transient trombocytopenia and leukopenia can occur, but levels usually return to normal within 24 to 48 hr following a single hemoperfusion. Adsorption or activation of coagulation factors has also been observed rarely and may be clinically significant in patients with liver failure (Isik *et al.*, 2013).

7. The Future of Detoxification by Hemoperfusion

In summary, hemoperfusion is an extracorporeal treatment based on adsorption, historically reserved for the treatment of acute poisonings. Its use was popularized in the 1970s after several *in vitro* and animal experiments had demonstrated its efficacy, and it was even preferred over hemodialysis in the management of overdosed patients. With the advent of new and more efficient dialytic modalities, hemoperfusion is now less frequently performed in the western world. However, hemoperfusion still remains popular in developing countries (Ghannoum *et al.*, 2014).

8. References

(CDC) CfDCaP. (2007) Unintentional poisoning deaths — United States, 1999–2004. *Morb Mortal Wkly Rep* **56:** 93–6.

Bo L. (2014) Therapeutic efficacies of different hemoperfusion frequencies in patients with organophosphate poisoning. *Eur Rev Med Pharmacol Sci* **18:** 3521–3.

Bryczkowski C, Geib AJ. (2012) Combined butalbital/acetaminophen/caffeine overdose: Case files of the Robert Wood Johnson Medical School Toxicology Service. *J Med Toxicol* **8:** 424–31.

Cherukuri H, Pramoda K, Rohini D, *et al.* (2014) Demographics, clinical characteristics and management of herbicide poisoning in tertiary care hospital. *Toxicol Int* **21:** 209–13.

Ghannoum M, Bouchard J, Nolin TD, *et al.* (2014) Hemoperfusion for the treatment of poisoning: technology, determinants of poison clearance, and application in clinical practice. *Semin Dial* **27:** 350–61.

Ghannoum M, Gosselin S. (2013) Enhanced poison elimination in critical care. *Adv Chronic Kidney Dis* **20:** 94–101.

Gil HW, Kim SJ, Yang JO, *et al.* (2010) Clinical outcome of hemoperfusion in poisoned patients. *Blood Purif* **30:** 84–8.

Ishigaki S, Fukasawa H, Kinoshita-Katahashi N, *et al.* (2014) Caffeine intoxication successfully treated by hemoperfusion and hemodialysis. *Intern Med* **53:** 2745–7.

Isik Y, Soyoral L, Karadas S, *et al.* (2013) Effectivity of one session charcoal hemoperfusion treatment in severe carbamazepine poisoning. *Iran Red Crescent Med J* **15:** 749–51.

Juneja D, Singh O, Bhasin A, *et al.* (2012) Severe suicidal digoxin toxicity managed with resin hemoperfusion: A case report. *Indian J Crit Care Med* **16:** 231–3.

Khodayar MJ, Kiani M, Hemmati AA, *et al.* (2014) The preventive effect of atorvastatin on paraquat-induced pulmonary fibrosis in the rats. *Adv Pharm Bull* **4:** 345–9.

Liu MW, Su MX, Zhang W, *et al.* (2014) Protective effect of Xuebijing injection on paraquat-induced pulmonary injury via down-regulating the expression of p38 MAPK in rats. *BMC Complement Altern Med* **14:** 498.

Lozano R, Naghavi M, Foreman K, *et al.* (2012) Global and regional mortality from 235 causes of death for 20 age groups in 1990 and 2010: A systematic analysis for the Global Burden of Disease Study 2010. *Lancet* **380:** 2095–128.

Malhi N, Turnbull A, Tan J, *et al.* (2014) A national survey of marine biotoxins in wild-caught abalone in Australia. *J Food Prot* **77:** 1960–7.

Rahman MH, Haqqie SS, McGoldrick MD. (2006) Acute hemolysis with acute renal failure in a patient with valproic acid poisoning treated with charcoal hemoperfusion. *Hemodial Int* **10:** 256–9.

Roberts DM, Buckley NA, Mohamed F, *et al.* (2010) A prospective observational study of the clinical toxicology of glyphosate-containing herbicides in adults with acute self-poisoning. *Clin Toxicol (Phila)* **48:** 129–36.

Sanaei-Zadeh H, Valian Z, Zamani N, *et al.* (2011) Clinical features and successful management of suicidal digoxin toxicity without use of digoxin-specific antibody (Fab) fragments — is it possible? *Trop Doct* **41:** 108–10.

Shalkham AS, Kirrane BM, Hoffman RS, *et al.* (2006) The availability and use of charcoal hemoperfusion in the treatment of poisoned patients. *Am J Kidney Dis* **48:** 239–41.

Tyagi PK, Winchester JF, Feinfeld DA. (2008) Extracorporeal removal of toxins. *Kidney Int* **74:** 1231–3.

Villaveces A, Mutter R, Owens PL, Barrett ML. (2013) Causes of Injuries Treated in the Emergency Department, 2010: Statistical Brief #156.

Chapter 4

Application of Hemoperfusion in Artificial Kidney

Du Hongyan,* Xie Qingwu and Cai Jianhua

1. Introduction

The artificial kidney is an important application of hemoperfusion treatment. For more than 10 years of development, the combination of hemoperfusion (HP) and hemodialysis (HD), i.e. combined artificial kidney (HD + HP), has been gradually accepted in clinical applications and has become an ideal treatment for maintenance hemodialysis patients.

Before the emergence of the combined artificial kidney, maintenance hemodialysis was a standard treatment for uremic patients. However, pure hemodialysis is less effective than nature kidney, and accumulated uremic toxins can cause the long-term complications in dialysis patients. In order to improve the removal of uremic toxins, importance has been attached to combination therapies such as the combined artificial kidney.

The complementation of adsorption separation with membrane separation can remove uremic toxins more completely than previous treatments. Clinical applications have indicated that combined HD and HP treatments can remove uremic toxins more effectively and

---

* R&D Department, Jafron Co., Ltd., Address: No. 98, Technology Six Road, High-tech Zone, Zhuhai, Guangdong, China, E-mail: duhongyan@jafron.com.

improve the clinical symptoms of uremic patients, and is one of the most important advances in kidney disease treatments during the recent decade.

This paper will review the emergence and development of the combined artificial kidney from the pathophysiology of kidney disease, including the process and treatment modalities of the combined artificial kidney and the advances in relevant clinical studies.

2. Kidney Diseases and Artificial Kidney Technology

2.1. *An Overview of Kidney Disease*

The kidney is an important organ that maintains a balance in the internal environment, and its primary functions include eliminating the end products of protein metabolism, regulating body fluid, electrolyte, and acid-base balances, and secreting active substances, such as renin and erythropoietin.

Kidney disease can be divided into chronic kidney disease (CKD) and acute kidney injury (AKI), and most instances occur in the former group. CKD is the progressive loss in renal function, and has a high incidence rate worldwide. It is caused by primary glomerulonephritis, hereditary nephritis, secondary diabetic nephropathy and hypertensive nephropathy, and no effective treatment has been found to treat CKD (Wang Zhigang, 2010).

CKD is a worldwide public health problem. Data from the third National Health and Nutrition Examination Survey (NHANES III) showed that 11% of ordinary Americans over the age of 20 suffered from varying degrees of CKD. A cross-sectional study of CKD in China found that about 10.8% of the Chinese people over the age of 18 suffered from CKD (Luxia Zhang *et al.*, 2012).

About 1–2% of CKD patients develop end-stage renal disease (ESRD), also known as uremia. The National Kidney Foundation (NKF) Kidney Disease Outcomes Quality Initiative (KDOQI) Clinical Practice Guidelines for Chronic Kidney Disease divide CKD into five separate stages. These stages are based on glomerular filtration rate

(GFR), and stage 5 CKD is defined as end stage renal disease, with a GFR lower than 15 mL/(min·1.73 m^2).

The clinical manifestations of uremia are complex, and almost all physiological systems of the human body can be affected to varying degrees, including gastrointestinal symptoms, blood system symptoms (e.g. anemia and bleeding tendency), cardiovascular system symptoms, renal osteodystrophy, neuropsychiatric symptoms, respiratory symptoms, skin symptoms, water, electrolyte and acid-base imbalance, endocrine and metabolic disorders, and immune dysfunction (Cyrielle Almeras and Angel Argiles, 2009).

2.2. *Kidney Disease and Uremic Toxins*

These complex clinical symptoms of uremia are primarily caused by the large number of substances metabolized through the kidneys that cannot be removed from the body due to impaired kidney functioning. Studies have shown that the levels of more than 200 substances in uremic patients are significantly higher than those in people without kidney disease, and the substances closely related to the metabolic disorders in clinical manifestations of uremia are called uremic toxins.

According to studies by the European Uremic Toxin Work Group (EUTox), uremic toxins are divided into three types: small-molecule toxins, middle-molecule toxins, and protein-bound toxins (Vanholder R *et al.*, 2003; Vanholder R *et al.*, 2008; Vanholder R, Van Laecke S and Glorieux G, 2008; Vanholder R and Ziad A. Massy, 2009).

Small-molecule toxins are defined as small free molecules with molecular weights lower than 500 Daltons, such as urea (60), creatinine (113), and uric acid (168).

Middle-molecule toxins have a broad molecular weight range which is generally considered to be from 500 Daltons to 65,000 Daltons. Examples of middle-molecule toxins include the parathyroid hormone (9,225), β2-microglobulin (11,818), interleukin-6 (26,000), complement factor D (23,750), and leptin (16,000).

Protein-bound toxins are toxins with small molecular weights that bind with serum proteins due to charges or molecular configuration, so that their chromatography displays the properties of middle-to-large molecules. These toxins include phenol (94), spermine (202), and indoxyl sulfate (251).

Currently, the understanding of uremic toxins is far from being sufficient, and the number of newly discovered toxins is on the increase. In 2003, the European Uremic Toxin Work Group reviewed and analyzed 857 papers published between 1968 and 2002, and classified 90 uremic toxins. In 2008, this work group classified 25 new uremic toxins, adding to the complexity of studying uremic toxins.

However, as the number of relevant studies continues to rise, the relationships between these uremic toxins and clinical symptoms, as well as the pathogenetic mechanisms, are being more explicitly described.

Small-molecule toxins have been studied thoroughly, including the minerals, urea, and creatinine accumulating in the body. The use of creatinine, urea nitrogen and serum potassium concentrations as dialysis indicators, and urea clearance as a dialysis adequacy indicator, has become standard methodology in clinical applications.

B2-microglobulin (β2-MG), parathyroid hormone (PTH) and cytokines are the most thoroughly studied middle-molecule toxins. B2-MG is a protein that is the most directly related to dialysis-related amyloidosis, as well as an important indicator for predicting the mortality of hemodialysis patients. PTH has two well-defined roles: first, the concentration of PTH has been considered an important symbol for renal osteopathy treatment; second, PTH, calcium, phosphorus, and calcium-phosphorus products are independent risk factors for cardiovascular disease. Cytokines have also proved to be related to a great number of CKD complications such as atherosclerosis and calcification.

Indoxyl sulfate and p-cresol are the two most important protein-bound toxins. Considerable evidence suggests that these two substances are related to cardiovascular complications in CKD patients, and that indoxyl sulfate is also related to the progression of CKD.

Table 4.1. Middle-Molecule and Protein-Bound Toxins

Toxin	MW	Toxin	MW
Middle-Molecule Toxins			
Adrenomedullin	5,729	Hyaluronic acid	25,000
Atrial natriuretic peptide	3,080	Interleukin-1β	32,000
β2-microglobulin	11,818	Interleukin-6	24,500
endorphin	3,465	κ-Ig light chain	25,000
Cholecystokinin	3,866	λ-Ig light chain	25,000
Clara cell protein (CC16)	15,800	Leptin	16,000
Complement factor D	23,750	Methionine-enkephalin	555
Cystatin C	13,300	Neuropeptide Y	4,272
Degranulation inhibiting protein I[c]	14,100	Parathyroid hormone	9,225
Delta-sleep inducing peptide	848	Retinol-binding protein	21,200
Endothelin	4,283	Tumor necrosis factor-α	51,000
Protein-bound Toxins			
2-methoxyresorcinol	140	Melatonin	126
3-deoxyglucosone	162	Methylglyoxal	72
CMPF	240	N^{ε}–(carboxymethyl)lysine	204
Fructoselysine	308	p-cresol	108
Glyoxal	58	Pentosidine	342
Hippuric acid	179	Phenol	94
Homocysteine	135	P-OHhippuric axid	195
Hydroquinone	110	Putrescine	88
Indole-3-acetic acid	175	Quinolinic acid	167
Indoxyl sulfate	251	Retinol-binding protein	21,200
Kinurenine	208	Spermidine	145
Kynurenic acid	189	Spermine	202
Leptin	16,000		

Indoxyl sulfate is produced as indole which is metabolized in the liver. The protein-binding rate of indoxyl sulfate is as high as 93%, and the protein-binding rate of p-cresol is 94% (Wang Zhigang, 2010).

Clinical research has shown that removal of these substances can significantly improve survival rates in ESRD patients. Since the existence of each toxin corresponds to many symptoms, the goal is to simultaneously reduce the levels of all toxins to normal levels.

2.3. *Treatment of Kidney Disease with Artificial Kidney*

Injuries associated with CKD and some acute kidney diseases are often permanent, so patients must receive artificial kidney treatment or kidney transplants once they develop end-stage renal disease. Clinical studies suggest that these two therapies are not significantly different in the prognoses and mortality rates of patients.

Based on the association between uremic toxins and the clinical symptoms of uremia, artificial kidney treatment removes uremic toxins in order to relieve the symptoms of uremia. Maintenance dialysis is the primary treatment for uremic patients, as well as the earliest artificial kidney technology, and is used to treat about 70% of the patients that receive artificial kidney treatment.

An annual data report (ADR) published by the United States Renal Data System (USRDS) in 2011 found that a total of 370,274 patients in the United States were receiving hemodialysis by December 31, 2009. These patients were also known as maintenance dialysis patients because they received dialysis three times a week.

However, the high rate of dialysis in patients with kidney disease does not mean that it is the most effective treatment option. First, dialysis is only a maintenance treatment, and it is impossible to completely cure uremic patients whose kidneys are seriously injured using dialysis; second, dialysis only partially replaces detoxification functioning of the kidney, which is primarily caused by the complex constituents of uremic toxins. Hemodialysis can effectively remove small-molecule toxins but has limited effects on the removal of middle-molecule and protein-bound toxins that play an important role in uremia.

The constituents of uremic toxins, or their pathophysiological roles, have yet to be completely elucidated, and the conventional method of assessing dialysis adequacy, which uses small-molecule toxins (urea clearance) as the only measure, has been questioned.

Since dialysis adequacy cannot compare with kidney detoxification function, a lot of uremic toxins still accumulate in the human body, leading to a number of complications, such as amyloidosis and arteriosclerosis, which greatly impact survival rates in patients. Further, renal osteodystrophy, pruritus, and restless leg syndrome seriously affect patients' quality of life.

3. Hemoperfusion and Combined Artificial Kidney

3.1. *The Emergence of the Combined Artificial Kidney*

In order to enhance dialysis adequacy, researchers have made a great number of innovations in materials and clinical applications. Further, they have developed a series of new treatments such as hemofiltration, hemodiafiltration, high-flux dialysis, long time dialysis, hemoperfusion, and plasma adsorption (Sunny Eloot *et al.*, 2014).

These technologies can be roughly divided into membrane separation and adsorption separation. Membrane separation can effectively remove small-molecule toxins, but has a limited effect on middle-molecule and protein-bound toxins. Adsorption separation such as hemoperfusion found its mature applications in the field of kidney disease lately; however, it is highly efficient in adsorbing middle-molecule and protein-bound toxins.

Similar to the development of other hemoperfusion applications, the earliest application of hemoperfusion in the field of kidney disease used adsorbent activated carbon (George Dunea, 2012). Bock *et al.* (1920) found that activated carbon can remove nitrogenous substances such as urea and uric acid. Yatzidis *et al.* (1964) used activated carbon to conduct 20 clinical perfusion experiments for the first time, reporting that activated carbon hemoperfusion can remove creatinine, uric acid, glycosides, phenols, guanidino compounds and organic acids, and significantly improve lethargy and gastrointestinal symptoms in patients. After Chang (1966, 1969, 1975) invented collodion-coated activated carbon, the blood compatibility problem of activated carbon was solved, and collodion-coated activated carbon has been gradually applied to hepatic encephalopathy, and has been shown to be a great method to treat poisoned patients (Chang *et al.*, 1973; Chang, 1975; Fiona *et al.*, 2011).

The early design idea of hemoperfusion was to look for alternatives to dialysis. However, as an adsorbent, activated carbon can remove only uremic toxins and metabolites but not urea or water and electrolyte balance. As a result hemoperfusion was successfully used in combination with hemodialysis (HD+HP) (Chang 1975;

Chang *et al.*, 1982; Martin *et al.*, 1978). In the 1990s, hemoperfusion techniques using adsorbent resins as adsorbents were also used as a combined artificial kidney (HD + HP).

The so-called combined artificial kidney is the combined clinical use of multiple blood purification technologies, typically represented by "hemoperfusion + dialysis (HD + HP)," to treat uremia.

3.2. *The Development of the Combined Artificial Kidney*

The emergence and development of the combined artificial kidney is inseparable from studies that explore the structures and adsorption principles of adsorbents. Studies indicate that pore size and specific surface area are the primary structural parameters that affect the adsorption performance of porous adsorbent materials such as activated carbon and adsorbent resins. Pore size determines the adsorption spectrum of an adsorbent, while specific surface area determines the adsorption rate for a specific solute. Since activated carbon has a small pore size, it displays a limited ability to adsorb middle-molecule toxins.

Unlike activated carbon, adsorbent resins can be conveniently used to prepare macroporous adsorbents and improve the ability to adsorb middle-molecule toxins. The design of their skeletal structures or ligands can give adsorbent resins new binding forces, including hydrophobic binding force, electrostatic binding force, and antigen–antibody reaction, and these new properties expand the applicable scope of adsorbent resins.

In the field of kidney disease, designed adsorbent resins can effectively adsorb middle-molecule and protein-bound toxins, and can remove almost all types of uremic toxins when combined with dialysis.

From the end of the 20th century to the beginning of the 21st century, there were at least three types of resin hemoperfusion cartridges, including HA seriesHemoperfutor (Jafron Biomedical Co., Ltd., China), Lixelle (Kaneka Corporation, Japan), and betasorb (Cytosorbents Corporation, not released). Relevant product parameters are shown in Table 4.2.

In addition to the above commercial products, there is the vortex-flow plasmapheretic reactor (VFPR), which is currently studied in

Table 4.2. List of Hemoperfusion Cartridges Used for Combined Artificial Kidney

Product Name	HA130	Lixelle	Betasorb
Carrier	styrene/divinylbenzene microspheres	cellulose microspheres	styrene/divinylbenzene microspheres
Ligand	NA	cetylamine	NA
Coating material	collodion	NA	polyvinylpyrrolidone
Preservation solution	water for injection	Citric acid/Sodium citrate solution	—
Effects	molecular sieve effect and hydrophobic effect	molecular sieve effect and hydrophobic effect	molecular sieve effect and hydrophobic effect
Targets	middle-molecule and protein-bound toxins	β2-microglobulin and cytokines	middle-molecule and protein-bound toxins
Year of release	1997	1996	not released

Table 4.3. Comparison of Removal Effects of Different Hemopurification Treatments on Uremic Toxins

Removal Method	Small-molecule Toxins	Middle-molecule Toxins	Protein-bound Toxins
Low-flux dialysis	Strong	Poor to no	No
High-flux dialysis	Moderate to strong	Poor	Poor
HF	Moderate to strong	Poor	Poor
HDF	Strong	Slightly strong	Poor
HP	Varying	Strong	Strong
HD+HP	Strong	Strong	Strong
Bioartificial kidney	Strong	Strong	Unknown

laboratory settings. It is prepared by grafting anti-β2-MG antibodies onto agarose microspheres, and specifically binds with β2-MG based on antigen–antibody reaction (Eric A. Grovender *et al.*, 2004).

The Prometheus system developed by Fresenius also uses adsorbent resins as its critical component, and is primarily for artificial livers, but has also been reported to be used for kidney failure (Falko Brettschneider, 2013).

The emergence of these products has not only diversified hemoperfusion cartridges, but also made it promising to enhance dialysis adequacy and change uremia treatment as shown in the following section.

Chen Xiangmei *et al.* (2005) reviewed the uremic toxin removal efficiencies for different treatments. They found that the effective middle-molecule and protein-bound toxin adsorbing capabilities of the combined artificial kidney allow for the complete removal of uremic toxins, giving them obvious advantages over other treatments (Table 4.3).

3.3 *Treatment Modalities of the Combined Artificial Kidney*

Through more than 10 years of technological development and clinical practice, use of the combined artificial kidney has gradually formed into mature treatment modalities. Generally, the continuous use of the combined artificial kidney for 2–6 months could significantly reduce uremic toxin concentrations and improve clinical symptoms in patients.

Typically, HD and HP are used together weekly, i.e., a perfusion cartridge is applied in one out of the three treatments within a week. In real cases, the frequency of using a perfusion cartridge can also be determined by the patient's actual condition, with each usage limited below six hours.

During combined HD and HP treatment, it is advised to connect a hemoperfusion cartridge before a dialyzer. First, use the combination therapy at a blood flow velocity of 180–250 mL/min. After 120 to 150 minutes, remove the hemoperfusion cartridge and continue to dialyze for two hours in order to complete the treatment.

During HD + HP treatment, heparin or low molecular weight heparin is often first used to heparinize the hemoperfusion cartridge and to heparinize the whole body for anticoagulation (the usage and dosage vary between products).

Different from dialyzers, all current hemoperfusion cartridges are designed for disposable use. Adsorbent resins are more or less saturated within two hours of treatment. Absorbed uremic toxins are difficult to elute, and reuse can cause a huge risk.

4. The Clinical Application of the Combined Artificial Kidney

The improvements that result from hemoperfusion treatment based on adsorbent resins, especially from HD + HP, in dialysis adequacy have been gradually acknowledged over more than 10 years of clinical practice, and the treatment has become one of the standard treatments for uremic patients. In Japan and China, the use of hemoperfusion products for treating kidney failure has been successively covered by medical insurance.

Considerable research evidence suggests that adsorbent resins can effectively reduce the middle-molecule and protein-bound toxins in the blood of hemodialysis patients and significantly improve clinical symptoms related to sleep, pruritus, appetite, blood pressure, and joint pain, considerably enhancing uremic patients' quality of life.

4.1. *Removal of Uremic Toxins*

Shun-Jie Chen *et al.* (2011) reported the results of a prospective randomized controlled study using HD + HP treatment. They selected 100 maintenance hemodialysis patients and randomly divided them into two groups after the four-week run-in period. The patients in the HD + HP group ($n = 51$) were treated with maintenance hemodialysis twice a week and with HD + HP once a week; the patients in the HD group ($n = 49$) were treated with hemodialysis three times a week. The follow-up period was two years on average.

This study thoroughly investigated HD + HP treatment. The primary endpoint was patient death, and the secondary endpoints included conventional clinical indicators, leptin, hsCRP, IL-6, β2-MG, PTH, TNF-α, and the Index of Dimensions of Medical Outcomes Study 36-Item Short Form Health Survey (SF-36 Chinese edition).

The results indicated that the mortality rates of the HD + HP and HD groups were 12.77% and 31.82%, respectively, at the end of the two-year period, and that the mortality rate of the HD + HP group was always lower than that of the HD group throughout the two-year follow-up period.

After the two-year observation period was over, the levels of leptin, hsCRP, PTH, IL-6, β2-MG, and TNF-α in the patients in the HD + HP group decreased significantly, while the levels of those indicators in the patients in the HD group increased, as shown in Table 4.4.

Meanwhile, this study observed that the serum concentration, systolic blood pressure (SBP), diastolic blood pressure (DBP), heart rate (HR), cardiothoracic ratio, left ventricular mass index (LVMI), erythropoietin doses and types of antihypertensive drugs used in the patients treated with HD + HP were lower than those of the patients treated with HD ($P < 0.05$), while the hemoglobin concentration, left ventricular ejection fraction, and body mass index were higher than HD treated patients ($P < 0.05$), indicating that the patients treated with HD + HP obtained a better quality of life.

Table 4.4. Reductions in Levels of Main Uremic Toxins in Patients Before and After Study

	HD + HP	HD
Leptin	31.34%	−10.04%
hsCRP	20.58%	−19.38%
PTH	12.77%	−12.67%
IL-6	13.47%	−14.86%
β2-MG	13.88%	−8.32%
TNF-α	12.56%	−6.67%

This conclusion was also confirmed by the results of the SF-36 scores, which quantitatively assessed the patients' quality of life from the aspects of the patients' physical functioning, role physical, bodily pain, general health, vitality, social functioning, role emotional, and mental health. The results indicated that the aggregate scores of the patients treated with HD + HP reached 59.76 ± 19.46, while those of the patients treated with HD only reached 41.09 ± 15.52, showing that the patients treated with HD + HP had an obvious survival advantage ($P < 0.05$).

Indicators with the biggest differences in their values between the patients treated with HD + HP and the patients treated with HD included bodily pain (64.42 ± 27.54 vs. 44.31 ± 21.45), mental health (65.09 ± 20.24 vs. 55.23 ± 21.47), and general health (48.48 ± 18.29 vs. 40.43 ± 10.78), showing the main benefits of HD + HP to patients.

This confirmed that HD + HP treatment is superior to pure HD in the removal of middle-molecule toxins in patients' bodies and has potential advantages in improving MHD patients' quality of life and in increasing their survival rate.

Other relevant clinical research articles also led to similar conclusions. We summarized the studies in the 29 papers published in different magazines from 2004 to 2014, as shown in Table 4.5. All the included articles used controlled trials or randomized controlled trials.

Table 4.5. Study on Targets Adsorbed by HA130

Target	Molecular Weight	Adsorption Rate① (%)	Number of Cases In Literature (HD + HP Group/Control Group)	Literature
PTH	9225	42.07	223/181	Shun-Jie Chen *et al.*, 2011; Chen kai, Li han, 2010; Zhang Junli *et al.*, 2009; Wang Jingmei *et al.*, 2010; Zhao Jin *et al.*, 2011; Wang Jidong *et al.*, 2012; Xu Yan, Ling Kai, 2011; Zhang Chengguo *et al.*, 2007; Chen Li, 2012
2-MG	11800	21.77	146/114	Shun-Jie Chen *et al.*, 2011; Zhao Jingyu *et al.*, 2011; Chen Kai, Li Han, 2010; Wang Jingmei *et al.*, 2010; Chen Li, 2012
TNF-	51000	31.86	193/191	Shun-Jie Chen *et al.*, 2011; Zhu Zhengxi *et al.*, 2011; Zhao Bing Feng, 2008; Zhou Xiyu *et al.*, 2011; Hu Yangqing *et al.*, 2010; Zhao Huiyuan *et al.*, 2005; Ma Wenlu *et al.*, 2010
IL-6	26000	37.3	197/191	Shun-Jie Chen *et al.*, 2011; Zhao Huang *et al.*, 2010; Zhu Zhengxi *et al.*, 2011; Zhao Bing Feng, 2008; Zhou Xiyu *et al.*, 2011; Zhao Huiyuan *et al.*, 2005; Ma Wenlu *et al.*, 2010
CRP	115–140K	23.91	178/176	Shun-Jie Chen *et al.*, 2011; Zhu Zhengxi *et al.*, 2011; Zhao Bing Feng, 2008; Zhou Xiyu *et al.*, 2011; Hu Yangqing *et al.*, 2010; Zhao Huiyuan *et al.*, 2005

Leptin	16000	36.29	151/149	Shun-Jie Chen *et al.*, 2011; Song Jie *et al.*, 2010; Hu Yangqing *et al.*, 2010; Lou Tanqi *et al.*, 2004
neuropeptide Y	4272	22.68	50/50	Song Jie *et al.*, 2010
FGF-23	32000	9.4	31/17	Zhang Junli *et al.*, 2009; Miao Liying *et al.*, 2013
endothelin	4283	11.14	34/34	Ping Yehong *et al.*, 2011
AOPP$^{\beta}$	71-670K	16.02	40/40	Peng Kanfu *et al.*, 2006; Xu Bo *et al.*, 2013
IL-1$^{\beta}$	32000	14.72	15/15	Ma Wenlu *et al.*, 2010
IL-8	8359	40	24/20	Zhao Huang *et al.*, 2010
Hcy	135	21.59	34/34	Ping Yehong *et al.*, 2011
IS	251	24.86	19/19	Deng Dai *et al.*, 2014
PCS	108	23.98	19/19	Deng Dai *et al.*, 2014
HA	178	11.84	19/19	Deng Dai *et al.*, 2014

①: Adsorption rate represents the increase achieved by HD + HP relative to HD.

②: AOPP is an umbrella term for the oxidation products of a group of plasma proteins.

At least 15 common uremic toxins were effectively removed by hemoperfusion, or were removed more effectively when combined with dialysis or high-flux dialysis (Table 4.5). These uremic toxins not only include typical middle-molecule toxins such as PTH, β2-MG, leptin, and inflammatory factors TNF-α and IL-6, but also include typical protein-bound toxins such as Hcy, IS, PCS, and HA.

A summary of the above literature indicates that HA130 can adsorb middle-molecule toxins with molecular weight ranging from 4,000 to 51,000, as well as proteins with greater molecular weights, essentially covering the uremic toxins that cannot be completely removed by dialysis due to their molecular weight.

Some studies also analyzed the difference in Kt/V values between experimental and control groups (Lou Tanqi *et al.*, 2004; Miao Liying *et al.*, 2013). The results showed no significant difference, indicating that hemoperfusion cannot significantly enhance the removal of small-molecule solutes, that hemoperfusion and hemodialysis are complementary, and that the conventional method of using Kt/V to characterize dialysis adequacy is not sufficient.

The adsorption rate data show that HA130 is not highly selective about targets and exhibits excellent adsorption performance in a very broad adsorption spectrum, primarily because the forces of neutral macroporous adsorbent resins include both porous adsorption and hydrophobic action. According to the existing data, HA130 has good adsorption performance for the removal of middle-molecule toxins with molecular weight ranging from 8,000 to 30,000 and slightly poorer adsorption performance for the removal of protein-bound toxins. However, the cause has not been clear, especially because there is relatively little clinical data on protein-bound toxins.

It is worth noting that studies on protein-bound toxins found that the combined artificial kidney slightly improved the removal of hippuric acid with a low protein binding rate (30%), and significantly enhanced the adsorption rates for the removal of IS and PCS toxins with high protein binding rates (93% and 94%, respectively) when compared to hemodialysis, suggesting that adsorption may be competitive or has become saturated (Deng Dai *et al.*, 2014).

The middle-molecule toxins that have received the most attention include PTH, which is directly related to the disturbances in calcium-phosphorus metabolism leading to renal osteopathy, β2-MG, which is directly related to the dialysis associated amyloidosis, leptin, which is directly related to the nutritional status of hemodialysis patients, and the inflammatory factors that are directly related to the micro-inflammatory state in dialysis patients (TNF-α, IL-6, CRP).

The micro-inflammation caused by inflammatory factors has received increasing attention because it is related to multiple pathophysiological lesions and clinical complications in uremic patients. However hemodialysis could even cause increases in inflammatory factors (Boenisch O *et al.*, 2002). Therefore, high hopes have been pinned on adsorption separation technologies such as hemoperfusion. As shown by the statistics in Table 4.5, the combined artificial kidney can remove approximately 30% of inflammatory factors based on hemodialysis, which is important for alleviating inflammation in patients and significant for reducing a great number of clinical complications.

Similar reports also covered AOPP (Peng Kanfu *et al.*, 2006; Xu Bo *et al.*, 2013), leptin (Lou Tanqi *et al.*, 2004), and FGF-23 (Miao Liying *et al.*, 2013), where treatment using combined artificial kidney obtained good removal effects after no significant differences had been found before and after dialysis treatment.

Although most of these studies were based on the combination of low-flux dialyzers with hemoperfusion, with the gradual popularity of high-flux hemodialyzers (HFHD), an increasing number of studies have been conducted on HFHD + HP and have also shown increased toxin removal effects, consistent with theoretical predictions. Studies indicate that the protein-bound toxin removal effects for high-flux and low-flux dialysis were not significantly different, but a previous study found that the combined artificial kidney adsorbed protein-bound toxins (Deng Dai *et al.*, 2014).

Theoretically, the use of hemoperfusion based on any type of membrane separation technology can remove more toxins since the principles behind the two separation technologies are completely different. However, we have also noticed that relevant evidence is not sufficient, and further studies are required to confirm these findings.

We have also noticed that in many studies, researchers used the levels of relevant substances in non-diseased individuals as controls (Zhang Junli *et al.*, 2009; Zhao Huiyuan *et al.*, 2005; Ma Wenlu, *et al.*, 2010; Xu Bo *et al.*, 2013), and the results indicated that neither hemodialysis nor combined artificial kidney nor other artificial kidney could reduce typical middle-molecule toxins or protein-bound toxins to the level of healthy individuals; namely, relevant toxins still accumulate in the body but at slower rates. The molecular weight intercepted by high-flux dialyzers has reached 60 kD, and since further dialysis technology development is difficult (Sunny Eloot *et al.*, 2014), membrane permeability cannot be further increased (otherwise, albumin would extravasate). Adsorption separation technology with high specific surface area and high affinity represented by hemoperfusion is likely to be a major direction in the future.

Using Betasorb, Winchester *et al.* (2004) and Silberzweig J *et al.* (2003) also confirmed that HD + HP treatment can effectively improve the removal of middle-molecule toxins. In a clinical study conducted on 12 ESRD patients (eight males and four females aged 48 ± 9 on average), researchers compared the patients treated with HFD for one week and the patients treated with HFD + HP for one week. The results indicated that HFD treatment removed β2-MG at a rate of 44 mL/min, and HD + HP treatment removed β2-MG at a rate of 143 mL/min ($P < 0.001$), and 225 mg and 402 mg of β2-MG were removed respectively at a time. HD + HP treatment showed significant removal effects on β2-MG, leptin, angiogenin, retinol binding protein, IL-18, and complement factor D, with increases of 36.7%, 30.7%, 30.6%, 54%, 8.8%, and 34% respectively from dialysis. In addition, HD + HP treatment adsorbed very few white blood cells and blood platelets, as well as very little albumin.

Lixelle was initially designed simply for dialysis-related amyloidosis caused by the accumulation of β2-MG. However, it has been found to adsorb a series of middle-molecule substances, broadening its applications. A study on the adsorption spectrum of Lixelle indicated that Lixelle can adsorb proteins with low molecular weights ranging between 4,000 and 20,000 (Hidetoshi Kutsuki, 2005), including lysozyme, retinol-binding protein, insulin, interleukin

IL-1β, and IL-6 in addition to β2-MG. However, Lixelle has a weak ability to adsorb uremic toxins with molecular weight greater than 30 K. Nevertheless, Lixelle has a broad adsorption spectrum and can significantly improve uremic toxin removal effects when combined with dialysis.

Falko Brettschneider *et al.* (2013) and Meijers BK *et al.* (2008) used the Prometheus system for treatment using HD + HP. This device contains hydrophobic adsorbent and ion exchange resins and was initially an artificial liver product designed to treat severe liver disease. However, since the Prometheus system can remove toxins, researchers have studied its applications in treating kidney failure.

Research results indicate that the Prometheus system can effectively remove protein-bound toxins such as indoxyl sulfate and p-cresol. Four hours of treatment decreased the level of indoxyl sulfate by 64.36%, and the p-cresol level by 65.40%. By contrast, dialysis reduced the levels of indoxyl sulfate and p-cresol by only 38.89% and 32.51%, respectively. These results confirm that the binding of adsorbents with uremic toxins plays a crucial role in clinical treatment.

4.2. *Improvements in Clinical Symptoms*

Kidney functioning is fairly well understood, and the clinical symptoms of kidney failure patients are attributed to the accumulation of one or several uremic toxins in the body. Although the pathophysiological effects of many specific uremic toxins have not been fully clarified, enhancing dialysis adequacy and trying to remove excessive metabolic products such as uremic toxins from the human body are of great significance for improving clinical symptoms.

As indicated by a study, patients subjected to hemodialysis for four to 10 years suffer from poorer sleep, more severe joint pain, and poorer physical conditions than patients subjected to hemodialysis for less than four years, and patients subjected to hemodialysis for more than 10 years have significantly reduced quality of life. Therefore, the dialysis adequacy of maintenance hemodialysis patients has a significant effect on their quality of life, and adequate dialysis should result in high quality of life (Ma Zu *et al.*, 2004).

In view of this, a great number of clinical studies on the combination of hemoperfusion and hemodialysis have focused on improving clinical symptoms, showing that HD + HP can effectively improve complications associated with dialysis.

A meta-analysis based on 11 randomized controlled trials (RCT) using HA130 to improve the clinical symptoms of dialysis patients (Song Kewei *et al.*, 2010) led to the following conclusion: A comparison of the clinical symptoms of the 571 patients treated using HD + HP and pure HD indicated that hemoperfusion significantly alleviated the patients' pruritus, improved their sleep and appetite, controlled their bone disease and muscle weakness, and significantly improved the clinical symptoms and physiology of the dialysis patients.

In addition to the above four improvements in clinical symptoms, hemoperfusion has also proved to play a crucial role in reducing resistant hypertension and increasing nerve conduction velocity. Relevant literature is listed in Table 4.6; all the included articles used controlled trials or randomized controlled trials to conduct studies.

Pruritus and sleep problems have been the most thoroughly studied (Table 4.6).

Zhao Jin *et al.* (2011) used a scoring method to score pruritus and obtained relatively quantitative data. The highest possible score in this system was 40. The control trials indicate that the average scores of the HD + HP group before and after treatment were 27.5 ± 6.1 and 14.9 ± 3.2, respectively, with a decline of approximately 50%, while the average scores of the dialysis group before and after treatment showed no significant changes. Wang Jidong *et al.* (2012) used Sergo's method to score pruritus, and defined a decline of over 60% in the pruritus score as the effective remission of pruritus. Through four weeks of treatment, the remission rate for the HD + HP group was 86%, while that for the HD group was only 6.67%. In addition, the symptoms of a case which had been treated with HDF ineffectively, and two cases which had been treated with HFHD ineffectively, were remitted after they were treated with HD + HP. Xu Yan *et al.* (2011) used the same scoring method and obtained an effectiveness of 77.27% for the HD + HP group, and an effectiveness of

Table 4.6. Improvements in Complications Associated with HA130 in Maintenance Dialysis Patients

Clinical Symptoms	Number of Cases in Literature (HD + HP Group/ Control Group)	Literature
Pruritus	177/178	Liu Huanwen, 2009; Chen Kai, Li Han, 2010; Wang Jingmei *et al.*, 2010; Zhao Jin *et al.*, 2011; Wang Jidong, *et al.*, 2012; Xu Yan, Ling Kai, 2011; Zhang Chengguo *et al.*, 2007; Guo AH *et al.*, 2011
Sleep problems	138/129	Liu Huanwen, 2009; Chen Kai, Li Han, 2010; Wang Jidong *et al.*, 2012; Zhang Chengguo *et al.*, 2007; Chen Li, 2012; Guo AH *et al.*, 2011
Loss of appetite	108/99	Liu Huanwen, 2009; Chen Kai, Li Han, 2010; Wang Jingmei *et al.*, 2010; Wang Jidong *et al.*, 2012; Zhang Chengguo *et al.*, 2007
Resistant hypertension	78/69	Liu Huanwen, 2009; Feng Jianhua *et al.*, 2011; Wang Jidong *et al.*, 2012; Zhang Chengguo *et al.*, 2007
Joint pain	62/65	Wang Liufen, 2010; Chen Kai, Li Han, 2010; Wang Jingmei *et al.*, 2010
Nerve conduction	45/46	Chi Yanchun *et al.*, 2006; Chen Xueqin *et al.*, 2012

28.57% for the dialysis group. Other qualitative studies also led to similar conclusions, indicating that HD + HP treatment plays a significant role in remitting pruritus.

Chen Li *et al.* (2012) used the Pittsburgh Sleep Quality Index as a tool for measuring sleep quality to assess the effects of HD + HP treatment. The results indicated that the sleep quality score of the maintenance hemodialysis patients decreased from 10.71 ± 5.67 to 6.67 ± 3.35 (the total score was 21, and a higher score would indicate poorer sleep quality) with statistical significance ($P < 0.01$), indicating that HD + HP has a significant effect on improving the sleep quality of maintenance dialysis patients. Other non-quantitative assessment of sleep quality also led to similar conclusions.

According to the statistics of patients reported in the literature (Table 4.6), only 32.3% of the patients (32 in 99) treated with HD had significantly better sleep, while 67.6% of the patients (73 in 108) treated with HD + HP had significantly better sleep.

Studies on appetite improvements indicated that 57.8 ± 5.5% of the patients treated with HD had better appetite, while 80 ± 5.9% of the patients treated with HD + HP had better appetite.

When it comes to resistant hypertension, Feng Jianhua *et al.* (2011) found that after 12 weeks of HD + HP treatment, the average systolic pressure of the experimental group reduced from 174 ± 11 to 142 ± 9, the diastolic pressure reduced from 106 ± 7 to 92 ± 6, and the number of different types of anti-hypertensives taken reduced from 4.56 to 3.35; the systolic and diastolic pressures of the patients treated with HD during the same period showed no significant reductions, and the number of types of anti-hypertensives taken increased from 4.55 to 5.85, indicating a strong treatment effect of HD + HP. Other similar studies suggested that 68.5 ± 7.7% patients having resistant hypertension after HD + HP treatment achieved improvement, while only 44.1 ± 11.5% of the patients treated with HD showed improvement. Wang Liufen *et al.* (2010) used a quantitative method to study joint pain relief. In their model, joint pain was divided into five scores (0, 1, 2, 3, 4), where higher scores would indicate more severe symptoms. They found that the nine cases in the HD + HP group suffered from significantly less joint pain (0.9 ± 1.0) after being treated with hemoperfusion for one month and that the 12 cases in the HD group after being treated for one month had no significant difference in joint pain. Qualitative research has shown support to this conclusion. Chen Kai *et al.* (2010) found that 82% of the patients treated with HD + HP three times a week suffered from less joint pain, and only 43% of the patients treated with HD suffered from less joint pain.

Peripheral neuropathy is also a common clinical symptom in maintenance hemodialysis patients, and its core symptom is distal symmetrical peripheral polyneuropathy, which is primarily characterized by sensory disturbance. Nerve biopsy results indicate that 75% of uremic patients had peripheral neuropathy, and that the treatment effects of conventional hemodialysis were not satisfactory.

Chi Yanchun *et al.* (2006) studied the effects of HD + HP treatment on the conduction velocities of sensory nerves such as the median nerve, the tibial nerve, and the lateral popliteal nerve in maintenance hemodialysis patients. After two months of treatment, the results indicated that the nerve conduction velocities of the HD + HP group improved, while those of the HD group showed no significant improvements. This conclusion was further confirmed in a study by Chen Xuexun *et al.* (2012).

In addition, Yamamoto *et al.* (2011) also reported that HD + HP treatment reduced the amyloidosis in maintenance hemodialysis patients. In this study, 17 maintenance hemodialysis patients with amyloidosis used this adsorption column three times a week for one year, and indicators such as pinch strength, the visual analog scale for joint pain, and activities of daily living were measured every three months. After one year of column use, pinch strength increased from 5.9 ± 3.0 pounds to 7.2 ± 3.2 pounds ($P < 0.05$), and the visual analog scale for joint pain and activities of daily living score also improved.

Kutsuki *et al.* (2005) reviewed a series of studies relevant to Lixelle. The results indicated that in different studies, HD + HP treatment showed significant improvements in clinical symptoms and pointed out that these improvements were primarily due to the removal of β2-MG and cytokines by HD + HP treatment. As reported in six independent studies, six months of HD + HP treatment increased the activity of daily living (ADL), alleviated the joint pain of a total of 54 patients, and gradually improved various patient symptoms, such as nocturnal awakening, anchylosis, joint mobility, motor nerve latency, pinch strength, grasping power, and number of bone cysts.

A latest study by Gejyo *et al.* (2013) has further confirmed the above conclusion, indicating that HD + HP treatment is very effective in improving patients' clinical symptoms, both subjectively and objectively. By surveying the patients treated with HD + HP throughout Japan, this study collected the data retrieved from answers to subjective questions and objective physiological indicators of 345 patients from 134 organizations. These indicators included activity of daily living (ADL), joint pain, the times of nocturnal awakening, and other relevant topics. The patients consisted of 162 males and 184 females aged 62.9 ± 7.7 on average, and were treated with dialysis for 25.9 ± 6.2 years (ranging from 10 years to 38 years), and with HD + HP for 3.5 ± 2.7 years (ranging from nine months to 11 years). According to qualitative questionnaires by the patients, 84.9–96.5% of the symptoms were inhibited. Objective evidence from the doctors indicates that HD + HP treatment was effective for 72.8% of the patients, and was partially effective for 24.1% of the patients. Therefore, this study argued that HD + HP treatment can inhibit the progression of the clinical symptoms and amyloidosis of most patients.

Through the above studies, we can see that combined artificial kidney (HD + HP) treatment enhances the improvements made by conventional dialysis treatment in patients' clinical symptoms in many aspects, and is effective for most patients. These complications are chronic clinical symptoms that plague uremic patients and seriously affect their quality of life. Their improvements are of great significance to enhancing uremic patients' quality of life and confidence in life.

5. The Promise of Hemoperfusion and the Artificial Kidney

The effects and safety of HD + HP treatment have been extensively recognized for more than 10 years in clinical practice, and it has contributed to improving the survival rate and quality of life of uremic patients. However, combined artificial kidney technology is not perfect, and it currently faces two challenges.

First, the adsorption spectra of hemoperfusion have not been fully studied. Due to the detection limitation during clinical studies, few types of uremic toxins, only equivalent to half the types of middle-molecule and protein-bound toxins identified by EUTox, have been found to be removable by hemoperfusion according to evidence-based medicine research so far. Therefore, research should focus on exploring these spectra in more detail.

Second, the adsorption performance of adsorbents, the core of hemoperfusion, needs to be further improved. Data from previous studies suggest that current combined artificial kidneys cannot reduce uremic toxins to normal levels, indicating that the combined artificial kidney can delay the occurrence of dialysis complications but cannot prevent uremic toxins from accumulating. Therefore, designing and synthesizing adsorbents with better adsorption performance, broadening the adsorption spectra of adsorbents, improving the saturated adsorption capacity for specific toxins, and further increasing the clearance of middle-molecule and protein-bound toxins are still the primary directions for the application of hemoperfusion in artificial kidney.

According to our review, hemoperfusion has shown huge potential for removing uremic toxins. We believe that through the application of hemoperfusion products by health care workers and patients, there will be a greater number of more detailed clinical studies and reports that will further increase evidence in evidence-based medicine, propose better clinical treatments, and further enhance hemodialysis adequacy and patient quality of life.

We hope that this paper can provide medical workers and practitioners with a positive perspective. We believe that there will be more efficient and comprehensive combined artificial kidney therapy in the future, which is expected to better simulate the kidney's toxin removal functioning, more effectively eliminate the complications caused by dialysis inadequacy and improve patient quality of life.

References

Almeras C, Argiles A. (2009) The General Picture of Uremia. *Seminars in Dialysis* **22(4):** 329–333.

Boenisch O, Ehmke KD, Heddergott A, *et al.* (2002) Creactive-protein and cytokine plasma levels in hemodialysis patients. *J Nephrol* **15:** 547–551.

Brettschneider F. (2013) Removal of protein-bound, hydrophobic uremic toxins by a combined fractionated plasma separation and adsorption technique. *Artif Organs* **37(4):** 409–416.

Chang TMS. (1966) Semipermeable aqueous microcapsules ("artificial cells"): With emphasis on experiments in an extracorporeal shunt system. *Trans Am Soc Artif Intern Organs* **12:** 13–19.

Chang TMS. (1969) Removal of endogenous and exogenous toxins by a microencapsulated absorbent. *Can J Physiol Pharmacol* **47(12):** 1043–5.

Chang TMS, Gonda A, Dirks JH, Malave N. (1971) Clinical evaluation of chronic, intermittent, and short term hemoperfusions in patients with chronic renal failure using semipermeable microcapsules (artificial cells) formed from membrane-coated activated charcoal. *Trans Am Soc Artif Intern Organs* **17:** 246–252.

Chang TMS. (1972) Haemoperfusions over microencapsulated adsorbent in a patient with hepatic coma. *Lancet* **ii:** 1371–1372.

Chang TMS, Gonda A, Dirks JH, *et al.* (1972) ACAC microcapsule artificial kidney for the long term and short term management of eleven patients with chronic renal failure. *Trans Am Soc Artif Intern Organs* **18:** 465–472.

Chang TMS, Coffey JF, Barre P, *et al.* (1973) Microcapsule artificial kidney: Treatment of patients with acute drug intoxication. *Can Med Assoc J* **108:** 429–433.

Chang TMS. (1975) Microencapsulated adsorbent hemoperfusion for uremia, intoxication and hepatic failure. *Kidney Int* **7:** S387–S392.

Chang TMS, Barre P, Kuruvilla S, *et al.* (1982) Phase one clinical trial of a new composite artificial kidney: A single unit combining dialysis with hemoperfusion. *Trans Am Soc Artif Intern Organs* **28:** 43–48.

Chen K, Li H. (2010) The clinical observation of the combination of hemoperfusion with hemodialysis on the of uremia complication. *International Journal of Transplantation and Hemopurification* **8(6):** 35–37.

Chen L. (2012) The influence of combination of hemodialysis with hemoperfusion on the quality of sleep. *Chin J Blood Purif* **11(6):** 342–343.

Chen SJ, *et al.* (2011) Combination of maintenance hemodialysis with hemoperfusion: A safe and effective model of artificial kidney. *The International Journal of Artificial Organs* **34(4):** 339–347.

Chen XM, Chu QJ. (2005) The clearance effect of different therapeutic apheresis on the protein-bind toxin. *Chin J Blood Purif* **4(11):** 581–591.

Chen XQ, *et al.* (2012) The influence of different blood purification profile on the peripheral neuropathy in end-stage renal disease patients. *Chin J Blood Purif* **11(12):** 661–663.

Chi YC, *et al.* (2006) Clinical efficacy of hemoperfusion and hemodiafiltration on treating uremic peripheral neuropathy. *Chin J Blood Purif* **5(11):** 766–768.

Deng D, *et al.* (2014) Comparison of different blood purification modalities for removal of serum protein-bound uremic toxins in maintenance hemodialysis patients. *Chin J Blood Purif* **9(13):** 639–642.

Dunea G. (2012) Charcoal hemoperfusion: Promise unfulfilled. In: Ing TS, Rahman MA, Kjellstrand CM (eds), *Dialysis History, Development and Promise*. World Scientific Publishing Co., Singapore, pp. 437–440.

Eloot S, *et al.* (2014) Extracorpo real removal of uremic toxins: Can we still do better? *Seminars in Nephrology* **34(2):** 209–227.

Feng JH, *et al.* (2011) The observation of clinical efficacy of hemoperfusion in the uremic patients with refractory hypertension. *Chin J Hemorh* **21(1):** 71–73.

Garlich FM, Goldfarb DS. (2011) Have advances in extracorporeal removal techniques changed the indications for their use in poisonings? *Advances in Chronic Kidney Disease* **18(3):** 172–179.

Gejyo F, *et al.* (2013) Survey of the effects of a column for adsorption of b2-microglobulin in patients with dialysis-related amyloidosis in Japan. *Therapeutic Apheresis and Dialysis* **17(1):** 40–47.

Grovender EA, *et al.* (2004) Single-chain antibody fragment-based adsorbent for the extracorporeal removal of b2-microglobulin. *Kidney International* **65:** 310–322.

Guo AH, *et al.* (2011) Adsorptive dialysis for cleaning uremic middle molecular substances. *Journal of Clinical Rehabilitative Tissue Engineering Research* **15(12):** 2261–2264.

Hu YQ, *et al.* (2010) Effect of hemodialysis combined with hemoperfusion on inflammation and nutrition status in maintenance hemodialysis patients. *Chin J Blood Purif* **9(12):** 672–674.

Jourde-Chiche N, *et al.* (2009) Protein-bound toxins — update 2009. *Seminars in Dialysis* **22(4):** 334–339.

Kutsuki H. (2005) β2-Microglobulin-selective direct hemoperfusion column for the treatment of dialysis-related amyloidosis. *Biochim Biophys Acta* **1753:** 141–145.

Liu HW. (2009) The application and nursing to the patients in the treatment of maintenance hemodialysis through hemodialysis combined hemoperfusion. *Chinese Journal of Dialysis & Artificial Organs* **20(3):** 40–41, 46.

Lou TW, *et al.* (2004) Clinical investigation of different blood purification techniques on clearing serum leptin of chronic hemodialysis patients. *Chin J Blood Purif* **3(2):** 78–81.

Ma WL, *et al.* (2010) Efficacy of clinical application of various hemodialysis modilities of cytokines removal for uremic patients. *International Journal of Transplantation and Hemopurification* **8(1):** 40–44.

Ma Z, *et al.* (2004) Multi-center study of quality of life in hemodialysis patients. *Chin J Blood Purif* **3(7):** 380–384.

Martin AM, Gibbins JK, Oduro A, Herbert R. (1978) Clinical experience with cellulose-coated carbon hemoperfusion. In: Chang TMS (ed): *Artificial Kidney, Artificial Liver, and Artificial Cells.* Plenum, New York, pp. 143–152.

Meijers BK, Weber V, Bammens B, *et al.* (2008) Removal of the uremic retention solute p-cresol using fractionated plasma separation and adsorption. *Artif Organs* **32:** 214–219.

Miao LY, *et al.* (2013) The comparison of the different therapeutic apheresis on the FGF-23 clearance of the maintenance hemodialysis patients. *Chin J Nephrol* **29(2):** 150–151.

Peng KF, *et al.* (2006) The influence of therapeutic apheresis on the AOPP level of uremia patients. *Chin J Blood Purif* **5(7):** 386–387.

Ping YH, *et al.* (2011) Clinical effectiveness of blood perfusion for refractory hypertension of patients undergoing maintenance hemodialysis. *J Clin Nephrol* **11(6):** 264–266.

Silberzweig J, *et al.* (2003) Clinical evaluation of small molecular weight protein (SMWP) removal with a novel sorption device (Betasorb) combined with high-flux dialysis (HFD). *Abstr Am Soc Artif Intern Organs* **92:** 426–427.

Song J, Li H, Zhang XD. (2010) Influence of different blood purification treatment on the serum leptin and neuropeptide γ levels in patients with chronic renal failure. *Chin J Cell Mol Immunol* **26(11):** 1116–1118.

Song KW, *et al.* (2010) Effects of combined hemoperfusion-hemodialysis treatment on complications in chinese patients with chronic renal failure: A meta-analysis. *Chinese General Practice* **13(3):** 257–259.

Vanholder R, *et al.* (2003) Review on uremic toxins: Classification, concentration, and interindividual variability. *Kidney International* **63:** 1934–1943.

Vanholder R, Van Laecke S, Glorieux G. (2008) The middle molecule hypothesis 30 years after: Lost and rediscovered in the universe of uremic toxicity? *J Nephrol* **21:** 146–160.

Vanholder R, Van Laecke S, Glorieux G. (2008) What is new in uremic toxicity? *Pediatr Nephrol* **23:** 1211–1221.

Vanholder R, Massy ZA. (2009) Progress in uremic toxin research: An introduction. *Seminars in Dialysis* **22(4):** 321–322.

Wang JD, *et al.* (2012) Influence of various blood purification methods on skin itching in patients subject to maintenance hemodialysis. *J Clin Nephrol* **12(2):** 67–69.

Wang JM, *et al.* (2010) The clinical application of the combination of hemoperfusion with hemodialysis for chronic kidney failure. *J Clin Nephrol* **10(8):** 371–372.

Wang LF (2010) The influence of the hemoperfusion on the joint pain of maintenance hemodialysis patients. *International Journal of Transplantation and Hemopurification* **8(5):** 35–36.

Wang ZG. (2010) *Blood Purification*. 3rd ed. Beijing Science & Technology Press, Beijing.

Winchester JF, *et al.* (2004) Sorbents in acute renal failure and end-stage renal disease: Middle molecule and cytokine removal. *Blood Purif* **22:** 73–77.

Xu B, *et al.* (2013) Removal of serum advanced oxidation protein products by different purification methods and their effect on cardiovascular morbidity in patients with maintenance hemodialysis. *J Clin Nephrol* **13(2):** 60–63.

Xu Y, Ling K. (2011) Treatment study of hemoperfusion combined with hemodialysis in eliminating parathyroid hormone due to uremia. *J Clin Nephrol* **11(6):** 267–269.

Yamamoto Y, *et al.* (2011) Long-term efficacy and safety of the small-sized β2-microglobulin adsorption column for dialysis-related amyloidosis. *Therapeutic Apheresis and Dialysis* **15(5):** 466–474.

Zhang CG, *et al.* (2007) Treatment effect of different blood purification methods on pruritus in patients with uremia. *Chin J Blood Purif* **6(3):** 171–173.

Zhang JL, *et al.* (2009) Effect of hemoperfusion on the level of serum fibroblast growth factor 23 in hemodiaiysis patients. *Chinese Journal of Clinical Medicine* **16(4):** 594–596.

Zhang LX, *et al.* (2012) Prevalence of chronic kidney disease in China: A cross-sectional survey. *Lancet* **379(9818):** 815–822.

Zhao BF. (2008) The effect of hybrid artificial kidney on the micro-inflammation state of maintenance hemodialysis patient. *Chinese Journal of Dialysis and Artificial Organs* **19(3):** 14–16.

Zhao H, *et al.* (2010) Removal of humoral mediators and the effect on the survival of septic patients by hemoperfusion with neutral microporous resin column. *Therapeutic Apheresis and Dialysis* **14(6):** 596–602.

Zhao HY, *et al.* (2005) The influence of the hemoperfusion on the IL-6, TNF-α, CRP. *Chin J Blood Purif* **4(10):** 566–568.

Zhao J, *et al.* (2011) The observation of treatment effect of the combination of hemoperfusion with high-flux hemodialysis on the skin itch of maintenance hemodialysis patients. *J Clin Nephrol* **11(4):** 175–176.

Zhao JY, *et al.* (2011) The impact of different blood purification methods on residual renal function among maintenance hemodialysis patients. *Chin J Blood Purif* **10(2):** 71–73.

Zhou XY, *et al.* (2011) Effects of hemoperfusion using HA130 resin hemoperfusion apparatus on microinflammatory state in maintenance hemodialysis patients. *Journal of Clinical Rehabilitative Tissue Engineering Research* **15(21):** 3909–3912.

Zhu ZX, *et al.* (2011) The influence of various blood purification methods on microinflammatory state in patients with maintenance hemodialysis. *Chin J Blood Purif* **10(1):** 18–21.

Chapter 5

Fundamental Knowledge of Hemoperfusion and Plasmaperfusion

Kazuo Teramoto*

1. Introduction

1.1. *What is Hemoperfusion?*

Hemoperfusion is an extracorporeal form of treatment. In hemoperfusion, blood is continuously drawn from a patient to pass through a device for removing detrimental substances, and then returns to the patient. Its major uses include removing waste products from the blood in patients with renal failure and removing poisonous substances from the blood in emergency situations or in other chronic diseases. Besides artificial kidneys and other blood-purification devices, artificial heart and lung and collection of blood components are classified under hemoperfusion in a broad sense. However, this chapter mainly describes various devices for blood purification therapy. There are filter-type modules made of hollow fiber membrane and/or adsorbent-type modules such as activated charcoal.

1.2. *Usage of Module*

Hemoperfusion is applied to therapy of diseases difficult to remedy with medicine because it needs a special apparatus. As the blood

* Department of Pathology (Pathology and Disease Regulation), Shiga University of Medical Science, Seta Tsukinuwa-cho Otsu Shiga, 520–2192 Japan.

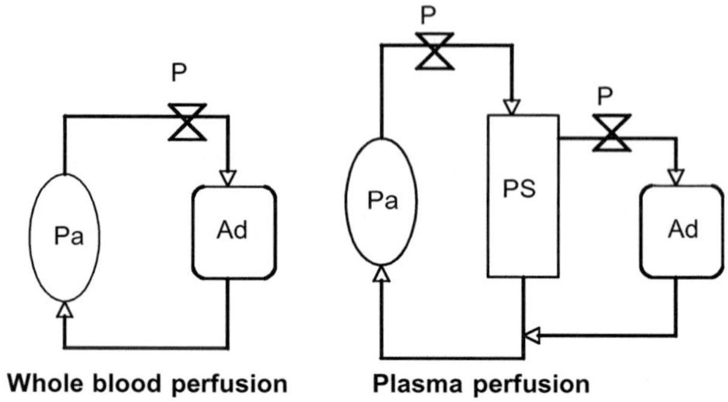

Fig. 5.1. Scheme of hemoperfusion method.

returns to the patient after it passes through module, the module for hemoperfusion should be absolutely safe for the body in all respects. The hemoperfusion module comprises of plastics that are foreign matter for body. Therefore, when blood passes through the module, coagulation proceeds on the surface without an anticoagulating agent. Therefore, anticoagulant is indispensable for prevention of clot formation. Heparin, Nafamostat mesylate, citric acid or argatroban are used as anticoagulant. Heparin is mainly used as anticoagulant for hemoperfusion in Japan. Nafamostat mesylate is used for an adsorbent adsorbing a lot of heparin, or for patients with bleeding tendencies.

Adsorbent-type hemoperfusion is carried out by a whole blood perfusion method or a plasma perfusion method using a plasma separation module (Fig. 5.1).

2. Selection of a Module with High Blood Compatibility

In hemoperfusion there are some problems shown below, because a hemoperfusion device is a foreign object for the body.

(1) Release of harmful substance from adsorbent
(2) Formation of clotting

(3) Hemolysis
(4) Activation of white blood cells

These problems must be overcome for safe hemoperfusion before application to therapy.

2.1. *Prevention of Release of Harmful Substances from Adsorbent*

In terms of harmful substances, there are minute particles, heavy metal ions, organic solvents and other organic compounds used as synthetic reagents for preparation of adsorbents.

2.1.1. *Minute particles*

When minute particles are shed from an adsorbent, they may choke capillary vessels in the lung or brain to induce pulmonary or cerebral infarction. Therefore, we must avoid discharge of minute particles from the adsorbent. When porous macro-reticular ion exchange resins are applied to an adsorbent module, they should be thoroughly sieved to remove minute particles before use. As these resins are tough, there is low probability that fine particles fall away from its surface. However, when activated charcoals are used as adsorbents, they may release powder during use because they are very brittle, even though after thorough sieving. Therefore, they must be coated with a semipermeable membrane of hydrophilic polymer such as poly(2-hydroxyethyl methacrylate) or cellulose. They are used in plasma perfusion type for certain safety.

2.1.2. *Heavy metal ions*

It is well known that heavy metal ions are poisonous to central the nervous system (Chang LW, 1990). Their toxicity is considered to be due to blocking of calcium channel or inhibition of enzyme reaction.

Polyethylene terephthalate (PET) resin is widely found in medical use such as for adsorbents or filters, and it is prepared by polymerization of terephthalic acid and ethylene glycol in the presence of a catalyst like antimony or germanium compound. Poly(lactic acid) is a useful polymer in medical contexts because it is biocompatible and bioabsorbable. It is prepared by polymerization of lactic acid in the presence of Tin (II) 2-ethylhexanoate as catalyst. When using these polymers, it is essential to confirm no release of these metal ions from the device.

2.1.3. *Solvent and reagent*

When solvents and reagents that are used for adsorbent preparation remain in the adsorbent, they may be released into blood during hemoperfusion, causing harm to the body. In general, solvents tend to remain in the adsorbent when a solvent having similar chemical structure and polarity to the polymer is used for adsorbent preparation.

Besides the solvent, we should note the residue of the reagent and its decomposition products. Especially when the reagent is an aromatic compound — they tend to remain in the adsorbent, since aromatic compounds are slightly soluble in water. Even if they are insoluble in water, they may dissolve into blood by complex-formation with albumin. Polymeric compounds such as polyvinylpyrrolidone or polyethylene glycol monododecyl ether are also difficult to remove completely from the plastic mold by extraction, although they are soluble in water. This phenomenon is due to the tangles among polymer chains. Therefore, we should avoid such compounds as synthetic reagents, and when we use such polymers as reagents, we should note their release during hemoperfusion.

2.2. *Prevention of Thrombus Formation*

2.2.1. *Mechanism of blood coagulation*

Blood coagulation cascade: Blood is gifted to coagulate for prevention of hemorrhage when a blood vessel is damaged. When blood was exposed to collagen, which means vessel injury, Factor 12, Factor 11, Factor 9, Factor 10, Factor 5, prothrombin and fibrinogen in the coagulation cascade are activated successively and form thrombus (Fig. 5.2) (Israels *et al.*, 2006). Clotting time by this cascade refers to activated thromboplastin time (APTT). In practice, APTT is measured by adding an activating agent such as kaolin to plasma sample, and its normal value is 10 to 12 sec.

Blood also tends to coagulate when it comes into contact with foreign substances like plastics, because plastics of the module are an abnormal substance for body.

In hemoperfusion, when platelets or monocytes are stimulated, they release inflammatory cytokines that induce cell surface expression of tissue factor (TF) on endothelial cells.

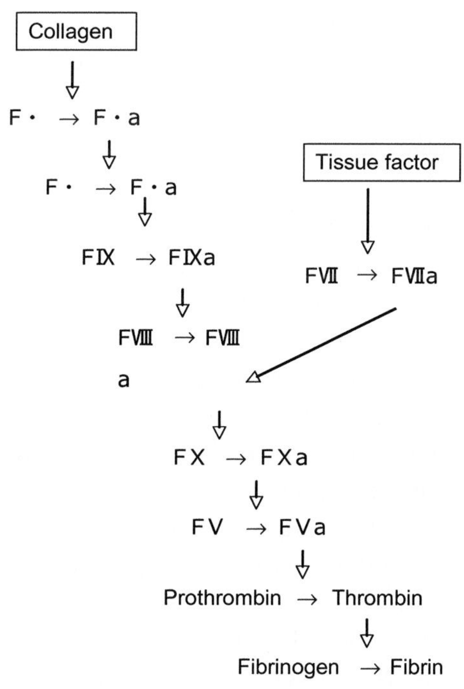

Fig. 5.2. Blood coagulation cascade.

TF is a 47-kDa protein expressed in various cells, and is one of the key initiators of coagulation cascades. TF progressively activates Factor 7, Factor 10, Factor 5, prothrombin and fibrinogen, and forms thrombus ultimately (Fig. 5.2). Although intact endothelial cells and monocytes do not express TF, they change to express TF under stimulation by various stimulants such as endotoxin, tumor necrosis factor-α, IL-1β, histamine, oxidized LDL or vascular endothelial growth factor (Steffel *et al.*, 2006). Incidentally, the soluble form of TF that is shed from cell surface molecule is observed in the blood of cancer patients or coronary disease patients. However, it is controversial whether these soluble TF generate clotting or not. Clotting time by this cascade is referred to as prothrombin time (PT). PT is actually measured by adding an activating agent such as kaolin to plasma sample, and its normal value is 20 to 40 sec.

Besides APTT and PT, activated whole blood clotting time (ACT) is also used for evaluation of coagulation ability of blood. ACT is measured by adding an activating agent such as kaolin to whole blood, and its normal value is 90 to 120 seconds. ACT is often used for monitoring at the bedside because it can be measured without centrifugation.

2.2.2. *Plastics as foreign object*: As noted above, plastics comprising the hemoperfusion device are considered to induce thrombus formation mainly by APTT cascade that is initiated by activation of Factor 12. Therefore, APTT usually becomes shorter on the surface of plastics. However, some adsorbents among plastics may adsorb several factors in this cascade, and in such case, APTT becomes longer.

On the other hand, in view of the cascade from TF, when blood flows through the device, platelets or monocytes adhere to the plastics surface. When these cells strongly attach to the plastics, they are activated and release pro-inflammatory cytokines to let endothelial cells express TF. Therefore, APTT becomes shorter in this case, but APTT becomes longer when the adsorbent adsorbs most of one of these coagulation proteins. Moreover, since anticoagulants such as heparin are usually fed during hemoperfusion, the relationship between coagulation time and adsorbent is complex.

2.3. *Prevention of Thrombus during Hemoperfusion*

As modules for hemoperfusion are made from artificial substances, anticoagulants such as heparin or nafamostat mesilate are used for prevention of coagulation.

Clinically available anticoagulants are as follows:

(1) Unfractionated heparin
(2) Low molecular weight heparin
(3) Nafamostat mesilate
(4) Argatroban
(5) Citric acid

Their half-life is one to two hours for heparin and eight minutes for nafamostat mesilate in the blood. Heparin is the first choice because it is effective, safe, and cheap. However, when too much heparin is added, blood becomes difficult to coagulate. Therefore, the amount of anticoagulant to be added is usually adjusted while APTT, PT and ACT are monitored.

In the case of hollow fiber membrane, there are few problems with clotting because the blood stream is smooth and it does not adsorb a lot of anticoagulant. On the other hand, usually, but not always, a cationic adsorbent of high capacity for the target molecule tends to adsorb a lot of heparin, and hydrophobic adsorbent tends to adsorb a lot of nafamostat mesilate. When the adsorbent adsorbs too much anticoagulant, the same amount of anticoagulant as it adsorbed should be supplemented to prevent coagulation. In this case it is not easy to control the coagulating time of blood.

2.4. *Features of Individual Anticoagulants*

Unfractionated heparin: There are unfractionated heparin, a low molecular weight heparin, nafamostat mesilate, sodium citrate, and Argatroban as anticoagulants. Unfractionated heparin was extracted from intestinal mucous membrane of pig or calf, and purified. It is a sort of glycosaminoglycan and a highly sulfated molecule of 3,000 to

20,000 dalton. It comprises glucosamine units and uronic acid units (Fig. 5.3).

Heparin tends to bind to various proteins. It forms a complex with antithrombin-3 in blood to inhibit activity of active form of Factor-10 and thrombin. Its half-life is 40 to 90 minutes in blood. ACT is used to monitor its concentration.

Low molecular weight heparin: Low molecular weight heparin (LMH) is prepared by fractination of heparin, and its molecular weight is 4,000 to 8,000 dalton. LMH forms a complex with antithrombin-3 and inhibits activity of active Factor-10a. LMH likes unfractionated heparin, but does not inhibit thrombin function. Therefore, LMH can be applied to a light bleeding tendency patients. Its half-life is about three hours. Its concentration can be monitored by anti-Factor-10a activity, but not by ACT.

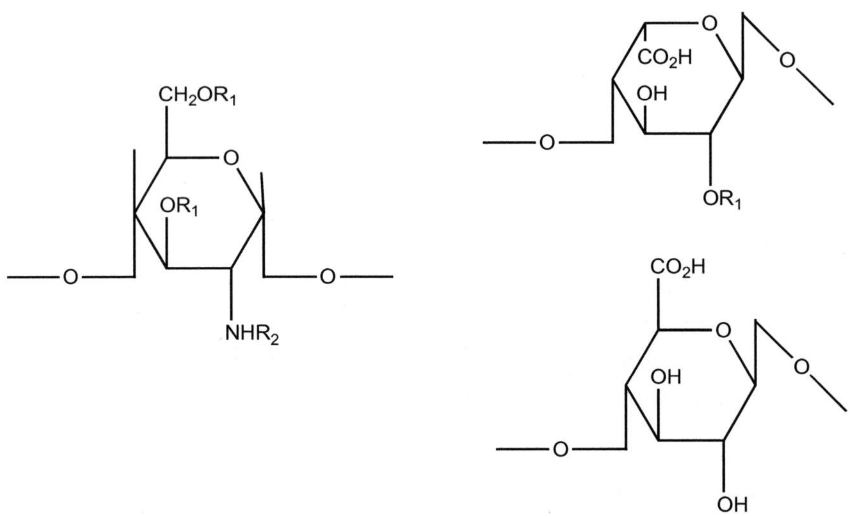

A = Glucosamine unit B = Uronic acid unite

R_1 = H or SO_3H, R_2 = SO_3H or $COCH_3$

Fig. 5.3. Structure of heparin: its components.

Nafamostat mesilate: Nafamostat mesilate, 539 dalton, is a low molecular weight inhibitor of protainase (Fig. 5.4). It inhibits active forms of Factor-12, Factor-10 and Factor-7, and platelet function to prevent coagulation. Its half-life is very short at eight minutes. Therefore, it can be used for patients with bleeding tendencies, although it is expensive compared to heparin.

In the case of dialysis, nafamostat mesilate (NM) passes through semipermeable membrane, but it becomes unable to pass through the membrane due to complex formation with albumin in blood. However, when NM was added just before the module, most of the NM is dialyzed before it binds to heparin. Therefore, a low molecular weight anticoagulant like NM is usually added at the back of module. Besides, NM is inappropriate to anticoagulant of charcoal column because an activated charcoal column tends to adsorb aromatic compounds such as NM. During hemoperfusion, the concentration of NM is monitored by ACT.

Argatroban: Argatroban, 526 dalton, is a water-insoluble arginine analog with strong anticoagulant activity (Fig. 5.4). It strongly inhibits fibrin formation by thrombin and suppresses platelet aggregation. It also blocks the contraction of blood vessels. Its half-life is about 30 min. It is applied to dialysis of antithrombin-3-lack patients because its anticoagulant property does not relate to antithrombin-3. It is appropriate to dialysis of heparin-induced thrombocytopenia

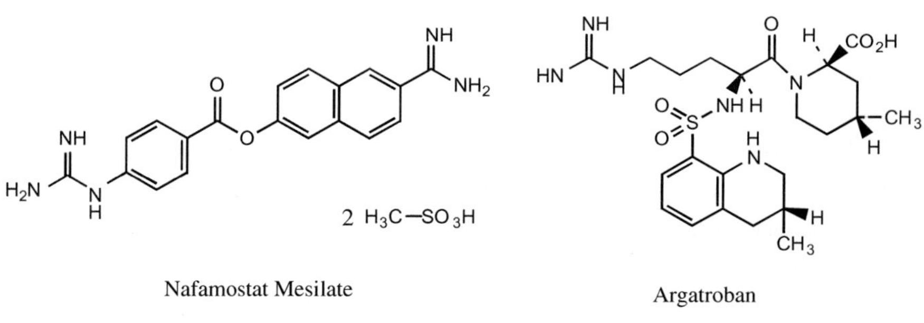

Fig. 5.4. Chemical structure of anticoagulant.

patients. As it is decomposed in the liver, it is not appropriate for patients with low hepatic function. It is usually used when other anticoagulants are unsuitable, because it is expensive.

Citric acid: Sodium citrate blocks thrombus formation by chelating calcium ions. Usually, it is used as an anticoagulant in preservation of plasma, platelet or blood stem cells. Occasionally, it is used as an anticoagulant for hemoperfusion.

2.5. *Activation of Complement*

The complement system is a part of the immune system and associated with activation of white blood cells during hemoperfusion. The basic functions are as follows:

(1) Assistance of phagocytosis of antigens
(2) Chemotaxis of macrophage and neutrophil
(3) Cell lysis of foreign cell
(4) Agglutination of pathogen

The complement system works as shown in Fig. 5.5. Its activation is typically triggered by antigen–antibody complexes or bacteria and bacteria-derived substances like endotoxin. C3b binds to the surface of pathogens, leading to greater internalization by phagocytes. C5a is an important chemotactic protein, helping to recruit macrophage and neutrophil. Both C3a and C5a have anaphylatoxin activity, directly triggering degranulation of mast cells as well as increasing vascular permeability. Therefore, they may cause an abrupt decline of blood pressure.

The complex consisting of C5b, C6, C7, C8, and C9 is called a membrane attack complex; it forms a transmembrane channel to cause osmotic lysis of the target cell like bacteria.

In hemoperfusion, if the module is contaminated with bacteria or its products, like endotoxin, it induces complement activation along the cascade starting from C3 in Fig. 5.5. A module made from

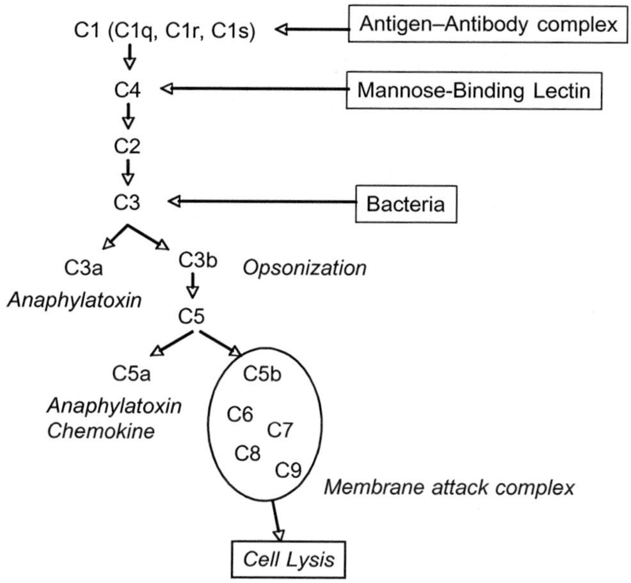

Fig. 5.5. Complement system.

natural substances like cellulose, especially, may be contaminated with endotoxin, and may then induce adverse effects such as fever or hypotension.

In the case of antibody-immobilized adsorbent columns, complement activation may occur by formation of antigen–antibody complex along the cascade starting from C1 in Fig. 5.5. Further, although the adsorbent is not an antibody-immobilized adsorbent, there is a probability that it induces complement activation by adsorbed proteins in the case of adsorption of immunoglobulin or immune-complex.

2.6. *Prevention of Hypotension by Module*

Anionic biomaterials such as dextran sulfate immobilized cellulose or tryptophan immobilized poly(vinyl alcohol) or poly(methacryl sulfonate) tend to activate high molecular weight kininogen and let it release bradykinin.

Bradykinin is nonapeptide Arg-Pro-Pro-Gly-Phe-Ser-Pro-Arg-OH that has antihypertensive property, and pain-producing activity. An increase of bradykinin induces hypotension, as shown in Fig. 5.6 .

Angiotensin converting enzyme inhibitor as antihypertensive inhibits decomposition of bradykinin to increase serum level of bradykinin as shown in Fig. 5.7. Therefore, when using an anionic module,

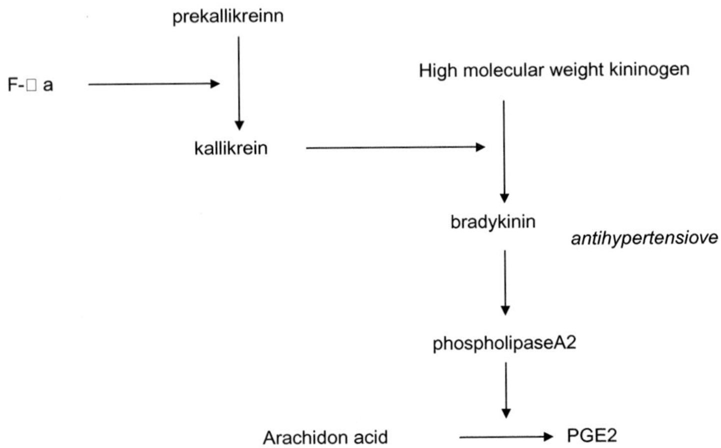

Fig. 5.6. Kinin-kallikrein system.

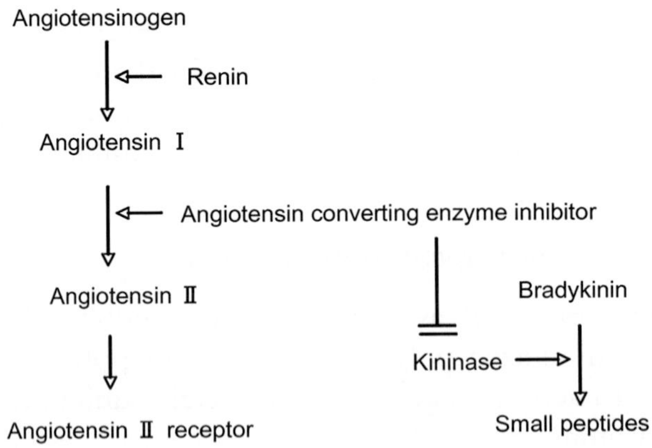

Fig. 5.7. Renin-angiotensin system.

we should consider whether or not the patient takes angiotensin-converting-enzyme inhibitors such as Captopril or Enalapril.

2.7. *Biocompatible Polymer*

It is desirable that components of the module including a blood circuit and column have high blood compatibility. Examples of biocompatible polymer are shown as follows (Jordan *et al.*, 2007):

(1) Polymer grafted with hydrophilic polymer chain
(2) Polymer with micro phase separation structure
(3) 2-Methacryloyloxyethyl phosphorylcholine polymer (MPC)
(4) Poly(2-methoxyethylacrylate)

Polymer grafted with hydrophilic polymer chain such as PEG-grafted or polyacrylamide-grafted polymer as shown in Fig. 5.8 is considered to be biocompatible because the hydrophilic layer and the soft long chain prevent adhesion of protein and platelet (Merrill and Salzman, 1983).

It has been reported that the polymer surface of microphase-separation-structure is biocompatible (He, 2014). Polymers that have a hydrophilic segment and a hydrophobic segment form microphase-separation-structures composed of a hydrophilic domain and a hydrophobic domain. Protein and platelet tend to adhere to a hydrophobic surface by van der Waals force, but do not adhere to hydrophilic surfaces. Therefore, the surface of the microphase-separation-structure does not

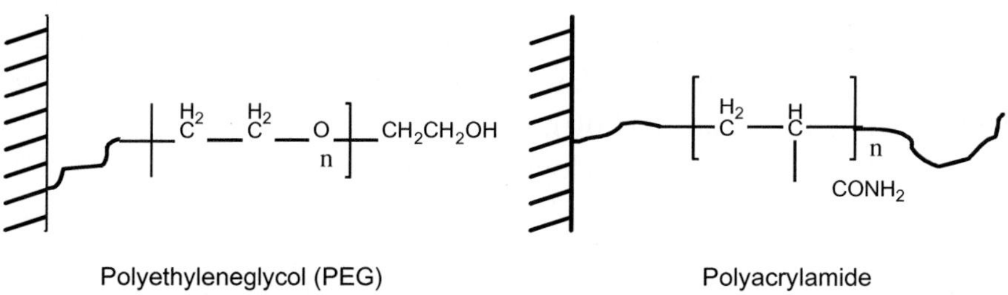

Fig. 5.8. Hydrophilic polymer grafted surface.

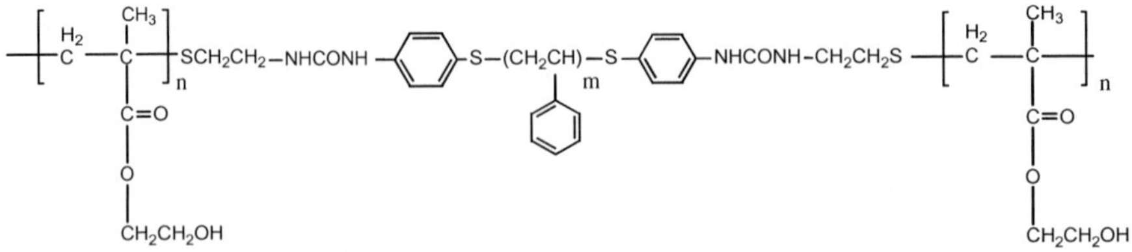

Fig. 5.9. Hydroxymethacrylate-styrene ABA block polymer.

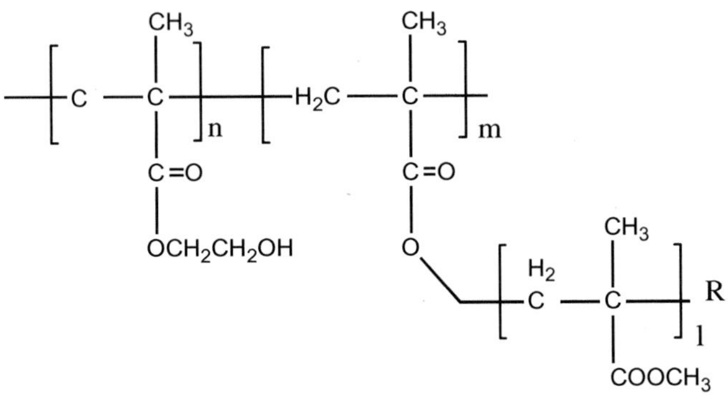

Fig. 5.10. PMMA-grafted PMMA/HEMA block copolymer.

adsorb proteins and platelets because a hydrophobic domain as a scaffold for adherence decreases in microphase-separation-structures.

The hydroxymethacrylate-styrene ABA block polymer shown in Fig. 5.9 is biocompatible because it forms a microphase-separation-structure (Okano, 1981).

The poly(methyl methacrylate)-grafted-poly(methyl methacrylate)/poly (hydroxyethyl methacrylate) block copolymer shown in Fig. 5.10 forms a microphase-separation-structure and is also biocompatible.

Poly(2-methoxyethylacrylate), shown in Fig. 5.11, is reported to be biocompatible (Saito, 2000).

Copolymer of 2-methacryloyloxyethyl phosphorylcholine (MPC), shown in Fig. 5.12, is reported to be biocompatible (Ishihara *et al.*, 1990). This is because it mimics a cell membrane.

Copolymer of n-dodecyl methacrylate and MPC (PMD), shown in Fig. 5.13, is reported to be highly biocompatible. The surface of

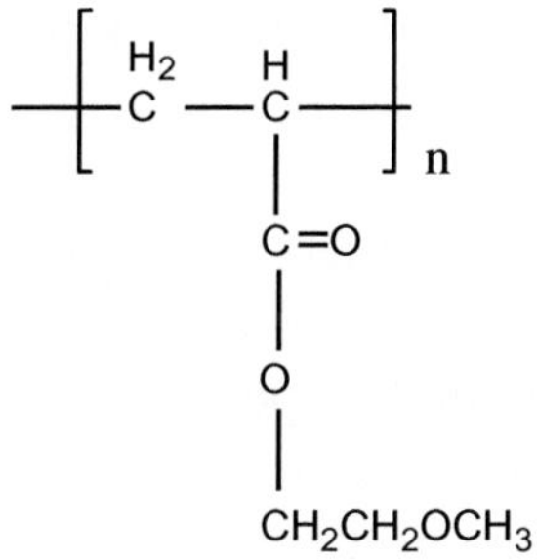

Fig. 5.11. Poly(2-methoxyethylacrylate).

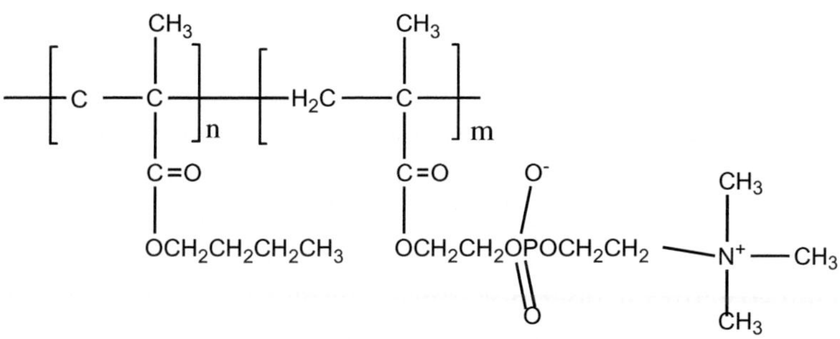

Fig. 5.12. MPC polymer.

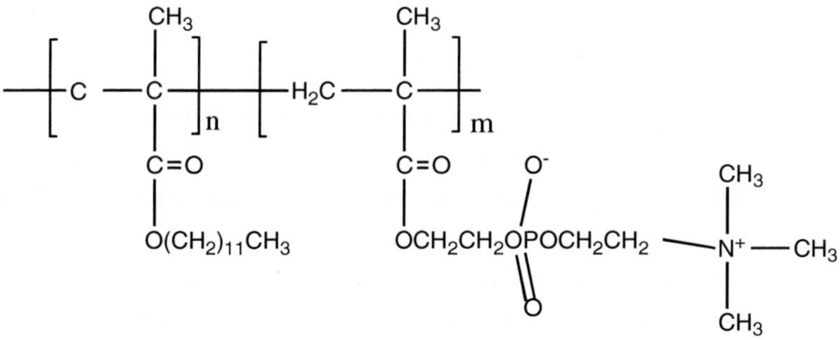

Fig. 5.13. Poly [MPC-co-n-dodecyl methacrylate].

PMD polymer does not adsorb protein. Fibroblast can proliferate on PMD surface, although it does not adhere to the surface (Katakura *et al.*, 2005).

These biocompatible polymers are available for coating of other plastics mold. However, these polymers are expensive and therefore general polymers were applied to perfusion circuit and column vessel. Thrombosis does not occur under smooth blood flow. Therefore, actually, polyvinylchloride, polypropylene, polystyrene, polyethylene terephthalate, polysulfone, and polycarbonate are being used for several parts of blood devices without problems.

2.8. *Suppression of Hemolysis*

Some plastics with particular chemical structures cause hemolysis. The chemical structure with a strong hydrophilic group and long alkyl group in a molecule tends to induce hemolysis. For example, chemical structure 1 in Fig. 5.14 causes hemolysis, but structure 2 does not. Structure 3, with a longer alkyl group, causes stronger hemolysis. Naturally structure 4 does not cause hemolysis. It is considered that the structure like surface-active agent disturbs red cell membrane and induces hemolysis.

2.9. *Suppression of Loss of Useful Substances*

In general, an adsorbent having high adsorbing capacity tends to adsorb other substances with the target molecule in the case of synthetic adsorbents. In particular, when the useful substances resemble the target molecule in polarity or molecular weight, they are adsorbed a lot with the target. When lost substances are valuable proteins such as albumin or coagulation factor, they are supplied by infusion of pooled plasma. However, it is best to avoid plasma infusion as far as possible, because it may contaminate the blood with unknown virus or germs. Therefore, it is essential to choose an adsorbent with as high specificity as possible.

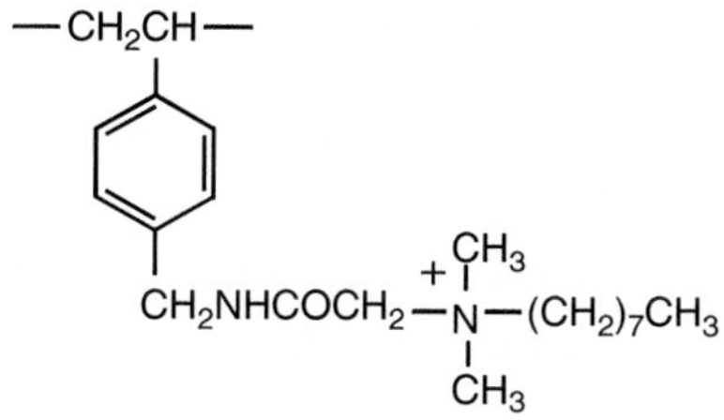

1

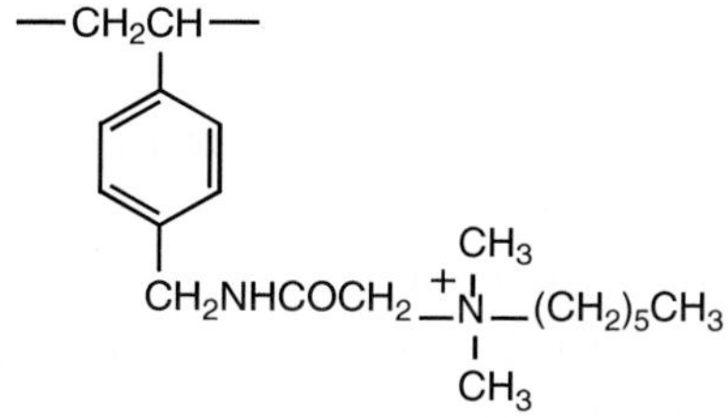

2

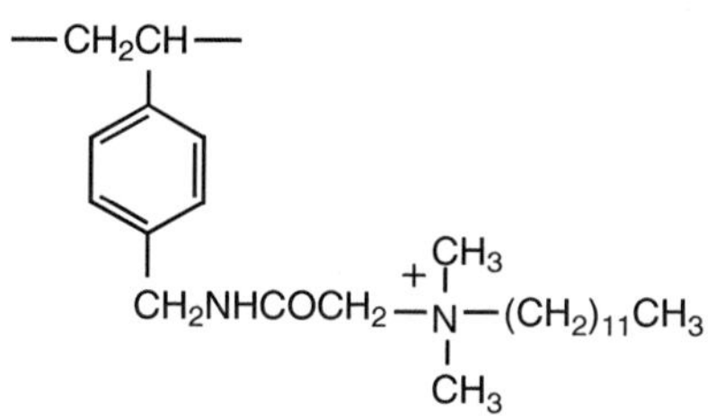

3

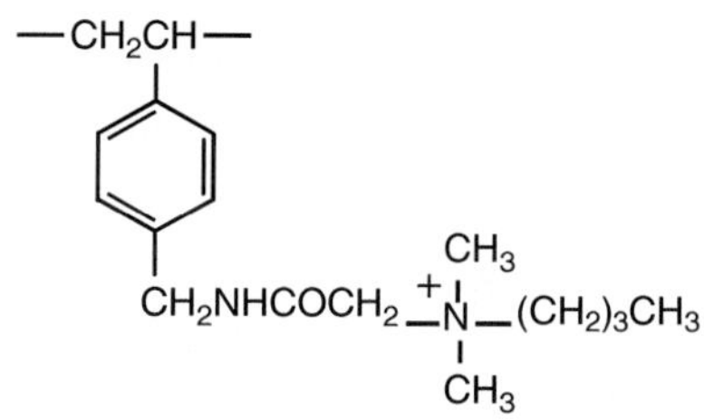

4

Fig. 5.14. Chemical structure of polystyrene derivatives.

3. Filter-type Module

3.1. *Hollow Fiber Membrane*

Various membranes having different sizes of pores are used for filter-type modules. Their shapes are almost hollow fiber. A bundle of hollow fibers are filled in the column to build the device. Although a size of pore limits a protein from penetrating, the separation is not sharp fundamentally, because protein particles are deformable and the membrane has many different pore sizes.

Generally, the hollow fibers are classified into three groups. First is the small-sized pore membrane for dialysis that passes low-molecular substances like urea, but not proteins. Second is the medium-sized pore membrane that passes albumin but not globulin. Third is the large-sized pore membrane that passes proteins but not cells. This type is used for plasma separation.

3.2. *Theory of Preparation of Hollow Fiber Membrane*

When a solution of water-insoluble plastic polymer is pushed out into water from a syringe, the polymer solidifies by phase separation to form fibrous material. In this process, when we use a syringe with a sheath and extrude polymer solution into water from the sheath, we can obtain hollow fiber. A coaxial double cylinder as shown in Fig. 5.15 is used as spinneret.

When the hollow fiber is developing, the outer and inner surfaces of the polymer layer solidify immediately with contact with water; on the other hand, the internal part solidifies gradually with penetration of water. Consequently, the outer and inner surfaces of hollow fiber are dense, and the inside of the membrane is porous. The density of these skin layers is dependent on the form of salts, the concentration of each salt in the coagulating solution, temperature and spinning speed. Therefore, the skin layer limits the molecule passing through the membrane, although almost all molecules in blood can pass the inside of membrane freely. In general, the inner surface is designed to

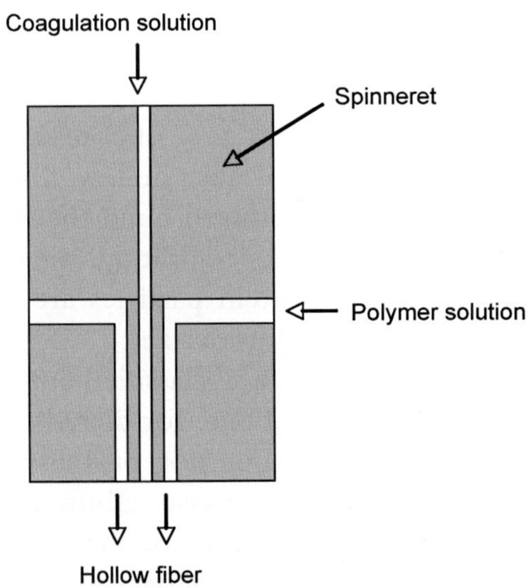

Fig. 5.15. Schematic diagram of spinneret for hollow fiber.

be denser than the outer surface because whole blood flows through the inner part of hollow fiber.

3.3. *Usage of Hollow Fiber Membrane*

A module of hollow fiber membrane is mainly used for dialysis of renal failure patients. The membrane is made from cellulose, poly-(acrylonitrile), ethylene-vinylalcohol copolymer, cellulose acetate, polysulfone, or poly(methylmethacrylate) by wet spinning.

As the blood stream in the hollow fiber is straight and as wide as 200 micrometers in diameter, blood flows smoothly. Moreover, the hollow fiber does not comprise of strong ionic groups such as sulfonic acid or quarternary ammonium group. Therefore, there are few clotting problems. As the hollow fiber membrane for dialysis comprises of small size of pore, it cannot dialyse large molecules such as beta-2 microgloblin from blood, although it can remove urea and creatinine. Therefore, there is a problem with beta-2 microglobulin being accumulated in patients receiving dialysis therapy for a long time.

Another use of hollow fiber module is separation of plasma from whole blood. For this use, hollow fiber membrane comprising of large-sized pores is used.

4. Adsorbent-type Module

4.1. *Adsorption Phenomenon*

There are three forces that work in adsorption: van der Waals force, hydrogen bond and ionic bond. The van der Waals force comprises London dispersion force, dipole–dipole interaction and quadrapole interaction. In adsorption of non-polar molecules, it is mainly the van der Waals force at work, although it is weak.

A hydrogen bond is the electrostatic attraction between hydrogen atoms and highly electronegative atoms such as fluorine, oxygen or nitrogen bearing lone pair. Hydrogen bonds vary in strength from weak to strong. In general, a hydrogen bond is five- to ten-fold stronger than the van der Waals force, although it is weaker than ionic bonding.

Coulomb's law, the following formula, represents electrostatic attractive forces;

$$F = k \times Q_1 \times Q_2 / d^2,$$

where Q_1 represents the quantity of charge on object 1, Q_2 represents the quantity of charge on object 2, and d represents the distance of separation between two objects. The symbol k is Coulomb's law constant, and depends on medium.

Adsorption by van der Waals forces is called physisorption, and adsorption by chemical forces such as the hydrogen bond and electrostatic attractive force is called chemisorption.

There are various interactions such as solute–solvent, solute–solute and solute–adsorbent interaction in aqueous solution (Fig. 5.16) (Otagiri, 2005). A solute–solute interaction is essential in blood because blood contains albumin and various glycoproteins in high concentrations.

For example, in the case of hydrophobic drugs like benzodiazepines that are insoluble in water and do not have any ionic group, they bind to serum protein such as albumin or alpha-1 acid glycoprotein in blood (Fig. 5.17). The binding is due to van der Waals forces. These binding are reversible, and equilibrium exists between the bound and unbound molecules. It is believed that the drug–protein complex releases drugs on the target tissue, because only unbound molecules are active.

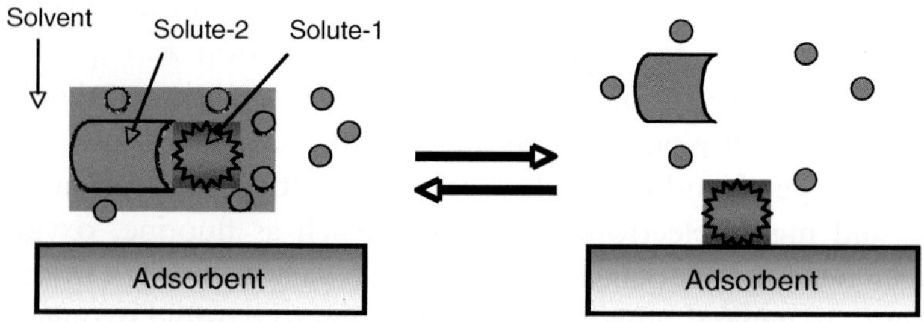

Fig. 5.16. Interaction between solutes and adsorbent.

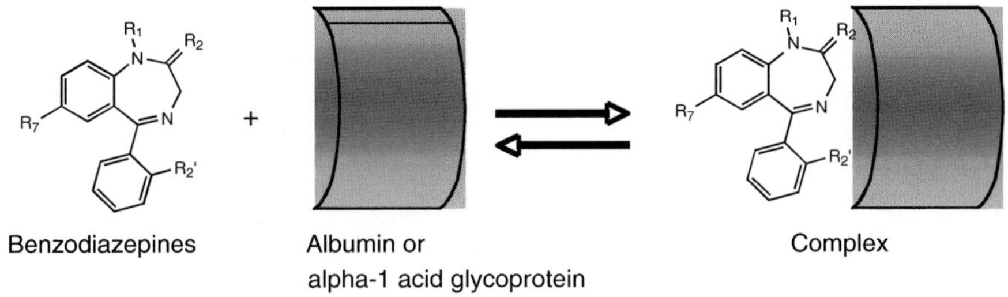

Fig. 5.17. Interaction of Benzodiazepines and serum protein in blood.

On the other hand, bilirubin exists in glucronized form and unglucronized form in blood (Fig. 5.18). Unglucronized bilirubin binds to the cationic part of albumin or alpha-1 acid glycoprotein, because it has two carboxyl groups. It is reported that human albumin has three drug binding sites, and the guanidine group of arginine in their sites play an important role in binding. Consequently, binding of bilirubin and albumin is strong because it is based on ionic force.

4.2. *Estimation of Adsorptive Capacity of Adsorbent*

The adsorptive capacity of an adsorbent should be measured in whole blood, plasma or serum because the target molecule may form a complex with serum proteins in blood, as described previously. The adsorption isotherm can be used in a routine manner for estimation of adsorptive capacity of an adsorbent. This is obtained as follows;

(1) Prepare five sealable tubes loaded with a known volume of serum or plasma that contains a target molecule.
(2) Add a known amount, a half amount, a quarter amount, one-eighth amount and one-sixteenth amount of adsorbent to the above tube.
(3) Shake the tube at the temperature of 37°C until equilibrium is established.
(4) Analyze the concentration of the target molecule after removal of the adsorbent, and obtain a decrease of the concentration.

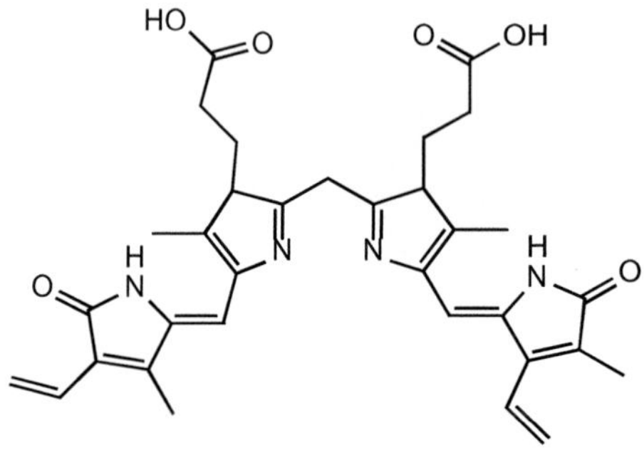

Bilirubin

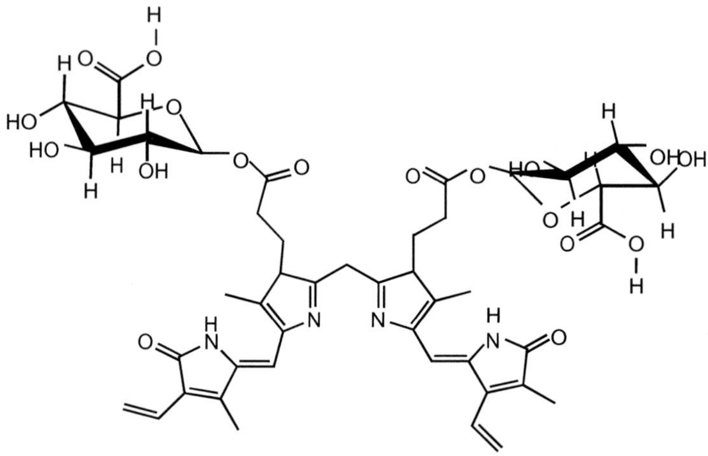

Bilirubin diglucuronide

Fig. 5.18. Chemical structure of bilirubin and bilirubin diglucuronide.

(5) Plot the concentration of the target molecule (Ci) as abscissa and the adsorbed amount of target molecule per weight of adsorbent (Wi) as ordinate. For analysis of adsorbing capacity, Freundlich equation (2) is useful in serum or plasma.

$$W = kC^{1/n}, \tag{2}$$

$$\mathrm{Log}\ W = \log k + (1/n) \log C,$$

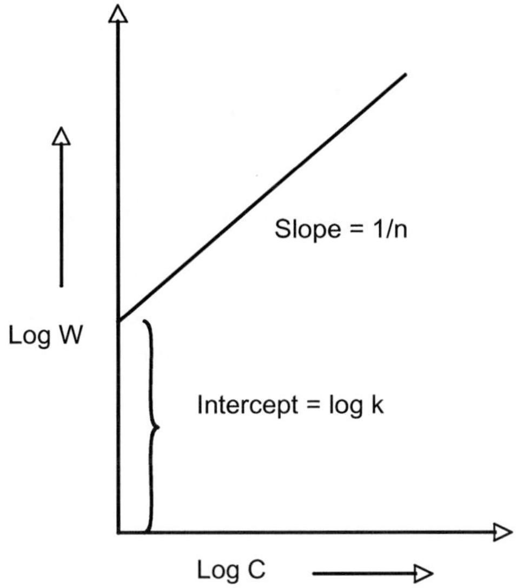

Fig. 5.19 Freundlich adsorption isotherm.

where k and $1/n$ represent adsorption constant, and C represents the concentration of the target molecule, and W represents the adsorptive capacity.

From Fig. 5.19, the slope $1/n$ represents affinity between the adsorbent and the target molecule, and the intercept log k represents basic adsorptive capacity of the adsorbent.

4.3. *Design of Adsorbent*

Hollow fiber membrane modules have been used for long time, and their safety is assured. Therefore, when the filter-type module cannot effectively remove harmful substances, an adsorbent-type module begins to be considered. Although the adsorbent with antibody as ligand shows high specificity of adsorption, it has some problems such as cost, sterilization, stability, storage, mass production, etc. Therefore, synthetic ligand is used for most adsorbent-modules, although its selectivity is inferior to that of antibodies. In order to raise adsorptive selectivity, various synthetic ligands were screened.

The value of log k in Freundlich adsorption isotherm gives a good guide to their screening. Selectivity is evaluated by comparison of k-value between target molecule and other components.

4.4. *Adsorbent for a Molecule of Low Molecular Weight*

Hemoperfusion column of activated charcoal column is applied to treatment of poisoning by agrichemicals or overdose of sleeping pills. These chemicals can be adsorbed effectively by activated charcoal, although they form a complex with protein such as albumin or alpha-1 acid glycoprotein in blood. Therefore, their complex forming force is considered to be the van der Waals force.

On the other hand, strong anion exchange resin that has a quaternary ammonium group (shown in Fig. 5.20) is applied to removal of bilirubin in jaundice. This is because bilirubin binds strongly to albumin or alpha-1 acid glycoprotein by electrostatic attractive force.

β2-microgloblin, a protein of 12 kDa, is a constituent of MHC class 1 and exists in the cell membrane of most of tissues. This small protein is concentrated in renal failure patients and activates immune cells. These cells release inflammatory cytokines and aggravate the symptoms.

Porous cellulose beads that immobilize hexadecylamine are used as adsorbents of β_2-microgloblin (Abe *et al.*, 2003). The adsorption is considered to be due to the van der Waals force.

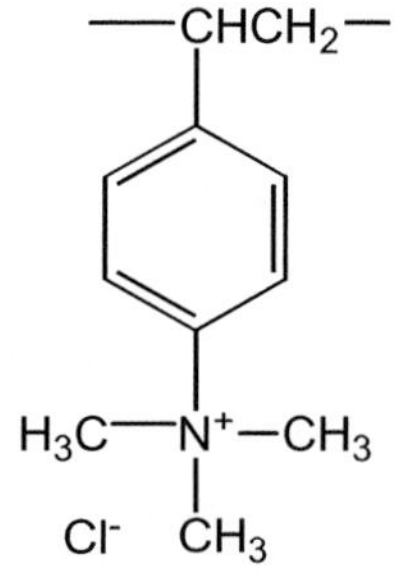

Fig. 5.20. Chemical structure of anion exchange resin.

4.5. *Adsorbent of Endotoxin from Gram-Negative Bacteria*

Endotoxin causes sepsis when it enters the peripheral blood because of serious injury or infection. Endotoxin is a lipopolysaccharide (LPS) that is one of the bacterial cell membrane constituents and is released from the membrane. LPS binds to toll like receptor 4 of monocytes or macrophage, and activates them. A large quantity of various cytokines is released by their cells, and causes cytokine storm to induce sepsis. The active site of LPS is lipid A (Fig. 5.21). Lipid A is harmful in the concentration of pico-gram order. Therefore, although Lipid A has two phosphoric acid groups as shown in Fig. 5.21, it is difficult to selectively remove such a low level of Lipid A from blood by using ion exchange resin because blood contains various ionic compounds at high concentration.

Polymyxin B (Fig. 5.22) is an effective antibiotic against Gram-negative bacteria and selectively binds to endotoxin and detoxifies it. Therefore, we chose Polymyxin B as the ligand for endotoxin adsorbent. Polymyxin B is cyclic peptide having five free amino groups

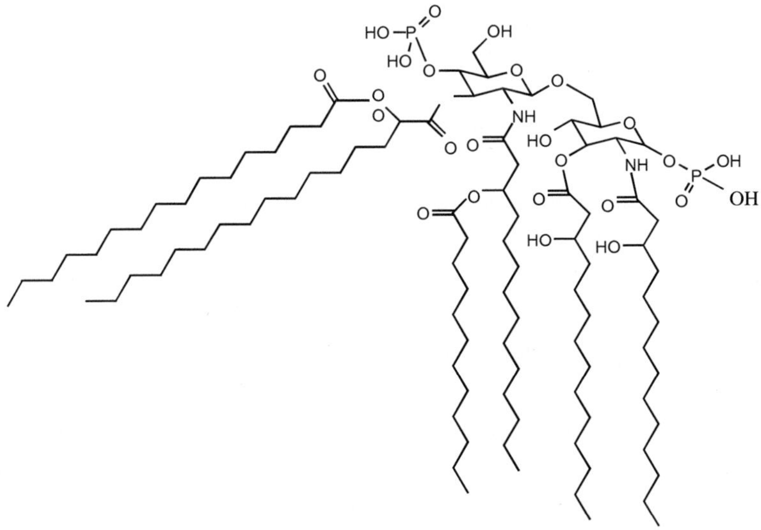

Fig. 5.21. Chemical structure of Lipid A.

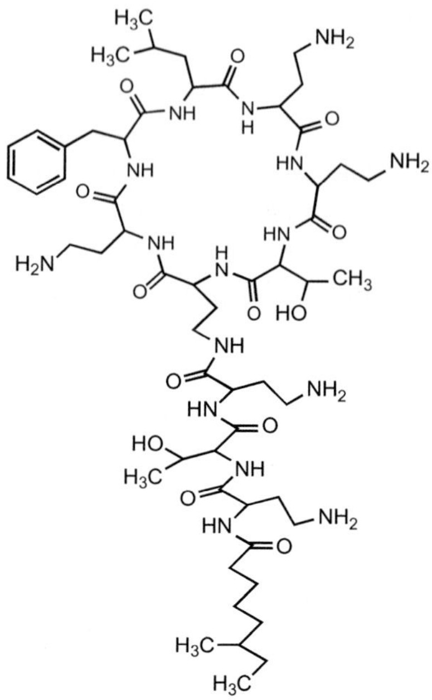

Fig. 5.22. Chemical structure of Polymyxin B.

derived from 2,4-diaminobutyric acid. We designed carrier for immobilization of Polymyxin B as follows;

Requirements of the carrier for endotoxin-remover are

(1) To be able to immobilize Polymyxin B with one of five free amino groups.
(2) To have low swelling ability for preventing nonspecific adsorption.
(3) To have low ionic density for preventing nonspecific adsorption.
(4) To have large surface area for high adsorbing ability.

Therefore, we developed chloroacetamidomethylated polystyrene fiber as the carrier. It is prepared by amidomethylation reaction of polystyrene/poly-propylene conjugated fiber with N-hydroxymethyl-2-chloroacetamide in the presence of sulfuric acid (Fig. 5.23). Obtained

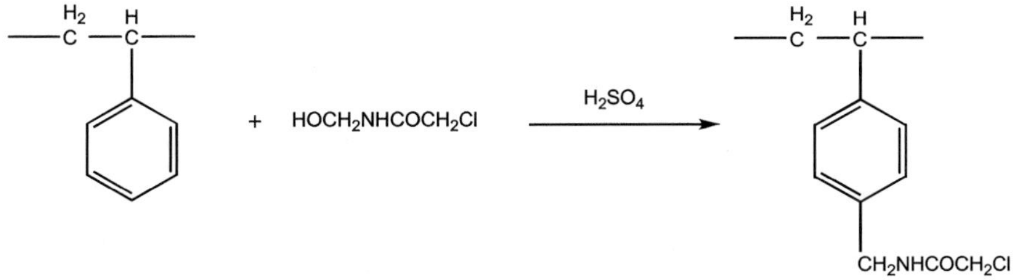

Fig. 5.23. Amidomethylation of polystyrene.

chloroacetamidomethylated fiber is reacted with Polymyxin B in aqueous solution, and gives Polymyxin B immobilized fiber (Fig. 5.24) (Kodama *et al.*, 1988; Hanazawa *et al.*, 2000).

From amino acid analysis of the adsorbent, we confirmed that one to two of five free amino groups in Polymyxin B are used for binding to polymer. This fiber shows low swelling property and, therefore, low nonspecific adsorption against serum proteins. This fiber shows endotoxin adsorbing capacity, where endotoxin was usually assayed by a biomedical method using Limlus amebocyte lysate. However, this assay cannot distinguish clearly between decomposition and adsorption of endotoxin. Therefore, we obtained adsorption isotherm between endotoxin and Polymyxin B immobilized fiber by using tetramethylrhodamine labeled LPS, and confirmed that it is adsorption (Teramoto *et al.*, 2002).

4.6. *Adsorbent of Lipoprotein*

Hyperlipoproteinemia causes cardiovascular diseases due to arteriosclerosis. When drugs are ineffective for hyperlipoproteinemia, hemoperfusion therapy is applied. As low density lipoprotein (LDL) has positive charge, the column of dextran sulfate immobilized cellulose beads is used for removal of LDL from blood. Plasma perfusion system is applied for this adsorption, because the complex of adsorbent and lipoprotein becomes gelatinous to raise the blood viscosity (Tani, 1996; Stefanutti *et al.*, 2012).

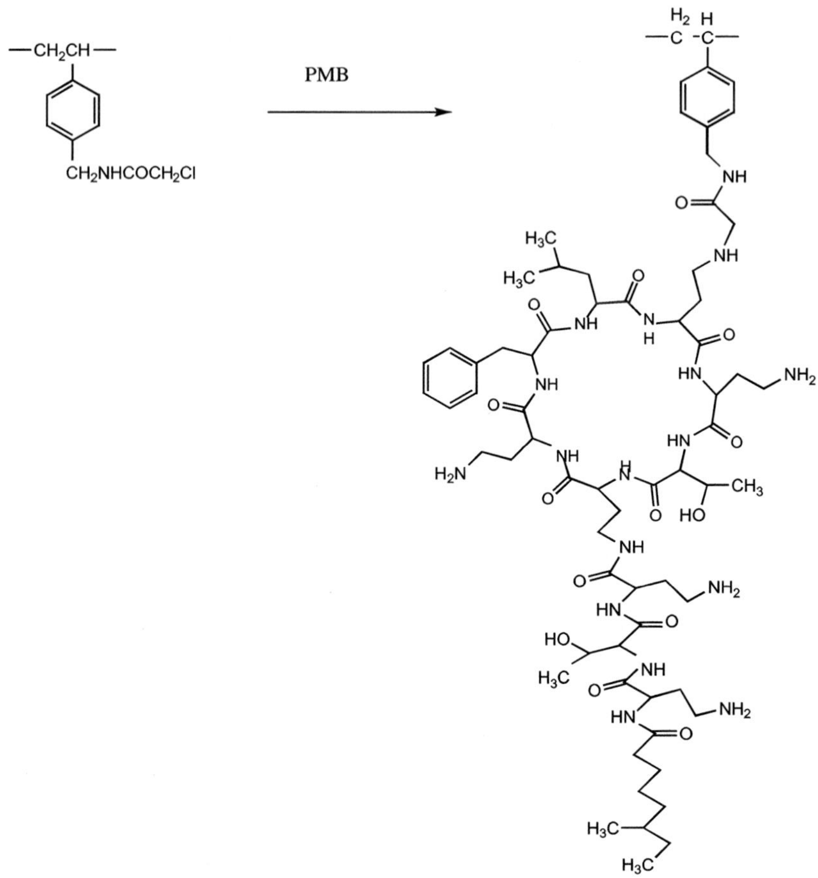

Fig. 5.24. Preparation of immobilized Plymixin B.

4.7. *Adsorbent of Protein Relative to Autoimmune Diseases*

A higher concentration of antibody or immune complex is observed in blood of patients with autoimmune diseases such as systemic lupus erythematosus, cryoglobulinemia, rheumatoid arthritis, Hasimoto's thyroiditis, or multiple sclerosis. This causes fever, arteriosclerosis, vasculitis, or renal failure. When drugs are ineffective on these diseases, hemoperfusion therapy is applied for removal of these immune related compounds such as immunoglobulin or immune complexes.

The antibody against acetylcholine receptor, DNA, or caldiolipin and these immune complexes are adsorbed by tryptophan or phenylalanine-immobilized beads (Fadule *et al.*, 1996). Their adsorptions are due to van der Waals force.

An adsorbent that immobilized protein A was developed (Iglesias *et al.*, 2004). Protein A is known to bind to Fc part of immunoglobulin G. It must be kept in mind that protein A binds may activate complement system to induce harmful effects. In addition, the adsorbent that immobilized antibodies such as anti-LDL or anti-IgG was also developed for removal of LDL or IgG. These adsorbents also should be considered in terms of whether they show harmful effects by complement activation or not.

4.8. *Adsorbent of White Blood Cells*

Inflammatory bowel disease such as ulcerative colitis or Crohn's disease is a sort of autoimmune disease, and an increase of activated polymorphonuclear leukocytes (PMN) and lymphocytes causes these diseases. Therefore, adsorbents of these cells were developed. One of them is cellulose acetate bead that decreases activated PMN by adsorption (Ohara *et al.*, 1997). Another is polyethylene terephthalate fiber that decreases activated PMN and lymphocytes by filtration method (Takenaka, 1996).

5. Future Perspective

Hemoperfusion has been used for therapy of renal failure, chemical poisoning, various autoimmune diseases, hypercholesterolemia and liver failure since 1970s. Hemoperfusion is applied to diseases that medicine cannot cure. Incidences of diabetes, heart diseases and cancer are increasing at present, but no clinically applicable hemoperfusion device has been developed for these diseases. Therefore, these diseases will become a target of hemoperfusion in the future.

5.1. *Adsorbent of Immunosuppressive Protein*

Cancer is the main cause of human deaths in advanced nations. In cancer patients, immunosuppressive proteins and immunoregulatory cells increase to create immunosuppressive circumstances that enhance cancer progress.

TGF-β, a representative immunosuppressive protein, is known to enhance tumor growth and metastasis in cancer. TGF-β exists in the form of a complex with latency associated peptides (LAP) in blood. This complex (LAP-TGF) is inactive in peripheral blood, but changes to active TGF-β on the target cells. LAP-TGF is a 100k Dalton molecule, while TGF-β is a 25k Dalton molecule. LAP has affinity for cationic adsorbents because it comprises phosphorylated carbohydrates. Polystyrene fiber with the fourth ammonium group is developed as an LAP-TGF adsorbent, and it shows anti-cancer effect on KDH-8 tumor-bearing rat (Yamamoto *et al.*, 2006).

6. References

Abe T, Uchita K, Orita H, *et al.* (2003) Effect of beta2-microglobulin adsorption column on dialysis-related amyloidosis. *Kidney International* **64:** 1522–1528.

Fadul JEM, Danielson Bo G, Wikstoem B. (1996) Reduction of plasma fibrinogen, immunoglobulin G, and immunoglobulin M concentrations by immunoadsorption therapy with tryptophan and phenylalanine adsorbents. *Artif Organs* **20(9):** 989–990.

Hanazawa K, Tani T, Oka T, *et al.* (2000) A new treatment for endotoxemia with direct hemoperfusion by Polymyxin immobilized fiber. *Therapeutic Apheresis* **4(2):** 142–145.

He Y, Xie D, Zhang X. (2014) The structure, microphase-separated morphology, and property of polyurethanes and polyureas. *J Mater Sci* **49:** 7339–7352.

Igresias J, D'Agati VD, Levine JS. (2004) Acute glomerulonephritis occurring during immunoadsorption with staphylococcal protein A column (Prosorba). *Nephrol Dial Transplant* **19:** 3155–3159.

Ishihara K, Aragaki R, Ueda T, *et al.* (1990) Reduced thrombogenicity of polymers having phospholipids polar groups. *J Biomed Mater Res* **24(8):** 1069–1077.

Israels S, Schwetz N, Boyar R, McNicol A. (2006) Bleeding disorders: Characterization, dental considerations and management. *J Canadian Dental Association* **72(9):** 827–8271.

Jordan SW, Chaikof EL. (2007) Novel thromboresistant materials. *J Vac Surg* **45:** 104A–115A.

Katakura O, Morimoto N, Iwasaki Y, Akiyoshi K (2005) Evaluation of 2-methacroyloxyethyl phosphorylcholine (MPC) polymer-coated dressing on surgical wounds. *J Med Dent Sci* **52:** 115–121.

Kodama T, Tani T, Hanazawa K, *et al.* (1988) Endotoxin detoxifying process. *United States Patent 4,661,260.*

Merrill EW, Salzman EW. (1983) Polyethylene oxide as a biomaterial. *ASAIO J* **6:** 1.

Ohara M, Saniabadi AR, Kokuma S, *et al.* (1997) Granulocytapheresis in the treatment of patients with rheumatoid arthritis. *Artif Organs* **21(9):** 989–994.

Okano T, Nishiyama S, Shinohara I, *et al.* (1981) Effect of hydrophilic and hydrophobic microdomains on mode of interaction between block polymer and blood platelets. *J Biomedical Material Res* **15(3):** 393–402.

Otagiri M. (2005) A molecular functional study on the interactions of drugs with plasma proteins. *Drug Metab. Pharmacokinet* **20(5):** 309–323.

Saito N, Motoyama S, Sawamoto J. (2000) Effects of new polymer-coated extracorporeal circuits on biocompatibility during cardiopulmonary bypass. *Artif Organs* **24(7):** 547–554.

Stefanutti C, Morozzi C, Perrone G, *et al.* (2012) The lipid- and lipoprotein-[LDL-Lp(a)] apheresis techniques updating. *G Chir* **33:** 444–449.

Steffel J, Luescher TF, Tanner FC. (2006) Tissue factor in cardiovascular diseases molecular mechanisms and clinical implications. *Circulation* **113:** 722–731.

Takenaka Y. (1996) Lymphocytapheresis. *Artif Organs* **20(8):** 914–916.

Tani N. (1996) Development of selective low-density Lipoprotein (LDL) apheresis system: immobilized polyanion as LDL-specific adsorption for LDL apheresis system. *Artif Organs* **20(8):** 922–929.

Teramoto K, Nakamoto Y, Kunitomo T, *et al.* (2002) Removal of endotoxin in Blood by Polymyxin B immobilized polystyrene-derivative fiber. *Therapeutic Apheresis* **6(2):** 103–108.

Yamamoto T, Ueda Y, Itoh T, *et al.* (2006) A novel immunotherapeutic modality with direct hemoperfusion targeting transforming growth factor-beta prolongs the survival of tumor-bearing rats. *Oncology Reports* **16:** 1277–1284.

Chapter 6

Hemoperfusion in Artificial Liver Support Systems

Lanjuan Li*

1. Introduction

1.1. *Metabolic Functions of the Liver*

The liver is a large, chemically reactant organ. It performs many important metabolic functions, including detoxification, transformation and synthesis of multiple substances; many of its functions are connected with other organs. This connectedness becomes especially clear when abnormalities arise in the liver. Liver functions include:

(i) Bilirubin metabolism
(ii) Metabolism of bile acids
(iii) Carbohydrate metabolism
(iv) Plasma lipids and lipoprotein metabolism
(v) Amino-acid metabolism, urea production
(vi) Dynamic removal of circulating substances
(vii) Synthesis of coagulation factor
(viii) Immunologic functions of the liver
(ix) Storage of vitamins
(x) Others

*State Key Laboratory for Diagnosis and Treatment of Infectious Diseases, the First Affiliated Hospital, School of Medicine, Zhejiang University, 79 Qingchun Road Hangzhou, China.

In clinical practice, some biochemical markers are used to assess liver function. For example, the biochemicals alanine aminotransferase (ALT), aspartate aminotransferase (AST), alkaline phosphatase and the serum bilirubin are markers of hepatobiliary disease. Additionally, the serum albumin level and prothrombin time are used to evaluate synthetic liver function.

1.2. *Definition of Liver Failure*

Liver failure is the inability of the liver to perform its normal synthetic, metabolic, excretory and biotransformation functions. Liver failure is usually manifested as coagulopathy, jaundice, hepatic encephalopathy and ascites. In terms of histopathological characteristics and the progression of disease, liver failure can be subdivided into four classes: acute liver failure (ALF), sub-acute liver failure (SALF), acute-on-chronic liver failure (ACLF), and chronic liver failure (CLF) (Organization Committee of 13th Asia-Pacific Congress of Clinical Microbiology and Infection, 2013). Liver failure is a devastating disease associated with high mortality.

Etiologies of liver failure are listed in Table 6.1.

In Eastern countries, hepatitis virus infections are the leading cause of liver failure. Drugs and hepatotoxic substances (such as

Table 6.1. Etiologies of Liver Failure

Common or Frequent	Less Frequent or Rare
Hepatitis viruses	Metabolic disorders
HAV, HBV, HCV, HDV, HEV	Wilson's disease, hereditary glycometabolism disorder, etc.
Other viruses	
Cytomegalovirus, Epstein-Barr virus (EBV), enterovirus, Herpes virus, etc.	Ischemia and anoxia
Drugs and hepatotoxic substances	Liver transplantation, partial hepatectomy, hepatic tumor
Acetaminophen, some herbs, etc.	
Alcohol	Conegenital biliary atresia
Bacteria and parasite infections	Others
Acute fatty liver of pregnancy	Trauma, radiation, etc.
Autoimmune liver diseases	

HAV: hepatitis A virus; HBV: hepatitis B virus; HCV: hepatitis C virus; HDV: hepatitis D virus; HEV: hepatitis E virus.

alcohol, chemicals, etc.) may also cause liver failure. However in Western countries, drug use is the primary cause for acute and subacute liver dysfunction, and long-term alcohol consumption often leads to chronic liver disease.

1.3. *Development of Artificial Liver Support System*

Artificial liver support therapy originates from natural processes of liver cell regeneration. It can eliminate noxious substances and temporarily function in place of the failed liver to create an environment for the regeneration of hepatocytes or liver transplantation. There are three types of artificial liver support systems: non-biological, biological and combined systems (Organization Committee of 13th Asia-Pacific Congress of Clinical Microbiology and Infection, 2013). Artificial liver support therapy could be employed as a means for the patients to recover native liver function, or as a bridge to transplantation.

The concept of the "artificial liver" was developed by Sorrentino in 1956 when he demonstrated that fresh liver tissue homogenate could produce urea from ammonium chloride and metabolize ketone bodies, barbiturate and salicylic acid.

Historically, most liver assistance techniques relied on hemoperfusion and dialysis. It was demonstrated that the powdered forms of activated charcoal that could strongly adsorb bilirubin had limited application due to their poor biocompatibility. This led to investigation of possible suitable adsorbents. Currently, hemoperfusion techniques used in artificial liver are mainly based on resin and activated charcoal. Resin includes ion-exchangers, high surface area adsorbents, polyacrylic ester, Chitosan, etc. Because of the coagulopathy associated with liver failure, blood should not be directly perfused unless the blood compatibility has been carefully adjusted using albumin coating (Chang, 1974).

Artificial liver support systems aim at the extracorporeal removal of water-soluble and protein-bound toxins associated with liver failure. However, the unspecific absorptive capacity of hemoperfusion is limited. Other blood purification technologies, including plasma exchange (PE), hemodialysis, hemofiltration, albumin dialysis, and especially hybrid techniques, are now involved in artificial liver support systems.

PE is a procedure that directly removes toxic substance by replacing toxic plasma with fresh healthy plasma. From the mid-1980s to late 1990s, Li Lanjuan (from the first affiliated hospital, School of Medicine, Zhejiang University, China), attempted using the PE-Centered method to detoxify patients with severe hepatitis and sustain them until they could regenerate their own liver or receive a transplant (Li *et al.*, 2006). With advances in blood purification technologies and a better understanding of the pathophysiology of hepatic failure, Li carried out a variety of extracorporeal liver support methods, such as plasma exchange with a small pore size plasma separator, combination of plasma exchange and hemoperfusion, combination of plasma exchange and hemofiltration, and paired plasma exchange filtration adsorption, etc. Li used different artificial liver methods based on the symptoms of individual patient, which is called Li non-bioartificial liver (LiNBAL) (Organization Committee of 13th Asia-Pacific Congress of Clinical Microbiology and Infection, 2013). Currently, LiNBAL has been widely used in clinical practice in China.

In Europe, there are several artificial liver support devices, such as the molecular adsorbent recirculating system (MARS), and the fractionated plasma separation and adsorption (FPSA), Biologic-DT system. These systems are based on hemoperfusion and other purification techniques.

MARS was introduced in 1993 (Gambro Lundia, Lund, Sweden). During MARS therapy, blood from the patient is pumped through a high-flux albumin-coated hemodialyser. Toxins (both albumin-bound and soluble) are transferred to the human albumin-enriched dialysate. The exogenous albumin dialysate is then regenerated in a closed loop by dialysis against a conventional dialysate and by adsorption through charcoal and anion-exchange resin columns (Wauters *et al.*, 2011).

Fractionated plasma separation and adsorption (FPSA) was implemented into a commercially available device in 1999 (Prometheus®, Fresenius Medical Care, Bad Homburg, Germany). Prometheus combines the treatment of fractionated plasma using adsorbents (neutral resin and an anion exchange resin) with a downstream high flux hemodialysis to remove water-soluble low molecular weight substances (Rifai, 2011).

However, the artificial systems are only able to provide detoxification functions of the liver, which is not enough in some cases. In the last two decades, improvements in hepatocyte isolation and culture techniques, a better understanding of hepatocyte–matrix interactions, an availability of new biomaterials and advances in tissue engineering techniques have resulted in the development of a new generation of extracorporeal liver assist devices — biological and combined biological artificial livers.

Biological and combined biological artificial livers not only have detoxification functions but also incorporate synthesis and biotransformation functions. It is likely that they are the direction of future development. They have partly completed Phase II/III clinical trials and shown their effectiveness on liver failure. Current research focuses on improving the safety and efficacy of these techniques.

1.4. *The Role of Hemoperfusion in Artificial Liver Support System*

Unlike the heart, lung or kidney, which have a single primary task, the liver either controls or affects almost every aspect of the metabolism, blood coagulation and fibrinolysis, the immune system, waste removal and most physiologic regulatory processes. In patients with liver failure, the range of toxic substances in the blood stream can be broad. It is perhaps not enough to remove these compounds using sorption therapy. More often, hemoperfusion is combined with other purification techniques in the management of liver failure.

For patients with cholestasis and hyperbilirubinemia, hemoperfusion is the only treatment, and bilirubin and bile acid are removed significantly from the blood. Common bilirubin adsorption columns include Medisorba BL-300 (Kuraray Medical Inc., Japan) (Nakaji *et al.*, 2002), Plasorba BR-350 (Asahi Medical, Japan), BS330 (Jafron Biomedical Co. Ltd, China) perfusion, etc. Plasorba BR-350 was investigated in five patients with intractable jaundice of various etiologies. Four liters of plasma were separated by membrane plasma separation (Plasmaflow OP-05) at a rate of 22.5 mL/min. The plasma was then perfused through the bilirubin adsorbent column. The concentration of total bilirubin was cut by 31–60%; total bile acids were reduced by 20–74% (Geiger *et al.*, 1992).

Generally, hemoperfusion has been combined with other purification techniques. In China, Li Lanjuan chooses appropriate methods according to the specific circumstances of each patient. For example, PE was added to hemoperfusion to enhance bilirubin removal. To patients complicated with hepatic encephalopathy, activated charcoal perfusion was added to plasma exchange (Li *et al.*, 2004). And now we are developing a device that includes plasma exchange, hemofiltration and bilirubin adsorption techniques. In Europe, MARS and FPSA, which is the combination of hemoperfusion and dialysis, are two commercially available extracorporeal liver support devices.

Hemoperfusion was not used in some methods of artificial liver support systems such as plasma exchange plus continuous hemodiafiltration (Sadahiro *et al.*, 2001), plasma exchange plus hemofiltration, and plasma exchange implementation with small membrane pore sizes (Li *et al.*, 2005). In designing the biological and combined biological artificial livers, charcoal columns are often placed before the hepatocyte bioreactor in the plasma circuit. For example, during HepatAssist BAL therapy, the patient's blood is first separated into plasma and cellular components using plasmapheresis. The separated plasma is directed into a high-flow plasma recirculation loop, where it first goes through the charcoal filter, and then through an oxygenator, a heater and a hollow fiber bioreactor inoculated with seven billion cryopreserved matrix-anchored porcine liver cells. By placing the charcoal column in the plasma circuit, porcine hepatocytes are likely to be protected from possible toxic effects of plasma in hepatic failure (Demetriou *et al.*, 2004).

In summary, it is clear that the application of hemoperfusion, especially in combination with other purification techniques, presents a potential aid to help the patient wait for a graft or the regeneration of hepatocytes. Up until now, non-biological artificial livers have been widely used in clinical practice, and their efficacy has been supported. In the biological and combined biological artificial liver field, current research focuses on improving its safety and efficacy.

2. Toxic Substances Accumulating in the Blood of Patients with Liver Failure

2.1. *The Pathophysiological Mechanism of Liver Failure*

According to clinical experience, it has been concluded that there are four phases during the development and progression of liver failure: the early phase, rise phase, platform phase and recovery phase. The most important pathological changes in the developmental process of liver failure are massive hepatocytes deaths, a large number of inflammatory cell infiltration and microcirculation obstructions, followed by elevated circulation of endotoxins and exacerbated immune liver injuries. Therefore, in accordance with clinical, pathological, and molecular biological changes, we believe that patients with liver failure have experienced three different injuries: direct hepatocytes damage, immune injury and endotoxemia injury.

In direct injury, viruses, drugs, hepatotoxic substances directly cause hepatocyte deaths.

In immune injury, hepatocyte death induces an immune response, resulting in immune injury and massive hepatocyte necrosis. In endotoxemia injury, after experience of the direct injury and immune injury, malfunction of liver detoxification capacity, intestinal barrier function and immune suppression induce endotoxemia and further accelerate the process of liver damage.

In the triple injury process, the noteworthy physiological and pathological changes are ischemia and hypoxia of the hepatocytes. This change is mainly caused by the immune damage, further exacerbates the immune injury, and plays a key role in the development of endotoxemia.

2.2. *Toxin Accumulation during Liver Failure*

During advanced liver failure, the liver cannot clear toxic metabolites from the portal vein, resulting in the direct release of toxic substances by necrotic liver tissue. The accumulation of toxic metabolites results in a series of serious clinical consequences. These toxic substances include ammonia, aromatic amino acids, fatty acids (mainly short-chain fatty acids), thiols, endogenous benzodiazepine-like substances, phenols, false neurotransmitters, γ-aminobutyric acid (GABA), middle molecules weight substances, bile acids, bilirubin, endotoxin, and cytokines. These water-soluble or protein-bound substances can cause local injuries such as cerebral edema, hepatic coma and suppression of liver regeneration, and systemic injuries such as systemic inflammatory response syndrome and multiple organ failure. Effectively removing these toxic substances is a fundamental goal of the application of artificial liver treatment in liver failure patients.

2.3. *Cytokines Released during Liver Failure*

Due to impaired immune function, intestinal bacteria translocation and infection, hepatitis patients often suffer from quite a high incidence of infection and endotoxemia (usually intestinal origin endotoxemia), mostly Gram-negative relative. Various inflammatory stimuli (e.g. endotoxins) can induce an excess of pro-inflammatory cytokines, such as TNFα, IL-1 and IL-6, by stimulating monocytes, macrophages and other effector cells. Studies have shown that patients with concurrent infection usually have a higher plasma TNFα level than that of non-infected patients, further increasing liver damage, leading to multiple organ dysfunction and even death. Therefore reducing the level of excess of cytokines in blood circulation, or inhibiting their activity, is an important aspect of the treatment of hepatitis and sepsis, preventing the occurrence of multiple organ failure, and reducing mortality.

During the past two decades, various research institutions have developed many biological engineering anti-cytokine formulations, such as the IL-1 receptor antagonist (IL-1Ra), the anti-TNFα monoclonal antibody (TNF-αMab), the anti-PAF antibody and the TNF soluble receptor. In animal experiments, these biological agents exhibit a good cytokine antagonistic effect and can effectively reduce the mortality of the experimental animals. However, clinical trials showed that these bioengineering formulations are not effective in reducing the mortality rate of sepsis patients. Instead, some experimental and clinical studies showed that hemoperfusion, plasma exchange and other artificial liver support therapies might reduce the level of circulating cytokines, which will be discussed in detail in the following text.

2.4. *Endotoxemia during Liver Failure*

Liver failure complicated with endotoxemia can present a range of clinical manifestations and complications, mainly in the form of fevers, disseminated intravascular coagulation, acute gastric bleeding, ascites and ascites infection, local Shwartzman reaction, hepatorenal syndrome, hepatoencephalopathy, metabolic disorder and shock. Endotoxins can act directly or indirectly on the body temperature regulating center in the hypothalamus, creating a rise in body temperature. Due to monocyte macrophage system damage or intestinal endotoxemia "blocking" of the monocyte macrophage system, procoagulant substances (e.g. endotoxins) and activated coagulation factors (such as factor IVa, Xa, VIa) cannot be effectively cleared, potentially causing DIC endotoxins to activate the bradykinin system that induces tissue ischemia. It also hinders cell metabolism, causing gastric bleeding and decreases in mucosal immunity.

2.5. *Characteristics of Toxic Substances in Hepatic Failure*

Toxic substances in liver failure can be divided into water-soluble and protein-bound toxins (Table 6.2). Albumin is the preferential binding protein. Albumin contains reversible binding sites for substances such as fatty acids, hormones, and trace metals. It has been proposed that albumin binding sites for these putative toxins become saturated in patients with liver failure, a consequence of decreased hepatic clearance, leading to an accumulation of unbound toxic substances and the development of organ dysfunction.

Indeed, the range of substances to be removed is broad and not completely identified. It is difficult to remove these compounds from the blood stream using adsorption therapy because of the limited, unspecific absorptive capacity of chemical adsorption. This also explains,

Table 6.2. Toxic Substances Accumulating in the Blood Due to Hepatic Failure

Protein-Bound Substances	Water-Soluble Substances
Bilirubin, un/conjugated	Ammonia
Bile acids	Gamma aminobutyric acid
Short-chain fatty acids	Aromatic amino acid#
Benzodiazepenes	Cytokines#
Mercaptans	Creatinine
Nitric oxide	
Indoyxlsulfate	
Copper	
Protoporphyrin	
Endotoxin	

Represents different classification in different documents (Hughes, 2002; Mitzer *et al.*, 2009)

at least to a certain extent, why no single detoxification measure, such as conventional hemoperfusion, hemodialysis, or hemofiltration, has achieved wide clinical use in the treatment of liver failure.

3. Adsorbent Materials used in Liver Failure Hemoperfusion

The adsorption mechanisms used for patients with liver failure include hydrophobic interactions, hydrogen bonding, electrostatic reactions, covalent bonding and chemical conversions, and several of these adsorption mechanisms can be implemented at the same time in one treatment. There are three kinds of absorbents: non-selective, selective and specific (Table 6.3). A typical material for the adsorption of blood perfusion comprises of activated charcoal and non-ionized or ionic resin. New pharmacological agents such as antiendotoxin monoclonal antibodies, antitumor necrosis factor antibodies, anti-IL-1 receptors, and anti-IL-6 antibodies are being

Table 6.3. Adsorbent Materials used in Clinical and Experimental Treatments

Non-selective adsorbent material	Activated charcoal: non-coated activated charcoal, activated charcoal coated Neutral resin: Amberlite XAD-7
Selective adsorbent material	Ionized resin: polystyrene derivative fibers fixed with polymyxin B (PMX-F), the anion exchange resin (plasorba BR-350), and modified anion exchange resin (DHP BR-350) Fiber materials: big pore fiber beads fixed with polymyxin B, the polyethylene amine bag coated big pore fiber beads, cationic modified fiber microspheres. Modified sepharose: Sephadex solid-phase column fixed with polymyxin B
Specific adsorbent material	Antibody coated microspheres detoxification system (MDS): anti-TNFα MDS, anti-IL-1β MDS

developed for the adjunctive treatment of sepsis. Toraymyxin (Toray Medical Co., Ltd., Japan) was developed as a selective adsorbent of endotoxins (Hanasawa *et al.*, 1989). Recent new adsorbents, such as polymyxinB-immobilized fiber, have been developed to primarily remove bacterial endotoxins, and some cytokines.

4. Removal of Toxic Substances by Hemoperfusion in Liver Failure Patients

4.1. *Bilirubin Adsorption*

Albumin-coated Amberlite XAD-7 is a blood compatible material that has been shown to remove bilirubin, phenols and middle molecular weight substances in patients with acute liver failure (Hughes *et al.*, 1979). *In vivo* hemoperfusion treatments run with hyperbilirubinemia animal models showed that a 40 mL resin column can remove approximately 20 mg of bilirubin in three hours. Hemoperfusion of female baboons (8–14 kg) showed no significant effect of the column on blood chemistry and coagulation factors, and repeated runs had no side effects (Sideman *et al.*, 1982).

Nikolaev VG *et al.* aimed to clarify the relationship between the extent of activation of synthetic granula and fibrous carbons, and the adsorption of bilirubin from protein solutions (Nikolaev *et al.*, 1991). They found a parallel increase in volume and specific surface area of micropores lead to a 4–5-fold increase of bilirubin adsorption from a 3% HSA solution. Carbon adsorbents synthesized for the elimination of bilirubin from protein-containing solutions can also be called deliganding adsorbents, since under some experimental conditions they eliminate other protein-bound ligands, viz. phenols by 97–99%, bile acids by 90–92% and sodium caprylate by 89–95%.

Resin adsorption capacity is much higher than that of activated charcoal. It has very strong adsorption effect of lipophilic substances and protein-containing substances. Cation exchange resin has better effect on ammonia removal, anion exchange resin is better at adsorption of bilirubin and organic anion, and neutral resin is good at adsorbing bilirubin, bile acid, free fatty acid and amide.

Gao L *et al.* used plasma collected from patients with chronic severe hepatitis flowing through BL-300 perfusion column in cycles with different pass-through volume (Gao *et al.*, 2008). The clearance rate of total bilirubin (TBil) and direct bilirubin (DBil) of the mixed plasma decreased gradually with the increase of total amount of pass-through. It was found TBil concentration decreased by 58.1%, DBil by 54.9%, total bile acid (TBA) by 86.9%, albumin by 7.5% and globulin by 11.7%, corresponding to the total adsorbed amount of 342.2 μmol, 292.3 μmol, 263.5 μmol, 7.0 g and 9.5g, respectively, when the total amount of pass-through reached 32.0 liters. However, ALB and GLB supplements are necessary for patients treated with artificial liver of this column.

4.2. *Endotoxin and Hemoperfusion*

Activated charcoal adsorption of endotoxins rely mainly on hydrophobic reactions. Bare activated charcoal can effectively remove endotoxins. Nolan JP *et al.* found that activated charcoal could remove 90% of the endotoxins in an isotonic saline solution within 15 minutes (Nolan *et al.*, 1975). However, blood (plasma) decreased the endotoxin clearance rate of activated charcoal both *in vitro* and *in vivo*.

Cholestyramine (Dowex 1-X2), a strongly basic anion-exchange resin, was used to test the endotoxin-binding characteristic. The resin to be tested was washed and swelled overnight, and 1 mg/ml of 51Cr-labeled endotoxin was added to the mixture, agitated and incubated at 37°C. The 1-X2 removed about 90% of the *E. coli* endotoxin from solution in 15 minutes. Increasing the mesh size appeared to offer more binding sites for each Dowex 1 resin. Activated charcoal also adsorbed about 90% of the endotoxin, but Amberlite XAD-2 (neutral resin) showed little binding capacity. Injection of a filtrate from unlabeled *E. coli* and *S. typhosa* resin-treated solution into rats demonstrated that both Dowex 1-X2 and activated charcoal prevented the transaminase rise noted in animals injected with untreated solution (Nolan *et al.*, 1975).

Amoureux MC *et al.* found a novel adsorption resin that removed 75% of endotoxins present in whole blood. This affinity resin displayed good hemocompatibility regarding the coagulation pathway. Minimal platelet, neutrophil and complement activation were observed. Additionally, there was no evidence of the consumption of coagulation factors or cell loss (Amoureux *et al.*, 2004).

4.3. *Other Absorbent and Endotoxemia*

Polymyxin B is a well-known antibiotic that inactivates endotoxins. However, it is a peptide that is not absorbed through the intestine, and is toxic to the central nervous system and the kidney. Therefore, to detoxify endotoxins, a biomaterial that fixes polymyxin B (toraymyxin) to polystyrene fibers was developed. Toraymyxin is nontoxic when used in direct hemoperfusion. In the Phase II and IIIa clinical studies, toraymyxin was safely used in patients with severe sepsis and/or septic MOF. Toraymyxin treatment significantly decreased the endotoxin level in both Phase II and IIIa studies (Hanasawa, 2002). PMX-direct hemoperfusion (PMX-DHP) was used to treat septic shock patients, and it was associated with elevations in mean arterial pressure and urine output, reduction in the mean dose requirement of vasopressor agents, and recovery from the shock state in all the patients. The PaO2/FIO2 ratio also showed significant improvement. Using the high-sensitivity endotoxin assay, endotoxin was detected in the blood of 20 out of the 24 patients (80%) before the PMX-DHP treatment, and a significant reduction in the endotoxin level was noted after the PMX-DHP treatment. The TNFα production capacity of whole blood, which was found to be lower in the septic shock patients than in healthy subjects, was significantly increased after PMX-DHP treatment. Elimination of ET by PMX-DHP in septic shock patients was confirmed by high-sensitivity endotoxin assays.

4.4. *Hemoperfusion and Cytokines Removal*

The Biologic-DTPF System combines the Biologic-DT hemodiabsorption system in series with the Biologic PF push-pull pheresis system, in which membranes separate plasma for direct contact between plasma proteins and the sorbents. Preliminary studies conducted in bovine serum albumin solution and in bovine plasma allowed charcoal and silica to be evaluated as adsorbents. Equilibrium binding experiments showed a high capacity of cytokine (IL-1 beta, TNFα) binding by powdered charcoal (70–90 ng/g). Kinetic binding studies in bovine plasma revealed relatively quick adsorption of IL-1 beta and IL-6 by charcoal, with the capacity range of 1.2–2.0 ng/g for tested cytokines (IL-1 beta and TNFα). Further laboratory studies with plasma have shown that powdered silica has an even greater binding capacity, up to 13 ng/g for TNFα depending upon particle size, and binds more rapidly for all tested cytokines than powdered charcoal. However, Cholestyramine is a more efficient sorbent for endotoxin removal than either charcoal or silica. *In vitro* tests using whole blood have demonstrated that the DTPF, with powdered charcoal as the sorbent, clears cytokines (TNFα, IL-1 beta, and IL-6) at 12.6–23.4 mL/min, bilirubin at 17.8–34.7 mL/min, and creatinine at 53.6–82.6 mL/min (Steczko *et al.*, 1999).

Direct hemoperfusion using a beta2-microglobulin-selective adsorbent column eliminates inflammatory cytokines and has a beneficial influence on the treatment of acute respiratory distress syndrome. Inflammatory cytokines such as interleukin (IL)-1 beta, IL-6, and soluble intercellular adhesion molecule 1 (sICAM-1) decreased significantly after the treatment. However, while IL-2 decreased significantly after the treatment, it had returned by the next treatment (Kono *et al.*, 2009).

Immobilized polymyxin-B fiber columns reduce cytokine plasma levels in patients with abdominal sepsis. Plasma concentrations (pg/mL) of IL-6, IL-10 and TNF-a were significantly lower after hemofiltration with a PMX fiber column. Survivors had higher IL-6 and lower IL-10 and TNF-a, pre-treatment plasma levels compared with deceased patients. Use of this technique was associated with lower norepinephrine support and an increased PaO2/FiO2 ratio (Zagli *et al.*, 2010). In another study, the serum levels of cytokines including IL-9, IL-12, IL-17, PDGF and VEGF were significantly decreased immediately after PMX-DHP ($P < 0.02$), and VEGF and IL-12 were most prominently reduced. In addition to PDGF and VEGF, IL-1β, IL-1ra, IL-8, IL-23, FGF basic, GM-CSF, IP-10, RANTES and TGF-β were eluted from used PMX-fibers. Improved pulmonary oxygenation after PMX-DHP treatment was highly correlated with the quantities of eluted VEGF. Removal of VEGF by PMX-DHP may contribute to the rapid improvement in oxygenation by suppressing vascular permeability in the lung (Oishi *et al.*, 2013).

4.5. *Remove Amino Acids through Hemoperfusion*

The effects of coated charcoal hemoperfusion on amino acids in plasma and CSF in grade III hepatic coma patients was evaluated with the galactosamine induced hepatic coma rat model. Hemoperfusion for one hour significantly reduced aromatic amino acids (AAA) and branched chain amino acids (BCAA) in plasma, and increased the BCAA/AAA molar ratio. When rats in grade III comas were treated with two consecutive one-hour hemoperfusions, the substantial reduction of plasma AAAs resulted in a significant decrease of AAA levels in CSF (Shi and Chang, 1984).

5. Application of only Hemoperfusion in Liver Failure Patients

5.1. *Bilirubin Adsorption*

The liver provides primary functions for homeostatic control, synthesis and clearance. If liver disease results in a loss of liver function below a critical level, homeostatic disequilibrium can results in a variety of clinical syndromes. Hyperbilirubinemia may result from isolated disorders of liver and biliary disease. Jaundice is a common sign of generalized hepatobiliary dysfunction. Liver diseases can be categorized as those in which the primary injury results from inflammation and hepatocellular necrosis, cholestasis, or a combination of the two. Cholestatic disorders can be further subdivided into those due to mechanical obstruction of the bile duct flow and those resulting from intrahepatic cholestasis. It is important to note that chronic cholestasis from any cause may result in the intrahepatic accumulation of compounds such as bile acids, which may themselves be toxic, causing hepatocellular injury. Plasma perfusion has become an effective means of treatment of hyperbilirubinemia.

Scientists have studied various perfusion materials to increase the adsorption capabilities of various toxins. Preliminary studies have shown albumin-coated Amberlite XAD-7 to be blood compatible. Hemoperfusion using an albumin-coated Amberlite XAD-7 column was performed in 19 patients that were in comas due to fulminant hepatic failure. The clinical procedure was well tolerated, and the study found good blood compatibility. No significant changes were observed in beta-thromboglobulin, screen filtration pressure, plasma electrolytes, calcium, protein or albumin. The total plasma bilirubin fell by a mean of 24 μmol/L, with a reduction in 21 of the 25 perfusions studied of up to 104 mμmol/L during perfusion. Mean plasma levels of total bile acids were 137 ± 19 μmol/L and 115 ± 16 μmol/L before and after four hours hemoperfusion, respectively. The amount of bile acids recovered by elution of the resin column was over three times greater than that apparent from the change in plasma levels. Sephadex G-25 column chromatography of material eluted from the resin column showed various peaks were removed, including substances in the middle molecular weight range (1000–5000 daltons). Of the patients treated, eight (42%) survived to leave hospital (Bihari *et al.*, 1983).

Xiang H *et al.* studied 94 different causes of chronic liver diseases in patients with severe jaundice that were treated with plasma perfusion by HB-H-6 resin. The TBil, DBil and indirect bilirubin (IBil) values were 387.80 ± 183.08 μmol/L, 238.66 ± 103.52 μmol/L, and 127.23 ± 62.00 μmol/L before treatment, and 291.80 ± 135.58 μmol/L, 183.10 ± 76.29 μmol/L, and 92.85 ± 54.25 μmol/L after HB-H-6 resin plasma perfusion treatment. The level of the TBil, DBil and IBil eliminated by HB-H-6 resin were 23.68% ± 9.14%, 21.54% ± 9.90%, and 27.09% ± 16.84% respectively (Xiang *et al.*, 2011).

Song *et al.* (2013) studied the therapeutic effect of plasma absorption on patients with hyperbilirubinemia. Sixty-five patients were randomly divided into two groups (treatment group and control group). After treatment with plasma absorption, the liver functions of the patients were ameliorated. The prothrombin time activity (PTA) two weeks and four weeks after treatment increased significantly ($P < 0.05$), and TBIL level significantly decreased ($P < 0.01$). The total effective rate of treatment and control group after four weeks was 39.40% and 12.50% respectively; there were significant differences between two groups ($P < 0.05$). The total effective rate after eight weeks in the treatment group was 66.67%, and 40.63% in the control group ($P < 0.05$). This study demonstrated that the liver function of patients with hyperbilirubinemia could be significantly improved by plasma absorption, which could prevent disease progression and significantly improve the clinical cure rate. It is particularly suitable for patients those unfit for the treatment with plasma exchange.

It was also found that combination of two kinds of adsorbents will increase the adsorption effect. Chen *et al.* (2012) reported 10 patients with acute-on-chronic liver failure treated with plasma perfusion by disposable hemoperfutor (HA-330) and bilirubin adsorption column (BS-330) (Fig. 6.1). Clinical symptoms and changes in physical signs were observed during the treatment. Statistically significant differences ($P < 0.05$) were observed in the TBil, DBil, ALT, AST, alkaline

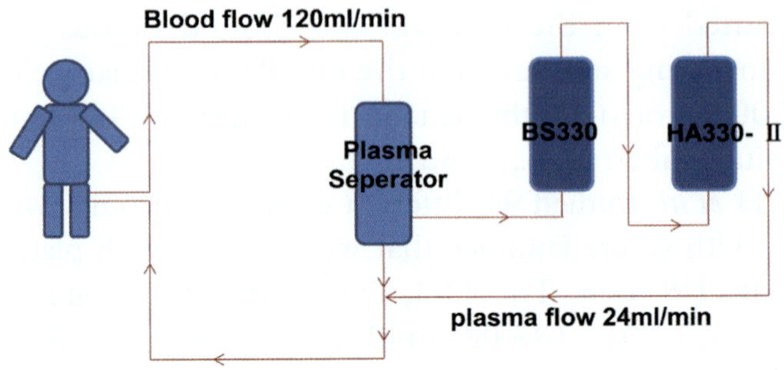

Fig. 6.1. Hemoperfusion model with two combined resins.

phosphatase, TBA, albumin and globulin serum levels before and after treatment. In contrast, no statistically significant differences ($P > 0.05$) were observed in the serum concentrations of potassium, sodium, creatinine and urea nitrogen, or in prothrombin time, activated partial thromboplastin time and aminogram changes. Detailed changes in the serum biochemistry and clotting function before and after treatment are shown in Table 6.4.

5.2. *The Effect of Hemoperfusion on Hepatic Coma Patients*

Hepatic encephalopathy is a reversible impairment of cognitive function or level of consciousness in patients with liver disease or portal-systemic shunts. Hepatic encephalopathy may be related to high blood ammonia, false neurotransmitters, increasing aromatic amino acids and imbalance of branched chain amino acids and aromatic amino acid. Hemoperfusion was very efficient in removing molecules of up to 5,000 molecular weight (Chang, 1972; Chang and Migchelsen, 1973; Chang and Lister, 1980; 1981), as well as protein-bound molecules. Hemoperfusion resulted in the recovery of consciousness in a grade IV hepatic coma patient (Chang, 1972). In 1974, 22 patients with fulminant hepatic failure, who deteriorated to grade IV comas despite full supportive therapy, were treated with repeated periods of hemoperfusion through columns containing activated charcoal (Gazzard *et al.*, 1974). The procedure was well

Table 6.4. Changes in Serum Biochemistry and Clotting Function after Treatment with Two Kinds of Adsorbents

	Pre-Treatment	Post-Treatment	*P* value
ALB (g/L)	33.54 ± 3.22	28.45 ± 1.61	0.000
TBil (μmol/L)	423.42 ± 202.10	284.53 ± 134.32	0.000
DBil (μmol/L)	265.11 ± 120.49	205.53 ± 105.73	0.000
TBA (μmol/L)	206.72 ± 84.09	163.00 ± 62.88	0.001
ALT (U/L)	176.42 ± 280.11	153.79 ± 254.59	0..033
AST (U/L)	118.84 ± 82.19	93.79 ± 59.79	0.005
AKP (U/L)	117.42 ± 4.57	112.68 ± 33.06	0.030
APTT (S)	50.58 ± 12.68	87.50 ± 47.45	NS
PT (S)	17.80 ± 8.84	23.56 ± 16.69	NS
K^+ (mmol/L)	3.96 ± 0.39	4.03 ± 0.46	NS
Na^+ (mmol/L)	131.65 ± 7.36	132.12 ± 7.72	NS
Cr (μmol/L)	57.25 ± 12.15	60.75 ± 17.41	NS
Bun (mmol/L)	4.77 ± 1.65	4.73 ± 1.75	NS

ALB: albumin; TBil: total bilirubin; DBil: direct bilirubin; TBA: total bile acid; ALT: aminotransferase; AST: aspartate aminotransferase; AKP: alkaline phosphatase; APTT: activated partial thromboplastin time; PT: prothrombin time; Cr: creatinine; Bun: urea nitrogen.

tolerated clinically. Eleven of the patients regained consciousness, and 10 left the hospital. The column extracted most amino acids from the plasma, and during perfusion, arterial concentrations of phenylalanine, tyrosine, and methionine-amino acids known to be involved in the pathogenesis of the encephalopathy fell significantly.

In 1988, 137 patients with fulminant hepatic failure were entered into two controlled trials of charcoal hemoperfusion (O'Grady *et al.*, 1988). In trial A, 75 patients with grade 3 encephalopathy were randomized to receive five or 10 hr of hemoperfusion daily. Overall survival rates for the two groups were similar (51.3% vs. 50.0%), as was the frequency of major complications including cerebral edema and renal failure. In trial B, 62 patients with established grade 4 encephalopathy were randomized into a non-perfusion group, or to have 10 hr of hemoperfusion daily. Overall survival rates for the two groups were again similar (39.3% and 34.5%, respectively). In both trials, there was a significant relationship between survival and etiology that was independent of the use or duration of hemoperfusion. Thus, percentage survival for the acetaminophen-overdose cases was 52.9%, for hepatitis A 66.7%, for hepatitis B 38.9%, for presumed non-A, non-B hepatitis 20%, and for halothane or drug reaction 12.5%.

Previous study showed that hemoperfusion for the treatment of liver failure has the following characteristics:

1. The liver is a complex organ with many other functions besides detoxification. The consciousness recovery rate is higher, but not the survival rate.
2. The survival rate is different in patients with different etiologies, not only because of the differential responses to the treatment, but also the different prognoses of original diseases.
3. The biochemical index in the patients improved significantly, especially the concentration of hydrophilic substances.
4. The liver is a complex organ with various vital functions in synthesis, detoxification and regulation. Therefore its failure constitutes a life-threatening condition, manifesting as coagulopathy, jaundice, hepatic encephalopathy (HE) and ascites. Hemoperfusion is a justifiable treatment for liver failure. However, hemoperfusion alone cannot fully support liver function or treat complications such as hepatorenal syndrome.

6. Li Non-Bioartificial Liver (LiNBAL)

6.1. *The Development of LiNBAL*

The initial stages of the artificial liver support system (ALSS) in China began in the late 1970s to mid-1980s, during which activated charcoal and other absorbents were studied. The efficacy of hemoperfusion was investigated in animals with acute liver failure. Hemoperfusion was also carried out to detoxify patients poisoned by drugs and toxic substances.

From the mid-1980s to late 1990s, Li Lanjuan attempted using the plasma exchange (PE)-Centered ALSS method to detoxify patients with severe hepatitis and sustain them until they could regenerate their own liver or receive a transplant. ALSS markedly improved the survival rate of patients with severe hepatitis. The efficacy of ALSS was first analyzed in 1997. In that study, 45 patients received ALSS therapy (the "ALSS group"), and 41 patients receiving other liver therapies made up the control group. The ALSS group included seven patients with acute and subacute severe hepatitis, and 38 with severe chronic hepatitis. The control group had five acute and subacute severe hepatitis patients, and 36 severe chronic hepatitis patients. The six-month survival rates for the ALSS group were 71.4% (5/7 acute and subacute hepatitis patients) and 44.7% (17/38 chronic hepatitis patients). In contrast, six-month survival in the control group was 20% (1/5 acute and subacute hepatitis patients), and only 5.6% (2/36 chronic hepatitis patients) (Li *et al.*, 2006).

In the mid-1990s, the use of ALSS was limited to several large hospitals in China. Li Lanjuan held ALSS training courses each year beginning in 1998, and ALSS was applied in more and more hospitals. According to unpublished data collected by the Artificial Liver and Liver Failure Group of the Chinese Society of Infection, ALSS therapy has now been carried out in approximately 200 hospitals in 31 provinces across China. An international symposium on liver failure and artificial liver technology was first held in China in 2001, and it has been held every two years since then. In 2002, the Artificial Liver and Liver Failure Group of the Chinese Society of Infection established guidelines for artificial liver therapy.

With advances in blood purification technologies, and a better understanding of the pathophysiology of hepatic failure, Li carried out a variety of extracorporeal liver support methods, such as plasma exchange with a small pore size plasma separator (Li *et al.*, 2005), combination of plasma exchange and hemoperfusion, combination of plasma exchange and hemofiltration, and paired plasma exchange filtration adsorption, etc. Li used different artificial liver methods based on the symptoms of individual patients (Table 6.5), which is called the Li non-bioartificial liver (LiNBAL). For example, when patients had

Table 6.5. Individualized Treatment with Li Non-Bioartificial Liver

Specific Circumstances of Patients	Methods
Drugs and poison-induced liver failure	PE + HF, PE + HD, or PE + hemoperfusion
Serious infection-induced liver failure	PE + HF
Early-stage liver failure induced by viral hepatitis	PE
Middle-stage liver failure induced by viral hepatitis	PE + HF
Liver failure patients with hepatic encephalopathy	PE + hemoperfusion, PE + HF or PE + CHDF
Liver failure patients with renal failure	PE + HF or PE + CHDF
Liver failure patients with water and electrolyte disturbance	PE + HF
Cholestasis	Bilirubin adsorption

PE: plasma exchange; HF: hemofiltration; HD: hemodialysis; CHDF: continuous hemodiafiltration.

hepatic encephalopathy, Li performed plasma exchange in combination with hemoperfusion. For patients with hepatorenal syndrome, Li chose plasma exchange and hemofiltration or CHDF. When the balance of water or electrolytes was disturbed, Li used plasma exchange and hemofiltration (Organization Committee of 13th Asia-Pacific Congress of Clinical Microbiology and Infection, 2013). Sometimes, more than three methods were used together for one patient.

6.2. *Effect of LiNBAL on Patients with Severe Viral Hepatitis: A Study of 400 Cases*

In 2004, Li Lanjuan analyzed 400 patients with severe viral hepatitis treated with LiNBAL in her hospital (Li *et al.*, 2004). The treatment group consisting of 400 viral severe hepatitis patients was treated with LiNBAL at the First Hospital of College of Medicine, Zhejiang University from 1995 to 2003. The control group consisted of 400 severe viral hepatitis patients that were treated at the same hospital from 1986 to 1994. LiNBAL therapy was carried out two or three times in the first or second week, then one time each week till the patient's condition was stable.

Table 6.6. Result of Liver Function Test before and after ALSS Therapy (Mean ± SE)

Liver Function	Pre-Treatment	Post-Treatment	*P*
ALT (U/L)	123.35 ± 281.32	53.15 ± 94.21	< 0.001
AST (U/L)	126.84 ± 115.25	63.71 ± 58.45	< 0.001
ALP (U/L)	127.97 ± 66.74	81.45 ± 39.62	< 0.001
TBil (μmol/L)	511.36 ± 192.81	257.38 ± 123.48	< 0.001
ChE (U/L)	2572.58 ± 2236.95	3119.24 ± 1812.62	0.001
γ-GT (U/L)	42.38 ± 53.85	20.91 ± 25.58	< 0.001
TBA (μmol/L)	256.36 ± 48.69	119.42 ± 49.37	< 0.001
PT (s)	31.8 ± 7.2	23.6 ± 6.8	< 0.001

Nearly 90% patients experienced an improvement in symptoms such as fatigue or abdominal distention after each treatment. Results of liver function tests improved significantly in all subjects. The serum ALT, AST and TBA levels declined significantly ($P < 0.001$), the level of serum TBIL decreased from 511.36 ± 192.81 μmol/L to 257.38 ± 123.48 μmol/L ($P < 0.001$), while the prothrombin time decreased from 31.8 ± 7.2s to 23.6 ± 6.8s (Table 6.6).

According to this retrospective analysis, the survival rate for chronic severe hepatitis patients who received LiNBAL was 43.4% (157/362), compared to 15.4% (55/358) of such patients treated with conventional medication. For patients with acute and subacute severe hepatitis, the survival rate was 78.9% (30/38) for those treated with LiNBAL, and 11.9% (5/42) for those treated with conventional medications. There were significant differences in the survival rate between the two groups ($P < 0.001$).

6.3. *Paired Plasma Exchange Filtration Adsorption*

There are some coupled plasma filtration adsorption (CPFA) trials abroad reported as potential treatment for SIRS, sepsis shock, MODS (Bellomo R *et al.*, 2003; Nalesso F, 2005; Mao HJ *et al.*, 2009). However, Livigni S *et al.* (2014) conducted a multicenter, randomized, non-blinded trial that compared CPFA with standard care in the treatment of critically ill patients with septic shock. It showed that CPFA did not reduce mortality in patients with septic shock, nor did it positively affect other important clinical outcomes. But the subgroup analysis suggested that CPFA could reduce mortality when a high volume of plasma is treated. Maggi U *et al.* (2013) reported that following liver transplantation in two patients, the first experienced early allograft dysfunction and the other hyperbilirubinemia linked to chronic rejection. After three cycles of CPFA, the bilirubin promptly decreased in both cases. Each cycle of treatment lowered the bilirubin of our patients by 40%.

As we know, plasma exchange requires a large amount of fresh frozen plasma, but plasma supply is in shortage. Bilirubin adsorption combined with hemofiltration can clear a portion of the toxins, but it will influence blood coagulation function in liver failure patients with a high risk of bleeding. Li Lanjuan found CPFA was good at clearing inflammatory mediators in the liver failure, but cannot correct coagulopathy. Thus, Li developed a new kind of artificial liver treatment called paired plasma exchange filtration adsorption (PPEFA). After the low volume plasma exchange, plasma adsorption parallel with plasma filtration continues. It can not only supply coagulation factors but also remove small molecular substances and re-establish the hydro-electrolyte, acid-based equilibrium. The team used this system to treat liver failure patients. It improved the clinical prognosis while plasma exchange successively with coupled plasma filtration adsorption. The experience showed one round of treatment needed 500–1,000 mL of plasma, which was much less than plasma exchange alone. Total bilirubin reduced to about a half. Additionally, patients are able to tolerate the treatment well, and there is no significant influence on blood cell and coagulation function. Further studies still need to be carried out to determine if the cure rate will improve compared with conventional plasma exchange. The treatment mode of paired plasma exchange filtration adsorption and a Pipeline connection diagram are demonstrated in Figs. 6.2 and 6.3.

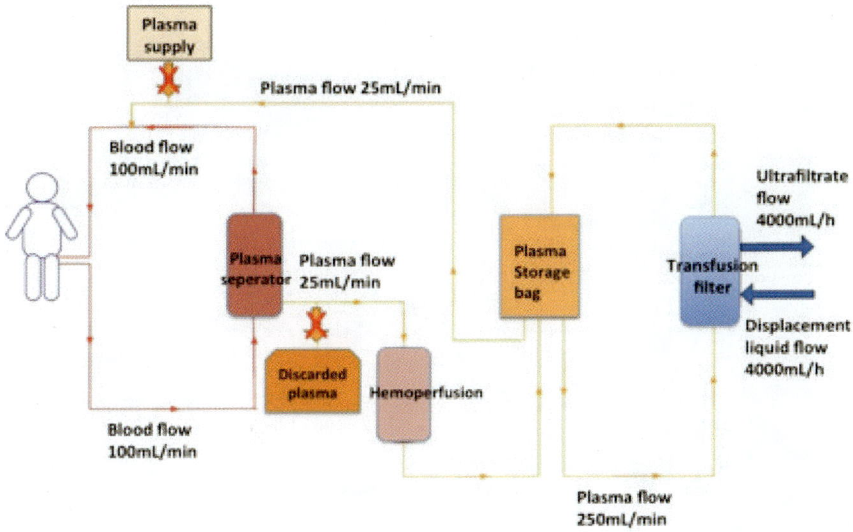

Fig. 6.2. Diagram of paired plasma exchange filtration adsorption.

6.4. *LiNBAL Combined with Liver Transplantation in the Treatment of Patients with Acute-on-Chronic Liver Failure*

Liver transplantation is the most effective therapy for advanced acute-on-chronic liver failure. However, liver transplantation is not always possible because of donor shortage. So LiNBAL is applied as a bridge to liver transplantation (Xu *et al.*, 2013). From January 2001 to December 2009, 171 patients with HBV related ACLF underwent liver transplantation in the First Affiliated Hospital, School of Medicine, Zhejiang University. Of the 171 patients, 115 received 247 sessions of LiNBAL treatment prior to liver transplantation (LiNBAL-LT group) and the other 56 received emergency liver transplantation (LT group). The Model for End-Stage Liver Disease (MELD) scores were 31 ± 6 and 30 ± 7 in the LiNBAL-LT group and the LT group. LiNBAL treatment resulted in improvement of liver function and better tolerance to liver transplantation. The average level of serum TBil before liver

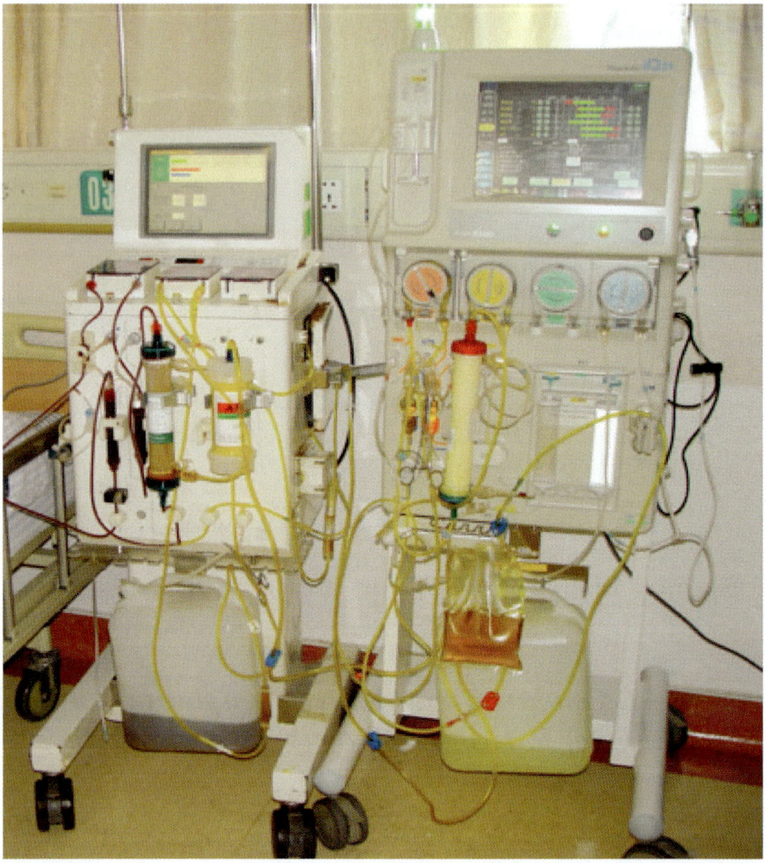

Fig. 6.3. Pipeline connection demonstration of paired plasma exchange filtration adsorption.

transplantation was lower than that before the first time of LiNBAL treatment (Table 6.7). The median waiting time for a donor liver was 12 days (2–226 days) from the first run of LiNBAL treatment to liver transplantation. Compared to the LT group, the beneficial influences of LiNBAL on intraoperative blood loss and endotracheal intubation time were also observed in the LiNBAL-LT group. The one-year and five-year survival rates in the LiNBAL-LT group and LT group were 79.2% and 83%, 69.7% and 78.6%.

Table 6.7. Liver Function and MELD Score before LiNBAL and before Liver Transplantation

	TBil (μmol/L)	ALT (U/L)	AST (U/L)	TBA (μmol/L)	MELD Score
Pre-first LiNBAL	557 ± 195	127 ± 113	182 ± 152	246 ± 98	31 ± 6
Pre-LT	476 ± 169	105 ± 80	165 ± 207	144 ± 108	29 ± 9
P value	< 0.05	> 0.05	> 0.05	< 0.05	> 0.05

Abbreviations: Pre-first LiNBAL, before the first run of LiNBAL; Pre-LT, before liver transplantation.

7. Mars System

MARS was designed by Drs. Stange and Mitzner from Rostock University in Germany based on the principle of protein ligand binding. The MARS system was developed in 1993, and clinically used in Europe in 1999 (Mitzner *et al.*, 2000) (Fig. 6.4); by now more than 20 units have been used in China.

7.1. *MARS System Composition*

MARS is designed using the liver detoxification and albumin transport mechanisms, and the combination of existing mature blood purification technologies such as hemodialysis and hemoperfusion. The MARS system mainly contains three parts: membrane separation, albumin recycling and low flow dialysis. MARS can remove protein-bound toxins and water-soluble toxins, reduce intracranial pressure and improve kidney function, helping prevent brain edema, HRS and MODS.

Membrane separation, albumin recycling and low flow dialysis work together. Free water-soluble toxins in the blood of patients are transferred to albumin regeneration cycle by the concentration gradient and then transferred to the low-flow dialysis fluid and removed. The whole process is a passive downhill process. After albumin-binding toxins (ABT) in the patient's blood cross the MARS membrane, they

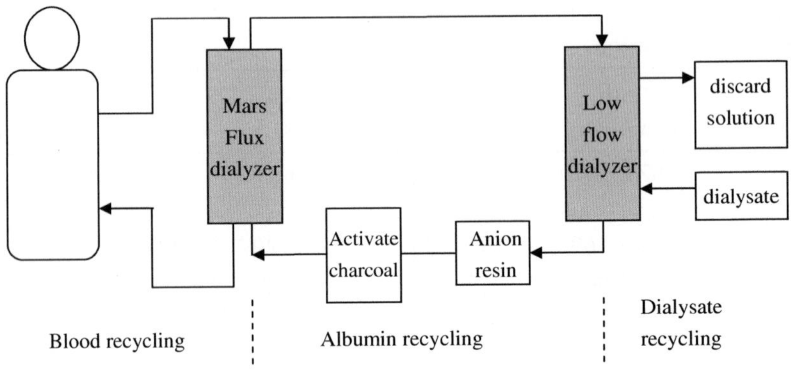

Fig. 6.4. Circulation pattern diagram of MARS system.

immediately combine with albumin in the albumin regeneration cycle, and then separate with albumin when flowing through the adsorption column and anion exchange resin column, where ABT remains in the adsorption column. This process is repeated again and again as toxins in the blood of patients continue to be cleared, and thus plays an alternative role in the detoxification of normal liver function (Fig. 6.5).

7.2. *MARS Complications and Countermeasures*

During MARS treatment, the patient's blood is not completely isolated from the adsorption material, but due to circulating albumin that acts as a detoxification intermediary, the whole process is safe and stable. The complications are significantly reduced compared to other treatments like hemodialysis. However complications such as disequilibrium syndrome, hypotension, dialysis allergic reactions, bleeding, hemolysis, air embolism, fever and muscle spasms may occur in MARS treatment. Currently, the most common clinical complication is mild thrombocytopenia in patient's peripheral blood. However, this does not cause bleeding complications, and platelet transfusion is not necessary. The main preventive measures include:

- shortening the first treatment time, generally no more than six hours;
- thoroughly rinsing dialyzer before treatment;

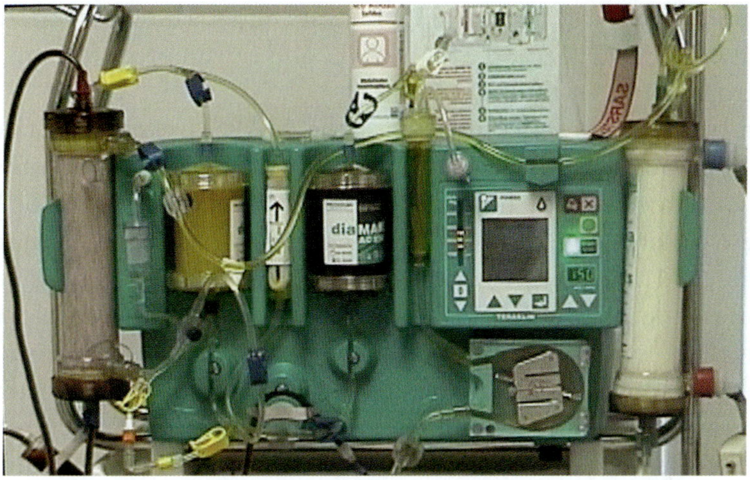

Fig. 6.5. MARS clinical setup.

- strictly abiding sterile and instrument instructions;
- closely observing patient vital signs and equipment operation, treating in time, discontinuation treatment when necessary.

7.3. *The Clinical Applications of MARS*

The first MARS clinical study occurred in 13 chronic severe hepatitis patients when conventional therapies were ineffective. Before treatment, average ALT was 106.5 U/L, AST was 422.75 U/L, serum cholinesterase was 22.7 U/L, serum TBil exceeded 342 μmol/L, serum prothrombin time measured using the quick method was less than 50%, all patients had stage II or above hepatic encephalopathy, and two or more common complications appeared at the same time or successively, such as hepatorenal syndrome, infection, and bleeding. After an average of 5.5 (2 to 14) MARS treatments, nine patients survived and four died, giving a survival rate of 69%. The Child–Pugh score of all patients (including deaths) decreased from 12.9 (Child grade C) to 9.69 (Child grade B). The degree of hepatic encephalopathy dropped from 2.82 to 0.75, serum TBIL decreased from 425.46 to 324.66 μmol/L, blood ammonia decreased from 97.55 to 57.11 μmol/L, serum cholinesterase increased from 22.7 to 30.18 μmol/L, and serum prothrombin time rose from 43.42 to 60.1%.

Inspired by the effectiveness of the MARS treatment, Stange J *et al.* (Stange *et al.*, 2000) retrospectively analyzed 26 patients with acute or chronic severe hepatitis who were treated with MARS. Most of these patients had a history of cirrhosis, in which the clinical manifestations include elevated transaminases, cholestatic and coagulation disorders, the presence of ascites, infection, hepatic encephalopathy, hepatorenal syndrome, and systemic inflammatory response syndrome and other complications. When conventional treatment was ineffective, MARS treatments were applied. After a single six-hour MARS treatment, serum TBil was reduced by 16 to 53%, and bile acids decreased by 10 to 90%. Of the 26 patients, 17 patients survived and nine died. MARS treatment improved the patient's prothrombin, antithrombin III and albumin were also significantly improved, and these improvements lasted in discharge.

Previous studies have indicated that MARS treatment has a positive effect on clinical manifestations and prognoses on type I hepatorenal syndrome (HRS) patients. Mitzner SR (Mitzner *et al.*, 2000) reported the clinical result of MARS treatment on patients with hepatorenal syndrome. They randomly divided 13 patients with hepatorenal syndrome on the basis of chronic liver disease into two groups: the first group ($n = 8$) received MARS and hemofiltration (HDF) treatment; the second group ($n = 5$) only received HDF treatment. Before treatment, the Child–Pugh score of the two groups was greater than 12, the mean serum albumin was less than 25 g/L, the average TBil was about 470 μmol/L, and the average prothrombin time activity (PTA) was about 32%. Additionally, these patients had hypotension, oliguria, multiple organ failure and other clinical manifestations. The primary endpoint was patient survival after 30 days. Arterial blood pressure increased, and the average survival time extended in type I HRS patients after MARS treatment. The average survival time in the MARS group was 25.2 ± 34.6 days, and 4.6 ± 1.8 days in the HDF group ($P < 0.01$). Seven-day survival in the MARS group was 67%, while that in the control group was 0%. However, the increase of urine was not statistically significant between both groups; four of the eight patients in MARS group had increases in urine, while HDF group did not show this performance.

At the European Association for the Study of the Liver annual congress held in Vienna in 2010, information was published in summary form about a multi-center controlled study on MARS (named MARS RELIEF). The research recruited 189 acute-on-chronic liver failure patients, and the results showed that compared with the medical group, MARS treatment significantly improved bilirubin, creatinine, and hepatic encephalopathy staging, but there was no significant difference in mortality at 28 days between the two groups (medical group mortality 40%, MARS group mortality of 41.2%, $P = 0.88$) (Mitzner, 2011).

In 2012, Vaid *et al.* (2012) published a meta-analysis that compiled data on MARS studies which showed that MARS had a definite effect in reducing the level of total bilirubin, hepatic encephalopathy degree, but showed no improvement in mortality (OR 0.91; 95% CI 0.64, 1.31; $P = 0.62$). The author admitted that they had little statistical power due to the small number of cases, infeasibility of the double-blind studies, and differential liver failure criteria, and recommended expanding the number of patients and clinical centers that used MARS treatment.

8. Prometheus System

The fractionated plasma separation and adsorption system (FPSA) was established in 1999 in Austria (Falkenhagen *et al.*, 1999). The system connects hemoperfusion and dialysis in series. First, plasma was filtered by a polysulfone membrane filter with a protein filter coefficient of 0.89, adsorbed by a 1–2 adsorber, and then the whole blood was dialyzed by a high flux dialyzer in order to simultaneous bind to large molecular protein toxins and middle small molecule water soluble toxins. Fresenius Company added the PFSA extension component on the base of a 4008 H Hemodialysis Machine, and the product was

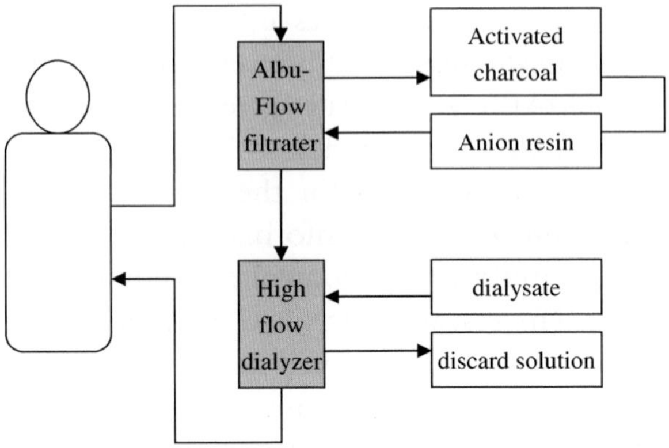

Fig. 6.6. Circulation pattern diagram of Prometheus.

named Prometheus system (Santoro *et al.*, 2006). The treatment component consists of protein separator with a screening coefficient of 0.6 (Albuflow), neutral adsorption resins (Prometh 01), anion resin adsorber (Prometh 02) and high-throughput dialyzer (F60S). The Prometheus system comprises of three loops: blood circulation, and albumin filtrate cycle and dialysis cycle (Fig. 6.6). Filter coefficient of AlbuFlow polysulfone membrane filter: β2-MG 1.0, albumin 0.6, IgG < 0.3, fibrinogen < 0.1, IgM < 0.01, relative molecular weight cut off 250,000. Albumin molecular weight was 68,000 and can pass the filter, but fibrin molecular weight was 340,000, which cannot pass the filter. Albumin solution and carrying toxins in blood were first separated by the AlbuFlowfilter, then perfused through the adsorber. After toxin adsorption of the two adsorbers, purified albumin solution was drained into patient's blood through high flux dialysis. No exogenous albumin was needed to supplement during the entire treatment process.

Prometheus is based on plasma direct adsorption and high throughput dialysis, while MARS is based on indirect albumin adsorption and indirect low-flow dialysis. Due to differences in design, Prometheus can remove albumin-bound toxins and water-soluble substances more effectively than MARS. Randomized crossover controlled trials on Prometheus and MARS have confirmed that clearance and reduction rate of albumin-bound and water-soluble products was higher with Prometheus than with MARS treatment.

In 2001 Kramer *et al.* (2001) published a controlled clinical study of FPSA in which HE was an end point. After six hours of FPSA treatment, 10 patients in the treatment group showed significantly improved HE levels and somatosensory evoked potentials of N70 peak latency. These improvements were not found in five control patients. The control data suggests that FPSA can rapidly improve neurophysiological dysfunction during severe HE. In 2003, Rifai *et al.* (2003) reported the first clinical treatment using Prometheus. There were 11 ACLF patients who accepted the treatment and nine cases had HRS. Nine patients accepted two Prometheus treatments; the average treatment time was 5 ± 1 hours, and each treatment interval was at least 4 hours. The 30-day survival rate was 36%, and three patients survived to be discharged from the hospital. The main adverse reactions were pipeline coagulation and transient hypotension.

Several published series of reports about FPSA system mainly included ACLF patients, some of which were complicated with HRS, and the vast majority of research data remained in the evaluation of biochemical, cytokines, hemodynamic and other indicators. No controlled studies regarding ultimate survival have been published yet. Grodzicki *et al.* (2009) reported the treatments of acute liver failure in 52 cases using the Prometheus 4008H FPSA system. The mean number of necessary treatments was 2.41 times, and the average treatment time was 6.3 hours. Chemical indicators such as plasma ammonia, bilirubin, ALT, AST, and creatinine, and BUN significantly improved after the treatment. Thirty-three cases received liver transplantation, and 28 patients ultimately survived, including 22 patients who underwent liver transplantation, and six cases of self-healing. There were 24 cases of death: 13 patients died while waiting for transplants, and 11 patients died after liver transplantation. The authors believe that the Prometheus system is good as a bridging method before liver transplantation. In ongoing HELIOS trials at European multi-center randomized trial sponsored by the Fresenius Company, approximately 200 cases of acute-on-chronic liver failure(ACLF) patients receive Prometheus or standard medical treatment. Kribben *et al.* (2012) published a preliminary report in which acute-on-chronic liver failure patients were randomly divided into a FPSA group ($n = 77$) and a standard medical therapy group (SMT group, n = 68). The Prometheus group accepted 8–11 treatments within three weeks, and the minimum treatment times were 4 hours. The FPSA group has a 28-day survival rate of 66%, while the SMT group's 28-day survival rate was 63% ($p = 0.70$). The 90-day survival rate of the FPSA group was 47%, compared to 38% in the SMT group ($P = 0.35$). No statistically significant different survival rates between the two groups at 28 days and 90 days were observed. However, pre-defined subgroup analysis found that in patients with a MELD score greater than 30 (a total of 48 cases, 24 cases in each group), 28-day survival rates were 42 and 57% in the SMT group and FPSA group, respectively, and 90-day survival rates were 9 and 48% in SMT groups and FPSA group ($P = 0.02$), respectively, indicating that the Prometheus system can improve survival rates.

9. Simple Albumin Dialysis (SAD)

The SAD system is designed based around the strong binding capacity of albumin on bilirubin and other toxins, thereby allowing blood bilirubin dissociation from protein, followed by transfer to the other side of the dialyzer membrane. The system uses a continuous renal replacement therapy (CRRT) device, albumin bicarbonate dialysate and albumin-impermeable ordinary synthetic membranes high flux dialyzers, online closed loop continuous hemodialysis to clear liver-derived protein-bound toxins. The method is simple and low cost, and some results show that the treatment effects of SAD are comparable to the MARS system.

10. Continued Albumin Purification System (CAPS) and Protein Adsorbent Recirculating System (PARS)

According to the basis of albumin dialysis, there have been many treatment systems similar to MARS, including CAPS and PARS.

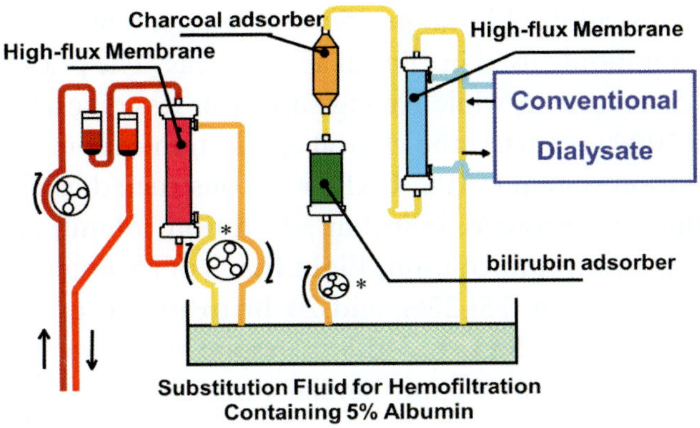

Fig. 6.7. Treatment pattern diagram of CAPS.

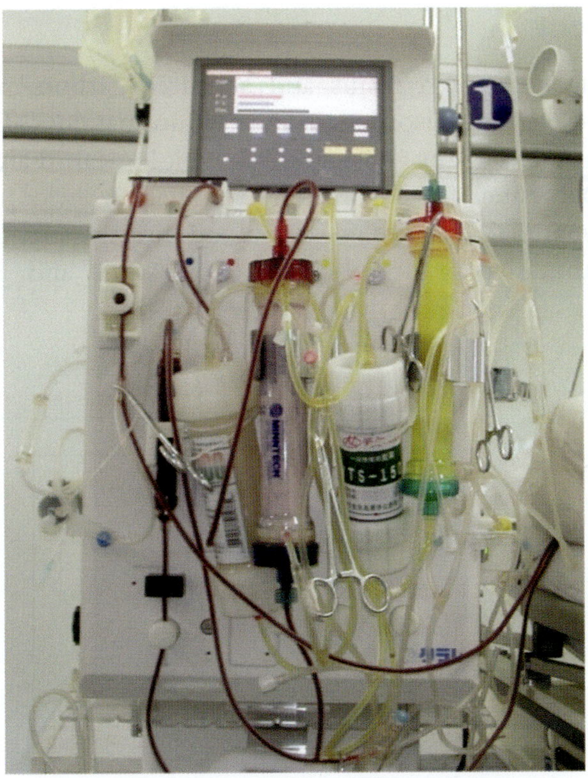

Fig. 6.8. Clinical application of CAPS.

These systems adopt blood purification equipment, dialyzers and adsorbers commonly used in clinical applications. The prices are lower than MARS treatment, while having similar or less efficacy to MARS. Abe *et al.* (2004) reported the clinical application of continuous albumin purification system CAPS. CAPS was constructed using triacetate film, 5% albumin dialysate, bilirubin adsorption column and an activated carbon adsorption column. Eight hours of treatment has shown to reduce bilirubin by 15–25%, and 24-hr treatment can adjust liver and kidney function (Figs. 6.7–6.8). Despite the lower efficacy, the CAPS and PARS systems apply similar principles as MARS and are available for clinical use.

11. Biologic-DT System and Biologic-DTPF System

Direct contact between resins and activated carbon may lead to severe blood clotting disorders in liver failure patients. The Indiana HemoCleanse company designed the biological dialysis system (Biologic-DT system), an adsorption hemodialysis device in which the adsorption material does not contact blood directly (Fig. 6.9). In this device, quantitative branched-chain amino acids and glucose are

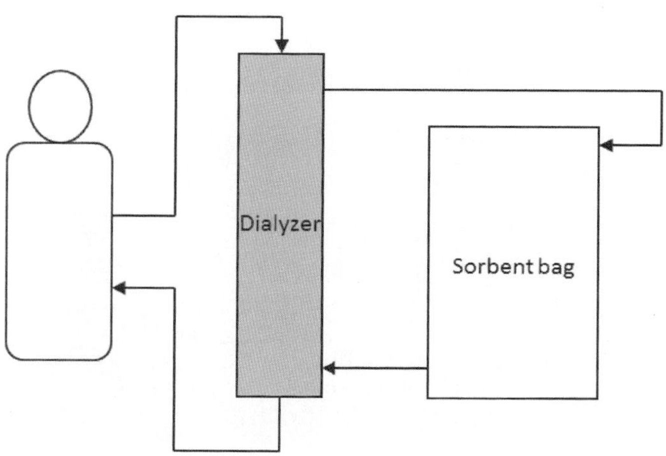

Fig. 6.9. Diagram of Biologic-DT system.

pre-loaded to the powdered activated carbon and cation exchange resin and then added to the dialysate plates of the dialysis machine, preventing blood–adsorbent contact and negating the need for an anticoagulant. The surface area of powdered activated carbon particle used in the system is up to 300,000 square meters, with a long time unsaturated absorption effect. Because the cut-off value of the molecular filtration membrane is low (100–5000D), it is more effective to remove small molecular toxins such as ammonia, phenol and other small molecules during liver failure than macromolecules like endotoxins, cytokines and bilirubin.

Ash *et al.* (2000) used this system to treat 15 patients with acute liver failure (11 cases needed respiratory support before treatment, 11 cases with renal failure). The average treatment time was four h/d, and the level of consciousness improved after treatment, among which four cases recovered, four patients recovered and underwent liver transplantation, and the remaining seven patients died. No white blood cells and platelet loss occurred after treatment. Hughes *et al.* (1994) conducted a randomized controlled study of Biologic-DT system on 10 cases of liver failure patients (five cases in the control group and five in the treatment group). After treatment the platelet count dropped from $163 \pm 34 \times 10^9$ to $101 \pm 1.3 \times 10^9$/L, plasma fibrinogen reduced from 0.53 ± 0.09 to 0.31 ± 0.08g/L, and blood activated clotting time extended from 190 ± 17 to 223 ± 22 s. However, there was no significant decrease in blood ammonia. Only one case survived in the treatment group, while three cases survived in the control group. Biologic-DT showed limited improvement in the treatment of severe acute alcoholic hepatitis. Ten patients with encephalopathy (grade II–IV) were entered into the study, five received treatment with the Biologic-DT for six hours daily for three days, and five received conventional treatment as controls. The system was biocompatible with no adverse effects on blood pressure or platelet counts, factor V, fibrinogen or anti-thrombin III. No bleeding episodes were observed, even with the use of small doses of heparin. After three days, blood ammonia was lower in the Biologic-DT treated patients than in the controls, although blood lactate was higher. There were slight increases in plasma TNF and IL-8 during treatment, possibly as a result of activation of white cells in the extracorporeal circuit (Ellis *et al.*, 1999). Although Ash *et al.* believed the Biologic-DT system had a stronger dialysis efficiency for ammonia, γ-aminobutyric acid and other water-soluble substances. Hughes *et al.* reported that the removal of metabolites by this treatment was limited, with no significant effect on blood ammonia levels, and further developments of the system will be needed for liver failure patients (Hughes *et al.*, 1994). Based on these inconsistent results, it is difficult to make any evaluations of the Biologic-DT system at present. As in other blood perfusion systems, the ultimate therapeutic effects (survival rate) are poor, but in this adsorbent-based dialysis system, blood does not contact the adsorbent directly, which can reduce the side effects compared to the conventional method. The system recently replaced its membrane types and increased endotoxin and albumin-bound macromolecular toxin clearance.

To improve the limitations of the Biologic-DT system in clearing macromolecules, a Biologic-DTPF system was developed and used in the treatment of acute liver failure and sepsis. Preliminary results of a clinical trial with the Biologic-DTPF System in the treatment of hepatic failure improved the ability to remove bilirubin and other strongly protein-bound toxins and showed some improvements for toxin removal. However, no further results have been reported since 2000.

The Biologic-DTPF System combines hemodiabsorption techniques (the Biologic-DT System with dialysis against powdered sorbent) with push-pull sorbent-based pheresis (the Biologic-PF System with powdered sorbent surrounding plasma filters). The DTPF system, also known as the plasma filtration detoxification system, includes hollow fiber plasma filtration units (e.g. PF system). The hole through the membrane filtration device is 0.5μm, and by changing the transmembrane pressure, plasma can directly come in contact with the powdered carbon or silicon adsorbent to remove the protein-binding and large-molecular-weight toxins. Further, the filtration membrane can effectively prevent membrane polarization of red blood cells and carbon powder adhesion. At blood flow rates of 200 mL/min, the system clears creatinine and aromatic amino acids at 120–160 mL/min, unconjugated bilirubin at 20–40 mL/min, and cytokines at 15–25 mL/min. Ash S.R. conducted a study of the DTPF System in treatment of patients with hepatic failure with Grade 3 or 4 encephalopathy and respiratory and kidney insufficiencies. Treatment appeared to be safe, and there were no significant hematologic changes. Physiologic changes include improved blood pressure and encephalopathy, as well as stable urine output. Chemical changes include decrease in plasma levels of bilirubin, aromatic amino acids, ammonium, creatinine, and interleukin-3 (IL-1beta) (Ash *et al.*, 2000).

12. The Application of Hemoperfusion in Bioartificial Liver

The liver is a complex organ that carries out a variety of functions and results in a multitude of clinical symptoms during liver failure. Non-biological artificial livers were developed in the last several decades to replace some failing liver functions (Carpentier *et al.*, 2009). However, non-biological artificial livers can replace only the detoxification liver function, though all systems can decrease toxin concentration and improve clinical status of the patient. However, these systems cannot replace the regulation or synthesis capacity. This lack of technology will lead to the development of bioartificial liver support devices that integrate biological components that contain liver cells with or without other non-biological components. The bioartificial liver will not only have detoxification capabilities, but will also replace the function of regulation and synthesis (Pless, 2010). Some attempts have successfully replaced liver functioning in liver failure patients. Those systems, in contrast to purely artificial systems, incorporate metabolically active cells to supply synthetic functions, as well as detoxification. The selection of ideal cell resources and the design of optimal cellular environments in bioreactors are the main issues in this field of research.

A variety of bioartificial liver systems have been developed in the past few decades (Gerlach *et al.*, 2008; Pless, 2010). The safety and efficacy of several systems have been supported in clinical studies. Bioartificial livers work to select cell resources and supply optimal environments for hepatocytes that were used in the bioartificial liver. There are two modes of perfusion used in bioartificial livers. One mode is the using of an adsorption column in the extracorporeal circuit in order to provide an optimal *in vivo* cell culture environment in the bioartificial liver system. Another is the perfusion of microencapsulated cells in the bioreactor of the bioartificial liver system.

12.1. *Hemoperfusion Enrolled as the Non-Biological Component*

Plasma from patients with liver failure has been shown to be cytotoxic to hepatocytes, so the charcoal adsorbent column in clinical extracorporeal circuit was used to reduce the level of toxic substances.

Some kinds of bioartificial liver system, such as the extracorporeal liver assist device (ELAD), HepatAssist, have done clinical trials which contain the charcoal absorbent column. ELAD was a bioartificial liver system that used the human hepatoblastoma cell line C3A as the cell resource. C3A cells were inoculated into the extracapillary space of one or two hollow fiber cartridges and were separated from the patient's blood by a filter to make the transfer of cells impossible. In order to let C3A in the bioreactor achieve optimal environmental conditions for expression of specific hepatocyte functions, perfusion of charcoal adsorber were integrated into the circuit. The ELAD system was constructed as follows (Ellis *et al.*, 1996): Venous access was gained from the double-lumen dialysis catheter into either femoral vein or the internal jugular vein. Two to four ELAD cartridges were used in the clinical circuit to provide the theoretical functioning hepatocyte mass of the liver. Blood was perfused through these cartridges in series at the rate of 150–200 mL/min. For anticoagulation, the initial heparin dose of 5,000 IU was given, followed by the continuous infusion to achieve an ACT (activated clotting time) of 200–250 s. Ultrafiltrate passing over the cells was formed at the rate of 400–900 mL/hr when negative pressure was applied across the membrane. Then, the fluid was passed through a 1-micron filter and returned to the patient. One perspective-controlled study was performed to evaluate ELAD in acute liver failure patients.

Twenty-four patients were divided into two groups. Group I was composed of 17 patients with a potentially recoverable lesion, while seven patients listed for transplantation were enrolled as Group II. These patients were randomly allocated to the ELAD group or control. The median time for ELAD hemoperfusion was 72 hr (range 3–168 hr). The result showed good biocompatibility, with no significant platelet consumption, and stable haemodynamics. The result showed that the deterioration of encephalopathy grade was more frequent in the control group, in seven of the 12 (58%) patients, than in the ELAD group, which occurred in three of the 12 patients (25%). In Group I, where survival for the ELAD cases was seven out nine (78%), there was a higher survival rate than in the controls, in which six of the eight (75%) survived. As for the Group II cases, one of the three (33%) survived in ELAD-treated patients, while only one of the four (25%) survived in the control patients. Both of the survivors underwent liver transplantation. For additive function test, ELAD improved galactose elimination capacity after six hours of therapy.

HepatAssist, another bioartificial liver system, was constructed as follows: 5–7 $\times$ 10^9 porcine hepatocytes were applied to systems that were previously cryopreserved. The hollow fiber bioreactor is similar to ELAD, resembling a dialysis cartridge and hollow fibers. The porcine hepatocytes were inoculated into the extracapillary space, and the patients' plasma was ultrafiltrated through the hollow fibers (Demetriou *et al.*, 2004; Detry *et al.*, 1999). Similar to the ELAD system, a charcoal adsorber serves to provide additional detoxification functioning and decrease the toxin load to the hepatocytes. Then, one membrane oxygenator was used to supply oxygen.

HepatAssist was the first system to complete its phase III clinical trial (Demetriou *et al.*, 2004). A total of 171 patients (86 in control and 85 in BAL) were enrolled in this clinical trial. Patients with fulminant/subfulminant hepatic failure and primary nonfunction after liver transplantation were included. Several factors were analyzed: survival rate at 30 days, disease severity, disease etiology, liver transplantation, and time to transplant. The results showed a 71% survival rate at 30 days for the BAL group and a 62% survival rate for the control (P = 0.26). Once primary nonfunctioning patients were excluded, survival rate was 73% for BAL group, and 59% for the control group (n = 147; P = 0.12). As for the entire patient population, there was no significant difference in survival rate between the two groups when survival was analyzed accounting for confounding factors (risk ratio = 0.67; P = 0.13).

However, survival rate in the fulminant/subfulminant liver failure patients was significantly higher in the BAL group than in the control group (risk ratio = 0.56; P = 0.048). This prospective, randomized, controlled clinical trial showed HepatAssist was safe and can improve survival rates in fulminant/subfulminant liver failure patients.

12.2. *Hemoperfusion Enrolled as the Biological Part*

The idea to enhance the effectiveness of the artificial liver by integrating the biological component first arose in the 1950s. Although the first attempt was achieved using whole liver perfusion, it was extremely complex because the whole liver is not easily available. In the past 30 years, biological components have focused on the isolated hepatocytes for extracorporeal liver support. Nearly all clinically applied bioartificial liver support systems are hollow fiber based. However, there are a variety of experimental systems that employ different means of bioreactor surroundings for the cells. An option for providing the three-dimensional environment is the microencapsulation of hepatocytes. One of the most widely used encapsulation substances is alginate gel. The cells can reform aggregates within the hydrophilic gel. Bead size is crucial to survival of the center cell in the beads because the supply of nutrients and oxygen is limited by diffusion. The cell-containing spheres may be utilized in fluidized-bed bioreactors or rotary systems mimicking microgravity (Pless, 2010).

One potential fluid system has been evaluated *in vitro* (Yu *et al.*, 2009). Microencapsulated HepLL cells were grown at high yield in large-scale roller bottles while free cells cultured in roller bottle spinners served as controls. The mechanical stability and permeability of the alginate-chitosan (AC) microcapsules were investigated, along with the growth, metabolism and functions of the encapsulated

HepLL cells. The results showed the microcapsules withstood the shear stress induced by high agitation rates. The microcapsules were permeable to albumin, but prevented the release of immunoglobulins. Cultures in roller bottles of immortalized human hepatocytes microencapsulated in the AC improved cell growth, ammonia elimination, albumin synthesis and lidocaine clearance. The results showed encapsulated HepLL cells might be cultured on a large scale in roller bottles. This makes them possible candidates for use in cell-based liver assistance therapies.

Then construction and evaluation of a choanoid-fluidized bed bioreactor (CFBB) based on microencapsulated immortalized human hepatocytes was done (Yu *et al.*, 2014). Encapsulated hepatocytes were placed in a constructed CFBB and circulated through Dulbecco's Modified Eagle's Medium (DMEM) for 12 hr, through exchanged plasma for 6 hr, and then compared with encapsulated cells cultivated under static conditions in a spinner flask. Levels of ALT and albumin were used to evaluate the CFBB during media circulation, whereas levels of ALT, TBil, and albumin were used to evaluate it during plasma circulation. Mass transfer and injury of hepatocyte were evaluated by comparing the results from the two experimental conditions. In addition, the viability and microstructure of encapsulated cells were tested in the different environments. The viability of encapsulated cells accounting for 84.6 ± 3.7% in CFBB plasma perfusion was higher than the 74.8 ± 3.1% in the static culture group ($P < 0.05$) after six hr. ALT release from cells was 29 ± 3.5 U/L and 40.6 ± 3.2 U/L at 12 hr ($P < 0.01$) in the CFBB medium circulation and static medium culture groups, respectively. Albumin secretion from cells was 234.2 ± 27.8 μg/1 × 107 cells and 167.8 ± 29.3 μg/L × 107 cells at six hr ($P < 0.01$), 274.4 ± 34.6 μg/L × 107 cells and 208.4 ± 49.3 μg/1 ×107 cells ($P < 0.05$) at 12 hr, in the two medium circulation/culture groups, respectively. Furthermore, ALT and TBil levels were 172.3 ± 24.1 and 236.3 ± 21.5 U/L ($P < 0.05$), 240.1 ± 23.9 and 241.9 ± 31.4 μmol/L ($P > 0.05$) at 6 hr in the CFBB plasma perfusion and static plasma culture groups, respectively. There was no significant difference in albumin concentration between two experimental plasma groups at any time point. The microstructure of the encapsulated hepatocytes kept healthier in the CFBB group compared with the static culture group after six hours of plasma perfusion. This indicates that the CFBB can function as a bioreactor for use in a bioartificial liver. The efficacy of this novel bioreactor is promising for the study of liver failure.

References

Abe T, Shono M, Kodama T, *et al.* (2004) Extracorporeal albumin dialysis. *Ther Apher Dial* **8:** 217–222.

Amoureux MC, Rajapakse N, Hegyi E, *et al.* (2004) Endotoxin removal from whole blood by a novel adsorption resin: Efficiency and hemocompatibility. *Int J Artif Organs* **27:** 480–487.

Ash SR, Steczko J, Knab WR, *et al.* (2000) Push-pull sorbent-based pheresis and hemodiabsorption in the treatment of hepatic failure: Preliminary results of a clinical trial with the BioLogic-DTPF System. *Ther Apher* **4:** 218–228.

Bellomo R, Tetta C, Ronco C. (2003) Coupled plasma filtration adsorption. *Intensive Care Med* **29:** 1222–1228.

Bihari D, Hughes RD, Gimson AE, *et al.* (1983) Effects of serial resin hemoperfusion in fulminant hepatic failure. *Int J Artif Organs* **6:** 299–302.

Carpentier B, Gautier A, Legallais C. (2009) Artificial and bioartificial liver devices: present and future. *Gut* **58:** 1690–1702.

Chang TMS. (1972) Haemoperfusions over microencapsulated adsorbent in a patient with hepatic coma. *Lancet* **ii:** 1371–1372.

Chang TMS. (1974) Platelet-surface interaction: effect of albumin coating or heparin complexing on thrombogenic surfaces. *Can J Physiol Pharmacol* **52(2):** 275–285.

Chang TMS, Lister C. (1980) Analysis of possible toxins in hepatic coma including the removal of mercaptan by albumin-collodion charcoal. *Int J Artif Organs* **3:** 108–112.

Chang TMS, Lister C. (1981) Middle molecules in hepatic coma and uremia. *Artif Organs* **4(Suppl):** 169–172.

Chang TMS, Migchelsen M. (1973) Characterization of possible "toxic" metabolites in uremia and hepatic coma based on the clearance spectrum for larger molecules by the ACAC microcapsule artificial kidney. *Trans Am Soc Artif Intern Organs* **19**: 314–319.

Chen J, Huang J, Chen Y, *et al.* (2012) A clinical study on the treatment of severe hepatitis by a combined artificial liver. *Hepatogastroenterology* **59:** 2273–2275.

Demetriou AA, Brown RS Jr, Busuttil RW, *et al.* (2004) Prospective, randomized, multicenter, controlled trial of a bioartificial liver in treating acute liver failure. *Ann Surg* **239:** 660–667; discussion 667–670.

Detry O, Arkadopoulos N, Ting P, *et al.* (1999) Clinical use of a bioartificial liver in the treatment of acetaminophen-induced fulminant hepatic failure. *Am Surg* **65:** 934–938.

Ellis AJ, Hughes RD, Wendon JA, *et al.* (1996) Pilot-controlled trial of the extracorporeal liver assist device in acute liver failure. *Hepatology* **24:** 1446–1451.

Ellis AJ, Hughes RD, Nicholl D, *et al.* (1999) Temporary extracorporeal liver support for severe acute alcoholic hepatitis using the BioLogic-DT. *Int J Artif Organs* **22:** 27–34.

Falkenhagen D, Strobl W, Vogt G, *et al.* (1999) Fractionated plasma separation and adsorption system: A novel system for blood purification to remove albumin bound substances. *Artif Organs* **23:** 81–86.

Gao L, Ji F, Shen Y, *et al.* (2008) The effect of BL-300 perfusion column on plasma purification for patients with chronic severe hepatitis. *Chin J Blood Purif* **11**: 615–617 (In Chinese).

Gazzard BG, Portmann BA, Weston MJ, *et al.* (1974) Charcoal haemoperfusion in the treatment of fulminant hepatic failure. *Lancet* **1:** 1301–1307.

Geiger H, Klepper J, Lux P, *et al.* (1992) Biochemical assessment and clinical evaluation of a bilirubin adsorbent column (BR-350) in critically ill patients with intractable jaundice. *Int J Artif Organs* **15:** 35–39.

Gerlach JC, Zeilinger K, Patzer JF. (2008). Bioartificial liver systems: Why, what, whither? *Regen Med* **3:** 575–595.

Grodzicki M, Kotulski M, Leonowicz D, *et al.* (2009) Results of treatment of acute liver failure patients with use of the prometheus FPSA system. *Transplant Proc* **41:** 3079–3081.

Hanasawa K. (2002) Extracorporeal treatment for septic patients: New adsorption technologies and their clinical application. *Ther Apher* **6:** 290–295.

Hanasawa K, Tani T, Kodama M. (1989) New approach to endotoxic and septic shock by means of polymyxin B immobilized fiber. *Surg Gynecol Obstet* **168:** 323–331.

Hughes R, Ton HY, Langley P, *et al.* (1979) Albumin-coated Amberlite XAD-7 resin for hemoperfusion in acute liver failure. Part II: in vivo evaluation. *Artif Organs* **3:** 23–26.

Hughes RD, Pucknell A, Routley D, *et al.* (1994) Evaluation of the BioLogic-DT sorbent-suspension dialyser in patients with fulminant hepatic failure. *Int J Artif Organs* **17:** 657–662.

Hughes RD. (2002) Review of methods to remove protein-bound substances in liver failure. *Int J Artif Organs* **25:** 911–917.

Kono K, Toda S, Hora K, *et al.* (2009) Direct hemoperfusion with a beta2-microglobulin-selective adsorbent column eliminates inflammatory cytokines and improves pulmonary oxygenation. *Ther Apher Dial* **13:** 27–33.

Kramer L, Gendo A, Madl C, *et al.* (2001) A controlled study of sorbent suspension dialysis in chronic liver disease and hepatic encephalopathy. *Int J Artif Organs* **24:** 434–442.

Kribben A, Gerken G, Haag S, *et al.* (2012) Effects of fractionated plasma separation and adsorption on survival in patients with acute-on-chronic liver failure. *Gastroenterology* **142:** 782–789.

Li LJ, Yang Q, Huang JR, *et al.* (2004) Effect of artificial liver support system on patients with severe viral hepatitis: A study of four hundred cases. *World J Gastroenterol* **10:** 2984–2988.

Li LJ, Zhang YM, Liu XL, *et al.* (2006) Artificial liver support system in China: A review over the last 30 years. *Ther Apher Dial* **10:** 160–67.

Li LJ, Liu XL, Xu XW, *et al.* (2005) Comparison of plasma exchange with different membrane pore sizes in the treatment of severe viral hepatitis. *Ther Apher Dial* **9:** 396–401.

Livigni S, Bertolini G, Rossi C, *et al.* (2014) Efficacy of coupled plasma filtration adsorption (CPFA) in patients with septic shock: A multicenter randomized controlled clinical trial. *BMJ Open* **4:** e003536.

Maggi U, Nita G, Gatti S, *et al.* (2013) Hyperbilirubinemia after liver transplantation: The role of coupled plasma filtration adsorption. *Transplant Proc* **45:** 2715–2717.

Mao HJ, Yu S, Yu XB, *et al.* (2009) Effects of coupled plasma filtration adsorption on immune function of patients with multiple organ dysfunction syndrome. *Int J Artif Organs* **32:** 31–38.

Mitzner SR, Stange J, Klammt S, *et al.* (2000) Improvement of hepatorenal syndrome with extracorporeal albumin dialysis MARS: Results of a prospective, randomized, controlled clinical trial. *Liver Transpl* **6:** 277–286.

Mitzner SR. (2011) Extracorporeal liver support-albumin dialysis with the Molecular Adsorbent Recirculating System (MARS). *Ann Hepatol* **10(Suppl 1):** S21–S28.

Nakaji S, Hayashi N. (2003) Bilirubin adsorption column Medisorba BL-300. *Ther Apher Dial* **7:** 98–103.

Nalesso F. (2005) Plasma filtration adsorption dialysis (PFAD): A new technology for blood purification. *Int J Artif Organs* **28:** 731–738.

Nikolaev VG, Sarnatskaya VV, Sigal VL, *et al.* (1991). High-porosity activated carbons for bilirubin removal. *Int J Artif Organs* **14:** 179–185.

Nolan JP, McDevitt JJ, Goldmann GS, *et al.* (1975) Endotoxin binding by charged and uncharged resins. *Proc Soc Exp Biol Med* **149:** 766–770.

O'Grady JG, Gimson AE, O'BrienCJ, *et al.* (1988) Controlled trials of charcoal hemoperfusion and prognostic factors in fulminant hepatic failure. *Gastroenterology* **94:** 1186–1192.

Oishi K, Mimura-Kimura Y, Miyasho T, *et al.* (2013) Association between cytokine removal by polymyxin B hemoperfusion and improved pulmonary oxygenation in patients with acute exacerbation of idiopathic pulmonary fibrosis. *Cytokine* **61:** 84–89.

Organization Committee of 13th Asia-Pacific Congress of Clinical Microbiology and Infection. (2013) 13th Asia-Pacific Congress of Clinical Microbiology and Infection Consensus Guidelines for diagnosis and treatment of liver failure. *Hepatobiliary Pancreat Dis Int* **12:** 346–354.

Pless G. (2010). Bioartificial liver support systems. *Methods Mol Biol* **640:** 511–523.

Rifai K, Ernst T, Kretschmer U, *et al.* (2003) Prometheus — a new extracorporeal system for the treatment of liver failure. *J Hepatol* **39:** 984–990.

Rifai K. (2011) Fractionated plasma separation and adsorption: Current practice and future options. *Liver Int* **31(Suppl 3):** 13–15.

Sadahiro T, Hirasawa H, Oda S. (2001) Useful of plasma exchange plus continuous hemodiafiltration to reduce adverse effects associated with plasma exchange in patients with acute liver failure. *Crit Care Med* **29:** 1386–1392.

Song S, Chen R, Chen M. (2013) Application of plasma absorption on patients with hyperbilirubinemia. *Biomedical Engineering and Clinical Medicine* **1:** 27–29 (In Chinese).

Stange J, Mitzner SR, Klammt S, *et al.* (2000) Liver support by extracorporeal blood purification: A clinical observation. *Liver Transpl* **6:** 603–613.

Steczko J, Ash SR, Blake DE, *et al.* (1999) Cytokines and endotoxin removal by sorbents and its application in push-pull sorbent-based pheresis: The BioLogic-DTPF System. *Artif Organs* **23**: 310–318.

Wauters J, Wilmer A. (2011) Albumin dialysis: Current practice and future options. *Liver Int* **31(Suppl 3):** 9–12.

Xiang H, Han T, Du Z, *et al.* (2011) Clinical scope analysis of plasma perfusion using HB-H-6 resin in patients with chronic liver disease and severe jaundice. *Biomedical Engineering and Clinical Medicine* **5:** 456–460 (In Chinese).

Xu X, Liu X, Ling Q, *et al.* (2013) Artificial liver support system combined with liver transplantation in the treatment of patients with acute-on-chronic liver failure. *PLoS One* **8:** e58738.

Yu CB, Pan XP, Yu L, *et al.* (2014). Evaluation of a novel choanoid fluidized bed bioreactor for future bioartificial livers. *World J Gastroenterol* **20:** 6869–6877.

Yu CB, Lv GL, Pan XP, *et al.* (2009). In vitro large-scale cultivation and evaluation of microencapsulated immortalized human hepatocytes (HepLL) in roller bottles. *Int J Artif Organs* **32:** 272–281.

Zagli G, Bonizzoli M, Spina R, *et al.* (2010) Effects of hemoperfusion with an immobilized polymyxin-B fiber column on cytokine plasma levels in patients with abdominal sepsis. *Minerva Anestesiol* **76:** 405–412.

Chapter 7

Sorption Therapy with the Use of Activated Carbons: Effects on Regeneration of Organs and Tissues

V.G. Nikolaev*

Summary

This article is devoted to the indirect effects of medicinal use of carbon sorbents, consisting in pathomorphism, i.e. an alteration of natural course of some diseases under the influence of sorption therapy and leading mainly to acceleration of regeneration in injured organs and tissues. Based upon a number of examples from experimental and clinical therapy with the use of activated carbons for blood purification (hamosorption), their peroral administration (enterosorption) and topical placement on wounds and burns (application therapy), the author demonstrates the fact that the capability to enhance and accelerate the regeneration processes in injured organs and tissues is a feature of all three types of sorption therapy. The potential uses of sorption therapy include possible uses for treatment of acute radiation sickness, chronic glomerulonephritis, biliary and portal

---

* Dedicated to my Teachers, Prof. R.E. Kavetsky, Prof. T.M.S. Chang, Prof. Yu.M. Lopukhin. R.E. Kavetsky Institute of Experimental Pathology, Oncology and Radiobiology of National Academy of Science of Ukraine, Kyiv, Ukraine. Volodymyr G. Nikolaev, 45 Vasilkovskaya Str., 03022 Kyiv Ukraine, phone/fax + 38 044 257 90 55, e-mail: aosiepor@yandex.ua.

cirrhosis, severe forms of leptospirosis, skin psoriasis, and psoriatic arthritis, idiopathic dilated cardiomyopathy, chronic kidney failure, as well as myeolo-, nephro-, cardio- and gonadotoxic side effects of intense cancer chemotherapy. Other areas include local application for speeding up skin regeneration in partial-thickness burns, as well as combined influence of enterosorption and local sorption improving the healing of diabetic ulcer.

1. Introduction

Since the pioneer works of Yatzidis (1963, 1976), Dunea and Kolff (1965), Hagstam *et al.* (1966) who have used activated charcoal produced from natural raw material for direct hemoperfusion (DHP, hamosorption), it became evident that despite the potential of hamosorption, the most critical deterrents to the wide clinical use of this method are high thrombogenecity of the sorbents, and generation of large amounts of carbon microparticles along with frequent development of dangerous pyrogenic reactions and the significant loss of blood cells. This problem has been solved via coating of external surface of carbon hemosorbents with ultrathin polymeric films which improve their hemocompatibility and decrease the risk of emersion of microparticles into patient's blood (Chang, 1969). Also, special attention has been focused on chemical refining of activated carbons for the reduction of pyrogenic reactions (Chang, 1969; Andrade *et al.*, 1971). At the same time polymeric coatings have significantly aggravated adsorptive characteristics of the sorbent and cut short the part of spectrum of withdrawn toxins with molecular weight higher than 1,000 Da (Denti *et al.*, 1975). Here the single exception was the technology proposed by Prof. Chang (1969, 1976b, 1978b) using ultrathin coatings in the form of ACAC (albumin-collodion coated activated charcoal) coatings. This ultrathin coating does not cover the macropore entrances on the surface of hemosorbent. Instead, the ultrathin membrane coats the surface of the internal channels of the adsorbent. This allows for ease of entry of molecules into the adsorbent channels to be more readily adsorbed. Unfortunately, most commercial devices did not use this technology, which was excellent for the preservation of sorptive properties of carbon matrix. Instead of ultrathin coating, most of them use much thicker coating. Unlike a few of the commercial devices with ultrathin membrane coating, most of them have only moderate capacity to remove toxic metabolites.

In the 1970s, a group of Moscow scientists headed by Prof. Yu.M. Lopukhin carried out aggressive surface modification of uncoated activated carbons of natural origin resulting in low thromboresistanse. They were able to use this for low-volume hemocarboperfusion in the treatment of mechanical jaundice, biliary cirrhosis, hereditary hypercholesterolemia, obstructive vascular diseases of extremities, alcoholic delirium, hypertoxic schizophrenia, bronchial asthma, psoriasis, systemic lupus erythromatosis, pemphigus, cold urticaria and some other immune-dependent pathologies. They observed that despite the small quantity of toxins removed from 0.5–0.75 CBV they could observe clinical effectiveness.

In 1973 we began systematical studies in the field of sorption therapy. In contrast to other researchers of that period who had been using "modified" technical activated charcoal for hamosorption, we placed our bets on carbon adsorbents specially produced for DHP from synthetic resins. These hemosorbents, for example, carbons of SCN mark, possessed a regular spherical shape, bulk weight of 0.35–0.4 g/cm^3, specific sorption area of 1,200–1,600 m^2/g and a smooth external surface that is hemocompatible without additional polymeric coatings (Nikolaev and Strelko, 1979). Somewhat later we developed synthetic carbon hemosorbents intended for purification of strongly protein-bound toxins from blood, as well as cytokines, bacterial toxins, and some hydrolases (Sarnatskaya *et al.*, 2002; Nikolaev *et al.*, 2011). These sorbents with bulk weight less than 0.22 g/cm^3 and specific surface area of 2,000–3,000 m^2/g were named "deliganding" and are presently known as HSGD (HemoSorbent Granulated Deliganding). The coating of carbon matrixes with native DNA or heparin has allowed us to produce combined hemoimmunosorbents for the treatment of autoimmune diseases (Snezhkova *et al.*, 1992, 2015, 2011).

Along with this, in our laboratory there has been developed the method of peroral administration of massive doses of carbon enterosorbent (enterosorption) and the method of usage of the highly activated fibrous carbons (more than 2,000 m^2/g) for treatment of wounds and burns (Nikolaev, 1990; Sakhno *et al.*, 1991; Eretskaya *et al.*, 1991), which became the second and third methods of sorption therapy.

In the present work, using some experimental and clinical illustrations, we attempt to demonstrate the correctness of at least two statements:

1. DHP with the use of synthetic carbon and combined hemosorbents is completely capable of causing significant modification of the course of many types of pathologies.
2. All three methods of sorption therapy possess an ability to potentiate and accelerate the regeneration processes in affected organs and tissues.

2. Hemosorption (DHP)

The powerful effect of hemocarboperfusion on the regenerative potential of injured organs and tissues has been clearly proven by the results of experiments in therapy of acute radiation sickness performed in 1976–77 and openly published only 15 years later (Nikolaev *et al.*, 1993). According to these results, survival of dogs upon homogenous external roentgen irradiation at a dose of 5.25 Gy increased from 3.2% to 68.4% and 62.4% after low-volume (0.5–1 CBV) hamosorption performed 2 or 24 hours after irradiation, respectively.

As it follows from Fig. 7.1, such results are based on a sharp increase of bone marrow regenerative potential. For example, at normal state the division rate of myelocytes expressed via total mitotic index value, is close to 15‰, but just in six hr after irradiation it decreases to 4‰, and then at 14th day — to a critical 2‰. In contrast, in dogs treated with hamosorption two hr after irradiation, the total mitotic index of bone marrow cells at six hr was the same as normal value; in 14 days it was significantly better. This will be described in detail in the next chapter.

Partial regeneration in patients with latent form of chronic glomerulonephritis who underwent a single hemocarboperfusion session

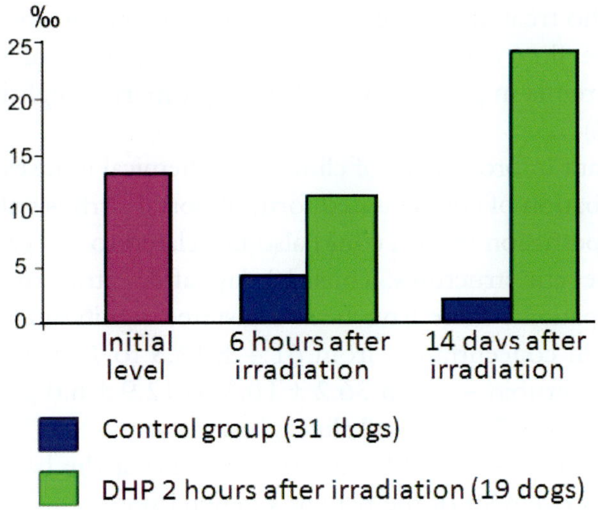

Fig. 7.1. Mitotic index of dog bone marrow after lethal irradiation in dose 5,25 Gy.

Table 7.1. Influence of Single Hemocarboperfusion on Course of Chronic Glomerulonephritis (n = 5)

Period Indexes	Before Hemocarboperfusion	After 25 Days
Creatinine, mmol/L	0.207 ± 0.02	0.134 ± 0.01*
Glomerular filtration, mL/min	78.9 ± 14.0	101.3 ± 4.07*
Proteinuria, g/day	0.513 ± 0.08	0.071 ± 0.08*

* – $P < 0.05$.

could explain significant improvement of renal functional indexes (Table 7.1).

As shown in Table 7.1, despite the small number of patients, a statistically significant decrease of serum concentration of creatinine and increase of glomerular filtration at the background of drastic fall of proteinuria grade was observed (Nikolaev *et al.*, 1985). Interestingly, based on analogous dynamics of the indexes, first of all, creatinine concentration, in 2004, i.e. 20 years later, the group of Japanese scientists who treated uremic patients with carbon enterosorbents has declared the first worldwide (in their opinion) case of therapeutic reverse of uremia in patients with chronic glomerulonephritis (Sanaka *et al.*, 2003).

Significant improvement of clinical-biochemical indexes in patients with exacerbation of compensated form of portal cirrhosis after a single hemocarboperfusion session could also be related to the partial regeneration of hepatic structures. Thus, 10 days after extracorporeal procedure in 48 patients there have been registered significant decreases of total bilirubin concentration from 61.4 ± 12.3 to 18.8 ± 6.6 μM/L, conjugated bilirubin — from 36.2 ± 10.5 to 12.9 ± 6.0 μM/L, transaminases — by 2-2.2-fold, alkaline phosphatase — by 1.3-fold, and C-reactive protein — 2-fold (Nikolaev, 1984a) at the background of essential improvement of the patient's general state.

As follows from Table 7.2, a single hamosorption session on SCN carbons resulted in objective improvement in only three out of nine patients with exacerbation of severe skin psoriasis, with no case of complete remission. However, if the carbon matrix was coated with a layer of native thymus DNA, complete remission was observed in 15 patients out of 27 (Snezhkova *et al.*, 1992). Complete remission is defined as total regeneration of skin integrity in place of psoriatic patches. Additionally, DNA-containing carbon adsorbents were found to be very effective for the treatment of psoriatic arthritis.

These observations clearly illustrate the fact that a reasonably selected biospecific ligand enhances the pro-regenerative action of the carbon hemosorbent.

Table 7.2. Clinical Effect of Hemoperfusion on DNA-Coated and Uncoated Hemosorbents in Patients with Exacerbation of Severe Forms of Psoriasis Vulgaris (Double-Blind Study)

	Clinical Effect, %			
Sorbent, Number of Patients	**Complete Remission**	**Objective Improvement**	**No Effect**	**Aggravation of Symptoms**
SCN + DNA, $n = 27$	55.5	37.0	7.5	0
SCN, $n = 9$	0	33.3	55.5	11.2

In many cases, deliganding HSGD hemosorbents seem to be clinically more effective than convenient synthetic activated carbons alone.

As follows from Table 7.3, the usage of hemosorbents resulted in a 5.6-fold decrease of child mortality due to multi-organ failure caused by "death cap amanita" poisoning. This is because of sorbent-induced enhancement of the regeneration processes in the tissues of the liver, kidneys and heart of the victims (Sarnatskaya *et al.*, 2007).

Figure 7.2 demonstrates the dynamic decrease of bilirubin concentration in a patient with a severe form of leptospirosis after hemoperfusion

Table 7.3. Usage of Hemosorbents SCN and HSGD in Treatment of Children with Death Cap Amanita Poisoning

Type of Hemosorbent	Number of Patients	Number of Lethal Outcomes	Lethality, %
Hemosorbents SCN	29	17	58.6
Hemosorbents HSGD	19	2	10.5

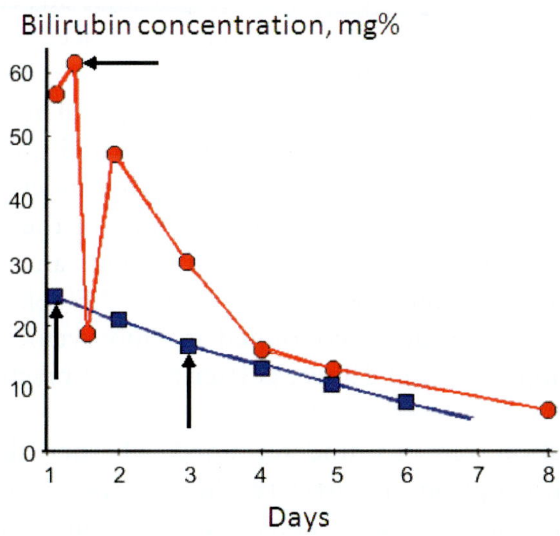

Fig. 7.2. Changes of bilirubin concentration in blood plasma of patient with biliary cirrhosis(▪) and patient with leptospirosis (•) after hemocarboperfusion onto conventional and deligandizating carbonic adsorbent, respectively.

on deliganding HSGD hemosorbents. For comparison, there is also presented the curve which shows the changes of this parameter after hamosorption on conventional SCN carbons. As one can see, in contrast to the procedure with the use of SCN, concentration of bilirubin actively adsorbed by the HSGD hemosorbent sharply falls, and again increases via its inflow from depot (rebound-effect). Still taking into account the bilirubin concentration decrease between the 3rd and 9th days after DHP, both types of uncoated synthetic carbon hemosorbents possess an expressed pro-regeneratory potential which is possibly more expressed in deliganding hemosorbents. (Table 7.3).

The result of the use of combined hemoimmunosorbents on the basis of highly active HSGS matrix coated with native DNA or its charge analog heparin for the treatment of dilated cardiomyopathy is also important (Todurov *et al.*, 2011). According to these data, two to four hemoimmunosorption sessions performed in 35 patients for two to three weeks resulted in an average 2.2-fold increase of ejection fraction and decrease of end-diastolic and end-systolic left ventricular volumes by 25 and 40% respectively in patients with heart failure of stage IV. Significant improvement of morphologic structure of myocardium and drastic increase of technetium-99 accumulation in its problem regions were detected.

The two cases described in Table 7.4 are of special interest; they are related to the results of sorption therapy of dilated cardiomyopathy developed as a late complication in patients who underwent anticancer therapy with rubomycinum — an antibiotic of the anthracycline line. After three weeks of therapy in both patients an extremely low ejection fraction has increased by 1.94- and 2.5-fold respectively. Characteristically, this index continued its improvement even after treatment termination and 10 weeks later reached practically normal values.

The question remains undetermined whether the described *effects* of hamosorption are *preserved* in the case of use of coated activated carbons. Some cues could be found in an old work of Odaka *et al.* (1976), who combined dialysis with hamosorption on coated activated carbons and detected the slower decrease of residual renal functions compared to this in the case of conventional

Table 7.4. Two Cases of Successful Usage of HSGD Hemosorbents with Immobilized Heparin for the Treatment of Dilated Cardiomyopathy (DCMP) in Patients Who in the Past Underwent Chemotherapy with Rubomycinum

Case Number	Time Interval Between Chemotherapy and Hamosorption (Years)	NYHA Functional Class	Initial Ejection Fraction (%)	Ejection Fraction After 3-Week Treatment (%)	Ejection Fraction 10 Weeks Later (%)
1	6	IV	18	35	44
2	4	IV	16	40	52

hemodialysis. As indirect proof of this one could consider the results recently described by Vogl *et al.* (2014) devoted to regional therapy of inoperable liver tumors by intra-arterial administration of high doses of strong radiomimetic — alkylating cytostatic melphalane, with its following depletion from outflowing blood with the use of coated hemosorbents of deliganding type effectively removing of protein-bound anticancer drugs (Vogl *et al.*, 2014). Despite significant melphalane quantity which remained in systemic circulation of the patient due its overflow out of perfusion circuit, hematologic reactions toward this procedure were moderate enough, what could be explained by analogy with the above-mentioned example of sorption therapy of acute radiation sickness, by myeloprotective action of hamosorption procedure performed in parallel with radiomimetic injury. Another argument for possible preservation of the effect of DHP-related-*pathomorphosis* upon the usage of carbon hemosorbents is the fact that modern deliganding activated carbons even in coated state are significantly better in terms of sorption properties than uncoated synthetic carbon hemosorbents of SCN type which served to obtain the main data on secondary effects of DHP.

3. Enterosorption

This method, based on the oral administration of large quantities of specially synthesized adsorbents, was proposed by us in the early 1980s (Nikolaev *et al.*, 1982; Nikolaev, 1984b, 1990). The fine fraction of SCN carbons with granule size of 0.15–0.4 mm were used for enterosorption initially. In recent times in the clinics there are new generations of carbon enterosorbents in the form of micronized powdered and fibrous highly active carbons with water as a binder (Nikolaev *et al.*, 2013). Apart from this, many other sorbents administered by oral route were also called "enterosorbents" (Enterosorption 1991; Nikolaev *et al.*, 2005).

Among many applications of enterosorption, its usage for renal-hepatic insufficiency caused by leptospirosis (Table 7.5) is of special interest because such treatment results in significant normalization of a number of biochemical and hematological indexes without the use of extracorporeal methods of detoxification, i.e. via quick enhancement of regenerative processes in patient's own detoxifying organs (Frolov *et al.*, 1986).

The next example of enterosorption pro-generative activity touches on the area of experimental oncology. Upon the development of Lewis lung carcinoma transplanted in hind leg of C57 Black mice, one could observe multiple degenerative changes in the morphology of livers and kidneys even though there was no metastasis to these organs (Kavetsky, 1977). As follows from Fig. 7.3, the use of enterosorbents significantly mitigates para-neoplastic manifestations in the tissues of liver and kidneys of tumor-bearing rats with evident predominance of regenerative processes versus degenerative ones (Sarnatskaya *et al.*, 2015) (Fig. 7.3).

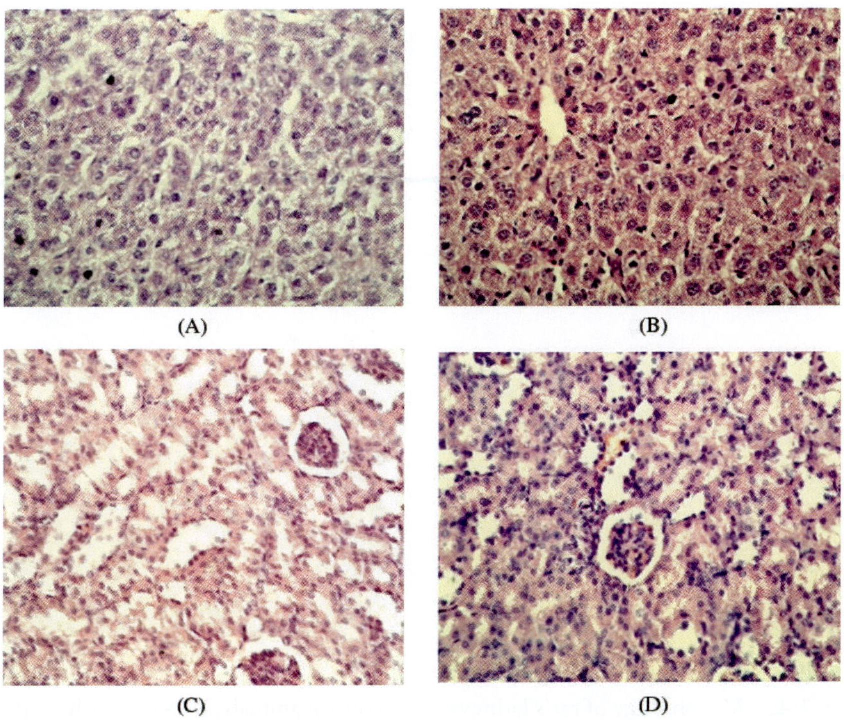

Fig. 7.3. Amelioration of morphology of liver (A,B) and kidneys (C,D) in Lewis carcinoma-bearing mice under the influence of enterosorption (B,D).

Similar types of systemic morphofunctional shifts were observed in the case of use of carbon enterosorbents as geroprotector in old rats what allowed to prolong their life span by 40–45% after initiation of the therapy (Frolkis *et al.*, 1984a, 1984b). With the use of light and electron microscopy, the decreased expression of morphological features of aging, and also increased synthesis of RNA and proteins in the cells of liver, kidneys and adrenal gland, amelioration of induction of the system of mixed oxidases and significant improvement of the indexes of lipid exchange were registered.

Figure 7.4 demonstrates the histological picture of kidney tissue of normal rat in comparison with that of animals with Guerin carcinoma transplanted in hind leg without therapy, or treated with cisplatin, or treated with cisplatin with simultaneous administration of the enterosorbent. As is evident from Fig. 7.4B, just Guerin carcinoma distant growth

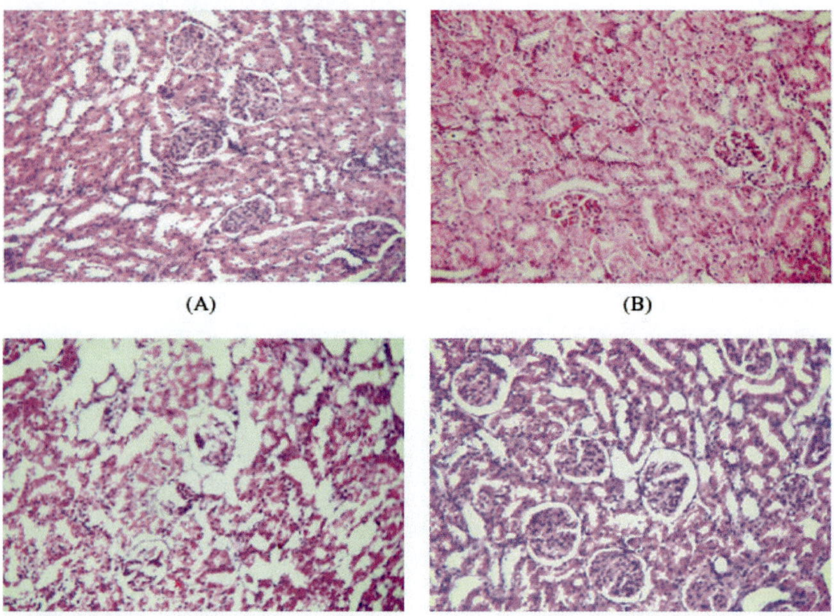

(A) (B) (C) (D)

Fig. 7.4. Morphology of rat's kidneys: A — control animals; B — rats with grafted Guerin carcinoma; C — rats with grafted Guerin carcinoma treated with cisplatin; D — rats with grafted Guerin carcinoma treated with cisplatin on the base of enterosorption.

caused significant changes of microscopic structure of kidneys demonstrates an expressed tubular cell dystrophy, formation of lumen and decreased glomerulus cell elements numbers in glomerulus which in sometimes are represented just by "naked" nuclei. In the majority of glomerulus one could observe hemorrhages of various degrees, in tubular lumen — small lesions with hemorrhages and hyaline inclusions.

Cisplatin caused notable disruption in renal tissue structure, alteration of anatomic region of glomerulus location which contained lower numbers of functional cells. There have been recorded dystrophic changes practically in all functional cell elements, tubular necrosis, and destruction of glomeruli and formation of large lumens between tubules. In the case of use of carbon enterosorbents there has been observed significantly better morphological picture of kidneys in tumor-bearing animals treated with cisplatin-based therapy. There were no erythrocytes and haemorrhages in glomerules, and its high cellularity has been registered. In the tubular lumen there were significantly lower numbers of erythrocytes, haemorrhages were rare. The large majority of cells in tubules are of normal morphological structure, the quantity of lumens between tubules and their sizes were lower (Sachno *et al.*, 2013). This example shows the pro-regenerative effect of enterosorption upon platinate-related kidney injury.

The fundamental action of enterosorption on the state of the injured kidney is illustrated in Table 7.6 by the results of the study of the effect of enterosorbent AST-120 similar to our old SCN-3M on gene expression profile in uremic kidney in rats (Aoyama *et al.*, 2003).

Table 7.5. Influence of Conservative Therapy with Use of Enterosorption on Laboratory Indexes in 15 Patients with Leptospirosis

Index	Before Treatment	7 Days After Beginning of Therapy
Total bilirubin, μmol/L	306.0 ± 24.2	140.0 ± 12.5
Creatinine, mmol/L	0.6 ± 0.02	0.2 ± 0.05
Urea, mmol/L	24.6 ± 3.5	14.5 ± 2.8
Leukocytes, $\times 10^9$/L	16.5 ± 1.5	9.5 ± 1.3
Platelets, $\times 10^9$/L	98.6 ± 2.5	135.5 ± 16.2

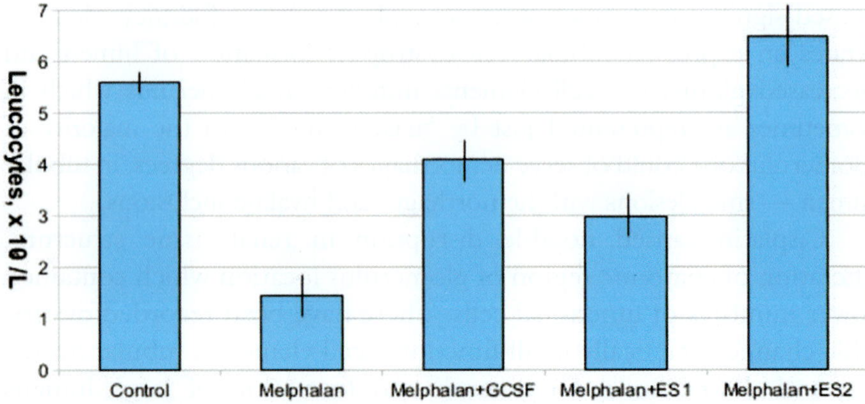

Fig. 7.5. Influence of enterosorption (C1 and C2) and granulocyte colony stimulating factor (GCSF) on the peripheral leucocytes count in the rats seven days after melphalan intravenous injection (3 mg/kg).

Table 7.6. Effects of Oral Adsorbent on Gene Expression Profile In Uremic Rat Kidney: cDNA Array Analysis

N / Conditions	From 343 Studied Renal Genes	
	Number of Upregulated Genes	Number of Downregulated Genes
Uremic kidney	139	45
Attenuation of expression level changes under the influence of AST-120	43	21

Functional categories: TGF-beta-1, intracellular signaling, transcription, translation, channels and transporters, proteases and its inhibitors.

It is necessary to note that the data on pro-regenerative action of enterosorption toward uremic kidney was obtained 20 years earlier by the group of St. Petersburg scientists headed by S.I. Riabov (Riabov *et al.*, 1987; Ryss *et al.*, 1996), who showed that in 44% pre-dialysis patients with renal failure of stages III–IV the oral administration of large doses of synthetic activated SCN carbons resulted in the decrease

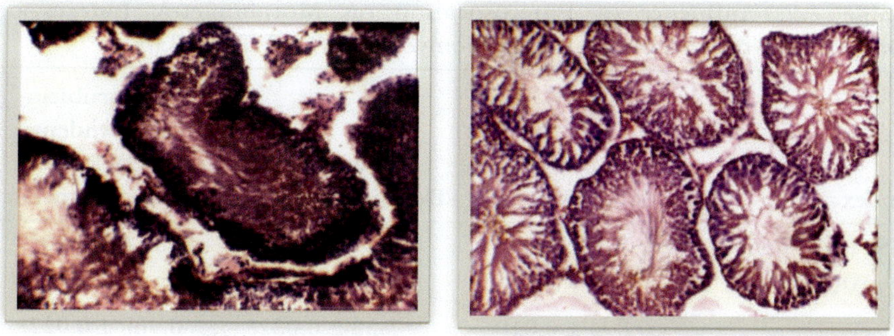

(A) (B)

Fig. 7.6. Gonadotoxic effect of melphalan on the morphology testiculus tissue (A) and gonadoprotective influence of enterosorption (B).

of average creatinine concentration from 0.84 ± 0.06 to 0.59 ± 0.04 μmol/l, and urea from 35.3 ± 9.0 to 23.1 ± 9.5 mmol/l.

As follows from Fig. 7.6, the deep leucopenia induced by introduction of alkylating radiomimetic preparation melphalane could be effectively corrected by treatment with granulocyte colony-stimulating factor (GCSF) and to a somewhat lower degree by administration of enterosorbent C1 with bulk weight of 0.28 g/cm^3. At the same time, carbon enterosorbent C2 with bulk weight of 0.2 g/cm^3 which differs from C1 first of all by twice higher special mesopore area, demonstrates much more expressed myeloregenerative effect compared to recombinant GCSF. Enterosorbent C2 exerts analogous pro-regenerative action toward testicles of rats (Fig. 7.6) treated with melphalane.[44] Here it's important to mention that C2 enterosorbent or GCSF did not worsen the therapeutic response of Guerin carcinoma on melphalane-based chemotherapy but demonstrated optimal myeloprotection when both were used together (Shevchuk *et al.*, 2014, 2015).

Table 7.7 shows that administration of carbon enterosorbents prevents the degenerative process in rat myocardium which results in the development of doxorubicin-dependent dilated cardiomyopathy (Shevchuk, 2014). Such a conclusion could be reached on the basis

Table 7.7. Influence of Enterosorption on Morphometric Indexes of Doxorubicin-dependent Dilated Cardiomyopathy (DCMP) in Rats

Index	Intact Rats	Doxorubicin-Dependent DCMP	Doxorubicin-Dependent DCMP After Enterosorption
ESLVWA, mm^2	118.0 ± 4.45	168.4 ± 7.63*	132.6 ± 3.06**
ESRVWA, mm^2	132.2 ± 6.27	134.8 ± 6.54	137.8 ± 4.88
PI	0.894 ± 0.010	1.25 ± 0.032*	0.965 ± 0.022**

ESLVWA — endocardial surface left ventricular wall area (ESLVWA);
ESRVWA — endocardial surface right ventricular wall area (ESRVWA);
PI — planimetric index (ESLVWA/ESRVWA).

of the significant decrease of endocardial surface left ventricular wall area in animals treated with enterosorbent and significant normalization of corresponding planimetric index. The obtained result certainly is in accordance with the above-mentioned data on hamosorption therapy of DCMP which developed as a late complication of cancer therapy with anthracycline antibiotics.

4. Application-Sorption Therapy

The majority of application sorbents for therapy of wounds and burns are represented by fibrous carbon materials with different structure and capacity (tissues, felts, free fibers). The diameter of these fibers is a few microns. Our concept of sorption-application therapy proposes that the bandage should contain large quantities of carbon materials with specific area of internal surface of 2,000 m^2/g and higher. Examples include «Borisfen» or ACFM «Dnepr» with useful properties (Table 7.8) of elasticity, hydroscopicity, hemostatic and bacteriostatic properties, high adsorptive and kinetic indices, ability to fix workable proteolytic enzymes or to fix and release different drugs, etc. These properties are important in their use in the therapy of different types of wounds and burns, decubitus, trophic ulcers, etc.

As an example, let us demonstrate an influence of carbon dressing on the rate of skin regeneration after experimental partial-thickness

Table 7.8. Sorption Capacity of Carbon Fibrous Activated Materials from Different Manufacturers

Material and Manufacturer	Pore Volume by Benzene, cm^3/g	Methylene Blue, mg/g (m.w. 373.9 D)	Vitamin B_{12}, mg/g (m.w. 1355.0 D)	Unconjugated Bilirubin, mg/g* (m.w. 584.7 D)
FM-10, Chemviron USA-Great Britain	0.456	60.0	0	0
FM-50, Chemviron USA-Great Britain	0.384	60.0	0	0.495
FM-100, Chemviron USA-Great Britain	0.343	24.0	0	0.483
C-TEX 20, MAST, Great Britain	0.759	297.1	11.8	0.3
ACF, Byelorussia	0.674	359.14	17.72	3.05
Borisfen, Ukraine	1.15	720.14	142.8	3.7

*from 3% human serum albumin solution.

burn (II–IIIA degree) in rats. In this experiment[a] devoted to the use of carbon dressing in the first minutes after the burn, the difference in the course of burn wound underneath of this dressing or gauze could be clearly observed just in 24 and 48 hr after the trauma (Fig. 7.7). Figure 7.8B shows the practically complete absence of epithelial layer in the control as well as notable signs of dermal damage 24 hr after burn trauma. Upon the use of carbon dressing there has been registered the presence of properly expressed layer of basal cells, the structural details of which evidence on active regeneration of the damaged structures (Fig. 7.8C). Consequently, the healing of skin under the carbon dressing occurs in significantly shorter period under the gauze (Fig. 7.8C) and burned surface in the experimental group in contrast to the control one (Fig. 7.8D) on the 7th day already possesses the morphological structure of normal skin, sometimes with the features

[a]The data were obtained by our laboratory research team Dr. L.A. Sakhno (L.A. Sakhno, V.N. Maslenny, A.S. Sidorenko, V.G. Korotich).

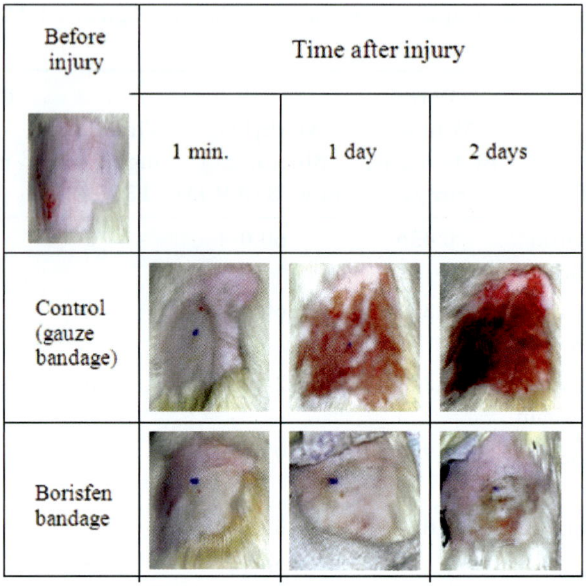

Fig. 7.7. Influence of BORISFEN carbonic bandage on the early dynamic of thermic injury of II-IIIA grade in the rats. Isobaric water steam exposition — 1 sec, area of injury — 6,3 cm^2.

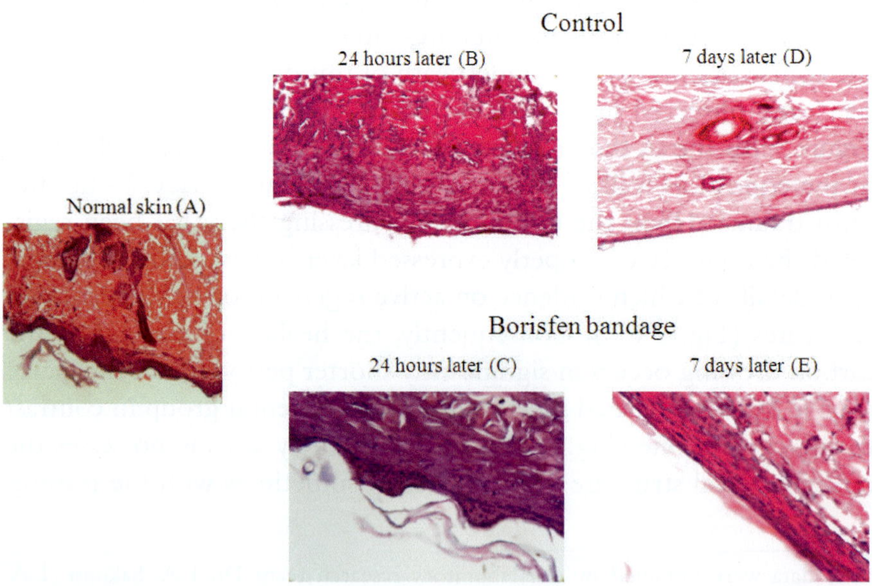

Fig. 7.8. Morphologic structure of skin 1 and 7 days later thermic injury of II-IIIA grade in the rats.

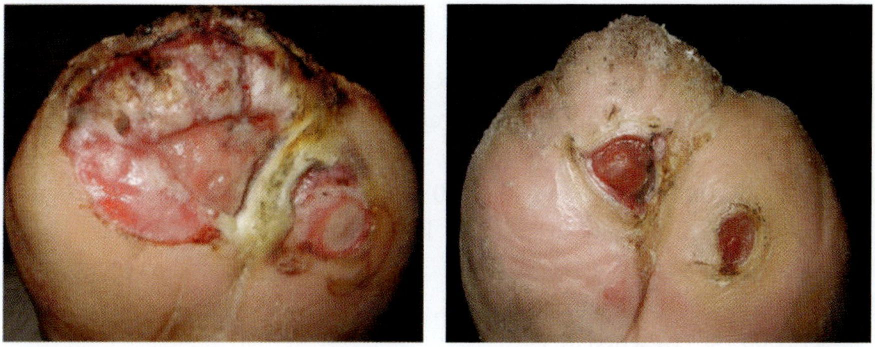

(A) (B)

Fig. 7.9. Example of the quick regeneration of skin in the case of interactable diabetic ulcer under influence of application therapy in combination with enterosorption (A — beginning of the treatment, B — 8 weeks after).

of moderate hyper-regeneration or with the presence of delicate, easily detached eschar (Fig. 7.8E). Apart from theoretical validity, such observations are of certain practical importance and demonstrate that it could be reasonable to include carbon dressings in medicine chests to provide prompt assistance upon domestic burns, especially burns in children.

The impressive effect of diabetic ulcer healing under the influence of carbonic bandage can be seen in Fig. 7.9. In this case, application sorption therapy was combined with enterosorption having a beneficial effects on diabetic patients (Bodnar *et al.*, 1983, 1984, Bodnar and Pristupiuk, 1986).

Conclusion

The ability to enhance and accelerate the regenerative process in affected organs and tissues is characteristic of all three types of adsorption therapy, based on the usage of uncoated synthetic hemosorbents, different types of activated charcoals for peroral consumption and highly activated carbonic fibrous materials (Kavetsky 1977; Nikolaev *et al.*, 1982, 2005, 2013; Bodnar *et al.*, 1983, 1984; Nikolaev, 1984b, 1990; Frolkis *et al.*, 1984a, 1984b; Frolov *et al.*, 1986; Bodnar and Pristupiuk, 1986; Riabov *et al.*, 1987; Enterosorption, 1991; Ryss *et al.*, 1996; Aoyama *et al.*, 2003; Sachno *et al.*, 2013; Shevchuk, 2014; Shevchuk *et al.*, 2014, 2015; Sarnatskaya *et al.*, 2015).

References

Andrade JD, Kunitomo K, Wagenen R. (1971) Coated adsorbents for direct blood perfusion. HEMA — activated carbon. *Trans ASAIO* **17:** 222–228.

Aoyama I, Enomoto A, Niwa T. (2003) Effects of oral adsorbent on gene expression profile in uremic rat kidney: cDNA array analysis. *Am J Kidney Dis* **41:** S8–14.

Bodnar PN, Bul'da ID, Nikolayev VG, Pristupiuk AM. (1983) Patent USSR #1209225. (In Russian.)

Bodnar PN, Nikolaev VG, Pristupiuk AM, Shcherbitskaia EV. (1984) Enterosorption in treating diabetes mellitus patients. *Vrach Delo* **4:** 79–81. (In Russian.)

Bodnar PN, Pristupiuk AM. (1986) Clinical evaluation of enterosorption in the treatment of patients with diabetes mellitus. *Problemi Endokrinilogii* **23:** 7–10. (In Russian.)

Chang TMS. (1969) Removal of endogenous and exogenous toxins by a microencapsulated absorbent. *Can J Physiol Pharmacol* **47(12):** 1043–5.

Chang TMS. (1976b) Microcapsule artificial kidney. Including update preparative procedures and properties. *Kidney Int* **10:** 218–224.

Chang TMS. (1978b) Artificial cells for artificial kidney, artificial liver and detoxification. In: Chang TMS (ed.), *Artificial Kidney, Artificial Liver and Artificial Cells.* Plenum Press, New York, London, pp. 57–72.

Denti E, Luboz M, Tessore V. (1975) Adsorption characterization of cellulose acetate coated charcoal. *J Biomed Mater Res* **9:** 143–150.

Dunea G, Kolff WJ. (1965) Clinical experience with the Yatzidis charcoal artificial kidney. *Trans Amer Soc Artif Inst Organs* **11:** 179–182.

Enterosorption. (1991) Belyakov NA (ed.), Center of sorption technology, Leningrad. (In Russian.)

Eretskaya EV, Sakhno LA, Nikolaev VG. (1991) Application sorption: Experience in clinical use and prospects of development. *Biomat Art Cells Art Org* **19**: 129–145.

Frolkis VV, Nikolaev VG, Bogatskaya LN, *et al.* (1984a) Enterosorption as a method for old animal life span prolonging. *Phisiol Jurn* **30(1):** 30–38. (In Russian.)

Frolkis VV, Nikolaev VG, Bogatskaya LN, *et al.* (1984b) Enterosorption in prolonging old animal lifespan. *Exp Gerontology* **19:** 217–225.

Frolov AF, Pletnyov VM, Lenartovich LS, *et al.* (1986) Complex use of methods of sorptional detoxication in the treatment of leptospirosis. *Vrachebnoe Delo* **4:** 102–105. (In Russian.)

Hagstam KE, Larsson LE, Thysell H. (1966) Experimental studies on charcoal Haemoperfusion in phenobalbital intoxication and uremia including histopatological findings. *Acta Med Scand* **180(5):** 593–610.

Kavetsky RE. (1977) *Tumor-Body Interaction*. Naukova Dumka, Kyiv. (In Russian.)

Nikolaev VG. (1984a) *Hemocarboperfusion in Experiment and Clinic.* Naukova Dumka, Kiev. (In Russian.)

Nikolaev VG. (1984b) Enterosorption. In: Chang TMS, Lin HB (eds.), *On Hemoperfusion and Artificial Organs.* Academic Publishers, China, pp. 87–99.

Nikolaev VG. (1990) Peroral application of synthetic activated charcoal in USSR. *Biomat Art Cells Art Org* **18:** 555–568.

Nikolaev VG. (1990) Peroral application of synthetic activated charcoal in USSR. *Biomat Art Cells Art Org* **18:** 555–68.

Nikolaev VG, Ryabov SI, Pyrig LA, *et al.* (1985) Unconventional methods of activated charcoal application in chronic renal disease therapy. *Nephrologia Mexicana* **6:** 148.

Nikolaev VG, Pinchuk LB, Umansky MA, *et al.* (1993) Early experimental studies on hemoperfusion as a treatment modality for acute radiation Disease. *J Artif Org* **17:** 1–4.

Nikolaev VG, Sarnatskaya VV, Mikhalovsky SV, *et al.* (2011) Deliganding carbonic adsorbents for simultaneous removal of protein-bound toxins, bacterial toxins and cytokines. In: Mikhalovsky S, Khajibaev A (eds), *Biodefence: Advanced Materials and Methods for Health Protection.* Springer Science, Business Media, B.V, pp. 289–305.

Nikolaev VG, Strelko VV. (1979) *Hamosorption on Activated Charcoals.* Naukova Dumka, Kiev. (In Russian.)

Nikolaev VG, Strelko VV, Korovin YuF *et al.* (1982) Theoretical foundations and practical application of enterosorption method. In: *Sorption Methods of Detoxification and Immunocorrection in Medicine,* pp. 112–114. Kharkov.

Nikolaev VG, Andreychyn MA, Bardakhivskaya KI, *et al.* (2013) Practical recommendations of the use of granulated carbon enterosorbent "Carboline." In: Nikolaev VG, Andreychyn MA (eds.) — IEPOR, 2013. — 16 p.

Nikolaev VG, Mikhalovsky SV, Gurina NM. (2005) Modern enterosorbents and mechanisms of their action. *Efferentnaja Terapiya* **11:** 3–17. (In Russian.)

Odaka M, Kobajash H, Tabata Y, Soma M. (1976) Three hour maintenance dialysis combined with direct hemoperfusion and hemodialysis. *Kidney Int* **2:** 197.

Riabov SI, Tsiura VI, Lukichev BG, *et al.* (1987) Reduced experimental blood creatinine as affected by the SKN enterosorbent. *Urol Nephrol (Mosk)* **2:** 49–52. (In Russian.)

Ryss ES, Riabov SI, Lutoshkin MB. (1996) A comparative evaluation of the efficacy of the clinical use of SKN-4M-, SKT-6A- and polifepan-type sorbents in treating patients with chronic kidney failure (clinical and experimental studies). *Ter Arkh* **68**: 39–43. (In Russian.)

Sachno LA, Yurchenko OV, Maslenniy VN, *et al.* (2013) Enterosorption as a method to decrease the systemic toxicity of cisplatin. *Exp Oncol* **35:** 45–52.

Sarnatskaya VV, *et al.* (2015) Effect of enterosorption on some biochemical and morphological manifestations of paraneoplastic syndrome induced in C57 black mice with transplanted Lewis lung carcinoma LLC/R9. *Exp Oncol* (in press). (In Russian.)

Sarnatskaya VV, Lindup WE, Walther P, *et al.* (2002) Albumin, Bilirubin and Activated Carbon: New Edges of an Old Triangle. *Artificial Cells, Blood Substitutes, and Immobilization Biotechnology* **2:** 113–127.

Sakhno LA, Nikolaev VG, Lozinskaya IA. (1991) Basic studies of local adsorption in burn treatment. *Biomat Art Cells Art Org* **19(1):** 111–128.

Sanaka T, Akizawa T, Koide K, Koshikawa S. (2003) Clinical analysis of renoprotective responding patients administrated with oral adsorbent in chronic renal failure secondary to chronic glomerulonephritis. *Ther Apher Dial* **7:** 269–78.

Sarnatskaya VV, Yushko LA, Sakhno LA, *et al.* (2007) New approaches to the removal of protein-bound toxins from blood plasma of uremic patients. *Artif Cells Blood Substit Immobil Biotechnol* **35:** 287–308.

Snezhkova EA, Kolyadenko VG, Nikolaev VG. (1992) DNA-coated carbon adsorbents experimental assessment and results of severe psoriasis treatment. *Biomat Art Cells Art Org* **20(5):** 1201–1221.

Shevchuk OO. (2014) Personal communication.

Shevchuk OO, Posokhova KA, Todor IM, *et al.* (2015) Combination of enterosorption with granulocyte colony-stimulating factor as a prospective approach for prevention of the alkylating drugs side effects. *Exp Onc* (in press).

Shevchuk OO, Posokhova KA, Sidorenko AS, *et al.* (2014) The influence of enterosorption on some haematological and biochemical indices of the normal rats after single injection of melphalan. *Exp Oncol* **36:** 94–100.

Snezhkova EA, Tridon A, Nikolaev VG, *et al.* (2015) Binding capacity of heparin immobilized on activated charcoal for DNA antibodies. *Bullet Experim Oncol Med* (in press). (In Russian.)

Snezhkova EA, Nikolaev VG, Tridon A, *et al.* (2011) Some reasons for development of heparin containing carbonic haemosorbents. *Efferent and Pisico-him Med* **3:** 3–5. (In Russian.)

Todurov BM, Kovtun GI, Mikhajlova AV, Druzhina AN (2011) Hemoimmuno- precipitation — a new approach in the treatment of dilated cardiomyopathy. *Sertseva Nedostatnist* **1:** 72–78. (In Russian.)

Vogl TJ, Zandos S, Scholtz JE, *et al.* (2014) Chemosaturation with percutaneous hepatic perfusions of melphalan for hepatic metastases: Experience from two european centers. *Fortschr Rontgenstr* **186:** 937–944.

Yatzidis H. (1963) A convenient hemoperfusion micro-apparatus over charcoal for the treatment of endogenous and exogenous intoxication. Its use as an effective artificial kidney. *Proc Eur Dial Transpl Assoc* **1:** 83–87.

Yatzidis H, Oreopoulus D. (1976) Early clinical trials with sorbents. *Kidney Int* **10:** 215–217.

Chapter 8

Adsorptive Treatment of Acute Radiation Sickness: Past Achievements and New Prospects

O.O. Shevchuk,* E.A. Snezhkova,† K.I. Bardakhivskaya,† V.G. Nikolaev†

Abstract

Until today the dangerous power of ionizing radiation remains untamed. In the case of a radiation accident, medical management is a challenge. Recent accidents (like the Fukushima nuclear power plant catastrophe) have shown that we need new approaches to the medical treatment of acute radiation sickness.

The hematopoietic system is one of the most sensitive and critical organs for radiation-induced health impairments. Modern techniques including stem cell transplantation and cytokine treatment to restore

*Horbachevsky Ternopil State Medical University, Ternopil, Ukraine.
Shevchuk Oksana O., Ternopil, Majdan Voli, 1, I. Horbachevsky Ternopil State Medical University, Ukraine 46001, phone + 380979006522. shevchukoo@tdmu.edu.ua
†R.E. Kavetsky Institute of Experimental Pathology, Oncology and Radiobiology of National Academy of Science of Ukraine, Kyiv, Ukraine.

hematopoietic functions are still not sufficiently effective for the treatment of acute radiation sickness.

That is why the use of sorption therapy in this field is considered promising. It was shown in 1976 that hemosorption on activated synthetic carbons significantly increased survival in animals which underwent external X-ray irradiation at the dose of 525 Rad (from 3.2% in control group to 68.4% in dogs treated with sorption). It is important to note that similar myeloprotective effect of enterosorption was shown in experiments and in clinical trials during the treatment of hematologic complications of intense cancer therapy.

On the models of cytostatic neutropenia caused by single injection of melphalan in doses 3 and 4 mg/kg, new carbon granulated enterosorbent with bulk weight of 0.18 g/cm^3, total specific surface of 2,162 m^2/g and mesopore area of 565 m^2/g (C2) caused significant elevation of leucocytes compared to classical sorptive materials.

Enterosorbent C2 optimized by its porous structure in combination with officinal preparation of granulocyte colony stimulating factor demonstrates much better protective effect in regard to hematologic toxicity and oxidative stress indexes than each of these preparations alone. Additive effect of carbon enterosorbents and granulocyte colony stimulating factor could be explained by the different mechanisms of their myeloprotective action and influence on different chains of prooxidative-antioxidative homeostasis. Taking into account the radiomimetic character of damaging action of melphalan, one could conclude on the perspective of combined use of carbon enterosorbents and myeloprotective cytokines for treatment of acute radiation syndrome.

NPP — nuclear power plant
ARS — acute radiation sickness
G-CSF — granulocyte colony-stimulating factor

Introduction

The development of the nuclear industry, accidental discharges at nuclear power plants (NPP), diagnostic and treatment, wider use of ionizing radiation in different technologies are constantly confronting us with the problem of radiation injuries, the severity of which is mostly related to the degree of bone marrow damage. Modern strategy of therapy of bone marrow syndrome is based on the use of hemostimulating therapy with myelotropic cytokines, but such treatment has some disadvantages. Thus, the problem of therapy of acute radiation injuries is still open both in its theoretical and practical aspects. Here we present the results and perspectives for the use of sorption therapy for the treatment of acute radiation sickness and cytostatic myelodepression.

Radiation Injuries Today

Acute and chronic radiation injuries became an important medico-social problem in Ukraine and worldwide after the accident at the Chernobyl Nuclear Power Plant in 1986 (Davis *et al.*, 2006; Mettler *et al.*, 2007; Belyi *et al.*, 2010). More recently, after the earthquake in Japan in March 2011, there was another global accident — at Fukushima NPP. Possible terrorist acts or military conflicts involving the use of atomic and radiological weapons could contribute to potential risk of radiation injuries as well. Also, radiation therapy is routinely used in oncology: approximately 50–70% of patients with malignant tumors are treated with radiotherapy (Shadad *et al.*, 2013; Hauer-Jensen *et al.*, 2014).

The action of ionizing radiation on an organism causes biological effects which depend on exposure and absorbed doses and also on the duration of such action, which initiates mutations in DNA, and the processes of apoptosis, cell necrosis and oncogenesis (Demina, 2007; Rzeszowska-Wolny *et al.*, 2009; Popov *et al.*, 2012). Simultaneously due to the excessive peroxidation the multiplicity of altered molecules are accumulated and lead to significant enhancement of the primary radiation effect. Ionizing radiation acts more potently on younger cells with the lowest differentiation. The severest damage is done to lymphoid organs (lymphatic nodes, spleen, thyroid gland), bone marrow, testicles, ovaries and gastro-intestinal mucosa. Parenchymal organs, namely liver, adrenal glands, kidneys, salivary glands and lungs are more radioresistant. Similar effects occur in the case of chemical compounds referred to as radiomimetics, among which the best known representatives are some anticancer preparations (Levis *et al.*, 1963; Povirk, 1996).

In any case, when there is exposure to ionizing radiation either radiomimetic agents, the main factor which determines the survival of the victims is the degree of injury to bone marrow and gastro-intestinal tract (Hauer-Jensen *et al.*, 2008; Manuylov *et al.*, 2012; Ho *et al.*, 2014).

In the last few decades, researchers have focused their attention on the study of possible ways to correct both early and long-term distant consequences of irradiation and chemo-radiotherapy. There are additive toxicity when ionizing radiation is combined with toxic action of xenobiotics, thermal burns, mechanical trauma, etc. Despite significant improvement of medical equipment for more precise and targeted irradiation of tumors, the development of effective means for enhancement of bone marrow cell proliferation (granulocyte colony stimulating factors, erythropoietin, interleukin-11, etc.), the problems of defence from and extenuation of negative consequences of ionizing radiation and radiomimetics remain very real.

Sorption Therapy of Acute Radiation Sickness

Sorption therapy creates a basis for the removal of toxic endogenous or exogenous compounds from body fluids. The most widely used types of sorption detoxification are purification of blood or its components (hemosorption), oral administration of massive doses of sorption materials (enterosorption), and also application-sorption therapy of wounds and burns.

The first results on sorption therapy of acute radiation sickness (ARS) were presented in 1976 by the group of Kyiv scientists headed by V.G. Nikolaev and L.B. Pinchuk. In this study 69 dogs underwent external X-ray irradiation at the dose of 525 Rad (5.25 Gy). The authors showed that hemosorption on activated synthetic carbons significantly increased animal survival from 3.2 % in the control group to 68.4 % in dogs treated with sorption two hours after the radiation injury, and to 62.4 % in animals treated with extracorporeal procedure 24 hours after irradiation (Nikolaev *et al.*, 1993). These results were re-tested and developed in the frame of a special closed program with research institutions of the Ministry of Health and the Ministry of Defense of the USSR (Chertkov *et al.*, 1996, 1999), and formed the basis for generation of instructional materials for early (before 36 hr) hemosorption therapy of mild-severe and severe radiation injuries (Vernigorov and Chertkov, 1983). Unfortunately, during the accident at Chernobyl NPP, due mainly to organizational circumstances, this time period was not upheld. Moreover, there could have been reluctance to use such complicated method of sorption therapy for radiation sickness as extracorporeal blood purification. If at that time, another more accessible method was available, such as enterosorption, the number of casualties would have been lower. Apart from hemosorption, for the treatment of consequences of massive external irradiation other methods of efferent (removing) medicine could be used, in particular, exchange plasmapheresis. This method produced moderate but significant increase of animal survival from 20% to 60%, mitigation of ARS and decrease of endogenous intoxication level typical for this pathology (Legeza 1994). Despite the result being inferior compared to the results of hemosorption, the study was valuable because in parallel with plasmapheresis an active "false" control was used which included all stages of extracorporeal procedure except plasma exchange itself. Since in these conditions no positive effects were registered, it demonstrated the leading role of detoxification in the observed curative effect of plasmapheresis in ARS. In another study (Muravskaya *et al.*, 1991) definite myeloprotective effect of enterosorption was demonstrated with the use of fibrous activated carbons in patients with lymphogranulomatosis who were treated with subtotal irradiation by eradicate scheme. It is important to note that analogous myeloprotective effect of enterosorption was present in experiments and in clinical trials during the treatment of hematologic complications of intense cancer therapy (Nonatskaya *et al.*, 1989). Such conjugated positive influence of enterosorption on activated carbons upon external irradiation and chemotherapy could be explained by radiomimetic properties of many anticancer preparations.

Myelostimulating Cytokines for Treatment of Bone Marrow Syndrome in Acute Radiation Sickness

The central point of the modern concept for treatment of ARS and acute hemopoietic syndrome is the use of myelostimulating cytokines, namely granulocyte (G-CSF)- and granulocyte-macrophage colony stimulating factors (Dainiak *et al.*, 2011) which reduce the duration of critical neutropenia and improve ARS prognosis on experimental animals and in clinical trials (Di Carlo *et al.*, 2011; Reeves, 2014). It is important to note that large G-CSF doses promote the survival and improve hematologic indexes of animals treated with combined gamma-neutron irradiation (Li *et al.*, 2011), but the FDA supposes that there are no infallible methods for effective irradiation counteraction (Singh *et al.*, 2012). So, this forces researchers to look for new methods of ARS therapy; in particular situations, the pegylated forms of G-CSF and other peptide myelostimulators, for example, recombinant interleukin-12 are used (Drouet and Herodin, 2010; Gluzman-Poltorak *et al.*, 2014).

Colony stimulating factors are also widely used for prophylaxis and therapy of febrile neutropenia caused by intense cancer chemotherapy (Mhaskar *et al.*, 2014; Pfeil *et al.*, 2014). A shortened period of neutropenia, decreased intensity and duration of fever and lower requirements for antibiotic use were observed. The pegylated forms of G-CSF, such as pegfilgrastim and lipegfilgrastim, are preferred (Ratti and Tomasello, 2014). However, these preparations are expensive; therefore the use of myeloid cytokines for cancer chemotherapy seems to be restricted (Barnes *et al.*, 2014).

Possibilities of Enterosorption upon Cytostatic Myelodepression

An influence of porous structured carbon enterosorbents on manifestations of hematological toxicity, functional indexes of liver and kidneys and the prooxidant-antioxidant activity balance have been studied on a rat model after single administration of melphalan (phenylalanine derivative of nitrogen mustard) at the dose of 3 mg/kg (Shevchuk *et al.*, 2014). It has been shown that carbon granulated enterosorbent (C1) with bulk density of 0.28 g/cm^3, specific surface of 1,719 m^2/g and mesopore area of 239 m^2/g has significantly lower protective activity compared to the sorbents with bulk weight of 0.18 g/cm^3, total specific surface of 2,162 m^2/g and mesopore area of 565 m^2/g (C2) when given at the same dose of 1 ml/200 g of rat weight. Leukocyte counts decreased from 5.24 ± 0.29 to 1.44 ± 0.14 at day 8 after melphalan administration. Counts in experiments with entersorption were 2.98 ± 0.24 and 6.48 ± 0.37 in animals treated with C1 and C2 respectively. Absolute neutrophil counts which decreased from 1.37 ± 0.09 to 0.56 ± 0.05 were maintained at the levels of 1.23 ± 0.14 and 2.12 ± 0.21 × 10^9/L at use of C1 and C2 respectively. Serum blood transaminases, alkaline phosphatase, total bilirubin and creatinine did not differ significantly in the case of use of alkylating agent, or enterosorbent administration. At the same time serum concentrations of urea, hydrophilic and hydrophobic fractions of medium molecular weight compounds which were significantly increased at day 8 after melphalan administration, more prominently decreased toward normal values in the case of use of C2 sorbent. Simultaneously the indexes of prooxidant activity which were significantly higher in animals after melphalan administration, decreased to their normal level in the group of animals treated with C2 enterosorbent. Reverse dynamics have been observed for the indexes of antioxidant activity which were suppressed by the alkylating agent. Significant normalization was observed in the group treated with C2 enterosorbent.

At the next work of this cycle (Shevchuk *et al.*, 2015), sorbent C2 was used with preparation of G-CSF filgrastim on a model of cytostatic neutropenia caused by administration of higher dose of melphalan (4 mg/kg). Here filgrastim and enterosorbent provided approximately comparable protective effect to leukocyte counts and drastic increase in the case of simultaneous use of both preparations. At day 8 after melphalan administration, absolute neutrophil count dropped from 1.37 ± 0.09 to 0.63 ± 0.05 and was then restored to the level of 0.79 ± 0.06 and 1.05 ± 0.12 for use of enterosorbent C2 and filgrastim, respectively. Both preparations, at day 8 after melphalan administration, increased this index to $1.86 \pm 0.16 \times 10^9$/L. In contrast to filgrastim, which did not alter significantly platelet counts, the use of C2 has shown a trend for elevation of this index. Combined use of the preparations resulted in significant normalization of platelat count.

The indices of prooxidant activity decreased significantly and nearly equally in the groups C2 and C2 + filgrastim, while being altered in different directions in animals protected with filgrastim only.

The indices of antioxidant activity, which were significantly suppressed in the melphalan-treated group, demonstrated just the trend to normalization after filgrastim administration, while the use of enterosorbent and combined use of the preparations provided significantly better effects.

So, in conclusion, enterosorbent C2 in combination with granulocyte colony stimulating factor demonstrates better protective effect for hematologic toxicity and oxidative stress than each preparation alone. The additive action of carbon enterosorbent and granulocyte colony stimulating factor could be explained by the different mechanisms of their myeloprotective action and influence on different chains of prooxidant-antioxidant homeostasis. Taking into account the radiomimetic character of damaging action of melphalan, the combined use of carbon enterosorbents and myeloprotective cytokines for the treatment of acute radiation syndrome is prospective. According to practical aspect, such a combination seems to be more simple and comfortable to use than extracorporeal hemocarboperfusion at early terms after irradiation.

Acknowledgements

This work was supported by the grant: "Problems of safety in nuclear power generation industry" from the Ministry of Education and Science of Ukraine.

The authors express their sincere gratitude to academician V.G. Baryakhtar for fruitful discussion.

References

Barnes G, Pathak A, Schwartzberg L. (2014) Pharmacoeconomics of granulocyte colony-stimulating factor: A critical review. *Adv Ther* **31(7):** 683–95.

Belyi D, Kovalenko A, Bazyka D, Bebeshko V. (2010) Non-cancer effects in acute radiation syndrome survivors in Ukraine. *Health Phys* **98(6):** 876–884.

Chertkov KS, Andrianova IY, Andrushchenko VN, Vernigorova LA. (1996) Experimental development of combined treatment for acute radiation sickness. In Berger; Horn. 9 international congress of the International Radiation Protection Association; Vienna (Austria), pp. 129–131.

Chertkov KS, Andrianova IE, Andrushchenko VN, *et al.* (1999) Advances and prospects of a combined therapy of acute radiation injury in experiments. *Radiats Biol Radioecol* **39(5):** 563–567.

Dainiak N, Gent RN, Carr Z, *et al.* (2011) First consensus for evidence-based management of the hematopoietic syndrome resulting from exposure to ionizing radiation. *Disaster Med Public Health Prep* **3:** 202–212.

Davis S, Day RW, KJ Kopecky, *et al.* (2006) Childhood leukaemia in Belarus, Russia, and Ukraine following the Chernobyl power station accident: Results from an international collaborative population-based case–control study. *Int J Epidemiol* **35:** 386–396.

Demina EA. (2007) The study of dose dependence of malignant tumors in the group of liquidators of consequences of Chernobyl catastrophe. *Problems of Radiat Med and Radiobiol* **13:** 146–150. (In Ukrainian.).

Di Carlo AL, Maher C, Hick JL, *et al.* (2011) Radiation injury after a nuclear detonation: Medical consequences and the need for scarce resources allocation. *Disaster Med Public Health Prep* **Suppl 1:** S32–44.

Drouet M, Herodin F. (2010) Radiation victim management and the haematologist in the future: Time to revisit therapeutic guidelines? *Int J Radiat Biol* **86(8):** 636–648.

Gluzman-Poltorak Z, Vainstein V, Basile LA. (2014) Recombinant interleukin-12, but not granulocyte-colony stimulating factor, improves survival in lethality irradiated nonhuman primates in the absence of supportive care: Evidence for the development of a frontline radiation medical countermeasure. *Am J Hematol* **89(9):** 868–879.

Hauer-Jensen M, Kumar KS, Wang J, *et al.* (2008) Intestinal toxicity in radiation — and combined injury: significance, mechanisms, and countermeasures. In: Larche RA (ed.), *Global Terrorism Issues and Developments.* Nova Science Publishers, Incv, pp. 61–100.

Hauer-Jensen M, Denham J, Andreyev J. (2014) Radiation enteropathy — pathogenesis, treatment and prevention. *Nat Rev Gastroenterol Hepatol* **11:** 470–479.

Ho IK, Cash BD, Cohen H, *et al.* (2014) Radiation exposure in gastroenterology: Improving patient and staff protection *Am J Gastroenterol* **109:** 1180–1194.

Legeza VI, Abdul' IuA, Chigareva NG, *et al.* (1994) Experimental study of therapeutic efficacy of plasmapheresis in acute radiation injury. *Radiats Biol Radioecol* **34(4–5):** 657–662.

Levis AG, Spanio L, De Nadai A. (1963) Radiomimetic effects of a nitrogen mustard on survival, growth, protein and nucleic acid synthesis of mammalian cells in vitro. *Exp Cell Res* **31(1):** 19–30.

Li M, Yu ZY, Xing S, *et al.* (2011) High dose granulocyte colony-stimulating factor enhances survival and hematopoietic reconstruction in canines irradiated by 2.3 Gy mixed fission neutron and gamma ray. *Zhongguo Shi Yan Xue Ye Xue Za Zhi* **19(4):** 991–8.

Manuylov AM, Bolokov MS, Gurmikov BN. (2012) The role of barrier function of gastrointestinal tract in clinical practice. Review of literature. *Novie Technologii* **4:** 302–327. (In Russian.)

Methodical recommendations on the use of hamosorption in clinical conditions for treatment of acute radiation sickness. Vernigorov LA, Chertkov KS (eds.), Moscow, 1983, 10 p. (In Russian.)

Mettler FA Jr, Gus'kova AK, Gusev I. (2007) Health effects in those with acute radiation sickness from the Chernobyl accident. *Health Phys* **93(5):** 462–469.

Mhaskar R, Clark OA, Lyman G, *et al.* (2014) Colony-stimulating factors for chemotherapy-induced febrile neutropenia. *Cochrane Database Syst Rev* **10:** CD003039.

Muravskaya GV, Nikolaev VG, Sergeev VP, *et al.* (1991) Enterosorption in oncotherapy. *Biomater Artif Cells Immobilization Biotechnol* **19(1):** 167–174.

Nikolaev VG, Pinchuk LB, Umansky MA, *et al.* (1993) Early experimental studies on hemoperfusion as a treatment modality for acute radiation disease. *Artif Organs* **17(5):** 362–365.

Nonatskaya LV, Plotnikov VM, Nikolaev VG. (1989) Decrease of hematotoxicity of anticancer preparation upon enterosorption. *Exp Oncol* **23(11):** 71–73. (In Russian.)

Pfeil AM, Allcott K, Pettengell R, *et al.* (2014) Efficacy, effectiveness and safety of long-acting granulocyte colony-stimulating factors for prophylaxis of chemotherapy-induced neutropenia in patients with cancer: a systematic review. *Support Care Cancer* Oct 7.

Popov D, Jones J, Maliev V. (2012) Radiation toxins — effects of radiation toxicity, molecular mechanisms of action, radiomimetic properties and possible countermeasures for radiation injury. In: Nenoi M (ed.), *Current Topics in Ionizing Radiation Research*. InTech, pp. 215–242.

Povirk LF. (1996) DNA damage and mutagenesis by radiomimetic DNA-cleaving agents: bleomycin, neocarzinostatin and other enediynes. *Mutation Research/Fundamental and Molecular Mechanisms of Mutagenesis* **355(1–2):** 71–89.

Ratti M, Tomasello G. (2014) Lipegfilgrastim for the prophylaxis and treatment of chemotherapy-induced neutropenia. *Expert Rev Clin Pharmacol* **20:** 1–10.

Reeves G. (2014) Overview of use of G-CSF and GM-CSF in the treatment of acute radiation injury. *Health Phys* **106(6):** 699–703.

Rzeszowska-Wolny J, Przybyszewski WM, Widel M. (2009) Ionizing radiation-induced bystander effects, potential targets for modulation of radiotherapy. *Eur J Pharmacol* **625**: 156–164.

Singh VK, Ducey EJ, Brown DS, *et al.* (2012) A review of radiation countermeasure work ongoing at the Armed Forces Radiobiology Research Institute. *Int J Radia Biol* **88(4):** 296–310.

Shadad AK, Sullivan FJ, Martin JD, Egan LJ. (2013) Gastrointestinal radiation injury: prevention and treatment. *World J Gastroenterol* **19(2):** 199–208.

Shevchuk OO, Posokhova KA, Sidorenko AS, *et al.* (2014) The influence of enterosorption on some haematological and biochemical indices of the normal rats after single injection of melphalan. *Exp Oncol* **36(2):** 94–100.
Shevchuk OO, Posokhova KA, Todor IM, *et al.* (2015) Prevention of myelosuppression by combined treatment with enterosorbent and granulocyte colony-stimulating factor. *Exp Oncol* **37**(2): 135–138.

III

Selective Adsorbents

Chapter 9

Hemoperfusion Selective Removal

Tohru Tani*

Selective removal is one of the selective removal methods relating to apheresis technologies. Initially, apheresis represented only blood separation. And therapeutic apheresis means the treatment of blood modification, especially blood letting (drawing) or separation for blood exchange. Usually, the targeted disease of apheresis is unclear, so the substance for removal or modification is not clearly decided.

But due to technological development, the toxic substances or materials removed have been gradually specialized (Duke, 1984).

The selectivity of this method is composed of adsorption, fractionation, salt-amino acid co-precipitation and plasma separation. The adsorption method has the highest selectivity. The most specific removal methods are an adsorption from perfused plasma, which is called plasma perfusion, or from whole blood, which is called hemoperfusion. Hemoperfusion or direct hemoperfusion method is performed with the adsorption device in the whole blood. Plasma perfusion is performed with the adsorption device in the plasma flow that is separated by a plasma collector or plasma membrane separator.

Removal methods are categorized by targeted materials, cell, protein, amino-acid, drug, foreign materials, toxic substance, middle molecule and so on (Nakane and Choichiro, 2004).

---

* Program Professor of Bio-medical Innovation Center, Shiga University of Medical Science, Japan. tan@belle.shiga-med.ac.jp

Endotoxin derives from outside of the body and has the strongest lethal activities in natural products. Endotoxin is one of the components in the cell wall of Gram negative bacteria. Traditionally, this material has been thought to be one of the agents which cause septic shock.

Nowadays, endotoxin is thought to be related to metabolite dysfunction, obesity and oily food ingestion (Patrice and Jacques, 2007). Endotoxin removal from blood is promising for the improvement of diseases mentioned above.

Y. Zhi mentions the adsorbents for removal of endotoxin and T. Shimizu introduces the following information. The device (Toraymyxin) mentioned by T. Shimizu (I am one of the inventors) has been applied in clinical cases since 1993. This column is one of the representative adsorption devices aiming to selectively adsorb single agents. Toraymyxin® is usually applied in 90-minute sessions by direct hemoperfusion (PMX).

Over 2,500 papers have been published on the efficacies of PMX and applied diseases including double-blinded clinical trials in the US, France and China (Tani and Shoji, 2010).

Low density lipoprotein (LDL) apheresis is reported by M. Shibata. LDL apheresis has been applied to treat hypercholesterolemia, especially familial hypercholesterolemia (FH) with homozygotes and heterozygotes (Makino and Marada-Shiba, 2014).

LDL apheresis is usually applied using plasma exchange, double filtration plasmapheresis, LDL adsorption and heparin-induced extracorporeal precipitation (HELP). The HELP system is popularly applied in western world; in Japan, LDL adsorption is popular (Yokoyama and Hayashi, 1984). The center belonging the author and the company have developed the Liposorber® system, which has dextran sulfate as a ligand, in the form of porous beads. This system selectively adsorbs Apolipoprotein charged positive and Lipoprotein. LDL apheresis removes immunoglobulin at the same time. This pheresis treatment should be performed with over 2 weeks' interval.

Effects of LDL apheresis have been reported for the prevention of atherosclerosis in FH (hemofiltration) patients and volume reduction of xanthoma.

B2-microglobulin is famous as a causative agent of amyloidosis and carpal tunnel syndrome in HD (hemodialysis) patients (Abe *et al.*, 2003). This column is used in the circuit of HD. This adsorbent has been reported to remove cytokines as well. The efficacies by removal of B2-microglobulin have been reported, such as improvement of symptoms from amyloidosis and suppression of bone cyst formation (Kuragano *et al.*, 2011). J. Kazama *et al.* report B2-microglobulin removal methods and clinical results in their chapter.

Cytokine removal treatments are reported by S. Oda. His paper mentions the adsorption method using hemofiltration (HF) or hemodialysis (HD) membrane. S. Oda has used continuous HF (CHF) to treat severe disease, severe sepsis or multi-organ failure (MOF) patients. They applied Poly-methel meta-acrilate (PMMA) membrane column for CHF, and found this membrane adsorbed a high amount of IL-6 cytokine. The author thinks PMMA membrane column contributes to IL-6 removal from patients with severe hyper-cytokinemia (Hirasawa and Oda, 2012).

A similar finding in HD membrane AN69 is reported by Thomas (Thomas and Moriyama, 2011). This system is also able to adsorb cytokines (Dahaba *et al.*, 2002) and was applied on severe sepsis or septic shock under Japanese national insurance in 2014 (Hirasawa and Oda, 2012).

In this chapter, two new apheresis methods should be presented: the molecular adsorbent recirculating system (MARS) of Baxter and plasma filtration dialysis (PDF). PDF is presented by Dr. Eguchi. MARS was advocated by Jan Stange and Steffen Mitzner of Rostock University (Stange and Mitzner, 1993). This system is categorized in extracorporeal albumin dialysis (ECAD). The advantage of this system is to be able to adsorb not only aqueous toxin but also albumin-binding agent. The targeted disease of MARS is mainly acute hepatic failure (Laleman *et al.*, 2006). The disadvantage of MARS is that it requires albumin infusion and a complex circuit.

The PDF system aims for direct filtration of middle molecule agents using a membrane of which sieving coefficient cuts off middle molecules, because there has been no system selectively removing middle molecules. MARS also purposes to remove a part of middle molecule attached albumin. PDF is actually performed by slow PE and CHF. The targeted diseases for PDF include severe disease relating to middle molecules in the blood, severe sepsis, hepatic failure and severe pancreatitis (Eguchi, 2010).

References

Abe T, *et al.* (2003) Effect of β2-microglobulin adsorption column in treating dialysis-related amyloidosis. **64:** 1522–1528.

Dahaba AA, *et al.* (2002) Procalcitonin and proinflammatory cytokine clearance during continuous venovenous haemofiltration in septic patients. *Anaesth Intensive Care* **30:** 269–274.

Duke OK. (1984) *Therapeutic Apheresis — A critical look ahead.* ISAO Press, pp. 7–10.

Eguchi Y. (2010) Plasma dia-filtration for severe sepsis. *Contrib Nephrol* **166:** 142–149.

Hirasawa H, Oda S. (2012) Continuous hemodiafiltration with a cytokine-adsorbing hemofilter for sepsis. *Blood Purif* **34:** 164–170.

Kuragano T, *et al.* (2011) Effectiveness of 2-microglobulin adsorption column in treating dialysis-related amyloidosis: A multicenter study. *Blood Purif* **32:** 317–322.

Laleman W, *et al.* (2006) Effect of the molecular adsorbent recirculating system and Prometheus devices on systemic haemodynamics and vasoactive agents with acute-on-chronic alcoholic liver failure. *Crit Care* **10:** R 108.

Makino H, Marada-Shiba M. (2014) Therapeutic effect of and evaluation method for LDL apheresis in familial hypercholesterolemia. *J Jpn Soc for Apheresis* **33:** 129–134.

Nakane M, Choichiro T. (2004) Basics of hemoperfusion. *Nippon Rinsho* **62:** 22–28.

Patrice DC, Jacques A. (2007) Metabolic endotoxemia initiates obesity and insulin resistance. *Diabetes* **56:** 1761–1772.

Stange J, Mitzner S. (1993) A new procedure for the removal of protein bound drugs and toxins. *ASAIO J* **39:** M621–625.

Tani T, Shoji H. (2010) Extracorporeal removal of endotoxin: The polymyxin-B-immbilized fiber cartridge. In: Ronco C, Rosner MH (eds.), ***Endotoxiemia and Endotoxin Shock***. Karger, Basel, pp. 35–39.

Thomas M, Moriyama K. (2011) AN6: Evolution of the world's first high permeability membrane. ***Contrib Nephrol*** **173:** 119–129.

Yokoyama S, Hayashi R. (1984) Specific sorbent of apolipoprotein-B containing lipoproteins for plasmapheresis. ***Arteriosclerosis*** **4:** 276–282.

Chapter 10

Hemoperfusion Method for Removing Endotoxin

Zhi Yuan*

1. Introduction

Endotoxins are the major pathogenic factors causing septicemia (with multiple abscesses and a longer duration), which is also called sepsis. Endotoxins are widely present in the external environment and the human body. Even at very low concentrations, endotoxins within the environment can cause respiratory symptoms such as cough, asthma and phlegm (George *et al.*, 2001). Endotoxins are toxic in the human body and can cause fever, disseminated intravascular coagulation, severe hypotension, metabolic disorders, organ failure and other diseases (Shimizu *et al.*, 2006; Abbass and Moghazy, 2011). In severe cases, endotoxins can lead to death. About 750,000 people suffer from septicemia each year in the United States, with a mortality rate of 28–50%. In 2010, 7.7% of the deaths in the UK were septicemia-related (Beers, 2006).

The commonly-used clinical method for treating septicemia is the use of antibiotics, which can damage the outer membrane of the bacterial cell wall and in turn release endotoxins, further exacerbating the

*Institute of Polymer Chemistry, Nankai University, China.
Key Laboratory of Functional Polymer Materials of Ministry of Education and Institute of Polymer Chemistry, Nankai University. E-mail: zhiy@nankai.edu.cn

condition (Opal, 2007). More recently, septicemia treatment strategies have focused on activating the immune system, treating the condition via immunomodulation (Yamakawa *et al.*, 2011; Ranieri *et al.*, 2012). Drotrecogin-α is the only approved immunotherapy drug for severe septicemia (Ranieri *et al.*, 2012). However, in 2011 the drug was withdrawn from the market, generating suspicion in people about the prospects of immune-modulatory interference in sepsis treatment (Evangelos and Bourboulis, 2013). An alternative to directly removes toxic substances from the blood is hemoperfusion, a method that has opened the door for rapid removal of toxins in the blood.

2. Endotoxin and Septicemia

One hundred years ago, Pfeiffer (1858–1945) found that cholera bacteria were still potentially toxic even after being killed by high temperatures. He defined the toxins that have high thermal stability and play an important role in stabilizing bacteria as endotoxins (Pfeiffer, 1892). Endotoxins can be released during various stages of bacterial cell division, growth, and death. Endotoxins are characterized by small size and high thermal and chemical stability.

Endotoxins are a major component of the outer membrane of the cell wall of Gram-negative bacteria (Fig. 10.1), consisting of three parts (Fig. 10.2): an external variable O- polysaccharide side chain, the relatively conservative core region in the middle, and a very conservative inner lipid A, which is the origin of the name lipopolysaccharide (LPS)(Raetz, 1990; Raetz, 2008). Lipid A (Fig. 10.3) is connected to the outer membrane of bacterial cells, with the polysaccharide moiety exposed. Lipid A, the most conservative part in endotoxins is phosphorylated (Hou and Zaniewski, 1990), and is the active site of the endotoxin. The lipid A structures of different bacteria are very similar, including 3-hydroxy-β-1,6-glycosidic linked glucose disaccharides and 3-hydroxy substituents that are covalently linked. However, there are still minor differences between them.

Lipid A is the active site of endotoxins. Researchers studied the structure-activity relationships using original structures, derivatives and synthetic analogues of lipid A. Surprisingly, the activity of

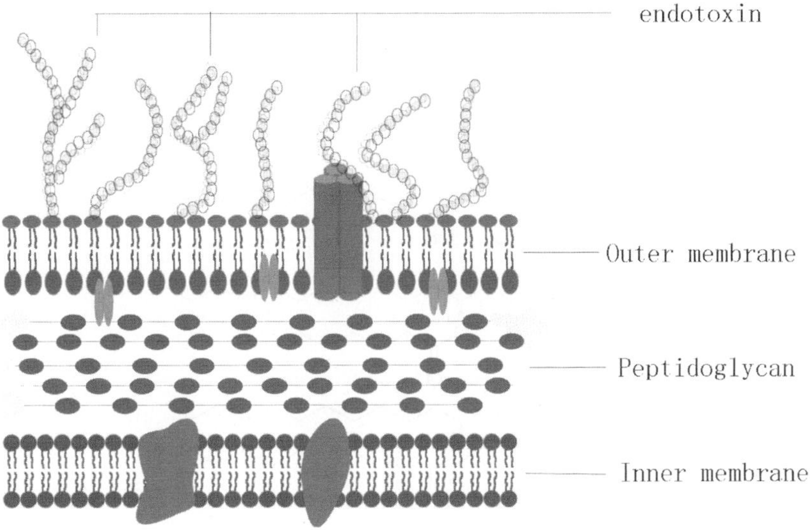

Fig. 10.1. Outer membrane structure of Gram-negative bacteria.

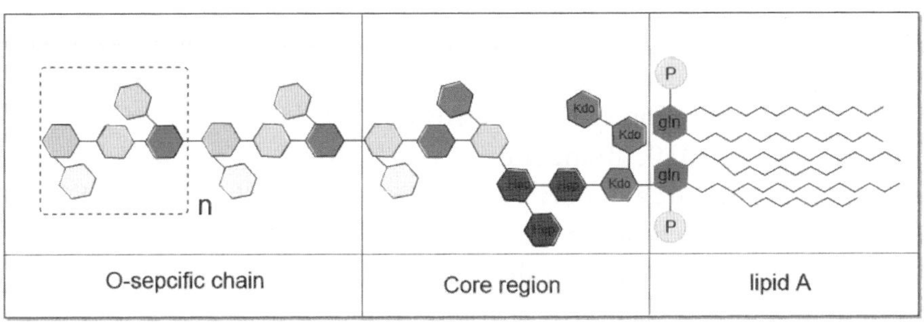

Fig. 10.2. Schematic diagram of endotoxin.

synthetic molecules is not higher than that of endotoxins, and any minor changes to lipid A will reduce endotoxin activity. Further, the local structure of lipid A has no effect on the human body. Thus the activity of endotoxins depends not only on the lipid A structural composition, but is jointly determined by the unique three-dimensional structure composed of the spatial factors, negative electric charge, and hydrophobic region (Schromm *et al.*, 1998).

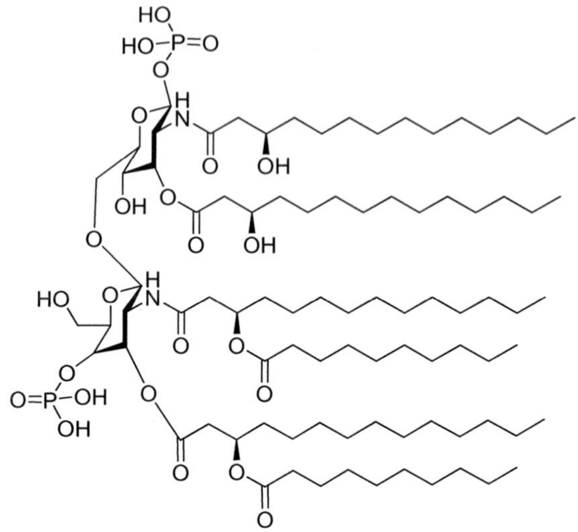

Fig. 10.3. Structure of lipid A.

Septicemia can occur in a variety of diseases such as extensive burns, acute obstructive cholangitis, obstruction, and cirrhosis. It can cause serious damage to the immune system, resulting in shock, multiple organ dysfunction syndrome (MODS) and multiple organ failure (Bhattacharjya, 2010; Brandenburg *et al.*, 2011). The 2003 US epidemiological research report mentioned that in the United States, 120,000 people died of septicemia, and this number is rising annually. Globally, septicemia causes 600,000 deaths each year, ranking 10th among fatal diseases, and the second among intensive care unit mortality, making it an important challenge for modern medicine (Martin *et al.*, 2003; Angus and Wax, 2001).

Septicemia is a pathophysiological disease that occurs when a large quantity of endotoxins are released by bacteria in blood or lesions. It can also be caused by the infusion of a large amount of liquid contaminated by endotoxins (Beers, 2006; Nakayama *et al.*, 2012). Therefore, the key to the treatment of septicemia lies in the timely and effective removal or destruction of endotoxins in the patient. Hemoperfusion, a direct method for clearing the toxins in blood streams, has achieved good results, and some products have been used in clinical practice.

3. Removal of Endotoxins by Adsorbents in Hemoperfusion

Compared to drug treatment, hemoperfusion does not cause bacterial death that releases endotoxins which therefore worsens the medical condition. Direct perfusion in patients with septicemia can effectively reduce patient mortality. The key to the treatment of septicemia with hemoperfusion depends on the removal of endotoxins, and the latter lies in the effectiveness of the adsorbent. Endotoxins are negatively charged, and the existence of large hydrophobic regions, which are important bases for designing adsorbents, will be described in the following sections.

3.1. *Activated Charcoal and Resin Adsorbents*

Cytokines (TNF, IL-1, IL-6, etc.) play an important role in the pathophysiological changes of septicemia. For example, when animals or human beings are given an injection of TNF, symptoms of septic shock will occur. Activated charcoal can remove endotoxins and cytokines (Bende and Bertoc, 1986; Gazzard *et al.*, 1974). Activated charcoal perfusion is safe, having a broad adsorption spectrum, and is widely used clinically. The adsorption principle of activated charcoal is based on its hydrophobicity, large specific surface area and pore structure. Using activated charcoal (DHP-1) to adsorb endotoxins in human plasma, the TNF and IL-6 removal rates have reached $17.2 \pm 6.2\%$ and $48.5 \pm 7.4\%$, respectively (Nagaki *et al.*, 1991). Odnopozov *et al.*, used activated charcoal in the hemoperfusion treatment of pigs that were endotoxin-shocked. All 14 porcines that underwent activated charcoal (SCN-IK) perfusion survived, and only 10 (out of 14 in the control group) survived in the control group (Odnopozov *et al.*, 1996).

While activated charcoal has strong adsorption capacities and has seen widespread use in clinical applications, its poor adsorption selection and blood compatibility affect its further application in clinical practice. Compared with activated charcoal, resin is a spherical polymer with high specific surface area. Its adsorption properties can be adjusted by changing its physical and chemical properties. Resins include non-polar and polar adsorbent resins. The non-polar resin contains no groups, such as styrene — diethylbenzene resin, while the polar resin is linked to a variety of polar groups, such as EA-1 resin, which has an amino group, is positively charged, and can adsorb negatively charged endotoxin molecules. The removal rates of TNF and IL-6 by neutral resin AmberliteXAD-7 can reach 32.5 ± 3.3% and 71.4 ± 3.8%, respectively (Yong *et al.*, 2012).

The adsorption capacity of neutral macroporous resin in the perfusion device depends primarily on the role of the molecule sieve formed by the three-dimensional network structures of the resin molecules, and the interaction between electric charge and hydrophobic groups. The current perfusion device used in clinical applications in China is the HA330 produced by the Jianfan company. The adsorbent is a macro-porous resin, polymerized from styrene/divinyl benzene. HA330 has a large adsorption capacity, high adsorption rate and high mechanical strength. It can effectively adsorb a variety of endogenous and exogenous toxins and inflammatory mediators, such as tumor necrosis factors, interleukin, endotoxins, and platelet-activating factor. The Yaoting Yu research team compared the adsorption capacity of NK-110, carbonized resin, and MET-1004 resin for the removal of TNF-α. Results showed that NK-110 and MET-1004 had a high adsorption capacity for TNF-α. (Jiao and YaoTing, 2001).

3.2. *Adsorbents with Polymyxin B and other Cations as Ligands*

In addition to activated charcoal and polystyrene resins, adsorbent with positively charged materials were developed for hemoperfusion and gradually used in clinical therapy. Polymyxin B (Fig. 10.4) is an antibiotic with a circular alkaline lipophilic peptide, produced by *Paenibacillus polymyxa* spores, and has an antibacterial effect on *Pseudomonas aeruginosa*, *Escherichia coli*, *Klebsiella pneumoniae* and other Gram-negative bacteria. Its mechanism for endotoxin removal is the electrostatic interaction between the positively charged polymyxin B and negatively charged phosphate groups in lipid A. Polymyxin B can damage the permeability of the outer layer or cell

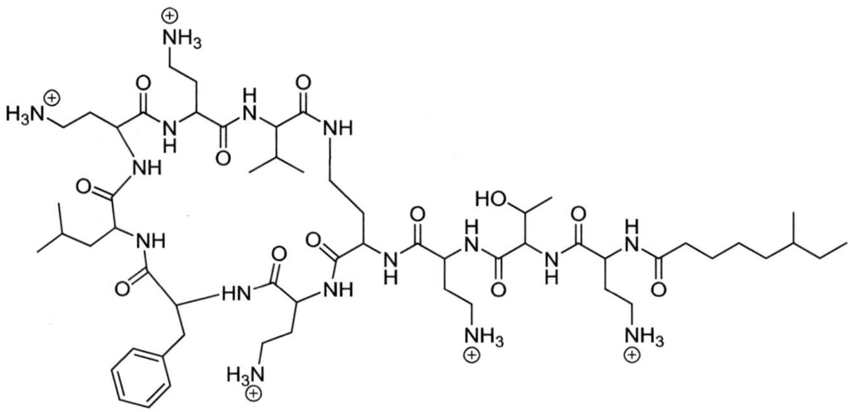

Fig. 10.4. Polymyxin B.

membrane of Gram-negative bacteria, and has antibacterial and endotoxin inactivation effects (Hirohike *et al.,* 1994). Polymyxin B has significant nephrotoxicity and neurotoxicity properties, and direct application can cause harm to the human body. However, if immobilized, it can be applied in hemoperfusion to treat septicemia.

Polymyxin B (PMX) immobilized on α-chloroacetamide-methyl polystyrene fibers was registered as a blood purification product (Toraymyxin) in Japan in 1994, and in the European market in 1998. Toraymyxin is a medical product developed by Toray Medical Co. Ltd as a whole blood perfusion adsorption column specifically used for toxin adsorption. It has been tested on mice (Cohen *et al.,* 1987), rabbits (Tetta *et al.,* 2000) and dogs (Shigui *et al.,* 1997), and has a higher survival rate than traditional drug treatment. Sixteen patients with sepsis were perfused by PMX; after two hours treatment, endotoxin concentrations dropped from 76 pg/mL to 21 pg/mL. After two weeks of treatment, nine of the 16 patients remained alive, and seven out of the nine patients were discharged from the hospital. (Hirohiko *et al.,* 1994). A septicemia clinical research team in Japan also obtained similar results (Kodama *et al.,* 1997). In six medical centers in Europe, the clinical treatment of 36 patients (17 received PMX treatment and 19 standardized treatment) showed that PMX had no side effects and could improve patients' cardiac and renal function (Jean-Louis *et al.*, 2005). During 2004–2007, 10 top hospitals in Italy applied early PMX treatment of septicemia (34 received conventional plus PMX treatment, while 30 received conventional treatment). The mortality rate for those receiving PMX treatment was reduced to 32%, compared to the 53% of mortality rate traditionally treated patients (Cruz *et al.*, 2009). There are other reports indicating that the best perfusion results were achieved when the oxygen consumption index is greater than 100 mL/min/m^2 (Hidehiko *et al.*, 2008). However, if polymyxin B leaks off the fiber column, there will be serious side effects. The fixation of other cationic groups, such as polyethyleneimine (PEI) (Mitzner *et al.*, 1993), diethylene diamine (DEAE) (Bengsch *et al.*, 2005), and amine (Ji Hong *et al.*, 2006; Yuan *et al.*, 2005), on the adsorbent can also significantly increase endotoxin adsorption rate. The adsorbent containing DEAE ligand was applied to 15 patients that were under intensive care with sepsis, and only 26.7% of them died during the 28-day treatment (Bengsch *et al.*, 2005). Yuan's research team studied the effectiveness of using a ligand with polymethyl methacrylate as the carrier, and for the first time computer simulation technology was used to design adsorbent material. (Yuan *et al.*, 2005). In this study, dimethylamine was found to have the best adsorption properties (Ji Hong *et al.*, 2006), and endotoxin adsorption can exceed 60%. It was also found that when there was a hydroxy in the β position of the end group (Table 10.1), there would also be hydrogen bonding, electrostatic and hydrophobic interactions between the adsorbents and

Table 10.1. Computer Analysis of Effect of α, β, ε-hydroxyl on Hydrogen Bonding

	Energy of Endotoxin Fragment (kcal/mol)	Energy of Endotoxin Fragment (kcal/mol)	Energy of Endotoxin Fragment (kcal/mol)	The Length of H Bond (Å)
α-OH	−14.591	19.382	−22.633	No
β-OH	−14.591	1.275	−23.875	1.942
ε-OH	−14.591	7.112	−17.956	No

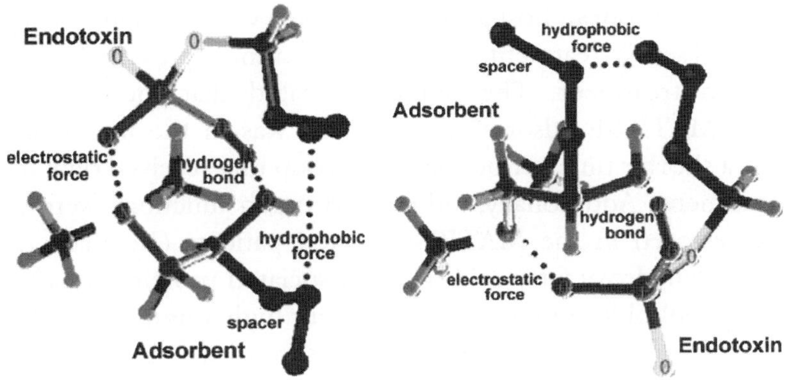

Fig. 10.5. Computer simulations of the interaction between adsorbent and endotoxin with the presence of hydroxyl at the β-position. Copyright permission from Elsevier (2005).

endotoxins, so that the adsorption effect was significantly enhanced (Ji Hong *et al.,* 2006). Computer simulations of the interaction between adsorbent and endotoxin are presented in Fig. 10.5.

3.3. *Immunosorbent*

The mechanism of immunosorbents is to immobilize an antigen or antibody on a carrier to adsorb pathogenic toxins in the blood. Immuno-adsorption treatment is highly selective in removing pathogenic factors, with almost no loss of useful components in plasma or whole blood of patients. This advance treatment effect avoids various adverse effects associated with infusion of plasma products (Deitch, 1992; Bone, 1996).

Since albumin in human serum has high affinity with lipid A and cytokines, it can be used to adsorb inflammatory cytokines and endotoxins in whole blood of patients with systemic infections. With macro-porous polymethylmethacrylate microspheres as a carrier and serum albumin as a ligand (MATISSE), the adsorbent prepared is satisfactory in removing endotoxins and cytokines. Preliminary studies on healthy volunteers and patients with sepsis showed that it is safe and has a good tolerance profile (Staubach *et al.,* 2003). Staubach conducted clinical research on 145 patients with G-bacterial infection sepsis in 31 European hospitals. The patients were randomly assigned to standard sepsis treatments or standard treatment plus the MATISSE adsorption treatment. The results indicated that the group that received MATISSE adsorption treatment was in the intensive care unit for a shorter time period than the group that received only standard treatment. Additionally, kidney and organ functions were more wholly restored in the MATISSE-treated patients (Reinhart *et al.*, 2004). Polyhydroxy fatty acid- (Pha) associated proteins (PhaP) and other hydrophobic compounds have a very high affinity for endotoxins (Jian *et al.*, 2011). Human lipopolysaccharide binding proteins (hLBP; rhLBP is the recombinant human lipopolysaccharide binding protein) in plasma are natural endotoxin receptors. Genes encoding hLBP mixed with PhaP in Pichia gs115 are expressed in producing a hybrid protein. The purified rhLBP-PhaP hybrid protein is fixed to polyhydroxybutyrate (Phb) particles, and the removal rate of endotoxins with rhLBP-Phap-Phb is more than 90%, with weak effects from the ionic strength but very strong effects from the pH changes (Jian *et al.*, 2011). There are also reports describing a strong binding capacity between lactoferrin and LPS (Maria *et al.*, 2012). The Yaoting Yu research team linked immune proteins IgG to spherical agarose gel, and found that its adsorption capacity may be related to the amino group and the carboxyl group of the IgG molecule and indolyl group in the tryptophan residues. In 3,000 U/mL TNF-α plasma, the maximum adsorption amount was 7,084.68 U/mL (Jiao *et al.*, 2005).

3.4. *Adsorbents with Amino Acids and Small Peptide Molecules as Ligands*

Linking of small molecules (amino acids and peptides) as ligands, directly or with a spacer to the matrix material, can improve endotoxin selection and adsorption rates. Yuan's research team linked lysine via hexamethylenediamine as a spacer on agar matrix. In an *in vitro* experiment, where the initial concentration of endotoxin was 10 EU/mL, the maximum adsorption capacity of the adsorbent was 253.1 EU/g (Tao *et al.*, 2004). Then they studied the adsorption properties of different spacers (ammonia, ethylenediamine and hemethylenediamine), and the hexamethylenediamine achieved the best removal rate (73.6%) in rabbits (Guanghui *et al.*, 2005). Other studies using amino acids as ligands were also conducted (Guang Hui *et al.*, 2005) and chitosan as the carrier was explored (Hui Yan *et al.*, 2005). Yaoting Yu's research team modified macroporous resin NK-110 with different amino acids and found that cysteine modified adsorbent was the best for adsorbing TNF-α, and static adsorption after 120 min could reach 7,683.30 μ/mL (Jiao *et al.*, 2001). The team also found that lysine had a good adsorption capacity with spherical cellulose as carrier (Hui *et al.*, 2004).

Wang Xiang's research team prepared an endotoxin adsorbent using chloromethylated resin as a base material and heparin as a spacer to link phenylalanine ligand (Xin Xin *et al.*, 2013). In *in vitro* perfusion experiments with aqueous solution and rabbit plasma, the endotoxin content was reduced to a low level within two hours, and the maximum removal rate was up to 79% in aqueous solutions and 72% in plasma. The influence on the content of other substances in the blood was also small (less than 10%). Chen Huanlin's team studied the use of polyvinylidene difluoride membrane to link various amino acids as ligand. Among them, serine had the highest removal rate (Kun, 2010). The research team subsequently found that the polysulfone affinity membrane, with serine as the ligand, was more efficient in removal of endotoxin (1.656 EU/cm^2) than the polyvinylidene difluoride membrane (Mo, 2012). Zheng Yongke *et al.* treated sepsis porcine using polyvinylidene fluoride — serine adsorption membrane in hemoperfusion, and the results indicated a significant reduction of oxidative stress and inflammation (Yong Ke and Qiu Ping, 2012).The Alteco endotoxin adsorption column immobilized polypeptides on the polyethylene column and was awarded the European certificate in 2005. The safety experiments with patients did not cause any complications (Blomquist *et al.*, 2009). A study in Finland found that five patients using the Alteco endotoxin adsorption column shortened their hospital stay and reduced their demand for cardiac drugs compared to the previous 10 control patients (Ala-Kokko *et al.*, 2009). Clinical trials conducted on patients directly compared the treatment efficiency by Alteco ($n = 32$) and PMX ($n = 14$). Results after 28 days of treatment showed that the PMX group's mortality was 14.2% and the Alteco group's was 31.3%. The treatment with PMX was slightly better than with Alteco (Rei *et al.*, 2013).

3.5. *Adsorbent with Supramolecular Compounds as the Ligand*

Cyclodextrins are a class of semi-natural cyclic oligosaccharides, which mainly include α-, β- and γ- cyclodextrins. Cyclodextrins and their hydrophilic derivatives can be combined with hydrophobic molecules to form inclusion compounds. It has been reported that γ- cyclodextrin polyurethane copolymer (cyclodextrin content of 14–20 mol %) can be used for the endotoxin removal in DNA solutions (Masayo *et al.*, 2013). After DNA solution containing LPS (15 EU/mL) was treated with the adsorbent, the remaining LPS in the solution was only 0.1 EU/mL, and 99% of the DNA was restored to a normal state. The key point of the adsorption is that the diameter of the hydrophobic LPS chains is about 0.7 nm and fits well into the inner diameter of the γ- cyclodextrin molecule cavity. Cyclodextrin is non-toxic and not expensive, but its inclusion of a variety of hydrophobic groups and molecules makes it less selective; it may also combine with other beneficial substances in the human body. It is believed that its selectivity can be improved with modification.

With the knowledge that helical peptide SC4 and βpep-25 with β-sheeting shape have some offsetting effects on endotoxins, Kevin H. Mayo *et al.*, designed and synthesized calixarene-based spiral/β-sheeting topological mimetics (Fig. 10.6), according to molecular size, surface amphiphilic nature, and structural features of endotoxins (Xuemei *et al.*, 2006). Even at very low LPS concentrations (10^{-8}M), it can bind with endotoxins, causing them to lose their activity. Nuclear Magnetic Resonance showed that the electrostatic and hydrophobic interactions occurred between the topological mimetics and lipid A of the LPS. In the series of calixarene topological mimetics that were designed, compounds A and B (Fig. 10.6) are the best. When a lethal dose of 600 μg of LPS was given to the mice, the survival rate

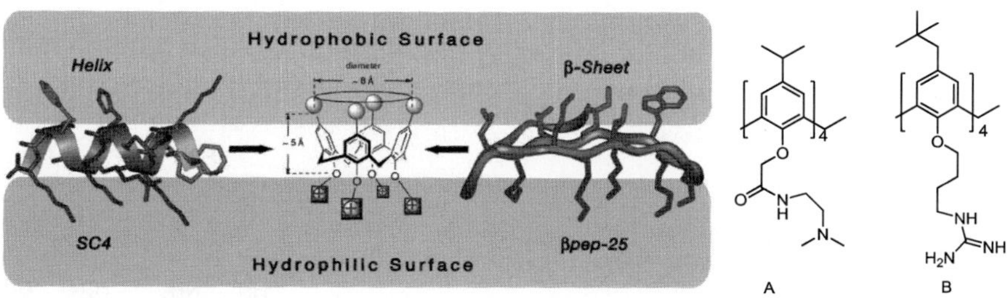

Fig 10.6. (Left) Schematic design, copyright permission from American Chemical Society (2006). (Right) The structure of compounds A and B.

reached 60% (A) and 40% (B) when injected with a 50 mg/kg dose of medication, while all mice in the control group died. The strong binding ability with LPS showed that this drug may be used as an adsorbent ligand to be applied in hemoperfusion for the removal of endotoxins.

3.6. *Adsorbents with Other Compounds as Ligands*

Pentamidine can be used for the treatment of kala-azar (leishmaniasis) and pneumocystis carinii echinococcosis. Pentamidine and lipid A on LPS have been shown to have a strong binding tendency, possibly because the distance between the two amidine groups matches the distance between the two anionic groups in lipid A (David *et al.*, 1994, 1995), and the binding force between pentamidine and LPS is enhanced through multivalent interactions. Laura *et al.* (Laura *et al.*, 2007), taking advantage of this feature of pentamidine, designed and synthesized a novel LPS adsorbent with pentamidine as the ligand and agarose as the carrier (Fig. 10.7). The maximum adsorption capacity of the adsorbent was 17 ± 2 μg (LPS)/mg (adsorbent), greater than that of agarose — polymyxin B adsorbent.

Lixelle adsorbent, an adsorption column formed by linking a hexadecyl chain onto porous cellulose, can be used to adsorb endotoxin. This method originated from the Lixelle system used to adsorb β2- microglobulin for treating dialysis amyloidosis (Tsuchida *et al.*, 2002). Lixelle can reduce the mortality rate in mice, and CTR

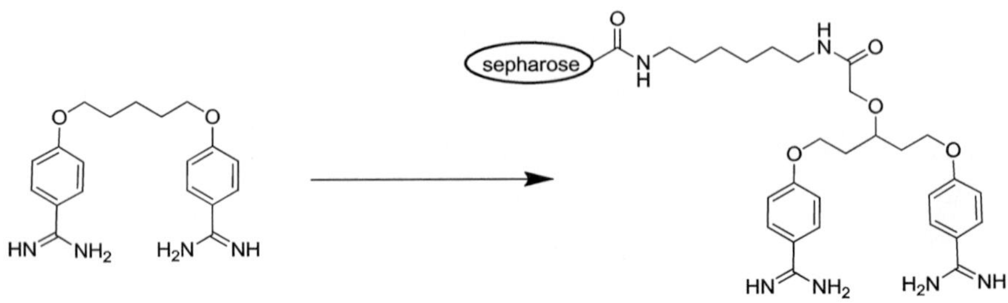

Fig. 10.7.

(a chemical compound that has a hexadecyl chain on porous cellulose) has the same effect (Tsuda and Taniguchi, 2006; Taniguchi *et al.*, 2006). However there are few clinical trials (Tsuchida *et al.*, 2002; Kono *et al.*, 2009), and further validation of results is needed.

It is reported that the allantoin crystals, derivatives of purine, can selectively adsorb endotoxins in aqueous solution with picomolar a nity (Vincent *et al.*, 2013). Atomic force microscopy and chemical inhibition tests have shown that endotoxin adsorption is not dependent on the hydrophobic and ionic interactions between the allantoin and endotoxin, but on interaction forces of hydrogen on the amide of the crystal. Allantoin crystals have a stronger binding capacity, which can reach 3×10^7 EU/g, than other known endotoxin-binding materials because of their small volume (500 nm) and large specific surface area. Although not sensitive to salt and hydrophobic compounds, the crystals are sensitive to amide compounds, such as formamide and urea. The two ureide structures of allantoin are capable of forming a three-dimensional hydrogen network structure. The author inferred that hydrogen bonding of the multivalent amide is the main driving force behind the adsorption of endotoxin by the allantoin crystals in an aqueous solution.

4. Conclusion

Hemoperfusion has been frequently used in clinical therapy. Endotoxins and the corresponding cytokines can be removed from the blood by hemoperfusion in order to ease and cure various diseases. Although the use of hemoperfusion to treat sepsis has a short history, this method has shown strong vitality. At present, the development of polymer adsorbents used in hemoperfusion treatment of sepsis have several characteristics: one is the gradual change of the source materials from mainly natural polymers to modified polymers; the other is the emphasis on improving blood compatibility, adsorption capacity and selectivity. It is believed that, with the development of chemistry, biology and other disciplines, the application of hemoperfusion in endotoxin removal will have pioneering prospects.

References

Abbass AN, Moghazy E. (2011) Factors affecting endotoxin removal from aqueous solutoions by ultrafiltration process. *Journal of Scientific & Industrial Research* **70:** 55–59.

Ala-Kokko T, Koskenkari J, Laurila J. (2009) LPS adsorber in abdominal septic shock [abstract A305]. *Crit Care* **13(suppl 1):** P280.

Angus DC, Wax RS. (2001) Epidemiology of sepsis: an update. *Crit Care Med* **29:** S109–S116.

Beers MH. (2006) *The Merck Manual of Diagnosis and Therapy*, 18th. ed. Merck Research Laboratories, Whitehouse Station, NJ.

Bende S, Bertoc L. (1986) Elimination of endotoxin from the blood by extracorporeal activated charcoal hemoperfusion in experimental canine endotoxin shock. *Cire Shock* **19:** 239–244.

Bengsch S, Boos KS, Nagel D, *et al.* (2005) Extracorporeal plasma treatment for the removal of endotoxin in patients with sepsis: clinical results of a pilot study. *Shock* **23:** 494–500.

Bhattacharjya, S. (2010) De novo designed lipopolysaccharide binding peptides: Structure based development of antiendotoxic and antimicrobial drugs. *Curr Med Chem* **17:** 3080–3093.

Blomquist S, Gustafsson V, Manolopoulos T, *et al.* (2009) Clinical experience with a novel endotoxin adsorbtion device in patients undergoing cardiac surgery. *Perfusion* **24:** 13–17.

Bone RC. (1996) Immunologic dissonance: a continuing evolution in our understanding of the systemic inflammatory response syndrome (SLRS) and the multiple organ dysfunction syndrome (MODS). *Ann Intern Med* **125:** 680–687.

Brandenburg K, Andrae J, Garidel P, *et al.* (2011) Peptide-based treatment of sepsis. *App Microbiol Biotechnol* **90:** 799–808.

Cohen J, Aslam M, Pusey CD, *et al.* (1987) Protection from endotoxemia: a rat model for plasmapheresis and specific adsorption with polymyxin B. *J Infect Dis* **155:** 690–695.

Cruz DN, Antonelli M, Fumagalli R, *et al.* (2009) Early use of polymyxin B hemoperfusion in abdominal septic shock: the EUPHAS randomized controlled trial. *JAMA* **301:** 2445–52.

David SA, Bechtel B, Annaiah C, *et al.* (1994) Interaction of cationic amphiphilic drugs with lipid A: implications for development of endotoxin antagonists. *Biochim Biophys Acta* **1212:** 167–175.

David SA, Mathan VI, Balaram P. (1995) Interactions of linear dicationic molecules with lipid A: structural requisites for optimal binding affinity. *J Endotoxin Res* **2:** 325–336.

Deitch EA. (1992) Multiple organ failure: pathophysiology and potential future therapy. *Ann Surg* **216:** 117–134.

Evangelos J, Bourboulis G. (2013) The failure of biologics in sepsis: Where do we stand? *International Journal of Antimicrobial Agents* **42S:** S45–S47.

Gazzard BG, Portmann B, Weston MJ, *et al.* (1974) Charcoal hemoperfusion in the treatment of fulminant hepatic failure. *Lancet* **1:** 1301–1307.

George CLS, Jin H, Wohlford-Lenane CL, *et al.* (2001) Endotoxin responsiveness and subchronic grain dustinduced airway disease. *Am J Physiol-Lung C* **280(2):** 203–213.

Guang Hui H, Hui Yan W, Tao L, *et al.* (2005) Preparation of adsorbents for endotoxin removal in hemoperfusion with amino acid as the ligand and investigations on their properties. *Chemical Journal of Chinese Universities* **26:** 1277–280.

Guanghui H, Tao L, Huiyan W, *et al.* (2005) Preparation of adsorbents for the removal of endotoxin. *Artif Cell Blood Sub* **33:** 227–237.

Hidehiko K, Takahiro M, Yuichiro S. (2008) Criteria for direct hemoperfusion with an immobilized polymyxin B fiber column based on oxygen metabolism. *Therapeutic Apheresis and Dialysis* **12(4):** 292–297.

Hirohike A, Masashi K, Tohru T, *et al.* (1994) Treatment of sepsis by extracorporeal elimination of endotoxin using polymyxin B-inimobilized fiber. *Am J Surg* **167:** 412–417.

Hou KC, Zaniewski R. (1990) Depyrogenation by endotoxin removal with positively charged depth folter cartridge. *J Parenter Sci Technol* **44:** 204–209.

Hui F, Jiao W, Yao Ting Y. (2004) Studies on adsorption performance of affinity adsorbent for bacterial endotoxins, *Chemical Journal of Chinese Universities* **25:** 1057–1059.

Hui Yan W, Guang Hui H, Mei Y, *et al.* (2005) Adsorbents for endotoxin based on chitosan matrix. *Chemical Journal of Chinese Universities* **26:** 680–682.

Jean-Louis V, Pierre-Francois L, Jonathan C. (2005) A pilot-controlled study of a polymyxin b-immobilized hemoperfusion cartridge in patients with severe sepsis secondary to intra-abdominal infection. *Shock* **23:** 400–405.

Ji Hong L, Mei Y, Hui Yan W, *et al.* (2006) Studies on adsorption mechanisms of adsorbents for endotoxin. *Chemical Journal of Chinese Universities* **27:** 1066–1069.

Jian L, Guanguan S, Mingliang Y, *et al.* (2011) Endotoxin removing method based on lipopolysaccharide binding protein and polyhydroxyalkanoate binding protein PhaP. *Biomacromolecules* **12:** 602–608.

Jiao W, Hui F, De-Ling K, *et al.* (2004) Studies on adsorption mechanism of TNFα with IgG ligand. *Chemical Research in Chinese Universities* **25(7):** 1277–1281.

Jiao W, Yao Ting Y, De Ling K. (2001) Adsorption of TNF-α onto the amino acid-modified NK-110 resin. *Chinese Journal of Biotechnology* **17:** 432–435.

Jiao W, YaoTing Y. (2001) Studies on adsorption of TNF α with macroporous resin. *Chemical Journal of Chinese Universities* **22:** 1813–1815.

Kodama M, Tani T, Hanasawa K. (1997) Treatment of sepsis by plasma endotoxin removal: hemoperfusion using a polymyxin-B immobilized column. *J Endotoxin Res* **4:** 293–300.

Kono K, Toda S, Hora K, *et al.* (2009) Direct hemoperfusion with a beta2-microglobulin-selective adsorbent column eliminates inflammatory cytokines and improves pulmonary oxygenation. *Ther Apher Dial* **13:** 27–33.

Kun X. (2010) A study on polyvinylidene fluoride grafted to amino affinity membrane and endotoxin removal. *Master Thesis of Zhejiang University.*

Laura AM, Mark SH, Roberta S, *et al.* (2007) Synthesis and application of a novel ligand for affinity chromatography based removal of endotoxin from antibodies. *Bioconjugate Chem* **8:** 559–566.

Maria ED, Mireyade G, Jesús SL, *et al.* (2012) Lactoferrin-lipopolysaccharide (LPS) binding as key to antibacterial and antiendotoxic effects. *International Immunopharmacology* **12:** 1–9.

Martin GS, Mannino DM, Eaton S, *et al.* (2003) The epidemiology of sepsis in the United States from 1979 through 2000. *N Engl J Med* **348:** 1546–1554.

Masayo S, Koji U, Kasane K, *et al.* (2013) c-Cyclodextrin–polyurethane copolymer adsorbent for selective removal of endotoxin from DNA solution. *Anal Biochem* **443:** 41–45.

Mitzner S, Schneidewind J, Falhenhagen D, *et al.* (1993) Extracorporeal endotoxin removal by immobilized polyethyleneimine. *Artif Organs* **17:** 775–781.

Mo Z. (2012) Polysulfone affinity membranes with serine as ligand and its endotoxin removal performance, *Master Thesis of Zhejiang University.*

Nagaki M, Hughes RD, *et al.* (1991) Removal of endotoxin and cytokines by adsorbents and plasma protein binding. *Int J Artif Orans* **14:** 43–50.

Nakayama M, Niki Y, Kawasaki T, *et al.* (2012) Enhanced susceptibility to lipopolysaccharide-induced arthritis and endotoxin shock in interleukin-32α transgenic mice through induction of tumor necrosis factor α. *Arthritis Res Ther* **14:** Article number R120.

Odnopozov VA, Laterre PF, Traystaman RJ, *et al.* (1996) Effect of hemocarboperfusion on organ blood flow and survival in porcine endotoxin shock. *Crit Care Med* **24:** 2021–2026.

Opal SM. (2007) The host response to endotoxin, antilipopolysaccharide strategies, and the management of severe sepsis. *Int J Med Microbiol* **297:** 365–77.

Pfeiffer R. (1892) Untersuchungen über das Choleragift. *Zeitschrift für Hygiene und Infektionskrankheiten* **11:** 393–411.

Raetz CR. (1990) Biochemistry of endotoxin. *Annu Rev Biochem* **59:** 129–170.

Raetz CR, Whitfield C. (2002) Lipopolysaccharide endotoxin. *Annu Rev Biochem* **71:** 635–700.

Ranieri VM, Thompson T, Barie PS, *et al.* (2012) Drotrecogin alfa (activated) in adults with septic shock. *N Engl J Med* **366:** 2055–64.

Rei S, Aleksandrova I, Kiselev V, *et al.* (2013) Selective adsorption of lipopolysaccharide in the complex treatment of patients with severe sepsis. *Critical Care* **17(Suppl 2):** P67.

Reinhart K, Meier-Hellmann A, Beale R, *et al.* (2004) Open randomized phase II trial of an extracorporeal endotoxin adsorber in suspected gram-negative sepsis. *Crit Care Med* **32:** 1662–68.

Schromm AB, Brandenburg K, Loppnow H, *et al.* (1998) The charge of endotoxin molecules influences their conformation and IL-6-inducing capacity. *J Immunol* **161:** 5464–5471.

Shigui L, Tobias R, Mcclure S, *et al.* (1997) Removal of endotoxin from recombinant protein preparations. *Clin Biochem* **30(6):** 455–463.

Shimizu T, Endo Y, Tsuchihashi H, *et al.* (2006) Endotoxin apheresis for sepsis. *Transfusion & Apheresis Sci* **35:** 271–282.

Staubach KH, Boehme M, Zimmermann M, *et al.* (2003) A new endotoxin adsorption device in Gram-negative sepsis: use of immobilized albumin with the MATISSE adsorber. *Transfus Apheresis Sci* **29:** 93.

Taniguchi T, Hirai F, Takemoto Y, *et al.* (2006) A novel adsorbent of circulating bacterial toxins and cytokines: The effect of direct hemoperfusion with CTR column for the treatment of experimental endotoxemia. *Crit Care Med* **34:** 800–06.

Tao L, Guang Hui H, Mei Y, *et al.* (2004) Hemperfusion adsorbents with lysine as ligand for removing of endotoxin, *Chemical Journal of Chinese Universities* **25:** 2284–2286.

Tetta C, Gianotti L, Cavaillon JM, *et al.* (2000) Coupled plasma filtration-adsorption in a rabbit model of endotoxic shock. *Crit Care Med* **28:** 1526–1533.

Tsuchida K, Takemoto Y, Sugimura K, *et al.* (2002) Adsorption of endotoxin by beta2-microglobulin adsorbent column (Lixelle): the new approach for endotoxinemia. *Ther Apher* **6:** 116–18.

Tsuchida K, Takemoto Y, Sugimura K, *et al.* (2002) Direct hemoperfusion by using Lixelle column for the treatment of systemic inflammatory response syndrome. *Int J Mol Med* **10:** 485–88.

Tsuda K, Taniguchi T. (2006) Effects of extracorporeal treatment with Lixelle on the mortality and inflammatory responses to endotoxin induced shock in rats. *Ther Apher Dial* **10:** 49–53.

Vincent V, Tim-Jang C, Rui-Jing C, *et al.* (2013) Amide-mediated hydrogen bonding at organic crystal/water interfaces enables selective endotoxin binding with picomolar affinity. *Appl Mater Interfaces* **5:** 4472–4478.

Xin Xin J, Xiang W, Jin Chun M, *et al.* (2013) Research of heparin-conjugating phenylalanine blood purification material to remove endotoxin. *Chinese Journal of Biomedical Engineering* **32:** 348–355.

Xuemei C, Ruud PMD, Irina N, *et al.* (2006) Topomimetics of amphipathic β-sheet and helix-forming bactericidal peptides neutralize lipopolysaccharide endotoxins. *J Med Chem* **49:** 7754–7765.

Yamakawa K, Fujimi S, Mohri T, *et al.* (2011) Treatment effects of recombinant human soluble thrombomodulin in patients with severe sepsis: a historical control study. *Crit Care* **15:** R123.

Yong C, Bin W, Guo Ye J. (2012) A comparative study of different anticoagulants effects on hemoperfusion in rabbit sepsis model. *Anhui Medical and Pharmaceutical Journal* **16:** 1232–1236.

Yong Ke Z; Qiu Ping X. (2012) Effect of hemoperfusion with absorbents using polyvinylidene difluoride-serine on intestinal injuries in porcine model with endotoxemia. *Chin J Crit Care Med (Electronic Edition)* **5:** 371–378.

Yuan Z, Yu M, Li J, *et al.* (2005) Endotoxin adsorbent using dimethylamine ligands. *Biomaterials* **26:** 2741–2747.

Chapter 11

Direct Hemoperfusion with β_2-microglobulin Adsorption Column for Patients with End-stage Kidney Disease

Suguru Yamamoto* and Junichro J. Kazama†

1. Introduction

β_2-microglobulin (β_2-m) is one of the middle-molecule uremic toxins and increases in chronic kidney patients undergoing dialysis treatment. An accumulation of β_2-m in dialysis patients induces onset of dialysis-related amyloidosis (DRA) as well as increase of all-cause mortality. Thus it is important to keep lower serum β_2-m level with sufficient removal of it with blood purification treatment. Direct hemoperfusion with use of β_2-m adsorption column, Lixelle, is effective treatment for removing β_2-m and improving osteo-articular symptoms related to DRA.

*Division of Clinical Nephrology and Rheumatology, Niigata University Graduate School of Medical and Dental Sciences, Niigata 951-8510, Japan.
†Division of Blood Purification Therapy, Niigata University Medical and Dental Hospital, Japan, Niigata 951-8520, Japan.

2. β_2-microglobulin in Patients with Chronic Kidney Disease

β_2-m is a polypeptide of 99 residues that has a molecular weight of 11.8 kDa. It forms the beta chain of the human leukocyte antigen (HLA) class I molecule, which are present on all nucleated cells. It has a well-known β-sandwich structure that involves a 7-strand β-pleated structure stabilized with a single disulphide bond (Cys25-Cys80) (Fig. 11.1). β_2-m secreted from cells is normally eliminated by the kidney via glomerular filtration and subsequent tubular catabolism with megalin (Orlando *et al.*, 1998).

Acute and chronic inflammatory diseases, such as multiple myeloma and lymphoma as well as chronic kidney disease (CKD), increase the production of β_2-m and show high serum level. On the other hand, proximal tubular injury increases β_2-m excretion in urine, and severe kidney damage induces the retention of β_2-m in serum due to impaired excretion from the kidney. Thus, serum level of β_2-m increases with impaired catabolism in proximal tubules as well as increased production in CKD patients, especially with no residual kidney function. The serum levels of β_2-m in patients undergoing hemodialysis are significantly higher compared to those in normal

Fig. 11.1. Structure of β_2-microglobulin (PDB 1JNJ) (Verdone *et al.*, 2002).

subjects (25–45 vs. 1–2 mg/L) (Gejyo *et al.*, 1986; Cheung *et al.*, 2006; Ikee *et al.*, 2008; Okuno *et al.*, 2009; Uji *et al.*, 2009). The continuous retention of β_2-m will lead to the development of DRA as well as higher mortality in patients undergoing long term-dialysis treatment.

3. Dialysis-related Amyloidosis

3.1. *Dialysis-related Amyloidosis as One of Systemic Amyloidosis*

Amyloidosis is defined as an insoluble protein fibril that is deposited mainly in the extracellular spaces of organs and tissues as a result of a sequence of changes in protein folding. Precursor proteins change their conformation to form amyloid fibrils, then the deposited amyloid induces organ damage with disease-specific conditions. In the nomenclature, DRA is defined as β_2-m-related (Aβ_2M) amyloid whose precursor protein is β_2-m accumulated in CKD patients undergoing dialysis treatment (Sipe *et al.*, 2014). Aβ_2M amyloid deposits in osteoarticular tissues mainly and leads to development of various disorders such as carpal tunnel syndrome (CTS), amyloid arthropathy, destructive spondyloarthropathy (DSA), and cystic bone disease, while it also deposits in organs, such as the intestine and heart with further disease progression. It is not clear the relation between serum β_2-m level and development of DRA, however, it is believed that to keep a high serum level of β_2-m during long-term dialysis treatment will induce development of DRA (Dember and Jaber, 2006). Recent progress of dialysis therapy has improved survival of patients with end-stage kidney disease, and DRA is still one of the serious complications for patients undergoing dialysis.

3.2 *Clinical Characteristics of Dialysis-related Amyloidosis*

The clinical risk factors of DRA are (i) long-term dialysis treatment, (ii) initiation dialysis treatment in old age, (iii) hemodialysis treatment with low purity dialysate, (iv) use of low-flux dialysis membrane (Davison, 1995). Recently it has been the trend to use high-flux dialysis membrane and high-purity dialysate in hemodialysis treatment, however, progress of dialysis treatment as well as treatment for other diseases will increase the number of patients who have dialysis treatment initiated at an older age and survive with long-term dialysis treatment, clinical risk factors for development of DRA.

Deposited $A\beta_2M$ amyloid induces various osteoarticular disorders, such as CTS, DSA, and cystic bone lesions (Kazama *et al.*, 2006). CTS is induced by the deposition of $A\beta_2M$ amyloid around synovium in carpal tunnel and the compression of the median nerve, and shows sensory disturbance and muscle weakness in hands. DSA is the defect of bone in the spine after deposition of $A\beta_2M$ amyloid. It is radiographically characterized by severe narrowing of the intervertebral disk space and erosions as well as cysts of the adjacent vertebral plates. DSA is mostly detected in the highly mobile areas, such as C5–C7 and L3–L5, and it shows sensory disturbance and muscle weakness in upper or/and lower limbs (Maruyama *et al.*, 1992). Cystic lesions occur in bones, such as carpal and femur, where $A\beta_2M$ amyloid deposits. Cystic bone lesions increase the risk of bone fracture. Thus patients with long-term dialysis treatment show various DRA-induced osteoarticular disorders. We researched over a four-year period 359 patients undergoing dialysis treatment who were admitted in our center for the treatment of their dialysis-related complications (Yamamoto *et al.*, 2008). DSA was a major cause of hospital admission in the patients undergoing dialysis therapy for 20 years or more, and the rate increased along with the increasing duration of dialysis therapy. The incidence rate of histories of surgeries for osteoarticular disorders related to DRA was 25.0%, 66.0%, and 77.8% in 20–24 years, 25–29 years, and 30 years or more after the initiation of dialysis therapy, respectively (Fig. 11.2). In the patients undergoing dialysis therapy for 30 years or more, the incidence rate of histories of surgeries for CTS, DSA, and joint arthropathy was 72.2%, 50.0%, and 22.2%, respectively, indicating they had analogous histories of surgeries for various osteoarticular disorders related with DRA.

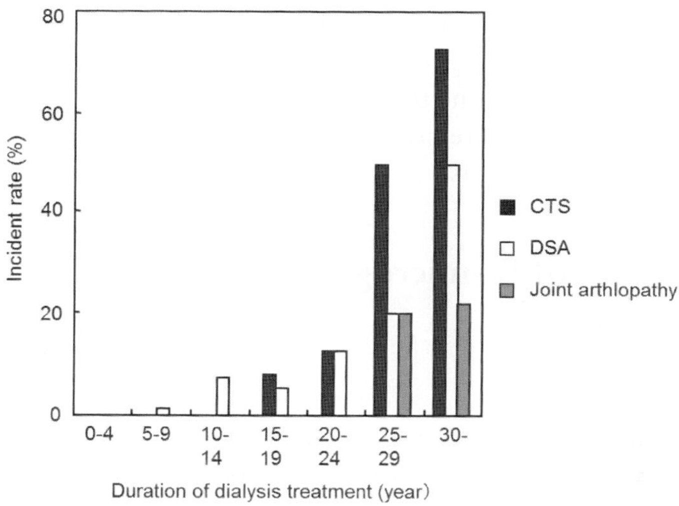

Fig. 11.2. Clinical manifestations of dialysis-related amyloidosis (DRA) in patients with chronic dialysis. The histories of DRA-rated osteoarticular disease, such as carpal tunnel syndrome, joint arthropathy and destructive spondyloarthropathy, increase with the duration of dialysis treatment, especially more than 30 years.

4. Motility

In the randomized Hemodialysis (HEMO) Study, the relationship between serum β_2-m levels or dialyzer β_2-m clearance and mortality over a period of 2.84 years was analyzed (Cheung *et al.*, 2006). In time-dependent Cox regression models, predialysis serum β_2-m levels were associated with all-cause mortality (Cheung *et al.*, 2006). Another study shows that higher serum β_2-m levels showed higher all-cause mortality and non-cardiovascular mortality during 3.33 years (Okuno *et al.*, 2009). In this study, there was no correlation between serum β_2-m and cardiovascular mortality. However, a cross-sectional study revealed that serum β_2-m is related to heart valve calcification, which is associated with carotid intima media thickness in dialysis patients (Ikee *et al.*, 2008). Furthermore, serum β_2-m levels were positively correlated with several cardiovascular risk factors, such as highly-sensitive C-reactive protein, troponin-T, myeloperoxidase, N-terminal pro-B-type natriuretic peptide and inversely correlated with prealbumin and ankle-brachial index (Kuragano *et al.*, 2010). Those reports suggest that high level of β_2-m is associated with mortality, partially due to cardiovascular disorder in patients with dialysis treatment. Thus removal of sufficient β_2-m may induce better survival in dialysis patients.

5. Removal of β_2-microgloblin with Blood Purification

5.1. *Overview*

To prevent DRA and improve mortality, removal of β_2-m by blood purification treatment is a main strategy in patients with chronic dialysis treatment (Kazama *et al.*, 2001). Practically, hemodialysis using high-flux dialyzer, hemodiafiltration, hemofiltration, and use of β_2-m adsorption column increase removal of β_2-m as compared to hemodialysis using low-flux dialyzer. Another strategy to reduce production of β_2-m is hemodialysis using biocompatible high-flux dialysis membrane and high-purity dialysate. In those methods, direct hemoperfusion using β_2-m adsorption column is most effective treatment for removal of β_2-m.

5.2. *β_2-m Adsorption Column*

Lixelle was developed for direct β_2-m adsorption from circulating blood, and has been used mainly for hemodialysis patients with progressive DRA. Lixelle contains porous cellulose beads at a mean diameter 460 mm in polycarbonate column (S-35) or a polypropylene column (S-25 and S-15) filled with a sodium citrate buffer. Column volumes are 350 ml, 250 ml and 150 ml in S-35, S-25 and S-15, respectively. The porous cellulose beads were designed to selectively adsorb β_2-m with appropriate pore size and hydrophobic interaction (Fig. 11.3). This adsorption column system is designed for direct hemoperfusion, and used in combination with a dialyzer, which is downstream of Lixelle via an exclusive connection tube (Fig. 11.4). When the blood from ESKD patients passed through the Lixelle S-35

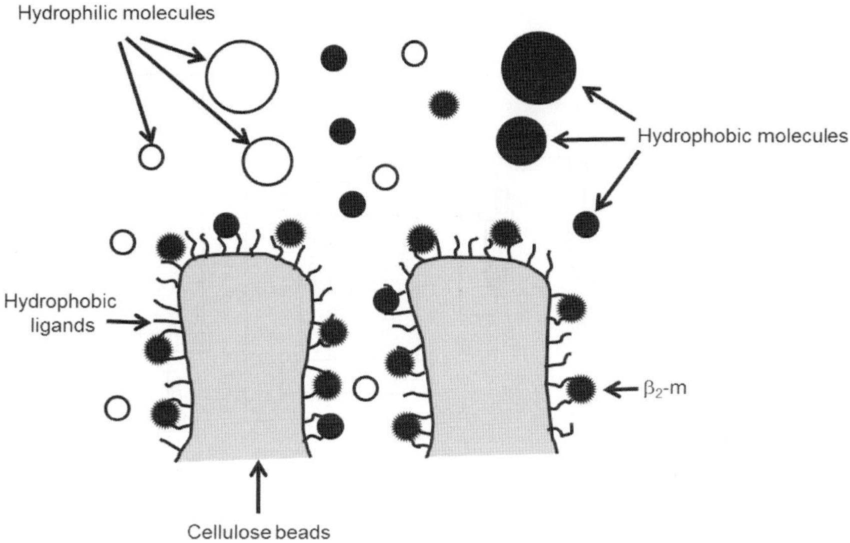

Fig. 11.3. Mechanism of Lixelle to adsorb β_2-m. Lixelle contains porous cellulose beads at a mean diameter 460 mm, and those are designed to selectively adsorb β_2-m with appropriate pore size and hydrophobic interaction.

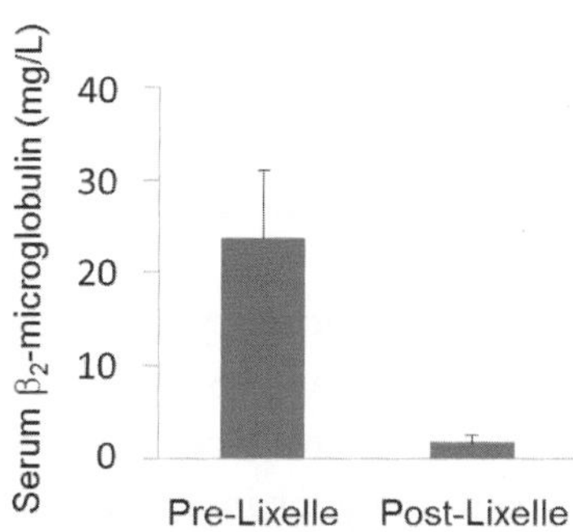

Fig. 11.4. Schematic representation of blood flow through the dialyzer and β_2-microglobulin adsorption column, Lixelle.

column, serum level of β_2-m was decreased from 26.3 ± 10.3 mg/L to 1.9 ± 1.2 mg/L (Fig. 11.5).

There are several clinical trials to show the effect of Lixelle on the reduction of β_2-m in patients undergoing hemodialysis (Gejyo *et al.*,

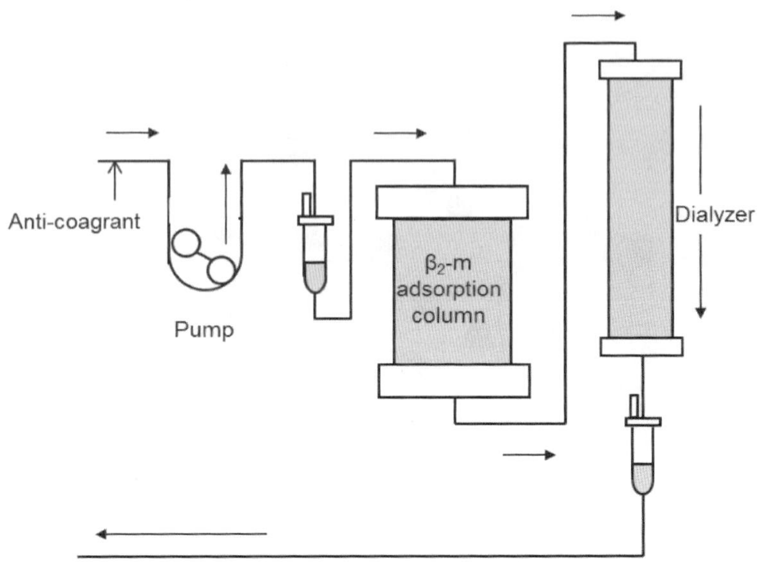

Fig. 11.5. Reduction of serum level of β_2-microglobulin by passing through Lixelle S-35 column.

1993, 1995, 2004; Nakazawa *et al.*, 1993; Homma *et al.*, 1995; Abe *et al.*, 2003). (Table 11.1) Lixelle column induces 60.0–78.9% of reduction of serum β_2-m, removes 157–300 mg of β_2-m, and the serum level of β_2-m after treatment reaches 6.8–13.5 mg/L (Gejyo *et al.*, 1993, 2004; Homma *et al.*, 1995). Some clinical studies showed reduction in the size of bone cysts when they are checked by X-ray (Homma *et al.*, 1995; Kuragano *et al.*, 2011). According to a prospective multicenter study, Lixelle column (S-35) that was placed in series with a polysulfone dialyzer increased serum β_2-m reduction in patients undergoing hemodialysis as compare to hemodialysis treatment without Lixelle (reduction rate: 74.1 ± 6.1 vs. 60.1 ± 6.3%).[20] In this study, six of 22 patients developed hypotension due to increase of priming volume in dialysis treatment. In this point, Lixelle S-15, minimum type of β_2-m adsorption column, decreased the serum level of β_2-m from 29.3 ± 9.6 mg/L to 24.7 ± 5.1 mg/L during one year of treatment without development of hypotension (Yamamoto *et al.*, 2011).

Table 11.1.

No. of References	No. of Patients	Type of Lixelle	Duration of Treatment	β_2-m at Entry	β_2-m at End	Improved Symptom
13	3	S-35	25–62 weeks	At entry: 28.4 ± 2.7–31.7 ± 5.1 before treatment	At end: 7.9 ± 0.8–10.4 ± 1.8 before treatment	Joint pain, joint mobility, frequency of nocturnal awakening, and sevierity of bone cyst
16	7	S-35	> 6 months	At entry: 29.5 ± 5.2 before treatment	At end: 24.4 ± 3.5 before treatment	Joint pain, Joint mobility, and frequency of nocturnal awakening
17	17	S-35	12 months	At entry: 34.5 ± 8.4 before treatment	At end: 28.8 ± 7.3 before treatment	Pinch strength, median motor terminal latency, VAS score, frequency of nocturnal awakening, and ADL score
18	22	S-35	24 months	Control at end: 11.9 ± 3.9 after treatment	Lixelle at end: 6.9 ± 2.1 after treatment	ADL score, pain score, and stiffness score
19	32					Bone cysts, ADL score, pain score, and frequency of nocturnal awakening
20	17	S-15	12 months	At entry: 29.3 ± 9.6 before treatment	At end: 24.7 ± 5.1 before treatment	VAS score, Pinch strength, and ADL score

Those clinical studies also show the improvements of DRA-related symptoms, such as joint pain and activity of daily living (Nakazawa *et al.*, 1993; Gejyo *et al.*, 1995, 2004, 2013; Homma *et al.*, 1995; Abe *et al.*, 2003; Yamamoto *et al.*, 2011). Gejyo *et al.* surveyed the clinical effect of Lixelle in ESKD patients with DRA. Of 345 patients with DRA, 56.2%, 35.1%, and 8.7%, patients felt that the treatment with Lixelle for 3.5 ± 2.7 years had improved, not changed, and worsened, and this result meant that Lixelle had inhibited the progression of DRA in 91.3% of patients (Gejyo *et al.*, 2013).

The Lixelle column has the property of adsorbing not only β_2-m but also several cytokines whose molecular size is from 4,000 to 20,000 Da (Kutsuki, 2005). Since use of Lixelle column induces the improvement of DRA-related clinical symptoms, the reasons for the effect may be reduction of some inflammatory cytokines as well as β_2-m.

The main adverse effect is fall in blood pressure during the treatment because of increase of blood volume in the extracorporeal circulation, and we should choose the size of Lixelle column bearing in mind the body size and blood pressure of patients.

6. Conclusion

A β_2-m adsorption column, Lixelle, has the property of reducing circulating level of β_2-m, which is related to the development of DRA and increased mortality in ESKD patients. This treatment has a beneficial effect on the activity of daily living, quality of life, and promises better survival for long-term hemodialysis patients.

References

Abe T, Uchita K, Orita H, *et al.* (2003) Effect of beta(2)-microglobulin adsorption column on dialysis-related amyloidosis. *Kidney Int* **64:** 1522–1528.

Cheung AK, Rocco MV, Yan G, *et al.* (2006) Serum beta-2 microglobulin levels predict mortality in dialysis patients: Results of the HEMO study. *J Am Soc Nephrol* **17:** 546–555.

Davison AM. (1995) beta 2-microglobulin and amyloidosis: Who is at risk? *Nephrol Dial Transplant* **10 Suppl 10:** 48–51.

Dember LM, Jaber BL. (2006) Dialysis-related amyloidosis: Late finding or hidden epidemic? *Seminars in dialysis* **19:** 105–109.

Gejyo F, Amano I, Ando T, *et al.* (2013) Survey of the effects of a column for adsorption of beta2-microglobulin in patients with dialysis-related amyloidosis in Japan. *Ther Apher Dial* **17:** 40–47.

Gejyo F, Homma N, Hasegawa S, *et al.* (1993) A new therapeutic approach to dialysis amyloidosis: Intensive removal of beta 2-microglobulin with adsorbent column. *Artif Organs* **17:** 240–243.

Gejyo F, Homma N, Suzuki Y, *et al.* (1986) Serum levels of beta 2-microglobulin as a new form of amyloid protein in patients undergoing long-term hemodialysis. *N Engl J Med* **314:** 585–586.

Gejyo F, Kawaguchi Y, Hara S, *et al.* (2004) Arresting dialysis-related amyloidosis: A prospective multicenter controlled trial of direct hemoperfusion with a beta2-microglobulin adsorption column. *Artif Organs* **28:** 371–380.

Gejyo F, Teramura T, Ei I, *et al.* (1995) Long-term clinical evaluation of an adsorbent column (BM-01) of direct hemoperfusion type for beta 2-microglobulin on the treatment of dialysis-related amyloidosis. *Artif Organs* **19:** 1222–1226.

Homma N, Gejyo F, Hasegawa S, *et al.* (1995) Effects of a new adsorbent column for removing beta-2-microglobulin from circulating blood of dialysis patients. *Contrib Nephrol* **112:** 164–171.

Ikee R, Honda K, Oka M, *et al.* (2008) Association of heart valve calcification with malnutrition-inflammation complex syndrome, beta-microglobulin, and carotid intima media thickness in patients on hemodialysis. *Ther Apher Dial* **12:** 464–468.

Kazama JJ, Maruyama H, Gejyo F. (2001) Reduction of circulating beta2-microglobulin level for the treatment of dialysis-related amyloidosis. *Nephrol Dial Transplant* **16 Suppl 4:** 31–35.

Kazama JJ, Yamamoto S, Takahashi N, *et al.* (2006) A beta-2 M amyloidosis and related bone diseases. *J Bone Miner Metab* **24:** 182–184.

Kuragano T, Kida A, Furuta M, *et al.* (2010) The impact of beta2-microglobulin clearance on the risk factors of cardiovascular disease in hemodialysis patients. *ASAIO J* **56:** 326–332.

Kuragano T, Inoue T, Yoh K, *et al.* (2011) Effectiveness of beta(2)-microglobulin adsorption column in treating dialysis-related amyloidosis: A multicenter study. *Blood purification* **32:** 317–322.

Kutsuki H. (2005) beta(2)-Microglobulin-selective direct hemoperfusion column for the treatment of dialysis-related amyloidosis. *Biochim Biophys Acta* **1753:** 141–145.

Maruyama H, Gejyo F, Arakawa M. (1992) Clinical studies of destructive spondyloarthropathy in long-term hemodialysis patients. *Nephron* **61:** 37–44.

Nakazawa R, Azuma N, Suzuki M, *et al.* (1993) A new treatment for dialysis-related amyloidosis with beta 2-microglobulin adsorbent column. *Int J Artif Organs* **16:** 823–829.

Okuno S, Ishimura E, Kohno K, *et al.* (2009) Serum beta2-microglobulin level is a significant predictor of mortality in maintenance haemodialysis patients. *Nephrol Dial Transplant* **24:** 571–577.

Orlando RA, Rader K, Authier F, *et al.* (1998) Megalin is an endocytic receptor for insulin. *J Am Soc Nephrol* **9:** 1759–1766.

Sipe JD, Benson MD, Buxbaum JN, *et al.* (2014) Nomenclature 2014: Amyloid fibril proteins and clinical classification of the amyloidosis. *Amyloid: the international journal of experimental and clinical investigation: the official journal of the International Society of Amyloidosis* **21:** 221–224.

Uji Y, Motomiya Y, Ando Y. (2009) A circulating beta 2-microglobulin intermediate in hemodialysis patients. *Nephron Clin Pract* **111:** c173–181.

Verdone G, Corazza A, Viglino P, *et al.* (2002) The solution structure of human beta2-microglobulin reveals the prodromes of its amyloid transition. *Protein Sci* **11:** 487–499.

Yamamoto S, Kazama JJ, Maruyama H, *et al.* (2008) Patients undergoing dialysis therapy for 30 years or more survive with serious osteoarticular disorders. *Clin Nephrol* **70:** 496–502.

Yamamoto Y, Hirawa N, Yamaguchi S, *et al.* (2011) Long-term efficacy and safety of the small-sized beta2-microglobulin adsorption column for dialysis-related amyloidosis. *Ther Apher Dial* **15:** 466–474.

Chapter 12

Endotoxin Adsorption: Direct Hemoperfusion with a Polymyxin B-Immobilized Fiber Column (Toraymyxin)

Tomoharu Shimizu, Masaji Tani, Yoshihiro Endo and Tohru Tani

1. Introduction

The prognosis of patients with endotoxemia and septic shock remains poor, despite the recent progress in surgical intervention and critical care. Endotoxins from gram-negative bacteria stimulate macrophages and other immune cells to produce various inflammatory mediators. These endotoxins play a major role in triggering toxic symptoms in septic shock (Mitaka and Tomita, 2011).

Polymyxin B is a well-known polycationic antibiotic that binds to and inactivates endotoxins. It prevents endotoxin-associated toxicity in humans; however, systemic administration is limited owing to its nephrotoxicity and neurotoxicity. Therefore, polymyxin B may be a strong candidate ligand for extracorporeal selective adsorption of circulating endotoxin in the blood. We developed the polymyxin B-immobilized fiber blood purification column (Toraymyxin®, Toray Medical Co., Ltd, Chiba, Japan) as a medical device to be used in conjunction with direct hemoperfusion to remove circulating endotoxins from the blood.

---

*Department of Surgery, Shiga University of Medical Science, Shiga, Japan.

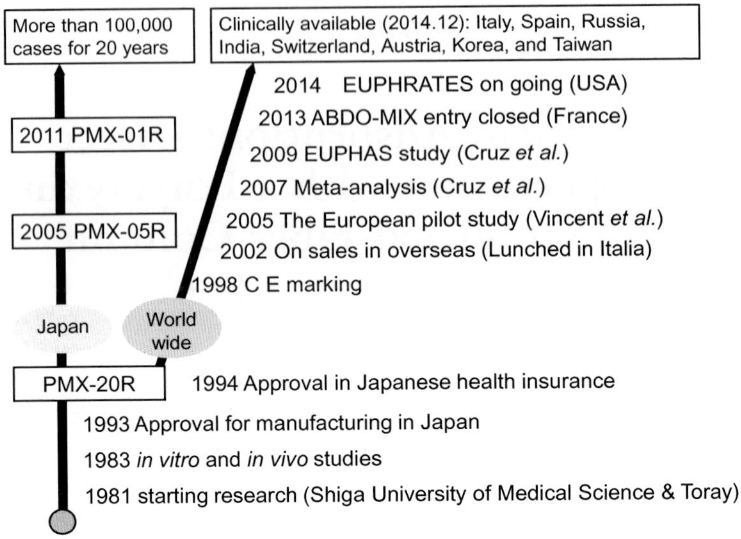

Fig. 12.1. History of the development of Toraymyxin. EUPHAS: Early Use of Polymyxin B Hemoperfusion in Abdominal Sepsis. ABDO-MIX trial: Effects of Hemoperfusion with a Polymyxin B Membrane in Peritonitis with Septic Shock. EUPHRATES trial: Evaluating the Use of Polymyxin B Hemoperfusion in a Randomized Controlled Trial of Adults Treated for Endotoxemia and Septic Shock.

The history of Toraymyxin is shown in Fig. 12.1 Polymyxin B-immobilized fibers were initially investigated in 1981 at the Department of Surgery, Shiga University of Medical Science as a collaboration with Toray Medical Co., Ltd. In 1982, preliminary results from this research were published; the results demonstrated that polymyxin B-immobilized fibers reduced endotoxin activity in saline solution, as measured using a semi-quantitative limulus amebocyte lysate (LAL) assay. *In vitro* and *in vivo* experimental studies have been ongoing since 1983.

A phase I clinical study was initiated in 1989. The first clinical report described 16 patients with septic multiple organ failure treated with direct hemoperfusion with Toraymyxin, and was published in 1994 (Aoki *et al.*, 1994).

In 1994, Toraymyxin was adopted by the Japanese National Health Insurance system to treat endotoxemia and septic shock. Since then,

Toraymyxin has been safely used in more than 100,000 cases in emergency and intensive care units in Japan. Three different columns are currently available for clinical use in Japan: PMX-20R for use in adults, PMX-05R for use in elderly patients or children, and PMX-01R for use in infants.

In 1998, Toraymyxin received CE approval in Europe. In 2005, the first randomized controlled trial (RCT) in Europe was published; the RCT showed that treatment with Toraymyxin is safe and improves cardiac and renal dysfunction due to sepsis or septic shock (Vincent *et al.*, 2005). In 2007, the findings of a meta-analysis of published literature demonstrated that Toraymyxin treatment decreases plasma endotoxin levels, thereby having favorable effects on mean arterial pressure, catecholamine use, partial pressure arterial oxygen and fraction of inspired oxygen ratio (PaO_2/FIO_2 ratio), and mortality (Cruz *et al.*, 2007). In 2009, the results of the second RCT in Italy, the Early Use of Polymyxin B Hemoperfusion in Abdominal Sepsis (EUPHAS) study, were published; the results showed that treatment significantly reduces sepsis-associated mortality.[5] Currently, two large RCTs are underway: the Evaluating the Use of Polymyxin B Hemoperfusion in a Randomized Controlled Trial of Adults Treated for Endotoxemia and Septic Shock (EUPHRATES) trial in the USA and the Effects of Hemoperfusion with a Polymyxin B Membrane in Peritonitis with Septic Shock (ABDO-MIX) trial in France. Toraymyxin is currently available for use in Italy, Spain, Russia, India, Switzerland, Austria, Korea and Taiwan (Fig. 12.1).

In this chapter, we review the development, clinical use, and efficacy of Toraymyxin, and comment on the current position of Toraymyxin for the treatment of severe sepsis and septic shock.

2. Development of Polymyxin B-Immobilized Fibers

2.1. *Polymyxin B-Immobilized Fibers*

Polymyxin B was covalently immobilized on the surface of polystyrene-derived polypropylene-reinforced conjugated carrier fibers to

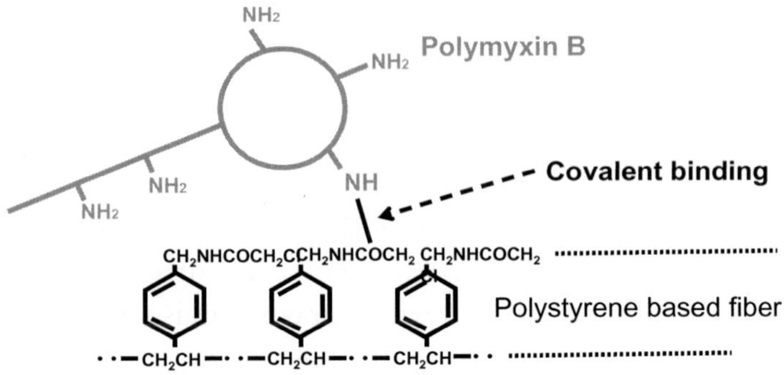

Fig. 12.2 Structure of a polymyxin B-immobilized fiber. Molecular conformation is shown. Polymyxin B covalently bound to polystyrene-based fibers. (Provided by Toray Medical Co., Ltd.)

enable selective adsorption of circulating endotoxins in the blood (Fig. 12.2). a-Chloroacetoamide methyl groups were chemically introduced into the polystyrene molecule to fix polymyxin B (Shoji *et al.*, 1998).

2.2. *Endotoxin Adsorption Capacity of Polymyxin B-Immobilized Fibers In Vitro*

When the polymyxin B-immobilized fibers or carrier fibers alone were incubated with synthetic lipid A *in vitro*, the polymyxin B-immobilized fibers, but not the carrier fibers, effectively adsorbed synthetic lipid A, as assessed using the LAL assay. These results indicated that polymyxin B is able to bind lipid A, even when it is immobilized (Fig. 12.3) (Shoji *et al.*, 1998). Endotoxin lipid A is located in the least variable region of lipopolysaccharide, and does not change from species to species or from strain to strain. The polymyxin B-immobilized fibers can adsorb various endotoxins with different O-side chains or chemotypes, indicating that immobilized polymyxin B binds specifically to the lipid A portion of the lipopolysaccharide (Fig. 12.4) (Shoji *et al.*, 1998).

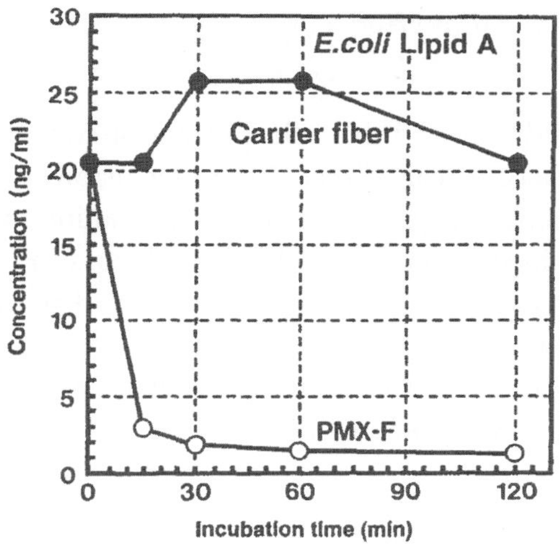

Fig. 12.3. Alterations in lipid A concentration by polymyxin B-immobilized fibers. Polymyxin B-immobilized fibers were incubated in bovine serum solution containing 20 ng/mL synthetic *Escherichia coli* lipid A under continuous stirring. Lipid A concentration was measured by means of a limulus amebocyte lysate assay. (Reproduced from Shoji *et al.* 1998)

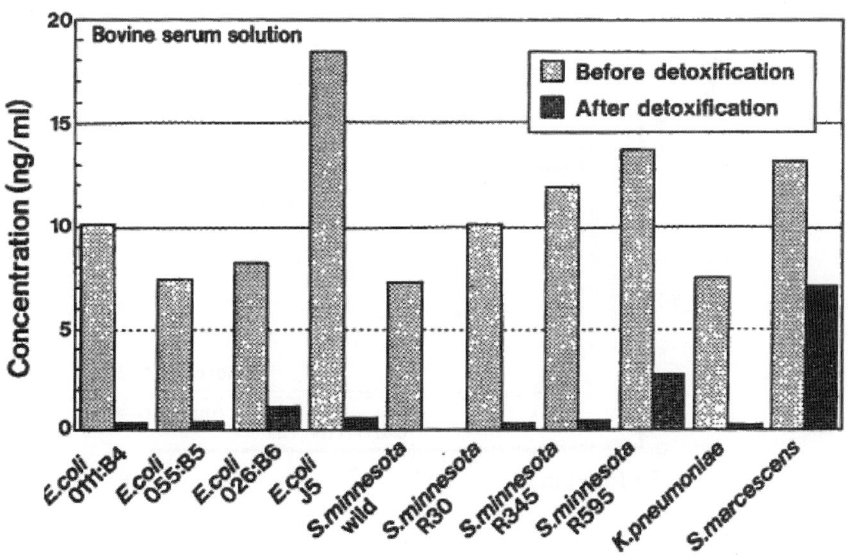

Fig. 12.4.

2.3. *Endotoxin Adsorption Capacity of Polymyxin B-Immobilized Fibers In Vivo*

A pre-clinical study showed that selective endotoxin removal from the blood by extracorporeal direct hemoperfusion with polymyxin B-immobilized fibers improved survival in a canine model following endotoxin administration (Fig. 12.5) (Hanasawa *et al.*, 1988). Extracorporeal hemoperfusion with polymyxin B-immobilized fibers resulted in increased survival in the canine endotoxemia model. There is therefore potential for the use of polymyxin B-immobilized fibers in humans (Hanasawa *et al.*, 1989).

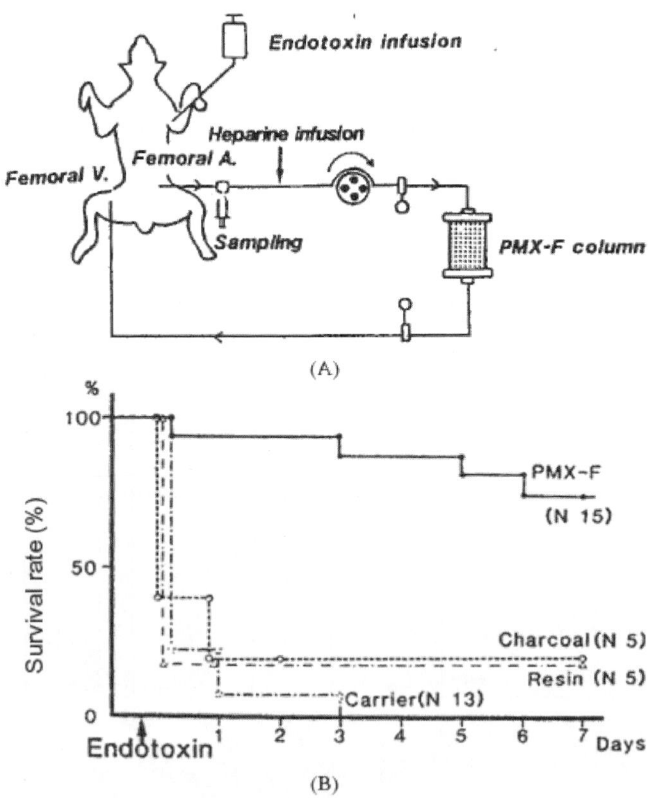

Fig. 12.5. Detoxification of endotoxin extracted from different species or strains of gram-negative bacteria using polymyxin B-immobilized fibers. (Reproduced from Shoji *et al.*, 1998)

3. Toraymyxin for Human Use

3.1. *Architecture and Specification of Toraymyxin*

Toraymyxin contains a knitted roll of polymyxin B-immobilized fibers. Use of a thin, fibrous carrier produces a hemo-adsorption cartridge with a large surface area that does not cause a large pressure drop in the blood flow compartment (Fig. 12.6). Blood flow through the Toraymyxin column is unidirectional and moves radially from the center to the outside of the polymyxin B-immobilized fibers, thereby improving adsorption capacity by ensuring a homogenous distribution of blood within the column (Fig. 12.7).

The endotoxin adsorption capacity of the Toraymyxin column was evaluated in a bovine serum perfusion model. Endotoxin-containing bovine serum solution (10 ng/mL endotoxin, total 1.5 L) was perfused through the Toraymyxin cartridge at a flow rate of

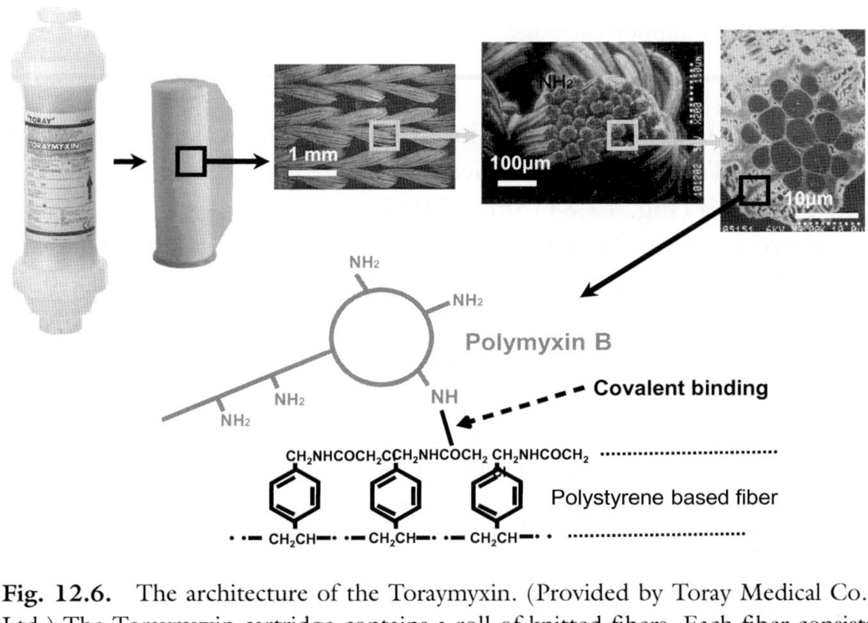

Fig. 12.6. The architecture of the Toraymyxin. (Provided by Toray Medical Co., Ltd.) The Toraymyxin cartridge contains a roll of knitted fibers. Each fiber consists of a bundle of ultra-fine fibers with a diameter of about 20 μm. The polymyxin B molecules are covalently bound onto the fiber surface, and therefore do not leak into the patient.

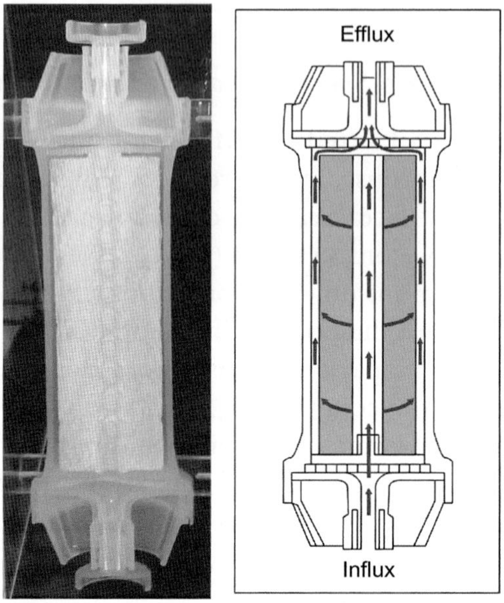

Fig. 12.7. The architecture of a Toraymyxin cartridge, and the schematics of blood flow inside the adsorption cartridge. (Provided by Toray Medical Co., Ltd.)

100 mL/min, and the change in endotoxin level was monitored by LAL assay. The endotoxin concentration reached equilibrium after 2 hr perfusion (Fig. 12.8) (Shoji, 2003).

There are three Toraymyxin columns currently available in Japan: PMX-20R (volume, 135 mL), PMX-05R (40 mL), and PMX-01R (8 mL) (Fig. 12.9). The first Toraymyxin column released was PMX-20R®, which was developed for the treatment of adults with septic shock (Shoji, 2003). The second column released was PMX-05R®, which was developed for use in pediatric or elderly patients with smaller circulating blood volumes or a body weight less than 40 kg (Nakamura *et al.*, 2007). Several reports have shown acceptable results using PMX-05R in pediatric and elderly patients with sepsis. In 2011, PMX-01R® was released for use in newborn or premature babies. Furthermore, several case reports describe the use of PMX-01R in patients with a body weight less than 1,000 g with severe sepsis.

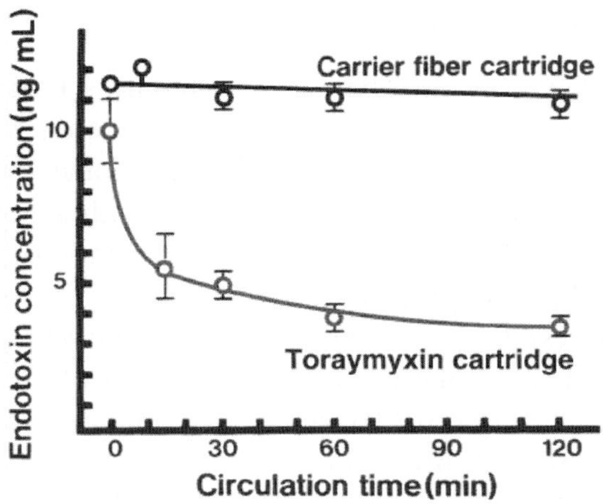

Fig. 12.8. Endotoxin adsorption capacity of the Toraymyxin cartridge, as compared to the carrier fiber cartridge alone. (Reproduced from Shoji H *et al.*, 2003) The bovine serum solution contained 10 ng/mL endotoxin, and was circulated through the Toraymyxin cartridge and carrier fiber cartridge.

	PMX-20R	PMX-05R	PMX-01R
length (mm)	224	133	133
Max. diameter (mm)	63	55	55
Shell diameter(mm)	49	40	25
Fiber weight (g)	56 ± 3	15 ± 2	4 ± 1
Blood volume (mL)	135 ± 3	40 ± 3	8 ± 2.5
rinsing volume (L)	4.0	2.0	0.5
Blood flow (mL/min)	80–120	20–40	8–12
Operation time (hr)	2.0	2.0	2.0

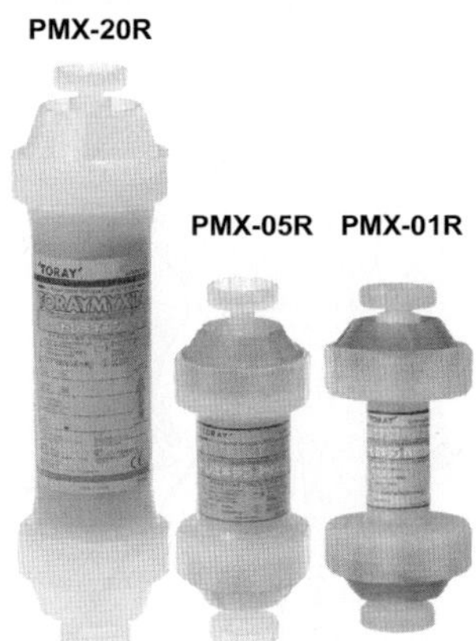

Fig. 12.9. Overview of the three Toraymyxin cartridges currently available for clinical use. (Provided by Toray Medical Co., Ltd.)

3.2. *Use of Toraymyxin in Hemoperfusion*

Prior to use, PMX-20R must be washed out with 4 L of physiological saline and primed with 1 L of physiological saline containing 20 mg of nafamostat mesilate or 1,000 U of unfractionated heparin. For patients with extremely severe conditions or with an extremely low blood pressure, an albumin solution containing anticoagulants can be used as the final priming solution.

Vascular access for hemoperfusion is usually via a central vein (internal cervical vein, subclavian vein, or femoral vein). The standard blood flow rate through the Toraymyxin column depends on the size of the column: 80–120 mL/min for PMX-20R, 20–40 mL/min for PMX-05R, and 8–12 mL/min for PMX-01R. An anticoagulant (20–30 mg/hr nafamostat mesilate or 40–60 U/kg/hr unfractionated heparin) should also be administered to prevent coagulation within the blood circuit (Fig. 12.10). Since coagulation dysfunction is

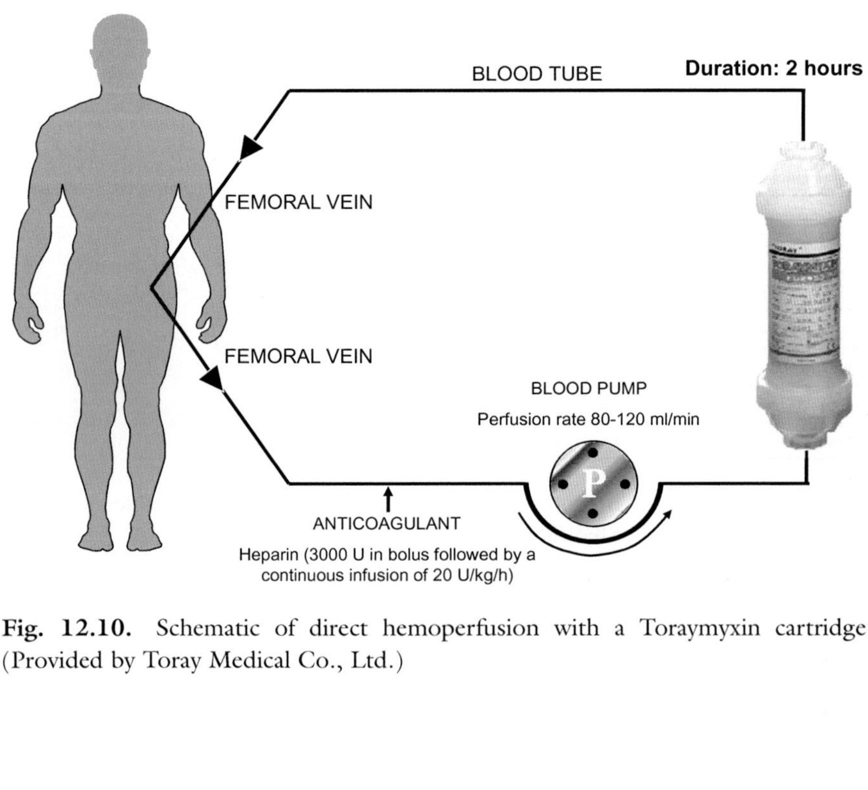

Fig. 12.10. Schematic of direct hemoperfusion with a Toraymyxin cartridge. (Provided by Toray Medical Co., Ltd.)

common in patients with septic shock, the short half-lives of these regional anticoagulants mean that they are safe for use during hemoperfusion. Nafamostat mesilate is the most commonly used anticoagulant for this purpose in Japan.

Since 1994, Toraymyxin has been used in over 100,000 cases in Japan. Despite adverse effects, including thrombocytopenia or hypotension, being reported, the incidence is rare, and there have been no serious adverse effects reported (Shoji, 2003).

4. Clinical Efficacy of Toraymyxin Therapy for Endotoxemia and Septic Shock

Aoki *et al.* published the first clinical report of Toraymyxin therapy for the treatment of severe sepsis or septic shock (Aoki *et al.*, 1994). Sixteen patients with septic multiple organ failure received 2 hr direct hemoperfusion with a Toraymyxin column in this phase I study. Toraymyxin therapy significantly decreased endotoxin levels from 76 pg/mL to 21 pg/mL. Furthermore, the patients' hyperdynamic state, a characteristic of endotoxic shock, resolved after treatment (Fig. 12.11) (Aoki *et al.*, 1994).

Tani *et al.* published the first prospective, multicenter, observational clinical trial of Toraymyxin for the treatment of patients with sepsis in Japan (Tani *et al.*, 1998). The survival rate of the Toraymyxin-treated group (54.0%) was significantly higher than that of the control group, which received standard therapies (36.4%). In the Toraymyxin-treated group, the mean plasma endotoxin concentration significantly decreased from 83.7 pg/mL before treatment to 56.4 pg/mL after treatment. Furthermore, the mean plasma endotoxin concentration decreased to 28.5 pg/mL the day after Toraymyxin therapy. Following Toraymyxin therapy, the mean plasma endotoxin concentration was lower in those who survived (18.8 pg/mL) than in those who died (88 pg/mL). Other parameters, including cardiac function, also improved after Toraymyxin therapy.

Nemoto *et al.* first demonstrated that the rate of overall survival in patients with sepsis significantly improved after polymyxin B

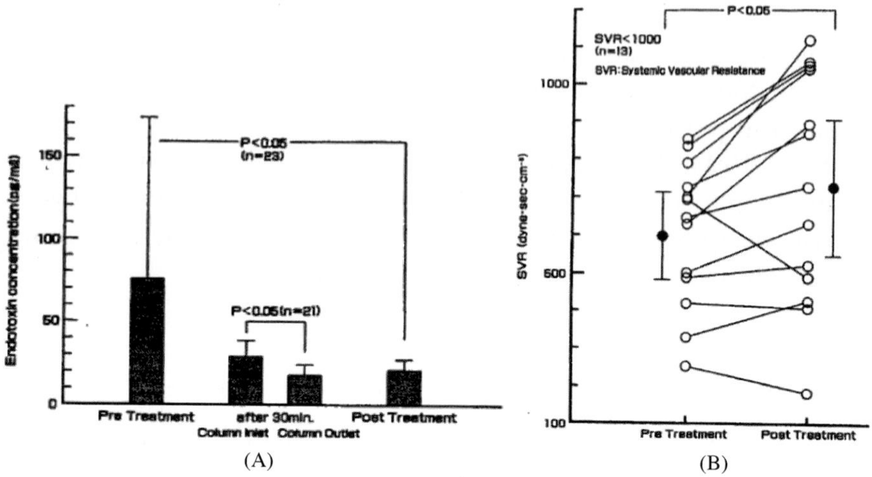

Fig. 12.11. Results from the first clinical trial of Toraymyxin in Japan. (Reproduced from Aoki H *et al.*, 1994) (A) Endotoxin concentration before and after Toraymyxin therapy. (B) Systemic vascular resistance (SVR) before and after Toraymyxin therapy. Toraymyxin therapy decreased the concentration of endotoxin in the blood and improved the hemodynamic status in patients with severe sepsis and septic shock.

treatment, compared to that after standard therapies (41% vs. 11%) in an RCT in 98 patients with sepsis in Japan (Nemoto *et al.*, 2001). Toraymyxin therapy improved the rate of survival in patients with an Acute Physiology and Chronic Health Evaluation II (APACHE II) score of <30, but not in patients with scores of >30, demonstrating that Toraymyxin treatment is most effective at improving patient outcomes when applied during the early stages of sepsis (Fig. 12.12).

In 2005, Vincent *et al.* published the results of a multicenter, open-label, pilot, randomized controlled study conducted in the intensive care units of six academic medical centers in Europe. This study was the first randomized controlled study of Toraymyxin conducted outside of Japan (Vincent *et al.*, 2005). Thirty-six postoperative patients with severe sepsis or septic shock secondary to intra-abdominal infection randomly received 2 hr Toraymyxin therapy (n = 17) or standard therapy (n = 19). Toraymyxin therapy was well tolerated, with no significant adverse effects. Patients treated with

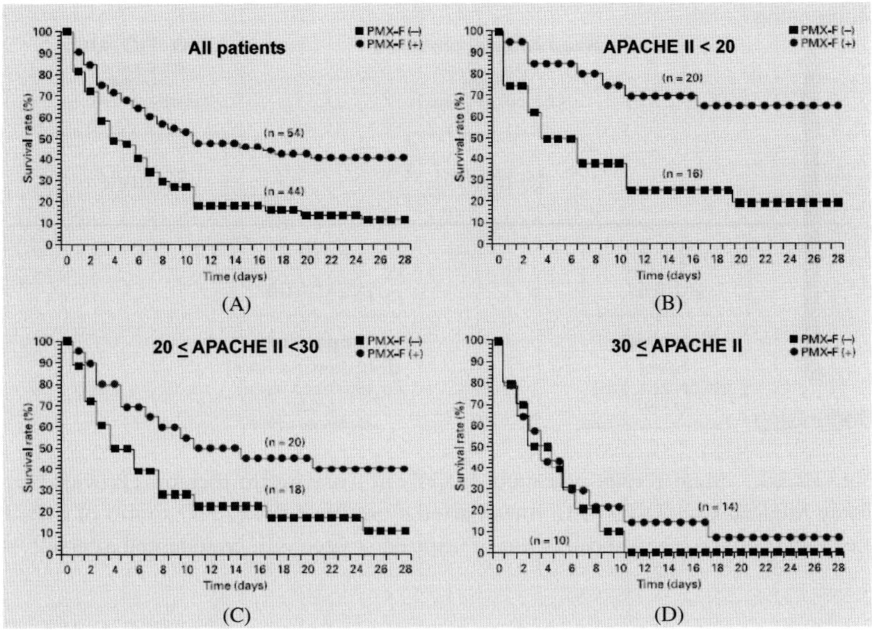

Fig. 12.12. Survival rate according to treatment group and patient severity, as measured by APACHE II score. (Reproduced from Nemoto H *et al.*, 2001) Toraymyxin therapy improved the rate of survival rate in patients with sepsis who had an APACHE II score < 30.

Toraymyxin therapy showed significant increases in cardiac index (P = 0.012 and 0.032 at days 1 and 2, respectively), left ventricular stroke work index (P = 0.015 at day 2), and oxygen delivery index (P = 0.007 at day 2), as compared to the controls. Furthermore, the need for continuous renal replacement therapy after study entry significantly reduced in the Toraymyxin-treated group (P = 0.043). These results indicate that Toraymyxin therapy is safe and improves cardiac and renal dysfunction due to sepsis or septic shock.

Cruz *et al.* published a systematic review of Toraymyxin therapy effectiveness for sepsis treatment (Cruz *et al.*, 2007). They included a total of 28 publications that reported at least one specified outcome measure for Toraymyxin therapy. In this meta-analysis of 1,425 patients (Toraymyxin therapy, 978 patients; conventional therapy, 447 patients),

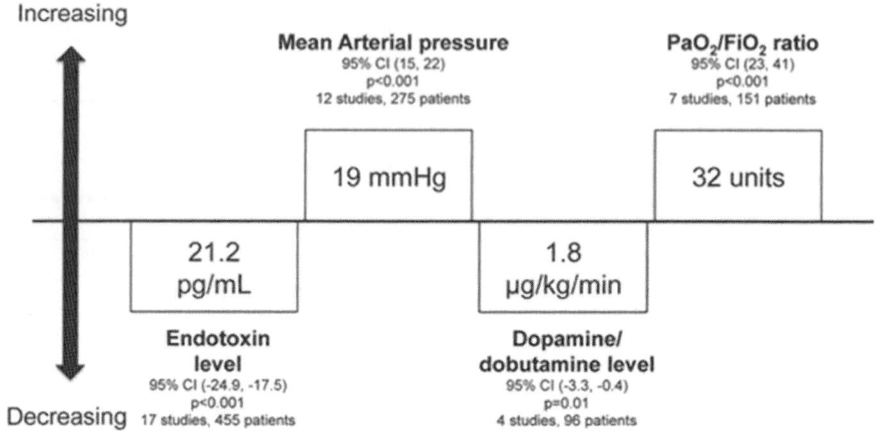

Fig. 12.13. Main results of a meta-analysis of Toraymyxin therapy. (Provided by Toray Medical Co., Ltd.) MAP: mean arterial pressure, PaO_2/FIO_2: ratio of partial pressure arterial oxygen and fraction of inspired oxygen, CI: confidential interval.

Toraymyxin therapy was associated with a significantly lower risk of mortality than that associated with conventional therapy (Toraymyxin, 33.5% vs. conventional treatment, 66.5%; risk ratio, 0.53; 95% confidence interval [CI], 0.43–0.65). A 33% to 80% reduction in plasma endotoxin levels was also observed. The large decrease in mortality was associated with an improvement in hemodynamic condition. After Toraymyxin therapy, mean arterial pressure significantly increased by 19 mmHg (mean increase, 26%; range, 14–42%) and dopamine/dobutamine dose was decreased by 1.8 µg/kg/min. Improvement in pulmonary function was also observed, as the PaO_2/FiO_2 increased by 32 units (95% CI, 23–41 units; $P < 0.001$) (Fig. 12.13).

Cruz *et al.* also published the Early Use of Polymyxin B Hemoperfusion in Abdominal Sepsis (EUPHAS) trial, which was a prospective, multicenter RCT conducted at the intensive care units of 10 Italian tertiary care hospitals in 2009 (Cruz *et al.*, 2007). Sixty-four patients with severe sepsis or septic shock who had undergone emergency surgery for intra-abdominal infections were enrolled in this study. Patients were randomized within 6 hr after open abdominal surgery to receive either conventional therapy (n = 30) or conventional therapy plus two 2 hr sessions of Toraymyxin therapy with 24 hr between sessions (n = 34). Compared to conventional therapy, Toraymyxin treatment significantly improved 28-day mortality (53%; 16/30 patients vs. 32%; 11/34 patients). In the Toraymyxin group, mean arterial pressure significantly increased from 76 mmHg (before treatment) to 84 mmHg (72 hr after treatment; $P = 0.001$), and vasopressor requirement (measured as inotropic score) significantly decreased from 29.9 (before treatment) to 6.8 (72 hr after treatment; $P < 0.001$). In contrast, in the conventional therapy group, mean arterial pressure (before, 74 mmHg; 72 hr after, 77 mmHg; $P = 0.37$) and inotropic score (before, 28.6; after, 22.4; $P = 0.37$) did not change significantly. Furthermore, the PaO_2/FIO_2 ratio significantly increased (before, 235; after, 264; $P = 0.049$) in the Toraymyxin-treated group, but not in the conventional therapy group (before, 217; after, 228; $P = 0.79$). Moreover, the sequential organ failure assessment (SOFA) score, which is an indicator of organ dysfunction severity, improved in the Toraymyxin-treated group, but not in the conventional therapy group (change in SOFA score, −3.4 vs. −0.1; $P < 0.001$). This was the first report showing an improved rate of survival in an RCT conducted outside of Japan.

The EUPHRATES trial, which is ongoing in the USA and Canada (Klein *et al.*, 2014), is a new trial designed to address the criticisms of previous studies. Circulating endotoxin levels in patients with septic shock are being evaluated by means of a novel endotoxin detection method called the endotoxin activity assay (EAA). A total of 360 patients with septic shock in which high EAA activity is detected will be enrolled and randomized to either Toraymyxin therapy or conventional therapy.

The ABDO-MIX trial is an ongoing study in France (Sato *et al.*, 2011). A total of 240 patients with peritonitis due to intestinal perforation were enrolled (enrollment period, 2010–2013), and the primary end point of the study was 28-day mortality. Data obtained from the trial is currently being analyzed, and the results are expected soon.

These ongoing RCTs are expected to produce compelling evidence that will change the positioning of blood purification techniques in the national and international guidelines for sepsis treatment.

5. Mechanism of Favorable Effects of Toraymyxin Therapy in Sepsis

5.1. *Removal of Endotoxin*

Toraymyxin was originally designed to specifically adsorb endotoxin. Tani *et al.* reported that Toraymyxin therapy significantly reduced plasma endotoxin concentrations from 83.7 ± 26.7 pg/mL to 56.4 ± 27.9 pg/mL after 2 hr of direct hemoperfusion with a Toraymyxin column (Nemoto *et al.*, 2001). The LAL assay, a standard endotoxin detection method, is prone to contamination, and lacks precision and accuracy, resulting in both false positive and false negative results. Indeed, our previous study showed that the endotoxin detection sensitivity of the turbidimetric LAL assay was very low (26.9%; 14 of 52 patients) in patients with severe sepsis and septic shock who required Toraymyxin therapy (Shimizu *et al.*, 2009). A new method for the detection and quantitation of endotoxin is therefore urgently needed.

Recently, Romaschin *et al.* developed the EAA, which detects endotoxins in whole blood using autologous neutrophil-dependent chemiluminescence (Romaschin *et al.*, 1998). The EEA has been approved by the United States Food and Drug Administration as a clinical method for the diagnosis of endotoxemia, and is currently being used in the ongoing EUPHRATES trial (Klein *et al.*, 2014).

A rapid LAL assay for detecting endotoxins by using a laser light-scattering particle-counting method, called endotoxin scattering photometry (ESP), has also been developed. Endotoxins bind to and activate factor C, a clotting enzyme, early in the LAL cascade, which ultimately results in the production of the insoluble protein coagulin. Coagulin spontaneously forms a polymer, which is detected in the turbidimetric LAL assay. Obata *et al.* reported that they were able to detect coagulin particles formed under agitation of the LAL cascade, and that the LAL cascade transfer period after endotoxin stimulation depends on the concentration of endotoxin before gelation using ESP (Obata *et al.*, 2008). The ESP method allows more rapid and sensitive detection of endotoxin compared with the turbidimetric LAL method.

Furthermore, compared to the standard turbidimetric LAL assay, the ESP method has been used to significantly discriminate between patients with sepsis and those with septic shock undergoing gastrointestinal emergency surgery (Shimizu *et al.*, 2013a). After prolonged (>2 hr) Toraymyxin therapy, the hemodynamic condition of patients with septic shock due to intra-abdominal infection improved. Interestingly, the reduction in plasma endotoxin concentration could be detected with ESP, but not with the standard turbidimetric method. We have demonstrated that endotoxin adsorption can occur, despite long-duration Toraymyxin therapy (Shimizu *et al.*, 2013b). We observed a reduction in endotoxin after ESP, even when using longer (>2 hr) Toraymyxin therapy. Thus, ESP appears to be sensitive enough to detect circulating endotoxin in the plasma of patients with septic shock who require Toraymyxin therapy.

5.2. *Other Potential Mechanisms of Toraymyxin Therapy*

Although Toraymyxin was originally developed for the adsorption of circulating endotoxin in the blood, recent studies have shown other potential mechanisms. Jaber *et al.* reported that it may be possible to use Toraymyxin to remove lipoteichoic acid (LTA) from peripheral blood (Jaber *et al.*, 1998). In an *in vitro* study, polymyxin B significantly suppressed LTA-induced tumor necrosis factor (TNF)-α production by peripheral blood mononuclear cells, suggesting that LTA may also be removed during direct hemoperfusion with polymyxin B, and that this removal may be due to LTA binding.

Anandamide and 2-arachidonyl glyceride, which are endogenous cannabinoids released by activated macrophages and platelets during endotoxic shock, play a crucial role in the induction of shock-related hypotension (Varga *et al.*, 1998). Wang *et al.* showed that anandamide was efficiently adsorbed by a polymyxin B-immobilized bead column *in vitro* (Wang *et al.*, 2000). Therefore, the adsorption of cannabinoids by polymyxin B may be an important mechanism to improve hemodynamic dysfunction following Toraymyxin therapy.

High-mobility group box 1 protein (HMGB1), a nuclear transcription factor, has also been implicated in delayed endotoxin lethality and systemic inflammation (Wang *et al.*, 2001). An experimental study demonstrated that serum levels of HMGB1 were lower in Toraymyxin-treated patients than in controls.The reduction in circulating HMGB1 levels may also contribute to the beneficial effects of Toraymyxin therapy in patients with sepsis (Hussein *et al.*, 2005; Sakamoto *et al.*, 2007; Abe *et al.*, 2011; Ueno *et al.*, 2011).

Other recent studies have reported a possible relationship between immune cells and the reduction in inflammatory mediators. Ono *et al.* demonstrated that the expression of HLA-DR, a major histocompatibility complex class II cell surface receptor, on monocytes, and CD16 on granulocytes, is markedly decreased in patients with septic shock. However, polymyxin B treatment increased leukocyte expression of these surface antigens (Ono *et al.*, 2004). The number of $CD16^+CD14^+$ monocytes and monocytic toll-like receptor 4 expression is dramatically increased in patients with severe infection, and recent studies have shown that Toraymyxin therapy effectively reduces the number of $CD16^+CD14^+$ monocytes and monocytic expression of toll-like receptor 4 in patients with septic shock (Tsujimoto *et al.*, 2004). Moreover, Ono *et al.* also demonstrated that the removal of surplus circulating $CD4^+CD25^+Foxp3^+$ regulatory T cells in patients with septic shock via hemoperfusion with Toraymyxin might represent a novel strategy for inducing recovery from sepsis-associated immunosuppression (Ono *et al.*, 2013). Tsuzuki *et al.* demonstrated immediate inhibition of NF-κB binding activity and suppression of TNF-α secretion after endotoxin neutralization with polymyxin B, regardless of whether the peripheral blood mononuclear cells were already producing TNF-α or not (Tsuzuki *et al.*, 2001). Reducing the number of circulating monocytes or modulating their function may contribute to improving pro-inflammatory responses following Toraymyxin therapy.

Although the evidence accumulated so far supports the clinical efficacy of Toraymyxin therapy, further evaluation is required to clarify the precise mechanisms of these beneficial effects. Nevertheless, in Japan, Toraymyxin therapy is widely accepted for the treatment of patients with septic shock. Furthermore, this useful clinical device is gaining acceptance as its use becomes more widespread.

References

Abe S, Hayashi H, Seo Y, *et al.* (2011) Reduction in serum high mobility group box-1 level by polymyxin B-immobilized fiber column in patients with idiopathic pulmonary fibrosis with acute exacerbation. *Blood Purif* 32(4): 310–316. [doi:10.1159/000330325]

Aoki H, Kodama M, Tani T, *et al.* (1994) Treatment of sepsis by extracorporeal elimination of endotoxin using polymyxin B-immobilized fiber. *Am J Surg* **167(4):** 412–417.

Cruz DN, Antonelli M, Fumagalli R, *et al.* (2009) Early use of polymyxin B hemoperfusion in abdominal septic shock: The EUPHAS randomized controlled trial. *JAMA* **301(23):** 2445–2452. [doi:10.1001/jama.2009.856]

Cruz DN, Perazella MA, Bellomo R, *et al.* (2007) Effectiveness of polymyxin B-immobilized fiber column in sepsis: A systematic review. *Crit Care* **11(2):** R47. [doi:10.1186/cc5780]

Hanasawa K, Tani T, Kodama M. (1989) New approach to endotoxic and septic shock by means of polymyxin B immobilized fiber. *Surg Gynecol Obstet* **168(4):** 323–331.

Hanasawa K, Tani T, Oka T, *et al.* (1988) Selective removal of endotoxin from the blood by extracorporeal hemoperfusion with polymyxin B immobilized fiber. *Prog Clin Biol Res* **264:** 337–341.

Hussein MH, Kato T, Sugiura T, *et al.* (2005) Effect of hemoperfusion using polymyxin B-immobilized fiber on IL-6, HMGB-1, and IFN gamma in a neonatal sepsis model. *Pediatr Res* **58(2):** 309–314. [doi:10.1203/01.PDR.0000169995.25333.F7]

Jaber BL, Barrett TW, Cendoroglo Neto M, *et al.* (1998) Removal of cytokine inducing substances by polymyxin-B immobilized polystyrene-derivative fibers during in vitro hemoperfusion of 10% human plasma containing Staphylococcus aureus challenge. *ASAIO J* **44(1):** 48–53.

Klein DJ, Foster D, Schorr CA, *et al.* (2014) The EUPHRATES trial (Evaluating the Use of Polymyxin B Hemoperfusion in a Randomized

Controlled Trial of Adults Treated for Endotoxemia and Septic Shock): study protocol for a randomized controlled trial. *Trials* **15:** 218.

Mitaka C, Tomita M. (2011) Polymyxin B-immobilized fiber column hemoperfusion therapy for septic shock. *Shock* **36(4):** 332–8.

Nakamura T, Kawagoe Y, Suzuki T, *et al.* (2007) Polymyxin B-immobilized fiber hemoperfusion with the PMX-05R column in elderly patients suffering from septic shock. *Am J Med Sci* **334(4):** 244–247. [doi:10.1097/MAJ.0b013e3180a5e8d8]

Nemoto H, Nakamoto H, Okada H, *et al.* (2001) Newly developed immobilized polymyxin B fibers improve the survival of patients with sepsis. *Blood Purif* **19(4):** 361–368; discussion 368–369. [doi:46966]

Obata T, Nomura M, Kase Y, *et al.* (2008) Early detection of the Limulus amebocyte lysate reaction evoked by endotoxins. *Anal Biochem* **373(2):** 281–286. [doi:10.1016/j.ab.2007.09.018]

Ono S, Kimura A, Hiraki S, *et al.* (2013) Removal of increased circulating CD4+CD25+Foxp3+ regulatory T cells in patients with septic shock using hemoperfusion with polymyxin B-immobilized fibers. *Surgery* **153(2):** 262–271. [doi:10.1016/j.surg.2012.06.023]

Ono S, Tsujimoto H, Matsumoto A, *et al.* (2004) Modulation of human leukocyte antigen-DR on monocytes and CD16 on granulocytes in patients with septic shock using hemoperfusion with polymyxin B-immobilized fiber. *Am J Surg* **188(2):** 150–156. [doi:10.1016/j.amjsurg.2003.12.067]

Romaschin AD, Harris DM, Ribeiro MB, *et al.* (1998) A rapid assay of endotoxin in whole blood using autologous neutrophil dependent chemiluminescence. *J Immunol Methods* **212(2):** 169–185.

Sakamoto Y, Mashiko K, Matsumoto H, *et al.* (2007) Relationship between effect of polymyxin B-immobilized fiber and high-mobility group box-1 protein in septic shock patients. *ASAIO J* **53(3):** 324–328. [doi:10.1097/MAT.0b013e3180340301]

Sato K, Maekawa H, Sakurada M, *et al.* (2011) Direct hemoperfusion with polymyxin B immobilized fiber for abdominal sepsis in Europe. *Surg Today* **41(6):** 754–760. [doi:10.1007/s00595-010-4504-9]

Shoji H. (2003) Extracorporeal endotoxin removal for the treatment of sepsis: Endotoxin adsorption cartridge (Toraymyxin). *Ther Apher Dial* **7(1):** 108–114.

Shoji H, Tani T, Hanasawa K, *et al.* (1998) Extracorporeal endotoxin removal by polymyxin B immobilized fiber cartridge: Designing and antiendotoxin efficacy in the clinical application. *Ther Apher* **2(1):** 3–12.

Shimizu T, Hanasawa K, Sato K, *et al.* (2009) Direct hemoperfusion with polymyxin-B-immobilized fiber columns improves septic hypotension and reduces inflammatory mediators in septic patients with colorectal perforation. *Langenbecks Arch Surg* **394(2):** 303–311. [doi:10.1007/s00423-008-0395-2]

Shimizu T, Obata T, Sonoda H, *et al.* (2013a) Diagnostic potential of endotoxin scattering photometry for sepsis and septic shock. *Shock* **40(6):** 504–511. [doi:10.1097/SHK.0000000000000056]

Shimizu T, Obata T, Sonoda H, *et al.* (2013b) The ability of endotoxin adsorption during a longer duration of direct hemoperfusion with a polymyxin B-immobilized fiber column in patients with septic shock. *Transfus Apher Sci* **49(3):** 499–503. [doi:10.1016/j.transci.2013.04.042]

Tani T, Hanasawa K, Endo Y, *et al.* (1998) Therapeutic apheresis for septic patients with organ dysfunction: Hemoperfusion using a polymyxin B immobilized column. *Artif Organs* **22(12):** 1038–1044.

Tsujimoto H, Ono S, Hiraki S, *et al.* (2004) Hemoperfusion with polymyxin B-immobilized fibers reduced the number of CD16+ CD14+ monocytes in patients with septic shock. *J Endotoxin Res* **10(4):** 229–237. [doi:10.1179/096805104225005814]

Tsuzuki H, Tani T, Ueyama H, Kodama M (2001) Lipopolysaccharide: neutralization by polymyxin B shuts down the signaling pathway of nuclear factor kappaB in peripheral blood mononuclear cells, even during activation. *J Surg Res* **100(1):** 127–134. [doi:10.1006/jsre.2001.6227]

Ueno T, Ikeda T, Ikeda K, *et al.* (2011) HMGB-1 as a useful prognostic biomarker in sepsis-induced organ failure in patients undergoing PMX-DHP. *J Surg Res* **171(1):** 183–190. [doi:10.1016/j.jss.2009.11.708]

Varga K, Wagner JA, Bridgen DT, *et al.* (1998) Platelet- and macrophage-derived endogenous cannabinoids are involved in endotoxin-induced hypotension. *FASEB J* **12(11):** 1035–1044.

Vincent JL, Laterre PF, Cohen J, *et al.* (2005) A pilot-controlled study of a polymyxin B-immobilized hemoperfusion cartridge in patients with severe sepsis secondary to intra-abdominal infection. *Shock* **23(5):** 400–405.

Wang H, Yang H, Czura CJ, *et al.* (2001) HMGB1 as a late mediator of lethal systemic inflammation. *Am J Respir Crit Care Med* **164(10 Pt 1):** 1768–1773. [doi:10.1164/ajrccm.164.10.2106117]

Wang Y, Liu Y, Sarker KP, *et al.* (2000) Polymyxin B binds to anandamide and inhibits its cytotoxic effect. *FEBS Lett* **470(2):** 151–155.

Chapter 13

Polyvinyl Alcohol Adsorbent in Hemoperfusion

Ou Lailiang*

1. Introduction

Poly(vinyl alcohol) (PVA) is a synthetic polymeric material with strong hydrophilicity, low toxicity, low cost, and good biocompatibility which was developed for nearly 50 years. PVA has become the largest synthetic water-soluble polymer in industrial output. The hydrophilic hydroxide groups in the molecular chain of PVA could participate in esterification and etherification, as well as in acetal (ketone) reactions. The wide scope of PVA applications is due to the ease with which various PVA functional groups can be connected in order to obtain new materials with different properties. These applications include the fields of industrial polymer chemistry, textiles, food, medicine, building, and cosmetics. The high blood compatibility and biocompatibility of PVA has resulted in its extensive application in biomedical fields (Zhang *et al.*, 2010; Bai *et al.*, 2010; Bai *et al.*, 2009; Hegazy *et al.*, 2004). In addition, crosslinking PVA

*The Key Laboratory of Bioactive Materials, Ministry of Education, College of Life Science, Nankai University, China.
Corresponding author: Ou Lailiang. E-mail: ouyll@nankai.edu.cn.

beads have been widely used to develop blood adsorbents, which have better mechanical strength than cellulose beads (Yu, 2014).

2. PVA Synthesis and Characteristics

2.1. *PVA Synthesis*

The structural formula of PVA is shown in Fig. 13.1.

PVA cannot be directly prepared by the polymerization of vinyl alcohol monomers, because vinyl alcohol is very unstable and cannot be existed freely. Generally, vinyl acetate is first polymerized into polyvinyl acetate, and then PVA is obtained by alcoholysis of polyvinyl acetate. The corresponding processes include the polymerization of vinyl acetate, the alcoholysis of polyvinyl acetate, and the recycling of acetic acid and methanol.

The molecular chains of PVA obtained from alcoholysis contain many hydrophilic and reactive hydroxyl groups, and the molecular structure is suitable for polymerization, hydrolysis, grafting reaction, and crosslinking modifications. PVA is water-soluble, flexible, smooth, wear-resistant, has many film-forming properties and can be used as an emulsification agent. The extent of these properties depends on the degree of polymerization, viscosity, water content and degree of alcoholysis.

In addition to linear PVA, crosslinked PVA are used as matrices in various functional materials. There are two techniques that can be used to synthesize crosslinked PVA. The first method crosslinks linear PVA using crosslinking agents (Zhang *et al.*, 1998), and the second method uses alcoholysis of the crosslinked poly (vinyl acetate) (Yao *et al.*, 1995). Glutaradehyde, epichlorohydrin, terephthaldehyde and formaldehyde are several common crosslinking agents used in the

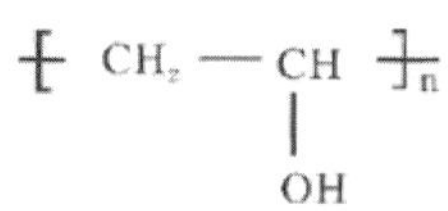

Fig. 13.1. The structural formula of PVA.

preparation of crosslinked PVA (Sun *et al.*, 2010). Glutaradehyde has many advantages over the others because of its high crosslinking speed, large crosslinking network and low toxicity.

Zhang *et al.* (1998) synthesized a crosslinked PVA gel using glutaraldehyde as the crosslinking agent through the reverse polymerization technique. PVA was first mixed with glutaraldehyde, and then hydrochloric acid was added to act as the catalyst. Next, paraffin oil with Span80 was added to the mixture and stirred at room temperature. Finally, the reaction was carried out at 70°C for 2 hr with appropriate stirring. At the end of the reaction, the crosslinked PVA beads were obtained after filtering, washing, drying, extraction and sieving.

Gu *et al.* (1993) reported a two-step process for the preparation of porous PVA beads crosslinked with epichlorohydrin (EP). In this two-step process, PVA aqueous solution and powered sodium hydroxide are mechanically stirred for 1 hr at 50°C. EP is introduced into the mixture for a pre-crosslinking, and then paraffin oil is added to the viscous mixture following the addition of Span 80. PVA beads were formed during the reverse suspension crosslinking reaction. During this process, it is very important to control the temperature, reaction time and stirring speed in order to avoid excessive crosslinking of the PVA solution. The synthetic scheme for the preparation of PVA beads is shown in Fig. 13.2, and the SEM photos of crosslinked PVA beads are shown in Fig. 13.3. Table 13.1 lists the pore parameters of the PVA beads.

2.2. *Biocompatibility of PVA*

The research and development of high-performance biomedical materials have become an important focal point in modern medicine. Medical materials, especially materials that come into direct contact with blood, should have good biocompatibility and mechanical properties. The blood compatibility of a normal human blood vessel is good because the endothelial cells contain large amounts of carbohydrates, such as sugar chains with hydroxyl groups. The structure of sugar chains rich in water is similar to aqueous gel (Gu *et al.*, 1993). PVA gels are a representative group of biocompatible polymers. They

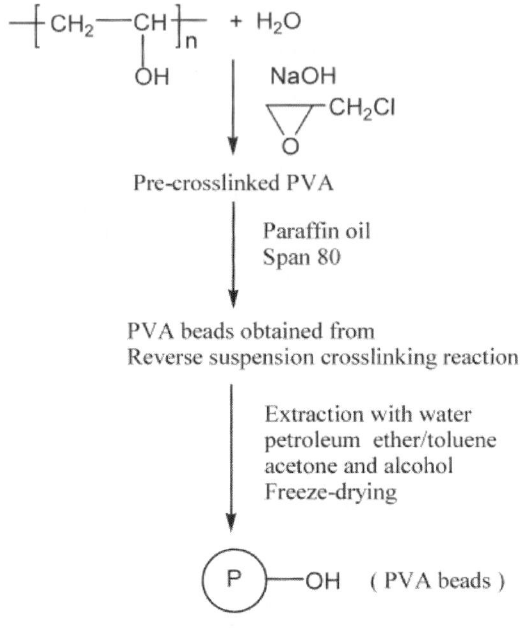

Fig. 13.2. Synthetic scheme for the preparation of PVA beads. (*With copyright permission from Gu et al. (1993)* monograph with Tianjin Science & Technology Translation & Publishing Corp.)

contain many hydroxyl groups with the physical properties of swelling and chemical crosslinking properties, which could effectively reduce the free energy at the interface between materials and blood, and improve the antithrombotic property of the materials.

3. Applications in Hemoperfusion

Generally, the adsorbents used to remove pathogenic blood toxins have been designed using porous resins that are linked with various bioactive ligands.

3.1. *PVA Adsorbents for Bilirubin Removal*

Bilirubin is a normal hemoglobin metabolite in senesced red blood cells. Bilirubin production disorders can result in high *in vivo*

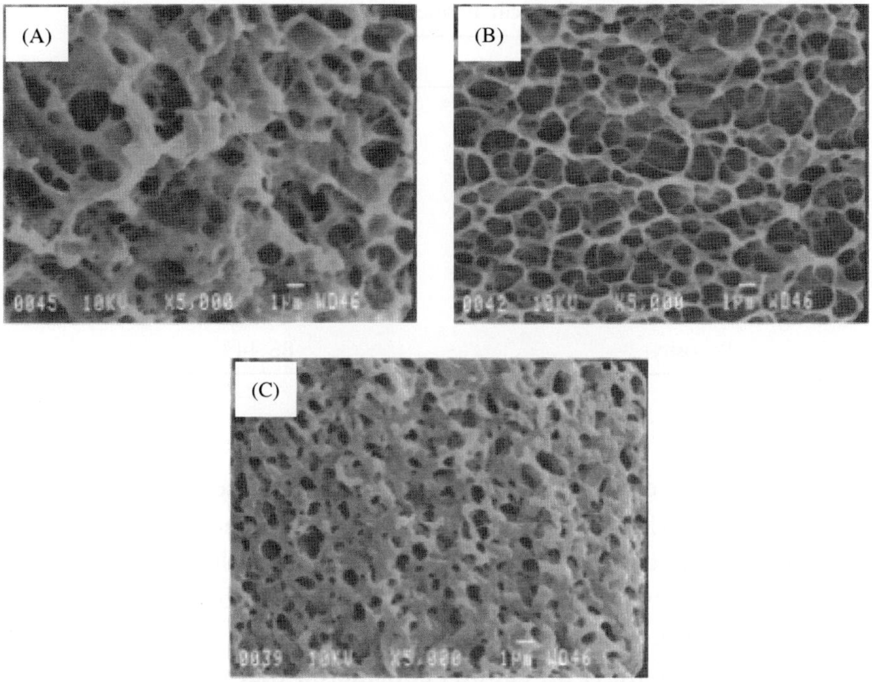

Fig. 13.3. SEM photos of crosslinked PVA beads. (A) PVA-6, (B) PVA-8, (C) PVA-10. (*With copyright permission from Gu et al. (1993)* monograph with Tianjin Science & Technology Translation & Publishing Corp.)

Table 13.1. Pore Parameters of the PVA Beads

Beads	Volume of Pores (Vp) (mL/g)	Porosity (φ)	Pore Size (μm)[a]
PVA-5	4.69	4.69	1–15
PVA-6	3.49	3.36	1–10
PVA-7	2.34	2.19	1–5
PVA-8	1.24	1.42	1–5
PVA-10	0.45	0.60	1–2

Volume of pores and porosity were the average of the differently sized beads.
[a]The pore size of the beads was estimated from their SEM images.

Table 13.2. Physical Properties of PVA Adsorbents

Parameters	Value
Mean diameter	150 μm
Apparent density	0.2654 g/mL
Skeletal density	1.183 g/mL
Specific surface area	195.64 m^2/g
Pore volume	3.2189 mL/g
Mean pore diameter	43.9 nm
Porosity	85.44%

accumulation levels that can result in various diseases (Lee *et al.*, 2002). Blood perfusion is the most promising technology for treating bilirubin-related disorders because of its efficacy, low cost and few side effects. Successful blood perfusion requires adsorbents with good blood compatibility, specific molecular recognition, and high adsorption capability. Wang *et al.* (2008) considered that in the preparation of a good adsorbent, the key point is the selection of matrix and ligands. They chose macro-porous PVA (Table 13.2) as the supporting matrix, and immobilized eight amine agents to the matrix as ligands for the adsorption of bilirubin. Among the eight ligands, trimethylamine (TMA), triethylamine (TEA), and 1,6-hexanediamine (HDA) yielded the highest bilirubin adsorption capacities, up to 80% of total bilirubin (Fig. 13.4).

Wang *et al.* (1998) studied the adsorption of crosslinked PVA gel to bilirubin. Their study showed that the crosslinking degree, particle size, adsorption temperature, and ion strength significantly influenced the adsorption efficiency. In addition, the particle size of PVA resin directly influences the adsorption performance. They also found that the adsorption rate and adsorption dynamic performance was higher in larger adsorbents than in smaller adsorbents. The main bilirubin binding sites of serum albumin are basic amino acid residues with positive charges that can form strong electrostatic interactions with negatively charged bilirubin carboxyl groups. Based on this, He *et al.*

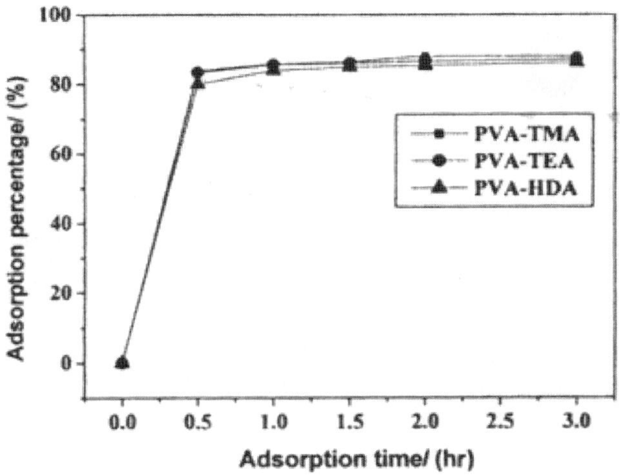

Fig. 13.4. Bilirubin adsorption kinetic curve of three amine modified adsorbents. (*With copyright permission from Wang et al. (2008)* monograph in *Chinese Journal of Dialysis and Artificial Organs.*)

(Zhang *et al.*, 1996) prepared a new adsorbent for bilirubin by immobilizing guanidine to crosslinked PVA gel, and found that the PVA immobilized with guanidine (PVA-G) had a high bilirubin adsorption capacity. The strong electrostatic force and hydrogen bonding were thought to be responsible for the high adsorption capacities of PVA-G (Fig. 13.5).

3.2. *PVA Adsorbents for TG and LDL Removal*

Triglycerides in the human body are composed of glycerol and long-chain fatty acids, and have important physiological functions in the body. However, when triglyceride content in serum exceeds a critical value, it will lead to high triglyceride diseases. Zhang *et al.* (2000) synthesized a new adsorbent for Triglyceride (TG) that used PVA as the matrix, was crosslinked by glutaraldehyde, and modified with taurine on the surface (Fig. 13.6). The adsorption effectively reduced the TG content and adsorbed little of other proteins. Additionally, the

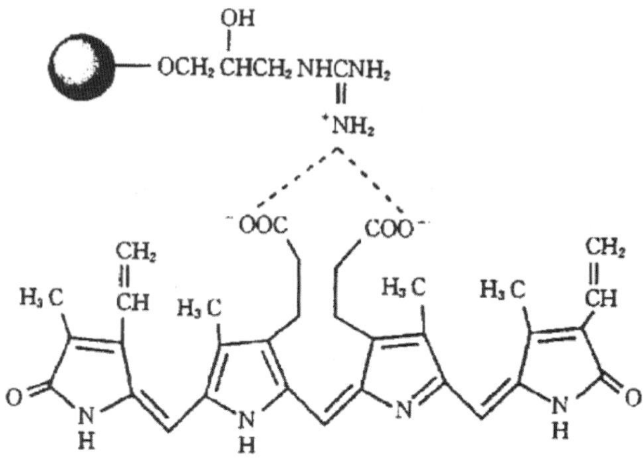

Fig. 13.5. Schematic diagram of the interaction between bilirubin and guanidinium. (*With copyright permission from Zhang et al. (1996)* monograph in *Chem J Chinese U.*)

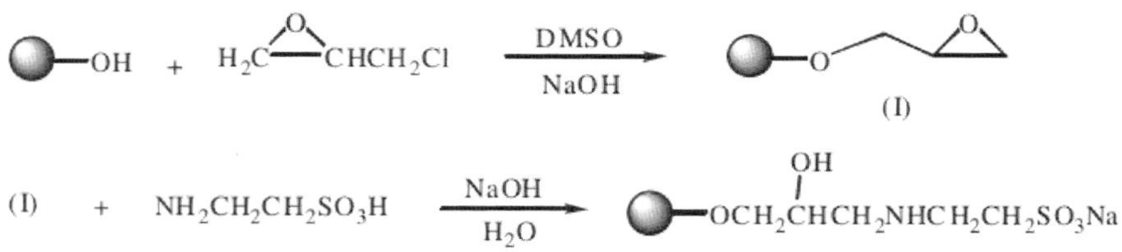

Fig. 13.6. The synthesis route of adsorbent crosslinked modified with taurine. (*With copyright permission from Zhang et al. (2000)* monograph in *Ion Exchange and Adsorption.*)

blood compatibility of the adsorbents was improved by introducing the anionic group to the surface.

Hyperlipidemia refers to the abnormal increase in plasma lipid content. Most patients with hyperlipidemia have high low-density lipoprotein (LDL) cholesterol levels. Therefore, it is very important for patients to reduce blood LDL content using blood perfusion. Xu *et al.* (2001) prepared a new LDL adsorbent by immobilizing polyethylene polyamine and a sulfonic acid group onto the crosslinked PVA beads (Fig. 13.7). The *in vitro* adsorption experiments showed this adsorbent had high LDL adsorption capacity, good adsorption

Fig. 13.7. The structure of LDL adsorbents.

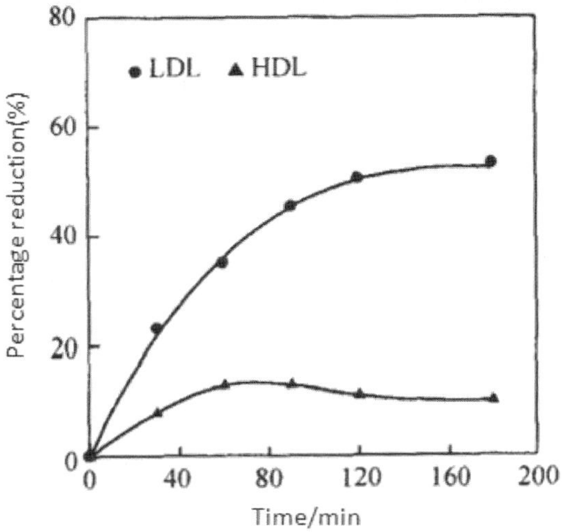

Fig. 13.8. Adsorption of LDL and HDL by immobilized polyethylene polyamine and a sulfonic acid group onto the crosslinked PVA beads. *(With copyright permission from Xu et al. (2001)* monograph in *Chinese Science Bulletin.*)

selectivity, low HDL adsorption (9%), and had fast adsorption rates (Fig. 13.8).

Li *et al.* (2004) synthesized a series of macroporous crosslinked PVA-based blood compatible adsorbents bearing carboxyl functional groups. Figure 13.9 shows the macroporous structure of the adsorbents, and the pore structure parameters are listed in Table 13.3. The LDL adsorption capacities from human plasma are shown in Table 13.4. The results indicated that different PVA-based adsorbents bearing the same carboxyl functional groups displayed different adsorption capabilities. The adsorption capacity depends on the type of the functional group that is immobilized onto the PVA beads. Moreover, the adsorbents bearing stronger acid functional groups

Fig. 13.9. SEM photos of the inner surface of PVA-based adsorbents. Crosslinking degree: (a) 19.7%; (b) 33.6%; (c) 40.6%. (*With copyright permission from Li et al. (2004)* monograph in *React Funct Polym.*)

Table 13.3. Pore Structural Parameters of PVA-Based Adsorbents

Sample	Crosslinking Degree (%)	ρ_a (g/mL)	ρ_s (g/mL)	S (m^2/g)	r (nm)	*P* (%)
1	19.7	0.5021	1.2590	28.0	171	60.1
2	26.5	0.4603	1.2803	29.7	187	64.0
3	33.6	0.4521	1.3014	42.5	135	65.3
4	40.6	0.4498	1.3102	46.8	124	65.7
5	46.6	0.4432	1.3345	58.2	103	66.8

ρ_a-apparent density; ρ_s-skeleton density; r-mean pore diameter, S-specific surface areas; P-porosity.

could adsorb more LDL than the adsorbents bearing relatively weaker acids. Further, the static electrostatic interaction between the functional group on the adsorbents and the LDL is considered to be the main driving force, because LDL is a negatively charged molecule under physiological condition.

In order to improve the LDL-C (low-density lipoprotein-cholesterol) adsorption capacity, Yu (Yu *et al.*, 2014) synthesized an amphiphilic adsorbent based on PVA containing cholesterol ligand and sulfonic dextran ligands. Prior to this work, all kinds of adsorbents for the removal of LDL based on PVA were hydrophilic. The dextran immobilized on the PVA beads not only reduced the steric hindrance as spacer, but also contained amphiphilic ligands. The results (Fig. 13.10) showed that the amphiphilic PVA adsorbents had a higher adsorption

Table 13.4. Adsorption of LDL onto Macroporous Crosslinked PVA-Based Adsorbents from Human Plasma

	Adsorption Capacity (mg/g)			
Adsorbents	**Tc**	**HDLc**	**LDLc + VLDLc**	**S**
$PVACH_2COOH$	0.78	0.24	0.54	2.25
PVA-HAD-CH_2COOH	1.32	0.4	0.92	2.30
PVA-DTPAA	1.16	0.42	0.74	1.76
PVA-HAD-DTPAA	1.74	0.53	1.21	2.28
PVA-TETA-DTPAA	1.54	0.38	1.16	3.05
PVA-MA	1.36	0.44	0.92	2.09
PVA-HAD-MA	1.16	0.34	0.82	2.41
PVA-TETA-MA	1.81	0.54	1.27	2.35
PVA-CF_2COOH	1.16	0.38	0.78	2.05
PVA-HAD-CF_2COOH	2.70	0.66	2.04	3.09
PVA-TETA-CF_2COOH	1.94	0.58	1.36	2.34

Tc, total cholesterol; VLDLc, cholesterol of very low density lipoprotein; LDLc, cholesterol of low density lipoprotein; TG, triglyceride; HDLc, cholesterol of high density lipoprotein. S represents adsorption selectivity and is defined as: (VLDLc + LDLc)/HDLc.

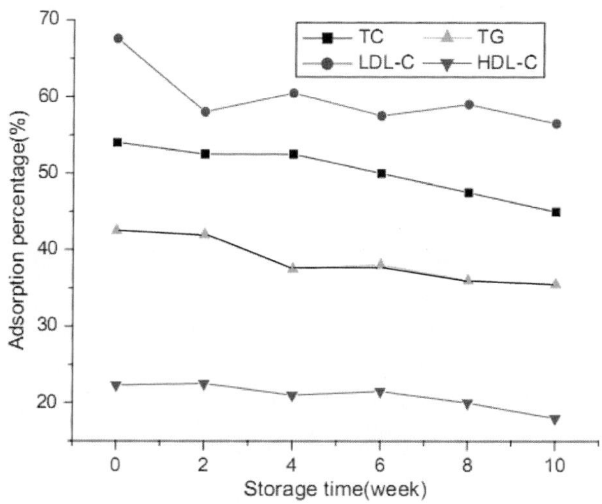

Fig. 13.10. Storage time versus the adsorption percentage. (*With copyright permission from Yu et al. (2014)* monograph in *Artif Cells Nanomed Biotechnol*, early online.)

capacity for total cholesterol (TC), LDL-C, triglyceride (TG), and a lower adsorption capacity for high-density lipoprotein-cholesterol (HDL-C); the adsorption percentage and adsorption capacity for TC, LDL-C, TG, and HDL-C were 54.4%, 67.6%, 42.5%, 10.4% and 4.02, 3.612, 2.154, 0.168 mg/g, respectively, and the ligand ratio of cholesterol to sulfonic ligands was 1.57:1.

Zhang (Zhang *et al.* 2004) synthesized two types of macroporous crosslinked PVA-based polyanion adsorbents by suspension polymerization. They found that two types of polyanion adsorbents had good adsorption properties and adsorption capacities for LDL, and the stronger acidity and larger number of functional groups lead to higher adsorption capacities and adsorption selectivity for LDL. These macroporous crosslinked PVA-based adsorbents could be applied in the treatment of patients with severe hypercholesterolaemia.

3.3. *PVA Adsorbents for the Removal of other Endogenous Toxins*

There are a large number of rheumatoid factors (RFs) in the blood of patients with rheumatoid arthritis (RA). Wang *et al.* (2009) prepared the adsorbents by immobilizing L-phenylalanine as ligands to the macroporous PVA beads for the removal of RF (Fig. 13.11). The parameters of studied adsorbents are listed in Table 13.5. The adsorbent could remove high levels of RFs from the blood of patients with

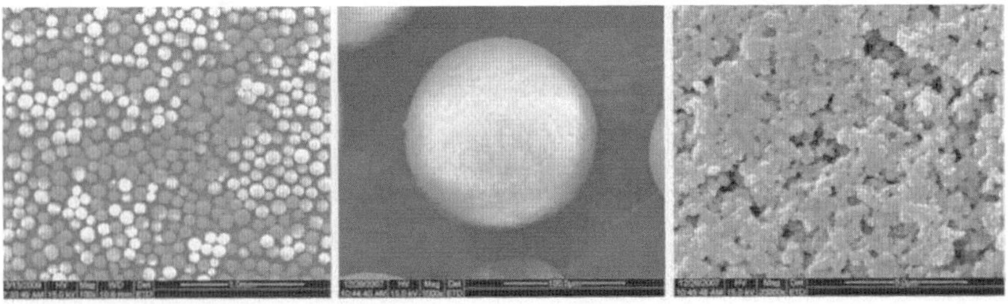

(A) (B) (C)

Fig. 13.11. SEM images of PVA beads: (a) ×100; (b) ×1000; (c) ×2000. (*With copyright permission from Wang et al. (2009)* monograph in *Artificial Cells, Blood Substitutes, and Biotechnology.*)

Table 13.5. Physical Properties of PVA Beads

Parameter	Data
Surface Area (S)/(m^2/g)	36.49
Skeletal density (d_s)/(g/mL)	1.6265
Apparent density (d_a)/(g/mL)	0.4242
Pore volume (V_p)/(mL/g)	1.74
Porosity (P)/%	73.9
Diameter (D)/μm	150
Pore size (r)/nm	1000
Water content (W_Q)/%	80
Swelling ratio (S_w)	4.17

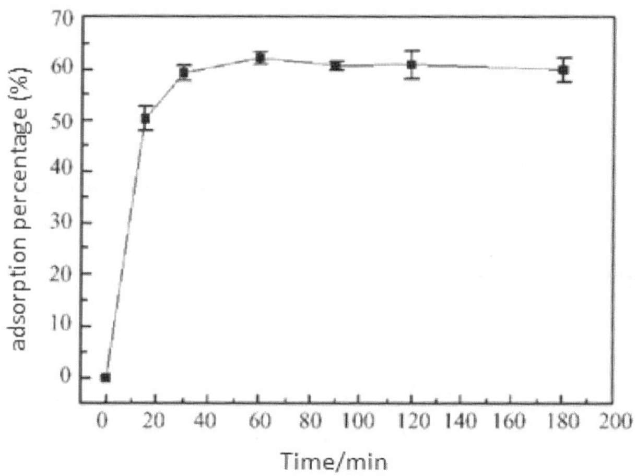

Fig. 13.12. Adsorption kinetic curve of adsorbents for RFs. (*With copyright permission from Wang et al. (2009)* in *Artificial Cells, Blood Substitutes, and Biotechnology.*)

RA at an adsorption capacity of 750 IU/mL with an excellent adsorption percentage (about 60%) (Fig. 13.12).

Recently, adsorbents for sepsis were prepared by incorporating hexane diamine onto macroporous PVA microspheres. And the adsorbents had high adsorption efficiencies for TNF-α (72%), IL-6 (91%) and endotoxins (96%) (Wang *et al.*, 2011; Yu, 2013).

4. Influencing Factors of PVA-based Adsorbent on Adsorption Performance

The process of adsorbing targeting molecules is very complex. There are many factors that influence adsorption performance, including the crosslinking degree of adsorbents, particle size of the adsorbent beads, the ligand immobilized to the adsorbents, temperature and ion strength.

1. *Crosslinking Degree of Adsorbents*

The crosslinking degree influences adsorption performance. Generally speaking, there is a negative relationship between adsorption capacity and crosslinking degree. In the literature by He (Zhang *et al.*, 1998) in 1998, it was found that with increasing amount of crosslinking agent, the adsorption rate and adsorption dynamic characteristics decreased to differing degrees. There are two reasons to explain this phenomenon. First, increasing the crosslinking agent results in a decreasing pore size in the PVA gel, preventing the diffusion of bilirubin molecules into the PVA beads. Second, increasing the crosslinking agent accordingly increases the hydrophobicity of the crosslinked PVA, preventing the bilirubin molecules from fully contacting the adsorbents.

Another study by Li (Li *et al.*, 2004) found that the adsorption capacity for LDL decreased rapidly when the crosslinking degree of PVA-based adsorbents exceeded 33.6% (Fig. 13.13). This is likely because the diffusion process plays a very important role in the adsorption process. Since the pore diameter of PVA-based adsorbents decreased with the increasing of the crosslinking degree, the diffusion of LDL into highly crosslinked adsorbents would be more difficult than that into slightly crosslinked adsorbents.

In addition, Zhang (Zhang *et al.*, 2004) studied the adsorption of LDL onto PVA-based adsorbents immobilized with sulfonic acid groups. This study found that the adsorption percentage of both LDL and HDL decreased with increasing crosslinking degrees, and that the adsorption selectivity of polyanion adsorbents for LDL also decreased.

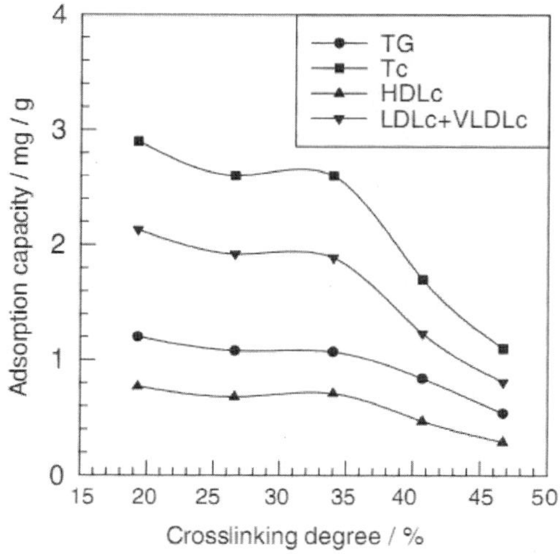

Fig. 13.13. The effect of crosslinking degree on the adsorption capacity of lipoproteins onto PVA-based adsorbents from human plasma (TG, in μmol/g). (*With copyright permission from Li et al. (2004)* monograph in *React Funct Polym.*)

These decreases are due to the reduced sulfonic acid group concentrations on the adsorbents with higher crosslinking degrees, resulting in the decreased adsorption selectivity for LDL.

2. *Ligand Immobilized to Adsorbents*

The adsorption capacity of the PVA-based adsorbents modified with ligands also depends on the number of present functional groups. Yu (2014) synthesized type II amphiphilic PVA adsorbents to study how the amounts of immobilized ligands affect the adsorption property. The initial TC, LDC-C, TG and HDC-C concentrations in the plasma of hyperlipidemia patients were 46.4, 178, 169 and 52 mg/dL, respectively. The TC and LDL-C adsorption capacities and adsorption percentages on the amphiphilic PVA adsorbents increased incrementally with immobilized ligands. The adsorbents had optimal adsorption properties when the amount of immobilized ligands reached 6.5 mg/dl. The adsorption percentages for TC, LDL-C and TC were 54.4, 67.6 and 42.5%, respectively (Table 13.6).

Table 13.6. Coupling Yield versus Adsorption Capacity

Coupling Yield (mg/mL, gel)	TC Adsorption Percentage (%)	Adsorption Capacity (mg/mL)	LDL-C Adsorption Percentage (%)	Adsorption Capacity (mg/mL)	TG Adsorption Percentage (%)	Adsorption Capacity (mg/mL)	HDL-C Adsorption Percentage (%)	Adsorption Capacity (mg/mL)
0	7.43	0.183	8.28	0.147	9.1	0.154	5.03	0.027
1.68	24.9	0.614	35.8	0.637	22.7	0.384	13.6	0.071
3.2	28.8	0.711	42.4	0.754	25.7	0.434	16.2	0.084
4.08	38.2	0.942	47.9	0.854	32.6	0.551	20.2	0.105
5.42	50.0	1.233	58.5	1.041	36.6	0.618	21.0	0.109
6.5	54.4	1.340	67.6	1.204	42.5	0.718	22.3	0.116

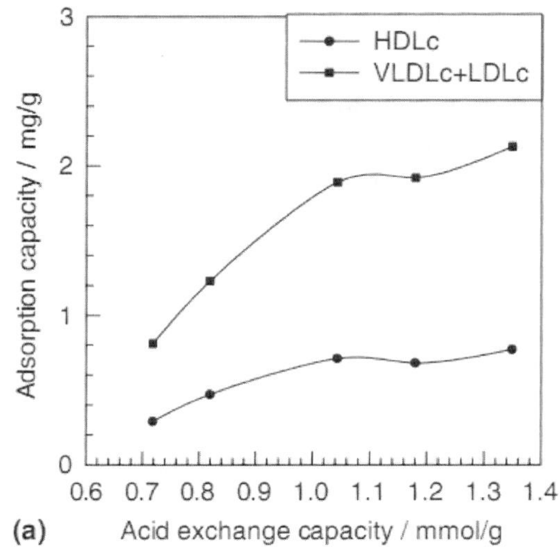

Fig. 13.14. The acid exchange capacity dependence of LDL adsorption on PVA-based adsorbents. (*With copyright permission from Li et al. (2004)* monograph in *React Funct Polym.*)

As shown in Fig. 13.14 (Li *et al.*, 2004), more functional adsorbent ligands resulted in a higher LDL adsorption capacity from human plasma under the employed experimental conditions. However there was no linear relationship between adsorption capacity and ligand amount.

3. *Particle Size of Adsorbent*

As the particle size of adsorbents decreases, the specific surface area will increase, affecting the adsorption properties of adsorbents (Xu *et al.*, 2001). The influence of particle size in sulfonated PVA-based adsorbents on the adsorption capacity and selectivity for LDL is listed in Table 13.7. The adsorption capacity for LDL showed clear increases, while HDL adsorbing capacity only showed slight increases with increasing particle size. This is because LDL is mainly adsorbed onto adsorbents through electrostatic interactions, and increasing the specific surface area of adsorbents enhances the contact between LDL

Table 13.7. Influence of Adsorbent Particle Size on LDL and HDL Adsorption

Particle Size	Concentration before Adsorption/ $mmol{\cdot}L^{-1}$		Adsorption Capacity/%	
	LDL	HDL	LDL	HDL
20~30	4.27	1.36	45	13
30~50	4.27	1.36	51	10
50~100	4.27	1.36	54	8

Table 13.8. Influence of Bead Size on Adsorption Efficiency of TG in Serum

PVA Beads Size (meshes)	20	60
Taurine amount on PVA (mmol/g)	0.86	0.93
TG levels in serum after adsorption (mg/dL)	59.3	56.2
The adsorption percentage of TG in serum (%)	61.7	63.7

C_0^{TG} = 154.7 mg/dL

and adsorbents, thus increasing the LDL adsorption capacity. Conversely, HDL is mainly adsorbed onto adsorbents through distribution effects, and because of the preferential adsorption for LDL, the free space for HDL is reduced. Therefore, the HDL adsorption capacity decreases to some degree. In other words, decreasing the particle size of adsorbents will help to improve the adsorption capacity and selectivity for LDL.

He *et al.* (Zhang *et al.*, 2000) found that the TG adsorption efficiency increased with the decreasing of particle size of beads for crosslinked PVA adsorbent beads modified with taurine (Table 13.8). The TG molecule does not contain ionic groups, thus the adsorption of TG onto PVA adsorbents with taurine ligands is mainly achieved by diffusion. The larger the surface area of the adsorbent beads is, the more chance of contact with TG the adsorbent has, and the better the performance of the adsorption kinetics. However, larger adsorbent particle sizes can immobilize fewer TG molecules, resulting in lower adsorption rates. Nevertheless, the particle size of the adsorbents used in blood perfusion could not be too small, causing a large flow resistance, increasing the perfusion time and influencing the clinical use of the adsorbents.

4. *Temperature*

Temperature could also affect the adsorption capacity of macroporous polymeric adsorbents. However, the influence of temperature on the adsorption process is complex, and the influence of different adsorbents for different targeting molecules is inconsistent. Investigation into the temperature dependence of LDL adsorption onto PVA-based adsorbent immobilized with trifluoroacetic acid is shown in Fig. 13.15 (Li *et al.*, 2004). The LDL adsorption process on the adsorbent is an exothermal process, and lowering the temperature is beneficial for LDL adsorption.

In vitro studies into bilirubin adsorption on crosslinked PVA gel revealed that the adsorption rate increased with increasing temperature (Fig. 13.16). The diffusion speed of bilirubin molecules entering

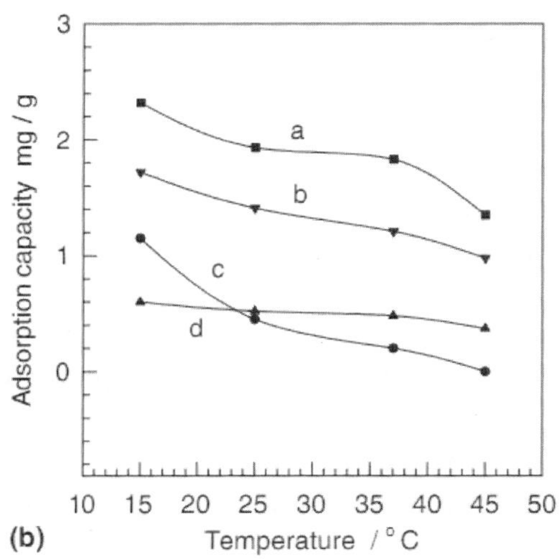

Fig. 13.15. The temperature dependence of LDL adsorption on PVA-based adsorbents: a. Tc; b. LDLc + VLDLc; c. TG (in μmol/g); d. HDLc. (*With copyright permission from Li et al., (2004)* monograph in *React Funct Polym.*)

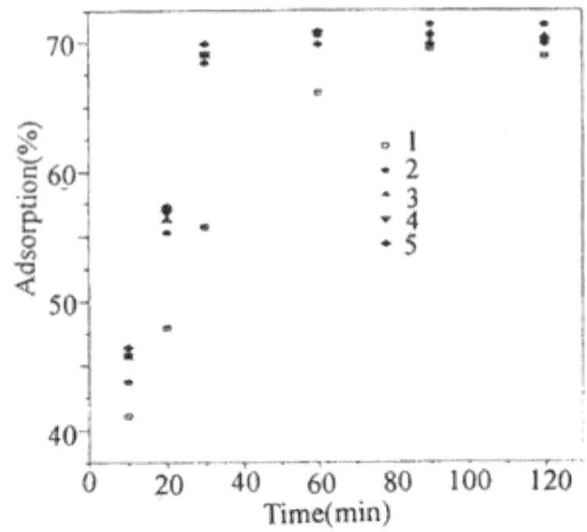

PVA8(20mesh); Bilirubin concentration 150mg/L;

1, 20 ℃; 2, 25 ℃; 3, 30 ℃; 4, 35 ℃; 5, 40 ℃

Fig. 13.16. The effect of temperature on adsorption.

into adsorbent resins increased with increasing temperature. Additionally, the swelling properties of the crosslinked PVA increased, resulting in a reduction in diffusion resistance of bilirubin molecules. However, once the adsorption equilibrium was reached, the adsorption capacity decreased with the rising temperature. This is because the PVA bilirubin adsorption of is dependent on hydrogen bonds between PVA and bilirubin molecules, which are weakened by increasing temperatures.

In a word, multiple factors could affect the adsorption capacity and selectivity of PVA-based adsorbents. Although the relationship between adsorption capacity and the crosslinking degree, particle size, ligand immobilized on adsorbents and temperature still remains unclear, the present work revealed that lowing the crosslinking degree and particle size, enlarging the pore size of the PVA matrix, and increasing the amount of the immobilized ligands would be beneficial for increasing the adsorption capacity and selectivity of PVA-based adsorbents.

The excellent properties of PVA indicate that PVA-based adsorbent is a promising adsorbent in blood perfusion and has potential for clinical treatments.

References

Bai X, Ye ZF, Li YF, Zhou LC. (2010) Preparation of crosslinked macroporous PVA foam carrier for immobilization of microorganisms. *Process Biochem* **45:** 60–66.

Bai X, Ye ZF, Qu YZ, *et al.* (2009) Immobilization of nanoscale Fe-o in and on PVA microspheres for nitrobenzene reduction. *J Hazard Mater* **172:** 1357–1364.

Brown MS, Goldstein JL. (1986) A receptor-mediated pathway for cholesterol homeostasis. *Science* **232(4746):** 34–47.

Gu HQ, Xu GF. (1993) *Biomedical Materials* (M). Tianjin Science & Technology Translation & Publishing Corp., Tianjin

Hegazy ESA, El-Salmawy KM, El-Naggar AA. (2004) Recovery of heavy metals from aqueous solution by using radiation crosslinked poly(vinyl alcohol). *J Appl Polym Sci* **94:** 1649–1656.

Hobbs HH, Russell DW, Brown MS, *et al.* (1990) The LDL receptor locus in familial hypercholesterolemia: mutational analysis of a membrane protein. *Annu Rev Genet* **24:** 133–170.

Lee KH, Julia W, Lee M, *et al.* (2002) Predicting the decrease of conjugated bilirubin with extracorporeal albumin dialysis. *Liver Transplat* **8:** 591–593.

Li HT, Zhang YM, Chen XF, *et al.* (2004) Synthesis and adsorption aspect of crosslinked PVA-based blood compatible adsorbents for LDL apheresis. *React Funct Polym* **58:** 53–63.

Ma KW, Cai SX, Wang X, *et al.* (2008) Preparation of heparin-immobilized PVA and its adsorption for low-density lipoprotein from hyperlipemia plasma. *J Mater Sci- M* **19:** 3255–3261.

Sun T, Li YY, Zhang HC, *et al.* (2010) Molecular imprinting technology based on cyclodextrins. *Prog Chem* **22(05):** 888–897.

Wang WC, Hu YH, Ou LL, *et al.* (2009) Preparation and evaluation of an adsorbent for rheumatoid factors. *Chinese Journal of Biomedical Engineering* **28:** 561–566.

Wang WC, Zhang SN, Hu YH, *et al.* (2008) Preparation of aminated macroporous polyvinyl alcohol resins and evaluation for bilirubin adsorption. *Chinese Journal of Dialysis and Artificial Organs* **19(2):** 1–5.

Wang YJ, Yu YT. (2011) Development of resin adsorbents for blood purification at Nankai University in China. *Artif Cell Blood Sub* **39:** 92–97.

Xu JK, Lu L, Li HT, *et al.* (2001) The selective adsorption of sulfonated crosslinked PVA ball on low density lipoprotein. *Chinese Science Bulletin* **46(23):** 1958–1961.

Yao X, You ZY. (1995) Study on water-loss control property of partial alcoholysis polyvinylacetate. *Speciality Petrochemicals* **5:** 29–32.

Yu YT. (2013) Adsorbents in blood purification: From lab search to clinical therapy. *Chinese Sci Bull* **58:** 4357–4361.

Yu YT. (2014) Amphiphilic polyvinyl alcohol adsorbent for the removal of low-density lipoprotein. *Artif Cells Nanomed Biotechnol,* Early online: 1–7.

Zhang KS, Sun JT, He BL. (1996) Preparation of new adsorbent for bilirubin by immobilizing guanidine on crosslinked polyvinyl alcohol gel. *Chem J Chinese U* **20:** 965–968.

Zhang KS, Sun JT, He BL. (1998) Studies of the adsorption of bilirubin on crosslinked polyvinylalcohol gel. *Ion Exchange and Adsorption* **14:** 204–209.

Zhang KS, Sun JT, He BL. (2000) Adsorption studies of triglyceride in serum by polyvinyl alcohol modified with taurine. *Ion Exchange and Adsorption* **16:** 324–329.

Zhang Y, Li YF, Li X, Yang L, *et al.* (2010) Selective removal for Pb2 in aqueous environment by using novel macroreticular PVA beads. *J Hazard Mater* **181:** 898–907.

Zhang YM. (2004) Synthesis and adsorption aspect of crosslinked PVA-based blood compatible adsorbents for LDL apheresis. *React Funct Polym* **58:** 53–63.

Chapter 14

Ion Exchange Resin in Hemoperfusion

Yongjian Wang*

1. Introduction

1.1. *Ion Exchange Resin*

Ion exchange resins are usually composed of a crosslinked polymer matrix and a functional portion, the ion-active groups, that are relatively uniformly distributed throughout the structure. The matrix provides stability, insolubility and structure, which imparts itself nontoxicity, and can be constructed using organic or inorganic matter. The organic matrixes are usually synthetic polymers (e.g. polystyrene-divinyl benzene) or made of natural polymers (e.g. cellulose, dextran). The inorganic resins include zeolite, clays, etc. General classifications of ion exchange resins are given in Fig. 14.1 and Fig. 14.2.

The active ion groups determine the major properties of the ion exchange resins. Ions with the same charge can be exchanged in stoichiometrically equivalent amounts. The exchange reaction can be represented by the formula

$$\mathbf{R - A + B \rightleftharpoons R - B + A,}$$

*Key Laboratory of Bioactive Materials (Ministry of Education), College of Life Sciences, Nankai University, Tianjin 300071, China. Email: yongjian_wang@nankai.edu.cn

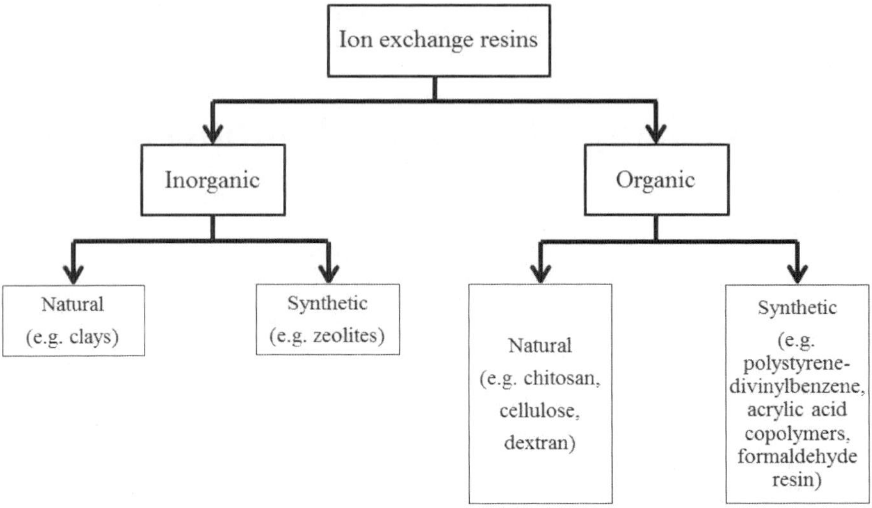

Fig. 14.1. A general classification of ion exchange resins according to polymer matrixes.

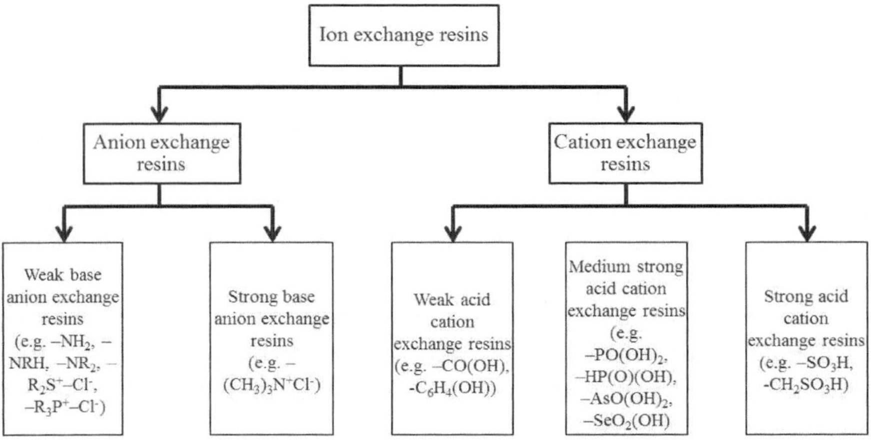

Fig. 14.2. A general classification of ion exchange resins according to active ion groups.

where R represents the matrix structure, A represents the exchangeable ion on the matrix, and B represents the substance that will be adsorbed. According to the properties of active ion groups, ion exchange resins can also be classified into cation exchange resins and anion exchange resins. Cation exchange resins include weak acid or strong acid cation exchange resins, with anionic functionalities and positively charged mobile ions. Weak acid cation exchange resins are primarily based on carboxylic acid groups, or their corresponding salts. They have a high affinity for hydrogen ions and a high capacity for alkaline earth metals. Strong acid cation exchange resins have strong acid functional groups, such as sulfonic acid groups or their corresponding salts. These resins have the ability of exchange cations or split neutral salts. Anion exchange resins include weak base or strong base exchange resins, with cationic functionalities and negatively charged mobile ions. Weak base anion exchange resins, such as those containing amino groups, are capable of high capacity sorption for strong acids. Strong base anion exchange resins functioning with strongly basic groups, e.g. containing quaternary ammonium groups, have the greatest affinity for weak acids such as carbonic acid.

Common uses of ion-exchange resins include water purification, metal recovery and separation, ion substitution, acid–base catalysis, as sensors or as solid electrolytes (e.g. in fuel cells, electrolyzers, electrodialysis devices), in chemical, food, and beverage industries, and in power, nuclear, semiconductor, and pharmaceutical industries (Barbaro and Liguori, 2009).

1.2. *Basic Features of Ion Exchange Resin for Use in Hemoperfusion*

1.2.1. *Exchange capacity*

Exchange capacity refers to the content of the exchangeable group in the resin, and is usually expressed as unit weight or volume (millimole per gram of resin, or millimole per milliliter of resin). The *in vitro* capacity of an ion exchanger resin can be determined by titration. The exchange capacity may limit the amount of substances that may be adsorbed onto a resin. Therefore, resins with high capacity will help in removing large amounts of toxin during hemoperfusion.

1.2.2. *Selectivity*

There is an equilibrium exchange reaction when ion exchange resins come in contact with a solution of anions or cations. Soluble ions may be removed from solution by exchange with the counter-ions adsorbed onto the resins. The sorption amount of soluble ions is also dependent upon the affinity of the ions for the ion exchange resins, as compared with the counter-ions of the resins. In general, the higher the valence of a cation, the greater its affinity. For ions of the same valence, affinity is generally directly related to molecular weight.

1.2.3. *Porosity*

Resin porosity determines the size of the species, molecule or ion that may penetrate into a resin matrix, as well as the rate of diffusion and exchange. In contrast to small organic ions, the penetration of large organic ions in the resin is slow, and these capacities are limited to the active ion sites that the organic ions can reach. Therefore, the porosity influences ionic selectivity. The porosity can be controlled by polymerization conditions.

1.3. *Hemoperfusion by Ion Exchange Resin*

Hemoperfusion is an effective extracorporeal technique that eliminates toxic substances from blood by an adsorptive mechanism during the passage of the blood through a resin. Hemoperfusion can be performed by either plasma perfusion (Fig. 14.3) or direct whole blood perfusion (Fig. 14.4). In plasma perfusion, plasma is first separated using a plasma separator to avoid accelerated coagulation in the column by direct perfusion; then the plasma is perfused through the column containing adsorption materials. Direct hemoperfusion can be used in cases where it is confirmed that coagulation is not stimulated in the column.

The first published application of hemoperfusion by an Amberlite ion exchange resin was performed by Muirhead and Reid in 1948 to remove uremic toxins in dogs (Muirhead and Reid, 1948). Pallotta

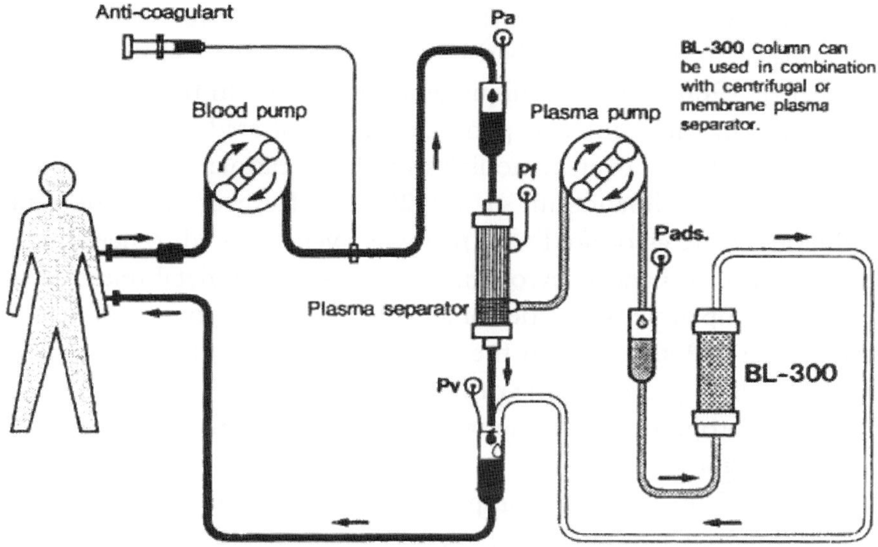

Fig. 14.3. Schematic diagram of plasma hemoperfusion (Nakaji and Hayashi, 2003). *(With copyright permission from John Wiley and Sons.)*

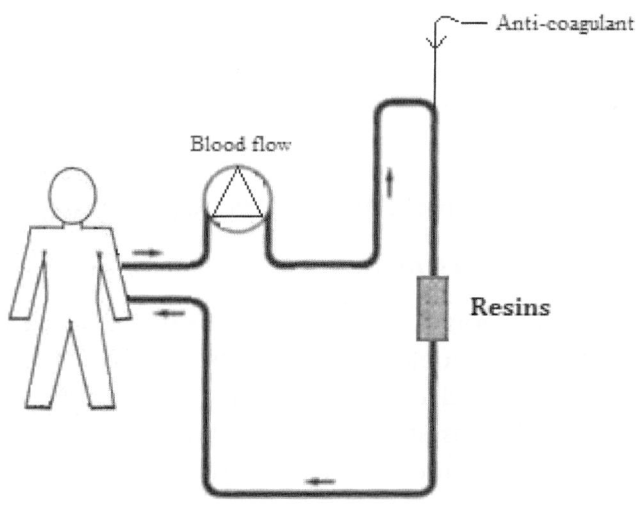

Fig. 14.4. Schematic diagram of direct whole blood hemoperfusion.

and Koppanyi used an anion exchange resin to remove salicylate and phenobarbital from phenobarbital-intoxicated dogs (Pallotta and Koppanyi, 1960). This technique was popularized in the 1970s after several *in vitro* and animal experiments demonstrated its efficacy, and was even preferred over hemodialysis in the treatment of drug overdosed patients (Ghannoum *et al.*, 2014). However, early sorbent techniques were associated with many problems related to hemolysis, electrolyte disturbances, pyrogenic reactions, heavy metal release and poor biocompatible properties of adsorbents such as thrombocytopenia. Chang (1966) reported the use of artificial cells to microencapsulate resins to prevent some of these problems.

Resin properties are important in determining the ability of hemoperfusion to eliminate toxins and poisons. Over the last 20 years, there have been great improvements in resin technologies that have promoted hemoperfusion development. Today, resins used in medical-device applications must undergo rigorous toxicologic and biocompatibility tests, and must guarantee no release of resin particles or leakages that could be harmful to the patient. Hemoperfusion by ion exchange resin is now widely used clinically for the treatment of liver failure, detoxification, removal of specific antibodies, low-density lipoprotein in hypercholesterolemia, and removal of mediators in inflammatory diseases such as sepsis.

2. Ion Exchange Resins in Hemoperfusion

2.1. *Ion Exchange Resins Used in Removing Ions and Drugs*

Hemoperfusion over ion exchange resins was first evaluated on small molecular ions. For example it has been used to treat hyperkalemia, which may produce electrocardiographic abnormalities, cardiac arrhythmias or cardiac asystole. It has previously been reported that hemoperfusion over a cation exchange resin is a simple, inexpensive and efficient method for lowering the concentration of serum potassium ions (Kessler, 1953; McLaughlin, 1959; Moore, 1961; Muirhead and Reid, 1948; Schechter *et al.*, 1958). Sandler *et al.* (1961)

reported that Dowex 50X8, a cation exchange resin, was suitable for use in patients. Perfusion over the resin had no deleterious effect on blood elements or coagulation (Sandler *et al.*, 1961). Besides the removal of potassium ions, Nealon and Ching (1962) demonstrated the resin could also remove ammonium in hepatic coma (Nealon and Ching, 1962).

Hemoperfusion using ion exchange resins can also remove ionic drugs from blood. It acts as a first aid treatment in patients who have overdosed on drugs when conservative treatment is not sufficient (Gelfand and Winchester, 1980). Studies have shown that cation exchange resins can remove alkaline drugs such as quinidine, quinine, mecamylamine, strychnine, morphine and chlorpromazine, and strongly basic anionic resins removed acidic drugs such as barbiturates, salicylates, bromides and iodides.

Hemoperfusion could be useful in neoplastic treatment. Giardina *et al.* (1993) reported an *in vitro* and *ex vivo* evaluation of methotrexate removal by anion exchange resins with different active groups (trimethylammonium, dimethylethanolammonium and dimethylammina) (Giardino *et al.*, 1993). The removal of methotrexate can help prevent several serious problems associated with the therapeutic application of high dose methotrexate, such as nephrotoxicity or hepatotoxicity. These results demonstrated that the resin with trimethylammonium group is effective at removing methotrexate from blood.

However, the nonspecific adsorption of calcium, magnesium and potasium, and hypotension are the major side effects of using ion exchange resins in drug overdose treatment. Faenze *et al.* (1992) tested an anion exchange resin, Dowex SAR, coated with methacrylic hydrogel which adsorbed HCO_3^- ions. In addition, the electrolytic balance was fully restored after 10 minutes of treatment (Faenza *et al.*, 1992).

Deng *et al.* (1997) designed a cation exchange resin to remove opioid drugs and anesthesia drugs by hemoperfusion, so as to reduce drugster dependency on drugs. The adsorption capacity for morphine is about 40.5 mg/g resin. The resin has little effect on red blood cells, white blood cells and platelets (Deng *et al.*, 1997).

2.2. *Ion Exchange Resin Used in Removing Bilirubin*

Bilirubin is formed as a result of the catabolism of hemoglobin from aged red blood cells in mammals (Ostrow, 1986), and is classified into two different states: one is an insoluble unconjugated state (indirect bilirubin), and the other is a soluble glucuronide-conjugated state (direct bilirubin). Indirect bilirubin is produced from hemoglobins broken by ageing or disease. Direct bilirubin is produced from indirect bilirubin in the liver cells by a reaction with UDP-glucuronyl transferase, and is transferred to the kidney or intestines, where it is finally decomposed. High concentrations of bilirubin can evoke hepatic or biliary tract dysfunction and permanent brain damage (Stocker *et al.*, 1987). Elevated concentrations of bilirubin in the blood of newborn infants suffering from unconjugated hyperbilirubinemia may affect neurological brain damage (kernicterus), morbidity and mortality. Mild hyperbilirubinemia is treated by phototherapy or by the administration of human serum albumin and drugs. Severe cases require exchange transfusion.

Hemoperfusion has become a promising technique to help the body eliminate excess bilirubin in blood. Bilirubin consists of an open chain of four pyrrole-like rings (tetrapyrrole), and includes two carboxyl groups, four secondary amines and two carbonyl groups (Bonnett *et al.*, 1976) (Fig. 14.5). In most cases, basic ion exchange materials are used to adsorb bilirubin.

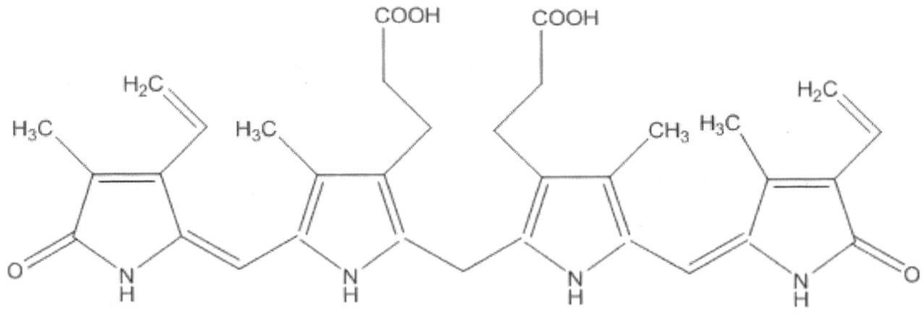

Fig. 14.5. The molecular structure of bilirubin.

2.2.1. *Ion exchange resin with quaternary aligand*

Strong basic anion exchange resins have a large number of quaternary ammonium functional groups. Bilirubin adsorption occurs by electrostatic interactions between the negative charge of the bilirubin carboxyl group and the positive charge of quaternary ammonium ion of the anion exchange resin.

Sideman *et al.* (1982) studied the application of anion exchange resins for the removal of bilirubin from blood, thus replacing exchange transfusion (Mor *et al.*, 1980; Sideman *et al.*, 1980; Sideman *et al.*, 1977; Sideman *et al.*, 1979). The anion-exchange resin used in this study for removal of plasma bilirubin did not alter the electrolyte composition, acid–base balance or the concentration of other measured metabolites (Lopukhin *et al.*, 1977). The resin particles can be made biocompatible by cross-linked albumin coatings to prevent platelet depletion and thrombocytopenia (Sideman *et al.*, 1981). Hemoperfusion in 8–14 kg female baboons showed no significant effect of the column on blood chemistry and coagulation factors and repeated runs had no ill effects (Sideman *et al.*, 1982).

The removal of toxic substances from the blood appears to be beneficial for patients with hepatic failures. Idezuki *et al.* (1978) developed a tertiary amine anion exchange resin, IONEX, in the late of 1970s (Idezuki *et al.*, 1978). Bile acid and bilirubin were removed by direct hemoperfusion with IONEX. *In vitro* tests showed that 67% of conjugated bilirubin, and 48% unconjugated bilirubin, were removed at the completion of the perfusion. Direct hemoperfusion in dogs revealed that the removal rate of bilirubin in the peripheral blood was about 49%, and about 67% for bile acid. In clinical applications, the average decreases in bilirubin and bile acid in peripheral blood of patients after direct hemoperfusion were 26% and 30%, respectively (Idezuki *et al.*, 1981). However, the IONEX adsorbs a large amount of heparin during hemoperfusion, thus limiting its use in clinical applications. Kanai *et al.* used a heparin-bound IONEX to avoid heparin adsorption. The results showed that it had a sufficient adsorption capacity for bilirubin and bile acids for reducing the blood level to a level similar to non-heparinized IONEX (Kanai *et al.*, 1985).

BR-601 is an anion exchange resin that has a high sorption capacity for bilirubin (Ouchi *et al.*, 1978). It can be used for removal of protein-bound bilirubin from critical care patients (Asanuma *et al.*, 1980). For patients with postoperative liver failure, the percent reductions in total and direct bilirubin, and in bile acids, were higher by BR-601 than by plasma exchange. Further, the coma grade was improved. However, the patients did not recover (Sakagami *et al.*, 1986).

A new bilirubin adsorbent resin based on styrene divinylbenzene, BR 350, was developed and investigated both experimentally and clinically by Morimoto *et al.* Plasma adsorption using a serial connection of BR 350 and noncoated charcoal was proceeded in order to achieve greater improvement in patients with postoperative hepatic failure. After *in vitro* perfusion of high bilirubin containing plasma through these columns for three hours, total bilirubin levels were drastically reduced to 21% of the pre-perfusion level in the serial connection of non-coated charcoal and BR 350. However, it remained high, above 40%, in the single use of each column. Total branched chain and aromatic amino acid levels were also drastically reduced in the serial connection of these columns to 50, 40, and 7%, respectively, while the total amino acid levels remained high at 87% in the single use of BR 350. Patients who received this treatment achieved an initial reduction of plasma total bilirubin and aromatic amino acids of 57 ± 6 and 84 ± 7%, respectively. Although the long-term prognosis for these patients was negative, this treatment resulted in improved clinical and laboratory findings. This plasma adsorption system could provide a possibility for an improved supportive therapy for hepatic failure, especially for patients with hepatic coma and hyperbilirubinemia (Morimoto *et al.*, 1989).

The efficacy of BR-350 for removal of bilirubin and bile acids in five patients with intractable jaundice of various origins was further investigated. The concentration of total bilirubin was cut by 31 to 60%, and total bile acids were reduced by 20 to 74%. Three patients recovered. Two patients died despite the bilirubin adsorption treatment. The effects of the adsorbent column on specific blood parameters, including the coagulation system, were measured. The data suggest that bilirubin adsorption should be examined further as a treatment for critically ill patients with intractable jaundice (Geiger *et al.*, 1992). But an unexpected decrease in plasma retinol (vitamin A) concentrations in a patient with type I Crigler–Najjar syndrome was found when BR-350 was used for the treatment of unconjugated hyperbilirubinemia (Ihara *et al.*, 1998).

Kuraray Medical Inc. developed a bilirubin adsorption column, Medisorba BL-300, that can be used for the treatment of hyperbilirubinemia (Nakaji and Hayashi, 2003). This column uses a porous anion exchange resin coated with hydrophilic 2-hydroxyethyl methacrylate copolymer (poly 2-hydroxyethyl methacrylate containing 0.5% glycidyl methacrylate as a comonomer) in order to improve blood compatibility. When the amount of coated PHEMA increases in the manufacturing process of BL-300 non-specific adsorption is suppressed. However, as the amount of coated PHEMA increases, the adsorption rate of the bilirubin decreases. Therefore, the suitable quantity of PHEMA coated on the adsorbent for BL-300 was chosen considering a balance of bilirubin adsorption performance and blood compatibility.

In vitro and clinical evaluations of BL-300 have been conducted by different groups. The average removal rate of total bilirubin was 52.2% in four treatments of two post-operative liver failure patients (Usami *et al.*, 1995). The removal rate was 57.3% for total bilirubin, 56.7% for direct bilirubin and 61.9% for bile acid for two patients with primary biliary cirrhosis and liver cirrhosis with jaundice (Suga *et al.*, 1990). Kabutan *et al.* conducted 16 treatments for patients with multiple traumas. The average removal rate of total bilirubin was 42.6%, and that of direct bilirubin was 39.9% (Kabutan *et al.*, 1991). Suzuki *et al.* performed 42 treatments for the patients with liver cirrhosis, primary biliary cirrhosis, and drug induced hepatitis. The average removal rate of direct bilirubin and indirect bilirubin was 37% and 36%, respectively (Suzuki *et al.*, 1991). These clinical data show that BL-300 is effective in the treatment of hyperbilirubinemia patients.

Wang *et al.* synthesized macroporous a poly-vinylpyridine resin with quaternary ammonium group. The removal of bilirubin by this

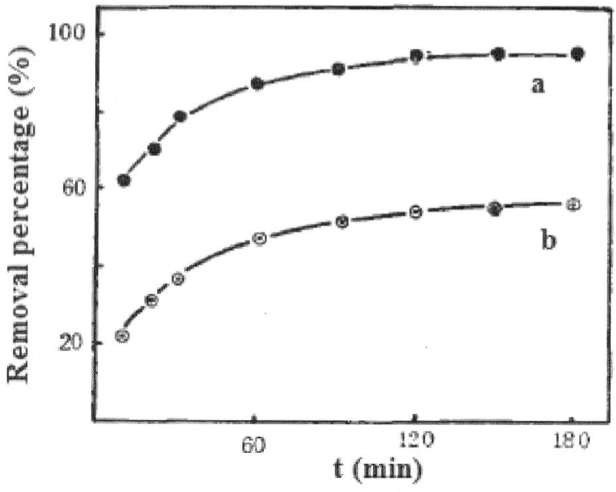

Fig. 14.6. Removal percentage of bilirubin by macroporous poly-vinylpyridine resins with quaternary ammonium groups (a) with tertiary amine groups (b). *Reproduced from Wang et al., 1989.* (*With copyright permission from* Acta polymerica sinica.)

resin was 91% during three hours of perfusion, which is higher than the removal by resin with tertiary amine group (Fig. 14.6). This result indicates that strong base group is more efficient than weak base in removal of bilirubin and is also the reason for the poor biocompatibility of the adsorbent (Wang *et al.*, 1989).

2.2.2. *Ion exchange resin with amino ligand*

Chitosan is a nontoxic, biocompatible, natural polysaccharide that has structural characteristics similar to glycosaminoglycans. It has high adsorption capacities for bilirubin and aromatic amino acids and can be used for the treatment of jaundice in dogs (Xu *et al.*, 1990). Chitosan resins with high amino group content have been shown to have high adsorption capacities (Xu *et al.*, 2003). Li *et al.* compared aminated chitosan resins with polystyrene divinyl benzene resins and found that aminated chitosan had higher adsorption efficacy for bilirubin in the plasma of severe hepatitis patients than polystyrene divinyl benzene resins (Table 14.1) (Li *et al.*, 2006).

Table 14.1. Adsorption Properties of Aminated Chitosan Resins and Polystyrene Divinyl Benzene Resins ($n = 10$)

Adsorbent	Adsorption Percentage of Total Bilirubin (%)	Adsorption Percentage of Direct Bilirubin (%)	Adsorption Percentage of Indirect Bilirubin (%)
Aminated chitosan resins	50.5 ± 3.4	57.0 ± 11.3	39.0 ± 7.2
Polystyrene divinyl benzene resins	30.1 ± 2.5	32.6 ± 3.0	27.6 ± 2.9

Reproduced from Li et al. (2006) (With copyright permission from BME & Clin Med.)

Cross-linked chitosan resin, prepared by crosslinking chitosan with glutaraldehyde, is highly stable. Cross-linked chitosan resins modified with di-, tri-, tetra-, and pentammines have a higher adsorption capacity for unconjugated and conjugated bilirubin than those with cross-linked chitosan resin due to the electrostatic and hydrophobic interactions (Yu *et al.*, 2000). Glutaraldehyde cross-linked chitosan resin can also be modified with poly(amidoamine) (PAMAM) dendrimers. The resin has satisfactory competitive abilities with BSA for bilirubin adsorption, though the adsorption percentage decreased in the presence of BSA (Wu *et al.*, 2011).

Poly(2-hydroxyethyl methacrylate) (PHEMA) resins are non-toxic and have good biocompatibility. The hydrophobic α-methyl groups and backbone of PHEMA impart hydrolytic stability to the polymer and support the mechanical strength of the polymer matrix. PHEMA resins used for bilirubin adsorption have been prepared by immobilizing amino groups on the adsorbent. The results showed that resins immobilizing hexanediamine and protonized hexanediamine had good adsorption efficiency and high adsorption capacity compared to PHEMA immobilizing triethylenetetramine. The adsorption of bilirubin by PHEMA adsorbent containing amino group could be attributed to the electrostatic interaction between amino group on adsorbent and carboxyl group of bilirubin, furthermore, the hydrophobic interaction and hydrogen bonding also played an important role (Lu *et al.*, 2003; Yuan *et al.*, 1999).

2.2.3. *Ion exchange resin with amino acid ligand*

Henning *et al.* synthesized polymer resins with amino acid ligands for bilirubin adsorption. They found that resins with either arginine or lysine ligands were capable of adsorbing bilirubin from aqueous solution (pH = 7.8), but those with other amino acid ligands showed no indication of sorption (Henning *et al.*, 1984). The amount of bilirubin adsorbed increased with increasing basicity of the amino acids in the resin. As the number of basic amino acids on the resin was increased from one to five, a 4.7-fold enhancement in the adsorption capacity was seen for arginine, while a 9.3-fold enhancement was obtained for lysine (Henning *et al.*, 1986). The resin containing (Arg)5(Ala)3 oligopeptides was able to retain adsorbed bilirubin in the presence of BSA (Henning *et al.*, 1987).

Poly-L-lysine coated chitosan beads were prepared for bilirubin binding. Such surface-modified chitosan beads exhibited high binding affinities for bilirubin in relation to activated charcoal and cholestyramine. The binding capacity was proportional to the amount of polylysine bonded to the chitosan beads. The hemolytic potential of all the poly-L-lysine modified beads is compatible with polystyrene control tubes. It appears that polylysine-modified chitosan may be an excellent sorbent for hemoperfusion due to its high binding affinity, capacity, and good blood compatibility (Chandy and Sharma, 1992). The polylysine coated polyamide resins have an improved binding affinity for bilirubin over the sorbents with lysine containing oligopeptide pendants, and a much higher affinity than cholestyramine. Additionally, replacement of poly-L-lysine with poly-D-lysine does not affect the adsorption capacity (Zhu *et al.*, 1990).

2.2.4. *Ion exchange resin with guanidine ligand*

Polyvinyl alcohol (PVA) is a widely used hydrophilic gel, and is compatible with blood. It contains large amounts of hydroxyl groups that enables it to be easily modified with various functional groups. A new adsorbent for bilirubin was prepared by immobilizing guanidine on a

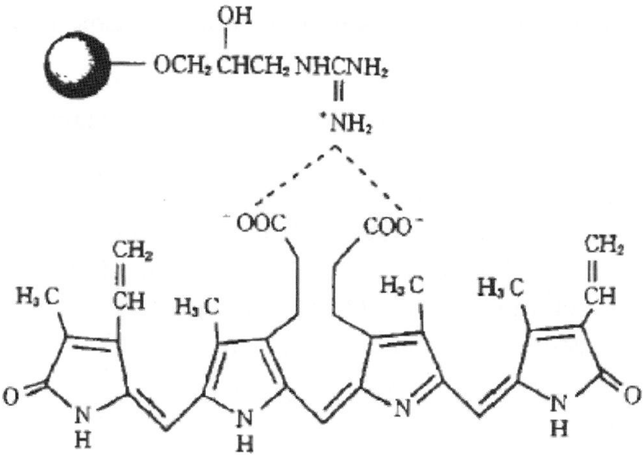

Fig. 14.7. Schematic diagram of the interaction between bilirubin and guanidine. (*With copyright permission from* Chemical Journal of Chinese Universities-Chinese.)

cross-linked polyvinyl alcohol resin (Fig. 14.7). This adsorbent has good adsorption performance for bilirubin in both albumin-free and albumin solutions. It was found that increasing guanidine could improve the adsorption performance, and the adsorption occurs by electrostatic interactions and hydrogen bonding (Zhang *et al.*, 1999).

2.2.5. *General discussion*

In most cases, basic ion exchange resins with primary amine, secondary amine and tertiary amine as ligands have been used for adsorbing bilirubin from plasma. Numerous natural polymers, like chitosan, agarose and agar as well as synthetic polymers, such as polystyrene divinyl benzene, ethylene glycol dimethacrylate, hydroxyethyl methacrylate, and acrylamide, have recently been investigated as sorbents for the removal of bilirubin directly from the plasma of patients suffering from hyperbilirubinemia. In general, the adsorption efficacy and hemocompatibility of the resins are related to the number and strength of the basic ligands, as well as the matrix of the resin.

2.3. *Ion Exchange Resin Used in Removing Low Density Lipoprotein (LDL)*

A high concentration of LDL in blood is one of the main factors causing hyperlipidemia, atherosclerosis and cardiovascular disease, and the reduction of LDL concentrations in blood can prevent or delay disease progression. Although diet and lipid-lowering drugs are common methods used to control LDL levels in clinical treatment, extracorporeal elimination of LDL is an effective treatment for familial hypercholesterolemia patients, and for those who cannot be sufficiently treated by diet and drugs.

Apolipoprotein B100 is the sole protein component of LDL, and has large amounts of alkaline amino acid residues such as lysine, arginine and histidine, on its surface. Therefore, cation ion exchange resins can be used to adsorb LDL via electrostatic interactions.

2.3.1. *Ion exchange resin with heparin ligand*

Heparin is a polyanionic polysaccharide containing a large number of carboxyl and sulfonic groups with a negative charge. It is the first ligand used for LDL adsorbent.

Lupien R. *et al.* prepared a LDL adsorbent by linking heparin to agarose beads. Reduction of plasma-cholesterol by the removal of LDL in an extracorporeal system was described as a possible approach for the treatment of familial hypercholesterolaemia (Lupien *et al.*, 1976). LDL was removed from the blood by their interaction with heparin in the presence of calcium ions. This technique was specific for the removal of LDL, because the concentration of high-density lipoproteins was not affected. The treatment was well tolerated by all three subjects (i.e., two hypercholesterolaemia patients and a normal volunteer), and there were no undesirable side effects. Several hematological parameters, clinical-chemistry tests, including serum enzymes, and immune-electrophoresis of plasma proteins were all unaffected by the treatment.

Other resins with heparin ligands have also been made. However, calcium ion should be added in plasma during the course of hemoperfusion by resin with heparin ligand, and should be removed to reduce the calcium concentration to its physiological concentration after hemoperfusion. The operation is cumbersome, thus this kind of resin has not been widely used.

2.3.2. *Ion exchange resin with dextran sulfate or sulfuric acid ligand*

Dextran sulfate has a structure similar to that of the LDL receptor, and seems to act as a type of pseudo-receptor (Tani, 1996). Low-molecular dextran sulfate (MW 4500) can selectively absorb apolipoprotein B with high affinity and low toxicity. Dextran sulfate was covalently bound to cellulose beads and used as the specific sorbent of apolipoprotein B-containing lipoproteins for plasmapheresis treatment (Mabuchi *et al.*, 1987; Yokoyama *et al.*, 1984; Yokoyama *et al.*, 1985). Further, hemoperfusion by this resin requires no additional calcium ions. The adsorption mechanism is a direct electrostatic interaction between dextran sulfate and the positively charged surface of apolipoprotein B-containing lipoproteins (LDL, very low density lipoprotein (VLDL), and lipoprotein (a) (Lp(a))). The adsorbent can be regenerated with 4.1% NaCl solution (Agishi, 1994) and reused. This LDL hemoperfusion device has been used clinically since 1988 (Bambauer *et al.*, 1997; Koga *et al.*, 1999; Mabuchi *et al.*, 1988).

Kobayashi *et al.* (2002) developed a new low-density lipoprotein adsorbent for use in whole-blood LDL-hemoperfusion. The adsorbent is made from dextran sulfate covalently bound to porous cellulose beads. The adsorbent has a large particle size (~250 μm). In whole blood perfusion tests, the adsorbent directly adsorbed atherogenic LDL cholesterol from whole blood and maintained the concentration of high-density lipoprotein cholesterol which is unchanged significantly. The adsorbent showed minimal side effects in terms of blood cell activation, complement activation, and blood cell loss, suggesting that it had excellent blood compatibility. In addition, the adsorbent showed mechanical stability and no hemolysis. The clinical trials for patients suffering from homozygous familial hypercholesterolemia, revealed that LDL apheresis could be continued safely without major technical problems for several years. Apheresis effectively reduced the serum levels of total and LDL cholesterol. Furthermore, it reduced Lp(a), which is not influenced by lipid-lowering drugs. The reduction of LDL cholesterol and Lp(a) may delay the progression of the atherosclerotic process, thereby reduce the risk of new episodes of coronary heart disease, thus extending the life expectancy of these patients (Fadul *et al.*, 1997).

Similar adsorbents can be prepared by immobilizing sulfuric acid groups on polymers. For example, sulfonated cross-linked dextran resins (Chen *et al.*, 1993), sulfonated macroporous cross-linked poly(glycidyl methacrylate) (Guo *et al.*, 2001), sulfonated hydroxyethyl cross-linked chitosan (Yu *et al.*, 1997), sulfonated cross-linked poly(vinyl alcohol) (Xu and He, 2002), and taurine immobilized cross-linked poly vinyl alcohol (Zhang *et al.*, 1999).

2.3.3. *Ion exchange resin with polyacrylate or carboxyl ligand*

Polyacrylate covalently bound to Fractogel was tested as a sorbent for lipoproteins from human plasma in a perfusion column. *In vitro* experiments indicated complete elimination of low-density lipoproteins from citrated plasma, no binding of high-density lipoproteins, fibrinogen, or the complement factors C3 and C4 were observed, and biocompatibility was demonstrated by extracorporeal plasma perfusion in sheep. The results indicated the potential use of polyacrylate-coated Fractogel for the removal of low-density lipoproteins from plasma *ex vivo* (Thies *et al.*, 1988).

In 1993, Bosch *et al.*, first described a new polyacrylate-based LDL adsorbent (DALI) that is directly applicable to whole blood (Bosch *et al.*, 1993), so that no cell/plasma separation is necessary. Whole blood hemoperfusion decreases the cost and complexity of the extracorporeal circuit. This adsorbent consists of anionic polyacrylate ligands that interact electrostatically with the positively-charged apoprotein B moiety of LDL, Lp(a) and VLDL, thereby retaining these atherogenic lipoproteins on the column. Furthermore, only lipoproteins can freely diffuse into the interior sponge-like structure of the adsorbent beads while cells are excluded (size exclusion effect); this significantly improves the hemocompatibility of the material.

High-density lipoprotein (HDL) can also penetrate the beads, but because the apo AI-coated HDL is not attracted to the ligand, it cannot be eliminated. Bosch *et al.* reported on the largest DALI database ever published. In a five-year follow-up, 12,291 DALI sessions were documented in 158 hypercholesterolemic atherosclerosis patients from 28 apheresis centers. This study supported the long-term efficacy, safety and ease of DALI apheresis handling. The results showed that the long-term therapy with DALI was safe, effective and selective as low-density lipoprotein cholesterol (LDL-C) and Lp(a) could be reduced to more than 60% per session in approximately 100 min treatment time, while high-density lipoprotein cholesterol decreased, and the incidence of acute adverse events were low (Bosch *et al.*, 2006). However, this column can be used for only one treatment.

Similar adsorbent can be prepared using immobilized carboxyl group as the ligand (Li *et al.*, 2004). Adsorbents bearing carboxyl as a functional group showed varying adsorption capability. The adsorption capacity depended on the type of the functional group that was immobilized on the macroporous cross-linked PVA beads. Adsorbents bearing stronger acids, such as trifluoroacetic acid, as the functional group could adsorb more LDL than the adsorbents bearing relatively weaker acids, such as chloroacetic acid, as the functional group (Table 14.2).

2.3.4. *Ion exchange resin with phosphate ligand*

A new low-density lipoprotein (LDL) adsorbent that uses phosphate group as the ligand was prepared by Yu *et al.* (2006) and Wang *et al.* (2008; 2009). *In vitro* adsorption assays showed that the adsorption

Table 14.2. Adsorption of Lipoproteins onto Macroporous Cross-Linked PVA-Based Adsorbents from Human Plasma

Adsorbent	TC (mg/g)	HDLC (mg/g)	LDLC+VLDLV (mg/g)
PVA-CF_2COOH	1.32	0.40	0.92
PVA-CH_2COOH	2.70	0.66	2.04

Reproduced from Li et al. (2004) (With copyright permission from Reactive & Functional Polymers.)

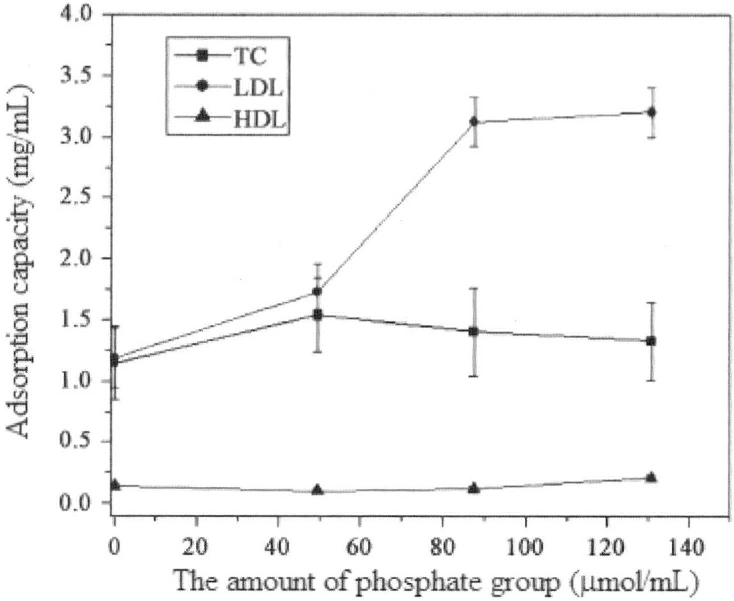

Fig. 14.8. Effect of amount of coupled phosphate on adsorption capacity. *Reproduced from Wang et al. (2008)* (*With copyright permission from* Chinese Journal of Biomedical Engineering.)

capacities of plasma lipoproteins were increased with the amount of phosphate group (Fig. 14.8). The pathogenic lipoproteins in the plasma, such as total cholesterol (TC), LDL and triglyceride (TG), could be removed significantly at rates of 42.9, 45.0 and 44.74%, respectively. Further, the reduction of HDL and other normal plasma components was very slight. Eight hyperlipidemia rabbits were treated *in vivo* with the PVA-phosphate adsorbent, and the removal of TC, LDL and TG was 45.03 ± 6.64%, 48.97 ± 9.92% and 35.42 ± 14.17%, respectively (Wang *et al.*, 2008; Wang, *et al.*, 2009).

2.3.5. *Ion exchange resin with amino acid ligand*

Amino acid was covalently attached to cellulose or polystyrene resins for the removal of LDL from plasma or blood. The results showed that resins with lysine or leucine ligands had a high adsorption capacity and selectivity for LDL (Chen *et al.*, 1992).

2.3.6. *Ion exchange resin with amphiphilic ligand*

Adsorbents with amphiphilic ligands for adsorption of low density lipoprotein (LDL) were prepared by Wang *et al* (2002; 2003). Resins were introduced by the sulfonic group, which acted as the hydrophilic moiety, and the cholesterol group, which acted as the hydrophobic moiety (Fig. 14.9). These adsorbents had a better selectivity in removing LDL-C, TC, TG compared to adsorbents which only contained hydrophobic or hydrophilic ligands (Table 14.3) (Cheng *et al.*, 2003; Wang *et al.*, 2003).

2.3.7. *General discussion*

Bambauer (2002) used four different adsorbents for patients suffering from familial hypercholesterolemia: cellulose with dextran sulfate ligand (Liposorber, Japan), polyacrylate coated polyacrylamide (Dali, Germany), sepharose coupled with specific antihuman apolipoprotein B-100 (immunoadsorbent) (Therasorb, Germany) and sepharose

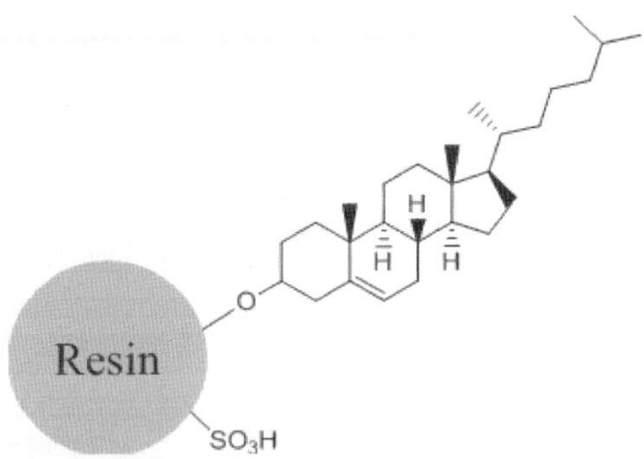

Fig. 14.9. Schematic diagram of cellulose adsorbent with amphiphilic ligands.

Table 14.3. Adsorption Properties of Adsorbents

Adsorbent	TC (mg/mL)	TG (mg/mL)	LDLC (mg/mL)
Amphiphilic ligand	1.836 ± 0.082	1.992 ± 0.056	1.519 ± 0.134
Cholesterol ligand	0.084 ± 0.011	0.0938 ± 0.009	0.036 ± 0.003
Sulfonic ligand	0.499 ± 0.011	0.458 ± 0.023	0.330 ± 0.008

(*With copyright permission from Biomaterials*)

coupled with special anti-lipoprotein (a) columns (immunoadsorbent) (Lipopak, Russia). Among the four treatments, an average reduction of 50.6% for total cholesterol, 52.2% for LDL, 64.3% for Lp(a), 43.1% for triglycerides, and an average increase of 10.3% for HDL was achieved. Severe side effects, such as shock or allergic reactions, were very rare (0.5%) in all treatments. At the end of the trials, there were no significant differences among the four treatments with respect to the clinical outcome experienced with the patients' total cholesterol, LDL, HDL, and triglyceride concentrations. However, to reduce high Lp(a) levels, the immunoadsorption treatment with special Lp(a) column (Lipobak) seems to be most effective: –59% (Lipobak) versus –25% (Liposorber), –25% (Therasorb), and –29% (Dali) (Fig. 14.10).

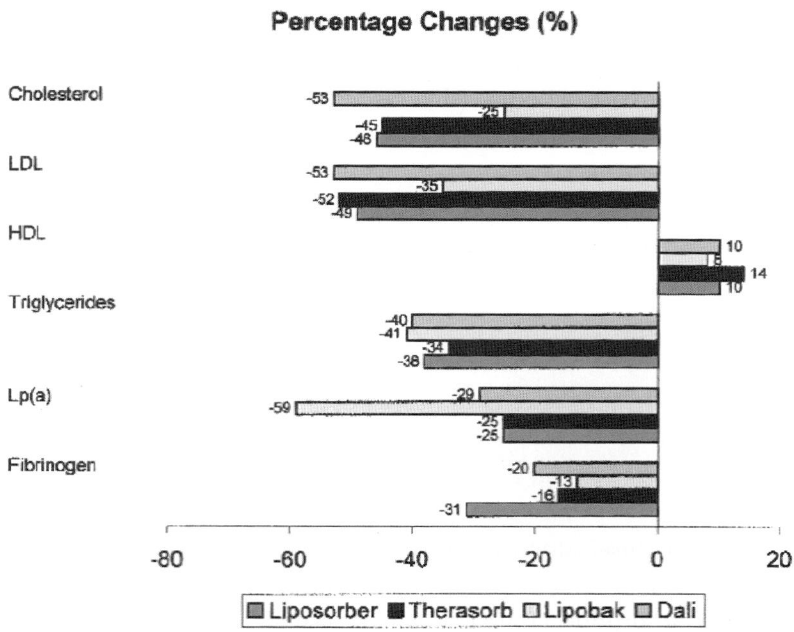

Fig. 14.10. The average percentage changes (%) in lipoproteins with four different adsorbents in LDL apheresis methods after a treatment period of 84.9 + 43.2 months. The average changes were calculated for both treatment periods [pretreatment time 1–6 years; LDL apheresis treatment time 1–9 years]. (*With copyright permission from Biomaterials.*)

The present data demonstrate that treatment with LDL apheresis of patients suffering from familial hypercholesterolemia resistant to maximum conservative therapy is very effective and safe, even in long-term applications (Bambauer, 2002). However, only the dextran sulfate and H.E.L.P. systems are currently approved for the use of familial hypercholesterolemia in the United States by the Food and Drug Administration (FDA) (Schwartz *et al.*, 2013).

2.4. *Ion Exchange Resin Used in Removing Endotoxin*

The biological effects of endotoxins on human beings and animals include affecting the structure and function of organs and cells, changing metabolic functions, raising body temperature, triggering coagulation cascades, modifying hemodynamics and causing shock, even at very low concentrations in the blood stream (Martich *et al.*, 1993). Because of the high endotoxin toxicity, the efficient removal of endotoxins from patient blood is very important in clinical medicine.

Endotoxins are lipopolysaccharides comprised of lipid A. The lipid A is responsible for the major bioactivity of endotoxins (Wang and Quinn, 2010). It contains long chain fatty acids and negatively charged phosphate groups. Anion exchange resins may interact with the negatively charged groups of lipid A so as to remove endotoxins.

2.4.1. *Ion exchange resin with polymyxin B ligand*

Polymyxin B is a cationic cyclic deca-peptide antibiotic agent, which has a strong bactericidal activity to Gram-negative bacteria. The action mechanism is attributed to the electrostatic interactions, between cationic antibiotic molecules with acidic phospholipids and the lipopolysaccharide (LPS) of the outer membrane of Gram-negative bacteria, which result in destroying the bacterial wall (Newton, 1956). Polymyxin B selectively binds to many kinds of LPS so as to recognize endotoxins from different origins. Hemoperfusion with polymyxin B-immobilized fiber columns can potentially interrupt the biological cascade of sepsis. This treatment is safe, and may improve cardiac and renal dysfunction due to sepsis or septic shock (Vincent *et al.*, 2005).

Cohen *et al.* (1987) immobilized polymyxin B on Sepharose 4B (PB-Sepharose). The columns removed 94% of 5 μg endotoxin *in vitro*. Plasma of rats was perfused over PB-Sepharose, which protected the rats from endotoxin-induced leukopenia, thrombocytopenia, and death (four of four survivors compared with none of the four controls) (Cohen *et al.*, 1987).

Direct hemoperfusion with polymyxin-immobilized fiber was carried out on dogs with *Escherichia coli*-induced sepsis. Endotoxin is thought of as the most important pathogenic factor in sepsis, making it an ideal potential therapeutic target. All of the dogs in the control group died within 18 hours after bacterial infusion. All of the dogs in the treated group survived for more than three days. One of the treated dogs survived for seven days and another for 14 (Hanasawa *et al.*, 1989).

Serious endotoxin intoxication can cause endotoxemia. Endotoxemia caused by cardiac surgery using cardiopulmonary bypass (CPB) can lead to systemic inflammatory response syndrome and deterioration of organ function. Ohki *et al.* evaluated the effectiveness of endotoxin removal with a polymyxin B-immobilized hemoperfusion cartridge in CPB on pigs. The results showed that the treatment reduced the inflammatory reaction caused by CPB, as well as cardiac and pulmonary dysfunctions (Ohki *et al.*, 2008).

Hemoperfusion by resins with polymyxin B ligands has been used in clinical applications (Novelli *et al.*, 2010a; Novelli *et al.*, 2010b). An early clinical report on polymyxin B hemoperfusion was published in 1994. Of the 16 patients with septic multiple organ failure that were treated with direct polymyxin B hemoperfusion, nine were alive two weeks after this therapy, and seven patients were discharged from the hospital (Aoki *et al.*, 1994).

Nakamura *et al.* reported that polymyxin B hemoperfusion is effective in reducing proximal tubular dysfunction for patients with severe sepsis, in part owing to the reduced plasma endotoxin levels (Nakamura *et al.*, 2004). Also, PMX-direct hemoperfusion might alleviate immunosuppression associated with severe sepsis (Kojika *et al.*, 2006).

In a preliminary study, the addition of polymyxin B hemoperfusion to conventional therapy significantly improved hemodynamics and organ dysfunction in a targeted population with severe sepsis and/or septic shock from intra-abdominal gram-negative infections. The survival of the polymyxin B group was 68%, higher than the 47% survival of the control group who underwent conventional treatment (Cruz *et al.*, 2009). Uric *et al.* retrospectively analyzed clinical profiles of 24 patients with septic shock (16 patients, gram-negative; eight patients, non-gram negative septic shock) who underwent direct hemoperfusion with polymyxin B-immobilized fiber. Their findings indicate that the improvement in hyperdynamic circulation was related directly to endotoxin removal by the polymyxin B-immobilized fiber column, and endotoxin has an important role in the development of hyperdynamic circulation in patients with gram-negative septic shock (Uriu *et al.*, 2002).

A recent study about polymyxin B-immobilized fiber column hemoperfusion in patients with septic shock found that after 24 hours of PMX treatment, the median radial plasma endotoxin concentration decreased from 16.48 to 1.857 pg/mL, and the concentration at the outlet circuit of the PMX column was further decreased to 0.779 pg/mL. Further, the median endotoxin removal rate was 74.4% (Mitaka *et al.*, 2014). The hemodynamic conditions of septic shock due to intra-abdominal infection were improved by continued duration of direct hemoperfusion (Shimizu *et al.*, 2013).

Taniguchi *et al.*, reported the efficacy of endotoxin adsorption therapy by polymyxin B hemoperfusion for patients with methicillin-resistant *Staphylococcus aureus* toxic shock syndrome who showed no improvement with conventional therapy. After hemoperfusion, the systolic arterial pressure significantly increased (from 89 to 125 mmHg, $P < 0.05$), patient white blood cell count decreased (from 17,640 to 10,090 /uL, $P < 0.05$), and sequential organ failure assessment scores decreased (from 13 to 9, $P < 0.05$). All patients recovered and were discharged from the ICU (Taniguchi *et al.*, 2013).

In another study for sepsis, polymyxin B-immobilized fiber columns added to conventional therapy significantly improved hemodynamics and organ dysfunction and reduced 28-day mortality in this targeted population (Cruz *et al.*, 2010).

2.4.2. *Ion exchange resin with amino acid ligand*

Fang *et al.* investigated the properties of cellulose beads linked with eight kinds of amino acids, and poly-L-lysine as ligand, to adsorb endotoxins in aqueous solution. The experimental results showed that arginine and lysine as ligands have the highest adsorption capacities of 182.0 and 160.0 pg/mL gel in 100.0 pg/mL endotoxin solution, respectively (Fig. 14.11). Adsorption isotherm of lysine as ligand showed that the endotoxin adsorption capacity increased linearly with increased endotoxin concentrations (Fang *et al.*, 2004b). They further evaluated the effectiveness of endotoxin removal by cellulose beads with lysine as ligand on New Zealand white rabbits with infectious symptoms. They found that the mean blood endotoxin concentration was reduced significantly (from 5.560.54 EU/mL to 0.410.26 EU/mL) after perfusion (Fig. 14.12), and liver function and renal function tests showed significant improvement. Other parameters, such as superoxide dismutase and malondialdehyde, were

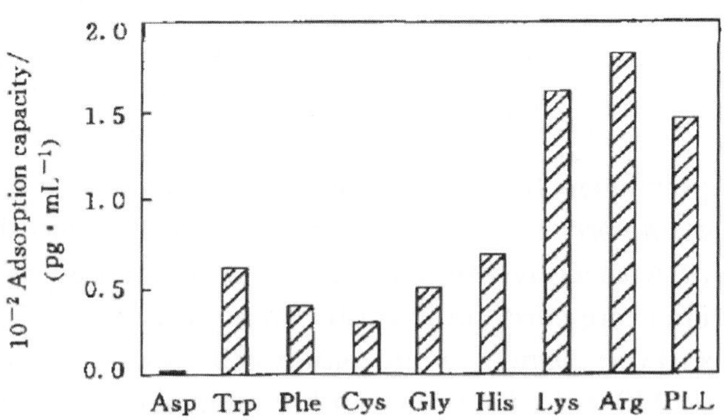

Fig. 14.11. Adsorption of endotoxins by cellulose beads with amino acids as ligand. (*With copyright permission from* Chemical Journal of Chinese Universities-Chinese.)

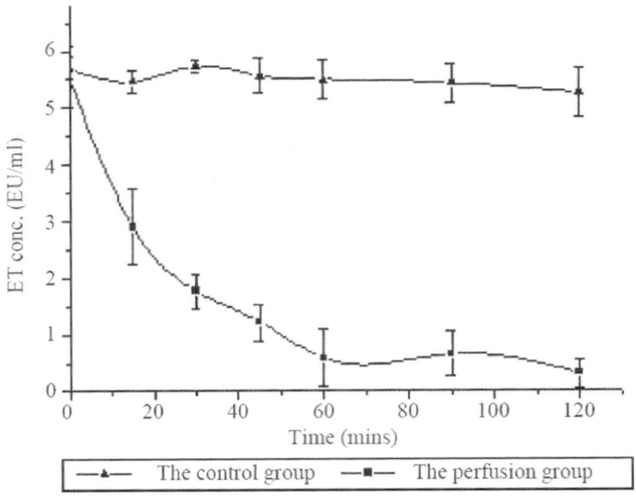

Fig. 14.12. Removal of ET from rabbit's blood by perfusion using Lys immobilized cellulose beads. (*With copyright permission from* Biomaterials.)

significantly improved. Moreover, the adsorbent had good mechanical strength, blood compatibility and cytotoxicity (Fang *et al.*, 2004a). Agar gel beads with lysine ligands also had good affinity to endotoxins, with an adsorption capacity of 253.1 EU/g, at an initial concentration of 10 EU/mL, and the adsorbent was blood compatible (Liu *et al.*, 2004).

Histidine was also used as a ligand to selectively remove endotoxins from serum (Zhang *et al.*, 2007). The results show that the endotoxin removal efficiency was more than 85%. The component content in serum before and after adsorption indicates that the nonspecific adsorption was almost negligible on this affinity adsorbent. The main function of the histidine ligand should be due to the alkalinity of imidazole, which forms a positively charged histidine at neutral pH, and can catch endotoxins by the formation of an anion–cation complex.

Another group also studied the effect of amino acids as ligands on the adsorption of endotoxins. A series of adsorbents based on amino acids ligands were prepared, and computer simulations were used to

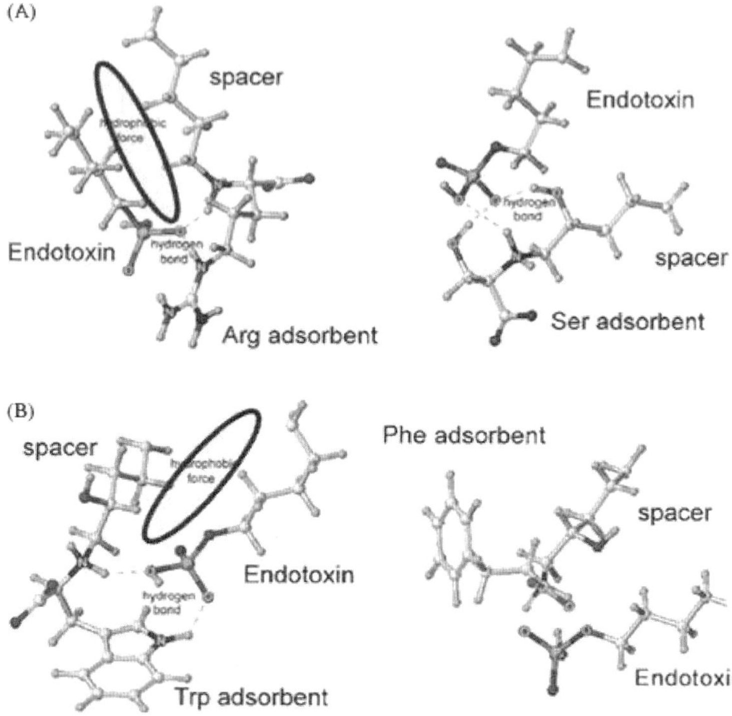

Fig. 14.13. Computer simulation of interaction models between endotoxin and ligand: the broken lines are hydrogen bonds formed between endotoxins and ligands, the black elliptical regions represent hydrophobic interaction. *Reproduced from Wei et al. (2007)* (*With copyright permission from* Journal of Chromatography B.)

study the interaction between the endotoxins and ligands (Fig. 14.13). The charge and polarity of amino acid ligands have a great influence on the adsorption of endotoxins. Electrostatic force plays a dominant role in the adsorption process, and the R group of the polar amino acid ligand makes it possible to form a hydrogen bond with an endotoxin molecule, which significantly enhances endotoxin adsorption. Further, arginine and serine were found to be better ligands for the adsorption of endotoxin than other amino acid ligands, and could effectively remove endotoxins from rabbit serum. The removal efficiency of adsorbents was up to 78% (Wei *et al.*, 2007).

2.4.3. *Ion exchange resin with other cationic or polycationic ligand*

A cross-linked chitosan matrix with divinyl amine ligands was prepared to adsorb endotoxins from plasma. This adsorbent was blood compatible and had a clearance of over 80% (Wang *et al.*, 2005a).

Bengsch *et al.* investigated the use of cellulose resin with diethylaminoethyl (DEAE) groups for reducing the plasma concentration of endotoxin in patients with severe sepsis. A significant reduction in plasma endotoxin levels from a median of 0.61 to 0.39 EU/mL was achieved ($P < 0.001$). Except for a transient and reversible increase of prothrombin time, no adverse events were observed in patients undergoing this adsorption apheresis treatment. The data show that the reduction of endotoxin by extracorporeal DEAE-cellulose-based plasma treatment may prove to be a promising therapeutic tool for patients suffering from bacterial sepsis and endotoxemia (Bengsch *et al.*, 2005).

An endotoxin adsorbent based on macroporous cellulosic beads with immobilized polyethylenimine (PEI) was prepared. The results showed that PEI, as well as polymyxin B, resulted in complete removal of endotoxins from plasma and water. However, the biocompatibility of the PEI absorber was superior to that of polymyxin B absorber (Mitzner *et al.*, 1993).

Weber *et al.* studied the removal properties for endotoxins by cationically modified cellulose adsorbents. Microspheres cationically modified through substitution with PEI or DEAE groups, and adsorbent manufactured by (physically) coating the cellulose matrix with PEI, exhibited a high adsorption capacity for endotoxins in human plasma, whereas activated charcoal and various anion exchange resins removed only small amounts of endotoxins under the same conditions (Weber *et al.*, 1995).

These adsorbents that use cationic or poly-cationic ligands have low production costs, high adsorption efficiency, and excellent biocompatibility. The adsorption process is primarily based on electrostatic interactions. With the adsorption of lipopolysaccharide by particles with lysine or amino groups, the adsorption properties were found to depend strongly on pK(a,app) but not necessarily on the

Table 14.4. Adsorptive Capacity for LPS and Apparent Dissociation Constant of Adsorbents

Adsorbent	$pK_{a, app}$	Amino-Group Content (meq g^{-1})	M_{lim}	Adsorptive Capacity for LPS (mg mL^{-1} wet adsorbent)	$K_{d,app}$ (M)
PMLG-NH_2	8.0	2.2	2,000	2.7	1.0×10^{-10}
PL/CMO = 90/10	7.4	5.3	8,000	1.8	2.6×10^{-10}
PL/CMO = 75/25	6.8	4.0	2,000	1.0	3.7×10^{-10}

(*With copyright permission from Journal of Chromatographia.*)

amino-group content. When pK(a,app) was increased, the LPS-adsorbing capacity was increased, and the apparent dissociation constant decreased (Table 14.4) (Sakata *et al.*, 2001).

2.4.4. *General discussion*

The endotoxin adsorption properties of various cationic ligands were compared. The results indicated that dimethylamine ligand had a higher adsorption capacity than amino, amide and hydroxy (Yu *et al.*, 2004). Electrostatic interaction is the main driving force during all adsorption events. Moreover, hydrogen bond and hydrophobic interaction between endotoxin and adsorbent could enhance their adsorption capacity (Li *et al.*, 2006; Yuan *et al.*, 2005).

Sato *et al.* (1993) evaluated the efficacy of hemoperfusion over polymyxin B immobilized fiber (PMX), and plasma perfusion over anion exchange resin (Resin), on experimental canine septic shock. The 24-hour survival rates of non-treated (Sham), PMX, and Resin groups were 0, 80, and 40%, respectively. In the Resin group, some parameters were significantly better than those of the Sham group, but the efficacy of this treatment was less than that of the PMX treatment. Hemoperfusion over polymyxin B immobilized fibers can detoxify circulatory endotoxin, resulting in improvement of systemic and organic disorders caused by sepsis (Sato *et al.*, 1993). However, polymyxin B is highly toxic to mammalian cells, as well as the kidney and nerve system. The patients must risk possible leaks of polymyxin B from the adsorber, and future development should focus on reducing or eliminating polymyxin B leakage.

Harm *et al.* (2014) consider that some commercial adsorbers do not have potential use in extracorporeal blood purification for endotoxin removal. They determined the endotoxin adsorption capacities of commercial adsorbers for endotoxin removal in buffer solution, protein solution, in serum and heparinized plasma; some of these were also characterized in whole blood. The tested lipopolysaccharides adsorbers were Toraymyxin (R) PMX-20R, Alteco (R) LPS Adsorber, DEAE-Sepharose, polymyxin B-Agarose, and EndoTrap (R) red. The results showed that DEAE-Sepharose was most effective in adsorbing lipopolysaccharides. Of the other tested endotoxin removal materials, only the polymyxin B-based adsorbers were able to reduce lipopolysaccharide activity. However, it was shown that the reduction in lipopolysaccharide activity was caused by desorbed polymyxin B that inactivated the endotoxins (Harm *et al.*, 2014).

Septic shock is not always triggered by endotoxins, and the removal of endotoxins alone does not lead to recovery in all septic shock patients (Cinel and Opal, 2009), however, there is clear biological rationale for endotoxin removal in the clinical treatment of severe sepsis and septic shock. The therapeutic effect of PMX-DHP has sometimes also been questioned. Rittirsch *et al.* (2008) reviewed the molecular interactions that occur during sepsis and suggested that hypercytokinemia is thought to play an important pathophysiological role, and the removal of cytokine should be considered as a therapeutic strategy in endotoxemia patients, septic patients and patients with many other critical illnesses, such as acute respiratory distress syndrome and severe acute pancreatitis.

3. Mechanism and Features of Ion Exchange Resins in Hemoperfusion

Hemoperfusion by ion exchange resin adsorbs toxins onto ion exchange resins so as to remove the toxins from blood. Adsorption by ion exchange resin relies on the principle that opposite electric charges attract one another, and the ionic functional groups of the resin interact with ions of opposite charge in solution. Toxic substances which are eliminated from blood should be ionic in solution. For example, bilirubin consists of two carboxyl groups, and has negative charges in blood that can be adsorbed by anion ion exchange resin.

Selective adsorption is mainly based on electric charge, and the selective adsorption of ion exchange resin is very important for hemoperfusion. The resin should only adsorb toxic substances, not other blood components. Molecules with opposite charges to the functional group are bound to the ion exchange resin; any neutral or like-charged molecules pass through the resin without binding. For example, Apolipoprotein B100, the sole protein component of LDL, has a large amount of alkaline amino acid residues, and is positively charged in blood. LDL can be adsorbed by cation ion exchange resin. Lipid A includes negatively charged phosphate groups. Anion exchange resins may interact with the negatively charged groups of lipid A so as to remove it from blood. Hence, the selection of ion exchange resin depends on the charge of the substance of interest. In general, for ions of the same charge, the higher the valence of an ion, the greater its affinity. For ions of the same valence, affinity is generally, with a few exceptions, directly related to molecular weight.

Selective adsorption is also associated with ionization degree. A weakly basic or acidic molecule that requires very low or very high pH for ionization will require the use of a strong ion exchange resin, because it functions at extreme pH levels. The carboxyl group of bilirubin is a weak acid group. For resins with amino acid as functional group, the adsorption capacity of bilirubin increases with the increasing basicity of the amino acids in the resin. For the separation of molecules with high charge from those with small charges, it is best to use weak exchange resins. If the charge differences between molecules are very small, weak exchange resins produce more resolution. In general, weak base anion exchange resins are capable of adsorbing strong acids at a high capacity. Strong base anion exchange resins have the greatest affinity for weak acids such as carbonic acid.

Ionic form also influences adsorption selectivity. The ionic form of an ion exchange resin refers to the counter-ion that is adsorbed onto the resin's functional group. This ion will be exchanged during adsorption. The strength of the interaction with a given resin varies for different counter-ions. The lower the selectivity of a counter-ion for the resin, the more readily it can be exchanged for another ion of like charge (for example, the toxin of interest).

Other factors that influence the adsorption include the matrix of resin and the spacer between the matrix and active ion. The matrix determines the accessibility of the adsorbed substance. For the removal of bilirubin in plasma of severe hepatitis patients, aminated chitosan had higher adsorption efficacy than polystyrene divinyl benzene resins (Li *et al.*, 2006). The spacer also influences the adsorption properties of resins. In general, the adsorbents with longer spacers have higher adsorption capacities than those with shorter spacers (Hou *et al.*, 2005; Yuan *et al.*, 2005), especially for the adsorption of large molecules. A new adsorbent was prepared by immobilizing polymyxin B on polystyrene resin via heparin spacer. The spacer reduces the steric hindrance during the adsorption process and improves the adsorption capacity of the resin (Wang *et al.*, 2013).

PAMAM has a branched structure and a large number of terminal functional groups, and it can be used as a new kind of spacer for adsorbents with high adsorption capacity (Wang *et al.*, 2005b). Yuan *et al.* prepared sulfonated cellulose resins bearing PAMAM as the spacer to link sulfonate groups. The results showed that the resin had a high amount of sulfonate groups, and high adsorption capacity for LDL (Yuan *et al.*, 2008).

4. The Key Issues of Ion Exchange Resins in Hemoperfusion

Currently, several ion exchange resins have been used in clinical applications to treat patients with drug intoxication, endotoxemia, hyperbilirubinemia, hyperlipemia, familial hypercholesterolemia, renal failure, autoimmune disease, and other diseases. It is selective, stable, safe and relatively inexpensive in hemoperfusion. However, there are still limitations in extending the applications of the technique. One of the main limitations is to develop high performance resins, including good biocompatibility, and enhanced capacity and selectivity for clearing poisons at lower costs and complication rates. The resin needs to be carefully considered from a physicochemical perspective, and/or for underpinning for the theory of disease.

References

Agishi T, Wood W, Gordon B. (1994) LDL apheresis using the Liposorber LA-15 system in coronary and peripheral vascular disease associated with severe hypercholesterolemia. *Curr Ther Res* **55**: 879–904.

Aoki H, Kodama M, Tani T, *et al.* (1994) Treatment of sepsis by extracorporeal elimination of endotoxin using polymyxin B-immobilized fiber. *Am J Surg* **167(4):** 412–7.

Asanuma Y, Malchesky PS, Smith JW, *et al.* (1980) Removal of protein-bound toxins from critical care patients. *Clin Toxicol* **17(4):** 571–81.

Bambauer R. (2002) Low-density lipoprotein apheresis: Clinical results with different methods. *Artif Organs* **26(2):** 133–139.

Bambauer R, Olbricht C, Schoeppe E. (1997) Low-density lipoprotein apheresis for prevention and regression of atherosclerosis: Clinical results. *Therapeutic Apheresis: Official Journal of the International Society for Apheresis and the Japanese Society for Apheresis* **1(3):** 242–8.

Barbaro P, Liguori F. (2009) Ion exchange resins: Catalyst recovery and recycle. *Chem Rev* **109(2):** 515–529.

Bengsch S, Boos KS, Nagel D, *et al.* (2005) Extracorporeal plasma treatment for the removal of endotoxin in patients with sepsis: Clinical results of a pilot study. *Shock* **23(6):** 494–500.

Bonnett R, Davies JE, Hursthouse MB. (1976) Structure of bilirubin. *Nature* **262:** 327–328.

Bosch T, Gahr S, Belschner U, *et al.* (2006) Direct adsorption of low-density lipoprotein by DALI-LDL-apheresis: Results of a prospective long-term multicenter follow-up covering 12 291 sessions. *Therapeutic Apheresis and Dialysis* **10(3):** 210–218.

Bosch T, Schmidt B, Blumenstein M, *et al.* (1993) Lipid apheresis by hemoperfusion: *In vitro* efficacy and *ex vivo* biocompatibility of a new low-density lipoprotein adsorber compatible with human whole blood. *Artif Organs* **17(7):** 640–52.

Bosch T, Schmidt B, Kleophas W, *et al.* (1997) LDL hemoperfusion: A new procedure for LDL apheresis: Biocompatibility results from a first pilot study in hypercholesterolemic atherosclerosis patients. *Artif Organs* **21(10):** 1060–1065.

Chandy T, Sharma C. (1992) Polylysine-immobilized chitosan beads as adsorbents for bilirubin. *Artif Organs* **16(6):** 568–76.

Chang TMS. (1966) Semipermeable aqueous microcapsules ("artificial cells"): With emphasis on experiments in an extracorporeal shunt system. *Trans Am Soc Artif Intern Organs* **12:** 13–19.

Chen B, Li C, Pan J, *et al.* (1992) A study of adsorbents for the adsorption of low density lipoprotein (LDL). *Ion Exchange and Adsorption* **8(3):** 234–238.

Chen B, Pan J, Tong M, *et al.* (1993) A study of adsorbents for the adsorption of low density lipoprotein (LDL). *Ion Exchange and Adsorption* **9(4):** 330–334.

Cheng Y, Wang S, Yu Y, *et al.* (2003) *In vitro*, *in vivo* studies of a new amphiphilic adsorbent for the removal of low-density lipoprotein. *Biomaterials* **24(13):** 2189–2194.

Cinel I, Opal SM. (2009) Molecular biology of inflammation and sepsis: A primer. *Crit Care Med* **37(1):** 291–304.

Cohen J, Aslam M, Pusey CD, *et al.* (1987) Protection from endotoxemia: A rat model of plasmapheresis and specific adsorption with polymyxin B. *J Infect Dis* **155(4):** 690–5.

Cruz DN, Antonelli M, Fumagalli R, *et al.* (2009) Early use of polymyxin B hemoperfusion in abdominal septic shock the EUPHAS randomized controlled trial. *JAMA* **301(23):** 2445–2452.

Cruz DN, de Cal M, Piccinni P, *et al.* (2010) Polymyxin-B hemoperfusion and endotoxin removal: Lessons from a review of the literature. In: Ronco C, Piccinni P, Rosner M (eds.), *Endoxemia and Endotoxin Shock: Disease, Diagnosis and Therapy. Contributions to Nephrology.* S. Karger AG Press.

Deng H, Guo X, He B. (1997) Preparation and structure of receptor adsorbent and its adsoption properties of anesthesia drugs. *Science in China (Series B)* **27:** 306–313.

Fadul J, Vessby B, Wikstrom B, *et al.* (1997) Treatment of homozygous familial hypercholesterolemia with low density lipoprotein apheresis: A 4 year follow-up study. *Artif Organs* **21(5):** 364–368.

Faenza S, Balestri M, Martinelli G, *et al.* (1992) Hemoperfusion with a new anion exchange resin corrects the metabolic alkalosis in pyloric stenosis: An experimental demonstration. *Int J Artif Organs* **15(11):** 677–80.

Fang H, Wei H, Yu Y. (2004a) *In vivo* studies of endotoxin removal by lysine-cellulose adsorbents. *Biomaterials* **25(23):** 5433–5440.

Fang H, Wei J, Yu Y. (2004b) Studies on adsorption performance of affinity adsorbent for bacterial endotoxins. *Chemical Journal of Chinese Universities-Chinese* **25(6):** 1056–1059.

Geiger H, Klepper J, Lux P, *et al.* (1992) Biochemical assessment and clinical evaluation of a bilirubin adsorbent column (BR-350) in critically ill patients with intractable jaundice. *Int J Artif Organs* **15:** 35–39.

Gelfand MC, Winchester JF. (1980) Hemoperfusion in drug overdosage: A technique when conservative management is not sufficient. *Clin Toxicol* **17(4):** 583–602.

Ghannoum M, Bouchard J, Nolin TD, *et al.* (2014) Hemoperfusion for the Treatment of Poisoning: Technology, Determinants of Poison Clearance, and Application in Clinical Practice. *Semin Dialysis* **27(4):** 350–361.

Giardino R, Fini M, Giavaresi G, *et al.* (1993) *In vitro* and *ex vivo* evaluation of methotrexate removal by different sorbents haemoperfusion. *Biomat Artif Cell Im* **21(4):** 447–54.

Guo X, Sun Y, Chen Y, *et al.* (2001) Study on synthesis of macroporous crosslinked poly(glycidyl methacrylate) and their adsorption properties for low density lipoprotein. *Chinese Journal of Biomedical Engineering* **20(1):** 17–22.

Hanasawa K, Tani T, Kodama M. (1989) New approach to endotoxic and septic shock by means of polymyxin B immobilized fiber. *Surg Gynecol Obstet* **168(4):** 323–31.

Harm S, Falkenhagen D, Hartmann J. (2014) Endotoxin adsorbents in extracorporeal blood purification: Do they fulfill expectations? *Int J Artif Organs* **37(3):** 222–232.

Henning D, Brown G, St-Pierre L. (1986) Polymer resins with amino acid containing pendants for sorption of bilirubin. II. Polyamide resins with various basic amino acids. *Int J Artif Organs* **9(1):** 33–8.

Henning D, Clas S, Brown G, *et al.* (1987) Polymer resins with amino acid containing pendants for sorption of bilirubin. III. Adsorption and desorption in the presence of albumin. *Biomater Artif Cells* **15(4):** 677–86.

Henning D, Lajoie G, Brown G, *et al.* (1984) Polymer resins with amino acid containing pendants for sorption of bilirubin. I. Comparison of Merrifield and polyamide resins. *Int J Artif Organs* 7**(4):** 209–14.

Hou G, Wang H, Liu T, *et al.* (2005) Preparation of adsorbents for the removal of endotoxin. *Artif Cells Blood Sub* **33(2):** 227–237.

Idezuki Y, Hamaguchi M, Hagiwara M, *et al.* (1978) Development of a new artificial liver support system: Treatment of hyperbilirubinemia by direct hemoperfusion using IONEX-column. *Artif Organs (Japan)* **7:** 1053.

Idezuki Y, Hamaguchi M, Hamabe S, *et al.* (1981) Removal of bilirubin and bile acid with a new anion exchange resin: Experimental background and clinical experience. *Trans Am Soc Artif Intern Organs* **27:** 428–433.

Ihara H, Shino Y, Hashizume N, *et al.* (1998) Decline in plasma retinol in unconjugated hyperbilirubinemia treated with bilirubin adsorption using an anion-exchange resin. *Journal of Nutritional Science and Vitaminology* **44(2):** 329–336.

Kabutan K, Kosaka F, Mizobuchi S, *et al.* (1991) Comparison of an anion-exchange resin with an uncoated powdered charcoal column in treatment by plasma perfusion. *Therapeutic Plasmapheresis* **9:** 335–338.

Kanai F, Takahama T, Onishi K, *et al.* (1985) Hemocompatibility of heparin-fixed IONEX. *Trans Am Soc Artif Intern Organs* **31:** 451–8.

Kessler BS, Liebler JB, Abraham JT, *et al.* (1953) Reduction of hyperkalemia by circulating blood through a cation exchange resin. *Proc Soc Exp Biol Med* **84:** 508.

Kobayashi A, Nakatani M, Furuyoshi S, *et al.* (2002) *In vitro* evaluation of dextran sulfate cellulose beads for whole blood infusion low-density lipoprotein-hemoperfusion. *Therapeutic Apheresis* **6(5):** 365–371.

Koga N, Watanabe K, Kurashige Y, *et al.* (1999) Long-term effects of LDL-hemoperfusion on carotid arterial atherosclerosis in familial hypercholesterolemic patients. *J Intern Med* **246:** 35–43.

Kojika M, Sato N, Yaegashi Y, *et al.* (2006) Endotoxin adsorption therapy for septic shock using polymyxin B-immobilized fibers (PMX): Evaluation by high-sensitivity endotoxin assay and measurement of the cytokine production capacity. *Therapeutic Apheresis and Dialysis* **10(1):** 12–18.

Li H, Zhang Y, Chen X, *et al.* (2004) Synthesis and adsorption aspect of crosslinked PVA-based blood compatible adsorbents for LDL apheresis. *Reactive & Functional Polymers* **58(1):** 53–63.

Li J, Yu M, Wang H, *et al.* (2006) Studies on adsorption mechanisms of adsorbents for endotoxin. *Chemical Journal of Chinese Universities-Chinese* **27(6):** 1066–1069.

Li T, Zhang J, Lu W, *et al.* (2006) A screening study *in vitro* on the adsorptive property of novel chitosan for plasma bilirubin and cytokines. *BME & Clin Med* **10(3):** 131–134.

Liu T, Hou G, Yu M, *et al.* (2004) Hemperfusion adsorbents with lysine as ligand for removing of endotoxin. *Chemical Journal of Chinese Universities-Chinese* **25(12):** 2284–2286.

Lopukhin UM, Molodenkev MN, Leykin UA, *et al.* (1977) Removal of plasma bilirubin by hemoperfusion in dogs. *Am J Gastroenterol* **68(4):** 345–353.

Lu L, Yuan Z, Shi K, *et al.* (2003) Studies on adsorption of bilirubin by crosslinking poly (2-hydroxyethyl methacrylate) adsorbents containing amino group. *Chemical Journal of Chinese Universities-Chinese* **24(3):** 454–458.

Lupien P, Moorjani S, Awad J. (1976) A new approach to the management of familial hypercholesterolaemia: Removal of plasma-cholesterol based on the principle of affinity chromatography. *Lancet* **1(7972):** 1261–1265.

Mabuchi H, Koizumi J, Michishita I, *et al.* (1988) Effects on coronary atherosclerosis of long-term treatment of familial hypercholesterolemia by LDL-hemoperfusion. *Contrib Infus Ther* **23:** 87–96.

Mabuchi H, Michishita I, Takeda M, *et al.* (1987) A new low density lipoprotein apheresis system using two dextran sulfate cellulose columns in an automated column regenerating unit (LDL continuous apheresis). *Atherosclerosis* **68:** 19–25.

Martich GD, Boujoukos AJ, Suffredini AF. (1993) Response of man to endotoxin. *Immunobiology* **187(3–5):** 403–16.

McLaughlin ED, Nealon TF, Gibbon JH. (1959) A cation exchange artificial kidney. *Trans Am Soc Artif Intern Organs* **5:** 8.

Mitaka C, Fujiwara N, Yamamoto M, *et al.* (2014) Polymyxin B-immobilized fiber column hemoperfusion removes endotoxin throughout a 24-hour treatment period. *J Crit Care* **29(5):** 728–732.

Mitzner S, Schneidewind J, Falkenhagen D, *et al.* (1993) Extracorporeal endotoxin removal by immobilized polyethylenimine. *Artif Organs* **17(9):** 775–81.

Moore RJ, Lehman TH, Hodges CV. (1961) Treatment of acute hyperkalemia utilizing extracorporeal perfusion of blood through cation exchange resin columns. *Surg Gyn Obstet* **112:** 67.

Mor LA, Sideman S, Mor L, *et al.* (1980) Analysis of hemoperfusion columns: Selective removal of protein-bound metabolites. In: Sideman S, Chang TMS (eds.), ***Hemoperfusion: Kidney and Liver Supports and Detoxification***, Part 1. New York: Hemisphere-McGraw-Hill.

Morimoto T, Matsushima M, Sowa N, *et al.* (1989) Plasma adsorption using bilirubin-adsorbent materials as a treatment for patients with hepatic failure. *Artif Organs* **13:** 447–452.

Muirhead E, Reid A. (1948) Resin artificial kidney. *J Lab Clin Med* **33:** 841–844.

Nakaji S, Hayashi N. (2003) Bilirubin adsorption column Medisorba BL-300. *Therapeutic Apheresis and Dialysis* 7**(1):** 98–103.

Nakamura T, Kawagoe Y, Matsuda T, *et al.* (2004) Effects of polymyxin B immobilized fiber on urinary N-acetyl-beta-glucosaminidase in patients with severe sepsis. *Asaio Journal* **50(6):** 563–567.

Nealon TF, Ching N. (1962) An extracorporeal device to lower blood ammonia levels in hepatic coma. *Trans Am Soc Artif Intern Organs* **8:** 226–8.

Newton BA. (1956) The properties and mode of action of the polymyxins. *Bacteriol Rev* **20(1):** 14–27.

Novelli G, Ferretti G, Poli L, *et al.* (2010a) Clinical results of treatment of postsurgical endotoxin-mediated sepsis with polymyxin-B direct hemoperfusion. *Transplantation Proceedings* **42(4):** 1021–1024.

Novelli G, Ferretti G, Ruberto F, *et al.* (2010b) Early management of endotoxemia using the endotoxin activity assay and polymyxin B-based hemoperfusion. In: Ronco C, Piccinni P, Rosner M (eds.), *Endoxemia and Endotoxin Shock: Disease, Diagnosis and Therapy. Contributions to Nephrology*. S Karger AG Press.

Ohki S, Oshima K, Takeyoshi I, *et al.* (2008) Endotoxin removal with a polymyxin B-immobilized hemoperfusion cartridge improves cardiopulmonary function after cardiopulmonary bypass. *J Surg Res* **145(1):** 74–79.

Ostrow JD. (1986) Bile pigments and jaundice. In: Ostrow JD (ed), Marcel Dekker Inc., New York & Basel.

Ouchi K, Piatkiewicz W, Malchesky PS, *et al.* (1978) An efficient, specific and blood compatible sorbent system for hepatic assist. *Trans Am Soc Artif Intern Organs* **24:** 246–9.

Pallotta AJ, Koppanyi T. (1960) The use of ion exchange resins in the treatment of phenobarbital intoxication in dogs. *J Pharmacol Exp Ther* **128:** 318–27.

Rittirsch D, Flierl MA, Ward PA. (2008) Harmful molecular mechanisms in sepsis. *Nature Reviews Immunology* **8(10):** 776–787.

Sakagami K, Miyazaki M, Matsuoka J, *et al.* (1986) Artificial liver support for postoperative hepatic failure with anion exchange resin (BR-601). *Acta Medica Okayama* **40(5):** 249–55.

Sakata M, Todokoro M, Kai T, *et al.* (2001) Effect of cationic polymer adsorbent pK(a) on the selective removal of endotoxin from an albumin solution. *Chromatographia* **53(11–12):** 619–623.

Sandler JL, Cohn HE, Clark JE, *et al.* (1961) An ion exchange resin kidney for use on human patients. *Trans Am Soc Artif Intern Organs* 7**(1):** 3–8.

Sato T, Orlowski JP, Zborowski M. (1993) Experimental study of extracorporeal perfusion for septic shock. *ASAIO J* **39(3):** M790–3.

Schechter DC, Nealon TF, Bibbon JH. (1958) An ion exchange resin artificial kidney. *Surg Forum* **9:** 110.

Schwartz J, Winters J, Padmanabhan A, *et al.* (2013) Guidelines on the use of therapeutic apheresis in clinical practice evidence-based approach from the writing committee of the American Society for Apheresis: The sixth special issue. *J Clin Apheresis* **28(3):** 145–284.

Shimizu T, Obata T, Sonoda H, *et al.* (2013) The ability of endotoxin adsorption during a longer duration of direct hemoperfusion with a polymyxin B-immobilized fiber column in patients with septic shock. *Transfusion and Apheresis Science* **49(3):** 499–503.

Sideman S, Mor L, Brandes JM. (1980) Sorption of bilirubin by anion exchange resins *in vitro*. *ASAIO J* **3:** 85.

Sideman S, Mor L, Rousseau I, *et al.* (1977) Removal of bilirubin from the blood of jaundiced infants. In: Kenedi RM, Courtney JM, Gaylor GDS, Gilchrist T (eds.), *Artificial Organs*. Macmillan Press, London.

Sideman S, Mor L, Brandes JM. (1979) Removal of bilirubin by hemoperfusion with ion exchange resins. *Trans Am Soc Artif Intern Organs* **25(1):** 497–501.

Sideman S, Mor L, Mihich M, *et al.* (1982) Resin hemoperfusion for unconjugated bilirubin removal. *Contributions to Nephrology* **29:** 90–100.

Sideman S, Mor L, Mordohovich D, *et al.* (1981) *In vivo* hemoperfusion studies of unconjugated bilirubin removel by ion exchange resin. *Trans Am Soc Artif Intern Organs* **27:** 434–438.

Stocker R, Yamamoto Y, McDonagh AF, *et al.* (1987) Bilirubin is an antioxidant of possible physiological importance. *Science* **235(4792):** 1043–1046.

Suga M, Tsuchida M, Arimura Y, *et al.* (1990) Basic and clinical evaluation of an anion-exchange resin column for treatment of jaundice. *Therapeutic Plasmapheresis* **8:** 214–218.

Suzuki Y, Morita T, Hasegawa H, *et al.* (1991) Clinical performance of an artificial liver support system in severe liver failure: Evaluation of four systems. *Therapeutic Plasmapheresis* **9:** 77–80.

Tani N. (1996) Development of selective low-density lipoprotein (LDL) apheresis system: Immobilized polyanion as LDL-specific adsorption for LDL apheresis system. *Artif Organs* **20:** 922–929.

Taniguchi T, Sato K, Kurita A, *et al.* (2013) Efficacy of endotoxin adsorption therapy (polymyxin B hemoperfusion) for methicillin-resistant Staphylococcus aureus toxic shock syndrome: A case report about five patients. *Minerva Anestesiologica* **79(7):** 758–761.

Thies K, Prigent S, Heuck C. (1988) Selective removal of low-density lipoproteins from plasma by polyacrylate-coated Fractogel *in vitro* and in experimental extracorporeal perfusion. *Artif Organs* **12(4):** 320–4.

Uriu K, Osajima A, Hiroshige K, *et al.* (2002) Endotoxin removal by direct hemoperfusion with an adsorbent column using polymyxin B-immobilized fiber ameliorates systemic circulatory disturbance in patients with septic shock. *American Journal of Kidney Diseases* **39(5):** 937–947.

Usami M, Nomura H, Nishimatsu S, *et al.* (1995) *In vivo* and *in vitro* evaluation of bilirubin removal in postoperative hyperbilirubinemia patients. *Artif Organs* **19**: 102–105.

Vincent JL, Laterre PF, Cohen J, *et al.* (2005) A pilot-controlled study of a polymyxin B-immobilized hemoperfusion cartridge in patients with severe sepsis secondary to intra-abdominal infection. *Shock* **23(5):** 400–405.

Wang F, Wang X, Xiong Y, *et al.* (2013) Study on preparation and property of a new adsorbent for endotoxin removal in blood purification. *J Biomed Eng* **30(3):** 635–40.

Wang H, Chen C, Tong M, *et al.* (1989) Synthesis of macroporous polyvinylpyridine resins and its removal of unconjugated bilirubin. *Acta Polymerica Sinica* **(2):** 171–176.

Wang H, Hou G, Yu M, *et al.* (2005a) Adsorbents for endotoxin based on chitosan matrix. *Chemical Journal of Chinese Universities-Chinese* **26(4):** 680–682.

Wang S, Yu Y, Cui T, *et al.* (2002) Cellulose amphiphilic adsorbent for the removal of low density lipoprotein. *Artif Cell Blood Sub* **30(4):** 285–292.

Wang S, Yu Y, Cui T, *et al.* (2003) A novel amphiphilic adsorbent for the removal of low-density lipoprotein. *Biomaterials* **24(16):** 2799–2802.

Wang T, Wang W, Wang L, *et al.* (2008) Phosphate-based affinity adsorbents for the removal of low density lipoprotein: Preparation and *in-vitro* adsorption capability test. *Chinese Journal of Biomedical Engineering* **27(1):** 132–136.

Wang W, Xie H, Sun L, *et al.* (2009) Macroporous poly(vinyl alcohol) microspheres bearing phosphate groups as a new adsorbent for low-density lipoprotein apheresis. *Biomedical Materials* **4(6):** 065007.

Wang X, Peter JQ. (2010) Endotoxins: Lipopolysaccharides of Gram-Negative Bacteria. In: Wang X, Quinn PJ (eds.), *Endotoxins: Structure, Function and Recognition*. Subcellular Biochemistry: Springer, 233 Spring Street, New York, NY 10013, United States.

Wang Y, Liu J, Duo J, *et al.* (2005b) New method for the preparation of adsorbent with high adsorption capacity. *Chinese Science Bulletin* **50(21):** 2432–2435.

Weber C, Henne B, Loth F, *et al.* (1995) Development of cationically modified cellulose adsorbents for the removal of endotoxins. *ASAIO J* **41(3):** M430–4.

Wei Z, Huang W, Li J, *et al.* (2007) Studies on endotoxin removal mechanism of adsorbents with amino acid ligands. *Journal of Chromatography B-Analytical Technologies in the Biomedical and Life Sciences* **852(1–2):** 288–292.

Wu L, Zhang Z, Guo Y, *et al.* (2011) Preparation of hexanediamine-contained PAMAM dendrimers supported on chitosan microspheres and their application in the adsorption of bilirubin. *Chemical Journal of Chinese Universities-Chinese* **32(6):** 1436–1444.

Xu G, Zou H, Qou B, *et al.* (1990) A study on a new adsorptive antidote for artificial liver-chitosamine. *J Biomed Eng* 7(2): 131–136.

Xu H, Xiao L, Li J, *et al.* (2003) Study on adsorption properties for bilirubin of novel aminated chitosan resin. *Ion Exchange and Adsorption* **19(6):** 489–495.

Xu J, He B. (2002) Preparation of a new adsorbent for binding low density lipoprotein. *Chemical Journal of Chinese Universities* **23:** 1421–1424.

Yokoyama S, Hayashi R, Kikkawa T, *et al.* (1984) Specific sorbent of apolipoprotein B-containing lipoproteins for plasmapheresis. Characterization and experimental use in hypercholesterolemic rabbits. *Arteriosclerosis (Dallas, Tex.)* **4:** 276–282.

Yokoyama S, Hayashi R, Satani M, *et al.* (1985) Selective removal of low density lipoprotein by plasmapheresis in familial hypercholesterolemia. *Arteriosclerosis (Dallas, Tex.)* **5:** 613–622.

Yu H, Fu G, Liu L, *et al.* (2006) Facile synthesis and adsorption properties of phosphonated cellulose beads for selective removal of low-density lipoprotein. *Chinese Chemical Letters* **17(9):** 1193–1196.

Yu M, Wang H, Liu T, *et al.* (2004) Hemperfusion adsorbents used in the removing of endotoxin. *Chemical Journal of Chinese Universities-Chinese* **25(8):** 1465–1467.

Yu Y, Gu H, He B, *et al.* (1997) The studies of adsorption properties of sulfonated hydroxyethyl crosslinked chitosan beads for low density lipoprotein in plasma. *Acta Polymerica Sinica* **(5):** 606–610.

Yu Y, He B, Gu H. (2000) Adsorption of bilirubin by amine-containing crosslinked chitosan resins. *Artif Cell Blood Sub* **28(4):** 307–320.

Yuan Y, Wang Y, Yu Y. (2008) Study of the LDL adsorbent with dendrimer PAMAM as the spacer. *Ion Exchange and Adsorption* **24(1):** 1–9.

Yuan Z, Wei B, He B, *et al.* (1999) Studies on the adsorption of bilirubin by adsorbents. *Chemical Journal of Chinese Universities-Chinese* **20(6):** 903–905.

Yuan Z, Yu M, Li J, *et al.* (2005) Endotoxin adsorbent using dimethylamine ligands. *Biomaterials* **26(15):** 2741–2747.

Zhang K, Sun J, He B. (1999) Preparation and property study of a novel adsorbent for low-density lipoprotein. *Acta Polymerica Sinica* **(6):** 662–667.

Zhang K, Sun J, He B. (1999) Preparation of new adsorbent for bilirubin by immobilizing guanidine on cross-linked polyvinyl alcohol gel. *Chemical Journal of Chinese Universities-Chinese* **20(6):** 965–968.

Zhang Y, Yang H, Zhou K, *et al.* (2007) Synthesis of an affinity adsorbent based on silica gel and its application in endotoxin removal. *Reactive & Functional Polymers* **67(8):** 728–736.

Zhu X, Brown G, St-Pierre L. (1990) Adsorption of bilirubin with polypeptide-coated resins. *Biomater Artif Cell* **18(1):** 75–93.

Chapter 15

LDL Adsorption

Hisashi Makino*, Tamiko Tamanaha*,
Mariko Harada-Shiba†

1. Introduction

Low-density lipoprotein (LDL) apheresis, which depletes circulating LDL cholesterol (LDLC) mechanically, has been developed as the treatment for refractory familial hypercholesterolemia (FH) patients.

FH is an autosomal dominant disorder caused by a mutation in the gene encoding the LDL receptor. Patients homozygous for FH have severe hypercholesterolemia (600–1,000 mg/dl), cutaneous and tendon xanthomas and premature atherosclerosis (Goldstein and Brown, 1982).

The development of HMG-CoA reductase inhibitors (statin) has enabled effective treatment for most of heterozygous FH (Illingworth, 1993). Indeed, the average age at coronary artery disease (CAD) onset was significantly higher after widespread use of statins compared to before October 1989, when statins were approved in Japan (Harada-Shiba *et al.*, 2010). However, patients homozygous for FH are resistant to these drugs because statins reduce LDLC levels mostly via an increase in the number of LDL receptors in the liver. Therefore,

*Division of Endocrinology and Metabolism, National Cerebral and Cardiovascular Center, Japan.
†National Cerebral and Cardiovascular Center Research Institute, Japan.

LDL apheresis is the only practical method to control LDL levels in patients homozygous for FH and in some other types of severe hypercholesterolemia such as autosomal recessive hypercholesterolemia. LDL apheresis is also effective for other disease such as peripheral artery disease (PAD) and focal segmental glomerular sclerosis (FSGS). Currently, plasma exchange, double filtration plasmapheresis, LDL selective adsorption, heparin-induced extracorporeal precipitation, and direct hemoperfusion are available as LDL apheresis therapy. Among them, LDL selective absorption is the most popular method.

2. Method of LDL Selective Absorption

The first trial of LDL apheresis in homozygous FH patients was performed as plasma exchange by DeGenne *et al.* (1967). This therapeutic approach provided improvement of coronary artery stenosis and withdrawal of xanthoma accompanied with LDLC reduction. However, plasma exchange has the disadvantage that various important substances such as immunoglobulin are removed along with LDLC. Therefore, today, plasma exchange is used only for FH patients aged < 10 years who cannot be treated with LDL absorption because of small capacity of extracorporeal circulation.

In the 1980s, dextran sulfate-coated cellulose beads were proved a potent specific sorbent of apolipoprotein B-containing lipoproteins (Yokoyama *et al.*, 1984). The early system of LDL absorption (LA-40 system) had a large column (400 ml) using this sorbent. However, this system could not remove LDLC adequately, since the column was saturated if the treated plasma volume was large.

Currently, LDL adsorption system (LA-15 system, Kaneka, Osaka, Japan) has two small columns whose volume is 150 ml. Each column is alternately reused during apheresis therapy by eluting adsorbed LDL on saturated column by 5% NaCl. This current system can treat a large volume of plasma and reduce the cardiovascular burden during apheresis therapy. Therefore, this system can be available for patients with cardiac dysfunction or small body mass, and today, the LA-15 system is widely used all over the world.

LDL absorption system promotes the production of bradykinin by activation of coagulation system since adsorption column has negative charge (Krieter *et al.*, 2005). Therefore, it is important to note that patients who received angiotensin-converting enzyme inhibitor must not be treated with this system.

3. Preventive Effects of LDL Apheresis on Atherosclerosis

3.1. *Mechanism of Preventive Effect on Atherosclerosis*

LDL adsorption therapy can remove not only LDL-C but also cell adhesion molecules such as intracellular adhesion molecule-1 and vascular cell adhesion molecule (Sampietro *et al.*, 1997; Utsumi *et al.*, 2007), and fibrinolytic factors such as fibrinogen and plasminogen activator inhibitor-1 (Maher *et al.*, 2009). LDL adsorption also can remove inflammatory cytokines such as tumor necrosis factor-a and interleukin-1 (Hovland *et al.*, 2010), and reduce reactive oxygen species production via suppression of NAPDH oxidase (Hara *et al.*, 2009). These factors play a key role in the progression of atherosclerosis. Thus LDL adsorption can prevent atherosclerosis dependent and independent of LDLC removal. LDL adsorption therapy was also reported to improve vascular endothelial function via induction of bradykinin and nitric oxide production (Tamai *et al.*, 1997).

LDL apheresis also improves atherogenic lipid metabolism in addition to direct removal of LDLC. In patients with peripheral artery disease, oxidized LDL was removed by LDL adsorption and removal rate of oxidized LDL was associated with the improvement of walking distance (Tamura *et al.*, 2013). We also found that small dense LDL-C which has strong atherogenic action was efficiently removed by LDL apheresis in FH patients. Furthermore, we reported that LDL apheresis could remove ApoC3 (Yuasa et al., 2014), which is an endogenous lipoprotein lipase inhibitor, and promote atherosclerosis progression (Kawakami and Yoshida, 2009). LDL adsorption could remove pro-protein convertase subtilisin/kexin type 9 (PCSK9) (Tavori *et al.*, 2013; Hori *et al.*, 2015). PCSK9 plays a pivotal role in lipid metabolism by enhancing endosomal and lysosomal degradation of LDL receptor in the liver (Lambert *et al.*, 2009). In addition, PCSK9 is involved in inflammatory process by upregulating pro-inflammatory gene expression (Tang *et al.*, 2012). Thus, LDL adsorption has various anti-atherosclerotic effect as well as LDL-C removal.

3.2. *Clinical effect of LDL apheresis and therapeutical target LDLC level*

3.2.1. *LDL apheresis in homozygous FH*

Homozygous FH patients are usually started with LDL apheresis therapy at about 10 years of age. Before LDL adsorption therapy, they are treated with plasma exchange therapy if possible. Figure 15.1 shows one case of homozygous FH treated with LDL apheresis. Total cholesterol level was above 600 mg/dL before introduction of LDL apheresis. On the other hand, the range of total cholesterol level was from 100 to 500 mg/dL after introduction of LDL apheresis. Thus,

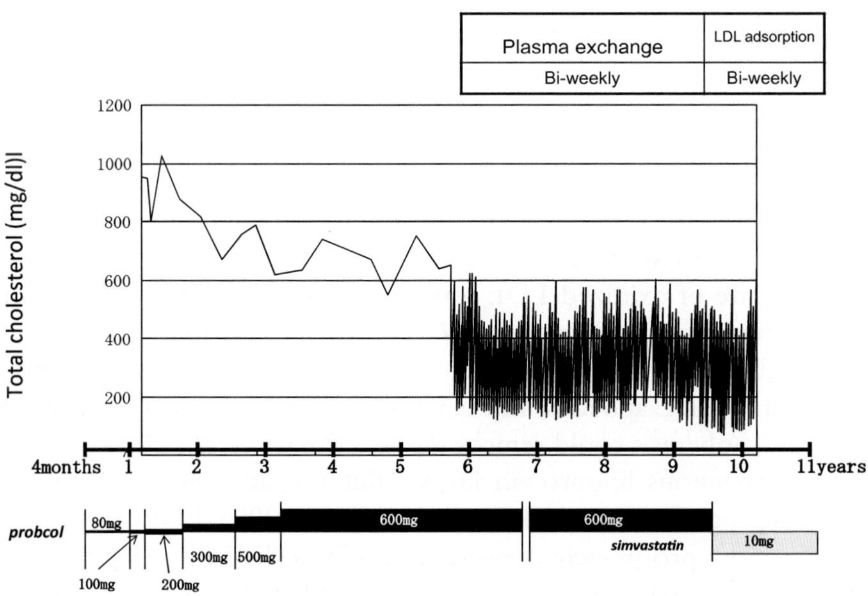

Fig. 15.1. Clinical course of homozygous hypercholesterolemia (Male) treated with LDL apheresis.

Table 15.1. Clinical Characteristics of Patients with Homozygous Familial Hyperlipidemia (Harada-Shiba and Makino, 2003)

Patient no.	**Age Span During Low Density Lipoprotein Apheresis Treatment**	**Cardiovascular Findings**
1 (F)	4–19 years old	Coronary artery stenosis has not occurred. Aortic wall is smooth and valve is normal.
2 (M)	6–13 years old	At 5 years, coronary artery and aorta were almost normal. At 12 years, thickening of the aortic wall and supravalvular stenosis (pressure gradient 70 mmHg).
3 (F)	9–12 years old	At 10 years, coronary artery stenosis was not observed. At 27 years, 100% stenosis of right coronary artery was apparent.
4 (F)	6–24 years old	Obstruction of left main trunk was found at age 6. The aortocoronary bypass surgery was carried out at age 10 and 15. 50% stenosis of graft was observed at age 24.
5 (M)	12–27 years old	Aortic wall became smoother once, but aortic regurgitation worsened
6 (F)	12–40 years old	Aortocoronary bypass was carried out at 18 years.
7 (M)	27–29 years old	At 29 years, patient died of acute myocardial infarction.
8 (F)	25–32 years old	At 31 years, patient had aortic valve stenosis and intact coronary artery. Patient died of myocardial infarction at age of 32.

the cholesterol level of this patient could not be controlled until the apheresis therapy was started.

We investigated the long-term effect of LDL apheresis in eight patients with homozygous FH (observation period was 3–20 years) (Table 15.1) (Makino and Harada-Shiba, 2003). Five patients were treated with LDL apheresis from childhood. Among these five patients, one was free from atherosclerotic disease, one revealed remission of atherosclerosis. The other three patients had supravalvular aortic stenosis or coronary artery stenosis, but atherosclerotic diseases in these three patients were relatively mild. On the other hand, three patients who started to be treated with LDL apheresis from adult had severe atherosclerosis. Two of these three patients died from myocardial infarction, and one had a coronary artery bypass before initiation of LDL apheresis. Græsdal A *et al.* also reported that patients who received LDL apheresis therapy from <10 years old revealed mild atherosclerotic change whereas two patients who started LDL apheresis as adults already had coronary artery disease before LDL apheresis initiation (Græsdal *et al.*, 2012). Thus, LDL apheresis should be initiated as early as possible in homozygous FH to prevent atherosclerotic disease progression. There is no international guideline of target LDLC level of LDL apheresis therapy. However, the recommendations of the HEART-UK LDL Apheresis Working Group is as following (Thompson *et al.*, 2010):

(1) An acute reduction in total cholesterol of $\geq 65\%$ or in LDL cholesterol of $\geq 70\%$ on average during each procedure.
(2) An interval mean total cholesterol of <270 mg/dl or LDL cholesterol <252 mg/dL (or decreases of $>60\%$ or $>65\%$, respectively from baseline values off all treatment).
(3) A baseline level of total cholesterol of <349 mg/dl or LDL cholesterol <329 mg/dL (or decreases of $>50\%$ or $>55\%$, respectively from baseline values off all treatment).

In Japan, it is common that total cholesterol level before apheresis therapy should be below 300 mg/dL and total cholesterol level after apheresis therapy should be below 100 mg/dL.

There are increased stresses on the cardiovascular system throughout pregnancy and particularly delivery. Both blood volume and cardiac output increase by 25 to 80% (Hankins *et al.*, 1985). The main point of concern has been the aggravation of coronary insufficiency of the mother, mainly due to these hemodynamic changes in pregnancy (Barss *et al.*, 1985). Therefore, during pregnancy, homozygous

FH and heterozygous FH with severe coronary artery disease should be treated by LDL apheresis to prevent further progression of coronary atherosclerosis and to tolerate the increased stress of delivery. High lipid levels may also affect placental vasculature and cause retardation of fatal growth (Beigel *et al.*, 1990; Kroon *et al.*, 1996). Therefore, homozygous FH and heterozygous FH with severe coronary artery disease should have LDL apheresis during the course of pregnancy to control their cholesterol level not only for their own sake but also for normal fatal growth. In any case, pregnancy and delivery in women with ischemic heart disease is hazardous and should be monitored very closely.

3.2.2. *LDL apheresis in heterozygous FH*

It is hard to assess the need for the use of LDL apheresis to treat heterozygous FH based on purely scientific grounds. However, there is some evidence of the usefulness of LDL apheresis for prevention of atherosclerosis progression of heterozygous FH. LDL-apheresis atherosclerosis regression study (LAARS) demonstrated that LDL apheresis significantly prevented coronary artery disease progression compared with only medication (Mabuchi *et al.*, 1998). The progression of coronary artery disease was well prevented especially in the case of LDLC level reduced to <100 mg/dL before apheresis, although mean LDLC level before apheresis was 249 mg/dL and mean LDLC level after apheresis was 105 mg/dL in LAARS (Matsuzaki *et al.*, 2002). Mabuchi *et al.* also reported that incidence of CVD in heterozygous FH patients who received LDL apheresis was less frequent than in patients who received only medication therapy in the Hokuriku-FH-LDL-Apheresis Study (Mabuchi *et al.*, 1998). Furthermore, LDL apheresis therapy could induce regression of atherosclerotic plaque of coronary artery LDL apheresis in heterozygous FH (Low Density Lipoprotein-Apheresis Coronary Morphology and Reserve Trial) (Matsuzaki *et al.*, 2002). In our observations, heterozygous FH patients who discontinued LDL apheresis therapy had poor prognosis compared with patients who continued apheresis therapy (Fig. 15.2), although patients

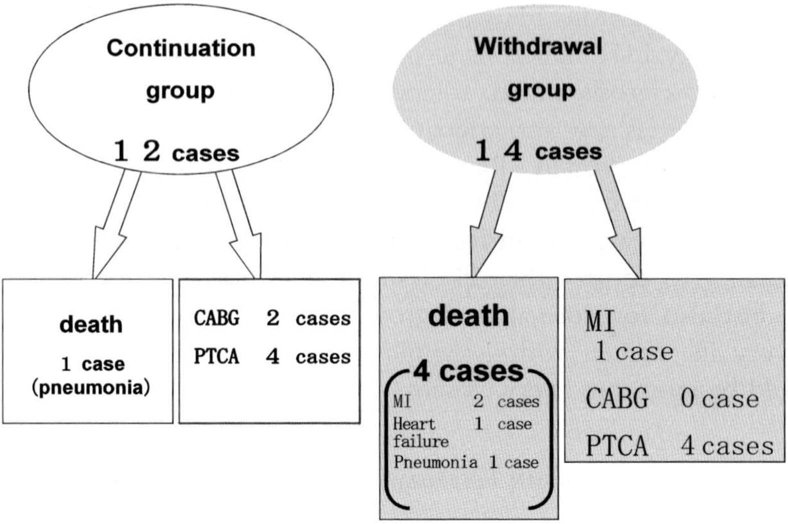

Fig. 15.2. Comparison of prognosis of heterozygous FH between LDL apheresis continuation group and LDL apheresis withdrawal group.

who discontinued LDL apheresis had almost the same level of LDLC compared with patients who continued LDL apheresis. These reports indicate that heterozygous FH patients with severe atherosclerotic disease require LDL apheresis therapy. Thompson *et al.* (2008) recommended that patients with heterozygous FH in whom there is objective evidence of continuing significant progression of coronary disease and whose LDL cholesterol remains 193 mg/dL or decreases by <40% despite combination drug therapy should be considered for LDL apheresis. International Panel on Management of Familial Hypercholesterolemia (2004) indicates that suggested efficacy targets for apheresis in FH heterozygotes are an interval mean LDL cholesterol of ≤100 mg/dL or a reduction of ≥60% from baseline values off treatment (Civiera, 2004).

3.2.3. *LDL apheresis in PAD*

Generally, for treatment of PAD, LDL apheresis is performed once a week for 10 weeks, and treating 2,000–5,000 mL of plasma for each session. Several reports showed efficacy of LDL apheresis on PAD. Peripheral arterial disease LDL apheresis multicenter study showed that LDL adsorption improved ankle brachial index (ABI) and maximum tolerated walking distance (Tsuchida *et al.*, 2006). Other report showed that the improvement of resting leg pain in PAD patients undergoing hemodialysis maintained at least four weeks after final apheresis probably by improvement of vascular endothelial function in addition to lipid reduction (Morimoto *et al.*, 2007). Thus LDL apheresis is good treatment option for PAD in case revascularization is impossible and medical management is not able to improve the symptom. However, characteristics of suitable PAD patients for LDL apheresis have been still unclear.

3.2.4. *LDL apheresis in FSGS*

LDL apheresis is reported to be effective for refractory nephrotic syndrome due to FSGS. Refractory nephrotic syndrome persist dyslipidemia, which exacerbates not only progression of atherosclerosis but also damages of glomeruli and renal tubules (Attman *et al.*,1997). Therefore, correction of dyslipidemia is important for the treatment of refractory nephrotic syndrome. LDL apheresis also can prevent progression of FSGS through enhanced response to immunosuppressant by amelioration of intracellular drug transport, removal of inflammatory cytokine, and removal of humoral factors responsible for nephrotic syndrome (Muso, 2014). Generally, LDL apheresis is carried out 12 times in three months.

4. Conclusion

The number of heterozygous FH patients who should be treated with LDL apheresis has been decreasing because of the therapeutic advance of lipid lowering agents, although homozygous FH must be treated with LDL apheresis. Furthermore, novel lipid lowering agents such as PCSK9 antibody (Roth *et al.*, 2012) and microsomal triglyceride transfer protein inhibitor (Cuchel *et al.*, 2013) are under development. Therefore, the number of patients who should be treated with

LDL apheresis and the frequency of LDL apheresis on each patient may further decrease in the future. However, LDL adsorption can strongly remove atherosclerotic lipoproteins. Moreover, LDL adsorption can remove other atherogenic factors. Indeed, previous studies indicate that LDL apheresis could improve the prognosis of FH compared to medication therapy. Therefore, LDL apheresis is still an important therapeutic option for FH patients with severe atherosclerosis.

References

Attman PO, Samuelsson O, Alaupovic P. (1997) Progression of renal failure: Role of apolipoprotein B-containing lipoproteins. *Kidney Int Suppl* **63:** S98–101.

Barss V, Phillippe M, Greene MF, Covell L. (1985) Pregnancy complicated by homozygous hypercholesterolemia. *Obstet Gynecol* **65:** 756–7.

Beigel Y, Hod M, Fuchs J, *et al.* (1990) Pregnancy in a homozygous familial hypercholesterolemic patient treated with long-term plasma exchange. *Am J Obstet Gynecol* **162:** 77–8.

Civiera F. (2004) For international panel on management of familial hypercholesterolemia. Guidelines for the diagnosis and management of heterozygous familial hypercholesterolemia. *Atherosclerosis* **173:** 55–68.

Cuchel M, Meagher EA, du Toit Theron H, *et al.* (2013) Efficacy and safety of a microsomal triglyceride transfer protein inhibitor in patients with homozygous familial hypercholesterolaemia: A single-arm, open-label, phase 3 study. *Lancet* **381:** 40–6.

de Gennes JL, Touraine R, Maunand B, *et al.* (1967) Homozygous cutaneotendinous forms of hypercholesteremic xanthomatosis in an exemplary familial case. Trial of plasmapheresis and 1402 heroic treatment. *Bull Mem Soc Med Hop Paris* **118:** 1377–1402.

Græsdal A, Bogsrud MP, Holven KB, *et al.* (2012) Apheresis in homozygous familial hypercholesterolemia: The results of a follow-up of all Norwegian patients with homozygous familial hypercholesterolemia. *J Clin Lipidol* **6:** 331–9.

Goldstein JL, Brown MS. (1982) *The Metabolic Basis of Inherited Disease*, 5th ed. MacGraw-Hill, New York.

Hankins GDV, Wendel GD, Leveno KJ. (1985) Myocardial infarction during pregnancy: A review. *Obstet Gynecol* **65:** 139–46.

Hara T, Kiyomoto H, Hitomi H, *et al.* (2009) Low-density lipoprotein apheresis for haemodialysis patients with peripheral arterial disease reduces reactive oxygen species production via suppression of NADPH oxidase gene expression in leucocytes. *Nephrol Dial Transplant* **24:** 3818–25.

Harada-Shiba M, Sugisawa T, Makino H, *et al.* (2010) *J Atheroscler Thromb* **17:** 667–674.

Hori M, Ishihara M, Yuasa Y, *et al.* (2015) Removal of plasma mature and furin-cleaved proprotein convertase subtilisin/kexin 9 by low-density lipoprotein-apheresis in familial hypercholesterolemia: Development and application of a new assay for PCSK9. *J Clin Endocrinol Metab* **100:** E41–9.

Hovland A, Hardersen R, Nielsen EW, *et al.* (2010) Hematologic and hemostatic changes induced by different columns during LDL apheresis. *J Clin Apher* **25:** 294–300.

Illingworth R. (1993) How effective is drug therapy in heterozygous familial hypercholesterolemia. *Am J Cardiol* **72:** D54–58.

Kawakami A, Yoshida M. (2009) Apolipoprotein CIII links dyslipidemia with atherosclerosis. *J Atheroscler Thromb* **16:** 6–11.

Krieter DH, Steinke J, Kerkhoff M, *et al.* (2005) Contact activation in low-density lipoprotein apheresis systems. *Artif Organs* **29:** 47–52.

Kroon AA, Aengevaeren WR, van der Werf T, *et al.* (1996) LDL-Apheresis Atherosclerosis Regression Study (LAARS). Effect of aggressive versus conventional lipid lowering treatment on coronary atherosclerosis. *Circulation* **93:** 1826–35.

Lambert G, Charlton F, Rye KA, Piper DE. (2009) Molecular basis of PCSK9 function. *Atherosclerosis* **203:** 1–7.

Mabuchi H, Koizumi J, Shimizu M, *et al.* (1998) Long-term efficacy of low-density lipoprotein apheresis on coronary heart disease in familial hypercholesterolemia. Hokuriku-FH-LDL-Apheresis Study Group. *Am J Cardiol* **82:** 1489–95.

Maher VM, Kitano Y, Neuwirth C, *et al.* (2009) Plasminogen activator inhibitor-1 removal using dextran sulphate columns. Evidence of PAI-1 homeostasis. *J Thromb Thrombolysis* **28:** 166–72.

Makino H, Harada-Shiba M. (2003) Long-term effect of low-density lipoprotein apheresis in patients with homozygous familial hypercholesterolemia. *Ther Apher Dial* **7:** 397–401.

Matsuzaki M1, Hiramori K, Imaizumi T, *et al.* (2002) Intravascular ultrasound evaluation of coronary plaque regression by low density lipoprotein-apheresis in familial hypercholesterolemia: The low density lipoprotein-apheresis coronary morphology and reserve trial (LACMART). *J Am Coll Cardiol* **40:** 220–7.

Morimoto S, Yano Y, Maki K, *et al.* (2007) Efficacy of low-density lipoprotein apheresis in patients with peripheral arterial occlusive disease undergoing hemodialysis treatment. *Am J Nephrol* **27:** 643–8.

Muso E. (2014) Beneficial effect of LDL-apheresis in refractory nephrotic syndrome. *Clin Exp Nephrol* **18:** 286–290.

Roth EM, McKenney JM, Hanotin C, *et al.* (2012) Atorvastatin with or without an antibody to PCSK9 in primary hypercholesterolemia. *N Engl J Med* **367:** 1891–1900.

Sampietro T, Tuoni M, Ferdeghini M, *et al.* (1997) Plasma cholesterol regulates soluble cell adhesion molecule expression in familial hypercholesterolemia. *Circulation* **96:** 1381–5.

Tamai O, Matsuoka H, Itabe H, *et al.* (1997) Single LDL apheresis improves endothelium-dependent vasodilatation in hypercholesterolemic humans. *Circulation* **95:** 76–82.

Tamura K, Tsurumi-Ikeya Y, Wakui H, *et al.* (2013) Therapeutic potential of low-density lipoprotein apheresis in the management of peripheral artery disease in patients with chronic kidney disease. *Ther Apher Dial* **17:** 185–92.

Tang Z, Jiang L, Peng J, *et al.* (2012) PCSK9 siRNA suppresses the inflammatory response induced by oxLDL through inhibition of NF-κB activation in THP-1-derived macrophages. *Int J Mol Med* **30:** 931–8.

Tavori H, Giunzioni I, Linton MF, Fazio S. (2013) Loss of plasma proprotein convertase subtilisin/kexin 9 (PCSK9) after lipoprotein apheresis. *Circ Res* **113:** 1290–5.

Thompson GR; HEART-UK LDL apheresis working group. (2008) Recommendations for the use of LDL apheresis. *Atherosclerosis* **198:** 247–55.

Thompson GR, Barbir M, Davies D, *et al.* (2010) Efficacy criteria and cholesterol targets for LDL apheresis. *Atherosclerosis* **208:** 317–21.

Tsuchida H, Shigematsu H, Ishimaru S, *et al.* (2006) Effect of low-density lipoprotein apheresis on patients with peripheral arterial disease. Peripheral arterial disease LDL apheresis multicenter study (P-LAS). *Int Angiol* **25:** 287–92.

Utsumi K, Kawabe M, Hirama A, *et al.* (2007) Effects of selective LDL apheresis on plasma concentrations of ICAM-1, VCAM-1 and P-selectin in diabetic patients with arteriosclerosis obliterans and receiving maintenance hemodialysis. *Clin Chim Acta* **377:** 198–200.

Yokoyama S, Hayashi R, Kikkawa T, *et al.* (1984) Specific sorbent of apolipoprotein B-containing lipoproteins for plasmapheresis. Characterization and experimental use in hypercholesterolemic rabbits. *Arteriosclerosis* **4:** 276–282.

Yuasa Y, Osaki T, Makino H, *et al.* (2014) Proteomic analysis of proteins eliminated by low-density lipoprotein apheresis. *Ther Apher Dial* **18:** 93–102.

Chapter 16

Treatment of Metabolic Diseases by Hemoperfusion

Xue, Jun*

1. Introduction

Lipid metabolism disorder is a major reason for the development of atherosclerosis, cardiovascular and cerebrovascular diseases. The close relationship between atherosclerosis and lipid metabolism disorder is confirmed by a large number of experimental studies that use animals, including plaque histopathology, tissue chemistry analyses, and epidemiological surveys. Moreover, numerous EBM studies have shown that the lipid-lowering therapy played an important role in the prevention of grade I and grade II cerebrovascular diseases.

Active diet control, appropriate physical activity and lipid-lowering drugs are effective in treating a majority of the lipid disorders in patients. However, these drugs sometimes fail to achieve target goals in patients with familial hypercholesterolemia. Moreover, some patients could not tolerate the lipid-lowering drugs and suffered from serious side effects. Meanwhile, some patients suffer from acute ischemic vascular disease and severe microcirculation with severe hyperlipidemia; for these patients rapidly decreasing lipids is very important. As a result, extracorporeal lipid-lowering therapies began

*Professor, Department of Nephrology, Huashan Hospital, Fudan University, Shanghai, P. R. China.

emerging in 1965, and the first application of plasma exchange therapy for the treatment of hyperlipidemia emerged in the late 1980s. Highly selective extracorporeal lipid-lowering therapies have made considerable progress with the rapid development of polymer materials science and engineering (Borberg H *et al.*, 2013). This chapter will give a brief introduction to the technical characteristics and clinical efficacy of these therapies.

2. The Technical Characteristics of the Extracorporeal Lipid Purification Therapy

2.1. *Non-Selective*

2.1.1. *Plasma exchange (PE)*

Plasma exchange was the first therapy used to treat hereditary lipid metabolism disorder, with 2 to 4.5 liters plasmapheresis for each treatment. However, plasma exchange not only removes harmful low-density lipoproteins (LDL), lipoprotein (a) [Lp (a)], cholesterol and other ingredients, it also removes the high-density lipoproteins (HDL), albumin, immunoglobulin and other beneficial components of blood. This makes it necessary to add a large amount of fresh plasma, often resulting in allergic reactions, cross-infection and other complications, so this non-selective extracorporeal treatment with lipid-lowering therapy is rarely used now.

2.2. *Semi-Selective*

2.2.1. *Double filtration plasmapheresis therapy (DFPP)*

Membrane filtration is called membrane different filtration (MDF) or cascade filtration (CF). The filtration removal range of the plasma protein is based on the different apertures of the plasma component separator (Stefanutti C *et al.*, 2012). The filtration system is illustrated using a flowchart in Fig. 16.1 (Japan Asahi Corporation Plasauto-iQ). The first filter is an ordinary plasma separator with a 0.2 μm pore that

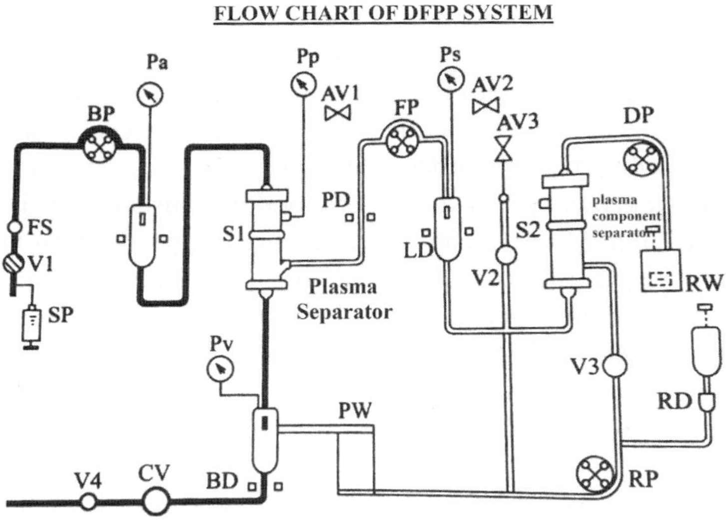

Fig. 16.1. Double filtration plasmapheresis. (1) The plasma and blood cells are separated in the plasma separator (S1). (2) Then large proteins, such as LDL, are separated from the plasma in S2 (the plasma component separator). (3) Supply albumin and saline based on individual needs (SP: syringe pump V1~V4: valve 1~valve 4 FS: flow sensor BP: blood pump Pa: arterial pressure S1: separator 1 Pp: plasma pressure AV1~AV3: air valve 1~air valve 3 PD: plasma detector FP: filtration pump Ps: second pressure LD: liquid detector S2: separator 2 DP: drain pump RP: return pump RW: return weight RD: return detector PW: plasma warmer Pv: veinal pressure BD: blood detector CV: clamp valve).

separates the plasma, red blood cells and other formed components within the blood. The second semi-selective column has a hollow fiber with a pore size of approximately 0.03 μm, and is the semi-selective column. This filter not only separates plasma macromolecules, such as LDL, very low density lipoproteins (VLDL), intermediate density lipoproteins (IDL) and fibrinogen, but also some beneficial components, such as HDL, immunoglobulins, hormones, and small albumin molecules and other beneficial constituents. The general treatment protocol is followed : (1) selection of the vascular access: the preferred right median vein → left elbow median vein; if the difference between the median cubital vein condition can be considered, such as deep vein femoral vein, internal jugular vein. (2) Blood flow: blood flow should be between 100–120 mL/min for an ideal plasma separation rate. However, most cubital vein patients do not meet the above requirements, according to the author's experience, and a blood flow of 60–90 mL/min is more viable. This results in an extension of the treatment time, but the effects are improved. (3) The separated plasma flow: generally 1/4 to 1/3 of blood flow, usually that is about 20~30mL/min. (4) Anticoagulation method: unfractionated heparin is generally used; the first dose is 2,000 to 3,000 units, then 500 to 1,000 units sustained per hour. The heparin dose may be increased in hypercoagulable or high viscosity patients. (5) Supplementary albumin, 30 g or the equivalent generation of plasma products.

In recent years, there have been some improved DFPP methods aimed at reducing the loss of small molecules, with an albumin recovery rate of 90%, removing the need for additional albumin. An improved DFPP, also known as pulsed flow cascade filtration, captured the beneficial small component by the waste plasma circulation flowing into second separator again and again (Legallais C *et al.*, 1996; Kardas F *et al.*, 2012). In this process, there is no waste solution, allowing the total amount of the filtered plasma to make it to the second separator all the time (Fig. 16.2, the Japanese company, Asahi Plasauto-iQ), and the cycle is repeated until the end of treatment to improve the retention efficiency of the albumin, HDL and other beneficial constituents. The remains in the plasma component separator and the circulation path of the waste are rich in lipid macromolecular protein particles, but there are less small molecules, such as albumin particles. The Japanese KURARAY company launched a dual filtration heat circulating lipid therapy that inserted the loop heating system in the separation recirculation, bringing the circulating plasma temperature to 42°C. This therapy method is beneficial for improving the separation performance in order to retain more albumin, HDL and other blood components. Another improved method of the plasma component separator is that it reverses the connection, the plasma flowing into the separator through original outlet and leaving through original inlet, so the conventional transmembrane flow has been opposite; this reverse filtration method actually expands membrane area to 1.7 times (see Fig. 16.3, Asahi,

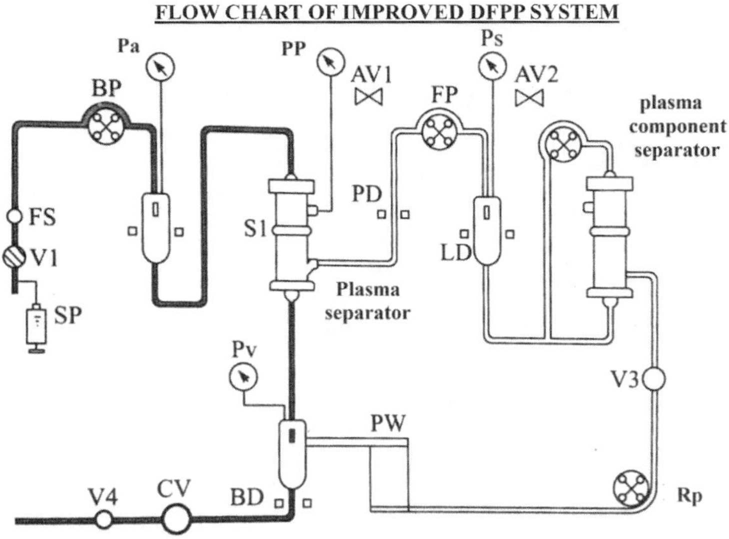

Fig. 16.2. Double filtration plasmapheresis (DFPP). (1) The plasma and blood cells are separated in the plasma separator (S1). (2) Then large proteins, such as LDL, are separated from the plasma in S2. (3) Supply albumin and saline based on individual needs. (SP: syringe pump; V1~V4: valve 1~valve 4; FS: flow sensor; BP: blood pump; Pa: arterial pressure; S1: separator 1 Pp: plasma pressure AV1~AV3: air valve 1~air valve 3; PD: plasma detector; FP: filtration pump; Ps: second pressure; LD: liquid detector; S2: separator 2; DP: drain pump; RP: return pump; RW: return weight; RD: return detector; PW: plasma warmer; Pv: veinal pressure; BD: blood detector; CV: clamp valve).

Japan Plasauto-iQ). For the modified method with the recycling process, as there is no specific loop of the waste except of residual liquid. At the end of the treatment, the plasma component separator and the pipe contain a large amount of lipoprotein, and only a small amount of beneficial particle, such as albumin.

2.3. *Highly Selective*

2.3.1. *Immunosorbent assay (immunoadsorption, IA)*

Based on antigen–antibody binding specificity, injected purified human LDL has been found to covalently bind with agarose in goats,

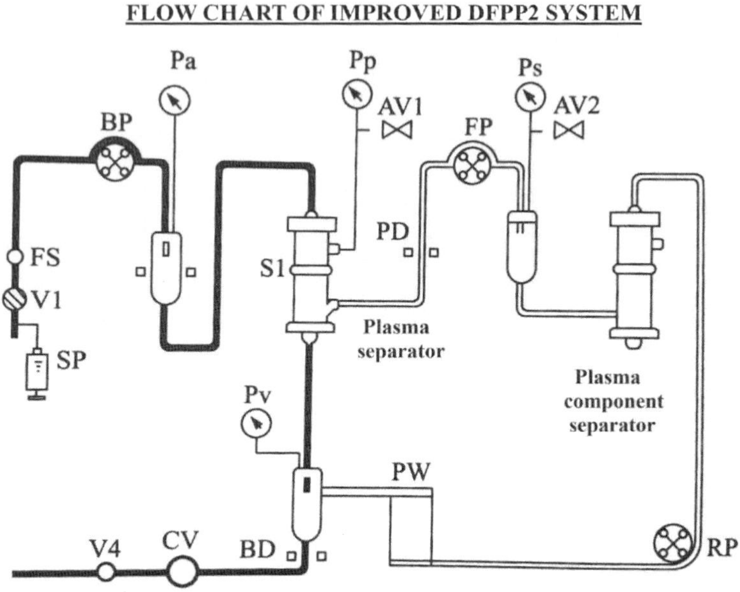

Fig. 16.3. Improved DFPP 2. (1) The plasma and blood cells are separated in S1. (2 Then large proteins, such as LDL, are separated from the plasma in a reversed transmembrane filtration way in the separator. (3) Supply albumin and saline to meet individual needs (SP: syringe pump V1~V4: valve 1~valve 4 FS: flow sensor BP: blood pump Pa: arterial pressure S1: separator 1 Pp: plasma pressure AV1~AV2: air valve 1~air valve 2 PD: plasma detector FP: filtration pump Ps: second pressure RP: return pump PW: plasma warmer Pv: veinal pressure BD: blood detector CV: clamp valve).

producing specific anti-LDL antibodies. These bound LDL particles can then be coupled to the surface of hydrogen bromide-activated Sepharose L1-4 to make an LDL-specific adsorption column. Further, the LDL adsorption column can be reused if it is washed with a saturated adsorption column buffer. Studies have shown that the efficiency and selectivity of the adsorption column remains unchanged, even after being used for more than 60 times (Jovin IS *et al.*, 2000; Julius U *et al.*, 2009). If the absorption column made of specific antibodies, such as monoclonal, polyclonal anti-apolipoprotein B (Apo-B), or anti-Lp (a) antibody, then the column could specifically bind to LDL, LDL Apo-B, VLDL, and Lp (a) extracorporeal. Such an adsorbent would have high bioaffinity and specificity. However, since it is a heterologous antibody, any antibodies that drop into the blood could easily cause allergic reactions. Therefore, specific antibodies may be the future of genetic engineering research, similar to products of the German company THERASORB.

2.3.2. *Dextran sulfate cellulose adsorption, DSA*

The covalent cross-linked dextran sulfate in porous cellulose has a structural form similar to the LDL receptor. When negatively charged, the surface can be combined with a specific positively charged LDL. So, the Dextran sulfate covalently binds to porous cellulose beads, forming multimeric complexes, and is then wrapped into the adsorption column. This therapy, applying the technique of physics chemistry affinity, is simple, stable, efficacious, and therefore widely used in the world. Figure 16.4 is a flowchart detailing the method (Japan Asahi Corporation Plasauto-iQ). According to patient condition, the therapy can process up to 5,000 ml of plasma. The Japanese KANEKA Company developed the DSA adsorption column in recent years, and now a whole blood perfusion of the adsorption column, which is applied in clinical setting (Otto C *et al.*, 2007).

2.3.3. *Heparin mediated extracorporeal LDL precipitation system (HELP) (Seidel D et al., 1996):*

Using isoelectric point of precipitation principles, mix heparin and acetate solution (pH = 4.85) at a 1: 1 ratio, raising the pH to 5.12, thus obtaining the isoelectric point of LDL. In such an environment, heparin maximum binding LDL, fibrinogen, VLDL and Lp (a) are nearly all deposited on the lipid deposition, while HDL, albumin and other beneficial components are hardly affected. The next step is to remove the heparin after complete adsorption of the heparin-anion exchange column, then, restore the physiological state including pH value and the electrolyte levels by bicarbonate dialysis, at last mix with separated red blood cells, and return the cleansed blood back to the body (Fig. 16.5, Germany B.Braun company Plasmat Futura). Each

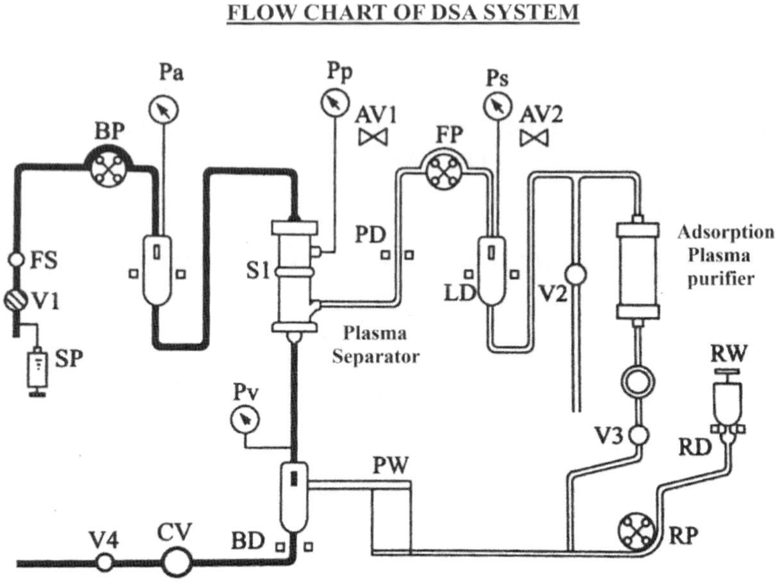

Fig. 16.4. DSA system. (1) The plasma and blood cells are separated in S1. (2) Then large proteins, such as LDL, are absorbed in the dextran sulfate cellulose column. (3) Supply albumin and saline to meet individual needs (SP: syringe pump; V1~V4: valve 1~valve 4; FS: flow sensor; BP: blood pump; Pa: arterial pressure; S1: separator 1; Pp: plasma pressure; AV1~AV3, air valve 1~air valve 3; PD: plasma detector; FP: filtration pump; Ps: second pressure; LD: liquid detector; S2: separator 2; DP: drain pump; RP: return pump; RW: return weight; RD: return detector; PW: plasma warmer; Pv: veinal pressure; BD: blood detector; CV: clamp valve).

therapy treats 2.5 to 3 liters of plasma; during the treatment due to its heparin was adsorbed to a column of anion exchange, so the amount of anticoagulant heparin is quite huge; the first agent is nearly 4,000 to 5,000 units, maintaining the 2,500–3,500 units per hour.

2.3.4. *Direct adsorption of lipoprotein from whole blood (DALI)*

Conventional lipid separation works to first separate the blood cells from the plasma. The DALI, a modified whole blood perfusion

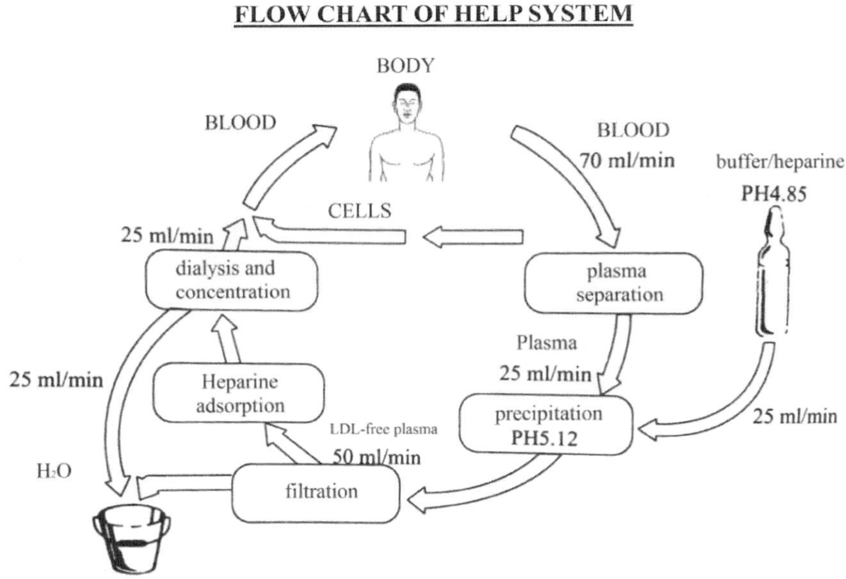

Fig. 16.5. (1) Mix the heparin and acetate mixture solution to reach a pH of 5.12, the isoelectric point of LDL. (2) Heparin maximum binding LDL, fibrinogen, VLDL and Lp (a), the mixture will be precipitated.

(hemoperfusion) technique, directly clears the LDL and Lp (a) from the whole blood (Bosch T *et al.*, 2000). The perfusion apparatus is made of polyacrylamide beads in a wrapped configuration. The polyacrylate ligands of positive charge bind with negatively charged LDL and Lp (a), resulting in the selective adsorption of the lipid composition (Tani N *et al.*, 1996), which has no effect on the blood cells and HDL. The particularity of the DALI system is in its anticoagulant technology, the mixture of 0.5 IU heparin and 0.375 mg citrate (ACD-A). Not only reaching the optimal anticoagulant effect, but also significantly avoiding the complement activation and hypocalcemia occurred. An operation flowchart is shown in Fig. 16.6 (German company Fresenius). The patient will feel better following treatment of 1.3 to 1.6 times of the whole blood volume.

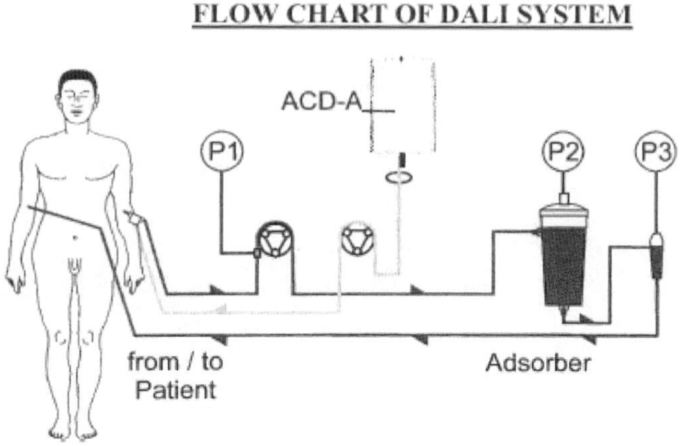

Fig. 16.6. DALI system. The LDL and Lp (a) are directly absorbed in the adsorber without the isolation of plasma and blood cells. (P1: arterial pressure P2: plasma pressure P3: second pressure ACD: acid citrate dextrose) (Dräger, Julius, Kraenzle *et al.* (1998))

3. Evaluation of Extracorporeal Blood Lipid Purification Treatments

Each previously described therapy has its pros and cons. However, simple plasma separation was eliminated due to its poor selectivity. The semi-selective double plasmapheresis therapy greatly improves the lipid component removal efficiency, but small amounts of albumin, immunoglobulin, high-density lipoprotein and other beneficial components are lost, necessitating the addition of albumin or plasma substitute products. The semi-selective therapy is widely used in clinical applications because of its low price and simple plasma processing capacity. The modified double plasma separation technique further reduces the loss of small proteins such as albumin, and no additional albumin is required. However, its operation is more complicated, especially the recycling, because it requires precise coordination with separate circulation speeds. The modified double plasma separation technique would be a cheap therapy option if it were well mastered.

The highly selective lipid purification therapy applies principles of biological or physical chemical affinity for the selective removal of unfavorable lipid compositions. Immunoabsorption assays apply biological affinity principles. However, the biggest drawback to this technique is that shedding antibodies into the blood could easily lead to an allergic reaction. In addition, there are considerable requirements associated with the preparation and preservation of specific antibodies, such as short hedges and cold storage. Furthermore, the strong acid buffer, as washing liquid in its system, has an impact on the protection of the environment, so this method has not been approved by the United States Drug and Food Control Commission (FDA).

The HELP, DSA and DALI systems are currently the most widely applied in clinical treatment, and are approved by the US FDA for therapy refractory hyperlipidemia. The technical character of the HELP system is the application of the physical and chemical affinity principles. The principle is that heparin in low pH environments will absorb a large quantity of the negative charges onto its surface, and then binds with the positively surface charged LDL, Lp (a), and fibrinogen closely, forming precipitates. The HELP system has the highest purification efficiency, and can reduce the LDL, Lp(a), and fibrinogen by 50% during every 3 liters plasma treatment. In addition, the HELP system has high biocompatibility. There was not a significant increase in the complement activation or inflammatory factors before and after the treatment. The best feature of the HELP system is that it did not activate the bradykinin system as in the DSA and DALI systems, so there is no need to stop taking angiotensin-converting enzyme inhibitors (ACEI). Moreover, there was no apparent hypotension, measure, vomiting, or facial flushing during treatment. However, the biggest drawback of the HELP system is that its operation is complex and requires a lot of material, so it is very expensive.

The DSA system simulates the unique structure of the LDL receptor, allowing it to absorb LDL using the covalently cross-linked dextran sulfate in porous cellulose. The biggest advantage of the DSA machine is that it is easy to operate and has good selectivity. However, the choice of absorbent material in the DAS system is a multivalent negatively charged substance. Contact with blood will produce the ABC (anion-blood contact) reaction that begins 15 minutes after cardiopulmonary bypass, and the patients will suffer hypotension and vocal cord edema. These responses in the patients are mainly a result of increased production of bradykinin. Thus, the patients should prohibit ACEI or ARB 24 or 48 hours before treatment.

Lipid perfusion with the DALI system can be applied to whole blood purification therapy. With its simple operation, fair biocompatibility and good results, clinical workers pay close attention to the therapy (Julius U *et al.*, 2007; Bosch T *et al.*, 2006). It may represent the future direction of lipids purification. There are two main disadvantages to this system. One is a relatively large extracorporeal circulation, and the second is the occurrence of ABC reactions in a small amount of the patients. In addition, the DALI system needs further clinical tests and verification, because it has only been used in a limited number of the cases.

4. Effect of Extracorporeal Blood Lipid Purification Treatment

4.1. *Regulate Plasma Lipids*

Several types of lipid-lowering extracorporeal therapies cleanse lots of harmful lipid component effectively, rapidly and safely, especially on LDL (Table 16.1).

A number of reports (Julius U *et al.*, 2008; Poli L *et al.*, 2006; Krebs A *et al.*, 2004; Borberg H *et al.*, 2009; Moriarty P *et al.*, 2010; Schuff-Werner P *et al.*, 1997) have revealed that severe lipid metabolism disorder in patients, after lipid purification LDL usually decreases more than 50%. Meanwhile, there are different degrees decline of Lp (a), triglycerides. The DSA, HELP, and DALI systems improve HDL preserving performance (Table 16.2).

Plasma cholesterol, LDL, Lp (a) and other lipid components dropped rapidly following treatment. Whether this will lead the rebounding of anabolism of LDL causes the concern of many scholars. Recent studies have shown that the stimulation of lipid-lowering does not raise cholesterol synthesis, except for LDL < 1~1.4mmol/L, where the cholesterol synthesis appears to increase (Bambauer R *et al.*, 2000).

Table 16.1. Decreased Effect of Plasma Lipid Separation on Plasma Components

	IA	DSA	HELP	DFPP
Heart	0	–3	–8	–13
Albumin	0	0	0	–2
TC	–74	–63	–64	–76
TG	–54	–34	–52	–62
LDL	–100	–99	–98	–100
HDL	–2	0	0	–15
VLDL	–50	–45	–55	–95
LP(a)	–95	–83	–90	–87
Apo-AI	–5	0	0	–19
Apo-B100	–100	–100	–100	–100
LDL	–3	–37	–88	–67
IgG	–1	0	–5	–17
IgA		–5	–5	–40
IgM	0	–9	–8	–91

(*Note*: these values represent the percentage decrease of the plasma component by lipid adsorption column or plasma component separator. TC, total cholesterol; LDL, low-density lipoprotein; HDL, high-density lipoprotein; VLDL, very low-density lipoprotein; Lp (a), lipoprotein (a); TG, triglycerides; Apo: apolipoprotein; Ig: immune globulin.)

Table 16.2. Change in Plasma Lipid Composition before and after Treatment

	IA	DSA	HELP	DFPP	DALI
Cholesterol	–50~–60	–50~–60	–50~–60	–50~–65	–50~–60
TG	–40~–50	–40~–50	–50	–40~–50	–10~–30
LDL	–60~–75	–60~–75	–60~–75	–60~–70	–60~–70
HDL	–10~–27	0~–10	0~–10	–10~–20	0~–5
Lp(a)	–40~–60	–40~–60	–40~–60	–40~–60	–50~–60

(LDL, low-density lipoprotein; HDL, high-density lipoprotein; Lp (a), lipoprotein (a); TG, triglycerides; IA, immunoadsorption; DSA, dextran sulfate cellulose adsorption; HELP, heparin mediated extracorporeal; LDL, precipitation; DFPP, double filtration plasmapheresis; DALI, direct adsorption of lipoprotein.)

Studies have also found that LDL anabolism temporarily decline after treatment, and the LDL receptor expression increases, while the Apo-B, Apo-AI and HDL metabolism secretions did not significantly change (Feillet C *et al.*, 1997; Frenais R *et al.*, 2001; Hovland A, 2012). Therefore, extracorporeal lipid-lowering therapy did not cause or aggravate lipid metabolism disorders.

4.2. *Improvement of Hemorheology*

Extracorporeal lipid-lowering therapy rapidly removes cholesterol, triglycerides, LDL, Lp (a), fibrinogen and other plasma macromolecule particles. These substances, especially the fibrinogen in plasma, aggravate the plasma viscosity. Therefore, after treatment, the hemorheological parameters improve significantly: the rate of red cell aggregation declines, erythrocyte functions are regulated and tissue oxygen supply increases, and microcirculation significantly improves. Data show that after DSA or DALI HELP treatments (Payhofer KG *et al.*, 2000), patient plasma fibrinogen decreased by 20 to 50%, high shear whole blood viscosity decreased by 10 to 25%, low shear viscosity decreased by 10 to 25%, plasma viscosity decreased by 10 to 20%, red blood cell permeability time (red cell transit time, RCTT) decreased from 20 to 25%, and blood tangential force-dependent platelet adhesion force decreased (Bosch T *et al.*, 2001). The HELP system especially, with its affirmative effect of decreasing fibrinogen, has a better performance in the microcirculation and tissue oxygenation aspects. In addition, extracorporeal lipid-lowering therapy also significantly influences the coagulation system; part of the clotting factor concentrations decrease, platelet aggregation decreases, and the hypercoagulable state improves. By analyzing proteomics (Spieker LE *et al.*, 2004), the decreased lipid purified protein components include three categories: (1) clotting factors and adhesion molecules (such as fibronectin), (2) blood rheology related proteins (mainly fiber fibrinogen), and (3) inflammatory factors (such as C-reactive protein). The main components of these proteins are cleansed by adsorption or restrictive filtration to achieve significant improvement of blood rheology and microcirculation.

4.3. *Significant Decrease of Oxidative Stress and Inflammation*

Oxidative stress and inflammation play an important role in the pathophysiology of the atherosclerosis process, and blood-membrane reacting in the extracorporeal circulation tends to induce oxidative stress. However, with the improved membrane materials and the use of appropriate anticoagulant, HELP, DSA, IA, and DALI systems have yielded very good biocompatibility (Dihazi H *et al.*, 2008; Russ M *et al.*, 2013). The HELP system has a special anion exchange column which absorbs the plasma of inflammatory cytokines, such as IL-6, IL-2, and TNF, and keeps these factors stable before and after treatment. Similarly, activate the DALI system does not complement and monocytes during the treatment. Meanwhile, with the oxidized LDL (ox-LDL) synchronous clearance, LDL cholesterol was enhanced antioxidant capacity after LDL-apheresis. The main reason is that the ratio of lipid to protein, linoleic acid and arachidonic acid content all decreased in LDL cholesterol. In contrast, a lot of antioxidants including VitE and plasminogen levels all increased, and some oxygen free radical scavenger activity also increased (Schettler V *et al.*, 2000). All the beneficial changes play a key role to prevent the formation of lipid-rich plaque. A study from Japan found that the patients' P-selectin, soluble intercellular adhesion molecule, high-sensitivity C-reactive protein and fibrinogen significantly lowered after DSA treatment (Hahnel D *et al.*, 1999; Utsumi K *et al.*, 2007). Particularly, shear stress-dependent platelet adhesion was decreased with the clearance of C-reactive protein and fibrinogen (Kobayashi S *et al.*, 2006). Therefore, LDL-apheresis not only clears the harmful lipids, large amounts of inflammatory cytokines and adhesion molecules, but also eliminated the adhesion activity of platelets, and inhibited atherosclerosis plaque from forming further. HDL has long been considered an anti-inflammatory factor, but some recent studies found that some of the components of HDL acted as pro-inflammatory factors. One study found that lipid purification decreased activity of pro-inflammatory HDL by 37% (Spieker LE *et al.*, 2006; Sheriff A *et al.*, 2014; Hovland A *et al.*, 2012). These findings further confirm the anti-inflammatory effects of lipid purification.

4.4. *Improvement of the Endothelial Function*

LDL-apheresis has been shown to improve endothelial function and vasodilation by significant decline of LDL, Lp (a) and other harmful lipid components. Mellwig's and other studies have found a significant improvement of myocardial perfusion after HELP treatment because it could improve coronary reserves and reduce coronary resistance. The authors believe this was due to improved vascular endothelial function (Sloan SR *et al.*, 2014). Cardiac magnetic resonance imaging after HELP therapy indicated myocardial perfusion significantly improved DSA treatment (van Buuren F *et al.*, 2012), due to the activation of bradykinin system, accompanied by increase of nitric oxide and prostaglandins plasma levels, and greatly improved endothelial function.

5. Clinical Application of Extracorporeal Blood Lipid Purification Treatments

Currently, there are two main options for the clinical application of LDL-apheresis: First is regular long-term treatment, and the second is short-term treatment. The former is mainly used to treat hereditary disorder of lipid metabolism in patients. The latter is mainly used for acute ischemic vascular disease with lipid metabolism or microcirculation disorders in patients.

5.1. *Indications (Bohl S et al., 2009; Schwartz J et al., 2013)*

5.1.1. *Familial hyperlipidemia with high LDL levels*

The US Food and Drug Administration (FDA) provides indications of long-term regular treatment:

(1) Familial hyperlipidemia (homozygous), LDL > 500 mg/dL.
(2) Familial hyperlipidemia (heterozygous), LDL ≥ 300 mg/dL.
(3) Familial hyperlipidemia (heterozygous), LDL ≥ 200 mg/dL and accompanied by myocardial infarction, unstable angina, coronary bypass surgery and other cardiovascular events.

5.1.2. *Phytanic acid storage disease (Refsum disease)*

Phytanic acid is a branched chain fatty acid that is derived exogenously from dietary sources. The inability to degrade phytanic acid results in its accumulation in fatty tissues, liver, kidney, myelin, and in lipoproteins in the plasma.

5.1.3. *Acute ischemic vascular disease with lipid metabolism or microcirculation disorder*

(1) Acute ischemic stroke
(2) Acute occlusive arteriosclerosis disease

5.1.4. *Acute pancreatitis with severe disorders of lipid metabolism*

LDL-apheresis benefits acute simple edema pancreatitis with dyslipidemia.

5.2. *Clinical Applications*

5.2.1. *Cardiovascular disease*

Familial hyperlipidemia was first treated in 1965, and since then, various extracorporeal lipid purification therapies have been improved on. Extracorporeal lipid purification therapy has become the main treatment method for familial hyperlipidemia, the class I, II level prevention of coronary artery disease in patients.

These patients need long-term regular treatment, and general treatment once every one to two weeks.

In recent years, many clinical studies have demonstrated that extracorporeal lipid purification techniques have good therapeutic effects for stabilizing plaque, reducing the acute coronary events, and preventing coronary restenosis after bypass surgery and percutaneous coronary angioplasty (PTCA). German scholar Schuff-Werner found that in two years of HELP regular treatment at weekly intervals in 50 familial hyperlipidemia patients, the numbers of regressive segments of coronary artery lesion were 1.8 times lower than that of progressive segments by digital coronary angiography analysis system. (Schuff-Werner P *et al.*, 1999; Bosch T *et al.*, 2005). Another group of small sample data showed that seven patients with familial hyperlipidemia altogether had 21 unstable plaques in the carotid artery, with only one case worsening, one worsening, 12 with no significant changes and eight regressive by the three-dimensional ultrasound examination after 6 to 12 months of regular HELP treatment. A 14-year-long retrospective study of coronary heart disease in 180 patients found that the average annual myocardial infarction events were 4.5 times in early 10 years without HELP, and then after two years HELP regular treatment that decreased to 1.5 times (Seidel D *et al.*, 1996).

A 7.7-year Japanese research study examined familial hyperlipidemia in 11 patients who were being treated with the DSA treatment system. This study used three-dimensional ultrasound to measure carotid intima and media thickness of the patients and found a 0.0023 mm decrease every year, while the control group had an average increase of 0.025 mm per year. In another Japanese L-CAPS study, 25 heterozygous familial hypercholesterolemic patients treated with LDL-apheresis and drugs and 11 patients treated with drugs underwent follow-up angiography 2.3 years later. One hundred and thirteen lesions were measured by quantitative angiography. When progression and regression were defined as a change in minimal lumen diameter of +/− 0.67 mm, in the apheresis group, two (8%) patients had progression, 19 (76%) stayed unchanged and four (16%) had regression, but in the control group seven (64%) patients had progression and four (36%) stayed unchanged. The frequency of regression or no change was significantly higher in the apheresis group than in the control group ($P < 0.004$) (Nishimura S *et al.*, 1999). Yamashita K *et al.* investigated restenosis rate after PTCA, and the result showed: the four-month rule DSA system treatment group restenosis rate was 29.4%, a little lower than the control group (47.1%) (Yamashita K *et al.*, 1998).

Other European and American multi-center clinical studies also showed that treatment using immunosorbent assays or by the regular DALI system, can effectively reduce the number of severe lipid metabolism disorder episodes and mortality in patients with myocardial infarction (Lui M *et al.*, 2014; Otto C *et al.*, 2007). Coronary angiography data also suggest that coronary stenosis improved to varying degrees. Similar results were obtained in the LAARS study (LDL-apheresis arteriosclerosis regression study). Extracorporeal lipid-lowering therapy can significantly improve the prognosis of patients with coronary heart disease and reduce the incidence of acute coronary events (Kroon AA *et al.*, 1996; Schuff-Werner P *et al.*, 1999; Weiss D R *et al.*, 2014).

Recent studies have found that the occurrence of acute coronary events was mainly due to instable plaque ruptures, and extracorporeal lipid-lowering therapy can rapidly reduce the lipid core, reducing pressure on the fibrous cap. Extracorporeal LDL-apheresis therapy can also lower blood flow shear stress on the fibrous cap of lipid-rich plaque and vascular endothelial damage, reduce lipid oxidation and improve endothelial function, reduce the hypercoagulable state, and inhibit platelet aggregation. Lipid purification therapy can play a stabilizing role and reduce the incidence of coronary events.

5.2.2. *Acute ischemic stroke*

Currently, in addition to thrombolytic therapy within six hours of an acute ischemic stroke, the primary treatments mainly include isovolemic haemodilution, anticoagulant and thrombolytic therapy. However, the effects of these treatments are not ideal. In a study that examined HELP therapy for treating acute ischemic strokes in 55 patients in one week intervals, the results showed that basic plasma fibrinogen was greater than 500 mg/dL, and that blood lipids and blood viscosity were at the upper limit of normal in all enrolled patients. Hemorheology significantly improved after treatment, such as high shear blood viscosity decreased by 17.4%, low shear blood viscosity decreased by 16.5%, plasma viscosity decreased by 17.7%, and some of the main artery of the brain blood flow increased by about 10%. Additionally, the brain microcirculation improved greatly; there was increased blood supply to the ischemic penumbra and significant improvements to the deformability of red blood cells. The red blood cell permeability time (red cell transmit time, RCTT) decreased by 18.5% and brain tissue oxygen supply increased by about 20%. Further, the HELP treatment was an improvement relative to the control subjects in Mathew scale, Mini-Mental State Examination, and activities of daily living test scores in the follow-up two weeks visit. These results support the hypothesis that the improved hemorheologic property of blood is an important factor in clinical recovery as well as basic neurological function (Walzl M *et al.*, 1994).

To evaluate the therapeutic effects of the HELP system in patients with acute ischemic stroke and probe into its possible mechanism, we enrolled 39 acute ischemic stroke patients with a therapeutic window of three to six days. The patients were randomly divided into two groups; the HELP treatment group included 26 patients: 17 males and 9 females aged 55–78 years. The control group consisted of 13 patients: 9 males and 4 females, aged 58–76 years. All patients routinely used isovolemic haemodilution therapy (Salvia + dextran), enteric-coated aspirin, heparin or low molecular weight heparin, if TC > 4.68mmol/L or TG > 1.7mmol/L, plus a statin or fibrate lipid-lowering drugs. After treatment, for all patients in the HELP group, TC, TG, LDL, Lp (a) and fibrinogen levels decreased by 30% to 50%, blood viscosity decreased by 10% to 20%, and HDL did not change significantly. At the three-month follow-up examination, the neurological function scores such as CSS (Chinese Stroke Scale) and ADL score (activities of daily living score) improved significantly compared to the control group, and blood viscosity had no significant rebound (Wu Duchu *et al.*, 1998).

In another of our studies, 10 patients with acute ischemic stroke were included and received HELP therapy in addition to low molecular weight dextran, salvia miltiorrhiza and aspirin. Matched with gender, age, European Stroke Scale (ESS) and fibrinogen, 20 patients with cerebral infarction treated with low molecular heparin were chosen as controls during the same period. We compared the efficacy and prognosis between the two groups. In order to clarify the effects of HELP treatment on endothelial function, human umbilical vein endothelial cells (HUVEC) were cultured with the serum of pre- and post-HELP therapy and control patients, then endothelial permeability and biomarkers of endothelial dysfunction or activation in the supernatant of incubated HUVEC, such as soluble thrombomodulin (sTM), soluble intracellular adhesion molecule-1 (sICAM-1) and monocyte chemoattractant protein-1 (MCP-1), were tested.

Twenty-one days after treatment, ESS scores in the HELP group (70.4 +/− 23.06) were higher than those in control group (60.7 +/− 18.94), but there was no statistical significance ($P > 0.05$). In the HELP group, the total efficacy rate reached 60% with ESS score (six cases of recovery and efficacy, four cases of inefficacy), while it was 40% in the control group (eight cases of recovery and efficacy, 12 cases of inefficacy), but unfortunately there was no significant difference between both groups using the chi (2)-test ($P > 0.05$). Interestingly, with the activities of daily living (ADL) score, the efficacy rate in the HELP group (60%) was markedly higher than that in the control group (20%) ($P < 0.05$). Furthermore, Pearson correlation analysis showed that the therapeutic window (the time from the patient's onset to receiving the therapy) was correlated to final ADL scores ($P = 0.044$) and the mean therapeutic window was 14.08 +/− 3.41 hr in patients who achieved efficacy criteria. After therapy, no difference was found in hemorheology, fibrinogen, blood lipid, oxidized low-density lipoprotein (oxLDL) or C-reactive protein (CRP) in the control group, while there was a significant decrease in the HELP group immediately after treatment ($P < 0.01$). But the hospital stay time was similar between the HELP group and the controls (23.11 +/− 10.65 vs. 21.53 +/− 8.73 days; $P > 0.05$) and 21 days later, the dimension of the largest cross-sectioned infarction area by CT scan did not significantly change for the two groups of patients ($P > 0.05$). When cultured with the patients' serum after HELP therapy, the concentration of sTM, sICAM-1, and MCP-1 in the supernatant of cultured HUVEC remained unchanged at 24 hr ($P > 0.05$), while it was remarkably higher when HUVEC was cultured with the serum of the controls and patients before HELP therapy ($P < 0.01$), except for sTM. The endothelial permeability was also improved after the HELP treatment. With the effects of improving hemorheology, decreasing acute phase reactive proteins such as CRP and fibrinogen, ameliorating endothelial cell function, the HELP system may be a novel therapy for acute ischemic stroke (Xue J *et al.*, 2007). Therefore, the outcomes of our studies are consistent with other studies. Of course, multi-center clinical research in this area is needed to clarify the indications for treatment.

5.2.3. *Acute occlusive arterial sclerosis*

According to the survey, half of the patients with arteriosclerosis obliterans were often associated with hyperlipidemia, mostly in the lower limb extremities of the arterial site. The indications of the lipid purification therapy for arteriosclerosis obliterans include the following four aspects: (1) arteriosclerosis obliterans was diagnosed more than Fontaine degree II; (2) after taking lipid-lowering drugs, lipid abnormalities were still present; (3) surgical treatment is difficult or impossible; (4) drug therapy had no significant effect. The most common applied lipid purification therapies are the DSA and HELP systems. The specific-LDL-apheresis treatment protocol is: twice a week for two weeks, then once a week for two weeks, then once every two weeks, a total of 10 sessions. LDL-apheresis for a single course of the treatment in about three months. Most patients began to experience an ant crawling feeling in the lower limp extremity after two to three treatments, which then turned into a variable temperature sensation, gradually extending throughout the whole body, and thereafter the patients' pain-free walking distance extended. Moreover, rest pain, numbness and intermittent claudication gradually reduced and disappeared. Gangrene disappearance and pulse of arterial dorsal pedis appearance occured in some patients at the end of treatment. Several studies have found that lipid purification can significantly improve patient limb perfusion, relieve pain, and promote the healing of ulcers (Ebihara I *et al.*, 2007; Weiss N *et al.*, 2012; Ramunni A *et al.*, 2012). Peripheral Arterial Disease LDL Apheresis Multicenter Study (P-LAS) confirmed the effectiveness of LDL apheresis on the improvement in physiological parameters such as ankle-brachial pressure index, maximum tolerated walking distance and clinical symptoms in patients with peripheral arterial disease (Tsuchida H *et al.*, 2006).

5.2.4. *Hormone resistance of nephritic syndrome*

Clinical studies by many Japanese scholars have similar results in the treatment of refractory nephrotic syndrome by LDL apheresis. Stenvinkel *et al.* conducted a 10-week long lipid lower treatment in the six patients with hormone-resistant nephrotic syndrome. The cholesterol, Lp (a), LDL, and urinary protein levels all decreased, and the serum albumin concentration increased (Stenvinkel P *et al.*, 2000). Some research also found that lipid lowering therapy could reduce plasma macrophages secretion of inflammatory cytokines and chemokines such as (MCP-1), inhibiting infiltration of inflammatory cells in glomerular, improve nephrotic syndrome, restore sensitivity to hormones, and slow glomerulosclerosis and tubulointerstitial fibrosis (Kobayashi S *et al.*, 2008). Furthermore, LDL apheresis could reduce serum levels of P-selectin, which is known to play an important role in the development of atherosclerosis as well as a reduction of serum C-reactive protein levels as standard biomarker of atherosclerosis. The possible mechanisms besides removal of toxic lipids are the reduction of the vasoconstrictive prostanoid and thromboxane A2 (TXA2) and an improvement in macrophage function evidenced by a significant amelioration of interleukin-8 production by lipopolysaccharide-stimulated peripheral blood mononuclear cells. It is intriguing to note that in terms of pharmacodynamics, LDL apheresis improves steroid and cyclosporine uptake into lymphocytes. In all, the lipid purification therapy improved prognosis of refractory nephrotic syndrome from the following aspects (Kobayashi T *et al.*, 2006): improved blood rheology, improved endothelial function, increased levels of nitric oxide and bradykinin, which reduces the level of oxidative stress and inflammation, increase in serum levels of vascular endothelial growth factor, and reduction of adhesion molecules on monocytes. The treatment effect is particularly significant for focal segmental sclerosis glomerulonephritis (Keith DS *et al.*, 2012) and the primary glomerulonephritis recurrence after transplantation.

5.2.5. *Other diseases*

Acute retinal ischemia, sudden deafness and acute pancreatitis are all the indications for LDL apheresis therapy. It can improve clinic symptoms and organic function, and aid disease recovery after LDL apheresis treatments. Over the years, there has been sufficient evidence recommending that this therapy should be the first choice for sudden deafness and acute retinal ischemia (Guerriero S *et al.*, 2009). Three weekly sessions of LDL apheresis were administered in 23 patients affected by nonarteritic ischemic optic neuropathy. Statistically significant reductions were achieved in all parameters: total cholesterol (44.6%), LDL cholesterol (54.6%), fibrinogen (60.9%), von Willebrand factor (38.6%), sE-Selectin (22.6%), sICAM-1 (14%) and sVCAM-1 (15.5%), each of which was correlated with an improvement in the mean deviation of the visual field. However, analysis of variance between the mean deviation improvement and the set of parameters taken together yielded highly significant results ($P < 0.0001$) (Ramunni A *et al.*, 2006). Sudden sensorineural hearing loss (SSHL) may be caused by a reduction of cochlear perfusion. LDL apheresis improves cochlear blood flow by acutely decreasing plasma cholesterol and fibrinogen. Remission rates of 217 patients with SSHL were analyzed retrospectively after single apheresis. All patients had been treated otherwise before without any improvement of hearing. 15% of all patients had complete remissions, whereas partial remissions were seen in 46%. No change of hearing threshold was seen in 33%, 2% worsened. Remission rates decreased from 70% for a time of two weeks between onset of SSHL and apheresis to 63% and 21% for six weeks and three months (Heigl F *et al.*, 2009).

6. Adverse Reactions of Extracorporeal Lipid Purification Therapy

The adverse effects of lipid separation are related primarily to extracorporeal circulation (Krebs A *et al.*, 2004; Hudgins LC *et al.*, 2008).

6.1. *Symptomatic Hypotension*

The incidence rate of symptomatic hypotension varies from report to report, but is generally anywhere from 3 to 6%. (Stefanutti C *et al.*, 2013; Hequet O *et al.*, 2010; Hovland A *et al.*, 2011), and may be associated with effective hypovolemia, vagus nerve disorders and poor cardiac function. Usually, the adverse reaction will not be serious. The incidence at our center is about 1.5%. It is quite important to actively monitor the blood pressure. Saline and colloidal solutions should be administered if the blood pressure drops.

6.2. *Allergic Reactions*

The incidence of allergic reaction is 0.5 to 3.6%. These reactions are often due to heterologous antibody or dextran sulfate molecules dropping into the blood in the IA or DSA treatments. With the improvement of these devices, plasma flows through the adsorption column in the pot before it enters the vein; the plasma travels through a special adsorption column that can almost completely absorb shedding particles. Therefore, this type of allergic reaction has been greatly decreased. Also, administering angiotensin converting enzyme inhibitors (ACEI) may cause hypotension, nausea, vomiting and other reaction, which may be related to excessive release of bradykinin. So it is recommended to withdraw these ACEI drugs before DSA treatments. No allergic reactions were reported in the 700 HELP treatment cases. Therefore we can conclude that the therapy is quite safe.

6.3. *Fever and Sepsis*

Fever is often related to the materials and methods, especially the bioaffinity adsorbent. Sepsis is often caused by exogenous contamination and improper operation.

6.4. *Non-specific Reactions*

Some patients will develop fatigue, weakness and discomfort after treatment (Stefanutti C *et al.*, 2013). This may be related to low blood sugar, changes in the environment and other factors.

6.5. *Hemolysis*

One study showed that the extracorporeal membrane might induce oxidative stress. This oxidative stress may lead to cellular abnormalities with clinical consequences, such as hemolysis and platelet hyperactivation (Amer J *et al.*, 2010). With the improvement of blood purification techniques and methodologies, such complications rarely occur.

6.6. *Anemia*

Clinical observations revealed an increased prevalence of iron deficiency anemia without chronic bleeding in patients treated with serial low-density lipoprotein (LDL) apheresis. This anemia is mainly caused by the significant removal of vitamin B12, transferrin, and ferritin (Bramlage CP *et al.*, 2010).

6.7. *Puncture Site Hematoma*

Puncture site hematoma is mainly related to poor vascular conditions, puncture technique and inappropriate compressing hemostasis. Especially in patients with direct arterial puncture, the complications of local hematoma are common. It will take a much longer time to press to stop blood, generally at least 30 minutes or more, and then bandage.

6.8. *Bleeding*

Bleeding is mainly related to the *vitro* anticoagulant, meanwhile coagulation factor concentrations decreased after LDL apheresis as well. Blood purification is prohibited for patients with bleeding tendencies.

7. Contraindications of Lipid Separation Therapy

The contradiction mainly includes two aspects: the first is the active bleeding or hemorrhagic tendency, the second is that the patients fail to tolerate the cardiopulmonary bypass procedure. Lipid separation therapy contraindications are acute hemorrhagic stroke, severe bleeding or high risk of bleeding peptic ulcer and other diseases, such as hypotension, acute myocardial infarction, and cardiopulmonary disease.

8. Conclusion: Clinical Comparison of *Vitro* Blood Lipid Purification Treatment

Because simple plasmapheresis has poor selectivity, it is necessary to supply a large amount of blood products to the patients, so the simple therapy has been replaced by other highly selective lipid separation therapies. Currently the most widely used clinical therapies are DFPP (or improved DFPP), IA, HELP, DSA and the DALI system (Table 16.3). Their lipid-lowering aspects have significant efficiency, biocompatibility and long-term toleration. Because the patients in the HELP system have a lower plasma fibrinogen, this system has advantages over other systems in hemorheology, and correcting microcirculation. The operation of IA, DSA, and the DALI systems are quite convenient. However, these three methods are more prone to cause allergic reactions than the HELP system, especially before taking ACEI therapy drugs. Although DFPP is slightly less selective than the latter four methods, it has a significant price advantage. Therefore, every type has its pros and cons, and should be applied by the practical clinical work.

Table 16.3. Comparison of Several Main Blood Fat Separation Methods

	DFPP	HELP	DSA	IA	DALI
LDL, Lp (a) clearance	+	+	+	+	+
Fibrinogen clearance	+	+	±	±	±
HDL preserve	±	+	+	±	+
Plasma protein	±	+	+	+	+
Safety	±	+	+	+	+
Vascular access	±	±	±	±	±

(LDL, low-density lipoprotein; Lp (a), lipoprotein (a); IA, immunoadsorption; DSA, dextran sulfate cellulose adsorption; HELP, heparin mediated extracorporeal LDL precipitation; DFPP, double filtration plasmapheresis; DALI, direct adsorption of lipoprotein)

References

Amer J, Frankenburg S, Fibach E. (2010) Apheresis induces oxidative stress in blood cells. *Ther Apher Dial* **14(2):** 166–71.

Bambauer R, Schid R, Latza R. (2000) Low density lipoprotein apheresis in treatment of hyperlipidemia: Experience with four different technologies. *Ther Apher* **4(3):** 213–217.

Bohl S, Kassner U, Eckardt R, *et al.* (2009) Single lipoprotein apheresis session improves cardiac microvascular function in patients with elevated lipoprotein(a): Detection by stress/rest perfusion magnetic resonance imaging. *Ther Apher Dial* **13(2):** 129–137.

Borberg H. (2013) The lower the better: target values after LDL-Apheresis and semi-selective LDL-elimination therapies. *Transfus Apher Sci* **48(2):** 203–206.

Borberg H. (2009) Comparison of different Lp (a) elimination techniques: A retrospective evaluation. *Transfus Apher Sci* **41(1):** 61–5.

Bosch T, Wendler T, Jaeger BR, *et al.* (2001) Improvement of hemorheology by DALI apheresis: Acute effects on plasma viscosity and erythrocyte aggregation in hypercholesterolemic patients. *Ther Apher* **5(5):** 372–376.

Bosch T, Gahr S, Belschner U, *et al.* (2006) Direct adsorption of low-density lipoprotein by DALI-LDL-apheresis: Results of a prospective long-term multicenter follow-up covering 12,291 sessions. *Ther Apher Dial* **10(3):** 210–218.

Bosch T, Lennertz A, Schmidt B, *et al.* (2000) DALI apheresis in hyperlipidemic patients: Biocompatibility, efficacy, and selectivity of direct adsorption of lipoproteins from whole blood. *Artif Organs* **24(2):** 81–90.

Bosch T. (2005) Therapeutic apheresis state of the art in the year 2005. *Ther Apher Dial* **9(6):** 459–468.

Bramlage, C.P., *et al.* (2010) Low-density lipoprotein apheresis decreases ferritin, transferrin and vitamin B12, which may cause anemia in serially treated patients. *Ther Apher Dial* **14(2):** 136–42.

Dräger LJ, Julius U, Kraenzle K, *et al.* (1998) DALI—the first human whole-blood low-density lipoprotein and lipoprotein (a) apheresis system in clinical use: procedure and clinical results. *European Journal of Clinical Investigation* **28(12):** 994–1002.

Dihazi H, Koziolek MJ, Söllner T, *et al.* (2008) Protein adsorption during LDL-apheresis: Proteomic analysis. *Nephrol Dial Transplant* **23(9):** 2925–2935.

Ebihara I, Sato T, Hirayama K, *et al.* (2007) Blood flow analysis of the head and lower limbs by the laser Doppler blood flowmeter during LDL apheresis. *Ther Apher Dial* **11(5):** 325–330.

Feillet C, Cristol JP, Michel F, *et al.* (1997) Cholesterol biosynthesis in normocholesterolemic patients after cholesterol removal by plasmapheresis. *J Clin Apheresis* **12(3):** 110–115.

Frenais R, Maugeais C, Ouguerram K, *et al.* (2001) Effect of low-density lipoproteins on apoliprotein AI kinetics in heterozygous familial hypercholesterolemia. *Metabolism* **50(6):** 635–639.

Gu Y, Xue J, Dong Q. (2002) The clinical research of HELP system on the systematic treatment of ischemia cardio-cerebrovascular disease. *China Blood Purification* **1(1):** 19–23.

Guerriero S, Giancipoli G, Cantatore A, *et al.* (2009) LDL apheresis in the treatment of non-arteritic ischaemic optic neuropathy: A 6-month follow-up study. *Eye* **23(6):** 1343–1344.

Hahnel D, Thiery J, Brosche T, *et al.* (1999) The role of plasmalogens in the enhanced resistance of LDL against copper induced oxidation after LDL apheresis. *Arterioscler Thromb Vasc Biol* **19(10):** 2431–2438.

Heigl F, Hettich R, Suckfuell M, *et al.* (2009) Fibrinogen/LDL apheresis as successful second-line treatment of sudden hearing loss: A retrospective study on 217 patients. *Atheroscler Suppl* **10(5):** 95–101.

Hequet O, *et al.* (2010) The first results demonstrating efficiency and safety of a double-column whole blood method of LDL-apheresis. *Transfus Apher Sci* **42(1):** 3–10.

Hovland A, Lappegard KT, Mollnes TE. (2012) LDL apheresis and inflammation–implications for atherosclerosis. *Scand J Immunol* **76(3):** 229–236.

Hovland A, Marcovina S, Hardersen R, *et al.* (2012) Three different LDL apheresis columns efficiently and equally reduce lipoprotein(a) concentrations in patients with familial hypercholesterolemia and small apolipoprotein(a) particles. *Transfus Apher Sci* **46(1):** 73–76.

Hovland A, *et al.* (2011) Patient tolerance regarding different low-density lipoprotein apheresis columns: frequent minor side effects and high patient satisfaction. *J Clin Lipidol* **5(1):** 45–9.

Hudgins LC, Kleinman B, Scheuer A, *et al.* (2008) Long-term safety and efficacy of low-density lipoprotein apheresis in childhood for homozygous familial hypercholesterolemia. *Am J Cardiol* **102(9):** 1199–1204.

Jovin IS, Taboski U, Muller-Berghaus G. (2000) Analysis of the long-term efficacy and selectivity of immunoadsorption columns for low density lipoprotein apheresis. *ASAIO J* **46(3):** 298–300.

Julius U, Frind A, Tselmin S, *et al.* (2008) Comparison of different LDL apheresis methods. *Expert Rev Cardiovasc Ther* **6(5):** 629–639.

Julius U, Parhofer KG, Heibges A, *et al.* (2007) Dextran-sulfate-adsorption of atherosclerotic lipoproteins from whole blood or separated plasma for lipid-apheresis–comparison of performance characteristics with DALI and Lipidfiltration. *J Clin Apher* **22(4):** 215–23.

Julius U, Tselmin S, Fischer S, *et al.* (2009) The Dresden Apheresis Center — experience with LDL apheresis and immunoadsorption. *Atheroscler Suppl* **10(5):** 12–16.

Kardas F, Cetin A, Solmaz M, *et al.* (2012) Successful treatment of homozygous familial hypercholesterolemia using cascade filtration plasmapheresis. *Turk J Haematol* **29(4):** 334–341.

Keith DS. (2012) Therapeutic apheresis rescue mission: Recurrent focal segmental glomerulosclerosis in renal allografts. *Semin Dial* **25(2):** 190–192.

Kobayashi S, Oka M, Moriya H, *et al.* (2006) LDL-apheresis reduces P-Selectin, CRP and fibrinogen — possible important implications for improving atherosclerosis. *Ther Apher Dial* **10(3):** 219–223.

Kobayashi S. (2008) Applications of LDL-apheresis in nephrology. *Clin Exp Nephrol* **12(1):** 9–15.

Kobayashi T, Ando Y, Umino T, *et al.* (2006) Complete remission of minimal-change nephrotic syndrome induced by apheresis monotherapy. *Clin Nephrol* **65(6):** 423–426.

Krebs A, Krebs K, Keller F. (2004) Retrospective comparison of 5 different methods for long-term LDL-apheresis in 20 patients between 1986 and 2001. *Int J Artif Organs* **27(2):** 137–148.

Kroon AA, Aengevaeren WR, Van der Werf T, *et al.* (1996) LDL-apheresis arteriosclerosis regression study (LAARS). *Circulation* **93(10):** 1826–1835.

Legallais C, Moriniere P, Fournier A, *et al.* (1996) Pulsed flow cascade filtration long-term experience in low density lipoprotein and lipoprotein a removal. *ASAIO J* **42(5):** m463–467.

Lui M, Garberich R, Strauss C, *et al.* (2014) Usefulness of lipid apheresis in the treatment of familial hypercholesterolemia. *J Lipids*, Epub Oct 19, 2014. [doi: 10.1155/2014/864317]

Moriarty P, *et al.* (2010) Comparison of different low density lipoprotein apheresis machines on brain natriuretic peptide levels in patients with familial hypercholesterolemia. *Ther Apher Dial* **14(1):** 74–8.

Nishimura S, Sekiguchi M, Kano T, *et al.* (1999) Effects of intensive lipid lowering by low-density lipoprotein apheresis on regression of coronary atherosclerosis in patients with familial hypercholesterolemia: Japan Low-density Lipoprotein Apheresis Coronary Atherosclerosis Prospective Study (L-CAPS). *Atherosclerosis* **144(2):** 409–17.

Otto C, Berster J, Otto B, *et al.* (2007) Effects of two whole blood systems (DALI and Liposorber D) for LDL apheresis on lipids and cardiovascular risk markers in severe hypercholesterolemia. *J Clin Apher* **22(6):** 301–305.

Payhofer KG, Barrett PH, Demant T, *et al.* (2000) Acute effects of low density lipoprotein apheresis on metabolic parameters of apolipoprotein B. *J Lipid Res* **41(10):** 1596–1603.

Poli L, Busnach G. (2006) Whole blood selective LDL-apheresis: A comparison of two different adsorbers. *Int J Artif Organs* **29(7):** 726–732.

Ramunni A, Brescia P, De Fino G, *et al.* (2012) Therapeutic apheresis in peripheral and retinal circulatory disorders. *Clin Res Cardiol Suppl* **7(Suppl 1):** 41–44.

Ramunni A, Ranieri G, Giancipoli G, *et al.* (2006) Is the efficacy of LDL apheresis in ischemic optic neuropathy linked to a reduction in endothelial activation markers? *Blood Purif* **24(4):** 405–412.

Russ M, Bedarf JR, Grosch-Ott S, *et al.* (2013) First level prevention instead of third level intervention—Review of research to improve biocompatibility and performance of capillary membrane apheresis in critically ill patients. *Artif Organs* **37(6):** 501–512.

Schettler V, Methe H, Schuff-Werner P. (2000) Acute effect of H.E.L.P. treatment on radical scavenging enzyme activities, total glutathione concentration in grannulocytes, and selenium in plasma. *Eur J Clin Invest* **30(1):** 26–32.

Schuff-Werner P, Schettler V. (1999) Plaque stabilization by LDL–apheresis. *Heart* **24(1):** 57–61.

Schuff-Werner P. (1997) Extracorporeal hemorheotherapy with selective plasma protein elimination. *Jpn J Apheresis* **16(1):** 25–30.

Schwartz J, *et al.* (2013) Guidelines on the use of therapeutic apheresis in clinical practice-evidence-based approach from the Writing Committee of the American Society for Apheresis: The sixth special issue. *J Clin Apher* **28(3):** 145–284.

Seidel D. (1996) H.E.L.P. apheresis therapy in the treatment of severe hypercholesterolemia: 10 years of clinical experience. *Artif Organs* **20(4):** 303–310.

Sheriff A, Schindler R, Vogt B, *et al.* (2014) Selective apheresis of C-reactive protein: A new therapeutic option in myocardial infarction? *J Clin Apher* **30(1):** 15–21.

Sloan SR, Andrzejewski CJ, Aqui NA, *et al.* (2014) Role of therapeutic apheresis in infectious and inflammatory diseases: Current knowledge and unanswered questions. *J Clin Apher.*

Spieker LE, Flammer AJ, Amacker N, *et al.* (2006) C-reactive protein influences shear stress-dependent platelet adhesion in patients with familiar hypercholesterolemia and coronary artery disease undergoing LDL apheresis. *Thromb Haemost* **96(4):** 540–542.

Spieker LE, Ruschitzka F, Badimon JJ, *et al.* (2004) Shear stress-dependent platelet function after LDL cholesterol apheresis. *Thromb Res* **113(6):** 395–398.

Stefanutti C, Morozzi C, Perrone G, *et al.* (2012) The lipid- and lipoprotein- (LDL-Lp(a)] apheresis techniques. Updating. *G Chir* **33(11–12):** 444–449.

Stefanutti C, Morozzi C, Di Giacomo S. (2013) Italian multicenter study on low-density lipoprotein apheresis Working Group 2009 survey. *Ther Apher Dial* **17(2):** 169–78.

Stenvinkel P, Alvestrand A, Angelin B, *et al.* (2000) LDL-apheresis in patients with nephrotic syndrome: Effects on serum albumin and urinary albumin excretion. *Eur J Clin Invest* **30(10):** 866–870.

Tani N. (1996) Development of selective low-density lipoprotein(LDL) apheresis system: Immobilized polyanion as LDL-specific adsorption for LDL apheresis system. *Artif Organs* **20(8):** 922–929.

Tsuchida H, Shigematsu H, Ishimaru S, *et al.* (2006) Effect of low-density lipoprotein apheresis on patients with peripheral arterial disease. Peripheral Arterial Disease LDL Apheresis Multicenter Study (P-LAS). *Int Angiol* **25(3):** 287–92.

Utsumi K, Kawabe M, Hirama A, *et al.* (2007) Effects of selective LDL apheresis on plasma concentrations of ICAM-1, VCAM-1 and P-selectin in diabetic patients with arteriosclerosis obliterans and receiving maintenance hemodialysis. *Clin Chim Acta* **377(1–2):** 198–200.

Van Buuren F, Kreickmann S, Horstkotte D, *et al.* HELP apheresis in hypercholesterolemia and cardiovascular disease: efficacy and adverse events after 8,500 procedures. *Clin Res Cardiol Suppl* **7:** 24–30.

Walzl M, Walzl B, Kleinert G, *et al.* (1994) LDL-precipitation and its influence on haemorheology and clinical symptoms in acute stroke and cerebral multi infarct disease. *Clin Hemorhest* **14(1):** 37–44.

Weiss DR, Franke D, Strasser EF, *et al.* (2014) von Willebrand factor, clotting factors, and clotting inhibitors in apheresis platelet concentrates. *Transfusion* **54(3):** 633–639.

Weiss N. (2012) A critical review on the use of lipid apheresis and rheopheresis for treatment of peripheral arterial disease and the diabetic foot syndrome. *Semin Dial* **25(2):** 220–227.

Wu DC, Zhang J, Lv CZ, *et al.* (1998) The preliminary treatment effect of HELP system with acute cerebral infarction, *Chinese Journal of Neural Mental Illness* **24(2):** 106–107.

Xue J, Dong Q, Han X, *et al.* (2007) Effects of HELP therapy on acute ischemic stroke and vascular endothelial cell function. *Ther Apher Dial* **11(3):** 171–176.

Yamashita K, Tasaki H, Tsuda Y, *et al.* (1998) Can aggressive lipid lowering using low-density lipoprotein apheresis prevent restenosis after percutaneous transluminal coronary angioplasty in patients with normocholesterolemia? *Ther Apher Dial* **2(3):** 210–7.

Zenti MG, Stefanutti C. (2011) Effects of selective H.E.L.P. LDL-apheresis on plasma inflammatory markers concentration in severe dyslipidemia: Implication for anti-inflammatory response. *Cytokine* **56(3):** 850–854.

Chapter 17

Cytokine Adsorbing Membrane: PMMA

Shigeto Oda,* Kenichi Matsuda,†
and Hiroyuki Hirasawa*

1. Introduction

1.1. *Development of the PMMA Membrane*

The polymethylmethacrylate (PMMA) membrane is a synthetic polymeric membrane developed by Basic Research Laboratories of Toray Industries, Inc. Mixing of isotactic PMMA and syndiotactic PMMA with different tacticities in an organic solvent generates a thermoreversible sol/gel. Through the application of this stereocomplex phenomenon, the PMMA membrane was prepared by immersing the hydrophobic matrix of PMMA in water to replace the organic solvent within the micropore space with water (Sakai and Tanzawa, 1978). Although originally developed as a reverse osmosis (RO) membrane, the PMMA membrane was developed for use as the first synthetic polymeric hollow fiber dialyzer for hemodialysis due to its insufficient rejection capacity for NaCl (Sakai, 2011).

*Department of Emergency and Critical Care Medicine, Graduate School of Medicine, Chiba University, Japan.

†Department of Emergency and Critical Care Medicine, School of Medicine, Yamanashi University, Japan.

Corresponding author: Shigeto Oda, MD, PhD, Address: 1-8-1 Inohana, Chuou-ku, Chiba City, Chiba 260-8670, Japan, TEL: +81-43-226-2353, FAX: +81-43-226-2371, E-mail: odas@faculty.chiba-u.jp.

The first commercial PMMA membrane dialyzers were the Filtryzer B1® and B2® series. They were granted marketing authorization in 1977 and used for the treatment of patients receiving maintenance dialysis. A newer product with improved water permeability and removal capacity for solutes, the BK series®, was subsequently developed and has been widely used as a high-performance membrane dialyzer.

1.2. *Features of the PMMA Membrane*

The PMMA membrane is a uniform membrane with a symmetric microporous structure throughout the matrix. It has a broader fractionation curve compared to other synthetic polymeric membranes, including a polysulfone membrane (Fig. 17.1). It has the advantage of a greater ability to remove middle-molecular-weight (represented by vitamin B_{12}) to high-molecular-weight substances (e.g. albumin) compared to the conventional cellulose membrane. Hakim *et al.* (1984) performed comparison of hemodialysis with cuprophane, cellulose acetate, and PMMA membrane dialyzers in patients with acute renal failure to investigate the effect of the dialysis membrane on

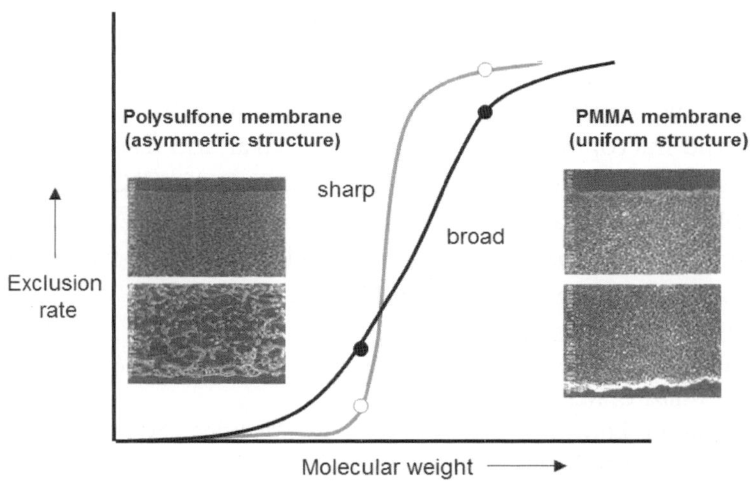

Fig. 17.1. Structure and removal characteristics of polysulfone and polymethylmethacrylate (PMMA) membranes.

complement activation and neutropenia. They reported that the PMMA membrane provided the best suppression of complement activation and improvement in neutropenia. From these results, the PMMA membrane came to be known to have an excellent biocompatibility among various dialysis membranes.

BK series with a larger membrane pore size were launched in 1984 and were intended for removal of substances with a higher molecular weight. In 1985, Gejyo *et al.* (1985) found that β_2-microglobulin (β_2MG) (molecular weight; 11,800) was a pathogenic substance of dialysis-related amyloidosis and reported that this substance was efficiently removed using a PMMA membrane dialyzer (Ono *et al.*, 1988). A subsequent long-term multicenter study in Japan demonstrated a significantly lower blood β_2MG level and a lower incidence of carpal tunnel syndrome and radiolucent bone cyst, conditions derived from amyloid deposition, in hemodialysis patients treated with a PMMA membrane dialyzer compared to those treated with other membrane dialyzers (Aoike, 1998; Arakawa, 1991). Interestingly, the β_2MG level in the dialysate was extremely low when a BK dialyzer was used for hemodialysis, and the primary mechanism of β_2MG removal was adsorption to the PMMA membrane itself (Yamada *et al.*, 1998). Sakai, one of the inventors of the PMMA membrane, later reminisced about this discovery as a series of serendipitous events (Sakai, 2011).

2. Development of Continuous Hemodiafiltration using a PMMA Membrane Hemofilter

In 1986, continuous hemofiltration (CHF) was introduced to the Department of Emergency and Critical Care Medicine, Chiba University Hospital (Chiba, Japan). At this time, PMMA membrane hemofilters were used. CHF was then replaced by continuous hemodiafiltration (CHDF), a combination of hemofiltration and hemodialysis, to enhance removal capacity for solutes. CHDF is now commonly performed in Japan as a renal replacement therapy for critically ill patients mainly in the intensive care unit (ICU).

In 1993, Bellomo *et al.* reported that various pro-inflammatory cytokines were removed by hemofiltration with dialysis. Pro-inflammatory cytokines are produced by immunocompetent cells in response to various insults, and high blood levels of these cytokines induce systemic inflammatory response syndrome (SIRS), characterized by fever or hypothermia, tachycardia, tachypnea, and leukocytosis or leukopeina (Shapiro and Gelfand, 1993). That is, SIRS is caused by hypercytokinemia and persistent SIRS triggers activation of the complement system, blood coagulation system, and neutrophils as well as injury to vascular endothelial cells, finally leading to multiple organ failure (MOF) (Wheeler and Bernard, 1999; Tracey, 2007).

In this context, our interest was focused on the fact that CHDF with a PMMA membrane hemofilter (PMMA-CHDF) produced effects other than renal support in various critically ill patients. More specifically, PMMA-CHDF performed in patients with severe acute pancreatitis (SAP) reduced abdominal pain and provided a rapid improvement in the inflammatory response. In addition, we often observed a rise in blood pressure and increase in urine volume in patients with acute renal failure soon after initiation of PMMA-CHDF. Based on these findings, we considered that the clinical effects of PMMA-CHDF might be ascribable to removal of some pathogenic substance(s) from the bloodstream. We therefore started an extensive clinical investigation into the effects of removal of various humoral mediators, including cytokines, by PMMA-CHDF (Hirasawa *et al.*, 1996).

3. Cytokine Removal with PMMA Membrane

3.1. *Investigation in Clinical Cases*

During PMMA-CHDF in various clinical conditions, blood samples were simultaneously collected from the inlet and outlet of the hemofilter column with concomitant collection of filtrate, in which various cytokines were measured. The results showed decreased blood levels of various cytokines at the outlet side of a hemofilter

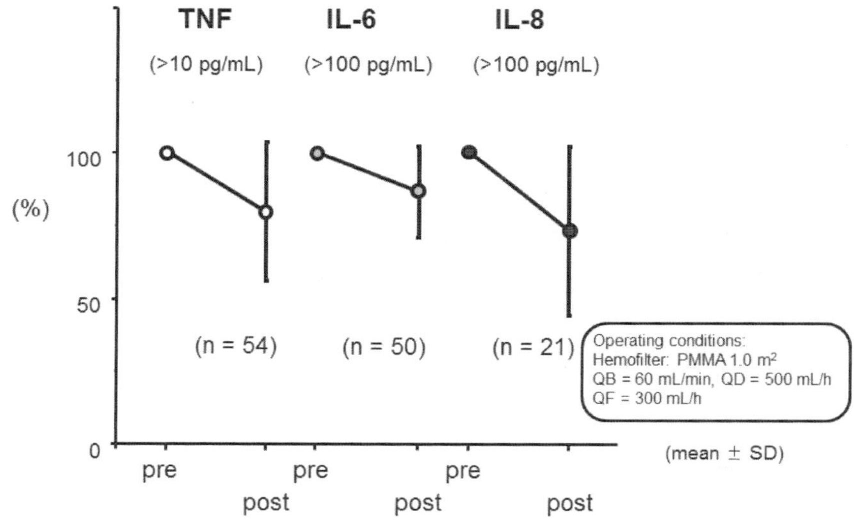

Fig. 17.2. Changes in blood cytokine levels from the pre- to the post-PMMA membrane hemofilter during continuous hemodiafiltration (CHDF).

TNF: tumor necrosis factor, IL: Interleukin, QB: blood flow rate, QF: filtration rate, QD: dialysate flow rate, CHDF: continuous hemodiafiltration, PMMA: Polymethylmethacrylate.

column (Fig. 17.2). On the other hand, almost no cytokines were detected in the filtrate, suggesting this decrease in blood cytokine levels may be caused by adsorption to the hemofilter membrane or deactivation/degradation during passage through the hollow fiber. Analysis of the relationship between cytokine clearance (calculated from blood level) and blood levels of cytokines revealed a significant positive correlation (Fig. 17.3). The higher cytokine clearance observed at higher blood levels indicated that cytokines were removed by adsorption to the PMMA membrane, as was the case with β_2MG (Matsuda *et al.*, 2001). We monitored changes in blood cytokine levels over time in patients undergoing PMMA-CHDF and found a significant decrease in blood cytokine levels three days after initiation of PMMA-CHDF in those who exhibited higher baseline cytokine levels but not in those with lower baseline levels (Fig 17.4). Furthermore, this decrease in blood cytokines during hemodiafiltration was not observed when a hemofilter

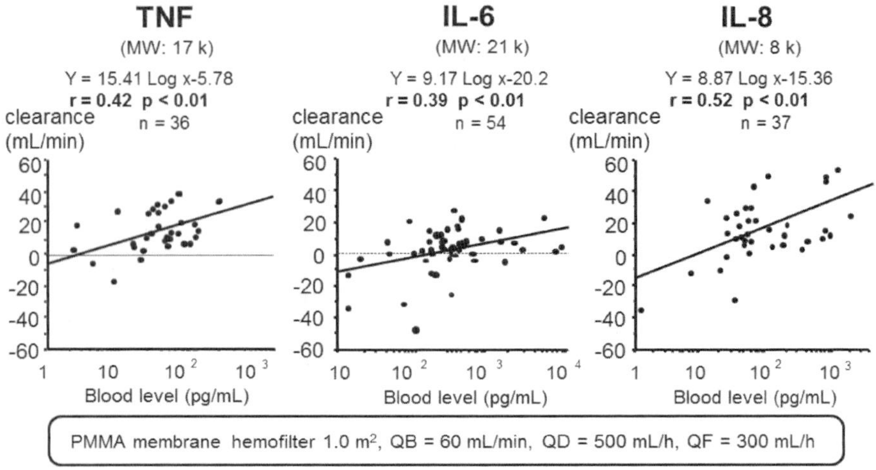

Fig. 17.3. Correlations between clearance and blood cytokine levels in patients treated with continuous hemodiafiltration with polymethymethacrylate membrane hemofilter (PMMA-CHDF).

PMMA: polymethylmethacrylate, CHDF: continuous hemodiafiltration, QB: blood flow rate, QD: dialysate flow rate, QF: filtration rate, TNF: tumor necrosis factor, IL: interleukin.

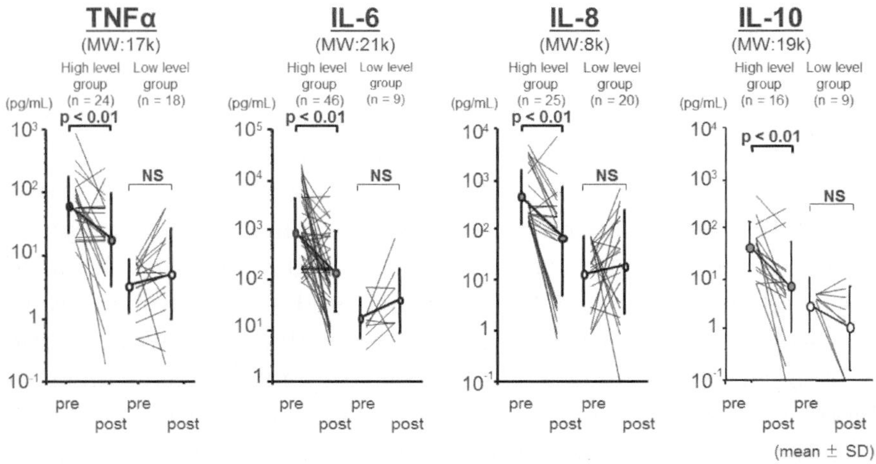

Fig. 17.4. Changes in blood cytokine levels after 3-day treatment of continuous hemodiafiltration with the polymethylmethacrylate membrane hemofilter (PMMA-CHDF).

TNFa: tumor necrosis factor-alpha, IL: interleukin, NS: not significant.

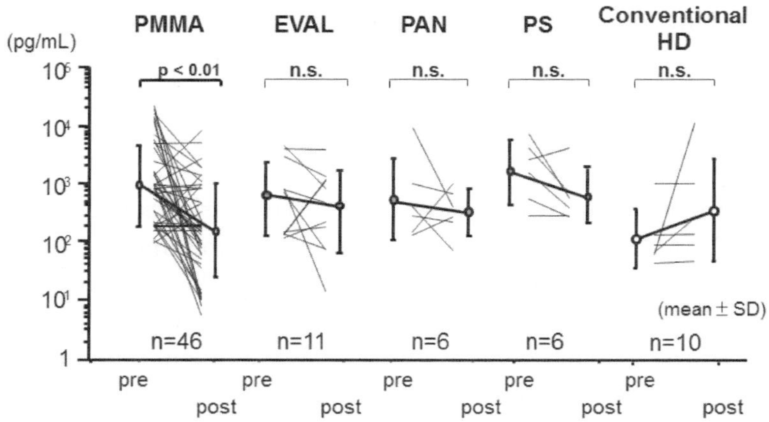

Fig. 17.5. Changes in blood interleukin-6 (IL-6) levels after 3-day treatment with continuous hemodiafiltration using various types of hemofilters and with conventional hemodiafiltration (HD).

PMMA: polymethylmethacrylate, EVAL: ethylene vinyl alcohol, PAN: polyacrylonitrile, PS: polysulfone, HD: hemodialysis.

made of a non-PMMA membrane material was used, suggesting this was a feature unique to the PMMA membrane hemofilter (Fig. 17.5). These results suggest various cytokines are efficiently adsorbed to a PMMA membrane hemofilter and removed from the bloodstream, thereby demonstrating the clinical effects of a PMMA membrane hemofilter to reduce blood cytokine levels.

Hakim *et al.* 1994 prospectively studied dialysis with PMMA membrane dialyzer compared to conventional cuprophane membrane dialyzer in acute renal failure (ARF) patients who required hemodialysis. The PMMA group (n = 37) showed significantly higher recovery rate from renal failure (62% vs 37%, $P = 0.04$) and higher tendency of survival rate (57% vs 37%, $P = 0.11$). Although survival rate was not significantly different between groups, the PMMA group showed significantly higher survival rate compared to cuprophane membrane group among patients with nonoliguric ARF (n = 20 for each group, 80% vs 40%, $P = 0.01$). They have attributed these differences in two membranes to biocompatibility. However, there is a possibility that the cytokine removal by the PMMA membrane resulted in these differences.

3.2. *In Vitro Study*

To investigate the cytokine adsorption characteristics and efficiency of the PMMA membrane, several *in vitro* experiments were performed. Figure 17.6 shows the efficiency of cytokine removal during concentration of ascites fluid containing a high concentration of IL-6 to 1/5 of the original volume using a PMMA membrane hemofilter (CH1.0N), a PMMA membrane dialyzer (BG2.1PQ), a cellulose triacetate (CTA) membrane hemofilter (UT1100S), and a polysulfone (PS) dialyzer (AEF-10). While IL-6 was scarcely detectable in the filtrate when a PMMA membrane hemofilter or dialyzer was used, it was detected in the filtrate using the UT1100S or AEF-10 dialyzers, suggesting IL-6 was removed by filtration in these membranes. However, the total IL-6 removal rate calculated from the IL-6 concentration in the concentrate for a PMMA hemofilter or

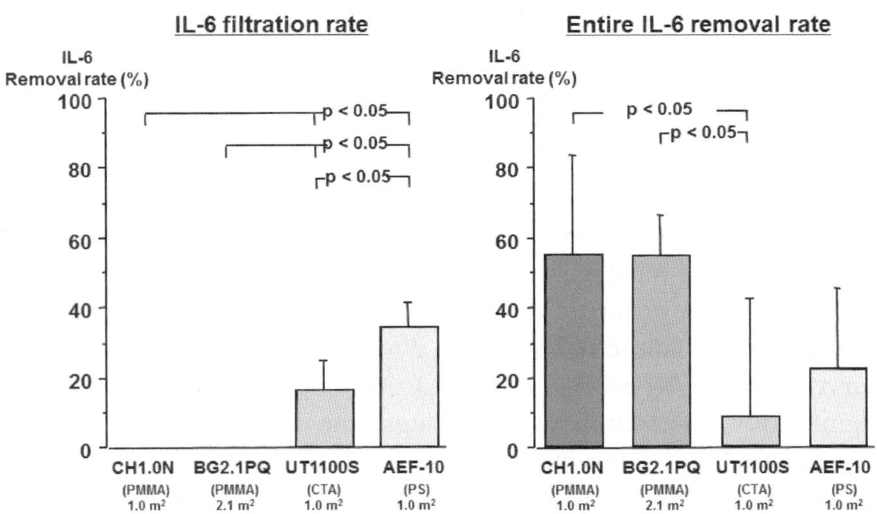

Fig. 17.6. Interleukin-6 (IL-6) removal rates for each hemofilter. Values are expressed as mean ± SD ($n = 6$).

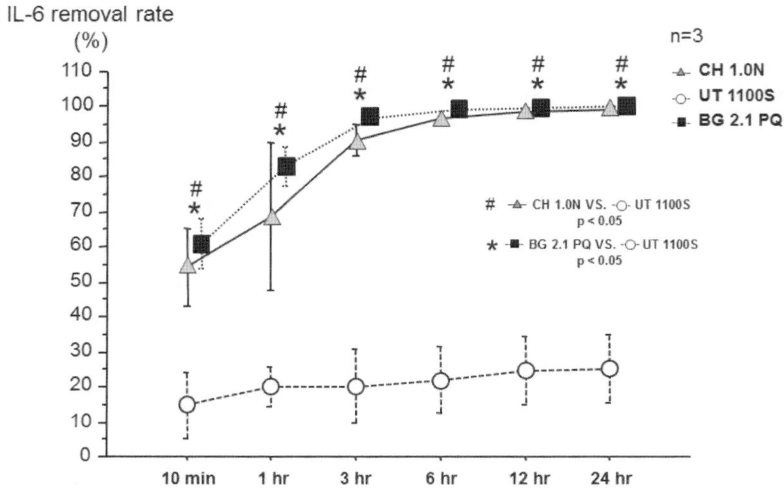

Fig. 17.7. Time-dependent changes in IL-6 removal by each hemofilter. Values are expressed as mean ± SD ($n = 6$).

dialyzer was significantly higher than those for devices with CTA or PS membrane, indicating that the PMMA membrane could remove IL-6 far more efficiently. Next, time-dependent changes in IL-6 removal rate from ascites fluid containing a high concentration of IL-6 were measured during simple recirculation with a pump, instead of one-way filtration (Fig. 17.7). While the IL-6 removal rate for a CTA membrane hemofilter increased to approximately 20% during the first one hour of recirculation and remained constant thereafter, the value steadily increased with time and reached approximately 100% three hours after the start of recirculation when a PMMA membrane hemofilter or dialyzer was used (Hirayama *et al.*, 2011). Thus, cytokine removal with a PMMA membrane hemofilter involved adsorption to the PMMA membrane, and this phenomenon occurred even with membranes of differing pore sizes.

To investigate the removal capacity of cytokines through a PMMA hemofilter in more detail, a perfusion experiment was performed: using 3 L heparinized porcine blood with continuous endotoxin (5 mg/L) injection to induce continuous cytokine production. Changes in the efficiency of cytokine removal by PMMA-CHDF were measured over 24 hr. The following operational conditions were employed: blood purifier, a PMMA membrane hemofilter (CH1.0N); blood flow rate (QB), 80 mL/min; filtration rate (QF), 300 mL/hr; and dialysate flow rate (QD), 1,000 mL/h. These operating conditions are commonly employed in routine clinical practice in our ICU and correspond to an effluent flow rate of 22 mL/kg/hr if a body weight of 60 kg is assumed: namely, they are not conditions for so-called "high volume hemofiltration." The mean clearance values thus measured for TNF-α, IL-6, IL-8, and HMG-B1 were 11.4, 20.9, 17.8, and 11.2 mL/min, respectively. Results also demonstrated that the cytokine-adsorbing effect of the PMMA membrane hemofilter was maintained during a prolonged CHDF session, although with a slight decrease observed after use for 24 hours. Furthermore, almost no cytokines were detectable in the filtrate. Thus, this experiment showed that cytokine removal was ascribed to adsorption to the PMMA membrane.

3.3. *Mechanism of Cytokine Adsorption using PMMA Membrane*

Tomisawa and Yamashita (2009) analyzed the adsorptive characteristics of a PMMA membrane dialyzer (BG1.6PQ) and a polyester polymer alloy (PEPA) membrane dialyzer (FLX-15GW) using chymotrypsinogen (molecular weight, 25,000) and albumin (molecular weight, 66,000) from the aspect of membrane technology. Although both dialyzers had comparable sieving coefficients for filtration of chymotrypsinogen, the PMMA membrane dialyzer exhibited a fractional adsorption for chymotrypsinogen approximately 20% higher than that measured with a PEPA membrane dialyzer. On the other hand, the fractional adsorption of albumin to a PEPA membrane dialyzer was constant regardless of QF (approximately 20%), while the PMMA membrane dialyzer yielded an increased value with the increase in QF and reached a plateau of 50–60% 10 hours after the start of the experiment. PEPA membranes have an asymmetric structure (as is the case with PS membrane), and adsorption to a PEPA membrane involves hydrophobic interactions. In contrast, adsorption to the

PMMA membrane may involve occlusion of protein molecules into open pores in a uniform and symmetric membrane structure.

4. Clinical Application of PMMA-CHDF

4.1. *Introduction of a Rapid Assay for IL-6*

Based on an idea that septic multiple organ failure is the result of hypercytokinemia, a clinical condition characterized by high blood levels of pro-inflammatory cytokines produced by immune-competent cells, we have aggressively introduced PMMA-CHDF for prevention and treatment of multiple organ failure, for cytokine modulation as a non-renal indication (Hirasawa *et al.*, 1996). However, knowing when to initiate PMMA-CHDF, how to evaluate the effect of cytokine removal, and when to terminate PMMA-CHDF proved difficult to define.

We therefore introduced a rapid assay for IL-6 based on chemiluminescent enzyme immunoassay (CLEIA) technology (Human IL-6 CLEIA Fujirebio®, Fujirebio, Tokyo, Japan) into the clinical laboratory of Chiba University Hospital in April 2000. This method provided almost real-time analysis of IL-6 blood levels within 30 minutes of blood sample collection. Figure 17.8 compares IL-6 blood levels measured with this assay system in ICU patients with various clinical conditions. While the IL-6 blood level observed in SIRS (due to trauma or burn) and sepsis without organ dysfunction was approximately 1,000 pg/mL, the values increased to several to tens of thousands pg/mL in severe sepsis with organ dysfunction and to tens to hundreds of thousands pg/mL in septic shock. Furthermore, as demonstrated in Fig. 17.9, the maximum IL-6 blood level significantly correlated with the maximum SOFA (Sequential Organ Failure Assessment) score often used as an index of organ dysfunction (Oda *et al.*, 2005a). The fact that IL-6 blood level is a good index of cytokine activation and closely correlates with the severity of organ dysfunction justified the use of IL-6 as an index for medical decision-making regarding the implementation of cytokine modulation. Based on these results, we perform critical care based on IL-6

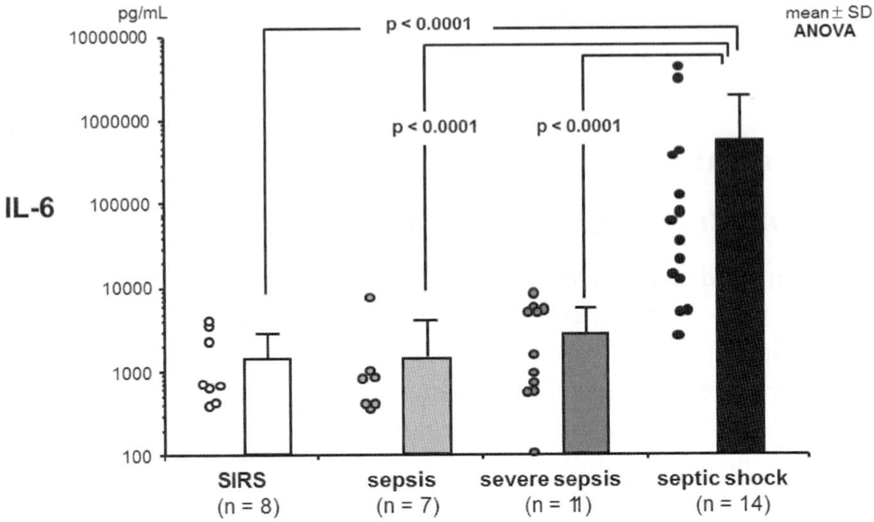

Fig. 17.8. Blood interleukin-6 (IL-6) levels in patients with SIRS, sepsis, severe sepsis, and septic shock on intensive care unit admission.

SIRS: Systemic inflammatory response syndrome.

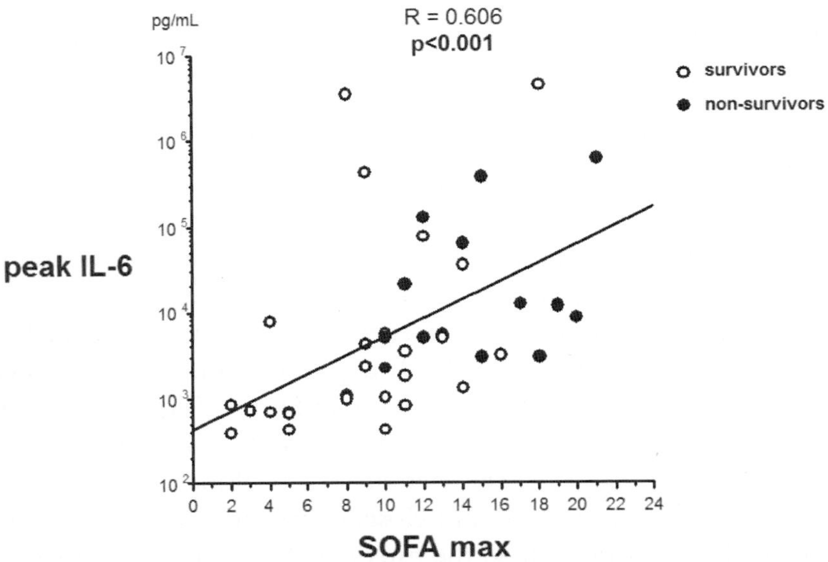

Fig. 17.9. Correlation between peak interleukin-6 (IL-60) blood level and maximum Sequential Organ Failure Assessment (SOFA) score during clinical course.

blood level as an index and aggressively apply cytokine modulation with PMMA-CHDF in patients with an IL-6 blood level of 10,000 pg/mL or higher.

4.2. *Efficacy on Severe Acute Pancreatitis (SAP)*

Severe acute pancreatitis (SAP), a typical example of SIRS not caused by infection, is often associated with development of multiple organ failure when aggravated and results in poor prognosis. Although the IL-6 blood level in SAP is not as high as in septic shock, persisting high IL-6 blood levels may lead to renal and respiratory failure. We therefore perform aggressive PMMA-CHDF in patients with SAP to prevent organ dysfunction. We have reported that early introduction of PMMA-CHDF rapidly reduced the IL-6 blood level and thereby prevented development of organ dysfunction (Oda *et al.*, 2002). Management of early SAP requires maintenance of organ perfusion with massive fluid administration, which may often cause an increase in intra-abdominal pressure (IAP) and lead to development of abdominal compartment syndrome (ACS), thereby further aggravating multiple organ failure. Early introduction of PMMA-CHDF with concomitant fluid management mainly using colloid solution successfully reduced interstitial edema and prevented development of ACS (Oda *et al.*, 2005b). This may be because cytokine removal prevents an increase in vascular permeability. The Japanese National Health Insurance has approved continuous hemodiafiltration for SAP patients regardless of the presence or absence of renal failure.

4.3. *Efficacy on Severe Sepsis/Septic Shock*

Even today, it is difficult for patients to survive severe sepsis/septic shock. As a part of the Surviving Sepsis Campaign launched in October 2002, the Surviving Sepsis Campaign Guidelines (SSCG) have been published and updated to standardize treatment for severe sepsis and septic shock (Dellinger *et al.*, 2004, 2013). Unfortunately, the current SSCG contain neither descriptions of nor countermeasures against hypercytokinemia, the underlying pathophysiology of severe sepsis/septic shock.

Based on our previous studies, we believe countermeasures against hypercytokinemia underlying organ dysfunction are important in the treatment of severe sepsis/septic shock and therefore adopted a therapeutic strategy involving measurement of IL-6 blood levels and early introduction of cytokine modulation by PMMA-CHDF. Particularly in patients with septic shock, PMMA-CHDF is performed in addition to early goal-directed therapy (Rivers *et al.*, 2001) recommended by SSCG for early improvement of tissue oxygen metabolism. As a result, a rapid increase in blood pressure and urine output was observed with a concomitant significant reduction in IL-6 blood level and a rapid decrease in blood lactate level (an index of tissue oxygen metabolism), as shown in Figs. 17.10 and 17.11. Consequently, of 43 patients with septic shock studied [mean Acute Physiology and Chronic Health Evaluation (APACHE) II score, 29.4; mean SOFA score, 13.7], 39 patients (90.7%) successfully

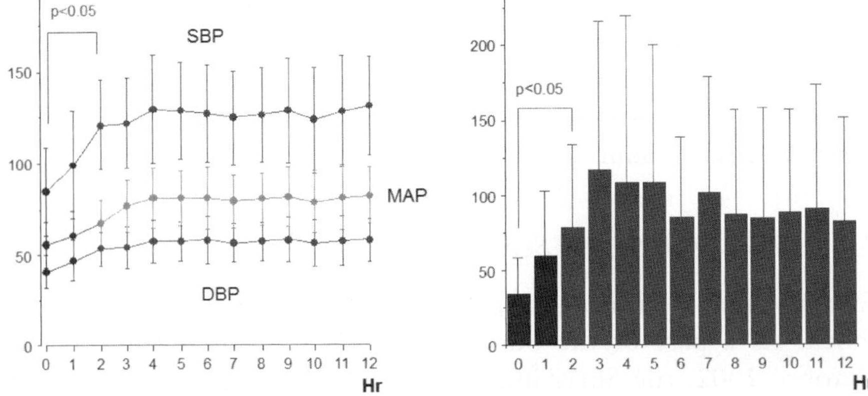

Fig. 17.10. Changes in blood pressure and urine output after initiation of continuous hemodiafiltration with the polymethylmethacrylate membrane hemofilter (PMMA-CHDF). Values are expressed as mean ± SD (n = 6). DBP: diastolic bllod pressure, SBP: systolic blood pressure.

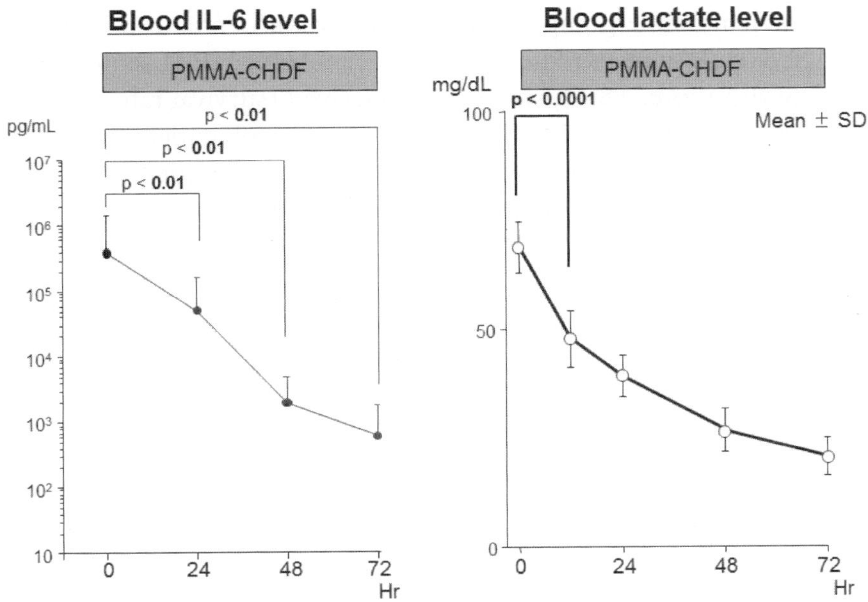

Fig. 17.11. Changes in blood interleukin-6 (IL-6) and lactate levels during continuous hemodiafiltration with the polymethylmethacrylate membrane hemofilter (PMMA-CHDF). Values are expressed as mean ± SD ($n = 6$).

recovered from shock with an excellent 28-day survival rate of 79.1% (Nakada *et al.*, 2008). Matsuda *et al.* (2009) reported that the survival rate in patients with septic acute kidney injury (AKI) receiving renal support by PMMA-CHDF was significantly higher than in those undergoing CHDF with a conventional polyacrylnitrile (PAN) membrane hemofilter, along with historical review of cytokine removal by blood purification therapy.

On the other hand, the survival rate in patients with refractory septic shock, a critical condition characterized by failure to recovery from shock despite aggressive treatment with conventional PMMA-CHDF and consequent early death, remains unacceptably low. To treat such patients, we have recently performed enhanced intensity PMMA-CHDF involving simultaneous operation of two extracorporeal CHDF circuits with two PMMA hemofilters to further enhance cytokine-removing capacity, after obtaining informed consent from each patient's family or next of kin. Although the baseline condition of 10 patients undergoing this new strategy with enhanced intensity PMMA-CHDF was extremely severe (mean APACHE II score, 40.2; mean SOFA score, 18.4), six of them survived (survival rate, 60%), a significantly better survival rate than 15 control patients (13.8%) matched for the severity of severe septic shock and undergoing conventional PMMA-CHDF (Matsumura *et al.*, 2012).

5. Future Perspective

Treatment of critical illness using hemofilters capable of adsorbing solutes has been studied mainly in Japan, and the clinical utility of such hemofilters (mostly PMMA membrane hemofilters) has been documented (Oda *et al.*, 2010). However, the clinical effects of a hemofilter that adsorbs cytokines (AN69 membrane hemofilter) have been reported in Europe and the United States (De Vriese *et al.*, 1999), and this therapeutic approach is expected to attract more and more attention in the future. We expect it will be widely used as an optional therapy for severe septic shock, in which it is difficult to achieve survival by a conventional therapeutic approach (Hirasawa *et al.*, 2012). Although no evidence to support the improvement of outcomes by CHDF for removal of humoral mediators is provided in The Japanese Guidelines for the Management of Sepsis (Oda *et al.*, 2014) published in November 2012, the implementation of CHDF for this particular purpose is recommended weakly at a low evidence level, admitting its potential to improve hemodynamics.

In Japan, a new PMMA membrane hemofilter (CH1.8W®) with an extended membrane area (from 1.0 m^2 to 1.8 m^2) and hollow fiber diameter (from 200 mm to 240 mm) for enhanced cytokine adsorbing capacity was marketed in June 2013. In addition, a cytokine-adsorbing AN69ST hemofilter (sepXiris®) was covered by the Japanese National Health Insurance for the first time for the treatment of severe sepsis/septic shock in July 2014, through a clinical trial conducted in Japan (Shiga *et al.*, 2014). We believe CHDF using cytokine-adsorbing hemofilters is a therapeutic approach to be disseminated not only in Japan but also worldwide.

References

Arakawa M, Niigata Research Program for β_2-microglobulin Removal Membrane. (1991) Long-term multicenter study on β2-microglobulin removal by PMMA BK membrane. *Neprol Dial Transplant* **(suppl 2):** 69–74.

Aoike I, Niigata Research Program for β_2-microglobulin Removal Membrane. (1998) Long term clinical experience with PMMA membrane. In: Ronco C (ed.), *Polymethymethacrylate. A Flexible Membrane for Tailored Dialysis.* Contrib Nephrol vol. 125, pp. 205–212. Karger, Basel.

Bellomo R, Tipping P, Boyce N. (1993) Continuous veno-venous hemofiltration with dialysis removes cytokines from the circulation of septic patients. *Crit Care Med* **21:** 522–526.

Dellinger RP, Cartlet JM, Masur H, *et al.*; Surviving Sepsis Campaign Management Guidelines Committee. (2004) Surviving Sepsis Campaign guidelines for management of severe sepsis and septic shock. *Crit Care Med* **32:** 858–882.

Dellinger RP, Levy MM, Rhodes A, *et al.* Surviving Sepsis Campaign Guidelines Committee including the Pediatric Subgroup. (2013) Surviving Sepsis Campaign: International Guidelines for Management of Severe Sepsis and Septic Shock: 2012. *Crit Care Med* **41:** 580–637.

De Vriese AS, Colardyn FA, Philippe JJ, *et al.* (1999) Cytokine removal during continuous hemofiltration in septic patients. *J Am Soc Nephrol* **10:** 846–853.

Gejyo F, Yamada T, Odani S, *et al.* (1985) A new form of amyloid protein associated with chronic hemodialysis was identified as β2-microglogulin. *Biochem Biophys Res Commun* **129:** 701–706.

Hakim RM, Fearon DT, Lazarus JM, Perzanowski CS. (1984) Biocompatibility of dialysis membranes: Effects of chronic complement activation. *Kidney Int* **26:** 194–200.

Hakim RM, Wingard RL, Parker RA. (1994) Effect of the dialysis membrane in the treatment of patients with acute renal failure. *New Engl J Med* **331**: 1338–1342.

Hirasawa H, Sugai T, Ohtake Y, *et al.* (1996) Blood purification for prevention and treatment of multiple organ failure. *World J Surg* **20:** 482–486.

Hirasawa H, Oda S, Nakamura N, *et al.* (2012) Continuous hemodiafiltration with a cytokine-adsorbing hemofilter for sepsis. *Blood Purif* **34:** 164–170.

Hirayama Y, Oda S, Wakabayashi K, *et al.* (2011) Comparison of interleukin-6 removal properties among hemofilters consisting of varying membrane materials and surface areas: An *in vitro* study. *Blood Purif* **31:** 18–25.

Matsuda K, Hirasawa H, Oda S, *et al.* (2001) Current topics on cytokine removal technologies. *Ther Apher* **5:** 306–314.

Matsuda K, Moriguchi T, Harii N, Goto J. (2009) Comparison of efficacy between continuous hemodiafiltration with a PMMA membrane hemofilter and a PAN membrane hemofilter in the treatment of a patient with septic acute renal failure. *Transfus Apher Sci* **40:** 49–53.

Matsumura Y, Oda S, Sadahiro T, *et al.* (2012) Treatment of septic shock with continuous HDF using 2 PMMA hemofilters for enhanced intensity. *Int J Artif Organs* **35:** 3–14.

Nakada T, Oda S, Matsuda K, *et al.* (2008) Continuous hemodiafiltration with PMMA hemofilter in the treatment of patients with septic shock. *Mol Med* **14:** 257–263.

Oda S, Hirasawa H, Shiga H, *et al.* (2002) Continuous hemofiltration/hemodiafiltration in critical care. *Ther Apher* **6:** 193–198.

Oda S, Hirasawa H, Shiga H, *et al.* (2005a) Management of intra-abdominal hypertension in patients with severe acute pancreatitis with continuous hemodiafiltration using a polymethyl methacrylate membrane hemofilter (PMMA-CHDF). *Ther Apher & Dial* **9:** 355–361.

Oda S, Hirasawa H, Shiga H, *et al.* (2005b) Sequential measurement of IL-6 blood levels in patients with systemic inflammatory response syndrome (SIRS)/ sepsis. *Cytokine* **29:** 169–175.

Oda S, Sadahiro T, Hirayama Y, *et al.* (2010) Non-renal indications for continuous renal replacement therapy: Current status in Japan. In: Suzuki H, Hirasawa H (eds.), *Acute Blood Purification*. Contrib Nephrol vol. 166, pp. 47–53. Karger, Basel.

Oda S, Aibiki M, Ikeda T, *et al.*; Sepsis Registry Committee of the Japnese Society of Intensive Care Medicine. (2014) The Japanese guidelines for the management of sepsis. *J Intensive Care* **2:** 55.

Ono T, *et al.* (1988) Clinical significance of a dialysis membrane that can remove β_2-microglobulin (β_2m). *Trans Am Soc Artif Intern Organs* **34:** 342–345.

Rivers E, Nguyen B, Havstad S, *et al.*; Early Goal-Directed Therapy Collaborative Group. (2001) Early goal-directed therapy in the treatment of severe sepsis and septic shock. *N Engl J Med* **345:** 1368–1377.

Sakai Y. (2011) Polymethylmethacrylate membrane with a series of serendipity. In: Saito A, Kawanishi H, Yamashita M (eds.), *High-Performance Membrane Dialyzers.* Contrib Nephrol, vol. 173, pp. 137–147. Karger, Basel.

Sakai Y, Tanzawa H. (1978) Poly(methymethacrylate) membranes. *J Appl Polym Sci* **22:** 1805–1815.

Shapiro L, Gelfand JA. (1993) Cytokines and sepsis: Pathophysiology and therapy. *New Horizons* **1:** 13–22.

Shiga H, Hirasawa H, Nishida O, *et al.* (2014) Continuous hemodiafiltration with a cytokine-adsorbing hemofilter in patients with septic shock: A preliminary report. *Blood Purif* **38:** 211–218.

Tracey KJ. (2007) Physiology and immunology of the cholinergic antiinflammatory pathway. *J Clin Invest* **117:** 289–296.

Tomisawa N, Yamashita CA. (2009) Amount of adsorbed albumin loss by dialysis membranes with protein adsorption. *J Artif Organs* **12:** 194–199.

Wheeler AP, Bernard GR. (1999) Treating patients with severe sepsis. *N Engl J Med* **340:** 207–214.

Yamada S, Kataoka H, Kobayashi H, *et al.* (1998) Identification of an erythropoietic inhibitor from the dialysate collected in the hemodialysis with PMMA membrane (BK-F) and its clinical effects. In: Ronco C (ed.), *Polymethymethacrylate. A Flexible Membrane for Tailored Dialysis.* Contrib Nephrol vol. 125. pp. 159–172. Karger, Basel.

Chapter 18

Clinical Aspects of Biospecific Sorbents

Kirkovsky V. Vasilevich*, Ryabzeva T.V.†

1. Introduction

Widely used modern medical techniques such as hemodialysis and plasmapheresis and their modifications, and hemoperfusion through uncoated carbon hemosorbents provide effective correction of metabolic and related functional disorders. Nevertheless, in the course of these manipulations of a patient's blood along with the removal of pathogenetic molecules, important physiological substances are also removed. Understanding of this fact has prompted some researchers to concentrate on the development of devices and methods of their application suitable for extraction from the blood of specific compounds which play a pivotal role in the formation of functional metabolic abnormalities in a number of diseases that are resistant to most modern medical therapy (Aasen and Kirulf, 1982; Lasson, 1984; Plate *et al.*, 1993, 1994; Kirkovsky, 1994, 2012).

Biospecific sorption is a modern method of efferent therapy for treatment of diseases associated with excessive activity of the endogenous proteolysis system and pathological accumulation of protein degradation products. Maximum metabolite specificity removal

---

* Belarusian State Medical University, Dzerzhinsky Ave 83, 220116, Minsk, Republic of Belarus. E-mail: kirkovskv.v@mail.ru

† Belarusian State Medical University, Dzerzhinsky Ave 83, 220116, Minsk, Republic of Belarus. E-mail: ta-yana@mail.ru

using this method is achieved through the use of highly specific interactions: protein–protein, enzyme–substrate, ligand–receptor, a peptide ligand, antibody–antigen, etc. Upon contact with a hemosorbent consisting of surfactant structures such molecules can be adsorbed (i.e. fixed on the surface hemosorbent) or absorbed (i.e. fixed in volume hemosorbent) to the latter. Depending on the nature of the sorbent, hemosorbtion can be nonselective or selective (biospecific).

Currently, as a sorbent basis for immobilizing medical supplies, specific ligand polymer hydrogels are widely used, in particular polyacrylamide gel. The polyacrylamide basis is chemically stable, biologically inert and admits the possibility of introducing a ligand to the polymerization step. The main disadvantages of the polyacrylamide matrix are the loss of ligand in gel volume, the possibility of toxic monomer penetration into the blood, and the incorporation of gel microparticles. Therefore, the main task of biochemical and extracorporeal medicine is the development and introduction of a new type of matrix without the above drawbacks. Currently, hemosorbents are being developed on polypropylene and polyethylene matrix modified acrylic acid.

2. Biospecific Anti-Proteinase Hemosorbent in Treatment of Acute Destructive Pancreatitis

Acute destructive pancreatitis (ADP) is one of the indications for application of hemosorbent with anti-proteinase activity. Pathogenesis of systemic disturbances in destructive pancreatitis is closely associated with a massive entry of pancreatic proteinases into the circulation followed by activation of blood proteolytic systems (Aasen and Kirulf, 1982; Lasson, 1984). The developing alteration of a balance between proteinases and their natural inhibitors plays the role of a trigger in molecular mechanisms of the endogenous intoxication syndrome, controlling gravity and the outcome of a disease. Because of this, the inhibitors of proteinases are the traditional components of ADP therapy. The rationale of polyvalent proteinase inhibitors (aprotinin and its analogues application) for such states in critically ill patients is well founded; however, because of their non-selectivity these drugs may damage the proteinase-mediated regulation of various physiological processes.

For a long time we are studying the efficiency of a biospecific antiproteinase hemosorbent containing ovomucoid from the whites of duck eggs, immobilized on the basis of polyacrylamide gel. There are also analogues for the polypropylene and polyethylene basis, modified by acrylic acid. That basis has a spherical shape with the diameter of 0.5 to 3 mm, after activation to form a layer with the attached ligand. Ovomucoid inhibits serine proteinases effectively. The specific activity of the sorbent was proved in experiments *in vitro* and on animals (Plate *et al.*, 1993, 1994). *In vitro* studies on model liquids containing serine proteinase have shown that biospecific hemosorbent with ovomucoid eliminates serine proteases in the amount of 1.7 mg/g. Experiments on animals (purulent peritonitis, acute radiation sickness and trypsin shock) proved the high effectiveness of this hemosorbent and developed container.

As shown by a comparative analysis of animal treatment, including a complex treatment method hemoperfusion through hemosorbent allowed reduce trypsin-like activity of plasma by 35–45% from the baseline. There were no negative trends in the pathological process during and after sorption (Vvedenskii *et al.*, 2009). Moreover, in animals with purulent peritonitis and radiation sickness, improvement was noted even in the process of hemosorption. It was a decrease in the severity of tachycardia, stabilization of blood pressure, and activation of intestinal peristalsis. Studying of biochemical indicators of all groups of animals showed that lowering plasma proteolytic activity after treatment provides a reduction of hypo-albuminemia expression and concentration of protein metabolic products and derivatives. The decrease of toxemia intensity after hemosorption indicates a significant decrease in the levels of leukocytosis and a neutrophilic shift. During the sorption process in the group of animals with experimental trypsin, shock reduction correlated with the stabilization of blood pressure. It is important to note that three procedures of hemosorption on biospecific antiproteinase hemosorbent including complex therapy resulted in a significant reduction in mortality in the groups of experimental animals.

Safety and efficacy of this hemosorbent confirmed three decades of clinical experience in patients with peritonitis of various origins, pancreatitis, extensive phlegmons and burn toxemia.

Here we present the results of clinical studies of this sorbent efficacy. Four hundred and thirty-one random patients with ADP were admitted to the surgical department in one of the Minsk hospitals in Belarus. The diagnosis of acute pancreatitis was based on clinical signs and proteolytic activity of plasma which reached 38.6 ± 11.4 mmol/C. l, and a raised plasma amylase level (≥ 1,000 U/L). The diagnosis was verified during laparoscopy or laparotomy. Hemosorption was performed after adequate correction of circulatory disturbances. Hemosorption was carried out for 80 ± 5 min at a blood flow rate of 60–80 mL/min after general intravenous heparinization (0.5 mg/kg). Hemosorption was carried out on extracorporal unit with ovomucoid (EU) and on uncoated charcoals (HS). The time intervals between successive hemosorption sessions (from 4 hr to 2 days) and the number of sessions was determined separately in each specific case, in accordance with the character and extent of generalization of the inflammatory process in the pancreas and abdominal cavity, gravity of endogenous intoxication, signs of acute organ system failure and therapeutic effect of the preceding hemosorption. The mortality was also analyzed in a group of 97 random patients with acute pancreatitis treated in Minsk hospitals before 1988 by a similar medicinal-surgical program, but with the use of hemosorptions (HS) on uncoated charcoals sorbents of the SKN type. Eighteen healthy donors served as control. It must be noted that patients of both groups were treated with intravenous infusion of the protease inhibitor from a bovine lung (5,000 U per body mass per day).

All studied patients were in a critical condition because of severe endogenic intoxication (SIRS) and circulatory disturbances. Patients had heavy abdominal disorders displayed, as dynamic ileus. In 88 patients there were signs of acute organ system failure. The special feature of biochemical alterations was the imbalance in the proteinase-inhibitor system (Table 18.1). This imbalance was manifested as a significant elevation of trypsin-like activity (TLA) in plasma ($P \leq 0.001$) and the decrease of α1-proteinase inhibitor (α1-PI) and α2-macroglobulin (α2-M) concentrations ($P \leq 0.001$). On the assumption of the total protein and albumin, and the increased level of urea and middle molecules (MM), there were also revealed distinct disturbances in the protein metabolism.

The dynamics of blood biochemical tests in the process of hemoperfusion are presented in Table 18.1. There is a significant reduction

Table 18.1. Biochemical Parameters of the Patient with ADP during Hemoperfusion on the Hemosorbent-containing Ovomucoid (EU) and on Uncoated Charcoals (HS)

Tests	Donor	Patients before HS	After HS	Before EU*	After EU	One Day after EU
TLA (mU/l)	14.7 ± 0.4	138.6 ± 10.5	110.9 ± 8.9	133.0 ± 25.0	83.23 ± 8.4	68.6 ± 7.6
a_1-PI (U/l)	24.6 ± 1.08	13.32 ± 1.2	16.99 ± 1.19	12.86 ± 2.6	13.6 ± 2.7	12.5 ± 1.98
a_2-M (U/l)	0.53 ± 0.02	0.39 ± 0.05	0.4 ± 0.06	0.43 ± 0.1	0.41 ± 0.12	0.42 ± 0.04
Total protein (g/l)	82.4 ± 1.6	59.6 ± 2.27	58.2 ± 2.8	69.1 ± 2.4	61.9 ± 2.3	60.59 ± 1.6
Albumin (g/l)	43.9 ± 0.96	26.6 ± 2.03	26.8 ± 0.7	25.3 ± 1.1	28.5 ± 1.3	26.1 ± 2.26
Urea (mmol/l)	4.3 ± 0.2	9.86 ± 0.8	9.3 ± 1.04	—	—	7.78 ± 0.74
MM (g/l)	0.53 ± 0.02	1.09 ± 0.07	0.93 ± 0.07	0.94 ± 0.21	1.08 ± 0.15	0.99 ± 0.06

of TLA both after blood perfusion through an extra corporal unit (−39.8 ± 9.6 mU/L, $P < 0.01$) and an HS session (−12.8 ± 4.7mU/L, $P < 0.05$). The hemosorbent containing ovomucoid has no distinct effect on the concentrations of total protein, albumin, urea, MM, a1-PI and a2-M. On the second day after EU, decreases in a plasma group average TLA (−32.8 ± 8.3mU/L, $P < 0.05$) and urea (−1.52 ± 0.74mmol/L, $P < 0.05$) were noted. The concentrations of total protein, albumin, a1-PI and a2-M were not significantly altered.

Within the next few days after hemosorption, in an overwhelming majority of patients, a reduction in the extent of SIRS and regression of functional and metabolic disorders were observed. This is displayed as an increase in mental and motor activities, a decrease in breathlessness, tachycardia and central venous pressure, and a rise in diuresis. During hemosorption, the peristaltic motions of the gastrointestinal tract appeared or became more intensive, and as a result loss of water and electrolytes through the gastric tube and flatulence were diminished. Involution of the inflammatory process in the pancreas was confirmed by a decrease in the plasma amylase level one day after hemoperfusion (−240 ± 35U/L, $P < 0.001$).

The positive therapeutic effect of the complex treatment involving the hemosorbent with ovomucoid was achieved in 405 patients and required two to seven hemosorption sessions. In 46 patients, rapid involution of signs of the destructive process in the pancreas was observed after a single hemoperfusion. The eleven of patients was caused by severe circulatory disorders developed in the first two days after admission. It must be noted that in the process of complex therapy there were no cases in which signs of acute organ system failure appeared or became more intensive. Any purulent complications in or around the pancreas were also not recorded. Thus, the lethality in this group of patients under study constituted 6.2%.

The mortality was also analyzed in a group of 97 random patients with acute pancreatitis treated in Minsk hospitals before 1988 by a similar medicinal-surgical program, but with the use of hemosorptions (HS) on uncoated charcoals sorbents of the SKN type.

The lethality in the group of patients treated by hemosorption on carbon sorbents was much higher (24.7%).

Unfavorable outcome of complex treatment with the biospecific hemosorbent containing ovomucoid usually registered in patients with severe inflammatory processes in the pancreas and retroperitoneal fat. Performing repeated biospecific hemosorption reduces the severity of functional and metabolic abnormalities. However, after a short time, they again increased and reached a baseline. Dissemination of purulent process to retroperitoneal fat, and in some cases severe erosive bleeding from vascular zone pancreas were the major adverse outcomes. In those cases where a comprehensive treatment for the proposed scheme starts on the first or second day of the onset of the disease, the pathological process was usually abortive in nature and patients were transferred to the general surgical hospitals.

High capacitance to the serine proteases (trypsin, plasmin, elastase, etc.) provides effective elimination of these substances at the volume of perfusion to 2.0 BCC adult patients. At the same time a clear positive trend recorded clinical and biochemical parameters of the pathological process in cases of radical surgical correction of that disease cause. As a result there is a reduction in heart rate, stabilization of the hemodynamic parameters, improvement in the motor-evacuation function of the gastrointestinal tract, increase in hourly urine output, and normalization of blood in the overall analysis. The concentration of middle molecule level in the after sorption period also reduced. The clinical effect of the hemosorbent with ovomucoid is accompanied by a decrease in plasma proteolytic activity and, as a result, in a reduction of the imbalance between proteases and inhibitors. Immediately after hemoperfusion, in spite of the trypsin-like activity decrease, the content of protease inhibitors in plasma retained its initial low level. This fact is understandable in the light of our knowledge on the rates of biosynthesis of these proteins.

Thus, our clinical experience indicates that sorbent with ovomucoid should be viewed as a tool of pathogenetic therapy in the acute destructive pancreatitis and other diseases associated with the impaired proteolytic system.

3. Hemoperfusion in Treatment of IgE-dependent Pathology

Today specific hemosorption is applied for treatment of allergic diseases. Traditional drug therapy for treatment of allergic patients is often ineffective. It is well known that the high level of IgE is the key element of pathogenesis of most allergic diseases. Considering this, the removal of IgE from the blood may lead to improving allergic patients' conditions. Taking the above said into account, a biospecific hemosorbent for selective and effective removal of IgE from the whole blood was developed. Below we present the results of our clinical experience with the use of this sorbent and propose the method of application of that hemosorbent to the patients resistant to the drug therapy with allergic and combined variants of bronchial asthma.

The specific antiIgE hemosorbent containing L-tryptophan, immobilized in the basis of polyacrylamide gel. For *in vitro* experiments to study sorption properties of that sorbent we used blood plasma obtained from patients who underwent plasmapheresis with a high level of IgE (500 IU/L and more). Plasma perfusion through the column with the hemosorbent was done, and we observed decreasing the IgE plasma level up to 40 % from starting.

Taking into account the obtaining of experimental data, we decided to use the developed hemosorbent with patients with allergic and combined bronchial asthma, because asthma was well known as a vivid example of allergic diseases. There were two groups of patients suffering from severe form of asthma with allergic and combined variants. Patients of the first group (basic) underwent the biospecific hemosorption on the sorbent for selective removal of IgE, along with the standard drug therapy. The patients of the second (control) group received the drug therapy only. Each patient in the first group underwent three procedures of hemosorption with 2–3 days interval. The vein-to-vein hemosorption per via mode by using peristaltic pump, anticoagulant — heparin (0.5 mg/kg), was performed. The perfusion volume was 1.5–2 of the whole circulating blood volume. There were no marked significant adverse effects of hemosorption in these patients.

Table 18.2. Dynamics Concentration of Immunoglobulins A, M, G in Patients with Asthma before and after Hemosorption

	Group 1 $n = 43$		Group 2 $n = 41$	
Immunoglobulins	**Before**	**After**	**Before**	**After**
IgA (g/l)	1.9 ± 0.1	1.8 ± 0.2	1.8 ± 0.1	1.7 ± 0.2
IgM (g/l)	2.3 ± 0.2	2.1 ± 0.3	2.2 ± 0.2	2.0 ± 0.1
IgG (g/l)	21.6 ± 2.1	19.3 ± 3.2	19.4 ± 1.6	18.9 ± 2.5

We studied the therapeutic effect of a biospecific hemosorbent with tryptophan on 43 patients with allergic and mixed forms asthma. All patients were resistant to conventional drug therapy, as well as to large doses of corticosteroids. The ongoing drug therapy included corticosteroids, inhaled B_2-agonists, methylxanthines, antispasmodics and anticholinergics.

Table 18.2 shows the results of the study of the dynamics concentration of immunoglobulins A, M, G in patients with allergic asthma and a comparison group (a group of patients with mixed asthma) before and after treatment.

As shown above, there was a slight decrease in IgA, IgM and IgG concentration after hemosorption. During the first session there was a decrease in concentration by an average of 31.1 ± 3.2% in hemosorption patients. However, by the time of the second session, in all patients of this group the levels of IgE increased. This is probably due to the access to the systemic circulation of membrane-bound immunoglobulin E. In the second session there was again a decrease of IgE by an average of 25.1 ± 2.2%. After the third session the IgE level was reduced by an average of 25 ± 2.4%. Thus, the concentration of IgE in blood plasma of patients with allergic asthma after treatment with a biospecific hemosorbent and tryptophan was reduced by 55% compared with the original.

Reducing the concentration of IgE clearly correlated with a positive clinical course of the disease and improvements of the respiratory function in patients with the allergic form of a disease (Table 18.3).

Table 18.3. Dynamics of Respiratory Indicator Function in Patients with Allergic Form of Disease before and after Treatment with Hemosorption

	Group	
Indicator (% dynamics of the original)	**Group 1** $n = 43$	**Group 2** $n = 41$
Lung vital capacity	+17.6 ± 3.3*	+7.5 ± 3.4
Forced expiratory volume in the first second	+35.6 ± 4.4*	+10.1 ± 3.4
Forced vital capacity	+23.8 ± 3.1*	+11.4 ± 4.7
Peak expiratory flow	+36.5 ± 3.9*	+19.4 ± 3.9
Total lung capacity	−135.5 ± 5.7*	−120.4 ± 8.1
Expiratory reserve volume	+ 36.9 ± 4.6**	+10.2 ± 5.8
Functional residual capacity	−71.9 ± 4.7	−60.6 ± 4.9
Bronchial resistance	−64.8 ± 5.3*	−44.6 ± 5.5

Note: "+"sign denotes an increase in the indicator mark, "–" marked reduction in the rates;
*–$p < 0.05$.
**–$p < 0.01$.

Clinical improvement was observed in 17 patients with allergic and mixed forms of asthma after the first hemosorption session. Decreased duration of asthma attacks, and decreased and less pronounced frequency of nocturnal asthma symptoms were observed in such patients. In connection with clinical improvement the dose of corticosteroids and B2-agonists was reduced in these patients. It should be noted that patients had rapid relief of concomitant manifestations of atopy. In particular, the 10 patients had minimized symptoms of allergic rhinitis, and 5-urticaria and 3-atopic dermatitis. For the remaining patients in this group the condition was without significant clinical dynamics. None of the study group patients experienced deterioration in the course of the disease.

As a result, the complex treatment of patients of the basic group, including the biospecific hemosorption on the sorbent for selective removal of IgE, allowed decreasing the IgE plasma level up to 50%. This decrease correlated with significant improvement of clinical and laboratory conditions of patients. There was no remarkable clinical and laboratory improvement in the patients of the second group. We can conclude that the developed biospecific hemosorbent may be widely used for treatment of patients with a severe form of allergic and combined bronchial asthma. This sorbent also has a good perspectives to be used in clinical situations where high IgE level plays a critical role. We have a successful experience with this sorbent application for treatment of patients with atopic dermatitis, polyvalent allergy, psoriasis. Of course, the severity and persistence of therapeutic effect of the therapy varied over a wide range. The final results depend on the initial concentration of immunoglobulin E, the severity of the pathological process and the possibility of detecting and preventing further exposure to the allergenic factor.

3.1. *Efferent Methods in Complex Treatment of Septic Pathology*

A distinct growth of diseases connected with sepsis is marked in the developed countries during the last few decades. Despite continuous perfection of intensive therapy methods and the presence of a constantly increasing spectrum of antibiotics, septic complications are the most frequent causes of death of patients in the intensive care units. According to a number of authors, septic complications in more than 50% of cases are caused by gram-negative bacteria. Clinical gram-negative infections are extremely fast developing, including such terrible complications as shock, poly-organic insufficiency, DVS, and respiratory distress syndrome of adults. According to the majority of authors, presence at a blood-groove lipid A, liberated at degradation or destruction of bacteria, conducts to initiation the cascade cytokine mediators, activation of cellular systems (monocyte-macrophages, endothelial cells, lymphocytes, granulocytes, fibroblasts), and as complement system and folding of blood. Endocrine and paracrine effects of cytokines and secondary mediators lead to vasodilatation and hypotension, decrease in warm emission, hyper activation of coagulation systems, damage of endothelium, and system adhesion of leukocytes leading to decreased perfusion and hypoxia.

That fact is obvious, that expressiveness of endotoxemia correlates with the severity and progression of gram-negative sepsis. High frequencies of septic complications and high lethality have led to an intensive search for new therapeutic methods. Nowadays, polymyxin B is recognized to be one of the effective remedies of struggle with endotoxemia, as a drug possessing the ability to connect and inactivate LPS both in bacterial membranes and in the free form. However it expressed neuro- and, especially, nephrotoxicity when applied in medical concentration polymyxin essentially limit its application in a kind injection forms.

The principal use of apheresis in treatment of sepsis is to remove toxic substances (bacterial endotoxin — lipopolysaccharide (LPS)) and restore normal organs functions. It is ideal to control the early phases of the inflammatory cascade when treating sepsis by removing microbial components, such as endotoxin. For the first time, G.W. Duff in 1982 showed *in vitro* sorption efficiency of the polymixin B-immobilized sorbent. In 1984, K. Hanasawa and co-workers developed PMX-F, a hemosorbent which consisted of covalent-binding of polystyrene fibers with polymyxin B. This could be used directly in extracorporal blood perfusion. In 1994, PMX-F sorbent was officially adopted for treatment of patients in clinics in Japan. Now, LPS-adsorbers produced in Sweden and Japan are applied in clinical practice worldwide (Tani *et al.*, 2010). An essential drawback of the above-mentioned sorbents is their rather low hemocompatibility and unacceptably high cost. There is an analogue made in Belarus, containing Polymyxin B, immobilized on the basis of polyacrylamide gel. This hemosorbent is much cheaper than foreign prototypes.

The therapeutic action of the polymyxin-immobilized hemosorbent was studied on two groups of experimental animals. The experiment was carried out on Chinchilla rabbits (weight from 2.5 to 3.5 kg). For the experiments the toxin LPS of the *E. coli* 0157:h7 ATCC®HB-10452™ (American Type Culture Collection, USA) at a dose of 0.5 mg/kg animal body weight, in a volume ratio of sterile saline to the toxin 2:1 was used. The solution was injected into the marginal ear vein of control animals for three minutes.

The basic hemodynamic data were controlled. In the experimental group, the dose of the toxin was incubated with the biospecific polymixin-immobilized hemosorbent for 60 minutes in a ratio of 1:2 hemosorbent–solution. Then, this solution was injected in the experimental animals.

Studies have shown that in response to the introduction of intravenous toxin solution hemodynamic instability was observed. It was reflected in a sharp decrease in blood pressure, tachycardia, disorders of external respiration, and convulsions. Against this background, 60% of the rabbits were pronounced dead. A different clinical picture was observed in the experimental group. Despite the presence of tachycardia in these animals, a significant decrease in blood pressure, which returned to normal without application of vasopressors an average after 100 ± 20 minutes. Death in this group of animals was not recorded.

Thus, the experimental data showed high efficiency of this hemosorbent in selective removal of the endotoxin from solution. Further medical tests of this hemosorbent were conducted.

We executed 72 sessions of vein-venous extracorporal hemoperfusion on the polymyxin-immobilized hemosorbent on 40 patients with abdominal sepsis (AS) peritonitis of various genesis (2001–2009). The level of endotoxin (lipopolysaccharide LPS) in blood was defined by a turbid-metric immunoassay. A turbid-metric immunoassay of LPS was made prior to and after hemosorption. To complex drug therapy of 40 patients with the abdominal sepsis, biospecific hemosorption on polymyxin-immobilized sorbents was added. All patients have been operated on in an emergency order. After operative intervention patients during the whole critical period were in the intensive therapy units where they were provided with complex treatment and necessary follow up.

Extracorporeal affinity sorption was carried out vein to vein with a peristaltic pump. Heparin was used as anticoagulant in an average dose of 80 ± 20 IU/kg. All patients in the postoperative period had a septic shock. Gram-negative bacteremia was revealed in 100% of cases. Monitoring of system hemodynamics indicators, gas structure of blood, general biochemical blood analyses and coagulogram were carried out. In all patients, hypotension demanding the application of dopamine in doses exceeding 5 mL/kg/min was observed. The majority of patients required respiratory support.

Carrying out during semi-hour perfusion of blood through the column with a hemosorbent with a speed of 50–70 mL/min allowed us to achieve the decrease in level LPS with 66.7 ± 3.1 pg/mL to 30.5 ± 2.2 pg/mL ($P < 0.01$). By the end of hemoperfusion, stabilization of system hemodynamics indicators (Table 18.4) was marked. It allowed us to reduce either the subsequent dosage or to cancel vasopressin drugs. In some patients the effect of hemodynamics stabilization was unstable; therefore PMX-hemosorption was carried out two to three times with an interval of 4–6 hours. In all patients by the end of hemoperfusion a positive dynamics in the general condition, the gas structure of blood, the general and biochemical blood analyses have been marked.

Analysis of the results shows that hemosorption on biospecific polymyxin-immobilized sorbents is effective in patients with abdominal sepsis caused by gram-negative flora. It is effective in eliminating endotoxin and leads to stabilization of hemodynamic parameters. So already in the process of hemoperfusion through this sorbent in most cases the increase of the minimum blood pressure while keeping pace administering drugs vasopressor action was observed. At the same

Table 18.4. Characteristics of Hemodynamics during Hemoperfusion on Polymixinimmobilized Sorbent

Characteristics	Before Hemosorption	After Hemosorption	*P*
Average arterial pressure (mm.m.p.)	75 ± 3	90 ± 1.6	< 0.01
Pulse pressure (mm.m.p)	42 ± 2	60 ± 1.5	< 0.01
Heart rate (impact/min.)	118 ± 5	101 ± 3	= 0.258
Central venous pressure (mm.m.p.)	8 ± 3	8 ± 4	= 0.76
Cardiac spike (l/min./м^2)	4.4 ± 1.0	4.6 ± 1.4	= 0.38
Index system vascular resistance (din-sec./см^2 × м^2)	1.41 ± 0.57	1.5 ± 0.5	= 0.02
Dopamine (mkg/kg/min.)	8 ± 0.05	3.2 ± 0.1	< 0.01

time, there is a decrease in heart rate, regression of impaired skin microcirculation, increased level of arterial blood saturation with oxygen, reduced levels of pro-calcitonin and pre-sepsin, including universal indicator severity of systemic inflammatory response syndrome — the so-called group of middle molecular compounds.

However, in most cases, after the completion of hemoperfusion on these biospecific hemosorbents, there was a marked increase in re-functional metabolic disorder that requires a restart in the complex therapy of methods of supporting extracorporeal correction of homeostasis, including extended techniques of renal replacement therapy and hemoperfusion using polymyxin-bound and uncovered coal hemosorbents.

In cases where a radical reorganization causes of the severe functional metabolic abnormalities is not possible, as in destructive pancreatitis or burn patients, the positive effect from the manipulation was of a rather short and unstable character. Moreover, a post-sorption period may substantially deepen the clinical manifestations of the systemic inflammatory response syndrome because of the improvement of microcirculation in the area of primary affect.

Studies have shown that the best clinical effect in this situation is achieved using a complex technique and involves the simultaneous perfusion through biospecific polymyxin-immobilized and anti-proteinase hemosorbents. In some cases, this therapy should be initiated in patients with severe purulent-septic processes and without signs of septic shock. The explanation for this may be the fact that such pathology plays an important role not only in aerobic but usually anaerobic microflora. We are sure it will be useful to register and use in clinical practice device, containing at least three components: polymyxin, ovomukoid and uncovered charcoal, which can improve the efficiency of treatment of such patients not only due to the elimination of toxins and excess gram-negative flora proteases, but also due to deligandization and transport environment by removing from the surface of albumin and blood cell membranes a number of hydrophobic compounds. The latter fact is of paramount importance, since the introduction of the cytoplasmic membrane main blood cells — red blood cell metabolic hydrophobic compounds and iatrogenic origin leads to gross violations of the transmembrane potential, and as a consequence, these cells sludging and frustration processes of blood perfusion in peripheral parts of the microcirculation. Our data on the nature of the therapeutic effect of these devices show their high efficiency.

One must admit that in the world literature there is no definite opinion on the effectiveness of devices and methods of this type in the complex therapy of seriously ill patients. Apparently, the mixed results of their application are largely dependent upon a variety of methods of their use. For example, many authors report a lack of effectiveness of hemoperfusion on hemosorbents containing polymyxin included in the complex therapy of patients with septic shock of various origins.

But our 30 years of experience with the extra- and intracorporal methods used to correct homeostasis in patients with severe functional and metabolic abnormalities on the basis of septic processes, as well as inflammatory and degenerative lesions of parenchymal organs, suggest that the beginning of such a combined therapy should be timely. This means that it is desirable to start hemosorption until the occurrence of functional insufficiency of organs of natural detoxification, mainly the liver, as this organ most heavily interferes with homeostasis. When the medical doctor's experience and intuition suggest a high probability of an unfavorable outcome of the pathological process, this justifies the inclusion of all available methods of assisted correction of homeostasis disorders. The therapy must begin immediately after adequate surgical correction of existing inflammatory disorders and correction of volemic state. The occurrence of acute liver failure on this background in almost all cases leads to adverse outcome of the process. In the event of acute renal dysfunction, the forecast is more optimistic. In modern medical practice, it is possible to use the extended techniques of renal replacement therapy.

It should be noted that the second prerequisite for optimal result of complex therapy is to conduct a comprehensive medical treatment in full. To potentiate the therapeutic effect it is necessary to use methods that can effectively influence the essential elements of the pathogenesis of this disease and its complications. Thus, in all patients with septic processes, no limit activation of the endogenous proteolysis occurs. In cases of destructive pancreatitis, this process plays a central role in the development of the pathological process. At generalization of infectious and inflammatory processes an important role have endotoxins, and in cases of septic shock special emphasis on the anaerobic microflora. With all of these diseases and their complications, a significant role is played by microcirculation disorders. In this regard, the rational use of columns containing hemosorbents ovomucoid, polymyxin and uncovered charcoal, along with membrane convection techniques such as CVVH give the best final result of treatment.

References

Aasen AO, Kirulf P. (1982) Studies on pathological plasma protolysis in patients with acute pancreatitis. A preliminary report. *Acta Chir Scand* **509:** 83–87.

Kirkovsky VV, Antiperovich OF, *et al.* (1994) Biospecific haemosorbents based on proteinase inhibitor. III. Biospecific antiptoteinase haemosorbent "Ovosorb" in complex therapy of acute destructive pancreatitis. *Biomaterials* **15:** 334–336.

Kirkovsky VV. (2012) Physico-chemical Methods of Homeostasis Correction. Moskva, Ruskij Vrach, 215p.

Lasson A. (1984) Acute pancreatitis in man. A clinical and biochemical study of pathophysiology and treatment. *Scand J Gastroenterol* **99:** 1–57.

Plate NA, Valuev LI, *et al.* (1993) Biospecific haemosorbents based on proteinase inhibitor. I. Synthesis and properties. *Biomaterials* **14:** 51–56.

Plate NA, Kirkovsky VV, *et al.* (1994) Biospecific haemosorbents based on proteinase inhibitor. II. Efficiency of biospecific antiproteinase haemosorbent "Ovosorb" in complex treatment of experimental generalized purulent peritonitis and acute destructive pancreatitis in dogs. *Biomaterials* **15:** 285–288.

Tani T, Shoji H, *et al.* (2010) Extracorporal removal of endotoxin: The polymyxin B-immobilized fiber cartridge. *Contrib Nephrol* **167:** 35–44.

Uriu K, Osajima A, Kamochi M, *et al.* (2002) Endotoxin removal by direct hemoperfision with an adsorbent column using polymyxin B-immobilized fiber ameliorates circulatory disturbance in patients with septic shock *Am J Kidney Dis* **39:** 937–947.

Vatasin AV. (2011) Selektivnaya adsorbziya endotoxina gramotrizatelnih bakterii pry hirurgicheskom sepsise. p. 19–29.

Vvedenskii DV, Kirkovskii VV, *et al.* (2009) Clinical efficacy of the biospecific hemosorbent "Liposorb". ***Vestnik Rossiiskoi Akademii Meditsinsikh Nauk/Rossiiskaja Akademija Meditsinsikh Nauk*** **10:** 40–43.

Chapter 19

Macroporous Polystyrene Adsorbent Resin and Its Application in Hemoperfusion

Ou Lailiang*

1. Introduction

Adsorbent resin, namely polymeric adsorbent, refers to a class of porous and highly cross-linked copolymers. It is a kind of polymer based on the characteristic of adsorption with the effect of concentration and separation of organic compounds. Mostly, the appearance of the adsorbent resins is small white particles with a diameter of less than 1 mm. After the development of porous networks towards the end of 1950s, it became necessary to distinguish these new materials from the conventional materials and thus terms like macroporous and macroreticular were introduced. Today, the terms porous, macroporous, microporous, macroreticular, permanently porous, sponges, heterogeneous were used to describe such materials (Flodin *et al.*, 1988). In general, macroporous copolymers refer to materials prepared in the presence of a pore-forming agent (diluent or porogen) to form a porous bead type resin.

*The Key Laboratory of Bioactive Materials, Ministry of Education, College of Life Science, Nankai University, China.
Corresponding author: Ou Lailiang, E-mail: ouyll@nankai.edu.cn.

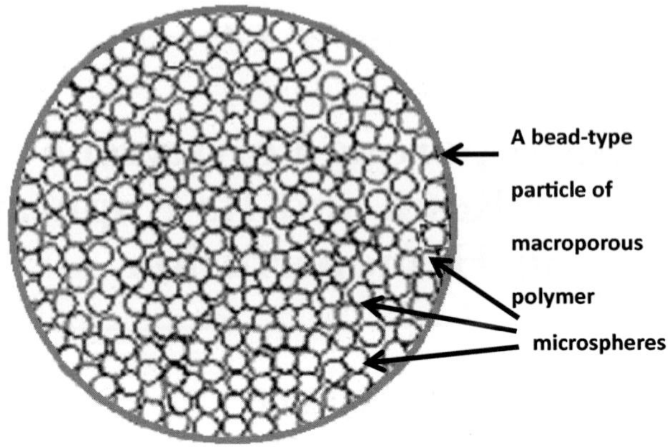

Fig. 19.1. Schematic representation of macroporous adsorbent resins.

Each macroporous polymer particle is composed of a large number of small microspheres, and comprised of a continuous gel phase and a continuous macroporous phase (Fig. 19.1). During the application of the adsorbents, the pores were filled with mobile phase, and the inner surface of the adsorbent resins is exposed to the solvent and solute molecules to capture the target molecules. Since the first successful synthesis of polymeric adsorbent with large pore structure and surface area in 1957, various synthetic adsorbents have come out in succession. Due to their outstanding advantages, including physical stability, active chemical properties, unique adsorption selectivity, mild desorption conditions, rather easy regeneration, long using cycle and cost savings, widespread applications of macroporous adsorbent resins have been used in many fields such as the chemical, food and pharmaceutical industries.

At present, one can say with confidence that the main type of polymeric adsorbent is represented by macroporous polystyrene resins. There are many definitions of macroporous polystyrene resin. Rabelo and Coutinho defined macroporous styrene-divinyl benzene (S-DVB) copolymers according to the kinetic data of their heptane uptake; those adsorbent resins that are able to adsorb more than 0.1 mL heptane per gram of copolymer within a short time (< 5 min) are called macroporous (Rabelo *et al.*, 1994). Porous polystyrene resins

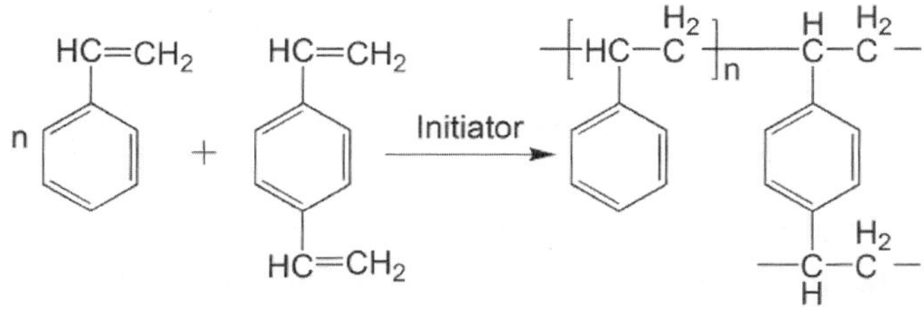

Fig. 19.2. Scheme of the synthesis of polystyrene resins. (*With copyright permission from Davankov et al., 1990* Monograph on Reactive Polymers.)

with network structure are generally obtained by copolymerization of styrene (S) with divinyl benzene (DVB) (Fig. 19.2). The adsorption mechanism of porous polystyrene adsorbent resins mainly depends on the hydrophobic and electrostatic interactions. Simplicity of synthesis, easy availability of initial materials, various physical structures and operational properties, have earned polystyrene networks much-deserved popularity (Davankov *et al.*, 1990). The aim of the present chapter was to overview macroporous polystyrene resin adsorbent and its application in hemoperfusion.

1.1. *History and Development of Polystyrene Adsorbent Resins*

The 1960s and 1970s were very important for the development of adsorbent resins. It was during this period that macroporous adsorbent resin, a very important and valuable product, was produced.

The beginning era of macroporous polystyrene resins is generally attributed to the product of Rohm & Haas Co., who synthesized porous styrene ion exchange resins with high specific surface area by suspension polymerization using S and DVB as monomers in 1962. One year later, the same company successively marketed a series of anion exchange resins (Amberlite MR) for water treatment. In 1964, a boron-specific adsorption resin, Amberlite XE 243, was developed which was synthesized by amination and chloromethylation of S-DVB copolymer and was recognized as the earliest development of macroporous adsorbent resin. On the basis of the above research, Rohm & Haas Co. developed various synthetic routes for preparing macroporous ion exchange resins and adsorbent resins with high specific surface area and pore volume. Then, for the first time, the company exploited several cross-linked polystyrene- and polyacrylate-based non-polar and moderately polar macroporous resins. In the late 1960s, Japan also developed and produced Diaion HP macroporous adsorbent resins. In addition, China started the research on macroporous adsorbent resin since 1970s, which was shortly put on production. Parameters of some macroporous polystyrene adsorbent resins are listed in Table 19.1.

In general, there are three developing periods of polystyrene type adsorbent. Polystyrene type polymers (S-DVB) were synthesized in

Table 19.1. Parameters of Macroporous Polystyrene Adsorbent Resins

Adsorbents	BET Surface Area (m^2/g)	Average Pore Diameter (nm)	Manufacturer
AmberliteXAD-2	≥ 300	18	Rohm&Haas (USA)
AmberliteXAD-4	≥ 750	10	
AmberliteXAD-16	≥ 800	15	Rohm&Haas (USA)
AmberliteXAD-761	150–250	60	Rohm&Haas (USA)
Diaion HP-20	≥ 600	40	Mitsubishi Chemical (Japan)
Diaion HP-21	≥ 570	16	Mitsubishi Chemical (Japan)
Diaion HP-2MG	≥ 500	40	Mitsubishi Chemical (Japan)
Sepabeads SP-825	≥ 1000	11	Mitsubishi Chemical (Japan)
Sepabeads SP-850	≥ 1000	7.6	Mitsubishi Chemical (Japan)
Sepabeads SP70	≥ 800	14	Mitsubishi Chemical (Japan)
Sepabeads SP700	≥ 1200	18	Mitsubishi Chemical (Japan)
Sepabeads SP207	≥ 630	21	Mitsubishi Chemical (Japan)
NKA-9	250–290	12–18	China
NK-110	≥ 500	16	China
HA	≥ 800	4–10	China
X-5	500–600	29–30	China
ADS-5	≥ 550	20–30	China

1934 and reported in 1935 by Staudinger and Heuer (Staudinger *et al.*, 1934). They were then converted into ion exchange resins one decade later (Alelio, 1945). During this period the gel-type resin was synthesized by conducting the crosslinking copolymerization of S with relatively small amount of DVB (generally below 8%) and yielded a product that swells in good solvents but does not dissolve in it (Davankov *et al.*, 2002). The degree of swelling can readily be regulated by adjusting the amount of DVB used during the synthesis. Swelling is an essential condition of the gel-type resin in order to function in the adsorption process. Since the outer surface area of the resin is negligibly small, the large interior surface area plays a decisive role to solute molecules in the swollen resin. That is why the first generation of polystyrene-type polymers is also described as resin with hidden porosity.

The second period of macroporous polystyrene started in 1952 with the first US patent by McBurney. Polystyrene with permanent porosity is the prominent characteristic of materials in this stage. In the patent it was stated that "A more porous resin is also formed if an inert solvent for the monomers which is immiscible with the suspending liquid is added to the monomers and after polymerization, is removed by leaching, drying or distilling from the hard polymerization particles" (McBurney, 1952). This method involved a suspension polymerization of a monomer mixture of S and DVB in the presence of an inert diluent (porogen), which is soluble in monomers but a poor solvent for the polymer. The permanent porosity results from micro-phase separation of the initially homogeneous mixture consisting of S, enhanced purity of DVB (above 8–15%) and a porogen that is miscible with monomers (Davankov *et al.*, 2002). In order to create macroporous structures, a phase separation must occur during the course of the crosslinking process (Okay *et al.*, 2000). The porogen after evaporation leaves voids in the structure that represent interconnected macropores. In fact, macroporous materials are opaque and display a total pore volume of about 1mL/g and an inner surface area of 20–300 m^2/g. In addition, the pore diameter of macroporous S-DVB copolymers lies in the range above 20 nm. One of the best commercially available adsorbent resins of the second generation of styrene-type copolymers is Amberlite XAD-4. Constant net volume in adsorption–regeneration cycles is an important advantage of these materials.

A substantially different network structure results from crosslinking polymerization in the presence of a porogen, which is a thermodynamically good solvent. Hyper-cross-linked polystyrene (HCLPS) resin represents a fundamentally new class of polymeric networks and forms the base for this period of polystyrene-type adsorbent. It could be prepared by an extensive post crosslinking of linear polystyrene in solution or S-DVB copolymers in highly swollen state with bi-functional compounds via the Friedel–Crafts reaction (Davankov *et al.*, 1990). As shown by Davankov and Tsyurupa (Sakodynsky *et al.*, 1972) who first introduced the hyper-crosslinked polymers, the copolymers of styrene with 37 wt. % of DVB were obtained by bulk polymerization in 1,2-dichloroethane in the presence of 1 wt.% of initiator (AIBN). There was no micro-phase separation because of the full compatibility of the porogen with polymer. Two main properties, i.e. the porous nature and high crosslinking degree, distinguished them from traditional S-DVB polymers. First, HCLPS is a resin with considerably small pore size, relatively high surface area (1,000 m^2/g) and porosity. In addition, the high degree of crosslinking is responsible for its high surface area. Second, in spite of high crosslinking density, the resin displays the same ability to be swollen in thermodynamically good solvents and in precipitating media (Sakodynsky *et al.*, 1972; Tsyurupa, 2006). Hyper-crosslinked polystyrene was first introduced in the early 1970s (Tsyrupa *et al.*, 1984), and the basic principle of the hyper-crosslinked polymers consists in the formation of a rigid highly solvated three-dimensional network. The excellent properties of HCLPS provide potential applications as an improved adsorbent for adsorbing organic vapors, recovering organic compounds from waste water, removal of trace amount of toxic metals, packing materials for various chromatography column and hydrogen storage.

2. Preparation and Properties of Macroporous Polystyrene Adsorbent Resin

Adsorbent resin is prepared from a mixture of S, DVB, and porogens (toluene and undecane). Different pore structures are formed by manipulating the ratios of the above ingredients (Roche, 2008). The synthetic methods of macroporous polystyrene resins can be classified into bulk, suspension, solution and emulsion polymerization. (Li *et al.,* 1999). However suspension polymerization is the most commonly used method in the synthesis of bead type adsorbent resin.

2.1. *Preparation of Adsorbent Resin by Suspension Polymerization*

Macroporous polystyrene adsorbent resin is prepared by suspension copolymerization of S, DVB in the presence of a diluent (porogen), initiator and suspending agent at 80 °C. Detailed studies concerning the suspension copolymerization of S with DVB were first reported by Kun and Kunin (Malovic *et al.,* 2007). The formation of porous structure was attributed to phase separation during the polymerization process. To illustrate the synthetic process, the technique is shown in Figs. 19.3 and 19.4 schematically. First, an oil phase

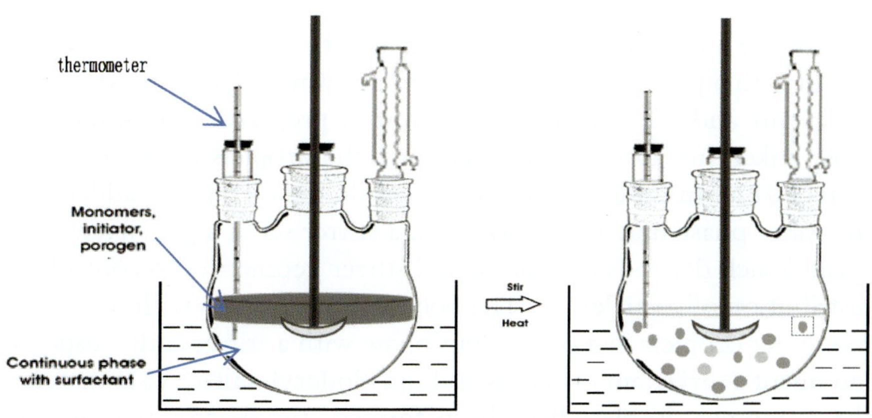

Fig. 19.3. Schematic diagram of suspension polymerization technique.

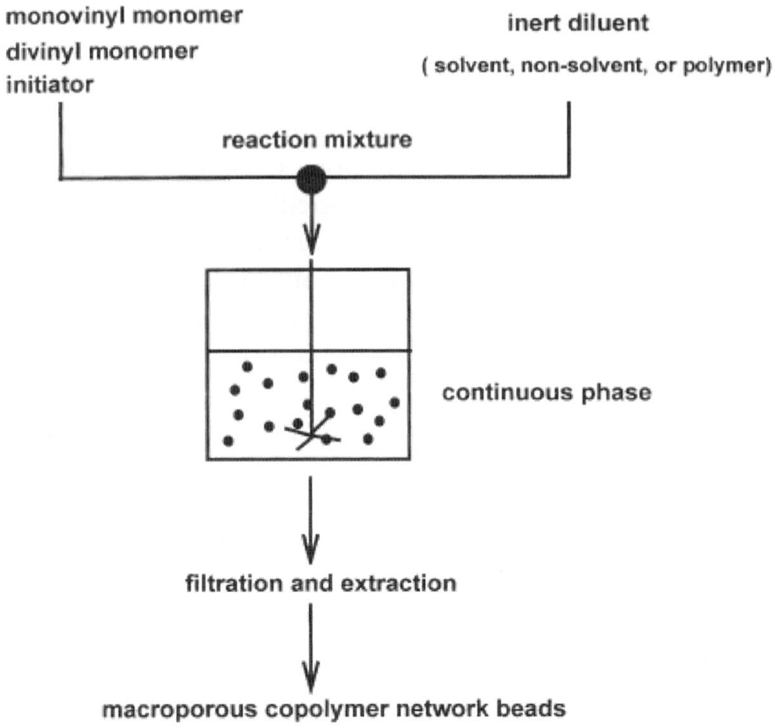

Fig. 19.4. Schematic representation of the suspension polymerization technique to produce macroporous copolymer network. (*With copyright permission from Okay, 2000 Monograph in* Prog Polym Sci.)

containing a vinyl and divinyl monomer with a free-radical initiator is mixed with a porogen. Then a continuous water phase with dissolved surfactant and/or polymeric stabilizers is prepared. The monomer, cross-linker, initiator and porogen form the droplets phase (Gokmen *et al.*, 2012). In order to achieve this result, the oil phase is added into the water phase under agitation with a stirrer set at a pre-determined speed which depends on reactor and stirrer geometries to control the distribution of particle size. The polymerization starts with dispersing monomer droplets in a continuous phase with a stirrer and a suspending agent (surfactant) such as sodium dodecyl sulfate (SDS). Once the monomer droplets have been formed, the reactor is heated to 80 °C gradually and polymerization takes place for 24 hr. The

Table 19.2. Characteristics of S-DVB Copolymer Beads

	Dilution Degree (%)				Surface
No.	Toluene	Heptane	Apparent Density (g/mL)	Swelling (%)	Area (m^2/g)
1	—	—	0.9890	31.0	54.5
2	25	—	0.9166	—	—
3	50	—	0.7500	74.0	59.0
4	100	—	0.6660	125.0	79.4
5	—	100	0.5960	51.0	74.1
6	25	75	0.5880	36.0	90.0
7	75	25	0.5830	45.0	92.3
8	50	50	0.5430	48.0	90.6

copolymerization and crosslinking reactions result in the formation of beads having a glassy, opaque, or milky appearance. Then the beads are extracted with a good solvent to remove the porogen or soluble polymers from the network (Okay *et al.*, 2000).

This synthetic method is widely applied in industry because of the low cost and upscaling possibilities, notwithstanding the fact that mult-dispersity is the main drawback of suspension polymerization.

R. Tank *et al.* (Gupta *et al.*, 2006) prepared S-DVB copolymer beads by suspension copolymerization of S and DVB using benzoyl peroxide as initiator, toluene, heptane or mixture as the porogen and magnesium hydroxide as suspending agent in water at 80 °C under nitrogen atmosphere. In the study, toluene, heptane, and their mixtures were incorporated as porogen and the effects of porogen on porosity of S-DVB copolymer were examined (Table 19.2).

2.2. *Post-crosslinking Preparation of Adsorbent Resin*

Another approach that will be briefly discussed in this section is called the post-crosslinking method or the Friedel–Crafts reaction method. This approach is based on the formation of extensive post-crosslinking between the phenyl groups on the polystyrene resins via Friedel–Crafts reaction (Davankov *et al.*, 1990). In the early 1970s, he used the Friedel–Crafts reaction to crosslink the phenyl and chloromethyl groups on the polymer chains and obtained the first hyper-crosslinked macroreticular polystyrene resin (Patterson *et al.*, 1971) (Fig. 19.5). The first step can be any heterogeneous polymerization such as suspension, emulsion, or precipitation polymerization, while the second step involves the Friedel–Crafts hypercrosslinking reaction where the beads are simply swollen in solution (dichloroethane) and heated in the presence of a catalyst ($FeCl_3$). The specific surface area of the resin could reach 1,000 m^2/g, (Fig. 19.6) (Ahn *et al.*, 2006). The commercial product is available from various manufacturers (Tsyurupa *et al.*,

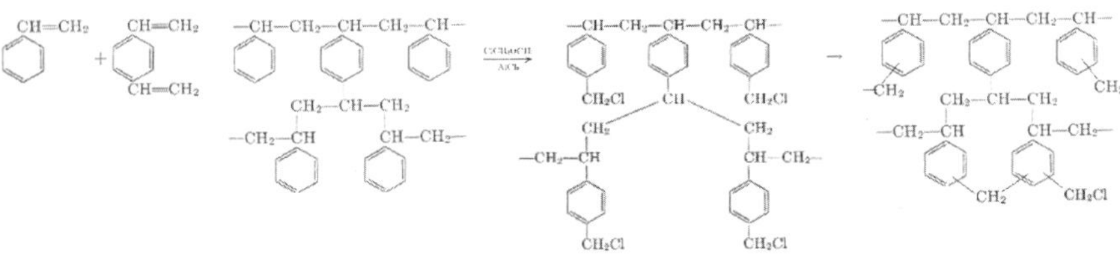

Fig. 19.5. Scheme of the synthesis of post crosslinking resin.

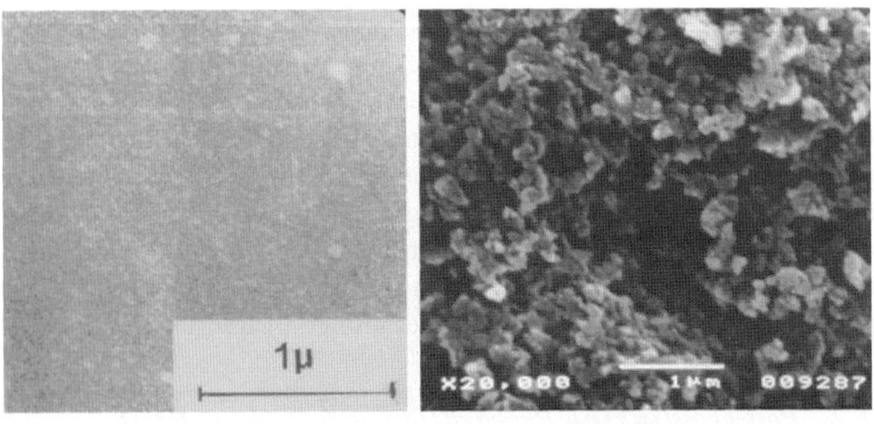

Fig. 19.6. Scanning electron micrograph of hyper-crosslinked networks obtained by post-crosslinking: (A) gel-type S-2%DVB copolymer, S_{app} = 1000 m^2/g (B) macroporous chloromethylated S-DVB copolymer, S_{app} = 1000 m^2/g. (*With copyright permission from Tsyurupa 2006 Monograph in* React Funct Polym.)

2006). Typical crosslinking agents which can be used are p-xylelenedichloride (XDC), 1,4-bis-chloromethyldiphenyl (CMDP) (Davankov *et al.*, 1974), monochlorodimethylether (MCDE) (Davankov *et al.*, 1976), Dimethylformamide (DMF) (Tsyrupa *et al.*, 1984), tris-(chloromethyl)-mesitylene (CMM), and bis-chloromethyl-1,4-diphenylbutane (DPB) (Tsyrupa *et al.*, 1984). The crosslinking reaction can occur between two polymeric chains or on the same chain between two phenyl groups.

In summary, post-crosslinking of polystyrene can be classified into the following four methods: Friedel–Crafts crosslinking reaction of linear polystyrene, Friedel–Crafts crosslinking reaction of gel-type S-DVB copolymers, post-crosslinking of macroporous low crosslinking S-DVB copolymers, and post-crosslinking of macroporous low crosslinking copolymers of S, DVB with polar monomers.

2.2.1. *Friedel–Crafts crosslinking reaction of linear polystyrene*

The post-crosslinking reaction of linear polystyrene (Fig. 5) can be carried out in dichloroethane or cyclohexane as solvent using $SnCl_4$ as catalyst, and bis-chloromethyl-1,4-diphenylbutane (DPB), 1,4-bis-chloromethyldiphenyl (CMDP), monochlorodimethylether (MCDE), 1, 2, 4, 5-benzene anhydride or dimethylformamide (DMF), as well as tris-(chloromethyl)-mesitylene (CMM) are usually used as crosslinking agents.

2.2.1.1. Friedel–Crafts crosslinking reaction of Gel-type S-DVB copolymers

In general, gel-type S-DVB copolymers with 0.3–2.0% of DVB preswollen in dichloroethane or nitrobenzene can be used as starting materials for subsequent transformations. The crosslinking reaction carried out in the presence of a crosslinking agent (such as MCDE, CMDP) and the hyper-crosslinked polystyrene adsorbent can be prepared. Kralova *et al.* described a method of preparation of permanently porous S copolymers by post-crosslinking of gel-type S-DVB copolymer according to the reaction scheme described in Fig. 19.7.

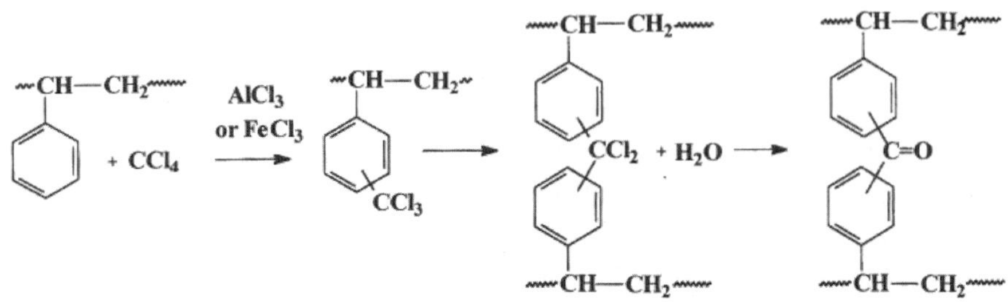

Fig. 19.7. Reaction scheme of post-crosslinking of gel-type S-DVB copolymers.

In the post-crosslinking reaction, tetrachloromethane was used as a crosslinking agent, aluminium or ferric chloride as a Friedel–Crafts catalyst and hyper-crosslinked resin with very high surface area was achieved (Hradil *et al.*, 1998).

2.2.1.2. Friedel–Crafts crosslinking of macroporous low crosslinking S-DVB copolymers

In a typical experiment, macroporous S-DVB copolymers with 3–8% of DVB were used as starting materials for subsequent post-crosslinking. The initial polymer was swollen adequately in dichloroethane or nitrobenzene. Hyper-crosslinked polystyrene with high porosity was obtained after addition of Friedel–Crafts catalyst and the crosslinking agent (CMME, carbon tetrachloride, thionyl chloride). The Friedel–Crafts catalysts include aluminium chloride, anhydrous zinc chloride, and tin tetrachloride. H-type polystyrene adsorbent resin was produced with this method by Nankai University Chemical Company.

2.2.1.3. Friedel–Crafts crosslinking of macroporous low crosslinking copolymers of S-DVB and polar monomers

This type of hyper-crosslinked polystyrene was prepared by the following procedure: (1) preparing the copolymer beads using S, a small amount of DVB and a third polar monomers such as methyl methacrylate, methyl acrylate, methylacrylic acid or acrylonitrile; (2) chloromethylation of the beads; (3) swelling the chloromethylated beads in the solvent such as nitrobenzene; (4) Friedel–Crafts crosslinking reaction in the presence of a catalyst such as anhydrous zinc chloride.

2.3. *Parameters of Macroporous Polystyrene Network*

Macroporous network is mainly characterized in the dried state. The following parameters are used for characterizing the pore structure:

Apparent Density d_0

$$d_0 = \frac{\text{mass of the bead}}{\text{volume of the bead}} \tag{1}$$

According to Eq. (1), the lower the parent density, the higher the porosity of the material.

True Density (or Skeletal ensity) d_2

$$d_2 = \frac{\text{mass of the bead}}{\text{volume of the polymer}} \tag{2}$$

According to Eq. (2), a material is homogeneous if its apparent density is equal to its true density.

Total Porosity P

$$P = \frac{\text{volume of pores}}{\text{volume of the beads}} \tag{3}$$

Total Pore Volume V_p

$$V_p = \frac{\text{volume of pores}}{\text{mass of the beads}} \tag{4}$$

2.3.1. *Specific surface area*

The specific surface area of the beads is one of the important performance parameters of porous resins, referring to the surface area of per gram of resin. It is often measured by nitrogen adsorption isotherms.

2.3.2. *Pore size and its distribution*

The average pore size and the distribution of a macroporous polymer can be measured by many methods, such as electron microscopy (Huxham *et al.*, 1992), size exclusion chromatography (Halasz *et al.*, 1978, 1980) and BET (Lowell, 1979).

The above parameters are crucially important in defining the porous nature of a resin.

2.4. *Effect of Porogen*

It has already been reported in many literatures that the solvating power of porogen has a critical effect on the pore structure of S-DVB copolymers. In general, porogen is an inert solvent that is miscible with the monomers and can swell the polymer formed. After polymerization, the porogen is removed from the network, leaving a porous structure within the crosslinked polymer beads. Therefore, the porogen acts as a pore-forming agent, which plays an important role in designing the porous structure of macroporous adsorbents. A number of porogens used in S-DVB copolymerization are listed in Table 19.3 (Okay *et al.*, 2000). Some of the porogens, such as toluene and cyclohexanol, are very good solvents for polystyrene. It is worth mentioning that cyclohexanol is one of the most efficient porogens for building up the highest porosity even at low amounts of DVB (Poinescu *et al.*, 1984; Okay *et al.*, 1986).

The solubility of polystyrene in inert porogen has a critical effect on the porous structure of macroporous polystyrene resin. Thermodynamically good solvents produce structures with small pores while poor solvents produce structures with large pores

Table 19.3. Porogens Used in Preparation of Macroporous S-DVB Copolymer Network

Porogen
Benzyl alcohol
Benzyl alcohol/various solvents
Cyclohexane
Cyclohexanol
Cyclohexanal/toluene mixtures
Cyclohexanone
Decane
Decane/toluene mixtures
Di-2-ethylhexylphosphoric acid
Di-2-ethylhexyl phthalate (DOP)
2-Ethyl-1-hexanoic acid
2-Ethyl-1-hexanol
2-Ethyl-1-hexanol/*n*-heptane mixtures
2-Ethyl-1-hexanol/toluene mixtures
Gasoline/toluene mixtures
Isoamyl alcohol/various solvents
n-Heptane
n-Heptane/ethylacetate
n-Heptane/isoamyl acetate
n-Heptane/tetraline mixtures
n-Heptane/toluene mixtures
n-Heptane/various mixtures
Pentanol
Poly(styrene-*co*-methyl methacrylate)/dibutyl phthalate
Polystyrene/2-ethyl-1-hexanol mixures
Polystyrene/dibutyl phthalate
Polystyrene/gasoline mixtures
Polystyrene/*n*-hexane mixtures
Polystyrene/toluene mixtures
Toluene
Tri-*n*-butylphosphate
1,2,3-Trichloropropane/2-ethyl-1-hexanol mixtures
2,2,4-Trimethyl pentane (isooctane)

(Poinescu *et al.*, 1988; Hilgen *et al.*, 1975). In summary, the following types of porogens are used to prepare macroporous S-DVB copolymer networks (Seidl *et al.*, 1967):

2.4.1. *Using good solvent as porogen (ν-induced syneresis)*

A good solvent is characterized by the Hildebrand solubility parameter (Milan *et al.*, 2005). Cyclohexanol, toluene, and dichloroethane are examples of thermodynamically good solvents or solvating porogens (SOL) for the preparation of porous polystyrene copolymers. Solvating porogens with good thermodynamic compatibility with the polystyrene network cause phase separation at a much later stage during the polymerization, and so form the microgel particles (Fig. 8B).

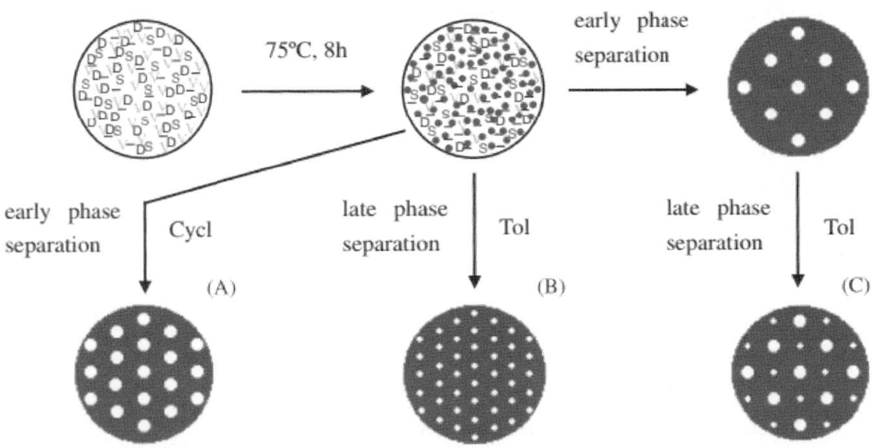

Fig. 19.8. Schematic representation of pore formation in poly(S-DVB) resins: (A) using a thermodynamically poor solvent (cyclohexanol, Cycl) as porogen (producing a relatively narrow pore size distribution in the macropore region); (B) using a thermodynamically good solvent (toluene, Tol) porogen (producing a relatively narrow pore size distribution in the micropore region); (C) using the mixture of tol and cycl as coporogens (inducing phase separation in two stages: an early separation in which the cycl phase separates to produce macropores and a late separation in which the tol phase separates to produce micropores). (*With copyright permission from Milan, 2005 Monograph in* J Sep Sci.)

As a result, addition of a good solvent produces small average pore diameter and therefore a considerable high specific surface area (50–500 m^2/g) and a relatively low pore volume, typically not exceeding 0.8 ml/g. The pore size distribution of the network is characterized by a large proportion of micro and mesopores. This type of pore formation is called ν-induced syneresis (Okay *et al.*, 2000).

2.4.2. *Using nonsolvent as porogen*

In contrast, the solvating porogens having poor thermodynamic compatibility with the polymer network such as n-heptane or alcohol cause an early phase separation and microgel formation, which results in a large pore volume (0.6–2.0 ml/g), a relatively large pore diameter and a rather low specific surface area (10–100 m^2/g) (Fig. 19.8A). The pore size distribution of the networks is characterized by a large proportion of meso and macropores (Okay *et al.*, 2000). This type of porogen is called a nonsolvating porogen (NONSOL) and the pore formation is called χ-induced syneresis.

Figure 19.9 illustrates the effect of solvating power of porogen on the pore size distribution of S-DVB copolymer beads with 10% DVB. It is well known that toluene is a good solvent for polystyrene, while cyclohexanol is a nonsolvating porogen for S-DVB copolymerization. Figure 19.9 indicates the number of meso and macropores (>20 Å) significantly increases as the solvating power of the porogen decreases (i.e. as the cyclohexanol content of the porogen increases). Moreover, the average pore size becomes larger and the pore size distribution shifts toward the larger pores in poor solvent interaction. This is a consequence of a change in the mechanism of pore formation from ν- to χ- induced synthesis (Erbay *et al.*, 1999).

2.4.3. *Using linear polymers as porogen*

Polymeric porogens such as linear polystyrene and poly (methyl methacrylate) will result in a pore volume up to about 0.5 ml/g, a specific surface area of 0.1 to 10 m^2/g and very large pores reaching

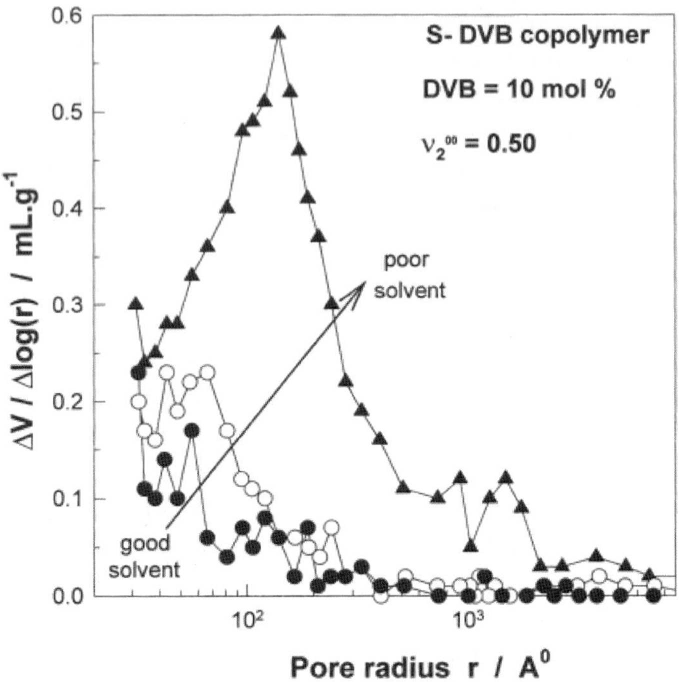

Fig. 19.9. Effect of solvating power of porogen on the pore size distribution of macroporous S-DVB copolymers. v200 = 0.50. DVB = 10%. Porogen = toluene (•), cyclohexanol/toluene (75/25 v/v) (○), and cyclohexanol (▲). (*With copyright permission from Erbay 1999 Monograph in* J Appl Polym Sci.)

the micrometer range. It is important to note that the amount and nature of polymeric porogen may either induce a porous or a nonporous hollow structure. In the order of good solvent, nonsolvent, and polymeric porogen, the specific surface area decreases below 1 m^2/g whereas the size of pores and total pore volume increases (Fig. 19.10) (Okubo *et al.*, 2002; 2000).

2.4.4. *Using oligomers as porogen*

Tsyurupa *et al.* (2006) considered that using oligomeric rather than polymeric porogens presented an opportunity to avoid some of the

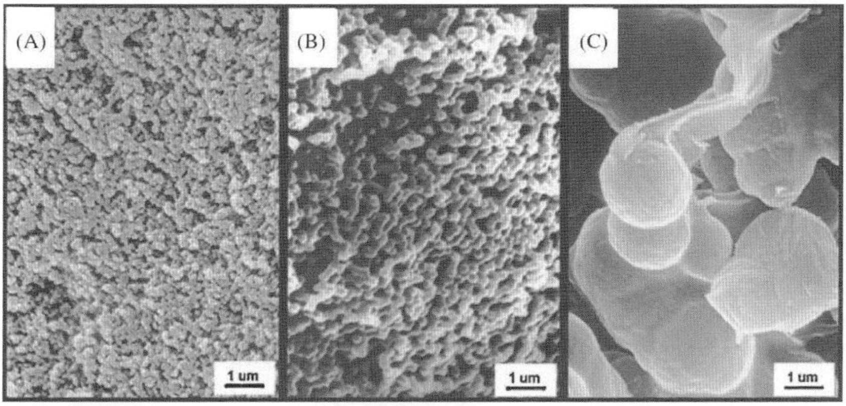

Fig. 19.10. SEM images of EGDMA-GMA beads prepared by using different porogens at 60% ratio: (A) toluene (good solvent), (B) dodecanol (nonsolvent), (C) PS solution in toluene (polymeric porogen, 15%). From left to right, pore size increases and total surface area decreases. (*With copyright permission from Okubo, 2002 Monograph in* Colloid Polym Sci.)

problems caused by the high molar mass of polymeric porogens. They postulated that for an appropriate oligomer with toluene as a solvent, the co-porogen might allow a double phase separation to occur during polymerization. As illustrated in Fig. 19.11A, the selection of oligomers as sole porogen in the polymerization of poly (DVB) result in an early phase separation, and yield a resin with macropores only. Using the oligomer as co-porogen with toluene induced phase separation in two stages (Fig. 19.11C), a bimodal pore size distribution might be achieved.

In addition, the amount of porogen is also an important factor that influences the properties of resins. Figure 19.12 and Table 19.4 (Gong, 2008) illustrate the effect of amount of heptane as porogen on the pore structure of S-DVB-MMA copolymer. The apparent density and true density of the polymeric adsorbent are inversely proportional to the percentage of porogen in monomers, but the pore volume (Fig. 19.13) and the surface area are directly proportional to it.

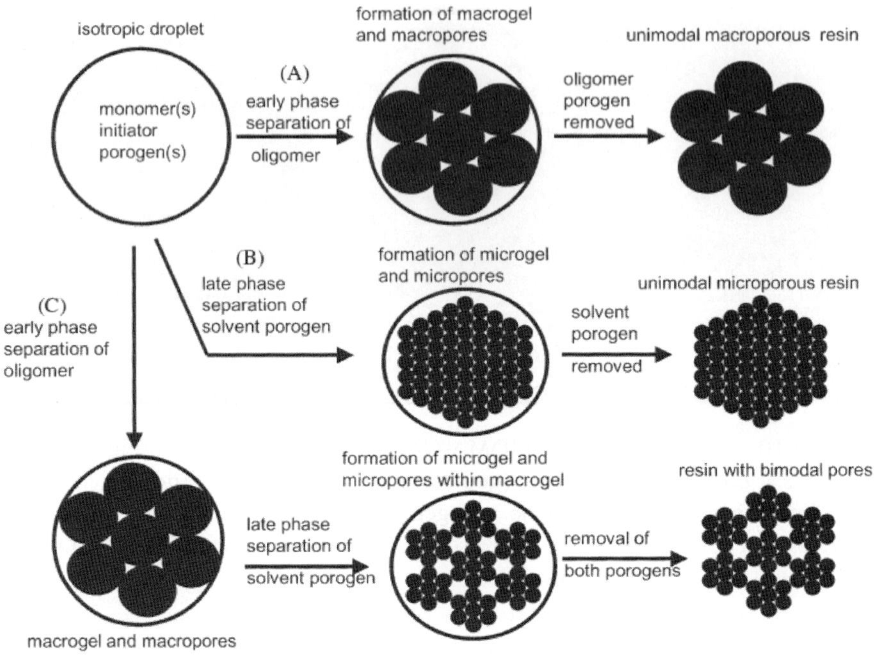

Fig. 19.11. Schematic representation of pore formation in poly(DVB) resins: (A) using an oligomer porogen (producing a relatively narrow pore size distribution in the macropore region); (B) using a thermodynamically good solvent porogen (producing a relatively narrow pore size distribution in the micropore region); (C) using a solvent and an oligomer as coporogens (inducing phase separation in two stages: an early separation in which the oligomer phase separates to produce macropores and a late separation in which the toluene phase separates to produce micropores). (*With copyright permission from Tsyurupa 2006 Monograph in* React Funct Polym.)

2.5. *Effect of Crosslinking Agent*

In addition to the effect of the porogen, the crosslinking agent in the polymerization mixture is another condition *sine qua non* for the formation of a macroporous structure. The percentage of the crosslinking agent present in the polymerization mixture is a very important characteristic that determines the extent of swelling and the structure of the final product (Milan *et al.*, 2005). In crosslinking polymerization of S-DVB in the presence of toluene as porogen, it was found

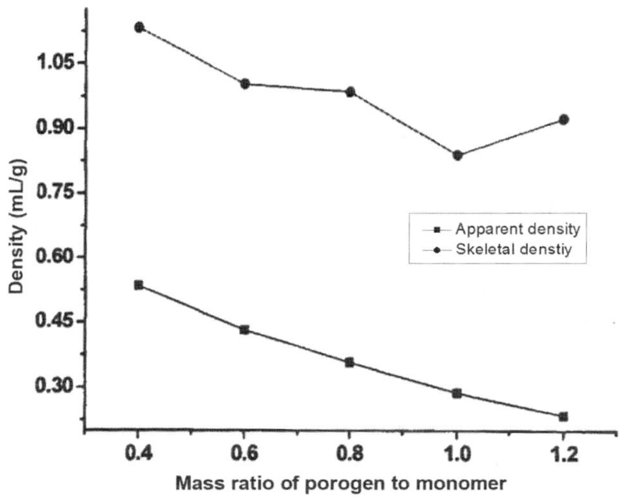

Fig. 19.12. Influence of porogen on density of resin. (*With copyright permission from Gong 2008 Monograph on PhD thesis of Jiangnan University.*)

Table 19.4. Pore Structure of 2%MMA Resins Prepared by Different Quantities of Pore Generating Agent

Amount of Porogen (g)	2	3	4	5	6
Apparent density	0.534	0.432	0.357	0.287	0.235
True density	1.132	1.002	0.984	0.839	0.923
Porosity	33.32%	44.03%	59.50%	61.75%	63.13%
Pore volume	0.453	0.786	1.546	1.864	1.897

that 30–60% of pendant vinyl groups are used by cyclization reactions, and an average of 100–800 cross-linkages are formed for one intermolecular cross-linkage (Okay *et al.*, 1995). As shown schematically in Fig. 19.14, the cyclization and multiple crosslinking reactions are mainly responsible for the formation of compact nuclei in S-DVB copolymerization.

Some of the crosslinking agents used in the polymerization of macroporous polystyrene and their thermal stabilities are listed in the Table 19.5. Among the S copolymers studied, the S-DVB copolymer

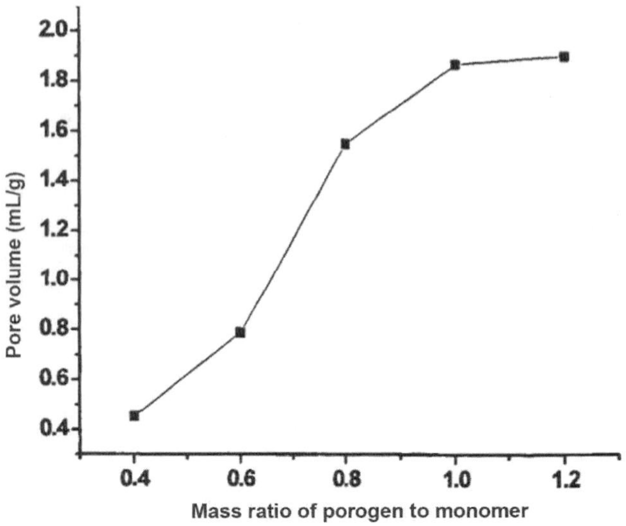

Fig. 19.13. Influence of porogen on pore volume of resin. (*With copyright permission from Gong 2008 Monograph on PhD thesis of Jiangnan University.*)

was found to be the most thermally stable, and the thermal stabilities are in the following order:

S-DVB>S-BADMA>S-EGDMA>S-NGDMA>S-BDDMA>S-HMGDMA. DVB is also the most common crosslinking agent used in synthesizing macroporous polystyrene networks.

In Figure 19.15, experimental data of total porosity data of S-DVB networks in the literature (Okay *et al.*, 1986; 1988) are collected and shown as a function of the crosslinking agent (DVB) concentration. It indicates that at a given concentration of monomer and porogen, raising DVB concentration causes an increase of porosity and then levels off gradually.

Figure 19.16 shows that the pore size distribution has a relationship with the concentration of the crosslinking agent at a fixed monomer concentration. The average pore diameter varies conversely with the surface area, but decreases with the increase of the crosslinking degree of network. The average pore diameter decreases sharply with the increase of the surface area and DVB content (Poinescu *et al.*, 1985).

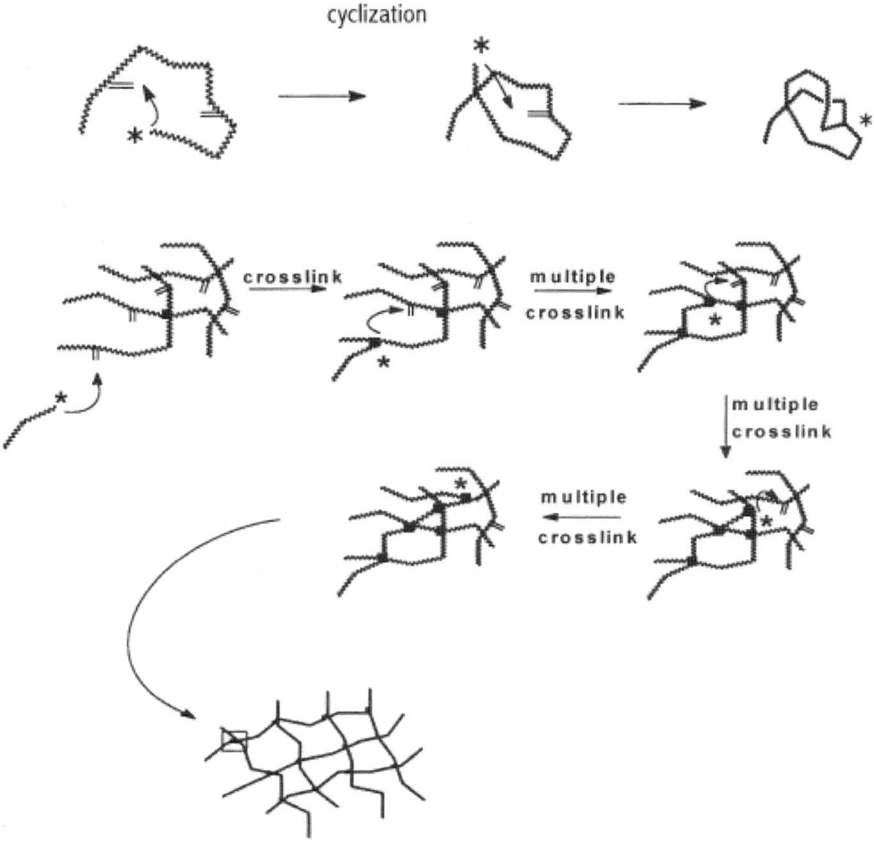

Fig. 19.14. Schematic representation of cyclization and multiple crosslinking reactions during FCC of vinyl/divinyl monomers leading to the formation of highly crosslinked regions in the final polymer network. • = crosslink, ╨ = pendant vinyl, * = radical center. (*With copyright permission from Okay et al., 1995 Monograph in* Macromolecules.)

In summary, varying the amount of DVB will produce macroporous resins with different crosslinking degree, and consequently the density, specific surface area and pore volume will also be greatly affected. As shown in Figs. 19.17, 19.18 and 19.19, the increase of crosslinking agent content causes the density of the polystyrene resin to decrease, however the pore volume and the specific surface area increased (Gong *et al.*, 2008).

Table 19.5. Thermal Degradation of Polystyrene Copolymer with Various Crosslinking Monomers

						Temperature (°C)			
Sample	S:crosslinking Agent	Crosslinking Agent	Compressive Strength (MPa)	Swelling (%)	IDT (°C)	10% Weight Loss	40% Weight Loss	60% Weight Loss	IPDT (°C)
1	80:20	EGDMA	44	44	210	324	365	408	407
2	80:20	NGDMA	43	48	170	277	345	365	405
3	80:20	BDDMA	17	40	180	281	362	387	388
4	80:20	HMGDMA	40	45	150	284	328	353	377
5	80:20	BADMA	20	54	225	317	370	395	419
6	80:20	DVB	>50	37	235	291	389	425	471

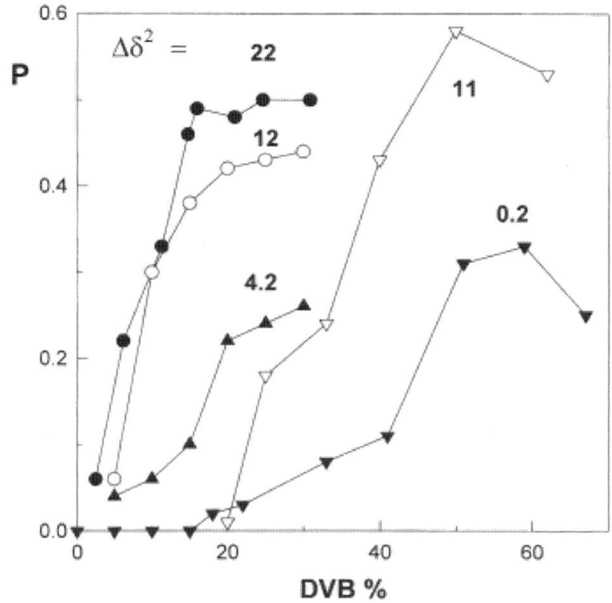

Fig. 19.15. The total porosity P of S-DVB copolymer networks shown as a function of DVB concentration. The difference between the solubility parameters of the porogen and the polymer is shown in the figure. (*With copyright permission from Poinescu et al., 1994 Monograph in* Macromolecular Symposia.) $v_2^{00} = 0.50$, porogen = cyclohexanol (•), toluene/cyclohexanol (1/3) (○), tolune/cyclohexanol (1/1) (▲), toluene (▼). v_2^{00}=0.40, porogen = toluene/gasoline (1/100) (V). (*With copyright permission from Erbay* et al., *1999 Monograph in* J Appl Polym Sci.)

3. Application of Macroporous Polystyrene Adsorbent in Hemoperfusion

Worldwide, around 1.8 million patients are undergoing treatment for end-stage renal disease (Grassmann *et al.,* 2004), of which about 1.4 million are on dialysis treatment and the rest are living with a functioning renal transplant. The molecules associated with dialysis treatments can be grouped according to their sizes (Table 19.6). Dialysis membranes are generally effective in removal of small molecules, which are usually water soluble. However, this is insufficient for removal of middle molecular weight molecules (MM) such as β2-microglobulin (β2M, MW 11.8 kDa).

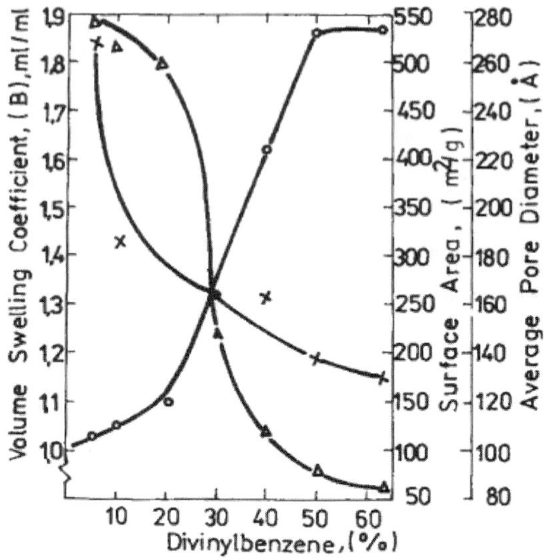

Fig. 19.16. Volume swelling coefficient (x), surface area (○), and average pore diameter (Δ) as a function of DVB content (porogen: 50 mL TBP/50g monomers). (*With copyright permission from Poinescu et al., 1985 Monograph in* Angew Makromol Chem.)

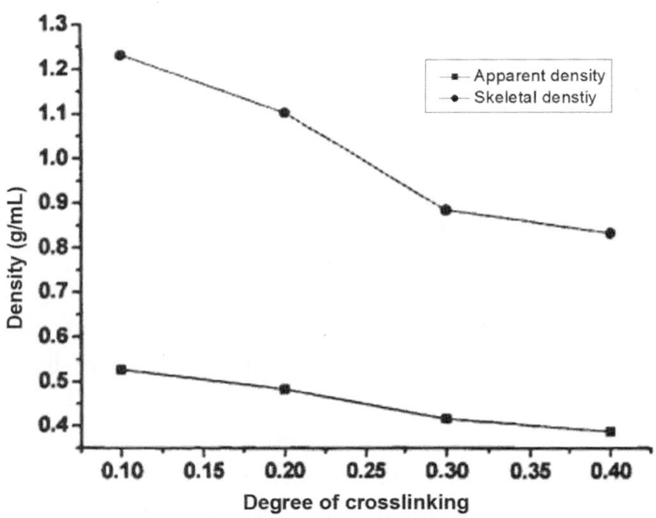

Fig. 19.17. Influence of crosslinking agent on density of resin. (*With copyright permission from Gong, 2008 Monograph on PhD thesis of Jiangnan University.*)

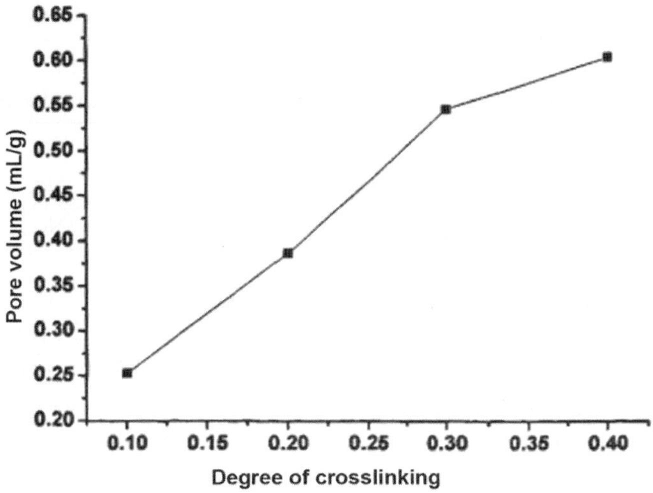

Fig. 19.18. Influence of crosslinking agent on pore volume of resin. (*With copyright permission from Gong, 2008 Monograph on PhD thesis of Jiangnan University.*)

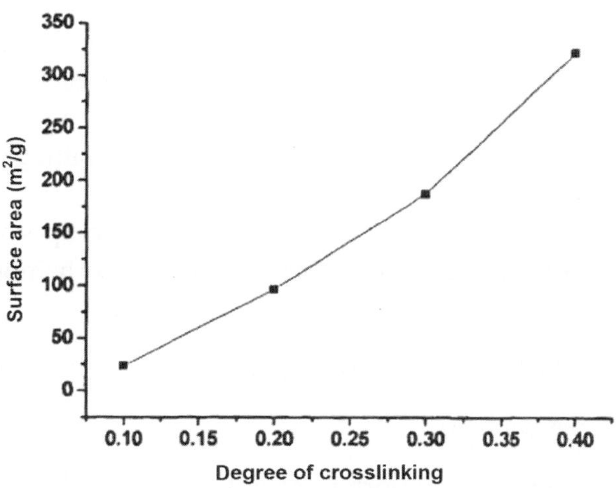

Fig. 19.19. Influence of crosslinking agent on surface area of resin (*With copyright permission from Gong, 2008 Monograph on PhD thesis of Jiangnan University.*)

Table 19.6. Toxin Removal According to Molecular Weight by Dialysis

Molecular Weight of Toxin	Size	Removed by Standard Hemodialysis
Low molecular weight	<500 Da	Yes
Middle molecular weight	500–60,000 Da	No
Large molecular weight	>60,000 Da	No

An alternative to hemodialysis is to bring the blood extracorporeally in contact with a biocompatible adsorbent (hemoperfusion, HP), which is now used in clinical practice. The use of adsorption treatments was reported in ancient Greece (Ghannoum *et al.*, 2014). The first published application of HP was performed by Muirbead in 1948 by using an ionic resin to remove uremic toxins in dogs (Muirhead *et al.*, 1948). The application of polymeric adsorption materials in blood perfusion was started in a study by Rosenbaum. In 1970, he used non-coated Amberlite XAD-2 resin as an adsorbent for blood perfusion. Animal experiment and clinical result indicated that the resin had a good therapeutic effect. Moreover, growing interest with in this result led to the updated development of adsorption-based systems for the selective removal of toxins from blood in treatment of sepsis, auto-immune diseases, liver and kidney failure, hyperlipidemia, endotoxemia, uremia and acute intoxication. Among the numerous adsorbent materials, many researchers have focused on macroporous S polymers, which are known to possess many advantageous adsorbent properties.

In this part, we introduce the application of macroporous polystyrene adsorbents in the removal of toxin molecules by hemoperfusion.

3.1. *Removal of Endogenous Toxins*

3.1.1. β_2 *Microglobulin*

The human kidney is an important excretory organ whose main functions are to remove superfluous water and toxic materials accumulating in the blood. Once kidney failure occurs, toxic materials, such as β_2 microglobulin (β_2M, M_r = 11800), accumulated in the blood, are considered to cause many health problems. Therefore, there is an urgent need to develop adsorbents that have high selective adsorption capacities for the removal of β_2M. Vadim Davankov (Davankov *et al.*, 2000) developed a biocompatible, hyper-crosslinked polystyrene type adsorbent for this purpose. In his study, a series of S-DVB copolymers were prepared with the total porosity varying between 1.0–1.7 cm^3/g, an apparent inner surface area of 550–800 m^2/g and a broad pore size distribution between 15–25 nm. Results indicated that the adsorbents removed 93–98% of β_2M and <5% of albumin and other essential proteins during *in vitro* tests with blood or plasma from patients of chronic kidney failure. Obviously, this is due to the size-exclusion effect and the difference in the hydrophobicity of the resins (Fig. 19.20).

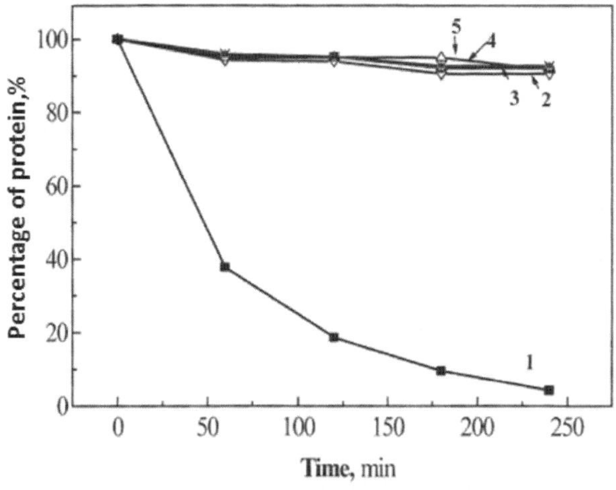

Fig. 19.20. Protein adsorption (selective) of hyper-crosslinked polymer in plasma. The polymer to plasma ratio was 1:20 (v/v). Temperature: 37°C. Initial concentrations of the proteins were as follows: (1) β2-microglobulin, 63.5 mg/L; (2) albumin, 3.15 g/dL; (3) pre-albumin, 30 mg/dL; (4) transferrin, 217 mg/dL and (5) total proteins, 6.2 g/dL. (*With copyright permission from Davankov, et al., 2000 Monograph in* J Chromatogr B.)

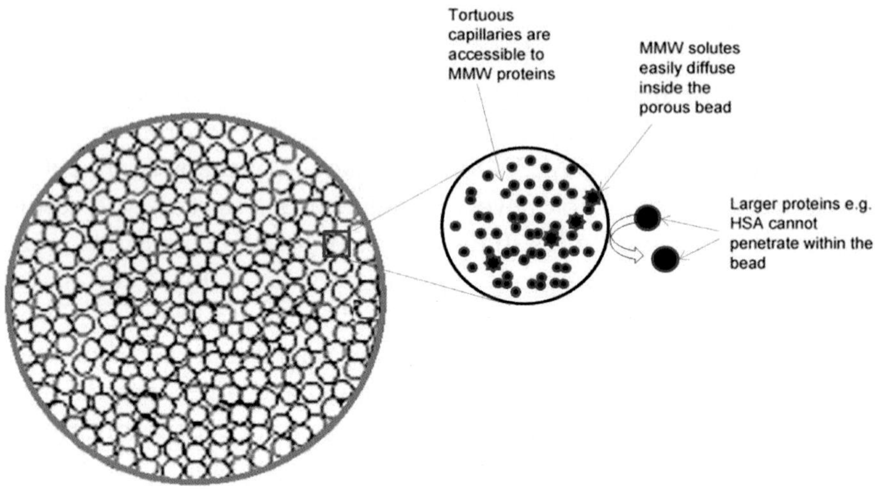

Fig. 19.21. Schematic diagram of a size-selective adsorbent bead with a nanoporous structure suitable for size exclusion of HAS while permitting smaller proteins, e.g. lysozyme, to diffuse within its pores.

3.1.2. *Lysozyme*

Malik (Malik *et al.*, 2009) developed a crosslinked polystyrene co-divinylbenzene adsorbent (PS:DVB 3:5, Tol:Nap 5:2) aiming to selectively remove middle molecular weight uremic toxins in the presence of competing solutes. Results show that polystyrene adsorbent could size-exclude serum albumin while allowing removal of smaller proteins (lysozyme) (Fig. 19.21).

3.1.3. *Cytokines*

There is increasing evidence that high levels of tumor necrosis factor (TNF-α) in serum are associated with a number of diseases, including septic shock syndrome, cachexia, AIDs and so on. Many studies have shown that hemoperfusion is one of the most effective methods for removal of pro-inflammatory cytokines such as interleukin-6 (IL-6) and tumor necrosis factor-a (TNF-α) (Davankov *et al.*, 2000; Morena *et al.*, 2003). Hyper-cross-linked S-DVB copolymer beads have a pore size distribution of 2 to 20 nm, which is suitable for removing

Table 19.7. Adsorption of TNF-α by Adsorbents

Adsorbent	Surface-to-Volume Ratio (m^2/g)	Pore Size (nm)	Polarity	Adsorption Capacity (ng/mL)	Adsorption Percentage (%)
NK-110	450–500	16.0	Weak	877.45	97.65
Carbonized resin	780–800	9.0	Weak	846.34	94.12
MET-104	334	17.6	Weak	888.90	98.63

molecules in the molecular weight range of 0.3 to 15.0 kDa, and restricting adsorption of larger molecules such as serum albumin (Davankov *et al.*, 2000).

Yu (Yu *et al.*, 2007) investigated the adsorption capacities of three macroporous resins, i.e. NK-110, carbonized resin and MET-1004, for the removal of TNF-α. Experimental results are listed in Table 19.7, which shows that MET-1004 and NK-110 have higher adsorption capacities than carbonized resin. Furthermore, the results also demonstrated that the pore size of macroporous resins have a strong impact on the adsorption capacity, that is, as the pore size increased, the adsorption capacities of the resin also increased.

3.1.4. *Bilirubin*

In acute liver failure, a range of toxic substances accumulate in the systemic circulation of the patient (Hughes *et al.*, 2002). Various macroporous resins were tailor-made by suspension polymerization of S and DVB with various amounts of porogens at Nankai University. Resins with different pore diameters and specific surface areas were studied for the adsorption capacity of unconjugated bilirubin (Table 19.8). The data show that NK110 resin having a pore size of 160 Å and specific surface area of 500 m^2/g displayed the highest adsorption capacity for bilirubin. This is because the large size of the unconjugated bilirubin molecule could diffuse freely through the large pores of the resin. However, Yaoting Yu from Nankai University pointed out that this does not mean that the larger the pore size, the higher the

Table 19.8. Surface Area and Pore Size of Resins versus Their Adsorption Capacities for Bilirubin

Resin	Average Pore Diameter (Å)	Average Pore Size (Å)	Specific Surface Area (m^2/g)	Adsorption Capacity (%)
X-5		152	483	79.3
NK-X-5110		160	500	87.5
A-6	≥120	125	330	41.0
A-7		250	200	28.0
D-14		500	38	22.0
Amberlite XAD-4		50	750	16.7
D-6	≤120	73	466	29.0
58-9		43	388	13.8
Amberlite XAD-2		90	300	12.0

adsorption capacity, because as the pore size increases the surface area will decrease, resulting in a decrease in adsorption capacity. Thus, there is a balance between the two parameters which should be optimized to obtain the highest adsorption capacity (Yu *et al.*, 2013). Based on the above findings, a slightly modified adsorbent resin was commercialized and used in clinics for the successful treatment of hyperbilirubinemia patients in whole blood hemoperfusion (Wang *et al.*, 2011).

Victoria Weber (Weber *et al.*, 2008) prepared two series of polystyrene divinylbenzene adsorbent resins with average pore sizes of 5–6 and 8–9 nm, respectively and investigated the effect of the pore size on the adsorption of bilirubin. Average particle sizes, pore sizes, and specific surface areas for the resins are listed in Table 19.9. Figure 19.22 shows typical electron micrographs of the polymers of large pore size series. As shown in Fig. 19.23 and Table 19.10, polymers L1-L3 efficiently adsorbed bilirubin, while S1-S3 displayed relatively poor bilirubin binding, which indicated that pores smaller than 5–6 nm are not accessible to bilirubin. As expected, polymers with small pores were found to be poor bilirubin adsorbents, while large pores were associated with high bilirubin adsorption.

Table 19.9. Pore Size and Surface Areas of Polystyrene-Divinylbenzene Resins

		Mesopore Volume (BJH Desorption) (cm^3/g)			Micropore Volume (HK) (cm^3/g)	
Sample (Average Size)	BET Surface Area (m^2/g)	2–50 (nm)	2–300 (nm)	MPD (nm)	0.5–2 (nm)	Average Pore Diameter (nm)
L3 (28 μm)	683	1.17	1.18	59.4	0.28	7.69
L2 (15 μm)	678	1.30	1.60	70.2	0.28	8.75
L1 (6 μm)	660	1.34	1.43	88.6	0.27	9.38
S3 (28 μm)	792	0.82	0.89	57.5	0.32	5.34
S2 (15 μm)	736	0.84	0.89	86.6	0.30	5.54
S1 (6 μm)	890	0.96	0.98	58.5	0.36	5.19

3.1.5. *Low Density Lipoprotein (LDL)*

The research and development of blood purification materials with good blood compatibility and high selectivity for LDL removal have important implications for the treatment of cardiovascular diseases. Heparin as a ligand was linked on macroporous polystyrene resin (XAD1600) to remove LDL selectively in hemoperfusion (Fig. 19.23). Experimental results show that the resin adsorbed 82.6 ± 2% of LDL and 7 ± 2% of HAS in a mixture of the two components (adsorbent (g): serum solution (ml) = 1:5). In addition, the resin adsorbed 82.4 ± 2% of LDL and < 10% of the total protein including albumin and globulin in the serum adsorption tests (adsorbent (g): serum solution (ml) = 1:5) (Tang *et al.*, 2012).

3.1.6. *Bile acid*

The total amount of bile acid in normal human serum is less than 1–2 mg/dL. In acute liver failure, metabolites include hydrophobic compounds (e.g. bile acids, bilirubin, hydrophobic amino acids), which are found to be bound to albumin with varying affinity. Consequently, removal of excess bile acid from the patient's blood is one of the

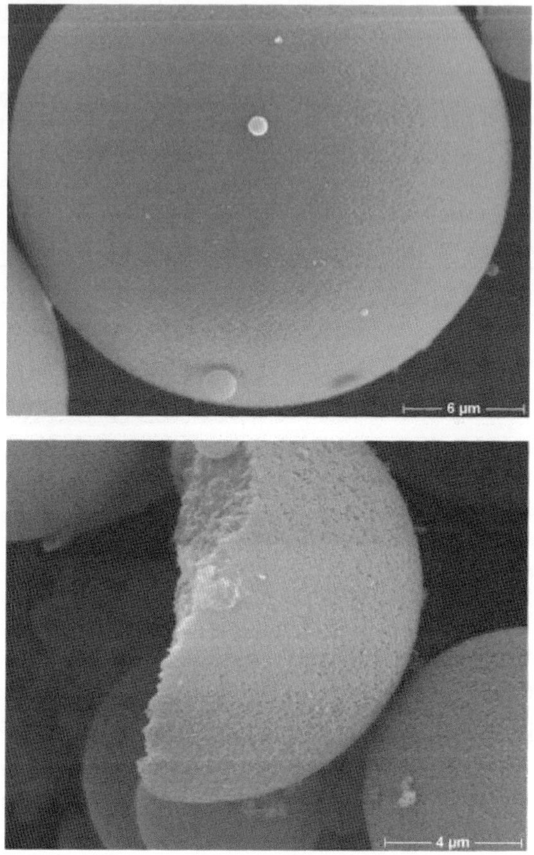

Fig. 19.22. Raster electron micrographs of adsorbent particles from the large pore-size series. (*With copyright permission from Weber et al., 2008 Monograph in* Biomacromolecules.)

important research topics for artificial livers. Binglin He investigated the adsorption ability of macroporous adsorbent resins for bile acid and found that macroporous resin NK110 is a new type of non-ionic adsorbent suitable for adsorption. Furthermore, he compared its adsorption behavior with Amberlite XAD-4 resin. The specific surface area, average pore size, and their influences on the adsorption are listed in Table 19.11. It was found that when the average pore size of the adsorbent is larger than 12 nm, the adsorption rate for bile acid increased with increasing surface area. Although the surface area of

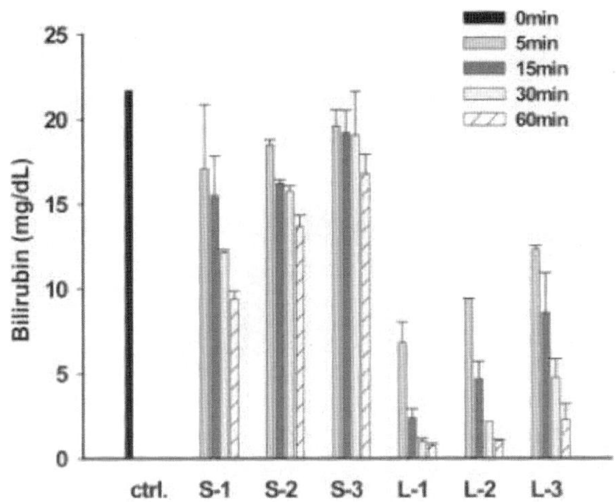

Fig. 19.23. Adsorption of bilirubin from plasma. Adsorption was tested *in vitro* using human plasma spiked with bilirubin (300 μM), and the ratio of adsorbent to plasma was 1:9 (vol/vol) for cholic acid. (*With copyright permission from Weber et al., 2008 Monograph in* Biomacromolecules.)

Table 19.10. Amount of Bilirubin Bound to Resins in Batch Experiments after 60 min (Weber *et al.*, 2008)

	Bilirubin Bound		Cholic Acid Bound	
Sample (Particle Size)	(μmol/mL Adsorbent)	(nmol/m²)	(μmol/mL Adsorbent)	(nmol/m²)
L3 (28 μm)	1.45	7.3	9.9	62.7
L2 (15 μm)	1.88	9.6	8.3	53.0
L1 (6 μm)	2.30	11.9	4.0	26.2
S3 (28 μm)	0.32	1.4	13.7	59.6
S2 (15 μm)	0.51	2.4	13.5	63.2
S1 (6 μm)	0.70	2.7	13.4	51.9

The BET surface area (Table 9) was used to calculate the amount of toxin bound per m²

Amberlite XAD-4 resin is 750 m^2/g, the pore size is only 5 nm so the adsorption rate is lower than NK110 resin. This result indicated that a resin adsorbing molecular compounds combined with a protein, should have pore size big enough, to allow macromolecules diffuse

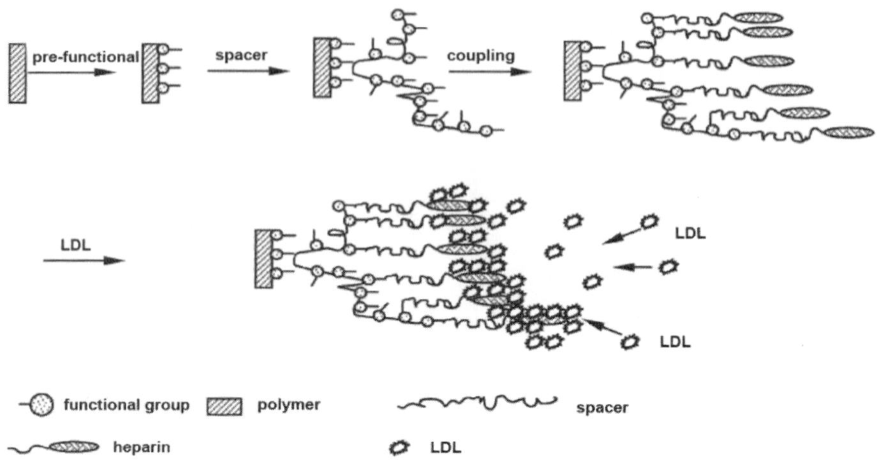

Fig. 19.24. The adsorption procedure of heparin modified polystyrene adsorbent for removal of LDL. (*With copyright permission from Tang et al., 2012 Monograph in* Journal of Medical Biomechanics.)

Table 19.11. Effects of Surface Area and Average Pore Size of Resins on Adsorption Rate

Resin	NK110	X-5	A-6	A-7	D-14	XAD-14
Surface area (m^2/g)	500	483	330	200	38	750
Average pore size (Å)	160	152	125	250	500	50
Adsorption rate (%)	93	87	80	68	45	75

freely in and out of the resin. The average pore sizes of A-7 and D-14 resins are 25 and 50 nm respectively, which are rather big, but their surface area are small (200 and 38 m^2/g, respectively), thus their adsorption rates are relatively low (68% and 45%). In other words, the specific surface area and average pore size are two essential factors that play an important role in the adsorption of adsorbents for substances with relative high molecular weights. In addition, the particle size of the resin also influences the adsorption for bile acid: the smaller the particle size, the greater the adsorption rate and capacity (Tong *et al.*, 1987).

Table 19.12. Clearance Rate (mL/min) of Amberlite XAD-4 and Activated Carbon for Drugs

	Adsorption Percentage (%)		
Drug	**Activated Carbon[1]**	**Activated Carbon[2]**	**Amberlite XAD-4**
Pentobarbital	64	68	100
Phenobarbital	76	55	96
Adalin	89	88	100
Methaqualone	81	80	100
Flurazepam	58	56	100
Diazepam	47	78	88
Imiramine	37	72	96
Doxepin	56	68	100
Promethazine	71	51	100
Promazine	60	40	100
Desoxyphenobarbital	96	65	100
Carbamazepine	96	95	99
Phenmetrazine	79	95	99
Phenmetrazin	78	85	100
Isopropylphenazone	70	79	100
Methylimidazole	72	80	100
Phenacetin	98	97	100
Acetaminophen	79	96	100
Acetylsalicylic acid	68	37	66
Propoxyphene	47	60	100
Pentazocine	96	86	100
Digoxin	30	46	99
Conqinine	62	52	98

[1]Coated by acrylic hydrophilic gel; [2]coated by cellulose

3.2. *Removal of the Exogenous Toxins*

Since the adsorbent resins were used in hemoperfusion for clinical treatments, after repeated practices and continuous improvements, they have become an effective and reliable method to treat and rescue patients with acute drug poisoning.

Table 19.13. Physical Properties and Adsorption Capacities of NK107

Resin	Density (g/mL)	Specific Surface Area (m^2/g)	Pore Diameter (Å)	Porosity (%)	Pore Volume (mL/g)	Particle Size (mm)	Adsorption Capacity (%)			
							Chorpro-mazine	Barbital	Methaq-ualone	Valium
NK107	0.37	1086	90	64	1.27	0.6–1.2	88.7	90.6	94.8	82.9

Table 19.14. Adsorption Capacity of Different Adsorbents for Hypnotics

	Adsorption Percentage of Hypnotics (%)				
Resins	Chlorpromazine Hydrochloride	Methaqu-Alonum	Antipsy-chotic	Phenobarbital Sodium	Barbital Sodium
NK-107	43.3	38.1	31.4	31.4	20.2
Amberlite XAD-4	37.0	31.6	19.8	18.2	9.5
Amberlite XAD-2	33.6	24.3	9.7	9.7	3.0
Activated carbon	44.7	35.5	27.6	27.6	34.3

Clinical trials have proved that Amberlite XAD-4 resin has a good adsorptive effect for drugs or poisons; see Table 19.12 (Yu, *et, al.* 1987).

Macroporous adsorbent resin NK107 made by the Nankai group with a high specific surface area of 1,070 m^2/g has a high adsorption capacity for small molecules such as drugs and chemicals (Table 19.13). *In vitro* experiments show that both adsorption capacity and adsorption rate of NK107 for drugs including barbital sodium, phenobarbital and methaqualone are higher than other resins, as shown in Table 19.14.

This resin is commercialized in China and used in hospitals all over the country, saving hundreds and thousands lives of suicidal or drug overdose patients. Furthermore, it is also widely used for detoxification treatment in the removal of endogenous pathogenic toxins in patients of various diseases by whole blood hemoperfusion.

References

Ahn JH, Jang JE, Oh CG, *et al.* (2006) Rapid generation and control of microporosity, bimodal pore size distribution, and surface area in Davankov-type hyper-cross-linked resins. *Macromolecules* **39:** 627–632.

Alelio GF. (1945) Production of synthetic polymeric compositions comprising sulphonated polymerizates of poly-vinyl aryl compounds and treatment of liquid media therewith. *United States Patent* 2,366,007.

Dan GP, Lu YZ, Liu CY, Wang LX. (1999) Preparation of styrene-divinyl benzene copolymer by two step polymerization. *Chemical Research and Application* **11:** 316–319.

Davankov V, Pavlova L, Tsyurupa M, *et al.* (2000) Polymeric adsorbent for removing toxic proteins from blood of patients with kindey failure. *J Chromatogr B* **739:** 73–80.

Davankov V, Tsyurupa M, Ilyin M, Pavlova L. (2002) Hypercross-linked polystyrene and its potentials for liquid chromatography: a mini-review. *J Chromatogr A* **965:** 65–73.

Davankov VA, Tsyrupa MP, Rogozhin SV. (1976) On factors determining the swelling capacity of crosslinked gels, II. *Angew Makromol Chem* **53:** 19.

Davankov VA, Tsyurupa MP, Rogozhin SV. (1974) Macronet isoporous gels through crosslinking of dissolved polystyrene. *J Polym Sci Symposium* **47:** 95.

Davankov VA, Tsyurupa MP. (1990) Structure and properties of hyper-crosslinked polystyrene the first representative of a new class of polymer networks. *Reactive Polymers* **13:** 27–42.

Davankov V, Pavlova L, Tsyurupa M, *et al.* (2000) Polymeric adsorbent for removing toxic proteins from blood of patients with kidney failure. *J Chromatogr B Biomed Sci Appl* **739:** 73–80.

Erbay E, Okay O. (1999) Pore memory of macroporous styrene-divinylbenzene copolymers. *J Appl Polym Sci* **71:** 1055–1062.

Flodin P. (1998) Structural aspects of porous gels. *Macromolecular Symposia* **22(1):** 253–268.

Ghannoum M, Bouchard J, Nolin TD, *et al.* (2014) Hemoperfusion for the treatment of poisoning technology, Determinants of Poison Clearance, and Application in Clinical Practice. *Semin Dialysis.* [doi: 10.1111/sdi.12246]

Gokmen MT, Prez FE. (2012) Porous polymer particles-A comprehensive guide to synthesis, characterization, functionalization and applications. *Prog Polym Sci* **37:** 365–405.

Gong J. (2008) Studies on Preparation and adsorption properties of polystyrene porous resin. PhD thesis, Jiangnan University.

Grassmann A, Gioberge S, Moeller S. (2004) ESRD patients in 2004: global overview of patient numbers, treatment and modalities and associated trends. *Nephrol Dial Transpl* **20(12):** 2587–2593.

Gupta DC, Beldar AG, Tank R. (2006) Suspension copolymerization of styrene and divinylbenzene: formation of beads. *J Appl Polym Sci* **101:** 3559–3563.

Halasz I, Martin K. (1978) Pore size of solids. *Angew Chem Int Ed* **17:** 901.

Halasz I, Vogtel P. (1980) Determination of morphological properties of swellable solids by size exclusion chromatography. *Angew Chem Int Ed* **19:** 24–28.

Hilgen H, DeJong GJ, Sederel WL. (1975) Styrene–divinylbenzene copolymers. II. The conservation of porosity in styrene–divinylbenzene copolymer matrices and derived ion-exchange resins. *J Appl Polym Sci* **19:** 2647–2654.

Hradil J, Kralov E. (1998) Styrene–divinylbenzene copolymers post-crosslinked with tetrachloromethane. *Polymer* **39:** 6041–6048.

Hughes RD. (2002) Review of methods to remove protein-bound substances in liver failure. *Int J Artif Organs* **25:** 911–917.

Huxham IM, Rowatt B, Sherrington DC, Tetley L. (1992) Molecular architectural changes in hydrated macroporous styrene-divinylbenzene resin sorbents revealed by transmission electron microscopy using image analysis. *Polymer* **13:** 2768–2777.

Li KY, Zhang JH, Xiang JF. (1999) The principle and technology of polymer synthesis. Science Press, Beijing.

Li L, Pan J, Yu Y. (2007) Development of sorbent therapy for multiple organ dysfunction syndrome (MODS). *Biomed Mater* **2:** R12

Lowell S. (1979) *Introduction to Powder Surface Area*. Wiley, New York.

Malik DJ, Webb C, Holdich RG, *et al.* (2009) Synthesis and characterization of size-selective nanoporous polymeric adsorbents for blood purification. *Sep Purif Technol* **66:** 578–585.

Malovic L, Nastasovic A, Sandic Z, *et al.* (2007) Surface modification of macroporous glycidyl methacrylate based copolymers for selective sorption of heavy metals. *J Mater Sci* **42:** 3326–3337.

McBurney CH. (1952) Resinous insoluble reaction products of tertiary amines with haloalkylated vinyl aromatic hydrocarbon copolymers. *United States Patent* 2,591,573.

Milan JB, Daniel H, Frantisek S. (2005) Methacrylate-based chromatographic media. *J Sep Sci* **28:** 1855–1875.

Millar JR, Smith DG, Marr WE, Kressman TRE. (1963) Solvent-modified polymer networks. Part I. The preparation and characterisation of expanded-network and macroporous styrene–divinylbenzene copolymers and their sulphonates. *J Chem Soc* 218–225.

Morena MD, Guo D, Balakrishnan VS, *et al.* (2003) Effect of a novel adsorbent on cytokine responsiveness to uremic plasma. *Kidney International* **63:** 1150–1154.

Muirhead EE, Reid AF. (1948) A resin artificial kidney. *J Lab Clin Med* **33:** 841–844.

Okay O, Kurz M, Lutz K, Funke W. (1995) Cyclization and reduced pendant vinyl group reactivity during the free-radical crosslinking polymerization of 1,4-divinylbenzene. *Macromolecules* **28:** 2728–2737.

Okay O. (1986) Heterogeneous styrene–divinylbenzene copolymers. Stability conditions of the porous structures. *J Appl Polym Sci* **32:** 5533–5542.

Okay O. (1988) Styrene-divinylbenzene copolymers. VI. Porosity formation in the presence of toluene-cyclohexanol mixtures as inert diluents. *Angew Makromol Chem* **157:** 1–13.

Okay O. (2000) Macroporous copolymer networks. *Prog Polym Sci* **25:** 711–779.

Okubo M, Konishi Y, Inohara T, Minami H. (2002) Production of hollow polymer particles by suspension polymerizations for ethylene glycol dimethacrylate/toluene droplets dissolving styrene-methyl methacrylate copolymers. *J Appl Polym Sci* **86:** 1087–1091.

Okubo M, Konishi Y, Minami H. (2000) Production of hollow polymer particles by suspension polymerizations for divinylbenzene/ toluene droplets dissolving various polymers. *Colloid Polym Sci* **278:** 659–664.

Patterson JA. (1971) Preparation of crosslinked polystyrene and their derivatives for use as solid supports or insoluble reagents. In: *Biochemical Aspects of Reaction on Solid Supports.* Academic Press, New York.

Poinescu IC, Beldie C, Vlad C. (1984) NMR linewidth study of a latex interpenetrating network. *J Appl Polym Sci* **29:** 23–34.

Poinescu IC, Popescu V, Carpov A. (1985) Synthesis of porous styrene divinylbenzene copolymers. *Angew Makromol Chem* **135:** 21–32.

Poinescu, IC, Beldie C. (1988) Styrene divinylbenzene copolymers. Influence of diluent mixture on matrix structure. *Angew Makromol Chem* **164:** 45–58.

Rabelo D, Coutinho FMB. (1994) Porous structure formation and swelling properties of styrene-divinylbenzene copolymers-effect of diluent nature. *Macromolecular Symposia* **84(1):** 341–350.

Roche I. (2008) Nanoporous polymeric adsorbents for blood purification. PhD thesis, Loughborough University Institutional Repository.

Sakodynsky KI, Panina LI. (1972) Polymeric sorbents in gase chromatography analysis. *Zh Analyt Khim* **27:** 1024.

Seidl J, Malinsky J, Dusek K, Heitz W. (1967) Makroporose styrol-divinylbenzol-copolymere und ihre verwendung in der chromato graphie und zur darstellung von ionenaustauschern. *Adv Polym Sci* **5:** 113–213.

Staudinger H, Heuer W. (1934) Uber hochpolymere verbindungen, 94, Mitteil,: Uber ein unlosliches poly-styrol. *Chem Ber* **67:** 1164–1172.

Tang JL, Wang X, Wang FF, Ji XX. (2012) Study on heparin modified macroporous polystyrene resin to selectively adsorb human low density lipoprotein. *Journal of Medical Biomechanics* **27:** 375–376.

Tong MR, Chen CZ, Yu YT, Ho P. (1987) Adsorption of cholic acid by macroporous adsorbent NK110. *Ion Exchange and Adsorption* **2:** 20–24.

Tsyurupa MP, Davankov VA. (2006) Porous structure of hyper-crosslinked polystyrene: state-of-the-art mini-review. *React Funct Polym* **66:** 768–779.

Tsyrupa MP, Lalaev VV, Davankov VA. (1984) Synthesis and some physic-chemical properties of macronet isoporous styrene polymers with cross-bridges of biphenylmethane type. *Acta Polym* **35:** 451.

Tsyrupa MP, Lalaev VV, Davankov VA. (1984) On reasons determining unusual properties of hyper-crosslinked styrene polymers. *Dokl AN SSSR* **279:** 156.

Wang YJ, Yu YT. (2011) Development of resin adsorbents for blood purification at Nankai University in China. *Artif Cell Blood Sub* **39:** 92–97.

Weber V, Linsberger I, Hauner M, *et al.* (2008) Neutral styrene divinylbenzene copolymer for adsorption of toxins in liver failure. *Biomacromolecules* **9:** 1322–1328.

Xu JH, Yuan XY, Huang JX. (2002) The study on solvent/emulsion polymerization to synthesize single-dispersed, micro-sized, cross-linked polystyrene. *Chinese Journal of Colloid & Polymer* **20:** 8–11.

Yu YT. (2013) Adsorbents in blood purification: From lab search to clinical therapy. *Chin Sci Bull* **58:** 4357–4361.

Yu YT, Chen CZ. (1987) Medical adsorbents and their development. *Journal of Biomedical Engineering* **4(3):** 211–221.

IV

Immunosorbent

Chapter 20

Immunoadsorbent in Hemoperfusion

Shenqi Wang*, Yaoting Yu†‡, and
Yanmiao Fan*

1. Introduction

Sorbent perfusion is a novel approach to blood purification which can specifically remove endogenous and exogenous pathogenic toxins from the blood of patients (Behim *et al.*, 1989). The technique involves passing whole blood or plasma of the patient through a cartridge filled with an adsorbent which can easily absorb the toxin or pathogenic molecules in the circulating blood. This approach has been developed in the past to be one of the most important areas in hemoperfusion, especially in clinical performances. The adsorbent is the core technique of the device and is composed of a carrier, spacer and ligand (Fig. 20.1a,b). According to selectivity, adsorbents can generally be classified as broad spectrum, affinity adsorbents and

*College of Life Science and Technology & Advanced Biomaterials and Tissue Engineering Center, Huazhong University of Science and Technology, 1037 Luoyu Road, Wuhan, P. R. China.

†The Key Laboratory of Bioactive Materials, Ministry of Education, Nankai University, Tianjin, China 300071.

‡Corresponding author: Yaoting Yu, Nankai University, College of Life Science, Weijin Road 94, Tianjin, China 300071. E-mail: yaotingyu@ nankai.edu.cn.

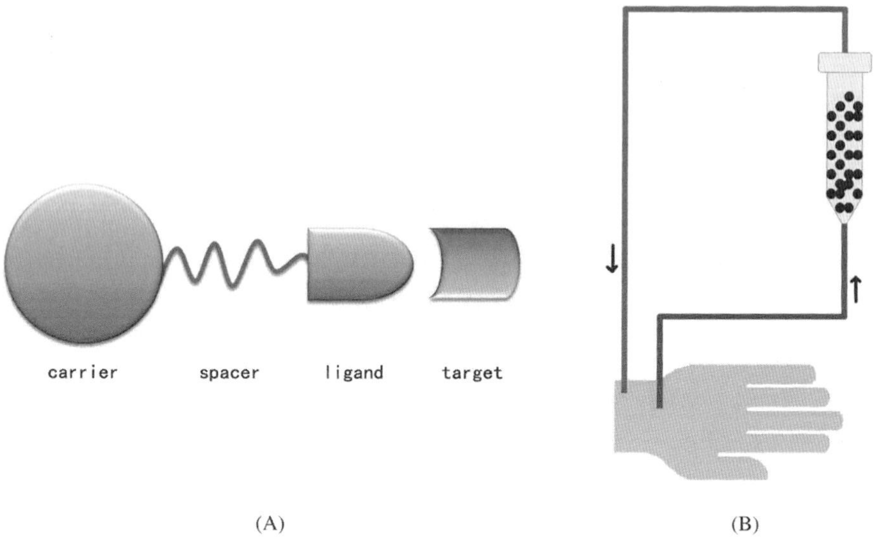

Fig. 20.1. Schematic diagram of sorbent-perfusion.

immuno-adsorbents, of which the latter has the highest selectivity (Lupien *et al.*, 1976; Sinitsyn *et al.*, 1990; Maaskant *et al.*, 1986; Kojima *et al.*, 1992).

In hemoperfusion, broad spectrum adsorbents such as activated charcoal can adsorb all kinds of pathogens existing in blood or plasma without a definite selectivity. An immunoadsorbent is defined as a preparation of antigen attached to a solid support or antigen in an insoluble form which absorbs homologous antibodies from a mixture of immunoglobulins (*Dorland's Medical Dictionary for Health Consumers*, 2007). It is based on the principle of antigen–antibody interaction, and is therefore highly specific. In general materials used for preparation of the carrier can be activated charcoal, natural or synthetic bioactive polymers in the form of beads, fibers and membranes. The pathogenic molecules in the blood of patients are absorbed by the adsorbent via hydrophilic (electro-static forces) or hydrophobic interactions. Macroporous resins usually show high adsorption capacities especially for the removal of high molecular weight or "middle molecule" toxins (Falkenhagen *et al.*, 1999; Terman *et al.*, 1977; Von Appen *et al.*, 1996). An antibody (Ab), also known as an immunoglobulin (Ig), is a large Y-shaped protein ("Y") produced by plasma cells that is used by the immune system to identify and neutralize foreign objects such as bacteria and viruses. The antibody recognizes a unique part of the foreign target, called an antigen (Murphy *et al.*, 2011; Litman *et al.*, 1993). Each tip of the "Y" of an antibody contains a paratope (a structure analogous to a lock) that is specific for one particular epitope (similarly analogous to a key) on an antigen, allowing these two structures to bind together with precision (Wikipedia public domain).

Yatzidis *et al.* used activated charcoal to treat patients with kidney failure in 1964. In order to prevent problems related to particle release and damage to blood cells, Chang modified the surface of activated charcoal with albumin-ultrathin polymeric membrane coating on the surface (1969) and treated patients with acute poisoning, renal failure and liver failure (Chang, 1975). Graf *et al.* were the first to show that immunoadsorbents could be employed to selectively remove antibodies from actively and passively immunized rabbits. Schenkein *et al.* (1971) extended Chang's albumin binding method (1969) and developed an extracorporeal immunoadsorbent system in which bovine serum albumin (BSA) was immobilized on agarose and proved to selectively remove BSA antibodies from the circulation. Because of the fragility of the supporting matrix which led to the releasing of fine particles causing thrombosis and poor biocompatibility properties of the adsorbent, new and stable carriers were developed. In order to solve the abovementioned problems, Chang introduced the concept of artificial cells, which are not a specific physical entity but an idea involving the preparation of artificial structures of cellular dimensions for possible replacement or supplement of deficient cell functions (Chang 1964 to 2007; Chang *et al.*, 1987; Chang and Poznansky, 1968; Chang and Malave, 1970).

Artificial cells have three significant characteristics: (1) The membrane of an artificial cell separates its contents from the outside environment; (2) the cell membrane is biocompatible, can be very thin, yet strong, allowing specified molecules to pass in and out, and has a large surface area; (3) artificial cells can contain the same biological materials as biological cells. All these characteristics solved the above problems in hemoperfusion. Next, Chang developed albumin-based artificial cells by binding albumin to the ultrathin collodion membrane of artificial cells, which improved their biocompatibility. Albumin coating was also applied to synthetic immunoadsorbents.

In the early '90s, adsorbents linked with antibodies or receptors used to treat autoimmune diseases were only classified as immunoadsorbent (Agishi, 1996). Following the progress of hemoperfusion, various ligands such as amino acids, sulfonic, phosphoric groups linked to polymers, synthetic peptides and protein A were developed and classified as ligands of immunoadsorbents. At the same time, clinical performances of immunoadsorbents had been developed for the treatment of autoimmune diseases, modification of hyper-acute renal xenograft rejection, removal of anti-HLA antibodies in transplant candidates and treatment of familial hypercholesterolemia with monoclonal antibodies to remove low-density lipoproteins. Then, a large number of immunoadsorbents were produced and commercialized (Terman, 1980; Terman *et al.*, 1979a; 1979b; Hakim *et al.*, 1990; Wingard *et al.*, 1991; Yang *et al.*, 2004). Chang's concept of the "artificial cell" played a significant role in the development of immunoadsorbents.

The aim of this chapter is to present a brief overview of immunoadsorbents in hemoperfusion, and the ligand's name is used to classify and describe the R&D of the adsorbents.

1.1. *Protein A as Ligand*

Protein A is a surface protein originally found in the cell wall of bacterium — Staphylococcus Aureus having a molecular weight of 56 kDa. Protein A is composed of five homologous Ig-binding domains that fold into a three-helix bundle. Each domain is able to bind proteins from many mammalian species, most notably IgGs (Graille *et al.*, 2000; Idusogie *et al.*, 2000). It binds the heavy chain within the Fc region of most immunoglobulins and also within the Fab region in the case of human VH3 family (Fig. 20.2).The immunoadsorbent of protein A is prepared by chemically linking the ligand to carriers such as agarose (sepharose) or silica beads via a spacer and is used for the

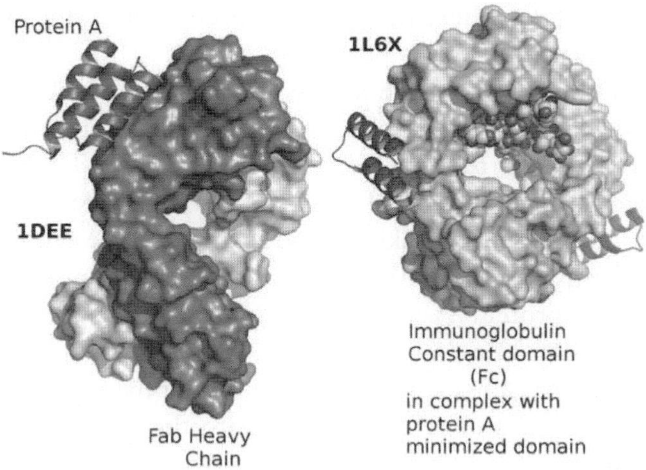

Fig. 20.2. Structure of a domain of protein A as a three-helix bundle binding to the heavy variable chain of a VH3 human Fab, left. Minimized protein A bound to Fc fragment of Rituximab. (Wikipedia, public domain.)

removal of immunoglobulins in autoimmune diseases (Alkana *et al.*, 2010; Samuelsson *et al.*, 1999; Freiburghaus *et al.*, 1988; Gjorstrup *et al.*, 1991; Matic *et al.*, 2000; 2001). Prosorba, the immunoadsorbent with protein A as ligand, was manufactured by Fresenius Hemocare (Redmont, CA, U.S.A.) and Immunosorba by Fresenius Hemocare (St. Wendel, Saarland, Germany). Both columns have been used clinically for the treatment of rheumatoid arthritis (Kutsuk *et al.*, 1998). It is important to point out that Prosorba immunoadsorbent encountered some unprecedented challenges and difficulties as a treatment in clinical therapy because commercial production of the column has been ceased by the FDA since 2006 (Ward *et al.*, 2011; Sanchez *et al.*, 2013), and its treatment was not available in recent years. The main problems encountered are as follows: A considerable amount of protein A could be released from the matrix, and such contamination would not be tolerated in clinical applications; it was shown that Protein A is a B-cell super-antigen and postulated to be evolved in staphylococci as a means to impair antibody-mediated defences in the host where the staphylococcus is invading (Goodyear *et al.*, 2003; Ward, 2011).

Super-antigens (SAgs) are a class of antigens which cause non-specific activation of T-cells resulting in polyclonal T cell activation and massive cytokine release (Llewelyn *et al.*, 2002). This cytokine is closely linked with induction of autoimmunity (Stauffer *et al.*, 2001), and the autoimmune Kawasaki Disease is known to be caused by SAgs infection (Arcus *et al.*, 2000). The contact of patient's blood with staphylococcal protein A may have an immunosuppressive effect caused by pharmacological mechanism rather than by an apheresis mechanism (Ward, 2011). Some leakage of Staphylococcus protein A (SPA) from the column (100–200 mg) into the patient's blood during apheresis has been described, and many of the patients formed anti-SPA antibodies (Sasso *et al.*, 2001). High cost of protein A adsorbent and low storage stability are also challenges for this adsorbent.

1.2. *Tryptophane as Ligand*

Tryptophane is one of the 22 standard amino acids that are essentially important in human diet. Due to the hydrophobicity property of the indole ring in the tryptophane molecule (Fig. 20.3), it is conducive to interact with acetylcholine receptor (AChR) and can be easily absorbed by the adsorbent. In 1982, Yamazaki *et al.* developed an adsorbent by linking tryptophane to PVA beads for the treatment of MG patients. Clinical performances demonstrated that the adsorbent

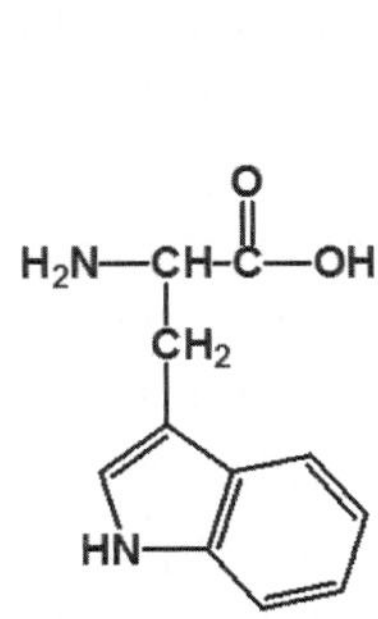

Fig. 20.3. Structure of tryptophan.

can effectively remove AChR from patient's plasma and alleviate MG syndrome. In 2005, Yu and Yang *et al.* synthesized immobilized tryptophan cellulosic bead type adsorbent and evaluated its adsorption capacity for binding AChR in whole blood. Experimental autoimmune myasthenia gravis (EAMG) rabbits were induced by Ta183-200 peptide, and extracorporeal whole blood hemoperfusion was conducted. Experimental results indicated that whole blood perfusion is an effective and safe approach in treating passive autoimmune myasthenia gravis by improving clinical manifestation, neuromuscular transmission function, enhancing the quantity of neuromuscular junction and reducing antibody titer. Bronchial asthma is a tracheal over-reacted disease stimulated by antigenic reaction, conducted by the antibody of immunoglobulin E. At present, the treatment of bronchial asthma is mainly by medicine; immunoadsorption by extracorporeal hemoperfusion is under investigation. In 2008, Wang and Yu *et al.* synthesized agar gel adsorbent with tryptophan as ligand for the removal of immunoglobulin E (IgE) in bronchial asthma. *In vitro* results indicated the adsorption percentage of IgE in whole blood was 70.73% (Wang *et al.*, 2008). Blood compatibility of the adsorbent was satisfactory and has a high potential to be used for whole blood purification in removal of IgE in bronchial asthma patients.

1.3. *Phenylanaline as Ligand*

Phenylanaline is an •-amino acid with the formula $C_6H_5CH_2CH(NH_2)COOH$ (Fig. 20.4). The non-polar characteristic of this essential amino acid makes it widely used to absorb hydrophobic proteins.

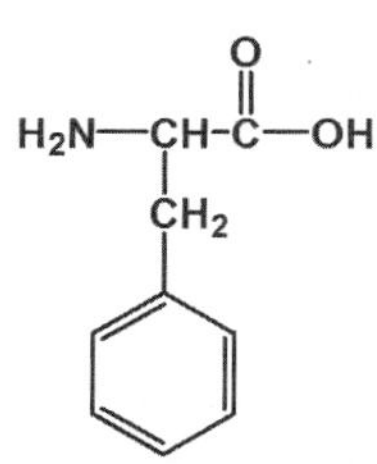

Fig. 20.4. Structure of phenylanaline.

In 1982, Yamazaki *et al.* developed Immusorba-PH-350 column by immobilizing phenylanaline on PVA beads and demonstrated that it could effectively remove autoimmune antibodies in the plasma of rheumatoid arthritis patients with good biocompatible properties. The adsorbent was commercialized and successfully used as a therapeutic treatment for rheumatoid arthritis. In 2006, Sugimoto *et al.* used Immunosorba-PH-350 column to treat Lupus Nephritis (LN). It could efficiently remove pathogenic substances, anti-DNA antibodies and immune complexes. In addition, concomitant immunoadsorption therapy with corticosteroid administration was shown to decrease disease activity and lower the overall corticosteroid requirement. Chen and Yang developed RA adsorbents by linking phenylanaline to agar beads which effectively removed IgGRF, IgMRF, IgARF from RA patient's plasma (2002). The adsorption percentage was 70%, 40% and 70%, respectively. In 2013, Jin and Wang synthesized chloromethylated polystyrene adsorbent with heparin as spacer, phenylanaline as ligand for the removal of endotoxin with better biocompatibility (Jin *et al.*, 2013).

1.4. *DNA as Ligand*

As shown in Fig. 20.5, the backbone of a DNA strand is composed of alternating phosphate and sugar residues (Ghosh *et al.*, 2003). The sugar moiety in DNA is 2-deoxyribose, which is a pentose (five-carbon) sugar. The sugars are joined together by phosphate groups that form phosphodiester bonds between the third and fifth carbon atoms of adjacent sugar rings (Wikipedia). Since Graf *et al.* demonstrated that immunoadsorbent could be used to remove antibodies from actively and passively immunized rabbits in 1969, many researchers began to develop immunoadsorbents suitable for extracorporeal circulation systems. In 1977, Terman *et al.* developed ssDNA based immunoadsorbent.

Single strand DNA (ssDNA) was dissolved in collodion solution and used to coat bead type charcoal via a procedure similar to Chang's (Chang *et al.*, 1966; 1972). Experimental results indicated that ssDNA was embedded in collodion-charcoal with no significant

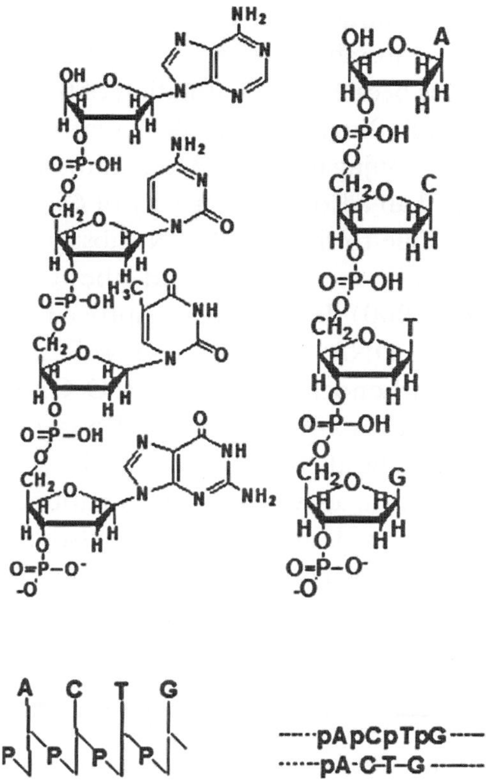

Fig. 20.5. Schematic structure of DNA.

release to the host. In later studies new and more stable carriers were developed by covalent linking of ssDNA on resin carriers which prevented ssDNA ligands from leaking, especially in sterilization and storage. Kato *et al.* (1996) developed ssDNA immunoadsorbent by using 1,1-carbonyldiimidazole as a spacer to link the hydroxyl group of DNA with the phosphate group in grafted poly(methacryloyloxyethyl phosphate). In another approach, he used poly-ion complex reaction between the anionic DNA and the cationic grafted poly (N,N-dimethyl-aminoethyl-methacrylate) chains to prepare the adsorbent. The adsorbent thus made exhibited a high adsorption capacity and specificity, but unfortunately most of the immobilized ssDNA ligand leaked out after autoclave at 115°C. They believed that leaking of DNA was due to instability of the poly-ion complex, but Zhu reported that this was because of denaturation of the DNA (Zhu *et al.*, 1999). Yihua Yu and Binglin He (1999) synthesized DNA-based immunadsorbent by conjugating DNA on chitosan beads via CNBr activation, which had an adsorption capacity of 65.33% for anti-DNA antibody in serum. Due to the unstable N-substituted iso-urea bond especially in autoclave sterilization and the toxicity of CNBr, Deling Kong *et al.* (2000) developed an improved DNA immunoadsorbent by coupling DNA on Sepharose 4FF via 5-norbornene-2,3-dicarboximido carbonochoronate (CL-CO-ONB), which had a better stability and was non-toxic. Deling Kong (2002) developed microparticle cellulose beads as the carrier and was able to enhance the adsorbent removal capacity of anti-DNA antibody. In an incubation of 20 mL of patient's plasma with 1 ml of adsorbent, 80% removal of anti-DNA antibody was achieved. In 1983, Yaoting Yu *et al.* developed carbonized resin beads from styrene-acrylonitrile copolymer by pyrolysis at high temperature. Anti-DNA antibody from calf thymus was then coated on the resin with collodion solution to form an immunoadsorbent. Whole blood from SLE patients was perfused through a cartridge containing the immunoadsorbent. Clinical reports and observations of 25 patients (Tables 20.1 and 20.2) showed that the pathogenic anti-DNA antibody was removed at an average of 56%, immuno-complexes were sharply reduced and symptoms were relieved. Whole blood hemoperfusion with the immunoadsorbent is effective (90%) for treatment of SLE (Yang *et al.*, 1988; Gao *et al.*, 1995; Kong *et al.*, 1998; Yu *et al.*, 1998; Zhu *et al.*, 1999; Yu *et al*, 2001; Yu *et al.*, 1998; Yu *et al.*, 1998) and the commercialized cartridges (HA280) are now used in hospitals all over China (Zhang *et al.*, 2012; Xu, 2013). Recently, Lokman Uzun *et al.* (2010) developed poly(hydroxyethyl methacrylate)-based affinity membrane for *in vitro* removal of anti-dsDNA antibodies and found that DNA incorporation could significantly increase the membrane's adsorption capacity up to 68×10^3 IU/g; plasma antibody level decreased from 875 to 144 IU/mL. All in all, PHEMA-based membranes carrying

Table 20.1. Clinical Report of 25 Patients

Case Number	Sex	Age (years)	Period of Illness (years)	Contents of anti-DNA Antibody (%)		Percentage Removal of Anti-DNA Antibody (%)	Immune Complexes (%)	
				B-HP	A-HP		B-HP	A-HP
1	F	29	2.5	52	19	63.5	+	—
2	F	32	3.5	65	16	75.4	++	±
3	F	25	3	51	26	49.0	++	—
4	F	40	3	49	28	42.8	+	—
5	M	29	3	43	23	46.5	++	—
6	M	31	1.5	60	15	75.0	+	—
7	M	40	2	45	24	46.7		
8	F	36	2	51	28	45.1	23	11
9	F	40	6	28 g/L	9.0 g/L	67.8	+	—
10	M	28	5	34.2 g/L	9.6 g/L	71.9	+	—
11	F	23	0.5	22.0 g/L	6.6 g/L	70.0	–	—
12	F	23	0.5	36.0 g/L	8.6 g/L	76.1	+	—
13	F	37	0.5	9.9	7.8	21.2	62 μg/mL	32 μg/mL
14	F	42	3	12.4	10.6	14.5	125 μg/mL	125 μg/mL
15	M	24	5	19.8 g/L	9.2 g/L	53.5	+	—
16	M	30	6	28.2 g/L	9.1 g/L	67.7	+	—
17	F	33	1	12.9 g/L	7.6 g/L	41.1	94 g/L	57 g/L
18	F	29	0.83	15.3 g/L	8.7 g/L	43.1	86 g/L	53 g/L
19	F	30	0.75	14.3 g/L	8.3 g/L	42.0	79 g/L	47 g/L
20	F	34	1	11.1 g/L	2.8 g/L	74.8	86 g/L	24 g/L
21	F	27	0.5	10.7	3.0	72.0	90 g/L	30 g/L
22	F	26	4	45	1	97.8		
23	F	30	0.67	54.1	37.7	30.3	55.6	9.0
24	F	24	0.5	+	—		+	—
25	F	22	3	++	±		++	—

B-PH: before hemoperfusion; A-HP: after hemoperfusion; (++), (+), (±), and (-) represent intense positive, positive, slightly positive, and negative, respectively.

With copyright permission by publisher.

Table 20.2. Clinical Observation of 25 Patients

Case Number	Symptoms before Hemoperfusion	Symptoms after Hemoperfusion	Follow-Up Observation
1	Facial butterfly erythema, oliguria, edema	Facial erythema obviously relieved, edema disappeared	Follow-up survey for 1 year, could take care of her living arrangements, took 20 mg/day prednisone orally
2	Facial butterfly erythema, high urine protein, leg edema	Symptoms disappeared	—
3	Facial edema with pigmentation, hypodynamia	Edema disappeared, pigments relieved, physical strength enhanced	Follow-up survey for 8 months, perfused twice, returned to work without administration of medicine
4	Hypodynamia, facial butterfly skin lesion	Physical strength enhanced, facial skin lesion relieved	—
5	Hypodynamia, sleepy, facial pigmentation, eyelid edema	Edema and pigments disappeared, physical strength enhanced	—
6	Facial butterfly erythema, oliguria, leg edema	Edema disappeared, facial condition turned to normal	Follow-up survey for 18 months, could take care of his living arrangements, took 20 mg/day prednisone orally
7	Eyelid and leg edema, general hypodynamia, sleepy	Edema relieved, physical strength enhanced	Follow-up survey for one year, could take care of his living arrangements, with administration of medicine
8	Facial pigmentation, leg edema, hypodynamia	Physical strength enhanced, pigments relieved, edema disappeared	—

9	Fever for 5 years, facial butterfly spots, percussion pain at kidney parts, tenderness at livers and ribs parts	Body temperature returned to normal, facial pigments relieved, percussion pain disappeared	Follow-up survey for 18 months, could do light work, took 10 mg/day prednisone orally
10	Fever and general aching pain for 5 years, waist pain, facial butterfly erythema	Normal body temperature, general aching pain and waist pain disappeared	Lived for 2 years after perfusion, died of uremia
11	Often fever, general aching pain, facial erythema	Fever and aching pain disappeared	Follow-up survey for 14 months, could work in the fields, took 10 mg/day prednisone orally discontinuously
12	Aching pain, hypodynamia, facial edema, erythema, fever	Aching and hypodynamia disappeared, facial edema relieved	Follow-up survey for 4 months, symptoms were normal
13	Facial erythema, hypodynamia, waist pain, hypertension	Facial erythema disappeared, waist pain and hypodynamia relieved, normotension	Follow-up survey for 2.5 years, took Chinese medicine orally, patient's condition was stable
14	Obvious rash, hypodynamia, leg and joint pain, hypertension	Hypodynamia and joint pain relieved, normotension	Follow-up survey for 2 years, could do housework, took Chinese medicine orally
15	Limb pain, joints of fingers and toes low-grade deformation	Limbs move about freely, peripheral circulation improved	—
16	General aching pain, fever, chest distress, expectoration	Joint pain and expectoration relieved, chest distress disappeared	Follow-up survey for 20 months, could do light work, patient's condition was better, took 10 mg/day prednisone
17	Legs edema, cardiectasis, bad appetite, pruritus cutanea	Edema relieved obviously, appetite improved	—

(*Continued*)

Table 20.2. (*Continued*)

Case Number	Symptoms before Hemoperfusion	Symptoms after Hemoperfusion	Follow-Up Observation
18	Rash, pruritus cutanea, bad appetite, waist ache, hypodynamia	Symptoms relieved	—
19	Legs hypodynamia slight edema, headache, hypertension	Symptoms relieved obviously	Follow-up survey for 2 years, could do light work, patient's condition was stable, took 15 mg/day prednisone
20	Rash, bad spirit	Better spirit, better recovery	—
21	General edema, headache, rash, joint pain, moderately sleepless	Spirit getting better, symptoms relieved, can sleep	—
22	Edema, oliguria, renal function deterioration	Edema disappeared, urine volume increased, symptoms improved	Follow-up survey for 18 months, could work, patient's condition was better, took 15 mg/day prednisone
23	General edema, joint pain, cardiectasis, cardiac insufficiency, palpitation, shortness of breath	Edema relieved, joint pain disappeared, can sleep	Follow-up survey for more than 2 years, could take care of her living arrangements, took 15 mg/day prednisone
24	Facial and general erythema, frequent heart failure	Erythema disappeared, heart failure relieved	Follow-up survey for more than 2 years, could do light work, took maintainable dosage of medicine orally
25	Coma, frequent tic, general edema, blood pressure 28/16 kPa, heart rate 110 bpm, obvious ascites, incontinence of feces and urine	Tic stopped, conscious, blood pressure 22/13 kPa, heart rate 90/bpm, can do light work	Follow-up survey for 2 years, put on hemodialysis because of renal failure, committed suicide because of economic and mental suffering

DNA are endowed with a good selectivity and a high adsorption capacity including commercially available benchmark.

1.5. *Dextran Sulfate as Ligand*

The electro-negativity of a ligand on an adsorbent plays an important role in absorbing bio-macromolecules such as LDL (Wang *et al.*, 2013). Flexible spacers with a certain length can effectively reduce the steric hindrance and improve its adsorption capacity. In 1987, Behm *et al.* linked dextran sulfate on cellulose beads which selectively removed total cholesterol, TG and LDL with a low adsorption capacity for HDL. The high adsorption capacity for LDL-C may be attributed to the electro-negativity of the sulfonic group on dextran sulfate molecule, and its backbone plays a flexible spacer role in the adsorption process (Fig. 20.6).

At present, dextran sulfate adsorbent is successfully used for clinical treatment of familial hyperlipidemia (Kobayashi *et al.*, 2002); Suzuki *et al.* (2000) reported that the same cartridge could efficiently

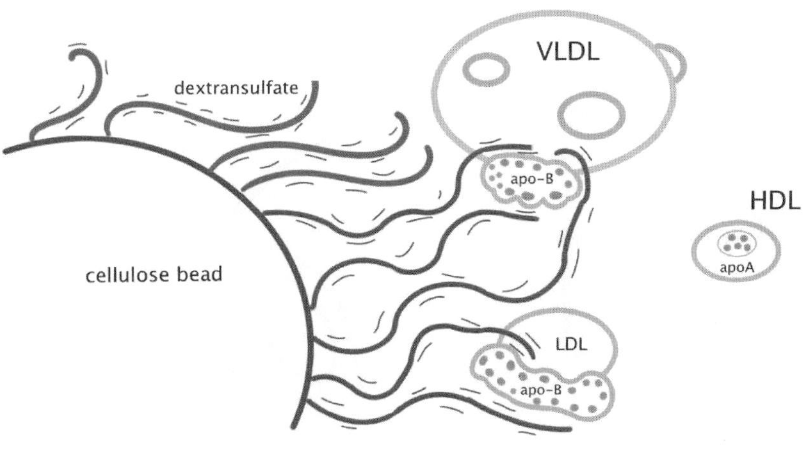

Fig. 20.6. Schematic diagram of the adsorption mechanism of dextran sulfate based adsorbent.

(*Source*: Kutsuki *et al.* (1998) Therapeutic selective adsorption of anti-DNA antibody using dextran sulfate cellulose column (Selesorb) for the treatment of systemic lupus erythematosus. *Ther Aperesis* **2:** 18–24, adapted. With copyright permission from Wiley Press.)

absorb both anti-Anx V antibodies and Anx V from serum in the treatment of SLE patients. Hidekaet *et al.* (1999) demonstrated that this type of adsorbent could also be used to remove a series of anti-nuclear antibodies (anti-SS-A, SS-B, Sm, Scl-70 and RNP antibody), therefore, it may also be applied to treat Bullous Pemphigoid and RA diseases (Braun *et al.*, 2002; Matsuki *et al.*, 1996; Ino *et al.*, 1997; Kutsuki *et al.*, 1998).

1.6. *Synthetic or Recombinant Peptide as Ligand*

Mutation studies (Bogan *et al.*, 1998; Clackson, 1995) indicated that only a fraction of hot spots on protein interfaces are responsible for most of the binding free energy of the protein complexes, and the hot spots compromise some short peptides (Fig. 20.7).

Therefore, if the hotspot of a certain protein is identified and confirmed, then their amino acids sequence (peptide), can be used as ligands to absorb certain proteins specifically. Based on this theory, peptides can be used as ligands in immunoadsorbents for

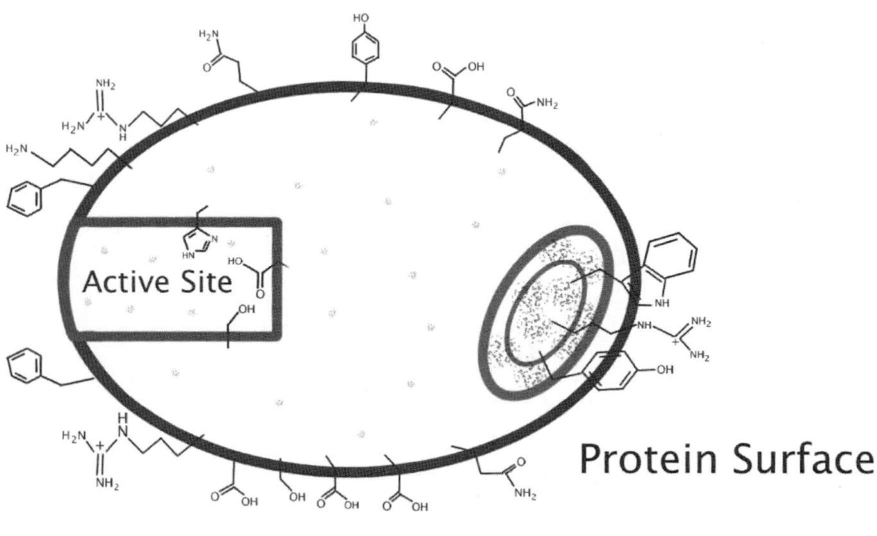

Fig. 20.7. Schematic diagram of hot spots on surface of protein. (*Source*: Li JH, PhD dissertation, 2004).

Lys-Lys-Gly-Trp-Lys-His-Trp-Val-Tyr-Tyr-Thr
183
-Cys-Cys-Pro-Asp-Thr-Pro-Tyr-Leu-Asp-Lys-Lys-Gly
200

Fig. 20.8. The structure of the segment 183–200 of alpha sub-unit.

Table 20.3. Removal Rate of Anti-AChR Antibodies and Plasma Proteins in Clinical Study

Anti-AChR Antibody	Removal Rate %
Blocking antibody	40.2
Binding antibody	12.4
Plasma protein	—
Total protein	5.8
Albumin	5.0
IgG	7.3
IgM	11.6
IgA	9.8

Nakaji S, Hayashi N. (2003) Adsorption Column for Myasthenia Gravis Treatment: Medisorba MG-50. *Therapeutic Apheresis and Dialysis* 7(**1**): 78–84. With copyright permission from Wiley Press.

hemoperfusion. Nakaji *et al.* (2003) found that the segment of 183–200 of the alpha sub-unit (Fig. 20.8) composing AChR is the site at which the blocking antibody attacks, and further discovered that torpedo alpha-183–200 peptide has a stronger affinity in blocking the antibody of MG patients than human alpha-183–200 peptide. Therefore they developed MG-50 immunoadsorbent with torpedo alpha-183–200 peptide as the ligand. Adsorption results indicated that MG-50 could effectively remove AChR specifically without loss of useful plasma components (Table 20.3).

Recent studies (Karlin, 2002; Tzartos, 2008) indicated that AChR consists of •, •, •, and • subunits (embryonic type) or •, •, •, and • subunits (adult type) in the stoichiometry of ~2•••(•). The subunits consists extracellular domain (ECD) of ~210 amino acids which carries the main epitopes for MG auto-antibodies. Based on this finding, Zisimopoulou *et al.* (2008) expressed the extracellular domains (ECDs) of all human muscle AChR subunits in *E. coli* and immobilized recombinant peptides on CNBr-Sepharose, then used it for immunoadsorption assays in MG sera. Adsorption results showed that this immunoadsorbent could induce AChR reduction in cell cultures and experimental MG rats, therefore immunoadsorbents based on ECD can be used as an efficient, antigen-specific therapy for MG. Tzartos *et al.* (2008) used ECD (expressed by yeast) as a ligand to remove AChR, and found that its adsorption capacity was higher than that extracted from *E. coli*.

For immunoadsorbents, on the one hand, they should have high specificity for immunoglobulin, on the other hand, they will not cause immunogenicity, which is essential for immunoadsorption treatment of autoimmune diseases. Kumiko Kawabata *et al.* (2014) developed a novel peptide-based immunoadsorbent for the removal of IgG in blood of rheumatoid arthritis patients. He grafted the IgG–Fc binding epitopes of Protein A onto a newly designed helix-loop peptide. Linear (linG7A5) and cyclic (cyG7A5) grafted peptides were synthesized to test their binding affinity and specificity. Results showed that peptide cyG7A5 had a high specificity for human IgG–Fc and no affinity to other plasma proteins, human serum albumin or fibrinogen. Therefore, it can be concluded that cyG7A5 based immunoadsorbent has a higher adsorption capacity and specificity for IgG than adsorbent with Protein A as ligand. In summary, because peptide linked immunoadsorbent has many advantages, e.g. high selectivity, high adsorption capacity, long shelf life, easy sterilization, low or no immunogenicity, therefore it has a high potential commercial interest in the future.

1.7. *LDL Antibody as Ligand*

Hemoperfusion is currently employed for the treatment of hyperlipidemia and serious cardiovascular disease when the reduction of low-density lipoprotein (LDL) in the patient's plasma appeared impossible to achieve via diet control or drug administration (Cao *et al.*, 2002; Sato *et al.*, 1996; Chen *et al.*, 1993). Adsorbents with LDL antibody

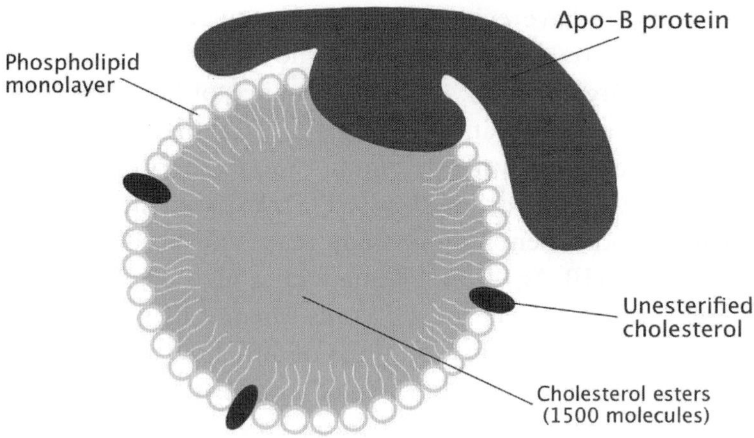

Fig. 20.9. Schematic diagram of low density lipoprotein.

as ligands can selectively absorb LDL because anti-ApoB100 antibody is located on the surface of the LDL molecule (Fig. 20.9), which is also responsible for the low adsorption of HDL and other plasma components.

Stoffel *et al.* constructed sepharose®-based immunoadsorption system with anti-apoprotein B-antibodies as ligands for the removal of LDL in the treatment of hypercholesterolemia (1981). This system has been used for more than 3,000 successful treatments in familial hypercholesterolemia patients by Borberg *et al.* (1988). Pokrovsky *et al.* linked anti-LDL on sepharose beads for the clearance of LDL from human plasma in 1990. In order to further improve the blood compatibility of the adsorbent, Yavuz *et al.* prepared a LDL adsorbent with poly(2-hydroxyethyl methacrylate--methacryloyl-amidophenylalanine) as the carrier and anti-LDL as the ligand. The adsorption capacity of LDL reached 16.0 mg/g in hypercholesterolemia human plasma.

1.8. *Anti-HBs as Ligand*

Hepatitis B virus (HBV) infection and its associated diseases are an important global problem — 75% of the world's population are living in areas where there are high levels of infection. More than two billion people worldwide have evidence of past or current HBV infection, and 350 million are chronic carriers of the virus (Uzun *et al.*, 2009; World Health Organization, 2008). Clearance of hepatitis B virus is conducive to treating hepatitis B. In 1978, Dodd *et al.* developed an immunoadsorbent by covalently binding mono-specific anti-hepatitis B (HBs) on controlled porous glass (CPG). Experimental results indicated that the adsorbent can strongly react and absorb hepatitis B surface antigen (HBsAg) from plasma with a high degree of specificity. It was estimated that the CPS adsorbent can specifically bind 32,000 passive hemagglutination units of anti-HBs per 100mg CPG beads. However, due to the high cost of anti-HBs, this type of immunoadorbent encountered some difficulties in clinical therapy. Because an imprinted adsorbent can be used multiple times, Uzun *et al.* synthesized a functional monomer via a reaction between methacryloyl chloride and l-tyrosine methyl ester (MAT), then a pre-organized complex was prepared with HBs and MAT in different ratios. The imprinted adsorbent can absorb HBsAb molecule 18.3 times more than anti-HAV antibody and 2.2 times more than IgE molecules as ligand.

1.9. *Immunoglobulin as Ligand*

The basic structure of immunoglobulins is illustrated in Fig. 20.10. Although different immunoglobulins can differ structurally, they all are built from the same basic units. The immunoglobulins are generally classified into five different types (IgG, IgM, IgA,IgD and IgE) according to differences of the amino acid sequences in the constant region of the heavy chains (Fig. 20.11). All immunoglobulins within a given class have very similar heavy chain constant regions. These differences can be detected by sequence studies or more commonly by serological methods.

Immunoglobulins can also be classified into subclasses based on small differences of amino acid sequences in the constant region of the heavy chains. All immunoglobulins within a subclass can have very similar heavy chains in the constant region.

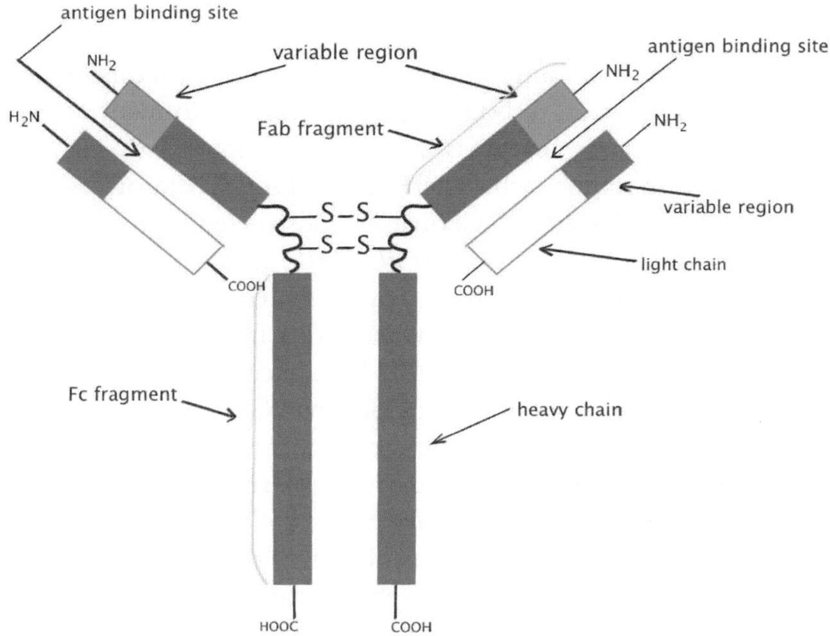

Fig. 20.10. Schematic diagram of immunoglobulin structure.

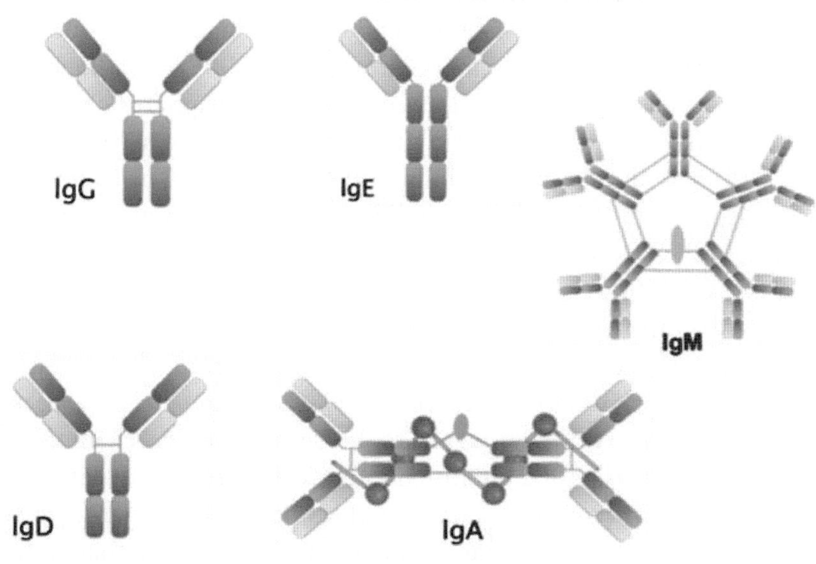

Fig. 20.11. Schematic diagram of five types of immunoglobulin.

1. IgG — Gamma heavy chains 2. IgM — Mu heavy chains 3. IgA — Alpha heavy chains 4. IgD — Delta heavy chains 5. IgE — Epsilon heavy chains

1.10. *IgG as Ligand*

IgG is the main iso-type antibody found in blood and extracellular fluid, which controls infection of body tissues, by binding to many kinds of pathogens, i.e. viruses, bacteria, and fungi (Junqueira & Carneiro, 2003). Based on unique characteristics of IgG, Therasorb Medical Systems (Unterschleissheim, Germany) constructed a Ig-Therasorb plasma perfusion column (Sueoka *et al.*, 2001) where polyclonal sheep anti-human IgG antibodies were covalently coupled to sepharose (Nakaji *et al.*, 2001). The polyclonal antibodies from sheep can specifically bind IgG (subclasses 1–4), IgM, IgA, circulating immune complexes, and fragments of immunoglobulins. Gaubitz *et al.* (2003) applied this column to treat various autoimmune diseases such as SLE and RA, etc. in Europe. The column can be regenerated by glycine buffer under control of an automatic adsorption-desorption device (ADA/Medicap, Ulrichstein). Therasorb is used to treat severe systemic lupus erythematosus and can effectively reduce pathogenic antibodies and disease activity under persistent, unchanged immunosuppressive treatments. The clinical efficacy of immunoadsorption is satisfactory (IgG > 97% was eliminated, IgM and IgA were reduced to 30%) and slightly superior to the published average percentage of responders in controlled plasmapheresis studies (Gaubitz *et al.*, 2003). In addition, no severe side effects occurred in the course of plasmapheresi, especially infections and allergic reactions of substitution fluids (Stummvoll *et al.*, 2005). It should be noted that the reusable twin pair columns of Ig–Therasorb (US$17,000) are significantly more expensive than the disposable PH-350 column (polyvinyl alcohol beads with phenylalanine as ligands, US$600) (Gaubitz *et al.*, 1998).

It was found that IgG can aggregate and form a soluble polymer at a certain temperature. In the course of aggregation, the configuration of IgG can transform some potential active sites into new active sites which cause IgG aggregates to be more reactive than that of single IgG (Fu *et al.*, 2002). Chunxiao Fu *et al.* synthesized an immunoadsorbent by linking heat-aggregated human IgG to agar gel beads for the treatment of rheumatoid arthritis (RA), which has a high adsorption capacity for RF and immunocomplexes (IC). The adsorption capacity for IgG, IgM, IgA was 74.8%, 84% and 87.3%, respectively. Almost at the same time, Jun Dai *et al.* (2001) synthesized a cellulose bead type RA adsorbent with heat-aggregated human IgG as ligand. Cellulose beads were epoxidized then coupled to heat-aggregated human IgG (HAH IgG). *In vitro*, the removal percentage of RF (IgG, IgM, IgA) was 69.6%, 70.3% and 77.0%, respectively. Following the work of Fu and Dai (Fu *et al.*, 2002; Dai *et al.*, 2001), Jiaqiang Chu (2003) used glycidyl 1, 4 butanediol as a spacer to synthesize the adsorbent, aiming to reduce the steric hindrance of the bulky reaction molecules. *In vitro* studies showed that the immunoadsorbent with glycidyl 1,4 butanediol as a spacer could absorb 60% of IgA in plasma, which was much higher than the CDI coupling method (35%). This is attributed to the longer spacer effect of glycidyl 1, 4 butanediol than CDI.

1.11. *IgE Antibody as Ligand*

Allergy is an immunologic system disorder, affecting approximately 20–30% of the general population and brings disability and productivity loss throughout the world (Casolaro *et al.,* 1996; Duarte *et al.*, 2006). Evidence demonstrated that IgE mediated allergic reaction in acute anaphylactic responses and allergic emergency cases (Bischoff *et al.*, 2007), extracorporeal perfusion treatment for selective removal of IgE can effectively alleviate allergic syndrome. Lebedin *et al.* (1991) developed an immunoadsorbent with anti-IgE as ligands by covalently immobilized monoclonal antibodies anti-IgE on Sepharose CL-4B. Seventeen clinical IgE apheresis treatments of five atopic asthma patients indicated that immunoadsorbent synthesized by Lebedin could effectively remove IgE from plasma. In addition, the adsorbents can also be used to treat allergic rheumatoid arthritis. Unfortunately, anti-IgE antibodies are expensive and difficult to purify (Sato *et al.*, 1989; Gorchakov *et al.,* 1991), which limited the application of the adsorbent. Many researchers tried to develop simpler and less expensive ligands to replace anti-IgE antibodies (Duarte *et al.,* 2006). In 2008, Laffer *et al.* synthesized IgE immunoadsorbent by linking monoclonal antibody 12 (mAb12) from IgE on sepharose beads which could efficiently remove IgE and IgE-bearing cells from peripheral blood streams (Laffe *et al.*, 2008).

1.12. *IgA as Ligand*

IgA nephropathy is a kidney disorder caused by deposits of IgA inside the glomeruli of the kidney. This chronic disease may progress over a period of 10 to 20 years and usually cannot be cured effectively. Immunoadsorbent can remove excess IgA by extracorporeal hemoperfusion. Jiaqiang Chu *et al.* (2004) activated agarose beads with epichlorohydrin then reacted with glutaraldehyde to introduce a spacer and finally linked sheep anti-human IgA antibody as the ligand. *In vitro* experiments showed that the adsorbent thus prepared could effectively remove excess IgA in plasma up to 74%.

1.13. *C1q as Ligand*

C1q is a 400kDa protein formed from 18 peptide chains in three subunits of six. Each six peptide subunit consists of a Y-shaped pair of triple peptide helices joined at the stem and ending in a globular non-helical head (Wikipedia) (Fig. 20.12). The unique C1q complex structure can be multivalent attached to the complement fixation sites of immunoglobulin. Systemic lupus erythematosus (SLE) is a typical autoimmune complex mediated disease characterized by flares, organ manifestations and auto-antibodies against nuclear antigens. Based on this principle, Gazitt *et al.* (1985) developed an immunoadsorbent using agarose-polyacrolein microsphere beads as carrier and Clq as ligand for the removal of circulating immune complexes in autoimmune diseases and neoplastic diseases. In 1999, Hiepe *et al.* synthesized an immunoadsorbent using C1q as multifunctional ligand to remove immune-complexes from plasma of SLE patients. C1q can selectively bind the Fc part of IgG and IgM via its globular region. Results showed that this type adsorbent could effectively remove IgG and IgM complexes, inflammatory proteins such as CRP, fibrinogen,

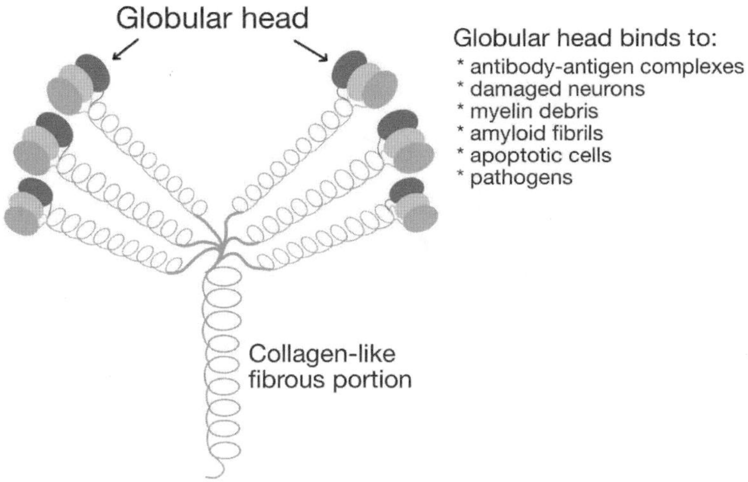

Fig. 20.12. Schematic structure of C1q.

fibronectin, DNA, C-reactive protein, nucleosome and anti-C1q-IgG auto-antibodies (Hiepe *et al.,* 1999). In addition, the European Consensus Lupus Activity Score (ECLAM) showed a statistically significant reduction of immune complexes in seven of eight patients. A decrease of circulating immune complexes (IgG) was observed in five of eight patients, C1q-bearing-IC was reduced significantly in seven of eight patients (Pfueller *et al.*, 2001).

1.14. *Heparin as Ligands*

Heparin, also known as unfractionated heparin, is a highly sulfated glycosaminoglycan which is widely used as an injectable anticoagulant (Fig. 20.13). It has the highest negative charge density of any known biological molecule (Wikipedia). Heparin is well known as a good biocompatible material and can be immobilized as an affinity ligand in protein purification and hemoperfusion. It can be immobilized on plastic surface by coating method and on various kinds of resins by covalently binding methods. Physical adsorption of heparin leaks off the resin surface easily. Chemical linking can make the heparin bind strongly to the resin surface (Huang *et al.*, 2011).

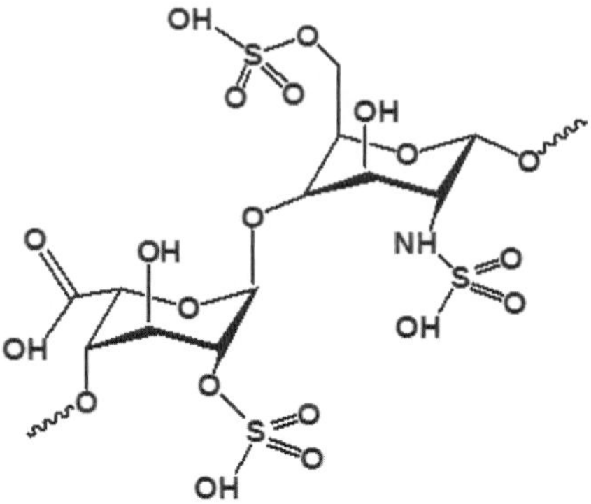

Fig. 20.13. Structure of heparin.

Huang *et al.* (2011) used an atmospheric pressure glow discharge (APGD) generator under ammonia/argon gas atmosphere to activate polysulfone (PSF) surface, then chemically bind heparin molecules in EDC/NHS solutions. ATR-FTIR and XPS results demonstrated that the surface has been modified and CA and TB colorimetric tests suggested an improvement in the hydrophilicity of heparinized PSF. Platelet adhesion tests showed good biocompatibility. In addition, heparinized PSF membrane has a good mechanical strength and absorbs LDL-C, but the adsorption capacity was too low (40 ng/cm^2). In 2012, Xiaodong Hou *et al.* developed a porous low density lipoprotein (LDL) adsorbent membrane by ^{60}Co γ-ray irradiation-induced grafting copolymerization of polypropylene (PP) non-woven fabric with acrylic acid, followed by immobilizing heparin covalently. Static adsorption and hemoperfusion tests demonstrated the adsorbent could efficiently and selectively remove LDL from human plasma. The adsorption capacity for LDL-C, TG, TC, and HDL-C was 33.3 ± 2.9 μg/cm^2, 202.4 ± 5.7 μg/cm^2, 64.9 ± 4.3 μg/cm^2 and 14.7 ± 1.9 μg/cm^2, respectively. The adsorption capacity of the adsorbent synthesized by Hou improved greatly, but its adsorption for HDL-C was also high. The above result may be attributed to the following factors: (1) The structure of human LDL shows that cholesterol is rich under the surface of phospholipid monolayer and in the cores of the LDL micro-particles. (2) Macropores of the adsorbent are conducive to enhance the adsorption capacity of free cholesterol and TG. (3) Synergistic effects of hydrophilicity and hydrophobicity can improve the specificity adsorption of the adsorbent (Wang *et al.*, 2008). Xiaodong Hou *et al.* (2014) prepared an amphiphilic macroporous polymeric membrane for the removal of LDL-C. Polypropylene (PP) nonwoven fibers were irradiated with acrylic acid under ^{60}Co γ-ray, followed by binding cholesterol via different length spacers. The amphiphilic (AMPA) membrane showed a high adsorption capacity for LDL-C and TG of 62.8 ± 3.8 μg/cm^2 and 132 ± 463.0 μg/cm^2, respectively. In addition, Ma *et al.* (2011) also synthesized some similar macroporous resin containing heparine as ligand.

1.15. *Poly-L-Lysine as Ligand*

The molecular structure of L-lysine and poly-L-lysine are shown in Fig. 20.14. As ligands, they are both positively charged and can easily react with DNA and other pathogenic molecules containing negative charges such as endotoxin. Poly-L-lysine has been widely used as a gene carrier because of its excellent DNA condensing ability (Senel *et al.*, 2003). In 2003, Senel *et al.* synthesized poly(2-hydroxyethyl methacrylate) membrane with poly-L-lysine as ligand. pHEMA membrane was prepared by UV initiated photo-polymerization and

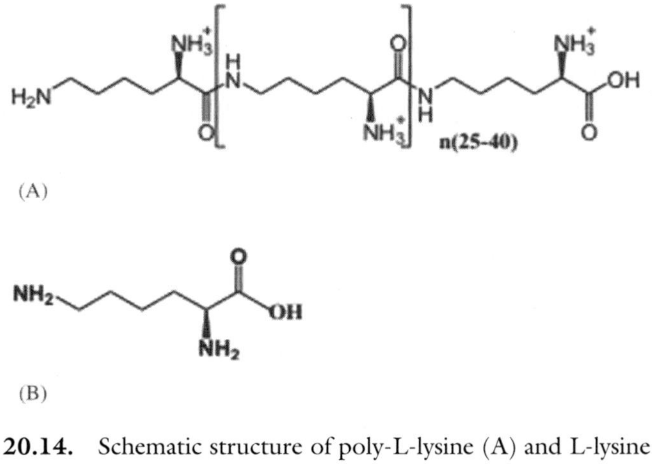

Fig. 20.14. Schematic structure of poly-L-lysine (A) and L-lysine (B).

poly-L-lysine was covalently linked on the membrane after activation with epichlorohydrin. Then DNA was ionically bond to the membrane (5,849 mg/m^2), which could be used to remove anti-DNA antibodies in autoimmune diseases.

1.16. *Polyanion as Ligand*

Since Behm *et al.* developed dextran sulfate based cellulose adsorbents to remove LDL in 1987, many researchers have tried different types of polyanion with various chemical structures to find the optimal ligand that is most suitable for the removal of LDL in human blood or plasma. Tani (2000) selected a series of synthetic polyanion compounds including polystyrene sulfonic acid, PVA sulfate, polyvinyl benzoic acid, polyacrylic acid, alginic acid and dextran sulfate to absorb LDL. He found that polyanions with sulfonate groups and hydrophilic backbones have a strong affinity for LDL, while the carboxyl group has a poor affinity for LDL. Therefore, dextran sulfate based cellulose adsorbent is regarded as one of the optimal adsorbent for LDL.

1.17. *4-Mercaptoethylpyridine as Ligand*

Mercaptoethylpyridine is a typical hydrophobic reagent under certain PH values (Fig. 20.15). Ren *et al.* (2009) found that MEP has a noncharged structure in neutral condition (pKa 4.8), but becomes positively charged when the pH is decreased to 4.8. This characteristic of the ligand enables the solid phase to bind proteins via hydrophobic interaction at neutral and low salt concentrations. (Corneillie *et al.*,

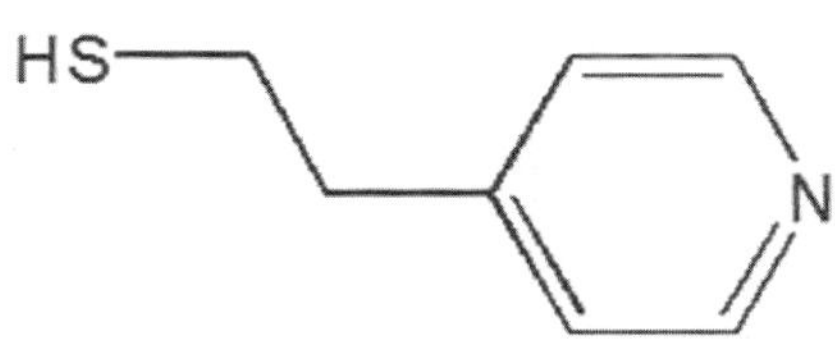

Fig. 20.15. Schematic structure of 4-mercaptoethylpyridine.

2004; Bak *et al.*, 2007; Dux *et al.*, 2006; Schwartz *et al.*, 2001). Ren *et al.* synthesized MEP-Sepharose adsorbents according to the method used by Burton and Harding (1998), with some modifications. The adsorbent could effectively remove RA factors and low nonspecific components from plasma. For detailed information about this type of immunoadsorbent, refer to Ren *et al.* (2009).

1.18. *Cell Membrane Proteins as Ligand*

Zenografts may help to alleviate the shortage of organs of human origin for transplantation (Lucchiari *et al.*, 1997). Hyper-acute vascular rejection remains the main barrier for the success of vascularized discordant xenografts. In the widely used plasmapheresis technique, adsorption is nonspecific; not only the pathogenic toxins are removed but antibodies are also depleted (Alexandre *et al.*, 1989). In order to improve the specificity, Lucchiari *et al.* (1997) synthesized Sepharose gel type beads as the carrier and porcine endothelial cell membrane or platelet membrane was used as the ligand. *In vivo* studies on animals showed that 97% of anti-porcine IgG antibodies and 96% of anti-porcine IgM antibodies were removed, respectively. At the same time, cytotoxic antibodies were also depleted by 85%. It should be noted that this type of immunoadsorbent is expensive, which may limit its further development and clinical trials.

1.19. *Avidin as Ligand for Cancer Therapy*

Extracorporeal immunoadsorption (ECAT) has been demonstrated as a new therapeutic method in radio-immuno targeting tumors, by removing excess radio-labeled monoclonal antibodies (MAbs) from blood, thus reducing the activity to radiosensitive organs and improving the tumor-to-normal-tissue uptake ratio (Wang *et al.*, 2002). Wang *et al.* (2002) used avidin-agarose adsorption column to remove radioactive substance in rat circulating blood, after administration of Indium^{111}–HMFG1–biotin. Animal experiment was conducted as follows: Forty-three rats of F1 breed of Brown Norway (BN) and Wistar Furth (WF) were used in the experiments. Thirty rats served as controls and were injected intraperitoneally with approximately 150 μg of non-biotinylated (n = 12) and biotinylated (n = 18) ^{111}In-HMFG1 (4.5–5.5 MBq per rat). Thirteen rats underwent ECAT starting 12 hours ($n = 11$) and 18 hours ($n = 2$) after intraperitoneal injection of ^{111}In-HMFG1-biotin. Adsorption results showed that the binding capacity of ^{111}In-HMFG1-biotin to avidin was high. Biotinylation did not enhance the excretion of HMFG1. When ECAT was employed, the WB and blood radioactivity were reduced by 35–40% ($P < 0.05$) and 75–86% ($P < 0.01$), respectively. After the completion of ECAT, the activity uptake in organs was significantly decreased, which suggested that radio-activity adsorption is beneficial to clear the excess radio labeled MAbs in the treatment.

1.20. *Receptor for AGE (RAGE) as Ligand*

Diabetic vascular complications are the leading causes of end-stage renal failure, acquired blindness, a variety of neuropathies, and accelerated atherosclerosis, which could account for the disabilities and high mortality rates found in diabetic patients (Yamagishi *et al.*, 2005). Long-term prospective studies have demonstrated that high serum concentration of advanced glycation end products (AGE) can predict cardiovascular diseases (CVD), deaths of diabetes patients and independent traditional CVD risk factors (Kilhovdet *et al.*, 2007; Nin *et al.*, 2011). Zhang *et al.* (2014) synthesized an immunoadsorbent for the removal of AGE by covalent binding of receptor AGE (RAGE) which was extracted from *Escherichia coli* and purified by chromatography onto agarose beads. Adsorption results demonstrated that RAGE-receptor-based immunoadsorbent can effectively remove AGE from human blood circulation.

1.21. *scFv as Ligand*

Fragment viable (Fv) is the smallest unit of immunoglobulin molecule which functions in antigen-binding activities. An antibody in single chain fragment variable (scFv) (Fig. 20.16) consists of variable regions

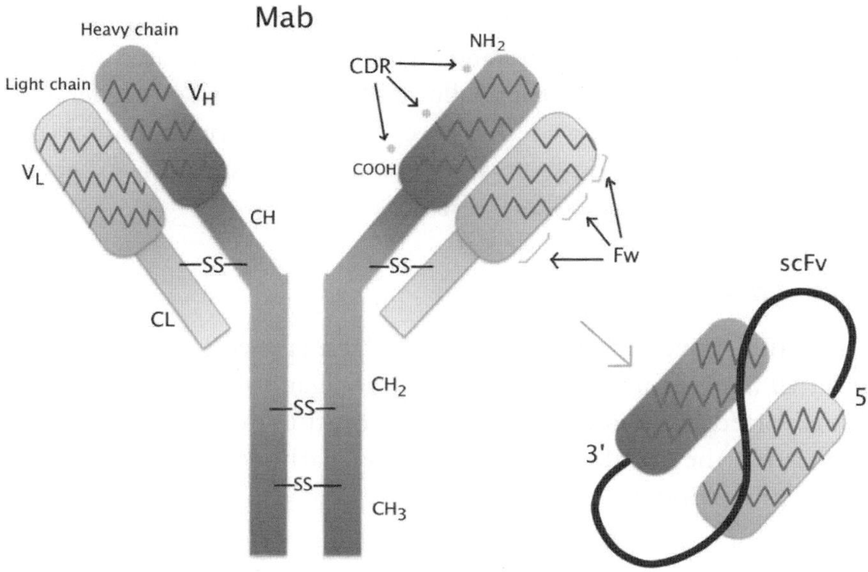

Fig. 20.16. Antibody model showing subunit composition and domain distribution along the polypeptide chains. Single-chain fragment variable (scFv) antibody generated by recombinant antibody technology appears in the shaded area.

of heavy and light chains, which are joined together by a flexible peptide linker. It can be easily expressed in functional form in *E. coli*, allowing protein engineers to improve the properties of scFv, such as increase of affinity and alteration of specificity (Griffiths *et al.*, 1998).

A key feature of dialysis-related amyloidosis (DRA) is the formation of amyloid fibrils consisting primarily of β_2-microglobulin (β_2m) and to a lesser degree glycated β_2m or advanced glycation end products (AGE-β_2m). Except for kidney transplantation, conventional kidney replacement therapies do not adequately address the removal of elevated concentrations of β_2-microglobulin or AGE removal from blood. In 2007, in order to construct a new type of immunoadsorbent for the treatment of chronic kidney disease (CKD), Daniels developed an anti-human β_2m single-chain variable region antibody fragment (scFv) and confirmed that it could specifically combine with β_2m. He then synthesized and immobilized scFv onto agarose beads and measured its binding capacity for β_2m, thermal stability, regeneration capacity, storage stability and sterility. Results indicated that the adsorbent thus made was thermally stable and could be regenerated. Adsorption results showed that it could quantitatively reduce the concentrations of AGEs, β-amyloid, etc. in plasma samples of diabetic patients with end-stage kidney disease.

1.22. *Other Ligands*

Beside the above mentioned ligands, which are widely used to construct all kinds of immunoadsorbents, the ligands in Table 20.4 were also reported in the literature. Static adsorption experiments demonstrated that they could effectively remove pathogenic toxins existing in patient's blood.

2. Mechanism and Features of Immunoadsorbent

An immunoadsorbent is the immobilization of an antigen on a solid matrix which binds selectively with the corresponding antibody. It is a specific interaction between the antigen and antibody. The determinant of the antigen epitope immobilized on the adsorbent can "recognize" well the binding site of the antibody or paratope which is located at the end of the arm of the Y-shaped structure antibody that contains various amino acid sequences (Fig. 20.17).

The determinant of the antigen generally fits well and binds closely with the binding site of the antibody which contributes to the high selectivity of the antigen–antibody interaction. The binding force between the antigen and the antibody is weak and generally involves non-covalent bonds such as electrostatic interactions, hydrogen bonds, van der Waals forces and hydrophobic interactions (Winchester *et al.*, 2003). Apart from the mechanism there are many important features of immunoadsorbents which are described as follows:

Specificity: High selectivity is a unique property or key parameter of an immunoadsorbent. To show the selectivity of the antigen–antibody reaction, Terman *et al.* (1976) established an extracorporeal

Table 20.4. Other Ligands used in Therapeutic Extracorporeal Plasmapheresis

Ligand	Biological Entity Absorbed	Selected Applications	Reference
Insulin	anti-insulin antibody		Burnouf *et al.*, 1998; Sawada *et al.*,1990
Factor VIII	anti-FVIII antibodies	Hemophilia A	Burnouf *et al.*, 1998; Sawada *et al.*, 1990
Factor IX	anti-FIX antibodies	Hemophilia B	Burnouf *et al.*, 1998; Sawada *et a*l.,1990
Oligosaccharide	Anti-blood type antibodies		Burnouf *et al.*, 1998; Sawada *et al.*, 1990
Anti-αfetoprotein antibodies	-αfetoprotein		Burnouf *et al.*, 1998; Sawada *et al.*, 1990
-αNCI Type IV collagen	Anti-αNCI; type IV	Goodpasture syndrome	Burnouf *et al;* 1998; Sawada *et al.*, 1990
Antithrombin III			Burnouf *et al.*, 1998; Sawada *et al.*, 1990
1α-Antitrypsin		Goodpasture syndrome	Burnouf *et al.*, 1998; Sawada *et al.*, 1990
Anti-human Protein C	Protein C		Burnouf *et al.*, 1998; Sawada *et al.*, 1990
Terminal trisaccharide A or B blood group antigen	Anti-A antibody Anti-B antibody	ABO-incompatible organ transplantation	Sanchez *et al.*, 2013
Hexadecyl group of resin	•-microglobulin	Dialysis-related Amyloidosis (DRA)	Suzuki *et al.*, 2003
von Willebrand factor		Willebrand disease	Burnouf *et al.*, 1992, 1998
Cu^{2+} (loaded porous chitosan particles)	IgG, IgM, IgA	Rhumatoid arthritis	Yue *et al.*, 2008;

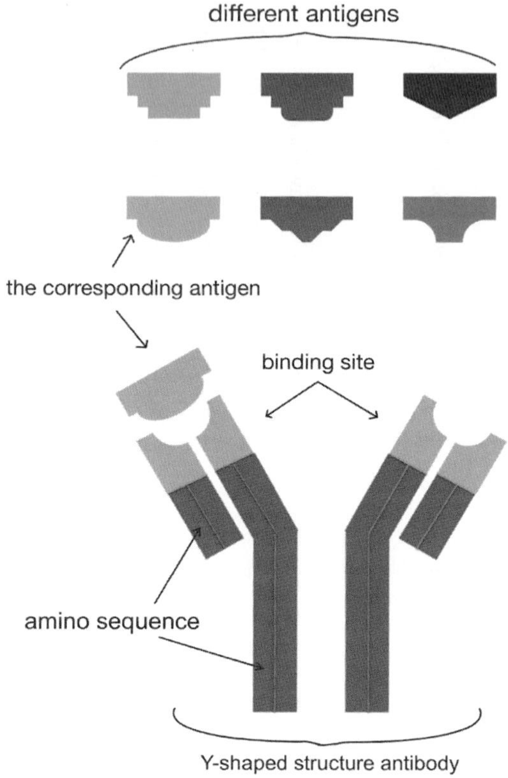

Fig. 20.17 Schematic diagram of antigen interacting with antibody.

circulating system in which ^{125}I-BSA was perfused over immobilized anti-BSA with γ-globulin as control. After two hours of perfusion, 89.6% greater uptake of ^{125}I-BSA than the control was achieved. Similar results were obtained by the same author in the immunoadsorption of DNA-antibodies by immobilized DNA in the extracorporeal system for the therapy of systemic lupus erythematosus. The principles of specificity of the antigen–antibody interaction are now widely used in clinical diagnosis and therapy. Although selectivity is the unique property of immunoadsorbents, there are several other important features or key parameters that governs the adsorption processes, namely surface area, pore diameter and adsorption kinetics.

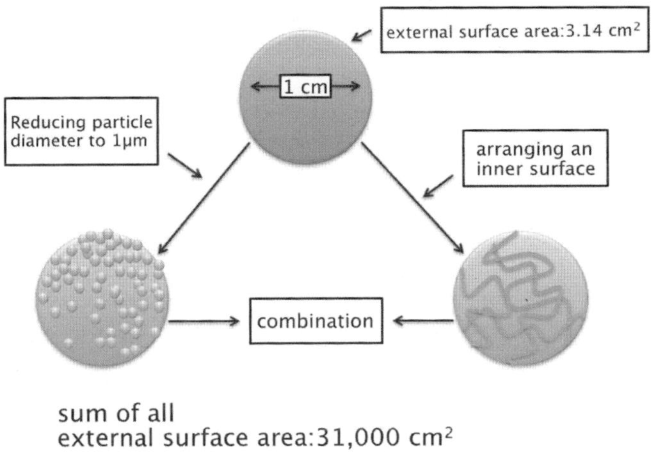

Fig. 20.18. Strategy for increasing the surface area of adsorption particles.

Source: Falkenhagen *et al.* (2002) Patient Safety Technology for Micro-adsorbent Systems in Extracorporeal Blood Purification. Artificial Organs **26(2):** 84–90, with copyright permission from Wiley Press.

Surface area: The surface area of an adsorbent remarkably affects its adsorption capacity. The capacity increases greatly as the area increases. It can be classified into external and internal areas. Falkenhagen *et al.* (2002) compared the external surface areas of large and small spherical particles (Fig. 20.18). Spheres of 1 cm in diameter display an outer surface area of 3.14 cm^2 whereas spheres of only 1 μm in diameter have 31,000 cm^2 at the same volume.

2.1. *Comparison of Surface Area for Small and Large-Size Particles*

Based on the external surface area, it is conceivable that small particles show a pronounced effect on the adsorption process, highly enhancing the adsorption capacity of an adsorbent. At the same time, the diffusion rate is much higher with small particles than with large particles (Fig. 20.19). The first clinical use of a micro-adsorbent system was designed by Ash (1992); this was commercialized as BioLogic DT. Principally, it is considered that extracorporeal blood purification system based on micro-adsorbent offers higher efficiency

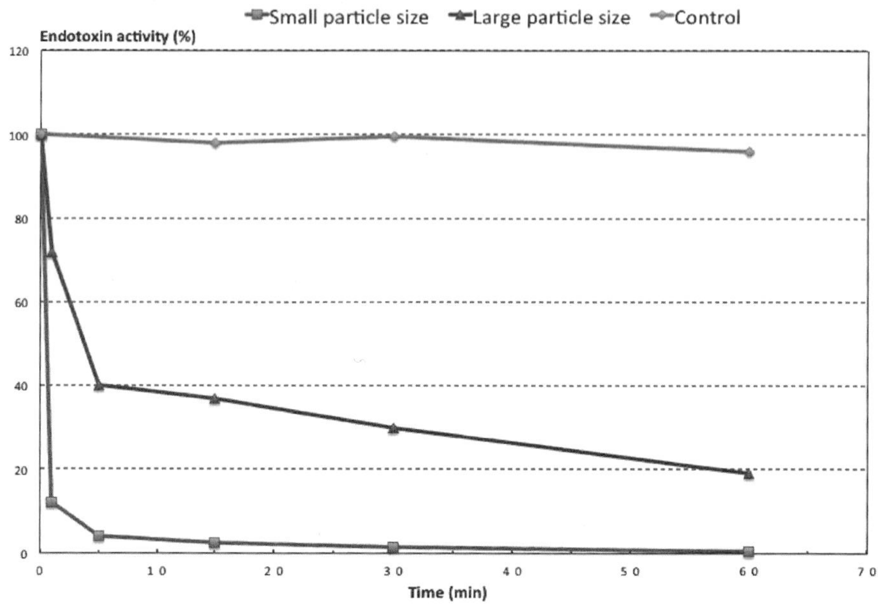

Fig. 20.19. Comparison of small (2-5μm) and large (150–300 μm) particle for endotoxin adsorption.
Source: Falkenhagen *et al.* (2002) Patient Safety Technology for Microadsorbent Systems in Extracorporeal Blood Purification. *Artificial Organs* **26(2):** 84–90, with copyright permission from Wiley Press.

and flexibility but is much more complicated than a conventional column-based adsorption system. Apart from the external surface area of a particle, the internal surface likewise contributes strongly to the overall surface area of a porous material, such as in macroporous resins. One gram of porous polystyrene beads (diameter ~ 300 μm) can be synthesized to have an internal surface area of more than 1,000 m^2 while its external surface area is less than 1 m^2. This type of adsorbent is used clinically for over-dose drug detoxification in China (Chen *et al.*, 1983). Therefore the adsorption capacity of a porous material has a greater advantage over non-porous substances. The immunoadsorbents mentioned in this review also have a high internal surface area, which contributes to their high removal capacities for pathogenic toxins.

2.2. *Pore Size*

Pore size of the adsorbent plays an important role where the adsorbates have different molecular weights. In general, carriers can be prepared to have three sizes of internal pores. In principle, mesopores have a pore diameter of 2–50 nm and can be used in detoxification hemoperfusion systems, where the toxin molecules are small molecular weight drugs such as chlorpromazine, barbital or methaqualone (MW $\leq$ 1kDa). The adsorbent used usually has a pore diameter of less than 10 nm (Chen *et al.*, 1983). Micropores with a pore diameter of 50–100nm can be used for targeting middle molecular weight uremic toxins (MW 4–40 kDa) like peptides and cytokines. A population of pores should be engineered which permits the entrance of the toxins freely into the interior of the adsorbent bead while excluding larger proteins such as albumin molecules (MW 66 kDa, molecular diameter 22.6 nm) (Davankov *et al.*, 2000) (Fig. 20.20). Macropores are pores

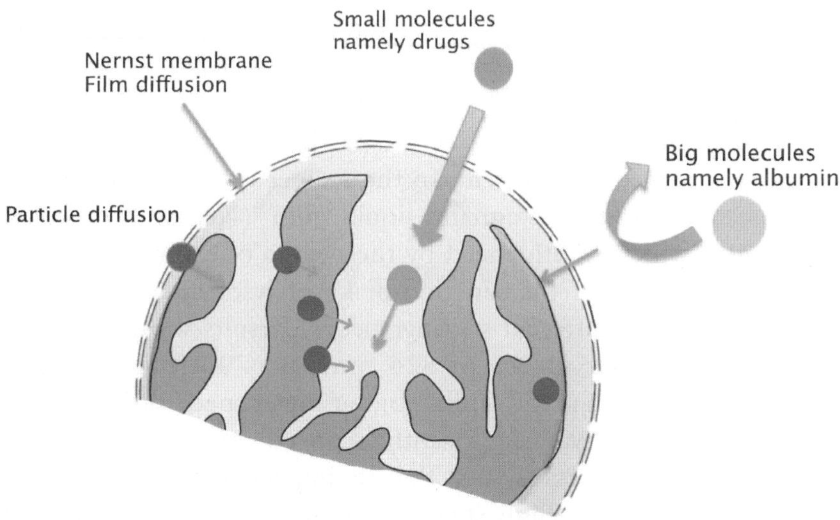

Fig. 20.20. Schematic representation of film diffusion, particle diffusion and penetration of small and big molecules into porous bead type adsorbent. Modified from *Fundamental and Clinical Aspects of Biomedical Engineering* (1988), Tianjin Science and Technology Press.

larger than 100 nm which can allow big molecules like immunoglobulins, LDL (MW 3 × 10^3 kDa and diameter about 22 nm) (Davankov *et al.*, 2000) and VLDL (diameter 30–80 nm) to diffuse freely in and out of the pores.

2.3. *Chemical Structure*

Functional groups on the chemical structure of the adsorbents highly affect their adsorption capacities and biocompatibility properties. For example, resins containing strong base groups have higher adsorption capacities than their corresponding weak base groups or resins without any functional groups. (Chen *et al.*, 1986). It is also well known that hydrophilic groups, namely amino or hydroxyl groups, can highly improve the biocompatible properties of an adsorbent.

2.4. *Adsorption Kinetics*

Another major parameter of the adsorbent is the adsorption rate, which can be characterized by Langmuir or Fruendlich adsorption isotherm. In principle, the adsorption rate is governed by two major factors: the diffusion rate of the molecules which arrive at the adsorption site located on the surface of the particle and the rate from the adsorption site into the particle. In the former case, diffusion of the solute occurs across a stagnant liquid film (Nernst membrane) (Fig. 20.20), which is influenced by the degree of external agitation. Roghe (2008) studied the kinetic uptake of lysozyme by a porous polystyrene resin at different stirring speeds. Results showed that the uptake at two higher speeds of 824 rpm and 700 rpm were both about 3 mg/L, which seem similar, but at lower speed of 84 rpm the uptake (2.2 mg/L) was significantly reduced, suggesting that the boundary layer thickness is such that its impact on the overall mass transfer is negligible at high stirring condition. The rate of particle diffusion is determined by the particle and pore size of the adsorbent. Generally, particle diffusion is the decisive step in the adsorption kinetics. Roche (2008) examined the influence of particle size on the mass transfer of protein adsorption. He found that the largest sized fraction (mean size 780 μm) showed considerable slower uptake of lysozyme (approximately 20%) than the 430 μm size fraction. The above reduction of particle size shortened the diffusion length of the protein within the adsorbent bead resulting in a lower uptake of lysozyme. As the diffusion path length is related to the radius of the particle, and the diffusion time is proportional to r^2, the time for the protein to diffuse from outside the particle to the center of the particle is four times less when the particle size is halved.

Kimmel (2011) examined the dynamics of cytokine (IL-6) adsorption within single sorbent particles (CytoSorb beads) using confocal laser scanning microscopy (CLSM). Results indicated that IL-6 adsorption dynamics could be modeled using classical adsorption/diffusion mechanisms, and a single model parameter, $\alpha = {}^{qmax}K/D$, can characterize interactions between IL-6 and CytoSorb beads, where $qmax$ and K stand for Langmuir adsorption isotherm parameters, and D stands for the effective diffusion coefficient within the porous matrix. After five hours of incubation with sorbent beads, IL-6 adsorption was limited to the outer 15 μm of the pore structure, which represents the use of less than 20% of available surface area for adsorption.

2.5. *Spacer Effect*

The adsorption capacity of adsorbents is known to be enhanced by the presence of a spacer molecule between the ligand and the matrix. Spacers, generally consist of relatively short aliphatic hydrocarbon chains (10 carbon atoms in length), which are effective at facilitating interactions involving small ligands and target solutes. To facilitate surface interactions between macromolecules or biological complexes, such as in immunoadsorption, the use of a much longer spacer, consisting of linear poly ethylene oxide (PEG) is proposed. Soltys *et al.* (2000) studied the adsorption capacity of apoB against anti-apoB IgG immobilized via PEG (1–2 mg/mL) was much greater than anti-apoB IgG immobilized directly (0.001 mg/mL) on the membrane surface. The diagrams below (Fig. 20.21) schematically show the spacers effect.

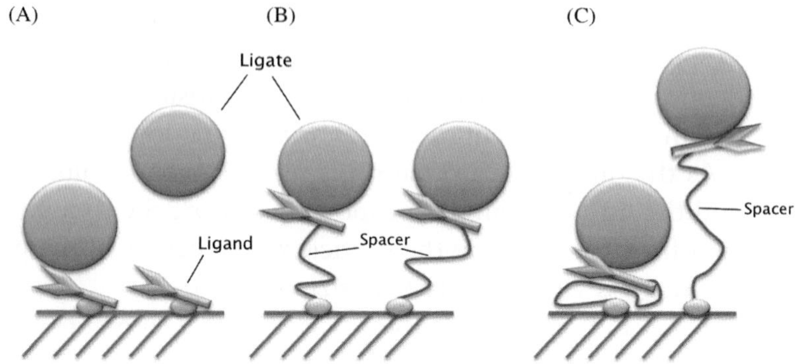

Fig. 20.21. Schematic diagram of spacer effect.

Note: The figure is not to scale for adsorption of LDL using a PEG spacer, in which case the diameter of the LDL ligate is 25 nm and the average end-to-end distance for a 17 kDa PEG spacer is 7.2 nm.

Source: Soltys PJ, Etzel MR. (2000) Biomaterials **21:** 37–48, with copyright permission from Elsevier Press.

Wang and Yu *et al.* (2008) designed and prepared cellulosic adsorbents with L-lysine as ligand and PEG having different molecular weights as spacers. Results showed that by increasing the molecular weight of PEG spacers from 1000Da to 6000Da, the average adsorption capacity of LDL was enhanced from 0.242 mg/mL to 0.903 mg/mL and the removal efficacy was remarkably increased (Table 20.5).

2.6. *Diseases Treated with Immunoadsorbents*

There are many advantages to treating diseases with immunoadsorbents as compared with plasmapheresis. The latter selectivity is low, causing losses of vitally important substances in the patient's blood, like clotting factors and fibrinogen. The amount of plasma exchanged is 2,000–3,000 mL while immunoperfusion can clear 9,000 mL of blood in one treatment. In plasmapheresis, plasma is discarded and fresh plasma needs to be infused into the patient, but in immmunoperfusion the plasma is returned. Since no plasma is supplemented, there is lower chance of infection. Immunosorption treatments have many advantages, but in the past, in spite of many types of adsorbents being investigated, only a few of them have been commercialized

Table 20.5. Adsorption Capacity of LDL–C by L-lysine Ligand with Different Molecular Weight Spacer

Mol. Weight of PEG Spacers (Da)	L-lysine Amount (mg/mL)	Average Removal Capacity of LDL (mg/mL)	Stoichiometric Capacity (mg LDL/mg L-Lysine)	Efficiency of Active Site (%)*
—	121.6	0.130 ± 0.013	0.001	1.018 ± 0.098
1,000	10.5	0.242 ± 0.017	0.023	21.947 ± 1.469
2,000	9.8	0.263 ± 0.013	0.027	25.556 ± 1.178
4,000	9.0	0.708 ± 0.017	0.078	74.917 ± 1.549
6,000	8.6	0.903 ± 0.003	0.105	100.000

Source: Wang *et al.* (2008) *Reactive & Functional Polymers* **68:** 261–267, with copyright permission from Elsevier Press.

(Table 20.6). Technological difficulties, mechanisms of diseases not being clarified, and economical aspects are the main reasons for this. Most of the diseases treated by immunosorption are autoimmune and familial hypercholestrolemia. Contraindications are severe decompensated heart failure, hypersensitive to proteins and unstable angina pectoris.

3. General Discussions

3.1. *Perspectives of Immunoadsorbent*

Immunoadsorption is an advanced technology which has grown rapidly in the past two decades and is used mainly for the treatment of diseases listed in Table 20.6. An ideal immunoadsorbent used as a therapeutic treatment in clinical performances should be composed of a well-defined carrier, a highly selective ligand and an appropriate length of spacer.

3.2. *Carriers*

The carrier of an immunosorbent is the solid support which links the ligand that binds the immunological pathogenic molecules in the blood of a patient. The material should be of medical grade

Table 20.6. Commercialized Immunoadsorbents

Manufacturer	Device Name	Carrier	Ligand	Selected Applications	Reference
Asahi Medical	Immusorba PH	PVA gel	Phenylalanine	many autoimmune diseases, for example, MG ,Guillain-Barre´ syndrome, and Fisher syndrome.	Nakaji *et al.*, 2001; Sato *et al.*, 1994
Asahi Medical	Immusorba TR	PVA gel	Tryptophan		Nakaji *et al.*, 2001
Kaneka	Selesorb	Cellulose	Dextran sulfate	treatment of systemic lupus erythematosus	Nakaji *et al.*, 2001; 153.Hashimoto *et al.*, 1991
Cypress Bioscience	Prosorba	Silica beads	Protein A	Rheumatoid arthritis	Nakaji *et al.*, 2001; Wiesenhuter *et al.*, 1994; Falson *et al.*,1999
Excorim	Immunosorba	Agarose beads	Protein A	MG and other immune diseases	Nakaji *et al.*, 2001; Braun *et al*, 1993
Kuraray	Medisorba MG	Cellulose beads	Acetylcholine receptor peptide	MG patients	Nakaji *et al.*, 1993, 2001; Oka *et al.*, 1993
Therasorb Medical Systems Unterschleissheim,	Ig-Therasorb	Cellulose beads	Sheep polyclonal antibodies	many kinds of autoimmune diseases	Nakaji *et al.*, 2001; Koll *et al.*, 1998
MiltenyiBiotec	TheraSorb TM- LDL	Sepharose	Anti-Apoprotein B100-antibodies	hyperlipidemia	Sanchez *et al.*, 2013
Kaneka	Liposorber VR D	Cellulose	Dextran sulfate	hyperlipidemia	Sanchez *et al.*, 2013
Pocard	LipopakVR	Agarose	Polyclonal antibodies to human ApoB	hyperlipidemia	Sanchez *et al.*, 2013; Bambauer *et al.*, 2012

Kuraray Medical Inc	Medisorba-MG 50	Cellulose	Peptide modified from alpha 183–200 of the torpedo acetylcholine receptor protein	MG	Sanchez *et al.*, 2013
Fresenius	MIRO	Polyacrylamide	Porcine C1q	SLE	Sanchez *et al.*, 2013
Fresenius	Globaffin	Sepharose	Peptide PGAM146	RA, acute kidney transplant rejection, Goodpasture's syndrome, RPGN, SLE, ITP, scleroderma, MG, FSGS, cryoglobulinemia, bullous pemphigoid, pemphigus vulgaris, Lambert-Eaton	Sanchez *et al.*, 2013
Fresenius	CoraffinV	Sepharose	Peptides PDCM349 and PDCM075	DCM	Sanchez *et al.*, 2013
Miltenyi Biotec	TheraSorbTM-Rheo	Sepharose	Penta-peptide(Gly-Pro-Arg-Pro-Lys)	Macular degeneration, diabetic complications, sudden hearing loss, venous leg ulcers, peripheral arterial disease	Sanchez *et al.*, 2013
Glycorex Transplantation AB	Glycosorb ABO ColumnVR	Sepharose	Terminal trisaccharide A or B blood group	ABO-incompatible organ transplantation	Sanchez *et al.*, 2013
Jiafron Biomedical Co., Ltd	HA280	Carbonized activated resin	Calf thymus DNA	Autoimmune diseases, such as SLE, pemphigus vulgaris, Lambert-Eaton, RA	Jiafron Biomedical Co., Ltd; Huang *et al.*, 2011; Su *et al.*, 2012

possessing good blood-compatible properties with minimal depletion of blood components. It is non-toxic, free from pyrogenic reactions, no carcinogenic or teratogenic effects. The carrier should present a large surface-to-volume area, and preferably be a porous polymer with controlled pore size and good mechanical strength in order to avoid embolism materials entering the circulating blood system of the patient. It should be storage stable, easily sterilized, activated and fabricated. Biomaterials suitable for preparing immunoadsorbent are shown in Table 20.7. For inorganic materials, only silica-based micro-beads have been successfully applied to clinical trials, because they are difficult to fabricate into beads and have poor biocompatibility properties. On the other hand, polymeric materials could be easily prepared into carriers having various shapes and sizes such as porous beads with micro-, meso- and macro- internal pore sizes, hollow fibers and membranes. It is obvious that a number of polymeric bead type adsorbents show unsatisfactory blood-compatible properties which need to be modified by being coated with a thin film of biocompatible biomaterial to improve their compatible properties.

3.3. *Ligand*

The ligand in an adsorbent plays the most important role in removal of pathogenic toxins from the patient's blood. A good immunoadsorbent, the ligand should be highly selective, firmly bound to the carrier with no leakage and no immunological activation side effect, have low nonspecific adsorption, be non-toxic, non-allergic, easily sterilized, good storage stability, easily available and cost effective. In the early stage, only antibodies were used as ligands, which caused immunological and technical problems especially difficulties in commercializing. The conventional method of linking antibody in a mild condition can prevent the denaturing reaction of the protein, but disordered orientation may cause a decrease of activity of the antibody (Alves *et al.*, 2013). In order to solve this problem, Kang *et al.* (2011) used the seed-conjugation method to prepare the immunoadsorbent with β_2-microglobulin-derived peptide (58–67) as ligand for the removal

Table 20.7 Biomaterials May Be Used in Hemoperfusion

Material	Repeated Unit	Main Fabrication Property	Activation Methods	Reference
Polydimethylsiloxane (PDMS)	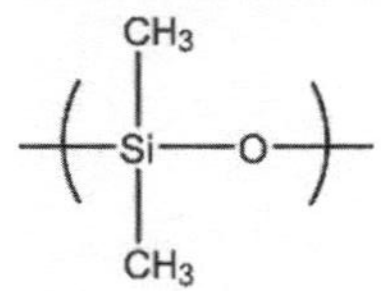	resin	silanization	Hermanson *et al.*, 1991,
Polyacrylate	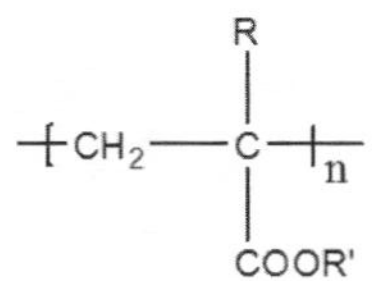	resin	N,N′-methylene bisacrylamide (MBA); glycerine	Hermanson *et al.*, 1991
Polystyrene, PS	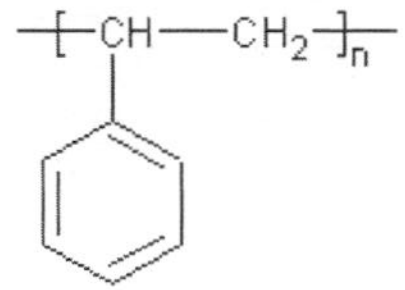	macroporous resin	sulfonation	Hermanson *et al.*, 1991
Poly(hydroxyethyl methacrylate) (PHEMA)	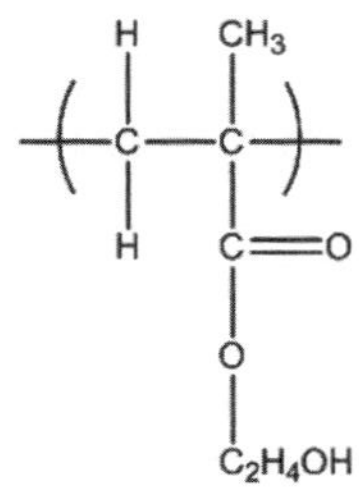	resin and membrane	sulfonation; hexamethylene diisocyanate	Hermanson *et al.*, 1991

(*Continued*)

Table 20.7. (*Continued*)

Material	Repeated Unit	Main Fabrication Property	Activation Methods	Reference
Polyacrylamide (PAM)	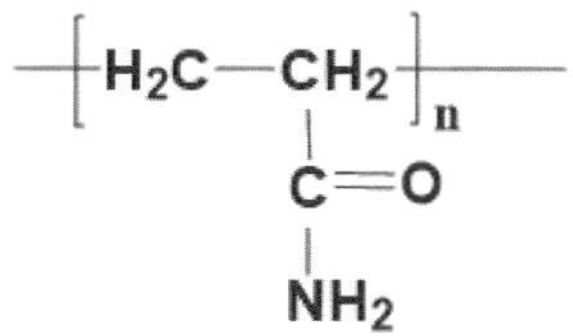	gel bead	sulfonation	Hermanson *et al.*, 1991
Controlled porous glass beads	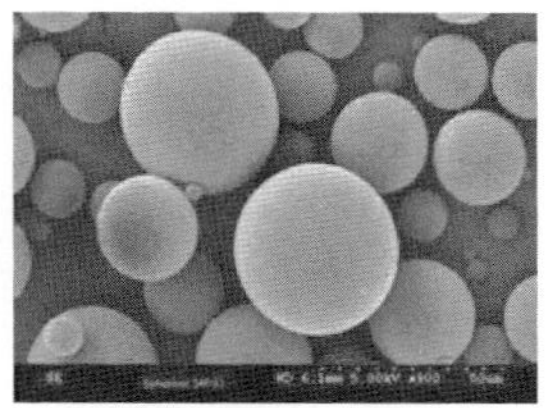	glass bead	silanization	Hermanson *et al.*, 1991
Polymerized activated carbon		resin	nitric acid	Hermanson *et al.*, 1991

(*Continued*)

Table 20.7. (*Continued*)

Material	Repeated Unit	Main Fabrication Property	Activation Methods	Reference
Polyvinyl acetate	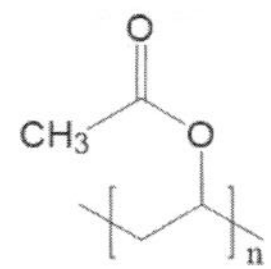	resin, membrane	3-aminophenylboronic acid	Hermanson *et al.*, 1991
Polyether sulfone	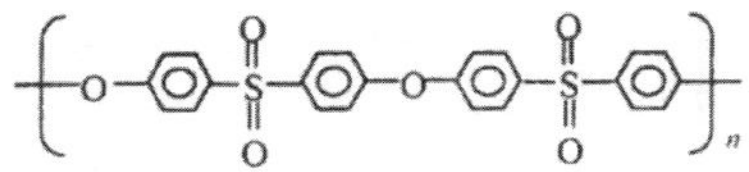	membrane, hollow fiber	sulfonation	Hermanson *et al.*, 1991
Dextran	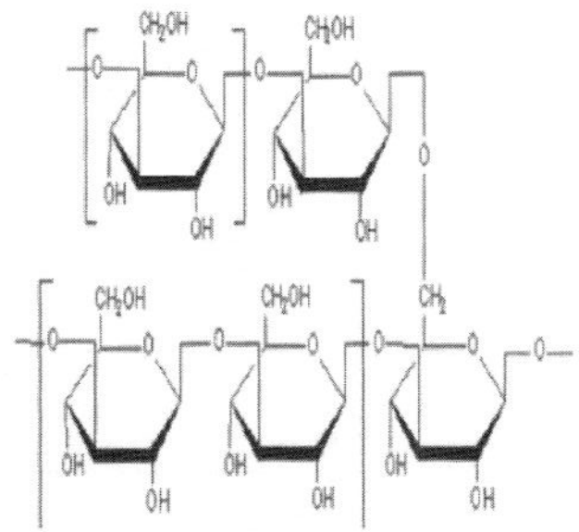	gel beads	cynanogen bromide; N-hydroxy succinimide esters(NHS); epoxy activation; hexamethylene diisocyanate	Hermanson *et al.*, 1991
Agarose	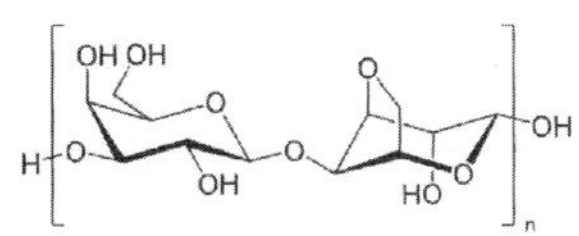	gel beads	cynanogen bromide; N-hydroxy succinimide esters(NHS); epoxy activation	Hermanson *et al.*, 1991

(*Continued*)

Table 20.7. (*Continued*)

Material	Repeated Unit	Main Fabrication Property	Activation Methods	Reference
Chitosan	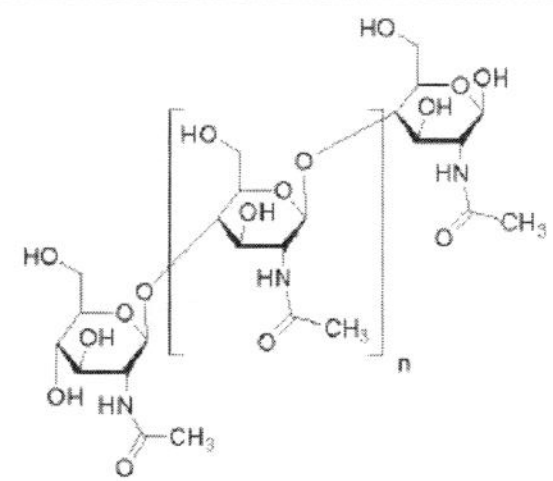	gel beads	cynanogen bromide; N-hydroxy succinimide esters(NHS); epoxy activation	Hermanson *et al.*, 1991
Cellulose	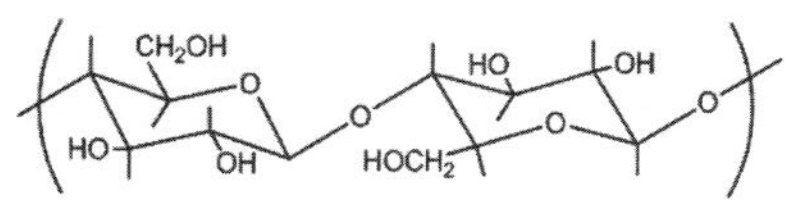	gel beads	cynanogen bromide; N-hydroxy succinimide esters(NHS); epoxy activation	Hermanson *et al.*, 1991
Glycosaminoglycans	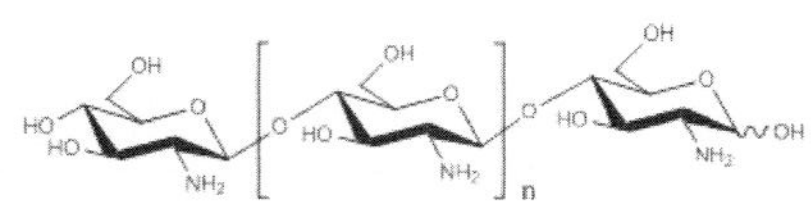	gel beads	cynanogen bromide; N-hydroxy succinimide esters(NHS); epoxy activation	Hermanson *et al.*,1991

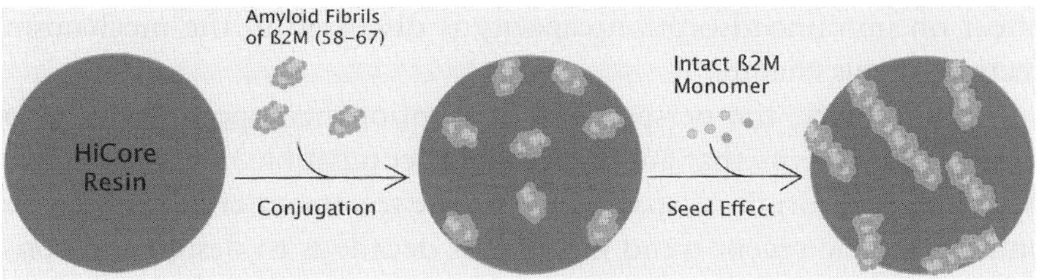

Fig. 20.22. Schematic diagram of seed-conjugation.

of intact β_2-microglobulin at neutral pH (Fig. 20.22). Adsorption results showed that β_2-microglobulin removal efficiency was improved, at the same time the method possesses high specificity and high binding capacity. The advantage of using peptide as ligand is that it has a linear structure, low molecular weight, high selectivity, low immunity, is more stable and easy to sterilize, and provides a simpler process for product production. From the theoretical point of view, deeper studies of peptide ligands will provide better understanding of the molecular recognition mechanism between the ligand and the target toxin molecule, so future trends should focus on peptide ligand studies.

3.4. *Spacer*

A spacer (arm) is a link that joins the ligand to the carrier in an adsorbent. Theoretically any small or big molecules can serve as the linker. However, most spacers are 10 atoms in length with active coupling functionalities on both ends. However, in immunoadsorption, often if not exclusively, the target molecules are macro-molecules like antibodies. Therefore the spacer serves to extend the ligand from the support, thus reducing the steric hindrance and allowing the much larger immunoglobulin or immuno-complex to orient for binding. Coupling via amine or carboxylate group are often used which can be conducted under mild experimental conditions. Some of the spacers used are diaminodipropylamine (DADPA), 6-aminocaproic acid (6-AC), 1,6-diaminohexane (DAH), ethylene diamine (EDA), carbodiimide and long chain of various molecular weight PEG. Spacer effect on immunoadsorption capacity is discussed in the mechanism section of this chapter.

In summary, the perspectives of sorption hemoperfusion are to develop adsorbents that are highly blood compatible, highly selective with a high adsorption capacity so that the treatment efficiency can be enhanced. The recent trend in the past decade is to design and construct medical devices to be as compact and efficient as possible. Therefore simple perfusion devices with highly efficient adsorbent, whole blood perfusion and home-care treatments are in great demand and recommended in the future.

3.5. *Social and Economic Impact*

There are multiple challenges for mandatory financed health care systems. All countries in the world, especially western countries, have to face the fact of ageing populations caused by higher life expectancies and lower birth rates (Zhu, 2010). In order to realize the above objectives, a number of new medical technologies need to be developed and be listed in the National Health Insurance System (NHS). Whether a new medical technology can be listed in NHS is determined by many important criteria. Although criteria are not equally designed in different countries, there are some common ones that should be considered:

(1) *Medical effectiveness*
Medical effectiveness is obviously the most important issue since it is directly beneficial to the recovery of the patient and is also a challenge to be listed in the NHS. Regenerative medicine, tissue engineering, nano-technology, biosensor, etc. are advance biotechnologies at the present stage. High sensitive, accurate and visible medical devices enable diagnosis at the molecular level. All these advanced technologies can promote the art of immunoadsorption and highly enhance its medical effectiveness.

(2) *Economic considerations*
Concerning economic aspects, some countries like the UK (and Germany to a smaller extent) partially base their decisions on

cost–outcome ratios like costs per life-year gained or costs per quality adjusted life-year (QALY) (Zhu, 2010). MARS has demonstrated that it has an immediate effect on life saving of patients with liver failure and other diseases, but its expenses are also very high. The direct medical costs per patient for initial hospitalization and one-year follow-up from a payer's perspective was €18,792 per patient and €9,638 for controls. The cost per life year gained was €29,719 (time horizon one year) (Hessel *et al.*, 2003). This cost and some recent therapeutic problems severely hindered MARS from becoming widespread clinical therapy and being listed in the worldwide National Health Insurance System. Berto *et al.* (2011) investigated cost-effectiveness of the treatment by polymyxin-B immobilized fiber (PMX-F) column with conventional medical therapy (CT) in the management of abdominal septic shock in Italy. Results showed that PMX-F-CT versus CT is a cost-effective intervention for treatment of severe sepsis/septic shock of abdominal origin based on the expected survival years (mean discounted PMX-F-CT 9.37 LY/patient, CT 4.92 LY/patient) and the expected mean cost (PMX-F-CT mean 59,922 EUR/patient, CT mean 42,712 EUR/patient). They believe that PMX-F-CT could be considered as a treatment in Italy and be listed in a National Health System hospital setting. The above cases suggest that medical effectiveness and expenses of prolonged patient's lifespan in the long run are two key factors affecting immunoadsorption therapy.

(3) *Social economic impact*

The third criteria "social economic impact" can also greatly affect the immunoadsorbent clinical therapy to be listed in the National Health Insurance System. Zhu *et al.* (2010) studied the socio-economic impact of systemic lupus in Hong Kong; she found that the average annual total cost was US$13,307 (2006) per patient. The direct costs dominated the total costs (62%), and the costs of inpatient care contributed 52% of the direct costs. Costs of SLE per subject are higher than those of other chronic diseases in Hong Kong. Due to most SLE patients being females and at working age, the disease affects their life and working quality so they have a significantly lower level of health-related quality of life (HRQoL) compared with the Hong Kong general population. In order to solve this problem, she suggested a new therapeutic approach should be adopted to treat SLE patients in the early stage, so SLE treatment could be listed in NHS. In China, hemoperfusion started in the early '80s; it was only in the past decade that it grew rapidly. Clinical performances developed from the initial detoxification to treatment of autoimmune diseases, renal failure, hyperbilirubinemia, hyperlipdemia, infectious disease, metabolic diseases, MODS, transplant rejection, and dermatonosus, etc. Whole blood perfusion is a characteristic treatment in China and the device (cartridge) cost is 1/5–1/3 of imported ones. This is because medical policy in China has from the beginning emphasized treatment efficiency, safety and cost effectiveness. In 2010, the Ministry of Health issued a revised National Standard for blood purification which provided a guide for hemoperfusion therapy. From then on, under the guidance of this new issue, hemoperfusion developed rapidly and the production outcome broke 0.3 billion RMB within the past three years. According to preliminary estimation, the market will reach 10 billion RMB per year (~1.7 billion US dollars) in the ongoing five years. Hemoperfusion has a great demand and market in China. During this occasion, the pioneer of hemoperfusion in China, Yu's research team received the Ho Leung, Ho Lee Prize and the National Prize of Progress in Science and Technology with Jafron Biomedical Co., which encouraged the R&D of sorption perfusion, thereby enhancing the health and life quality of the people.

References

Alexandre GPL, Gianello P, Latinne D *et al.* (1989) Plasmapheresis and splenectomy in experimental renal xenotransplantation. In: Hardy MA (ed.), *Xenografi.* Elsevier Science Publishers, New York, p. 259.

Alkan H, Berelib N, Baysal Z, *et al.* (2010) Selective removal of the autoantibodies from rheumatoid arthritis patient plasma using protein

A carrying affinity cryogels. *Biochemical Engineering Journal* **51:** 153–159.

Alvesa NJ, Mustafaoglu N, Bilgicer B. (2013) Oriented antibody immobilization by site-specific UV photo-crosslinking of biotin at the conserved nucleotide binding site for enhanced antigen detection. *Biosensors and Bioelectronics* **49:** 387–393.

Arcus VL, Proft T, Sigrell JA, *et al.* (2000) Conservation and variation in superantigen structure and activity highlighted by the three-dimensional structures of two new superantigens from Streptococcus pyogenes. *J Mol Biol* **299(1):** 157–68.

Ash S, Blare DE, Carr DJ. (1992) Clinical effect of a sorbent suspension dialysis system in treatment of hepatic coma (The Biologic DT). *Int J Artif Organs* **15:** 151-61.

Bak H, Thomas ORT. (2007) Evaluation of commercial chromatographic adsorbents for the direct capture of polyclonal rabbit antibodies from clarified antiserum. *J Chromatogr B* **848:** 116–130.

Bambauer R, Bambauer C, Lehmann B, *et al.* (2012) LDL-apheresis: Technical and clinical aspects. *Sci World J* 2012 Article ID 314283, 19 pages [http://dx.doi.org/10.1100/2012/314283]

Behim E, Klinkmann H. (1989) Selective and specific adsorbents for medical therapy. *Int J Artif Organs* **12:** 1–10.

Berto P, Ronco C, Cruz D. (2011) Cost-effectiveness analysis of polymyxin-b immobilized fiber column and conventional medical therapy in the management of abdominal septic shock in Italy. *Blood Purif* **32:** 331–340.

Bischoff SC. (2007) Role of mast cells in allergic and non-allergic immune responses: Comparison of human and murine data. *Nat Rev Immunol* **7:** 93–104.

Bogan AA, Thorn KS. (1998) Anatomy of hot spots in protein interfaces. *J Mol Biol* **280:** 1–9.

Borberg H, Gaczkowski A, Hombach V, *et al.* (1988) Treatment of familial hypercholesterolemia by means of specific immunoadsorption. *J CIin Apheresis* **4:** 59–65.

Bosch T, Gahr S, Belschner U, *et al.* (2006) Direct adsorption of low-density lipoprotein by dalildl-apheresis: results of a prospective Long-term multicenter follow-up covering 12 291 sessions. *Therapeutic Apheresis and Dialysis* **10(3):** 210–218.

Braun N, Risler T, *et al.* (1999) Immunoadsorption as a tool for the immunomodulation of the humoral and cellular immune system in autoimmune disease. *Ther Apheresis* **3:** 240–5.

Braun N, Junger M, Klein R, *et al.* (2002) Dextran sulfate (Selesorb) plasma apheresis improves vascular changes in systemic lupus erythematosus. *Therapeutic Apheresis* **6(6):** 471–577.

Burnoufa T, Goubranb H, Radosevicha M. (1998) Application of bioaffinity technology in therapeutic extracorporeal plasmapheresis and large-scale fractionation of human plasma. *Journal of Chromatography B* **715:** 65–80.

Burton SC, Harding DRK. (1998) Hydrophobic charge induction chromatography: Salt independent protein adsorption and facile elution with aqueous buffers. *J Chromatogr A* **814(1/2):** 71–81.

Cao NN, Yu YY, Wang M-Y, *et al.* (2002) In vitro study of a novel low density lipoprotein adsorbent. *Artif Cell Blood Sub* **30:** 53–61.

Casolaro V, Georas SN, Song Z, *et al.* (1996) Biology and genetics of atopic disease. *Curr Opin Immunol* **8:** 796–803.

Chang TMS. (1964) Semipermeable microcapsules. *Science* **146(3643):** 524–525.

Chang TMS. (1966) Semipermeable aqueous microcapsules ("artificial cells"): With emphasis on experiments in an extracorporeal shunt system. *Trans Am Soc Artif Intern Organs* **12:** 13–19.

Chang TMS. (1969) Removal of endogenous and exogenous toxins by a microencapsulated absorbent. *Can J Physiol Pharmacol* **47:** 1043–1045.

Chang TMS. (1971) The in vivo effects of semipermeable microcapsules containing L-asparaginase on 6C3HED lymphosarcoma. *Nature* **229:** 117–118.

Chang TMS. (1972) *Artificial Cells.* Thomas, Springfield, Illinois. (Free full text monograph available on www.medicine.mcgill.ca/artcell).

Chang TMS. (1975) Microencapsulated adsorbent hemoperfusion for uremia, intoxication and hepatic failure. *Kidney Int* **7:** S387–S392.

Chang TMS. (2005). Therapeutic applications of polymeric artificial cells. *Nature Review: Drug Discovery* **4:** 221–235.

Chang TMS. (2007) Monograph on "ARTIFICIAL CELLS: biotechnology, nanotechnology, blood substitutes, regenerative medicine, bioencapsulation, cell/stem cell therapy." World Scientific Publisher/Imperial College Press, 435 pages. (Free full text monograph available on www.medicine.mcgill.ca/artcell)

Chang TMS, Johnson LJ, Ransome OJ. (1967) Semipermeable aqueous microcapsules: IV. Nonthrombogenic microcapsules with heparin-complexed membranes. *Can J Physiol Pharmacol* **45:** 705–715.

Chang TMS, Poznansky MJ. (1968) Semipermeable microcapsules containing catalase for enzyme replacement in acatalsaemic mice. *Nature* **218:** 242–245.

Chang TMS, Malave N. (1970) The development and first clinical use of semipermeable microcapsules (artificial cells) as a compact artificial kidney. *Trans Am Soc Artif Intern Organs* **16:** 141–148.

Chen B, Pan JL, Tong MR, *et al.* (1993) A study of adsorbents for the adsorption of low density lipoprotein (LDL). *Ion Exch Adsorption* **9:** 330–334.

Chen CZ, Zhang QX, Yu YT. (1983) New uncoated adsorbent NK-107 of high surface area for treatment of acute intoxication, *Proceedings of the Fifth International Symposium on Hemoperfusion and Artificial Organs* by Chang TMS, He BL. China Academic Publishers.

Chu JQ, Yu YT, Zhang WH, *et al.* (2003) In vitro studies on the immunoadsorbent for removal of IgA in IgA-Neplropatly (II). *Chemical Journal of Chinese Universities* **24(4):** 636–638.

Chu JQ, Zhang WH, Yu YT, *et al.* (2004) In vitro studies of the immunoadsorbent for removal of IgA in IgA—nephropathy(III). *Chemical Journal of Chinese Universities* **25(8):** 1454–1457.

Clackson T, Wells JA. (1995) A hot spot of binding energy in a hormone receptor interface. *Science* **267:** 383–386.

Corneillie TM, Lee KC, Whetstone PA, *et al.* (2004) Irreversible engineering of the multi-element binding antibody 2D12.5 and its complementary ligands. *Bio-conjugate Chem* **15:** 1392–1402.

Dai J, Chen CZ, Yu YT. (2001) Preparation and properties of immunoadsorbent for rheumatoid factors. *Ion Exchange and Adsorption* **17(5):** 352–356.

Daniels CM. (2007) Developing and characterization of bioadsorbent for extracorporeal blood purification, PhD Dissertation.

Davankov V, Pavlova L, Tsyurupa M, *et al.* (2000) Polymeric adsorbent for removing toxic proteins from blood of patients with kidney failure. *J Chromatogr B Biomed Sci Appl* **739:** 73–80.

Dodda RY, Kobitaa JA. (1978) A novel immunoadsorbent: Use for the preparation of mono-specific antibodies to the hepatitis B antigen. *Journal of Immunological Methods* **20:** 117–129.

Duarte IS, Zollner RL, Bueno SMA. (2006) *In Vitro* evaluation of bio-specific and pseudo-bio-specific ligands aimed at extracorporeal treatment for immunoglobulin E removal. *Artificial Organs* **30(8):** 606–614.

Dux MP, Barent R, Sinha J, *et al.* (2006) Purification and scale-up of a recombinant heavy chain fragment C of botulinum neurotoxin serotype E in pichia pastoris GS115 protein Expr. *Purif* **45(2):** 359–367.

Falkenhagen D, Strobl W, Hartmann J, *et al.* (2002) Patient safety technology for micro-adsorbent systems in extracorporeal blood purification. *Artif Organs* **26(2):** 84–90.

Falkenhagen D, Schima H, Loth F. (1999) Arrangement for removing substances from liquids in particular blood. United States Patent, #5,855,782.

Falson DT, LaValley MP, Baldassare AR, *et al.* (1999) The prosorba column for treatment of refractory rheumatoid arthritis: A randomized, double-blind, sham-controlled trial. *Arthritis Rheum* **42:** 2153–9.

Fang H, Wei J, Yu YT. (2004) *In vivo* studies of endotoxin removal by lysine-cellulose adsorbents. *Biomaterilas* **25:** 5433–5440.

Freiburghaus C, Ohlson S, Nilsson IM. (1988) Extracorporeal systems for adsorption of antibodies in hemophilia A and B. *Methods Enzymol* **137:** 458–66.

Fu CX, Chen CZ, Yu YT. (2002) Heat aggregated IgG linked to agar gel — A new immunoadsorbent for rheumatoid arthritis. *Chinese Journal of Biomedical Engineering* **2(1):** 1–4.

Gao CL, Li BA, Chen CZ, *et al.* (1995) Clinical trials of immunoadsorbent in systemic lupus erythematosus therapy. *Artif Organs* **19:** 468–469.

Gaubitz M, Seidel M, Kummer S, *et al.* (1998) Prospective randomized trial of two different immunoadsorbers in severe systemic lupus erythematosus. *Journal of Autoimmunity* **11:** 495–501.

Gaubitz M, Schneider KM. (2003) Immunoadsorption in systemic lupus erythematosus: different techniques and their current role in medical therapy. *Therapeutic Apheresis and Dialysis* **7(2):** 183–188.

Gazitt Y, Margel Y, Lerner A. (1985) Development of a novel Clq immunoadsorbent for removal of circulating immunecomplexes: Quantitative isolation of hepatitis B virus surface antigen and immunecomplexes. *Immunology Letters* **11(1):** 1–8.

Ghosh A, Bansal M. (2003) A glossary of DNA structures from A to Z. *Acta Crystallogr D* **59(4):** 620–6.

Gjorstrup P, Berntorp E, Larsson L, *et al.* (1991) Kinetic aspects of the removal of IgG and inhibitors in hemophiliacs using protein A immunoadsorption. *Vox Sang* **61:** 244–50.

Goodyear CS, Silverman GJ. (2003) Death by a B cell superantigen: In vivo VH-targeted apoptotic supraclonal B cell deletion by a Staphylococcal toxin. *J Exp Med* **197:** 1125–1139.

Gorchakov VD, Sakodynskii KI, Lebedin YS, *et al.* (1991) Immunosorbents for clinical use: Ex vivo immunoglobulin E removal in allergy. *J Chromatogr* **563:** 166–71.

Graf MW, Uhr JW. (1969) Regulation of antibody formation by serum antibody I: Removal of specific antibody by means of immunoglobulin. *J Exp Med* **50:** 1175–1181.

Graille M, Stura EA, Corper AL, *et al.* (2000) Crystal structure of a Staphylococcus aureus protein A domain complexed with the Fab fragment of a human IgM antibody: Structural basis for recognition of B-cell receptors and superantigen activity. *Proc Natl Acad Sci USA* **97(10):** 5399–404.

Griffiths AD, Duncan AR. (1998) Strategies for selection of antibodies by phage display. *Current Opinion in Biotechnology* **9:** 102–108.

Hari P, Willi P, Sharma CP. (2000) Adsorption of human IgG on Cu2+-immobilized cellulose affinity membrane: Preliminary study. *J Biomed Mater Res* **50:** 110–13.

Hashimoto H, Tsuda H, Kanai Y, *et al.* (1991) Selective removal of anti-DNA and anticardiolipin antibodies by adsorbent plasmapheresis using dextran sulfate columns in patients with systemic lupus erythematosus. *J Rheumatology* **18:** 545–51.

Hermanson GT, A Krishna Mallia, Smith PK. (1991) Immobilized affinity ligand techniques.

Hessel FP, Mitzner SR, Rief J, *et al.* (2003) Economic evaluation and 1-year survival analysis of MARS in patients with alcoholic liver disease. *Liver International* **23(Suppl. 3):** 66–72.

Hidaka T, Suzuki K, Matzuki Y, *et al.* (2000) Evaluation of adsorption selectivity dextran sulfate bound cellulose beads for removal of anti-DNA antibodies. *Journal of Clinical Apheresis* **15:** 262–265.

Hiepe F, Pfüller B, Wolbart K, *et al.* (1999) C1q: A multifunctional ligand for a new immunoadsorption treatment. *Ther Apher* **3:** 246–51.

Hou XD, Zhang T, Cao AM. (2013) A heparin modified polypropylene non-woven fabric membrane adsorbent for selective removal of low density lipoprotein fromplasma. *Polym Adv Technol* **24:** 660–667.

Hou XD, Zhang T, Cao AM. (2015) Preparation of new amphiphilic macroporous nonwoven polymeric adsorbents aimed for selective removal of low-density lipoprotein from plasma. *Journal of Biomedical Materials Research B: Applied Biomaterials* **103(1):** 52–61.

Idusogie EE, Presta LG, Gazzano-Santoro H, *et al.* (2000) Mapping of the C1q binding site on rituxan, a chimeric antibody with a human IgG1 Fc. *J Immunol* **164(8):** 4178–84.

Ino N, Kamata N, Matsura C, *et al.* (1997) Immunoadsorption for the treatment of bullous pemphigoid. *Therapeutic Apheresis* **1(4):** 372–376.

Janeway C. (2001) *Immunobiology*, 5th ed. Garland Publishing. [ISBN 0-8153-3642-X] (Electronic full text via NCBI Bookshelf.)

Jin XX. (2003) The research of heparin connecting phenylanine blood purification materials to remove toxin, Master's thesis.

Junqueira L, Carneiro CJ. (2003) *Basic Histology*. McGraw-Hill.

Karlin A. (2002) Emerging structure of the nicotinic acetylcholine receptors. *Nat Rev Neurosci* **3:** 102–114.

Kato K, Ikada Y. (1966) Immobilization of DNA onto a polymer support and its potentiality as immunoadsorbent. *Biotechnology and Bioengineering* **51:** 581–590.

Kawabata K, Nagai H, Konishi N, *et al.* (2014) Peptide-based immunoadsorbents: Molecular grafting of IgG–Fc-binding epitopes of Protein A onto a de novo-designed helix-loop-helix peptide. *Bioorganic & Medicinal Chemistry* **22:** 1845–1849.

Kilhovd BK, Juutilainen A, Lehto S, *et al.* (2007) Increased serum levels of advanced glycation end products predict total, cardiovascular and coronary mortality in women with type 2 diabetes: A population-based 18 year follow-up study. *Diabetologia* **50:** 1409–17.

Kimmel JD. (2011) Characterizing cytokine transport in hemoadsorption beads used to treat sepsis. PhD dissertation, University of Pittsburgh.

Kobayashi A, Nakatani M, Furuyoshi S, *et al.* (2002) In vitro evaluation of dextran sulfate cellulose beads for whole blood infusion low density lipoprotein hemoperfusion. *Therapeutic Apheresis* **6(5):** 365–371.

Kojima S, Harada-Shiba M, Toyota Y, *et al.* (1992) Changes in coagulation factors by passage through a dextran sulfate cellulose column during low-density lipoprotein apheresis. *Int J Artif Organs* **15:** 185–190.

Koll RA. (1998) Ig-Therasorb immunoadsorption for selective removal of human immunoglobulin diseases associated with pathogenic antibodies of all classes and IgG subclasses, immune complexes, and fragments of immunoglobulins. *Ther Apheresis* **2:** 147–52.

Kong DL, Schuett W, Dai J. (2002) Development of cellulose-DNA immunoadsorbent. *Artificial Organs* **26(2):** 200–208.

Kong DL, Schuett W, Boeden HF, *et al.* (1999) Development of a DNA immunoadsorbent: Coupling DNA on Sepharose 4FF by an efficient activation method. *Artif Organs* **24(11):** 845–851.

Kong DL, Chen CZ, Lin EF, *et al.* (1998) Clinical trials of type I and in vitro studies of type II immunoadsorbents for systemic lupus erythematosus therapy. *Artif Organs* **22:** 644–650.

Kutsuki H, Takata S, Yamamoto K, *et al.* (1998) N. Therapeutic selective adsorption of anti-DNA antibody using dextran sulfate cellulose column (Selesorb) for the treatment of systemic lupus erythematosus. *Ther Apheresis* **2:** 18–24.

Kutsuki H, Takata S, Yamamoto K, *et al.* (1998) Therapeutic selective adsorption of anti-DNA antibody using dextran sulfate cellulose column (selesorb) for the treatment of systemic lupus erythematosus. *Ther Apheresis* **2(1):** 18–24.

Laffer S, Lupinek C, Rauter I. (2008) A high-affinity monoclonal anti-IgE antibody for depletion of IgE and IgE-bearing cells. *Allergy* **63:** 695–702.

Lebedin YS, Gorchakov VD, Petrova EN, *et al.* (1991) Ex vivo removal of IgE in atopic asthma by extracorporeal plasmoimmunoadsorption (EPIA): development of a clinical adsorbent. *Int J Artif Organs* **14:** 508–14.

Li J, Huang XJ, Ji J. (2011) Covalent heparin modification of a polysulfone flat sheet membrane for selective removal of low-density lipoproteins: A simple and versatile method. *Macromol Biosci* **11:** 1218–1226.

Li XH, Cheng Y, Yang L, *et al.* (2010) Extracorporeal whole blood immune-adsorption of passively transferred myasthenia gravis rabbits by cellulose-tryptophan column. *Artif Cell Blood Sub* **38:** 186–191.

Litman GW, Rast JP, Shamblott MJ, *et al.* (1993) Phylogenetic diversification of immunoglobulin genes and the antibody repertoire. *Mol Biol Evol* **10(1):** 60–72.

Llewelyn M, Cohen J. (2002) Superantigens: Microbial agents that corrupt immunity. *Lancet Infect Dis* **2(3):** 156–62.

Lucchiari N, Azimzadeh A, Wolf P. (1997) In vivo and in vtro optimization of depletion of IgM and IgG xenoantibodies by immunoadsorption using cell membrane proteins. *Artificial Organs* **21(4):** 278–286.

Lupien PJ, Moorjani S, Awad J. (1976) A new approach to the management of familial hypercholesterolaemia: Removal of plasma cholesterol based on the principle of affinity chromatography. *Lancet* **1:** 1261–1265.

Ma KW, Dai XZ, Feng SY. (2011) Heparinized polyvinyl alcohol to specifically adsorb low density lipoprotein from plasma. *Transfusion and Apheresis Science* **44:** 3–9.

Maaskant N, Bantjes A, Kempen HJM. (1986) Removal of low density lipoprotein from blood plasma using cross-linked, sulfated polyvinylalcohol. *Atherosclerosis* **62:** 159–166.

Matic G, Hofmann D, Winkler R, *et al.* (2000) Removal of immunoglobulins by aprotein A versus an antihuman immunoglobulin G-based system: Evaluation of 602 sessions of extracorporeal immunoadsorption. *Artif Organs* **24:** 103–7.

Matic G, Schutt W, Winkler RE, *et al.* (2000) Extracorporeal removal of circulating immune complexes: From non-selective to patient-specific. *Blood Purif* **18:** 156–60.

Matic G, Winkler RE, Tiess M, *et al.* (2001) Selective apheresis — time for a change. *Int J Artif Organs* **24:** 4–7.

Matsuki Y, Suzuki K, Kawakami M, *et al.* (1996) High-avidity anti-DNA antibody removal from the serum of systemic lupus erythematosus patients by adsorption using dextran sulfate cellulose columns. *Journal of Clinical Apheresis* **11:** 30–35.

Nakaji S, Oka K, Tanihara M, *et al.* (1993) Development of a specific immunoadsorbent containing immobilized synthetic peptide of acetylcholine receptor for treatment of myasthenia gravis. *Ther Plasmapheresis* **12:** 573–6.

Nakaji S, Hayashi N. (2003) Adsorption column for myasthenia gravis treatment: Medisorba MG-50. *Therapeutic Apheresis and Dialysis* **7(1):** 78–84.

Nin JW, Jorsal A, Ferreira I. (2011) Higher plasma levels of advanced glycation end products are associated with incident cardiovascular disease and all-cause mortality in type 1 diabetes: a 12-year follow-up study. *Diabetes Care* **34:** 442–7.

Oka K, Nakaji S, Tanihara M, *et al.* (1993) Development of specific immunoadsorbent containing immobilized synthetic peptide of acetylcholine receptor for treatment of myasthenia gravis. *Jpn J Artif Organs* **22:** 194–7.

Pfueller B, Wolbart K, Bruns A, *et al.* (2001) Successful treatment of patients with systemic lupus erythematosus by immunoadsorption with a C1q column: A pilot study. *Arthritis Rheum* **44:** 1962–3.

Pokrovsky SN, Adamova IY. (1990) Immunosorbents for LDL-apheresis. *Biomater Artif Cells Artif Organs* **18:** 623.

Ren J, Jia LY, Xu L. (2009) Removal of autoantibodies by 4-mercaptoethylpyridine-based adsorbent. *Journal of Chromatography B* **877:** 1200–1204.

Roghe L. (2008) Nanoporous polymeric adsorbent for blood purification. PhD dissertation, Loughborough University Department of Chemical Engineering.

Samuelsson G. (1999) What's happening? Protein A columns: Current concepts and recent advances. *Transfus Sci* **21:** 215–7.

Sanchez AP, Cunard R, Ward DW. (2013) The selective therapeutic apheresis procedures. *Journal of Clinical Apheresis* **28:** 20–29.

Sasso EH, Merrill C, Furst TE. (2001) Immunoglobulin binding properties of the Prosorba immunadsorption column in treatment of rheumatoid arthritis. *Ther Apher* **5:** 84–91.

Sato H, Watanabe K, Azuma J, *et al.* (1989) Specific removal of IgE by therapeutic immunoadsorption system. *J Immunol Methods* **118:** 161–8.

Sato Y, Agishi T. (1996) Low density lipoprotein adsorption for arteriosclerotic patients. *Artif Organs* **20(4):** 324–327.

Sawada K, Malchesky PS, Nose Y. (1990) Available removal systems: State of the art. In: Therapeutic hemapheresis in the 1990s. *Curr Stud Hematol Blood Transfus* **57:** 341–383.

Schenkein I, Bystryn JC, Uhr JW. (1971) Specific removal of in vivo antibody by extracorporeal circulation over an immunoadsorbent in gel. *J Clin Invest* **50:** 1864–1868.

Schwartz W, Judd D, Wysocki M, *et al.* (2001) Comparision of hydrophobic charge induction chromatography with affinity chromatography on protein A for harvest and purification of antibodies. *J Chromatogr A* **908:** 251–263.

Senel S, Bayramoglu G, Arıca MY. (2003) DNA adsorption on a poly-L-lysine-immobilized poly(2-hydroxyethyl methacrylate) membrane. *Polym Int* **52:** 1169–1174.

Shibuya N, Sato T, Osame M. (1994) Immunoadsorption therapy for myasthenia gravis. *J Neurol Neurosurg Psychiatry* **57:** 578–81.

Sinitsyn VV, Mamontova AG, Konovalov GA, *et al.* (1990) Apheresis of low density lipoproteins using a heparin-based sorbent with low antithrombin III binding capacity. *Atherosclerosis* **84:** 55–59.

Soltys JP, Etzel RM. (2000) Equilibrium adsorption of LDL and gold immunoconjugates to affinity membranes containing PEG Spacers. *Biomaterials* **21:** 37–48.

Stauffer Y, Marguerat S, Meylan F, *et al.* (2001) Interferon-alpha-induced endogenous superantigen. A model linking environment and autoimmunity. *Immunity* **15(4):** 591–601.

Stoffel W, Borberg H, Greve V. (1981) Application of specific extracorporeal removal of low density lipoprotein in familial hypercholesterolemia. *Lancet* **ii:** 1005–7.

Stummvoll GH, Aringer M, Smolen JS, *et al.* (2005) IgG immunoadsorption reduces systemic lupus erythematosus activity and proteinuria: A long term observational study. *Ann Rheum Dis* **64:** 1015–1021.

Sueoka A. (1997) Present status of apheresis technologies, part 3: Adsorbent. *Ther Apheresis* **1:** 271–83.

Sugimoto K, Yamaji K, Yang KS, *et al.* (2006) Immunoadsorption plasmapheresis using a phenylalanine column as an effective treatment for lupus nephritis. *Therapeutic Apheresis and Dialysis* **10(2):** 187–192.

Suzuki K, Satoh A, Hidaka T, *et al.* (2000) Brief report adsorption of anti-annexin V Using dextran sulfate bound cellulose beads. *Journal of Clinical Apheresis* **15:** 262–265.

Suzuki K, Shimazaki M, Kutsuki H. (2003) β_2-microglobulin-selective adsorbent column (lixelle) for the treatment of dialysis-related amyloidosis. *Therapeutic Apheresis and Dialysis* **7(1):** 104–107.

Suzuki K. (2000) The role of immunoadsorption using dextran-sulfate cellulose columns in the treatment of systemic lupus erythematosus. *Therapeutic Apheresis* **4(3):** 239–243.

Terman DS, Petty D, Harbeck R, *et al.* (1977) Specific removal of DNA antibodies in vivo by extracorporeal circulation over DNA immobilized in collodion charcoal. *Clin Immunol Immunopathol* **8:** 90–96.

Tetsuzo A. (1996) The current status of extracorporeal immunomodulation. *Transfus Sci* **17(3):** 439–444.

Tzartos SJ, Bitzopoulou K, Gavra I, *et al.* (2008) Antigen-specific apheresis of pathogenic autoantibodies from myasthenia gravis sera. *Ann NY Acad Sci* **1132:** 291–299.

Uzun K, Say R, Ünal S *et al.* (2009) Hepatitis B surface antibody purification with hepatitis B surface antibody imprinted poly(hydroxyethyl methacrylate-N-methacryloyl-l-tyrosine methyl ester) particles. *Journal of Chromatography B* **877:** 181–188.

Uzuna L, Yavuza H, Osmanb B, *et al.* (2010) Poly(hydroxyethyl methacrylate) based affinity membranes for in vitro removal of anti-dsDNA antibodies from SLE plasma. *International Journal of Biological Macromolecules* **47:** 44–49.

Von Appen K, Weber C, Losert U, *et al.* (1996) Microspheres based detoxification system: A new method in convective blood purification. *Artif Organs* **20:** 420–425.

Wang ZM, Garkavij M, Tennvall JG, *et al.* (2002) Application of extracorporeal immunoadsorption to reduce circulating blood radioactivity after intraperitoneal administration of indium-111–HMFG1–biotin. *Cancer* **94(4):** 1287–1292.

Wiesenhuter C, Irish B, Bertram J. (1994) Treatment of patients with refractory rheumatoid arthritis with extracorporeal protein-A immunoadsorption column: A pilot trial. *J Rheumatol* **21:** 804–12.

Winchester J, *et al.* (2003) Sorbents in acute renal failure and the systemic inflammatory response syndrome. *Blood Purification* **21:** 79–84.

World Health Organization. (2008) Hepatitis B, Fact Sheets, No. 204.

Xu XF. (2013) Clinical trials and nurse of hemoperfusion in dermatoses. *Journal of Youjiang Medical University for Nationalities* **35:** 433.

Yamagishi S, Imaizumi T. (2005) Diabetic vascular complications: Pathophysiology, biochemical basis and potential therapeutic strategy. *Curr Pharm Des* **11:** 2279–99.

Yamazaki Z, Fujimori Y, Takahama T. (1982) Efficiency and biocompatibility of a new immunoadsorbent. *Transactions of the American Society for Artificial Internal Organs* **28:** 318.

Yang L, Cheng Y, Yan WR, *et al.* (2004) Extracorporeal whole blood immunoadsorption of autoimmune myasthenia gravis by cellulose tryptophan adsorbent. *Artif Cell Blood Sub* **32:** 519–528.

Yang Y, Chen CZ. (2002) A new immunoadsorbent with phenylanaline as ligand for treatment of RA, Master's thesis, China.

Yang Y, Yu YT, Song JC, *et al.* (1988) A new DNA immune adsorbent for hemoperfusion in SLE therapy — A clinical trial. *Artif Organs* **12:** 444–448.

Yang Y, Yu YT, Song JC, *et al.* (1987) Preparation and clinical trial of a new DNA immune adsorbent for hemoperfusion in systemic lupus erythematosus (SLE) therapy. *Artif Organs* **11:** 334.

Yatzidis H. (1964) Research on extrarenal purification with the aid of activated charcoal. *Nephron* **72:** 310–312.

Yavuz H, Denizli A. (2003) Immunoaffinity beads for selective removal of cholesterol from human plasma. *Journal of Biomaterials Science, Polymer Edition* **14(5):** 395–409.

Yu YH, He BL. (1999) The preparation of immunoadsorbents and their adsorption properties for anti-DNA antibodies in SLE serum. *Reactive & Functional Polymers* **41:** 191–195.

Yu YT, Fu CX, Kong DL, *et al.* (1998). Immunoadsorbents for systemic lupus erythematosus and rheumatoid arthritis therapy. *Artif Cell Blood Sub* **26:** 94.

Yu YT, Chen CZ, Kong DL. (1998) Bead type cellulose based DNA immunoadsorbent. CN 98102354.1.

Yu YT, Chen CZ, Kong DL. (1998) Method for preparing a carbonized resin DNA immunoadsorbent. CN 98102355X.

Yu YT, Chen CZ, Kong DL. (2001) Method for preparing a carbonized resin DNA immunoadsorbent. US 6262172.

Yu YY, Zhang HM, Wang WC, *et al.* (2008) Selective Adsorbent for the removal of immunoglobulin E in bronchial asthma. *Artif Cell Blood Sub* **36:** 63–73.

Yue RZ, Liu T, Su BH. (2008) Preparation of Cu2+-loaded porous chitosan particles for immunoglobulin G adsorption. *Therapeutic Apheresis and Dialysis* **12(3):** 209–215.

Zhang C, Li B, Zhang XR, *et al.* (2012) Effect of hemoperfusion on serial cytokines of children with Henoch–Schonlein purpura. *Chinese Journal of Practical Medicine* **39:** 1–2.

Zhang YR, Lapidos KA, Gal-Moscovici A, *et al.* (2014) A receptor-based bio-adsorbent to target advanced glycation end products in chronic kidney disease. *Artificial Organs* **38(6):** 474–483.

Zhu BR, Iwata H, Kong DL, *et al.* (1999) Preparation of DNA-immobilized immunoadsorbent for treatment of systemic lupus erythematosus. *J Biomater Sci Polym Edn* **10:** 341–350.

Zhu Y. (2010) Socioeconomic impact of systemic lupus erythematosus in Hong Kong: Direct, indirect costs and health-related quality of life. PhD dissertation.

Zisimopoulou P, Lagoumintzis G, Poulas K, *et al.* (2008) Antigen-specific apheresis of human anti-acetylcholine receptor autoantibodies from myasthenia gravis patients' sera using escherichia coli-expressed receptor domains. *Journal of Neuroimmunology* **200:** 133–141.

Chapter 21

Natural Polysaccharide Adsorbent in Hemoperfusion

Shenqi Wang*,‡, Yaoting Yu†, and Yanmiao Fan*

1. Introduction

Polysaccharides are natural biomacromolecules composed of long chains of monosaccharide units bound together by glycosidic linkages, and following hydrolysis, yield the constituent monosaccharides or oligosaccharides. Polysaccharides range in structure from linear to highly branched (Varki *et al.*, 1999). Natural polysaccharides can be classified into different types based on the distinct properties of their monosaccharide building blocks. Polysaccharides can be amorphous, or even insoluble in water (Varki *et al.*, 2008, 1999). Natural saccharides are made of simple carbohydrates called monosaccharides, with the general formula $(CH_2O)_n$, where n is a number greater than or equal to three. Cellulose and chitin are examples of structural polysaccharides. Due to high biocompatibility and excellent processing, natural polysaccharides can be easily formed into

*Advanced Biomaterials and Tissue Engineering Center & School of Life Science and Technology, Huazhong University of Science and Technology, Wuhan, 430074, P. R. China.

†The Key Laboratory of Bioactive Materials, Ministry of Education, Nankai University, Tianjin, China 300071.

‡Corresponding author.

microbeads and membranes. Agarose (agar/agarose), cellulose, dextran and chitosan have been widely used to prepare adsorbents for use in hemoperfusion.

Chitosan, the deacylated derivative of chitin, is a natural polysaccharide that is extracted from the exoskeletons of crustaceans, or generated by fungal fermentation processes. Chitosan is a β-1,4-linked polymer of 2-amino-2-deoxy-d-glucose, allowing it to carry a positive NH_3^+ charge (Fig. 21.1; Liu *et al.*, 2010).

Dextran is a complex, branched glucan, composed of chains of various lengths (from 3 to 2000 kilodaltons). The linear chains consist of α-1,6 glycosidic linkages between glucose molecules, and branches formed from the α-1,3 linkages (Fig. 21.2; Lewis *et al.*, 2010).

Agarose is a linear polymer made up of repeating agarobiose units. Agarobiose is a disaccharide, consisting of D-galactose and 3,

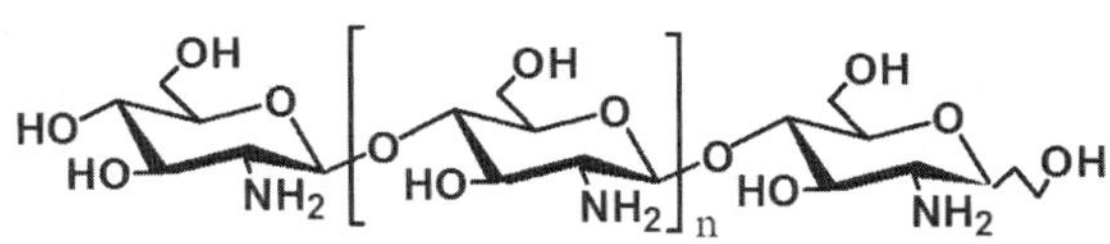

Fig. 21.1. Structure of chitosan.

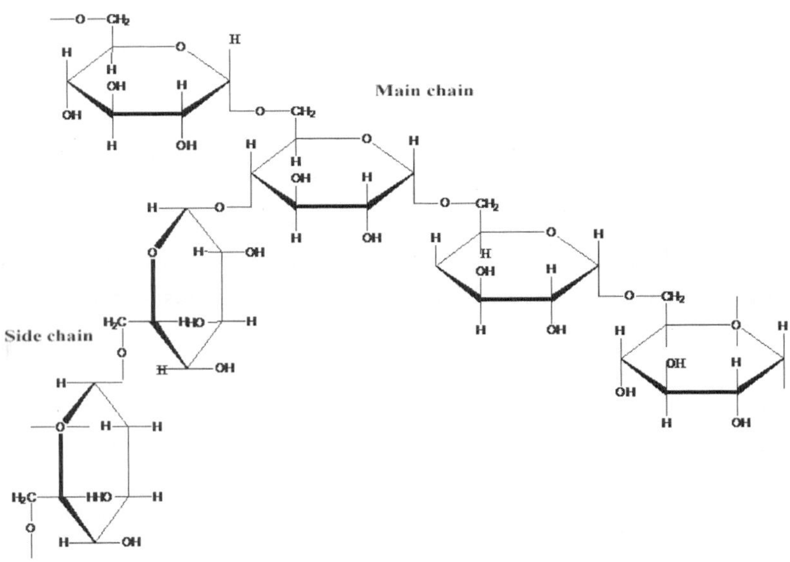

Fig. 21.2. Structure of dextran.

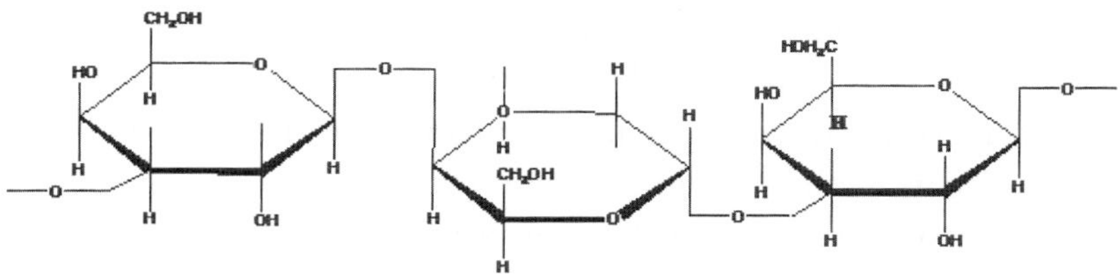

Fig. 21.3. Structure of agarose.

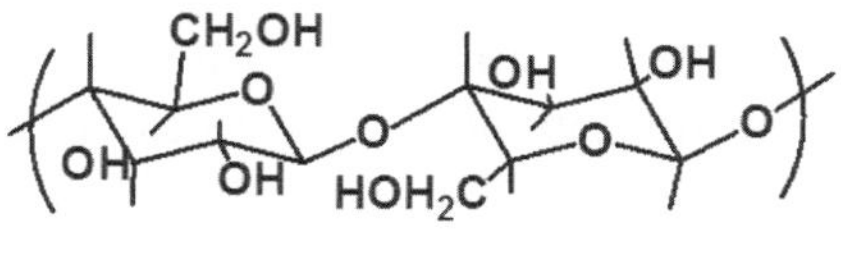

Fig. 21.4. Structure of cellulose.

6-anhydro-L-galactopyranose that is usually extracted from seaweed. Agarose is one of the two principal components of agar, and is purified from agar by removing agar's other component, agaropectin (Fig. 21.3; Wikipedia public domain).

Cellulose, a natural polymer, is an organic compound with the formula $(C_6H_{10}O_5)n$. It is a polysaccharide consisting of a linear chain of several hundred to many thousands of $\beta(1\rightarrow4)$ linked D-glucose units (Fig. 21.4).

In this chapter, we will introduce how to prepare, activate and evaluate natural polysaccharide beads and describe their adsorption capacities for the removal of pathogens in blood; moreover, we will introduce their applications in hemoperfusion in detail.

2. Preparation, Activation and Evaluation of Natural Polysaccharide Adsorbent in Hemoperfusion

2.1. *Preparation and Activation of Agarose (Agarose/Agar) Beads*

2.1.1. *Preparation of agarose (agarose/agar) beads*

To prepare the agarose beads, 400 mL of chloroform/toluene (chloroform to toluene ratio was 1:3.2) and 0.3 mL of Tween80 were mixed in a conical flask and heated to 60°C using a water bath. Next, 100 mL of 4% (wt%) agar solution was added, and an electronic stirrer was used to disperse the agar into uniform liquid beads. Following uniform dispersal, the reaction system was cooled to room temperature and was sieved to collect the agar beads. The mechanical strength of these agar beads was not high enough for further use, so they were crosslinked with EPI or glutaraldehyde before activation.

Agarose beads can be prepared using inverse-suspension crosslinking polymerization and emulsion techniques (Sun *et al.*, 2012). In this experiment, emulsion was used because it creates uniform porous agarose beads. The agarose was dissolved in boiling water to form a 6% agar solution for use in the water phase. A mixture of hexamethylene containing 0.3–4.0 wt% of Span-85 emulsifier and carbon tetrachloride containing 0.3–4.0 wt% of Tween emulsifier (v/v: 4:1) was used as the oil phase. The water phase was sprayed into the freezing oil phase under 0.5 MPa pressure of nitrogen gas at 60–65°C by a stainless steel container with a certain diameter nozzle to form uniform W/O emulsion. The emulsion was cooled and stirred for 10 minutes. The organic phase was discarded after one hour, and the water phase was collected and washed completely with water, 50% ethanol, and water again. Un-uniform sized agarose beads were then screened.

2.1.2. *Activation of agarose (agarose/agar) beads*

Crosslinked agarose is usually preferred over the non-crosslinked variety that is used in affinity applications. However, the crosslinking process results in the loss of approximately 30–50% of the activatable hydroxyl groups because they are consumed in the course of crosslinking the agarose beads (Hermanson GT *et al.*, 1992). Activation is used to efficiently link biospecific ligands to an insoluble matrix. Therefore, a ligand linked in a matrix should not suffer any adverse effects. Cyanogen bromide (CNBr), N-hydroxy succinimide esters (NHS), p-toluenesulfonyl chloride, sodium periodate, carbonyl diimidazole (CDI), epoxy chloropropane (EPI) and 2-fluoro-1-methylpyridinium tolune-4-sulfonate(FMP) are among

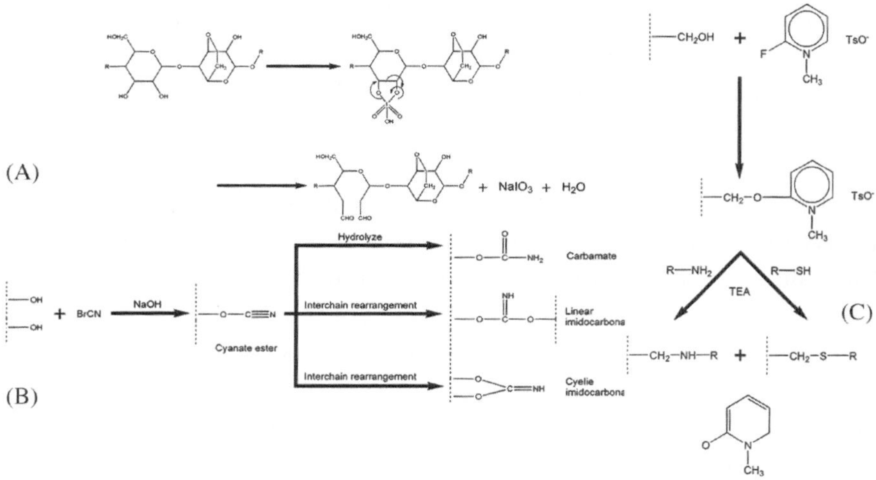

Fig. 21.5. Agarose activation mechanism by (A) Sodium periodate, (B) FMP and (C) CNBr.

some of the widely used compounds to activate agarose and agar (Fig. 21.5).

Although CNBr activation is the most commonly used method to create agarose beads, it has two significant disadvantages. First, this method can result in unstable *N*-substituted isourea bonds that can cause ligand leaking from the carrier. Second, this method can cause high nonspecific adsorption (Herman *et al.*, 1992). Recently, enzyme-mediated activation was also used to activate agarose beads due to its simple and mild reaction condition (Mendoza *et al.*, 2011).

2.2. *Preparation and Activation of Cellulosic Beads*

2.2.1. *Preparation of cellulosic beads*

One hundred grams of medical grade cotton was soaked in a 19% NaOH solution for three hours at room temperature. The cotton was squeezed, weighed, and then placed in a 1,500 mL conical flask at 25°C for three days. Fifty milliliters of carbon disulfide were added to the solution, which was then sealed and aged for five hours to convert the cellulose into a viscose solution. The viscose solution was then diluted to 1,000 mL with a 6% NaOH solution, to make a 10% cellulose viscous solution. In a reactor equipped with a stirrer, 800 mL of the mixture was stirred for 30 minutes at 300 rpm at room temperature. Then, 300 mL of 10% chlorobenzene, 200 mL of carbon tetrachloride and 2.0 g of potassium oleate were slowly stirred into the cellulose viscose solution for 30 min, until the liquid particles were uniformly dispersed. The temperature was then slowly raised to 90°C and kept constant for 2.5 hr, after which it was cooled to room temperature to solidify the liquid particles into resin beads. Cellulosic beads were filtered using 20–40 mesh and washed thoroughly with alcohol and distilled water to remove any impurities.

Compared to the gel cellulosic beads, macroporous beads can greatly enhance the adsorption capacity for middle and high molecules in the therapeutic embolization of meningiomas (Kai *et al.*, 2006; Hamada *et al.*, 1996a, 1996b; Kai *et al.*, 2007), and can be synthesized according to Wang *et al.* (2007) and Bai *et al.* (2006). A certain amount of pore-forming agent, such as calcium carbonate granules with an average diameter of about 0.2 mm, was added to a 10% viscous solution of cellulose, then mixed and dispersed to form cellulosic beads. After washing with diluted HCl solution to remove the pore-forming agent, various kinds of porous adsorbents can be prepared. Alternatively, macroporous cellulose beads can be prepared from a cellulose solution in ionic liquid by double emulsification techniques (Du *et al.*, 2010; Wang *et al.*, 2008).

Recently, cellulosic microspheres with a particle size below 5 μm have been adopted for use in blood purification (Thummemler *et al.*, 2011; Weber *et al.*, 2010), and can be an excellent matrix for the preparation of adsorbents for use in hemoperfusion.

2.2.2. *Activation of cellulose beads*

Epichlorohydrin is frequently used to prepare cellulosic adsorbents because it can easily activate cellulose (Fang *et al.*, 2004; Weber *et al.*, 2005; Zhou *et al.*, 1998). Ten grams of cellulosic beads was activated

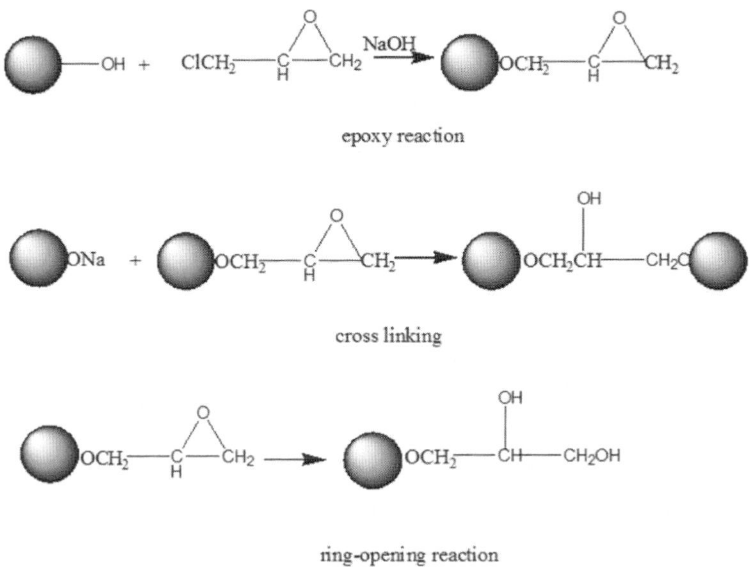

Fig. 21.6. Reaction activation of cellulosic beads. (With copyright permission from Intech Press.)

with 10 mL of epichlorohydrin in 20 mL 2 mol/L sodium hydroxide solution. The mixture was stirred at 40°C for 4 hr. Then, the epoxy-activated cellulosic beads were washed thoroughly with distilled water and further reacted with amino acids or proteins (Fig. 21.6). The concentration of sodium hydroxide solution used in the condensation reaction plays an important role in determining the number of activated epoxy groups linked to cellulose. This is attributed to the condensation and ring opening reaction of the epichlorohydrin molecule that competes in the reactions (Fig. 21.7).

2.3. *Preparation and Activation of Chitosan Beads*

2.3.1. *Preparation of chitosan beads*

Two methods were widely used to prepare chitosan microbeads. In the inversion suspension crosslinking polymerization method, formaldehyde was used as a pre-crosslinker to crosslink the chitosan

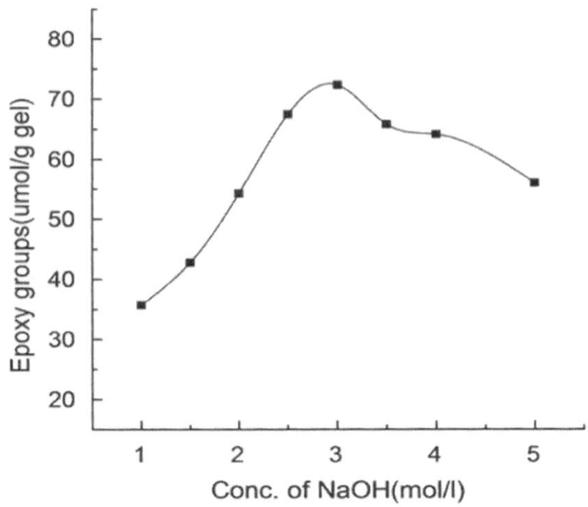

Fig. 21.7. Amount of cellulose epoxy groups versus NaOH concentration. (With copyright permission from Intech Press)

liquid beads dispersed in the solute and obtain EPI crosslinked chitosan microbeads at alkaline condition. Ten grams of chitosan were heated to 60°C, and 3 mL 30% (v%) H_2O_2 solution was added to the mixture and stirred for 30 minutes. Then, a certain amount of chitosan was dissolved into a 3% (wt%) solution of acetatic acid. Next, liquid paraffin and Span80 were added and stirred at 30°C for 30 minutes. The solution was then raised to 60°C, and a certain amount of formadehyde solution was added. The mixture was allowed to react for one hour, after which the pH of the solution was adjusted to 10. Next, a certain amount of EPI solution was added, and the solution was raised to 70°C. A 5% (wt%) solution of NaOH was dripped into the reaction system for four hours in order to maintain a pH value of 10. Finally, the chitosan microbeads crosslinked by EPI were collected by filtering the solution and washed in distilled water.

The phase-inversion method was used to prepare chitosan microbeads (Fu *et al.*, 2003). One gram of chitosan was dissolved into 19 mL of a 3% (wt%) acetic acid solution. The chitosan solution was then sprayed into an 8% (wt%) NaOH aqueous solution. Chitosan is insoluble in basic solutions, so this resulted in the formation of porous beads. The mean diameter of the chitosan beads was determined

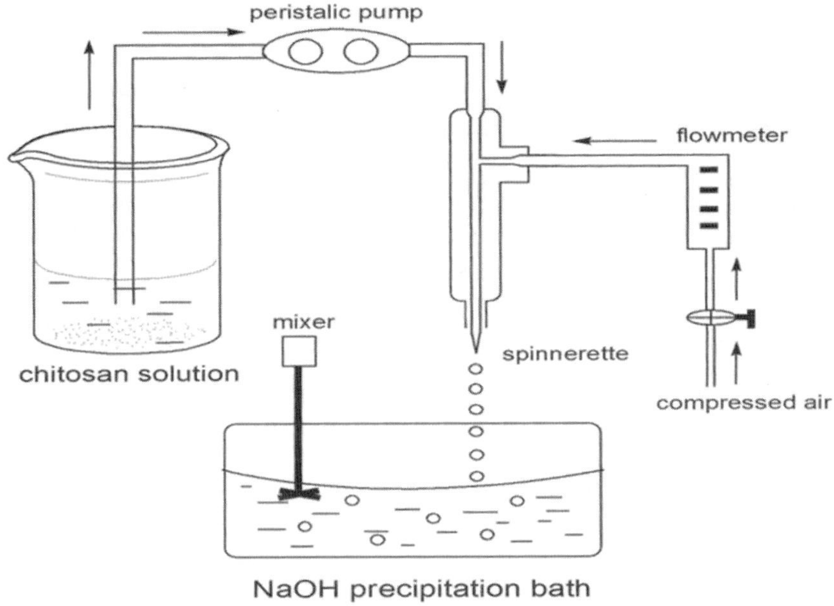

Fig. 21.8. Apparatus for preparation of chitosan beads.

using the diameter of syringe needle (Fig. 21.8), and they were generally 0.2–0.5 mm in diameter. The chitosan beads were then washed thoroughly with water to remove any impurities.

2.3.2. *Activation of chitosan beads*

Chitosan can be easily epoxidized by EPI under alkaline conditions because it has many active amine and hydroxyl groups (Fu *et al.*, 2003). Chitosan can be carboxylated by reacting it with chloro alkyl acid or glyoxylic acid (Hermanson *et al.*, 1992). It can also be acylated by reacting with saturated acid anhydrides (Jiang, 2001), etherificated by reacting it with alkaline chitosan and an etherification reagent, alkylated by reacting with sulfate or halogenated hydrocarbons (Hermanson *et al.*, 1992), and esterificated by phosphorus oxychloride, or by chitosan beads reacting with concentrated sulfuric acid. Acylation, carboxylation and esterification have been widely used to activate chitosan microbeads for use in hemoperfusion (Fig. 21.9).

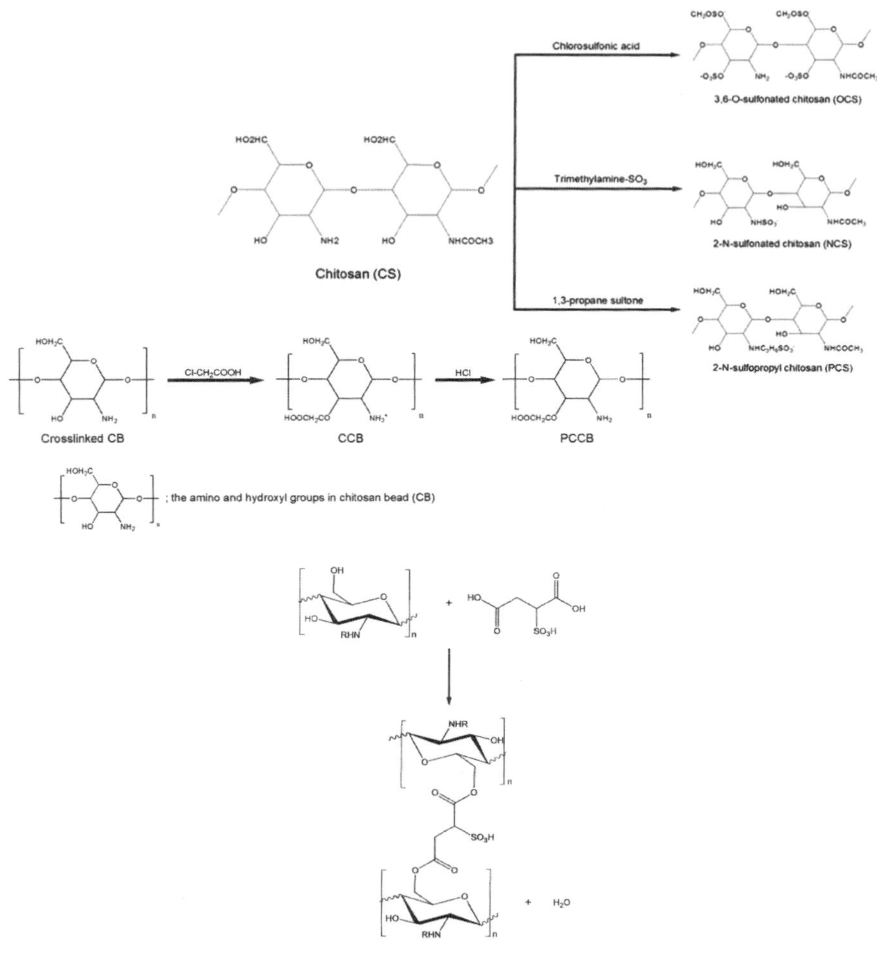

Fig. 21.9. The mechanism of chitosan activation by acylation, carboxylation and esterification.

2.4. *Preparation and Activation of Dextran Beads*

2.4.1. *Preparation of dextran beads*

Five grams of dextran were dissolved into 10 mL of distilled water using a magnetic stirrer. Then, 6 mL of EPI and 3 g of Span80 were uniformly mixed with 50 mL of liquid paraffin. The temperature was then raised to 70°C, and 4 mL 50%(wt%) solution of sodium hydroxide was slowly added. After two hours, the temperature was raised to 90°C for six hr. The solution was then cooled to room temperature to solidify the liquid particles into resin beads. Dextran beads were immersed in distilled water for 24 hr and then filtered using 20–40 mesh and washed thoroughly with alcohol and distilled water to remove any impurities.

2.4.2. *Activation of dextran beads*

The above activation methods used in cellulose, agarose, and chitosan beads can also be used to activate dextran beads.

2.5. *Evaluation of Natural Polysaccharide Beads*

Bead porosity and density were calculated using the following equations (Fang *et al.*, 2004; Bai *et al.*, 2006):

$$P = \frac{\rho_s \times Q}{\rho_s Q + (1 - Q)\rho_{H_2O}} \times 100\%, \tag{1}$$

$$D_P = \frac{wt_w}{V_w}, \tag{2}$$

$$Q = \frac{wt_w - wt_d}{wt_w} \times 100\%, \tag{3}$$

where P stands for porosity percentage, ρ_s stands for skeleton density, Q stands for water content, ρ_{H_2O} stands for water density, wt_w stands for the wet weight of the beads, wt_d stands for the dry weight of the beads, V_w stands for the volume of the wet beads, and Dp stands for packing density. Adsorption percentage and capacity were calculated using the following equations:

$$AP = \frac{[C]_B - [C]_A}{[C]_B} \times 100\%, \tag{4}$$

$$AC = ([C]_B - [C]_A) \times V_P, \quad (5)$$

where *AP* and *AC* stand for adsorption percentage and adsorption capacity, $[C]_B$ is the concentration before adsorption, $[C]_A$ is the concentration after adsorption, and V_p is the volume of plasma used during adsorption.

3. Typical Natural Polysaccharide Bioactive Bead Adsorbent for use in Blood Purification

3.1. *Agarose/Agar-based Adsorbent in Hemoperfusion*

3.1.1. *Agarose bead-based adsorbent for the treatment of IgA nephropathy*

In order to develop more cost effective and higher adsorption capacity Ig A adsorbents for the treatment of IgA nephropathy, Chu and Yu (2001) developed a series of agarose/agar-based IgA adsorbents and constructed an immunosorbent for the removal of IgA from patient serum by using CDI (N,N'-Carbonyldiimidazole) as an activator, agarose as a carrier and human immunoglobulin G(IgG) as a ligand. They studied the various factors affecting the immobilizing condition and adsorption capacity of IgA, and found that the immobilizing capabilities of IgG are optimal at pH values of 4.0 or 10.0 because IgG is easily denatured in weak acids and strong bases (Table 21.1). Therefore, they confirmed that the optimal pH for immobilization of IgG is 8.5.

The adsorption results indicated that the adsorption percentage of the adsorbent for IgA could reach 30–35%. Xing *et al.* (2003) prepared an immunoadsorbent that used amino acid as ligand for the removal of IgA. Adsorption experiments of the immunosorbent showed that it could remove lgA in serum (Table 21.2). Experiments showed that the immunosorbent was not toxic and had high

Table 21.1. Effect of pH on Coupling Efficiency of IgG

No.	pH Value	Coupling Efficiency (%)
1	4.0	68.2
2	5.0	56.3
3	6.0	52.2
4	7.0	41.5
5	8.0	54.2
6	9.0	59.3
7	10.0	67.2

Note: Initial concentration of IgG buffer is 2mg/mL; the ratio of agar to buffer solution is 1:1 (w:v); the solution reacted at 25°C for 2h. *Source*: Chu *et al.* (2002) *Ion Exchange and Absorption* **18(4):** 369–373.

hemocapability. Although its adsorption capacity is lower, it also can be used as immunosorbent of lgA nephropathy.

In order to improve the adsorption percentage for IgA, Chu *et al.* (2003) used glycidyl 1,4 butanediol to activate the agarose beads. Glycidyl 1,4 butanediol can easily couple with the human IgG, and has a long spacer that can effectively reduce the hindrance effect of IgG coupling reactions. *In vitro* experiments showed that the immunosorbent could adsorb 60% of IgA in plasma (Table 21.3). The agarose bead-based immunosorbents have higher selectivity for IgA. In order to improve the adsorbents' specific adsorption for IgA, Chu and Yu *et al.* (2004) constructed an immunosorbent with anti-human IgA as ligand. First, active epoxy groups were introduced into the agarose beads by reacting agarose with epoxy chloropropane; then the beads were reacted with ammonium hydroxide and glutaraldehyde to introduce a spacer, finally sheep anti-human IgA antibodies were linked to the spacer as a ligand. *In vitro* experiments showed that this adsorbent could effectively remove excess IgA in plasma, and the adsorption capacity reached 74%.

Table 21.2. Adsorption Efficiency of Agarose Carrier Coupling Amino Acids in Differing pH Conditions

		pH		
Ligand		**5.0**	**6.0**	**7.0**
Trp	Adsorption capacity (mg/0.34 mg)	109	70	124
	Adsorption efficiency (%)	34.71	22.29	39.49
Glu	Adsorption capacity (mg/0.34 mg)	52	95	53
	Adsorption efficiency (%)	16.56	30.25	16.88
Asp	Adsorption capacity (mg/0.34 mg)	98	56	83
	Adsorption efficiency (%)	31.21	17.83	26.43
Tyr	Adsorption capacity (mg/0.34 mg)	59	99	67
	Adsorption efficiency (%)	18.79	31.53	21.34
Phe	Adsorption capacity (mg/0.34 mg)	96	74	99
	Adsorption efficiency (%)	30.57	23.57	31.53
Lys	Adsorption capacity (mg/0.34 mg)	73	97	81
	Adsorption efficiency (%)	23.25	30.89	25.80

Source: Xing *et al.* (2003) *Ion Exchange and Adsorption* **19(3):** 241–246.

3.1.2. *Agar bead-based adsorbent for the treatment of rheumatoid arthritis*

Fu *et al.* (2002) synthesized an agar gel bead-based rheumatoid arthritis (RA) immunosorbent by linking aggregated IgG to agar gel beads. Experimental results showed that the adsorbents had a high adsorption capacity for rheumatoid factors (RF) and immunological complexes (IC) (Tables 21.4 and 21.5). In addition, the synthesized adsorbent was stable, compatible with blood, and its mechanical strength was suitable for hemoperfusion.

3.1.3. *Agar bead-based IgE adsorbents for the treatment of bronchial asthma*

Bronchial asthma is a tracheal over-reacted disease stimulated by antigens and non-antigens. The pathology and mechanism of bronchial

Table 21.3. Adsorption Time Interval and Immunoglobulin Adsorption Percentage (30°C)

	Adsorption Time/hr					Adsorption Time/hr			
Immunoglobulin	0.5	1	2	4	Immunoglobulin	0.5	1	2	4
IgE	11.90	13.90	14.86	15.28	IgM	4.95	5.86	6.21	6.85
IgG	18.25	20.66	21.05	21.36	IgA	58.4	59.7	61.3	63.7

Source: Chu *et al.* (2003) *Chem J Chinese Universities* **24:** 636–638.

Table 21.4. Adsorption Capacity for RA Factors

Amount of Immobilization (mg/g gel)	IgMRF		IgGRF		IgARF	
	Adsorption Capacity (IU/g)	Adsorption Percentage (%)	Adsorption Capacity (IU/g)	Adsorption Percentage (%)	Adsorption Capacity (IU/g)	Adsorption Percentage (%)
Control	795	25.2	810	23.4	300	13.3
2.19	2,310	73.3	1,695	49.0	765	34.0
2.65	2,445	77.6	1,863	47.0	1,200	53.3
4.00	2,640	84.0	2,580	74.8	1,965	87.3
5.33	2,520	80.0	2,160	62.6	1,890	84.0
6.35	2,490	79.0	1,800	52.2	1,140	50.7

Note: The initial concentration of IgMRF, IgGRF and IgARF were 1,050, 1,150 and 750 IU/mL.

Source: Fu *et al*. (2002) *Chinese Journal of Biomedical Engineering* **21(1)**: 1–4.

Table 21.5. Adsorption Capacity for IC

	IC	
Amount of Immobilization (mg/g gel)	**Adsorption Capacity (μg/g gel)**	**Adsorption Percentage (%)**
Control	49.5	12.0
2.19	199.5	48.4
2.65	239.3	58.0
4.10	297.0	72.0
5.33	297.0	72.0
6.35	260.7	63.2

asthma is still unknown. However, it is believed that asthma is a type I allergic disease that is controlled by the immunoglobulin E (IgE) antibody; therefore, the total level of IgE in patients with bronchial asthma is remarkably higher than normal levels (430 ng/mL). Sato *et al.* demonstrated in the 1980s that blood perfusion for asthma using immunosorbents can effectively alleviate bronchial arthritis. Many researchers began to develop and innovate adsorbents for treatment of bronchial asthma. Yu *et al.* (2008) developed an agar gel bead type adsorbent that contained succinic anhydride as spacer, and tryptophan as the ligand. They found that the spacer could enhance the adsorption capacity up to 70.73% (1309.403 IU/g agar-gel) which was higher than that without a spacer (53%) (Fig. 21.10).

The changes in IgG, IgA, IgM and C_3 and C_4 content in blood were evaluated before and after adsorption (Table 21.6). The maximal adsorption percentage for IgA was 26.17% (Table 21.6), and is much lower than IgE (70.73%), signifying that the adsorbent has a good selectivity for IgE.

The adsorption percentage for the leucocytes, erythrocytes and platelets was 27.3%, 2.4% and 6.9%. These satisfactory blood compatibility levels of the adsorbent show a high potential use in blood purification.

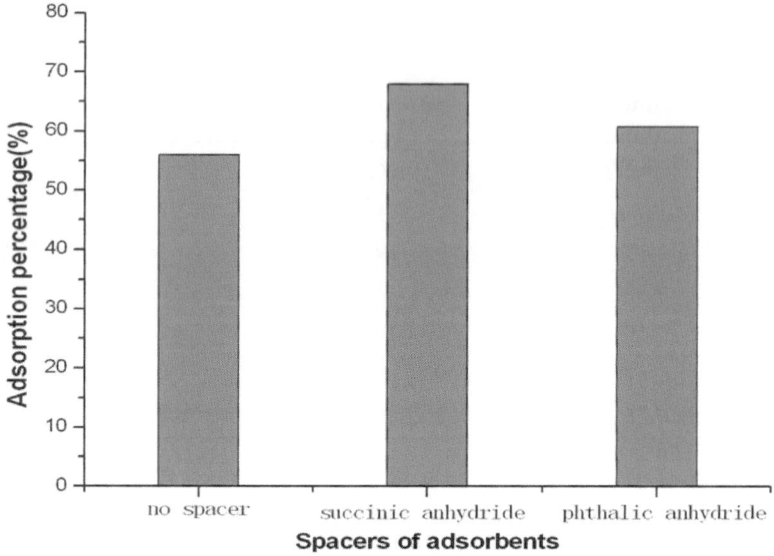

Fig. 21.10. Effect of spacers on adsorption capability. Adsorption dynamics study of adsorbent containing succinic anhydride spacer and tryptophan ligand.

(*Source*: Yu *et al.* (2008) *Artif Cell Blood Sub* **36:** 63–73, with copyright permission from Taylor & Francis Group Press.)

Table 21.6. Changes in IgG, IgA, IgM, C3, and C4 Before and After Adsorption

Adsorption Time (h)	0	0.5	1.0	1.5	2.0	Adsorption Percentage (%)
IgG	12.30	11.26	11.26	11.24	10.18	17.24
IgA	2.14	1.95	1.77	1.58	1.58	26.17
IgM	0.51	0.51	0.49	0.48	0.48	5.88
C_3	1.12	1.12	1.11	0.99	0.97	13.39
C_4	0.44	0.40	0.40	0.36	0.36	18.18

Source: Yu *et al.* (2008) *Artif Cell Blood Sub* **36:** 63–73, with copyright permission from Taylor & Francis Group Press.

3.1.4. *Agar bead-based adsorbents for the removal of TNF-α*

Wei and Yu *et al.* (2004) prepared an agarose gel-based TNF-α adsorbent with IgG as ligand because IgG can specifically bind to TNF-α with an adsorption capacity of 3,339.75–3,586.45 U/mL, and an adsorption percentage of 39%, which indicates that it can be used in the removal of endotoxins in human plasma. Moreover, Wei and Yu *et al.* (2004) investigated the adsorption mechanism of TNF-α with IgG ligands by modifying the IgG molecules using chemical reagents. The results showed that the amino group had a significant influence on the adsorption capacity, the carboxyl and indole groups had a negative effect, and the guanidine and imidazole groups had no effect on adsorption capacity. Furthermore, 1, 6-hexamethylene diamine (DAH) was used as a spacer to link the agar gel support and the IgG ligand, a favorable method for reducing the steric hindrance and elevating the adsorption capacity (Fig. 21.11).

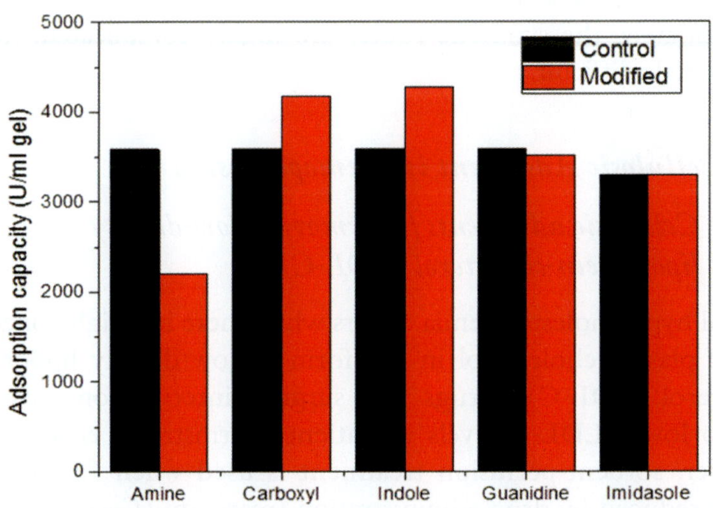

Fig. 21.11. Comparison of the modification effect between the different groups.
Source: Adapted from Wei *et al.* (2004) *Chemical Journal of Chinese Universities* 7: 1277–1281.

3.1.5. *Agar bead-based adsorbents for the removal of LDL*

Cao *et al.* (2001) studied the interaction mechanisms of human low-density lipoprotein (LDL), and very-low-density lipoprotein (VLDL) with cholesterol modified dextran (CHD) in serum, using turbidity and high performance liquid chromatography (HPLC). They found that CHD could induce self-aggregation of LDL and VLDL. Based on these results, Cao *et al.* (2002) prepared agar bead-based LDL adsorbents with cholesterol modified dextran as ligands. The ligand of the cholesterol modified dextran was synthesized by the following two steps: (1) synthesis of the cholesterol modified dextran, and (2) the introduction of the amino group on the dextran molecule. The synthesized ligand was then linked to the agar beads using glutaraldehyde. The amount of ligand linked to agar beads was found to greatly affect the adsorption capacity of LDL, TC, TG and HDL (Cao *et al.*, 2002). The adsorbent had optimal adsorption capacity for LDL, TC and TG when the amount of ligand immobilized on carriers reached 7.60 mg/mL gel (Table 21.7).

In addition, the adsorbent had higher selectivity and heat stability. Unfortunately, the synthetic routes are highly complicated, making clinical trials difficult.

3.2. *Cellulosic Adsorbent in Hemoperfusion*

3.2.1. *Cellulosic adsorbents for removing low-density lipoprotein-cholesterol (LDL-C)*

Familial hypercholesterolemia occurs when there is a high concentration of plasma cholesterol in the form of low-density lipoprotein-cholesterol (LDL-C). Drugs and surgical intervention have been used to lower LDL-C levels in patients (Armstrong *et al.*, 1983). However, sorbent-perfusion treatment is used when LDL-C levels are not reduced by drug administration treatment (Cao *et al.*, 2002; Sato *et al.*,1996; Chen *et al.*, 1993), and scientists have been engaged in developing different kinds of adsorbents to remove pathogenic substances since the late 1970s (Sreenivasan *et al.*, 1997; Bosch *et al.*, 1997, 1996).

Table 21.7. Amount of Ligand Attached on Carrier vs Adsorption Capacity

Amount of Ligands on Carrier (mg/mL gel)	TC Amount Adsorbed (mg) (%)	TG Amount Adsorbed (mg) (%)	LDL-C Amount Adsorbed (mg) (%)	HDL-C Amount Adsorbed (mg) (%)
Control	0.275 ± 0.06	0.231 ± 0.11	0.222 ± 0.01	0.041 ± 0.01
	7.4 ± 0.8	9.2 ± 4.4	8.09 ± 0.4	4.79 ± 1.2
1.68	0.956 ± 0.17	0.576 ± 0.07	0.921 ± 0.16	0.106 ± 0.01
	25.6 ± 4.5	23 ± 2.8	33.6 ± 5.8	12.4 ± 1.2
3.20	1.131 ± 0.14	0.651 ± 0.15	1.066 ± 0.04	0.126 ± 0.04
	30.3 ± 3.8	26 ± 6.0	38.8 ± 1.4	14.7 ± 4.7
4.08	1.413 ± 0.29	0.826 ± 0.14	1.281 ± 0.23	0.158 ± 0.03
	37.9 ± 7.8	33 ± 5.6	46.7 ± 8.4	18.5 ± 3.5
6.42	1.849 ± 0.23	0.927 ± 0.34	1.562 ± 0.19	0.163 ± 0.05
	49.6 ± 6.2	37 ± 13.5	56.9 ± 6.9	19.1 ± 5.8
7.60	2.010 ± 0.35	1.077 ± 0.22	1.806 ± 0.41	0.174 ± 0.06
	53.9 ± 9.4	43.0 ± 8.8	65.8 ± 14.9	20.3 ± 7.0

Note: The initial concentrations of TC, TG, LDL-C, and HDL-C were 248.6, 167, 183, and 57 mg/dL, respectively.
Source: Cao Ningning (2000) PhD dissertation, Nankai University.

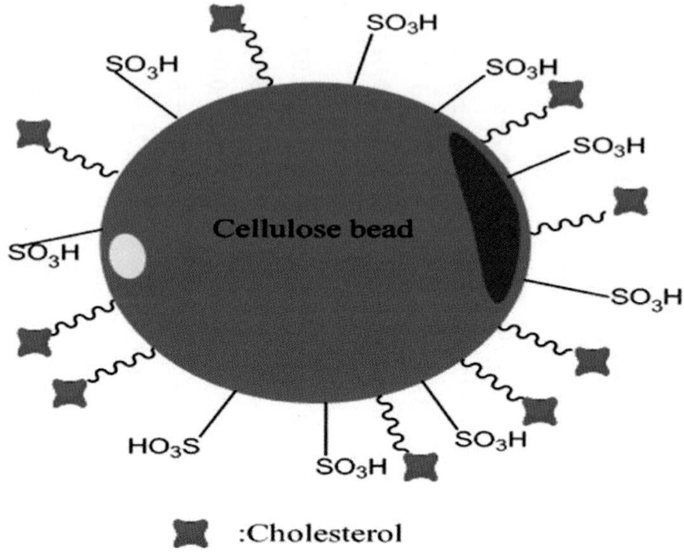

Fig. 21.12. Schematic structure of amphiphilic cellulose adsorbent. Adapted from: Wang SQ *et al.* (2002) *Artif Cells Blood Sub* **30:** 285–292, with copyright permission from Taylor & Francis Group Press.

Yu and Wang *et al.* (2003) developed a cellulosic adsorbent that linked amphiphilic ligand to adsorb LDL-C. Cellulose beads was reacted with cholesterol N-(6-isocyanatohexyl) carbamate in the presence of pyridine in DMSO at 80°C in order to introduce the hydrophilic moiety. This mixture was then reacted with chlorosulfonic acid in dimethyl formamide to introduce the sulfonic group (Fig. 21.12).

The effects of sulfonation and cholesterol grafting time on the swelling properties of the adsorbent were studied. Results showed that the sulfonation and grafting time of the cholesterol was three and five hours, respectively. The amphiphilic adsorbent had a high LDL-C adsorption capacity, without significantly adsorbing high-density lipoproteins. A rabbit model was constructed to test (Chapman *et al.*, 1980). Japanese white male rabbits were purchased from a local experimental animal institute, housed in a standard facility, and were fed standard chow and water for one week. The healthy rabbits were divided into a control group (group 1, $n = 6$) and a hyperlipiedemia

Table 21.8. Removal of Lipoproteins by Amphiphilic Adsorbent

Parameter	Before (mmol/L)	After (mmol/L)	Reduction (%)
TC	8.54 ± 1.01	3.33 ± 0.63*	61.20 ± 2.81
TG	1.845 ± 0.191	1.05 ± 0.153*	43.09 ± 2.43
LDL-C	3.619 ± 0.354	0.724 ± 0.07*	78.56 ± 0.147
HDL-C	0.216 ± 0.06	0.205 ± 0.057	5.09 ± 0.042

Note: $n = 6$,
Source: Wang SQ *et al.* (2003) *Biomaterials* **24:** 2189–2194, with copyright permission from Elsevier Press.

Table 21.9. Removal of LDL-C Subfraction by Amphiphilic Adsorbent

	Before Perfusion		After Perfusion		Change = (A-B)/B × 100%
	Mmol/L	%	Mmol/L	%	
LDL-1	0.51 ± 0.061	13.95 ± 2.1*	0.12 ± 0.046	15.1 ± 2.3	8.24 ± 2.6
LDL-2	1.65 ± 0.019	45.4 ± 5.3*	0.38 ± 0.043	53.4 ± 5.8*	17.36 ± 3.4
LDL-3	1.45 ± 0.052	40.1 ± 4.7*	0.23 ± 0.024	31.6 ± 3.6*	−21.20 ± 3.9

Note: $n = 6$
Source: Wang SQ *et al.* (2003) *Biomaterials* **24:** 2189–2194, with copyright permission from Elsevier Press.

group. Rabbits in the control group consumed water and 120–150 g of standard chow per day. In the hyperlipidemia model group, the rabbits were fed with standard chow supplemented with 0.5–1% cholesterol, 15% egg yolk and 5% animal oil. After eight weeks, the rabbits in the hyperlipidemia group were further divided into two groups, a group that received the same treatment, group No. 2 ($n = 6$), and a group that was treated with sorbent-perfusion ($n = 6$). The LDL-C levels decreased significantly after two hours of perfusion, indicating the adsorbent could effectively remove LDL-C (Table 21.8). Furthermore, the sorbent-perfusion treatment reduced the LDL-C subfractions, decreasing the risk for the development of atherosclerosis and myocardial infarction (Table 21.9).

Table 21.10. Effect of Coupled Phosphates on Adsorption Capacity

Sample		1	2	3	4
Amount of coupled phosphate (μmol/mL)		0	68.4	94.3	128.6
Adsorption capacity (mg/mL)	TC	0.893	0.735	0.939	1.586
	LDL	1.235	1.515	1.707	2.721
	HDL	0.134	0.039	0.058	0.063

Source: Wang T *et al.* (2008) *Chinese Journal of Biomedical Engineering* **27**: 132–136.

The authors also developed the adsorbent with lysine and phosphate groups as the ligand for the treatment of hyperlipidemia (Wang and Yu *et al.*, 2006). Compared to the amphiphilic adsorbent, the lysine adsorbent had a lower LDL-C adsorption capacity and total cholesterol, and a higher HDL-C. However, the phosphorylated cellulosic adsorbent had a higher LDL-C adsorption capacity, and a lower adsorption capacity for HDL-C (Table 21.10).

Haofeng Yu *et al.* (2007) synthesized PAA-grafted cellulosic adsorbent for the removal of LDL-C from human plasma. *In vitro* studies showed that this adsorbent could remove total cholesterol, TC, LDL-C, and triglyceride (TG) at levels of 5.55, 4.46, and 2.48 mg/mL, respectively. Unfortunately, it also removed 30% of HDL-C in the plasma.

3.2.2. *Cellulosic adsorbents for the removal of rheumatoid factors*

Rheumatoid arthritis (RA) is an immune disease that affects approximately 1% of the world population (Spector, 1990). Wang and Yu *et al.* (2004) covalently linked ss-DNA to cellulosic beads for the removal of rheumatoid factors. *In vitro* and *In vivo* studies showed that the adsorbent had a high IgMRF, IgGRF and IgARF adsorption capacities (Fig. 21.13), and was highly compatible with human blood.

Sorbent-perfusion using cellulose adsorbent, with ss-DNA as ligand, was conducted on 35 patients with RA. Forty-six patients were

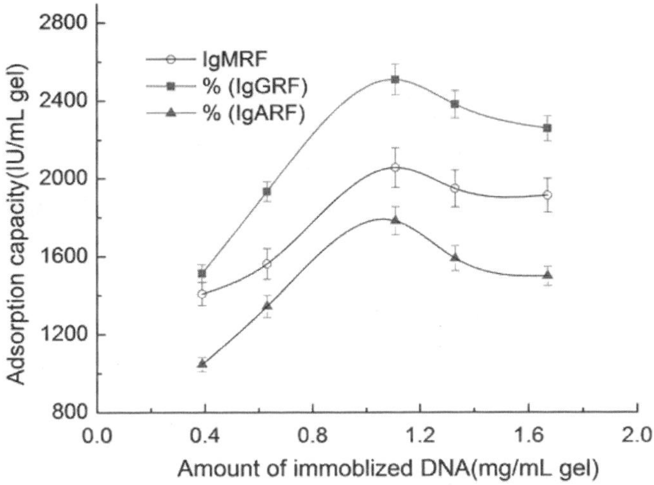

Fig. 21.13. Relationship between adsorption capacity and amount of immobilized DNA.

Source: Wang LY *et al.* (2004) *Chemical Research in Chinese Universities* **20:** 795–800.

hospitalized (Department of Internal Medicine, Zibo Central Hospital, Zibo, Shandong province) from December 1998–November 2000, and diagnosed with RA sickness. The extent of joint pain, swelling, morning stiffness, nodules under the skin, titer of rheumatoid arthritis factor (RF), value of immune-globulin and X-ray diagnosis of sick ankle were among the patient data collected. The patients were divided into several groups. Of the patients, 35 matched American Rheumatoid Arthritis Standard of 1987 (Arnett *et al.*, 1988).

RA patients were divided into four groups according to the functional ability of the diseased joint. Eleven patients had grade II joint sickness, 20 patients had grade III, and one patient was grade IV. The conventional treatment method was performed on 30 of the cases.

Following perfusion, the rheumatoid factor levels were reduced, and the platelet count showed no significant changes in 35 of the RA patients ($P < 0.05$) during treatment. After one week, joint pain, swelling, tenderness, and morning stiffness disappeared in 90 percent of the patients, and 80 percent of the abnormal indices recovered to a normal value. After two to three weeks, joint function in 82 percent of the patients was improved to grades II or I. Two to three months later, X-ray examinations showed that 80 percent of the bone matrix destruction was restored. Follow up tests (0.67–2.5 years) showed that the effective rate (97.14%, 81.25%) and total remission rate (82.86%, 59.38%) were all significantly improved, and patients had a higher rate of recovery when compared to routine therapy.

The results from the above clinical study suggest that sorbent-perfusion using cellulosic adsorbents could be an effective therapy for rheumatoid arthritis. The treatment is safe, because it does not cause hemolysis or pyrogenic side effects, is easy to sterilize and cost effective.

3.2.3. *Cellulosic adsorbents for removal of endotoxin*

Bacterial endotoxins (ET), or lipopolysaccharides (LPS), are components of the outer cell wall of gram-negative bacteria, and could play an important role in the pathogenesis of endotoxemia and septic shock (Raetz *et al.*, 1990; Callery *et al.*, 1991). Hui Fang and Yaoting Yu *et al.* (2004) synthesized an adsorbent for the removal of bacterial endotoxins by covalently immobilizing lysine onto cellulosic beads, and found that the adsorbent was biocompatible (Table 21.11). In order to evaluate the adsorbent's properties, rabbit models were constructed according to the following methods: Thirteen New Zealand white rabbits (weight 2.0–2.5 kg) were injected intravenously with 0.20 mg of LPS (*E. coli O55: B5, sigma*) to induce endotoxemia after being anaesthetized. The rabbits were classified into two groups, an LPS-treated group that was perfused using adsorbents ($n = 8$), and a control group ($n = 5$) that did not undergo perfusion. Sorbent-perfusion was conducted 1.5 hr after LPS administration using conventional perfusion techniques. Blood was drawn from the artery and returned to the vein using a peristaltic pump (Pharmacia-LKB). The perfusion was carried out at a rate of 5 mL/min for 2 hr, and the adsorbent showed a strong ET-binding capacity. After perfusion, the blood ET level decreased from 5.56 ± 0.54 EU/mL to 0.41 ± 0.26 EU/mL (Fig. 21.14). The results of liver function and renal function tests, as well as SOD and

Table 21.11. Biocompatibility and Toxicology Properties of the Adsorbent

Parameters Tested	Results
Hemolysis	1.1% (standard < 5%), (n = 3)
Platelet adhesion	18.76% (n = 3) (10% is excellent; 10–30% is good)
Whole body toxicology	Wt. of rat before test Wt. of rat after test 21.5 ± 0.61 (g) 24.4 ± 2.04 (g) (n = 5)
Allergic reaction	Grade 1, not higher than negative control (n = 10), none allergic
Cytotoxicology	$R^*_{\text{index of sample}}$ = 0/0 (n = 6), none toxic
Skin stimulation	PII** = 0.0-0.4, (N = 15), very mild toxic

$R^*_{\text{index of sample}} = R_{\text{index of extract}} - R_{\text{index of sample tested}}$ = 0.5/0- 0.5/0 = 0/0
PII** is the skin stimulation index.
Note: The biocompatibility and toxicology tests of the adsorbent were conducted by the Testing and Evaluation Research Centre of Biomedical Materials in Tianjin, China according to the Criteria of GB/T16886.5-1997 in correlation to ISO 10993.4:2002
Source: Fang H and Yu YT *et al.* (2004) *Biomaterials* **25:** 5433–5440, with copyright permission from Elsevier Press.

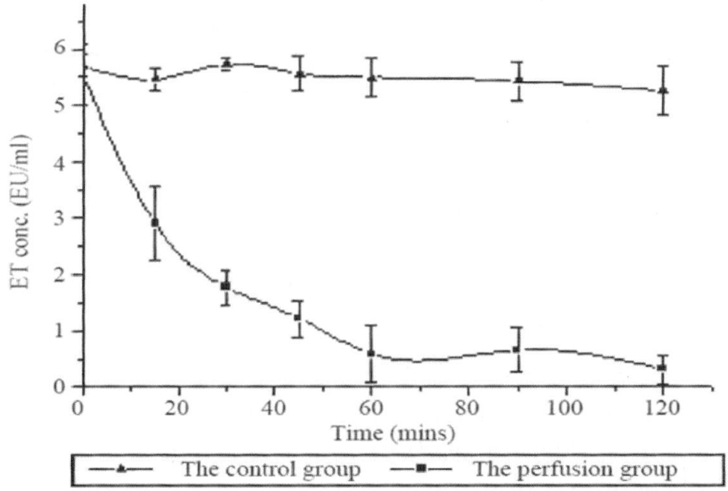

Fig. 21.14. Removal of ET from rabbit blood by perfusion using Lys immobilized cellulosic beads.

Note: EU is a measure of endotoxin content.
Source: Fang H and Yu YT *et al.* (2004) *Biomaterials* **25:** 5433–5440, with copyright permission from Elsevier Press.

Table 21.12. Liver and Renal Function Test Results

Parameter	Groups	0 min	60 min	120 min
SA	Perfusion	26.7 ± 2.4	24.5 ± 1.8	23.2 ± 2.3
(g/L)	Control	26.1 ± 3.7	22.0 ± 2.2	24.1 ± 4.6
TBL	Perfusion	0.04 ± 0.01	0.05 ± 0.02	0.05 ± 0.01
(mg/mL)	Control	0.05 ± 0.01	0.08 ± 0.03	0.09 ± 0.02
AST	Perfusion	15.3 ± 4.0	12.1 ± 3.2	14.0 ± 4.2
(U/L)	Control	17.9 ± 5.7	27.6 ± 6.9	35.7 ± 8.0
ALT	Perfusion	34.4 ± 9.5	25.9 ± 9.1	24.8 ± 8.5
(U/L)	Control	35.3 ± 9.9	34.5 ± 6.2	41.4 ± 1.8
AKP	Perfusion	113.1 ± 8.5	116.9 ± 19.2	118.1 ± 25.7
(U/L)	Control	109.4 ± 10.3	153.0 ± 17.3	195.5 ± 31.2
BUN	Perfusion	5.2 ± 0.6	4.9 ± 1.3	4.6 ± 0.6
(mmol/L)	Control	5.6 ± 1.1	7.3 ± 0.8	7.8 ± 0.7
CRK	Perfusion	61.8 ± 4.8	59.3 ± 7.5	70.9 ± 2.0
(umol/L)	Control	62.7 ± 4.2	80.4 ± 3.7	94.3 ± 4.7

Note: $n = 3$
Source: Fang H and Yu YT *et al.* (2004) *Biomaterials* **25:** 5433–5440, with copyright permission from Elsevier Press.

malondialdehyde (MDA) assays, showed that the septic symptoms were ameliorated with the removal of large amounts of ET in the blood. This removal also prevented further organ damage (Table 21.12; Figs. 21.15 and 21.16).

3.2.4. *Cellulosic adsorbents for the removal of anti-DNA antibodies in the treatment of systematic lupus erythematosus*

The abnormally high levels of anti-DNA antibodies and immune complexes in the plasma of systemic lupus erythematosus (SLE) patients can be removed by sorbent-perfusion. Yu and Kong (1998) synthesized

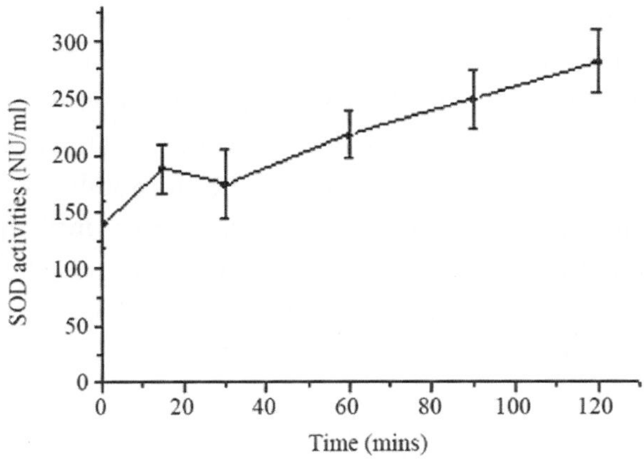

Fig. 21.15. Improved activities of serum superoxide dismutase (SOD) versus ET removal by perfusion.

Note: NU is a measure of SOD activity.

Source: Fang H and Yu YT *et al.* (2004) Biomaterials **25:** 5433–5440, with copyright permission from Elsevier Press.

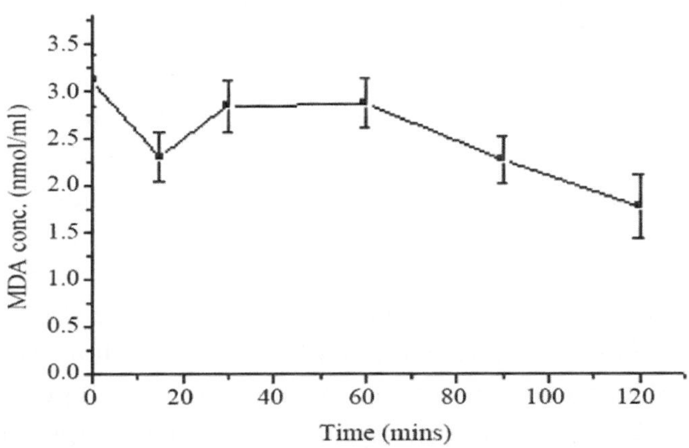

Fig. 21.16. Decrease in malondialdehyde (MDA) concentrations versus ET removal by perfusion.

Source: Fang H and Yu YT *et al.* (2004) *Biomaterials* **25:** 5433–5440, with copyright permission from Elsevier Press.

cellulose adsorbents using DNA as ligand for the removal of anti-DNA antibodies. The activation of cellulosic beads was conducted according to Kong *et al.* (1998). Two mL of activated cellulose beads were added to a 4.0 mL buffer solution containing 4.0 mg of DNA in a flask and stirred at 25°C for 20 hr on a shaker. The immobilized DNA beads were then washed with buffer solution and water until there was no DNA leakage detected in the rinse water using UV spectrometer at 260 nm. The obtained immunosorbent was stored at 4°C. *in vitro* adsorption tests showed that the DNA immunosorbent could remove 40–70% of anti-DNA antibodies from sheep plasma (Chu *et al.*, 2004). The maximum decrease in anti-DNA levels was 80% after 60 min of *in vitro* experiments. The high adsorption capacity shows that this method has high potential for use in clinical applications.

3.2.5. *Cellulosic adsorbent for the treatment of myasthenia gravis*

Myasthenia gravis (MG) is an autoimmune disorder characterized by a disturbance in neuromuscular transmission that results in muscle weakness. Yu and Yang *et al.* (2004, 2010) synthesized immobilized tryptophan cellulosic adsorbent and evaluated its adsorption capacity for binding acetylcholine receptors in the plasma of MG patients. Experimental autoimmune myasthenia gravis (EAMG) rabbits were induced by Ta183-200 peptide. Female rabbits weighing approximately 2 kg were injected intradermally at multiple sites with 500 μg of Ta183-200, which was emulsified with an equal volume of Freund's complete adjuvant. A booster injection of 500 μg Ta183-200 with Freund's incomplete adjuvant was administered after four weeks.

The rabbits underwent extracorporeal whole blood perfusions for two hours. Results showed no significant damage to the blood cells or changes in electrolyte concentrations. Whole blood sorbent-adsorption improved clinical manifestation and neuromuscular function of the EAMG rabbits (Table 21.13).

Neuromuscular transmission function has been evaluated by stimulating the deep peroneal nerve. The mean decrement of potentials evoked from the anterior tibial muscle, at three stimulation

Table 21.13. Grading of Clinical Manifestation pre-WBIA and 5 Days after WBIA

No. of Rabbits	Pre WBIA	5th Day after WBIA
1	+	0 – +
2	+	0
3	+	0
4	+	0
5	+	0
6	+	0
7	+ +	+
8	+	0

Note: 0 stands for the rabbits recovered normally after therapy.
(*Source*: Yang L *et al.* (2004) *Artif Cell Blood Sub* **32:** 519–528, with copyright permission from Taylor & Francis Group Press.)

frequencies in the therapeutic rabbit group, were determined. At 3Hz, the potential decreased from 21.87% to 17.87%, from 22.25% to 18.75% at 5 Hz, and from 24.37% to 23.25% at 10 Hz. Table 21.13 shows the changes of the electro-physiological features of EAMG rabbits after WBIA on the fifth (D5), and the eighth (D8) day after passive transfer. The same RNS was performed in the control rabbit group, but no reductions were found (Table 21.14).

The quantity of neuromuscular junctions can reflect neurotransmission function. The quantity of neuromuscular junctions per unit area (25 mm^2) of the therapeutic rabbit group was determined on the third day after WBIA (D8). The quantity of neuromuscular junctions per unit area increased from 9.825 ± 3.401 to 10.90 ± 2.879 ($P < 0.05$) after WBIA, which was higher than that of the control group ($P < 0.01$) (Fig. 21.17).

In conclusion, extracorporeal whole blood sorbent-adsorption is an effective and safe approach in treating the passive experimental autoimmune myasthenia gravis by improving clinical manifestation, neuromuscular transmission function, and enhancing the quantity of neuromuscular junctions and antibody titer.

Table 21.14. Comparison of RNS Reductions between Therapeutic and Control Groups

	Therapeutic Group (%)		Control Group (%)	
Frequency	**D5**	**D8**	**D5**	**D8**
3Hz	21.875 ± 3.226	17.875 ± 1.642	21.375 ± 2.615	21.125 ± 2.416
5Hz	22.250 ± 2.815	18.750 ± 1.388	21.875 ± 2.232	22.000 ± 1.772
10Hz	24.375 ± 1.685	23.250 ± 1.388	23.875 ± 1.126	24.125 ± 2.696

Source: Yang L *et al.* (2004) *Artif Cell Blood Sub* 32: 519–528, with copyright permission from Taylor & Francis Group Press.

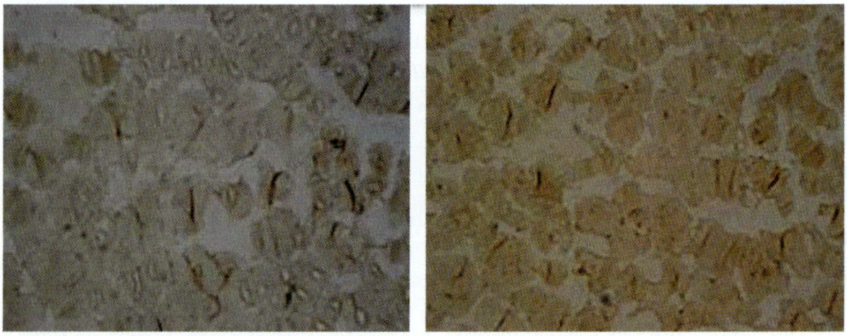

(A) (B)

Fig. 21.17. Neuromuscular junctions of the therapeutic rabbit group before and after WBIA on the third day (200).

(*Note*: (A) and (B) stand for the results before and after WBIA.
Source: Li XH *et al.* (2010) *Artif Cell Blood Sub* **38:** 186–191, with copyright permission from Francis Taylor Press.)

3.2.6. *Cellulosic adsorbents for the treatment of IgA nephropathy*

IgA nephropathy is one of the most common types of glomerulonephritis in many parts of the world, accounting for 15–20% of all glomerulo nephritis in western Europe, and 30–40% in Japan. However, it accounts for less than 10% in North America, and is characterized by elevated concentrations of IgA and its immune complexes (Harper *et al.*, 1999). It is conducive for alleviating disease by decreasing the concentration of IgA and its immune complexes in blood.

A cellulose bead based IgA immunosorbent was developed to deal with IgA nephropathy (Chu and Yu *et al.*, 2003). Cellulose beads were activated by epichlorohydrin, and reacting the activated cellulose beads with glutaraldehyde created aldehydic groups; Trp, Glu, Asp, Tyr, Phe and Lysine were selected as ligands and immobilized on the cellulose beads. Ethanolamine was used to block surplus aldehydic groups. Trp adsorbents had a higher adsorption capacity for IgA and its immune complexes (Table 21.15). Blood compatibility tests showed that the cellulose-based immunosorbent has a lower adsorption capacity for plasma proteins, total proteins, red blood cells and platelets, whose adsorption percentages are lower than 15%. Only the adsorption capacity for white blood cells was too high, reaching 26% after coming in contact with blood (Xing *et al.*, 2003). Acute toxicity tests demonstrated that the cellulose-based IgA immunoadsorbent is safe.

Table 21.15. Adsorption Efficiency of Cellulose Carrier Coupling Amino Acids in Different pH Conditions

		pH		
Ligand		**5.0**	**6.0**	**7.0**
Trp	Adsorption capacity (mg/0.34g)	136	121	94
	Adsorption efficiency (%)	43.31	38.54	29.94
Glu	Adsorption capacity (mg/0.34g)	106	60	89
	Adsorption efficiency(%)	33.76	19.12	28.34
Asp	Adsorption capacity (mg/0.34g)	66	95	63
	Adsorption efficiency (%)	21.02	30.25	20.06
Tyr	Adsorption capacity (mg/0.34g)	91	72	103
	Adsorption efficiency (%)	28.98	22.93	32.80
Phe	Adsorption capacity (mg/0.34g)	83	101	80
	Adsorption efficiency (%)	26.43	32.17	25.48
Lys	Adsorption capacity (mg/0.34g)	103	59	96
	Adsorption efficiency (%)	32.80	18.79	30.57

Source: Xing *et al.* (2003) *Ion Exchange and Adsorption* **19(3)**: 241–246.

3.2.7. *Cellulosic adsorbents for the removal of bilirubin*

Bilirubin is a pathogenic substance that causes jaundice and has been reported to be a neuronal and mitochondrial toxin (Roche *et al.*, 2004; Hass, 1999). Bilirubin contains many negative groups, i.e. carboxyl groups and carbonyl groups. Although plasma exchange has been successfully used to treat hyperbilirubinemia, there are possible risks of viral infection, or undesirable immune response. Therefore, a selective bilirubin adsorbent may be a safer, more effective treatment option.

Zhang and Gu (2008) designed eight types of cellulose-based adsorbent, each with different polyamines as ligand. The adsorbents with 1,6 hexamethylene diamine as ligands had the highest adsorption capacities for free bilirubin (Table 21.16).

3.3. *Chitosan-based Adsorbents in Hemoperfusion*

3.3.1. *Chitosan bead-based adsorbent for the treatment of MODS*

Multiple organ dysfunction syndrome (MODS) occurs when organ function is altered due to sepsis, septic shock or systemic inflammatory response syndrome (SIRS), and is the most common cause of death in intensive care units. Endotoxins (LPS) make up the cell walls of gram-negative bacteria and play an important role in the initiation and development of MODS. Therefore, the removal of endotoxins in blood is very important.

Li and Pan *et al.* (2007) constructed a porous chitosan microbead-based adsorbent for the removal of blood endotoxins. High porous chitosan microbeads were prepared using the phase-inversion method, and glutaraldehyde or EPI were used as the cross-linker. Then, three types of chitosan microbeads were prepared by immobilizing lysine, TEPA and HDA onto chitosan microbeads to function as ligands. Adsorption results showed that porosity and the ligand type can significantly affect the adsorption capacity for endotoxins (Table 21.17). The porosity is conducive for improving the adsorption capacity.

Table 21.16. Adsorption Performance of Amino-Functionalized Cellulose-Based Adsorbents in Simulated Serum Solution

Sample	Cel	Cel-TMA	Cel-TEA	Cel-ETA	Cel-HDA	Cel-DTA	Cel-TTA	Cel-TPA	Cel-EDA
Adsorption capacity mg/ml	0.199 ± 0.0305	0.277 ± 0.0058	0.153 ± 0.0065	0.101 ± 0.0111	0.647 ± 0.0088	0.298 ± 0.0116	0.316 ± 0.088	0.150 ± 0.0210	0.279 ± 0.0309
Adsorption percentage %	10.987 ± 1.6851	15.292 ± 0.3214	8.463 ± 0.3579	5.605 ± 0.6133	35.707 ± 0.4854	16.480 ± 0.6429	17.445 ± 0.4854	8.277 ± 1.1590	15.404 ± 1.7046

Note: Cel: cellulose bead; ethylenediamine (EDA); 1,6-Hexylenediamime (HDA); Diethylenetriamine (DTA); Triethylenetetramine (TTA); Tetraethylenepentamine (TPA); Ethanolamine (ETA); Trimethylamine (TMA); Triethylamine (TEA)
Source: Zhang (2008) Master's dissertation, Tianjin Medical University.

Table 21.17. Adsorption Capacity for Endotoxin and Chitosan Adsorbents with Different Porosities

Adsorbents	Chitosan (0.5% PEG)			Chitosan (1% PEG)		
	Ligand Amount	Adsorption Capacity	Adsorption Percentage (%)	Ligand Amount	Adsorption Capacity	Adsorption Capacity (%)
porosity Chitosan-lysine	142.08 μmol/g	1.757 EU/mL	87.9	172.37 μmol/g	1.868 EU/mL	93.4
porosity Chitosan-TEPA	1.2024 mmol/g	1.703 EU/mL	85.1	1.2465 μmol/g	1.770 EU/mL	88.5
porosity Chitosan-HDA	0.8098 mmol/g	1.676 EU/mL	83.8	0.8790 μmol/g	1.693 EU/mL	84.6

Source: Li *et al.* (2007) Master's dissertation, Nankai University, China.

Furthermore, lysine is easily immobilized on chitosan beads, and the adsorption capacity of endotoxins increased as the immobilized ligand increased.

3.3.2. *Chitosan bead-based adsorbents for the removal of LDL*

Liposorber has been shown to effectively clear LDL, TC and TG for the treatment of hyperlipidemia. Many researchers were working to develop chitosan-based LDL adsorbents because they are cost effective and easily activated. Fu and Li *et al.* (2004) developed chitosan bead-based adsorbents that used tryptophan as ligands for the selective removal of low density lipoproteins. They found that these adsorbents effectively removed LDL, TC and TG. Furthermore, they found that they had a high HDL adsorption capacity and an HDL adsorption percentage greater than 20%. Compared to the chitosan-based LDL adsorbents synthesized by Fu and Li *et al.* (2004), these had a higher performance and easier preparation route than other chitosan-based LDL adsorbents (Gamzazade *et al.*, 1997; Liu *et al.*, 2002; Sharma *et al.*, 1993; Yu and He, 1997; Yu *et al.*, 1997) In order to improve the adsorption capacity, Yu and Li constructed two types of chitosan-based LDL adsorbents that used sulfonic groups as the ligands (Yu and Li, 2007).

The sulfonic groups of type I adsorbent were introduced into the chitosan beads by directly sulfonating chitosan beads via sulfonic reagents at room temperature; the type II adsorbent's sulfonic groups were introduced after the different spacers were introduced into the adsorbents (Li, 2005). The results show that type II adsorbents have higher LDL-C adsorption capacities (60%) than type I ones (30%). This difference may be caused by several factors. Direct sulfonation might produce the sufficient number of sulfonic groups, but the LDL biomacromolecule has a diameter of approximately 25 nm (Lu, 2002); its steric hindrance may cause a reduction of the adsorption capacity (Wang *et al.*, 2013). In type II adsorbents, the spacers with a certain length can effectively reduce steric hindrance, which helps to improve sulfonic group efficiency, leading to higher LDL adsorption capacities.

3.3.3. *Chitosan bead-based adsorbents for the removal of bilirubin*

Bilirubin is a small molecule that contains many negative charged carboxyl groups. Chitosan contains some amine groups on its surface that are capable of adsorbing bilirubin. Li *et al.* (2005) synthesized a porous chitosan-based bilirubin adsorbent. The porous chitosan beads was prepared by inverse suspension polymerization. Then, EPI, EGDE and CNCl were used to activate the porous chitosan beads, and HAD was linked onto the beads. Of the three activation methods studied, the CNCl activation method resulted in porous chitosan beads with the highest adsorption capacity (Table 21.18).

It is worth noting that these adsorption results were obtained using bilirubin in PBS solution. Human plasma contains human albumin, which combines easily with free bilirubin. Therefore, when the above adsorbents were used to adsorb bilirubin from human blood, the adsorption capacities would likely to be lower than the adsorption capacity in PBS solution.

3.4. *Dextran-based Adsorbents in Hemoperfusion*

3.4.1. *Dextran-based adsorbent for the removal of LDL*

Wang and Yu *et al.* (2003) constructed a Dextran G-75 gel bead based amphiphilic adsorbent for the removal of low-density lipoproteins. Cholesterol groups were linked to Dextran G-75 by reacting

Table 21.18. Comparison of Different Porous Chitosan Bead Activation Methods

Adsorbent	Epoxy Groups (umol/g)	Adsorption Capacity (mg/g)	Adsorption Percentage (%)
CS-EPI-HDA	123.5	45.90	61.20
CS-EGDE-HDA	59.2	49.54	66.06
CS-CNCl-HDA		55.31	73.74

Note: Bilirubin concentration: 150mg/L; Bilirubin volume: 25 mL.
Source: Li Haiyan (2005), Master's dissertation, Nankai University, China.

Table 21.19. Adsorption Results of Amphiphilic Dextran Adsorbents

	TC		TG		LDL	
The Ratio of Cholesterol to Sulfonic Group	Adsorption Percentage mv ± sd (%)	Adsorption Capacity mv ± sd (mg/mL)	Adsorption Percentage mv ± sd (%)	Adsorption Capacity mv ± sd (mg/mL)	Adsorption Percentage mv ± sd (%)	Adsorption Capacity mv ± sd (mg/mL)
1:1.95	33.62 ± 3.24	0.507 ± 0.016	33.16 ± 11.31	0.496 ± 0.056	48.00 ± 3.27	0.455 ± 0.015
1:1.57	75.27 ± 4.47	1.836 ± 0.082	75.54 ± 2.80	1.992 ± 0.056	81.46 ± 8.88	1.519 ± 0.134
1:1.36	23.04 ± 1.68	0.723 ± 0.012	39.30 ± 2.966	0.557 ± 0.017	33.31 ± 4.59	0.760 ± 0.034
100% Cholesterol	3.77 ± 13.11	0.084 ± 0.011	5.10 ± 10.49	0.0938 ± 0.009	1.50 ± 8.09	0.036 ± 0.003
100% Sulfonic group	24.60 ± 2.14	0.499 ± 0.011	22.84 ± 4.94	0.458 ± 0.023	30.00 ± 2.38	0.330 ± 0.008

Note: mv ± sd: mean value ± standard deviation, N = 6;

Source: Wang *et al.* (2003) *Biomaterials* **24:** 2799–2802, with copyright permission from Elsevier Press.

with cholesterol N-(6-isocyanatohexyl) carbamate in the presence of pyridine in DMSO at 80°C. The mixture was then reacted with chlorosulfonic acid in formamide, which introduced the sulfonic group that created the hydrophilic moiety. The cholesterol ligand to sulfonic group ratio was found to greatly affect the adsorption capacity of LDL-C, TC, and TG (Table 21.19). When the cholesterol to sulfonic group ratio was 1:1.57, the optimal adsorption capacity for TC, TG and LDL-C reached 836 ± 0.082 mg/mL, 1.992 ± 0.056 mg/mL and 1.519 ± 0.134 mg/mL, respectively.

The non-specific adsorption of the amphiphilic adsorbent was tested for its adsorption of total plasma proteins. Results show that the total plasma protein adsorption was less than 10%, indicating that the adsorbent has a rather good selectivity. However, the Dextran G-75 based amphiphilic adsorbent has a weaker mechanical strength than the cellulose amphiphilic adsorbent for the removal of LDL.

4. Conclusion and Future Perspectives

Intensive research and development of natural polysaccharide adsorbents by sorbent-perfusion in blood purification have paved a path for the treatment of patients with autoimmune diseases, hyperlipidemia and inflammatory disorders. Animal experiments and clinical trials have shown that these methods are safe, efficient and cost effective. Therefore, they have high clinical potential for improving the living standards and survival rates of patients.

References

Adrian M, Roberto F-L, Jose MG. (2000) Essential role of the concentration of immobilized ligands in affinity chromatography: Purification guanidinobenzoatase on an ionized ligand. *J Chromatogr B* **740:** 211–218.

Armstrong VW, Windisch M, Wieland H, *et al.* (1983) Selective continuous extractacorporeal ekimination of low-density lipoprotein with heparin at acidic pH. *Trans Am Soc Artif Intern Organs* **29:** 323–327.

Arnett FC, Edworthy SM, Bloch DA, *et al.* (1988) The 1987 revised ARA criteria for rheumatoid arthritis rheum. *Arthritis and Rheumatism* **31:** 315–324.

Bai YX, Li YF. (2006) Preparation and characterization of crosslinked porous cellulose beads. *Carbohyd Polym* **64:** 402–407.

Behim E, Klinkmann H. (1989) Selective and specific adsorbents for medical therapy. *Int J Artif Organs* **12:** 1–10.

Bosch T. (1996) Low density lipoprotein hemoperfusion using a modified polyacrylate adsorber: *In vitro, ex vitro* and first clinical results. *Artif Organs* **204:** 344–345.

Bosch T, Schmidt B, Kleophas W, *et al.* (1997) LDL hemoperfusion — A new procedure for LDL apheresis: first clinical application of an LDL adsorber compatible with human whole blood. *Artif Organs* **209:** 977–982.

Callery MP, Kamei T, Mangino MJ, *et al.* (1991) Organ interactions in sepsis. Host defense and the hepatic-pulmonary macrophage axis. *Arch Surg-Chicago* **126:** 28–32.

Cao NN, Yu YY, Wang M-Y, *et al.* (2002) *In vitro* study of a novel low density lipoprotein adsorbent. *Artif Cell Blood Sub* **30:** 53–61.

Cao NN, Yu YT, Wang MY, *et al.* (2002) The interaction of human lipoprotein with cholesterol modified Dextran in Serum. *Chem J Chinese U* **23:** 1524–1528.

Chapman MJ. (1980) Animal lipoproteins: Chemistry, structure and comparative aspects. *J Lipid Res* **21:** 789–853.

Chen B, Pan JL, Tong MR, *et al.* (1993) A study of adsorbents for the adsorption of low density lipoprotein (LDL). *Ion Exchange & Adsorption* **9:** 330–334.

Cheng Y, Wang SQ, Yu YT, *et al.* (2003) *In vitro, In vivo* studies of a new amphiphilic adsorbent for the removal of low-density lipoprotein. *Biomaterials* **24:** 2189–2194.

Chu JQ, Chen JJ, Kong DL, *et al.* (2002) Studies on the immunoadsorbent for removal of IgA in IgA-nephropathy(I) *Ion Exchange and Adsorption* **18(4):** 369–373.

Chu JQ, Yu YT, Zhang WH, *et al.* (2003) *In vitro* studies on the immunoadsorbent for removal of IgA in IgA-neplropatly(II) *Chem J Chinese U* **24(4):** 636–638.

Chu J-Q, Zhang W-H, Yu Y-T, *et al.* (2004) *In vitro* studies of the immunoadsorbent for removal of IgA-nephropathy(III). *Chem J Chinese U* **25:** 1454–1457.

Dai J, Chen CZ, Yu YT, *et al.* (2001) Preparation and properties of immunoadsorbent for rheumatoid factors. *Ion Exchange and Adsorption* **17(5):** 352–356.

Du KF, Yan M, Wang QY, *et al.* (2010) Preparation and characterization of novel macroporous cellulose beads regenerated from ionic liquid for fast chromatography. *J Chromatogr A* **1217:** 1298–1304.

Falkenhagen D, Schima H, Loth F. (1999) Arrangement for removing substances from liquids, in particular blood. United States Patent #5, 855, 782.

Fang H, Wei J, Yu YT. (2004) *In vivo* studies of endotoxin removal by lysine-cellulose adsorbents. *Biomaterials* **25:** 5433–5440.

Fu CX, Chen CZ, Yu YT. (2002) Heat aggregated-IgG linked to agar gel — A new immunoadsorbent rheumatoid arthritis. *Chinese Journal of Biomedical Engineering* **21:** 1–4.

Hamada J, Ushio Y, Kazekawa K, *et al.* (1996) Embolization with cellulose porous beads I An experimental study. *AJNR Am J Neuroradiol* **17:** 1895–1899.

Hamada J, Kai Y, Nagahiro S, *et al.* (1996) Embolization with cellulose porous beads II clinical trial. *AJNR Am J Neuroradiol* **17:** 1901–1906.

Harris MJ, Struck CE, Case GM, *et al.* (1984) Synthesis and characterization of poly(ethylene glycol) derivatives. *J Polym Sci* **22:** 341–352.

Hass EL. (1999) Differentiation and diagnosis of jaundice. *AACN Clin Issues* **10(4):** 433–41.

Hou KC, Zaniewski R. (1991) Protein A immobilized cartridge for immunoglobulin purification. *Biotechnol Appl Bioc* **13:** 257–268.

Kai Y, Hamada J-I, Morioka M, *et al.* (2006) Clinical evaluation of cellulose porous beads for the therapeutic embolization of meningiomas. *AJNR Am J Neuroadiol* **27:** 1146–1150.

Kai Y, Hamada J-I, Morioka M, *et al.* (2007) Preoperative cellulose porous beads for therapeutic embolization of meninggioma: Provocation test and technical considerations. *Neuroradiology* **49:** 437–443.

Kojima S, Harada-Shiba M, Toyota Y, *et al.* (1992) Changes in coagulation factors by passage through a dextran sulfate cellulose column during low-density lipoprotein apheresis. *Int J Artif Organs* **15:** 185–190.

Kong DL, Chen CZ, Lin EF, *et al.* (1998) Clinical trials of type I and *in vitro* studies of type II immunoadsorbents for systemic lupus erythematosus therapy. *Artif Organs* **22:** 644–650.

Leckband DE, Kuhl T, Wang HK, *et al.* (1995) 4-4-20 anti-fliorescyl IgG Fab' recognition of membrane bpund hapten: Direct evidence for the role of protein and interfacial structure. *Biochemistry-US* **34:** 11467–11478.

Lewis SL. (2010) *Medical Surgical Nursing — Assessment and Management of Clinical Problems*, 8th ed. [ISBN 978-0323079150]

Li HY. (2005) Preparation of adsorbent based on porous chitosan beads for blood purification. Master's dissertation, Nankai Unoversity, China.

Li L. (2007) Preparation and adsorption properties of adsorbent for MODS. Master's dissertation, Nankai University, China.

Li XH, Cheng Y, Yang L, *et al.* (2010) Extracorporeal whole blood immune-adsorption of passively transferred myasthenia gravis rabbits by cellulose-tryptophan column. *Artif Cell Blood Sub* **38:** 186–191.

Liu W, Cao YL. (2010) Chapter 5: Scaffolds, In: *Tissue Engineering for the Hand*. World Scientific Press, Singapore.

Lupien PJ, Moorjani S, Awad J. (1976) A new approach to the management of familial hypercholesterolaemia: Removal of plasma cholesterol based on the principle of affinity chromatography. *Lancet* **1:** 1261–1265.

Maaskant N, Bantjes A, Kempen HJM. (1986) Removal of low density lipoprotein from blood plasma using cross-linked, sulfated polyvinylalcohol. *Atherosclerosis* **62:** 159–166.

Mendoza L, Mamo G, Flores N, *et al.* (2011) Laccase mediator system for activation of agarose gel: Application for immobilization of proteins. *Journal of Molecular Catalysis B: Enzymatic* **68:** 270–274.

Meng QQ, Wang JX, Ma GH, *et al.* (2009) Lyophilization of CNBr-activated agarose beads with lactose and PEG. *Process Biochemistry* **44:** 562–571.

Raetz CRH. (1990) Biochemistry of endotoxins. *Annu Rev Biochem* **59:** 129–170.

Roche SP, Kobos R. (2004) Jaundice in the adult patient. *Am Fam Physician* **69(2):** 299–304.

Sato Y, Agishi T. (1996) Low-density lipoprotein adsorption for arteriosclerotic patients. *Artif Organs* **20:** 324–327.

Sinitsyn VV, Mamontova AG, Konovalov GA, *et al.* (1990) Apheresis of low density lipoproteins using a heparin-based sorbent with low antithrombin III binding capacity. *Atherosclerosis* **84:** 55–59.

Spector TD. (1990) Rheumatoid arthritis. *Rheum Disclin North Am* **16:** 513.

Sreenivasan K. (1997) Synthesis and evaluation of cyclodextrin 2 hydroxyethy methacrylate copolymer as anovel adsorbent. *Polym Int* **42:** 22–24.

Soltys JP, Etzel RM. (2000) Equilibrium adsorption of LDL and gold immunoconjugates to affinity membranes containing PEG Spacers. *Biomaterials* **21:** 37–48.

Sun SJ, Tang YH, Fu Q, *et al.* (2012) Monolithic cryogels made of agarose–chitosan composite and loaded with agarose beads for purification of immunoglobulin G. *International Journal of Biological Macromolecules* **50:** 1002–1007.

Terman DS, Petty D, Harbeck Carr RI, *et al.* (1977) Specific removal of DNA antibodies *in vivo* by extracorporeal circulation over DNA immobilized in collodion charcoal. *Clin Immunol Immunopathol* **8:** 90–96.

Thummler K, Fischer S, Feldner A, *et al.* (2011) Preparation and characterization of cellulose microspheres. *Cellulose* **18:** 135–142.

Varki A, Cummings R, Esko J, *et al.* (2008) *Essentials of Glycobiology*, 2nd ed. Cold Spring Harbor Laboratory Press. [ISBN 0-87969-770-9]

Varki A, Cummings R, Esko J, *et al.* (1999). *Essentials of Glycobiology*. Cold Spring Harbor Laboratory Press. ISBN 0-87969-560-9.

Viswanathan N, C Sundaram S, Meenakshi S. (2009) Development of multifunctional chitosan beads for fluoride removal. *Journal of Hazardous Materials* **167:** 325–331.

Von Appen K, Weber C, Losert U, *et al.* (1996) Microspheres based detoxification system: A new method in convective blood purification. *Artif Organs* **20:** 420–425.

Wang DM, Hao G, Shi QH, *et al.* (2007) Fabrication and characterization of superporous cellulose bead for high-speed protein chromatography. *J Chromatogr A* **1146:** 32–40.

Wang HF, Li B, Shi BL. (2008) Preparation and surface acid-base properties of porous cellulose. *BioResources* **3:** 3–12.

Wang SQ, Li CM, Guo XJ, *et al.* (2008) Interaction study of LDL with charged ligands for effective LDL-C removing adsorbents. *React Funct Polym* **68:** 261–267.

Wang LY, Lu J, Yu YT. (2004) Interaction of rheumatoid factor with immobilized ss-DNA. *Chem Res Chinese U* **20:** 795–800.

Wang SQ, Guo XJ, Wang LY, *et al.* (2006) Effect of PEG spacer on cellulose adsorbent for the removal of low density lipoprotein cholesterol. *Artif Cell Blood Sub* **34:** 101–112.

Wang SQ, Yu YT, Cui T, *et al.* (2002) Cellulose amphiphilic adsorbent for the removal of low density lipoprotein. *Artif Cell Blood Sub* **30:** 285–292.

Wang T, Wang WC, Wang LY, *et al.* (2008) Phosphate-based affinity adsorbents for the removal of low density lipoprotein: Preparation and *in vitro* adsorption capability test. *Chinese J Biomed Eng* **27:** 132–136.

Wang YJ, Wang LY, Yu YT. (2005) Study on the rheumatoid arthritis adsorbent and the influence of spacers. *Chinese Journal of Biomedical Engineering* **24:** 486–491.

Weber V, Ettenauer M, Linsberger I, *et al.* (2010) Functionalization and application of cellulose microparticles as adsorbents inextracorporeal blood purification. *Macomol Symp* **294**-II: 90–95.

Weber V, Linsberger I, Ettenauer M, *et al.* (2005) Development of specific adsorbents for human tumor necrosis factor-α: Influence of antibody immobilization on performance and biocompatibility. *Biomacromolecules* **6:** 1864–1870.

Wei J, Fang H, Kong DL, *et al.* (2004) Studies on adsorption mechanism of TNFG with IgG ligand. *Chemical Journal of Chinese Universities* **25(7):** 1277–1281.

Xing KL, Chu JQ, Yu YT. (2003) Immunoadsorbent for IgA nephropathy—Study of absorb performance and hemocompatibility. *Ion Exchange and Adsorption* **19(3):** 241–246.

Yang L, Cheng Y, Yan WR, *et al.* (2004) Extracorporeal whole blood immunoadsorption of autoimmune myasthenia gravis by cellulose tryptophan adsorbent. *Artif Cell Blood Sub* **32:** 519–528.

Yang Y. (2002) Agarose resin for removal of rheumatoid arthritis factors. Master's dissertation, Nankai University, China.

Yatzidis H. (1964) Research on extrarenal purification with the aid of activated charcoal. *Nephron* **72:** 310–312.

Yu HF, Fu GQ, He BL. (2007) Preparation and adsorption properties of PAA-grafted cellulose adsorbent for low density lipoprotein from human plasma. *Cellulose* **14:** 99–107.

Yu HF. (2006) Synthesis and *in vitro* sorption properties of new adsorbents for low density lipoprotein from human plasma. PhD dissertation, Nankai University, China.

Yu YT, Zhang HM, Wang WC, *et al.* (2008) Selective adsorbent for the removal of immunoglobulin E in bronchial asthma. *Artif Cell Blood Sub* **36:** 63–73.

Zalipsky S. (1995) Chemistry of polyethylene glycol conjugates with biological active molecules. *Adv Drug Deliver Rev* **16:** 157–182.

Zhao YP, Huang MS, Wu W, *et al.* (2009) Synthesis of the cotton cellulose based Fe(III)-loaded adsorbent for arsenic(V) removal from drinking water. *Desalination* **249:** 1006–1011.

Zhang SN (2008) Preparation and evaluation of bilirubin adsorbent. Master's dissertation, Tianjin Medical University, China.

Zhou HM, Wang HR. (1998) *Chemical Modification of Proteins.* Beijing, Tsinghua University Press, 19 pp.

Chapter 22

Adsorbents for the Treatment of Autoimmune Diseases through Hemoperfusion

Jun Ren, Houliang Wei, Lingyun Jia*,†

1. Introduction

Autoimmune diseases, a family of more than 80 human diseases, are caused by defects in the immune system of the body causing it to attack its own organs and tissues. Such an immune disorder would result in damage to an individual's organs, tissues, or cells, and bring about a range of disabilities and deterioration of quality of life (Cooper and Stroehla, 2003). Many autoimmune diseases are hallmarked by the recognition of the self-antigens by antibodies, and these autoantibodies are regarded as pathogenic if they contribute directly to the development of the diseases (Eaton *et al*., 2007). The pathogenic mechanism of autoimmune diseases is not very clear, but still many studies indicate that, in general, autoimmune diseases may be caused by a combination of genes and environmental factors (Ellis *et al*., 2014). In addition, the immune response of men and women is also quite different. Women are more likely to develop an autoimmune disease.

*,† School of Life Science and Biotechnology, Dalian University of Technology, Dalian, 116023, P. R. China.
Corresponding author: lyj81@dlut.edu.cn, Tel.: +86-411-84706125; Fax: +86-411-84706125

Suppressing the destructive autoimmune response with medication is one of the approaches to treating autoimmune diseases. Immunosuppressive drugs are commonly used as they can reduce the overall immune response and thereby ameliorate the effects of these diseases. However, some patients do not respond to prolonged immunosuppression (Baggi *et al.*, 2008), and it can also cause severe side effects, such as reduced resistance to infection, therefore these drugs must be used with great caution. In recent years, much effort has been devoted to develop biologic agents, such as monoclonal antibodies, which can result in more targeted immunosuppression (Kao *et al.*, 2014). Most of these agents target a specific step in the inflammatory response of tissue-damaging. They can decrease T cells or B cells specifically, act on only activated T cells, inhibit particular cytokine mediators of inflammation, or block the recruitment and localization of lymphocytes to the target organ (Kao *et al.*, 2014; Yang *et al.*, 2014). Although these targeted drugs usually have fewer side effects, they may also increase the patient's vulnerability to infection (Knier *et al.*, 2014). Moreover, a therapy that works with one autoimmune disease will sometimes make another disorder worse (Carter and Zhao, 2010).

Another strategy for the treatment of autoimmune diseases is to remove the pathogenic autoantibodies through extracorporeal blood purification from patients' blood or plasma. Until two decades ago, therapeutic plasma exchange was the only procedure for antibody removal from the plasma. Plasma exchange could reduce autoantibodies by approximately 50–75% (Keller *et al.*, 1983). However, the most significant drawbacks of plasma exchange are the elimination of essential plasma components such as albumin and clotting factors, and the adverse effects of putting patients at high risk of allergic reactions and viral contamination, which result from the administration of the plasma substitutions (Hershko and Naparstek, 2005).

To overcome these problems, another alternative approach, immunoadsorption, has been introduced to remove pathogenic autoantibodies from the patient's plasma with a minimum of side effects which medication or conventional plasma exchange are known for (Samuelsson, 2001; Ullrich and Kuehnl, 2004). The blood or the separated plasma is extra-corporeally circulated through an immunoadsorption column containing adsorbents. The adsorbents are characterized with specific active ligands covalently bound to an inert and insoluble matrix (carrier material). The ligands are the carefully selected functional molecules which are capable of capturing the pathogenic autoantibodies. In this chapter, we chose to present and discuss commercial and recently developed immunoadsorbents for treatment of autoimmune diseases, with a special emphasis on the principles of the interaction between the ligands and autoantibodies.

2. Commercial Adsorbents for Autoimmune Diseases

Currently, there are several commercial immunoadsorption columns available for clinical use (Table 22.1). All the commercial adsorbents consist of a specific bioactive ligand and a type of gel bead as matrix material, such as cellulose, agarose, polyvinylalcohol, silica gel and polymethacrylate. According to the principle of the interaction between the ligand and the autoantibody, the adsorbents could be subdivided into three classes: non-selective, semi-selective, and highly-selective adsorbents.

2.1. *Non-Selective Adsorbents*

The non-selective adsorbent not only removes the autoantibody but also reduces the plasma levels of normal antibodies and many other substances like fibrinogen, albumin, and lipids at the same time, because of the non-specific interaction between ligands and plasma components. Two adsorbents of such type include the Selesorb column (Kaneka, Japan) and Immusorba column (Ashahi Medical, Tokyo, Japan).

In the Selesorb column, the negatively charged dextran sulfate (ligand) is linked to cellulose beads (matrix) (Kutsuki *et al.*, 1998). Anti-dsDNA antibodies, a group of anti-nuclear antibodies, are highly diagnostic of systemic lupus erythematosus (SLE). Despite DNA being an intracellular target, anti-dsDNA antibodies can exert pathogenic effects directly in the cell or indirectly via complement and/or

Table 22.1. Commercial Immunoadsorption Columns for Treatment of Autoimmune Diseases

Apparatus	Selesorb	Immusorba	Immunosorba	Prosorba	Ig-Therasorb	Medisorba MG-50	Miro	DNA230
Company	Kaneka, Japan	Asahi-Medical, Japan	Fresenius, Germany	Fresenius, Germany	Teterow, Germany	Kuraray, Japan	Fresenius, Germany	Jafron, China
Matrix	Cellulose bead	Polyvinylalcohol bead	Sepharose bead	Silica bead	Sepharose bead	Cellulose bead	Polyacrylate bead	Carbonized resin bead
Ligand	Dextran sulphate	Tryptophan, Phenylalanin	Protein A	Protein A	Polyclonal sheep anti-human Ig	torpedo α-183-200 peptide	C1q	dsDNA
Volume	150 mL	350 mL	62.5 mL	300 mL	300 mL	50 mL	300 mL	230 ml
Capacity	NA	Igs 2.4 g	1.2 g IgG	557 mg IgG	4 g IgG	NA	400 mg CIC	NA
Specificity	Igs Anti-dsDNA antibodies, lipids, and fibrinogen	Igs, fibrinogen and others	IgG, IgA, IgM	IgG, IgA, IgM	IgG, IgA, IgM	Anti-AChR antibodies	C1q-CIC, C1q-antibodies, anti-phospholipid antibodies	Anti-dsDNA antibodies
Regeneration	Yes	No	Yes	No	Yes	No	No	No

Fc receptor-mediated mechanisms (Giles and Boackle, 2013). The dextran sulfate enables electrostatic bonding with anti-dsDNA and anticardiolipin antibodies, but may also be effective in the treatment of hepatitis C virus (HCV-)-related cryoglobulinemia (Stefanutti *et al.*, 2003). It was also found that dextran sulfate only binds Fab, but not Fc of the anti-dsDNA antibodies (Kutsuki *et al.*, 1998).

The Immusorba column contains cross-linked polyvinyl alcohol gel beads as the matrix on which the hydrophobic amino acids tryptophan (TR) or phenylalanine (PH) are immobilized (Hirata *et al.*, 2003). *In vitro* studies have shown that Immusorba columns can bind several kinds of autoantibodies, including anti-AChR antibody, anti-ganglioside antibody, and anti-dsDNA antibody. In clinical trials, the Immusorba column has proven effective for a variety of neurologic disorders, such as MG and Guillain-Barré syndrome (Shibuya *et al.*, 1994). Its adsorption mechanism might be mainly hydrophobic interaction and partly ionic interaction that is based on the carboxylic group of the ligand (Hirata *et al.*, 2003).

Since both Selesorb and Immusorba demonstrate non-selective adsorption properties, they are applicable in more than one disease, depending on their capacity to interact with the pathogenic autoantibodies. However, they also cause reduction of plasma levels of various substances other than immunoglobulins, such as fibrinogen, albumin, and lipids (Braun *et al.*, 1999). Therefore, with regard to reducing the loss of other useful substances, adsorbents which possess a more specific interaction with antibodies are more promising for the treatment of autoimmune diseases.

2.2. *Semi-Selective Adsorbents*

Semi-selective adsorbents show relatively high affinity toward human antibodies from plasma proteins. The ligands of the commercial semi-selective adsorbents are staphylococcal protein A and sheep polyclonal anti-human immunoglobulin. Protein A is derived from staphylococcus and is a cell wall component that was found in *S. aureus* isolates more than 40 years ago (Matic *et al.*, 2001). Its capacity to react with the Fc fragment of antibodies renders it useful as a means for

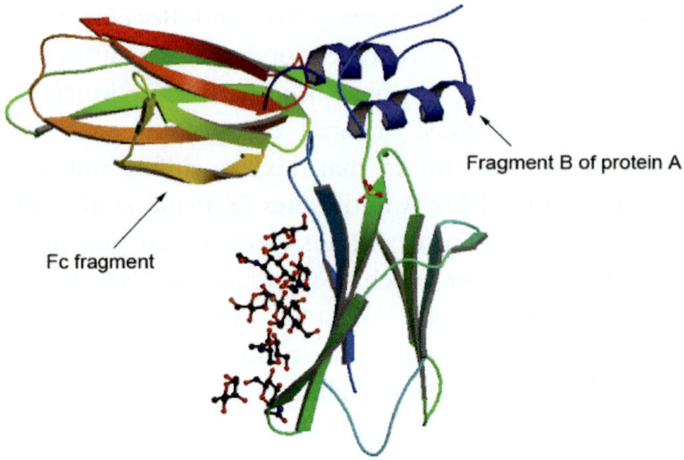

Fig. 22.1. Crystallographic refinement and atomic models of a human Fc fragment and its complex with fragment B of protein A.

extraction of immunoglobulins (Fig. 22.1) (Deisenhofer, 1981; Braun and Bosch, 2000). The most used adsorbents using staphylococcus protein A (SPA) as the ligand are Immunosorba, in which SPA is coupled to agarose beads, and Prosorba, in which silica is used as the matrix. These two columns have been approved by the FDA for the treatment of rheumatoid arthritis and idiopathic thrombocytopenic purpura (Prosorba) and hemophilia with inhibitors (Immunosorba).

For the Immunosorba column, SPA is immobilized on agarose (Sepharose®) through the cyanogen bromide activation of the beads. One Immunosorba column can remove about 1.2 g of IgG antibody at most (Braun and Bosch, 2000). A twin-column system is usually employed, in which two Immunosorba columns are used alternately in an adsorption/elution process during the treatment. The used column can be regenerated by citrate buffer solution (pH 2.3). A single treatment using Immunosorba columns can cause a considerable reduction in autoantibodies in the patient's blood due to the "unlimited" capacity of the columns and the large volume of plasma treated. A set of two Immunosorba columns can be re-used up to 20 times by the same patient.

Another commercial SPA-based adsorbent is the Prosorba column, which contains 200 mg of protein A immobilized within 123 g silica beads. One Prosorba column can bind about 0.6 g of IgG antibody at saturation (Braun and Bosch, 2000). Unlike Immunosorba, Prosorba is a single-column system which is not available for regeneration, so the capacity of Prosorba is limited to a maximum of 2,000 mL of processed plasma (Felson *et al.*, 1999). Although the restricted volume of processed plasma does not allow the significant removal of immunoglobulin or circulating immune complexes, this device has shown immune modulating effects. However, its mode of action remains somewhat obscure (Balint *et al.*, 1996; Bosch, 1996).

Apart from the FDA-approved indications, SPA-based immunoadsorption has been exploited for treatment of several other autoimmune diseases. In the treatment of SLE, Immunosorba can eliminate circulating immune complexes and anti-dsDNA antibodies with concomitant amelioration of inflammation (Braun, 1999, 2000). Possible beneficial effects for Goodpasture syndrome, Wegener's granulomatosis and myasthenia gravis have also been supported by some data (Benny *et al.*, 1999; Stegmayr *et al.*, 1999; Matic *et al.*, 2001). In most cases, SPA-based immunoadsorption is suggested for treating patients with life-threatening autoimmune diseases, especially in the case of autoimmune thrombocytopenic purpura patients who do not respond to corticosteroids, and in severe or drug-unresponsive rheumatoid arthritis.

However, several important issues should be addressed during treatment using SPA-based immunoadsorption. First, although the effectiveness of immunoadsorption has been proven in removing immunoglobulins and circulating immune complex from the blood, some inflammatory processes often occur in the tissue other than in the vascular bed. Therefore, simple removal of immunoglobulins from the blood is not able to eliminate the autoimmune process. Second, repeated treatment cycles with adequately processed plasma volumes are often used to prevent the redistribution of pathogenic autoantibodies. The almost complete removal of IgG may lead to a severe humoral immune deficiency, so that the patients are also exposed to the risk of complication resulting from infections. Third, patients suffering from autoimmune diseases may vary in their conditions. Although they may manifest similar symptoms, they could have different types of pathogenic factors, depending on the diversity of the autoantibodies. Since SPA-based adsorbents show a weak adsorption for IgG3, IgM and IgA, differences in the efficacy of the treatment may exist among different patients.

Compared with the SPA-based adsorption column, the Ig-Therasorb adsorption column (Therasorb Medical Systems, Unterschleissheim, Germany) can remove a larger spectrum of antibodies, including IgG (all subclasses), IgM and IgA (Jansen *et al.*, 2001). The Ig-Therasorb adsorption column is made of cellulose beads with sheep polyclonal anti-human immunoglobulin as the ligand. Like the Immunosorba system, Ig-Therasorb also employs a double-column system. Each column contains anti-human immunoglobulin immobilized into 150 ml of Sephroase beads, and one Ig-Therasorb column can eliminate approximately 4 g of IgG at saturation (Braun and Bosch, 2000). The efficacy of Ig-Therasorb has been demonstrated in several diseases, such as DCM.

2.3. *Highly-selective Adsorbents*

Unlike the non-selective and semi-selective adsorbents, highly-selective adsorbents can eliminate specific pathogenic autoantibodies. Medisorba MG-50 (Kuraray Medical Inc., Tokyo, Japan) is a highly-specific adsorption column to eliminate the anti-acetylcholine receptor (anti-AChR) antibody of MG (Oka *et al.*, 1993). The anti-AChR antibody belongs to the IgG subclass, which is implicated in the enhancement of AChR degradation and destruction of postsynaptic membrane through complement activation in MG (Vincent *et al.*, 2001). The adsorbent of this column is composed of cellulose beads as carrier material and covalent-bound synthetic peptide as a ligand that has a specific affinity to the pathogenic anti-AChR antibody. The amino acid sequence of the peptide is modified from the segment of α 183–200 of the torpedo AChR protein, and the segment is the acetylcholine binding site on AChR and the target site of anti-AChR antibody (Nakaji and Hayashi, 2003). Clinically, MG-50 specifically reduces blocking antibody titer and ameliorates MG symptoms. Nevertheless, the specificity of this technique, which is its major advantage, proved to be a significant drawback because the anti-AChR antibody is just one subclass among several others in the pathogenesis of MG. In addition, the rather expensive cost of the production of ligand peptide is also a problem (Nakaji and Hayashi, 2003).

Another adsorption column is Miro (Fresenius, St. Wendel, Germany) which is composed of polyacrylate as matrix and porcine C1q as ligand (Braun and Bosch, 2000). The principle of the C1q as the ligand is based on its specific interaction with circulating immune complexes. C1q, a component of the first complement component C1, is a glycoprotein with a molecular weight of 460 kDa and an isoelectric point of 9.3. There are 18 peptide chains folded together to form a collagen-like structure, which can bind the Fc-fragment of immunoglobulins at their globular C-terminus (Knobel *et al.*, 1975; Reid, 1976). Due to the weak affinity of soluble immunoglobulins to C1q, immunoglobulins bound in immune complexes lead to an increased affinity due to multi-point attachment (Duncan and Winter, 1988). This adsorbent could specifically remove C1q-circulating immune complex (C1q-CIC) and C1q-antibodies from the blood of patients suffering from *Lupus erythematosus* who often have deficient expression of C1q (Braun and Bosch, 2000).

DNA230 (Jafron, Zhuhai, China) is a DNA immunoadsorbent for hemoperfusion in SLE therapy. dsDNA was immobilized on macroporous spherical polymeric adsorbent (Yang *et al.*, 1988). The principle of this DNA immunoadsorbent is based on the interaction of dsDNA and anti-dsDNA antibodies.

3. Recent Developments on Adsorbents for Autoimmune Diseases

3.1. *Adsorbents Using Phema Membrane and Cryogel as Matrices*

The packed gel-beads in the traditional hemoperfusion column tend to cause the compaction of the bed under pressure, leading to the reduction of flow rates. An alternative adsorbent geometry would be microporous membranes in the form of flat sheets and hollow fibers (Braun and Bosch, 2000). Recently, a DNA attached poly(hydroxyethyl methacrylate) (PHEMA)-based microporous affinity membrane was prepared for selective removal of anti-dsDNA antibodies from SLE patient plasma *in vitro* (Uzun *et al.*, 2010). PHEMA membrane, one of the most widely used hydrophilic polymers in biomedical applications, was used as the matrix for its good blood compatibility. In order to further increase the blood compatibility of PHEMA, amino acid-based comonomer N-methacryloyl-l-alanine (MAAL) was introduced in the polymerization recipe to produce PHEMAAL membrane. PHEMAAL membrane had large pores in the range of 5–10 μm. The clotting time of PHEMAAL membrane increased, and loss of platelets and leukocytes of PHEMAAL membrane was very low compared with those of PHEMA membrane, indicating a better blood compatibility of PHEMAAL membrane. DNA loading on PHEMAAL membrane was 7.8 mg/g, and the PHEMAAL-DNA membrane could bind anti-dsDNA-antibodies in the range of $10–68 \times 10^3$ IU/g from SLE plasma.

Apart from the porous membrane, cryogels which have many macrospores provide another potential matrix due to their low pressure drops and lack of diffusion resistances compared with traditional packed-bed columns. Whole blood can be applied to cryogels without any pre-treatment. IgM-antibody (rheumatoid factor) removal from human plasma with supermacroporous poly(hydroxyethyl methacrylate) [PHEMA] cryogel carrying protein A has been evaluated (Alkan *et al.*, 2010). The PHEMA cryogel was prepared by bulk polymerization which was carried out in an aqueous solution of monomer frozen inside a plastic syringe (cryo-polymerization). After thawing, the PHEMA cryogel contains a continuous matrix with interconnected macropores. Pore volume in the PHEMA cryogel was 71.6%. Protein A molecules were covalently immobilized onto the PHEMA cryogel via cyanogen bromide (CNBr) activation. The supermacroporous structure of the PHEMA cryogel makes it possible to process blood cells without blocking the cryogel column. The maximum IgM-antibody adsorption capacity was 42.7 mg/g. IgM molecules could be repeatedly adsorbed and eluted without noticeable loss in the IgM binding amount.

3.2. *Adsorbents Using Mep As Ligand*

Although there are several reports on the improvement of the matrix geometry, the ligand is still the crucial factor affecting the performance of the adsorbent. The non-specific removal of useful substances from blood is the main problem of the commercial non-selective adsorbents. As for the selective adsorbents that use synthetic peptide or proteins, such as protein A, polyclonal anti-human immunoglobulin and C1q, as ligands, the high cost of the ligands and the adsorbents is an important issue, which may limit their widespread use. Besides, the lack of stability for cleaning and storage is also a major drawback for these protein-based adsorbents. These limitations have stimulated the search for a stable, simple and inexpensive ligand as an alternative to protein A and other protein ligands.

Our laboratory has recently reported the feasibility of 4-mercaptoethylpyridine (MEP, MW 139 Da), a simple and inexpensive synthetic compound, for autoantibody removal therapy (Ren *et al.*, 2009). MEP is a typical ligand of hydrophobic charge induction chromatography (HCIC), a new type of chromatography recently developed for antibodies purification in downstream bioengineering (Roque *et al.*, 2007). Important features of MEP-based HCIC stationary phase are the near physiological working condition and their remarkable ability to capture antibodies directly from complex feedstock. MEP has a non-charged structure under neutral conditions (pKa 4.8) but becomes positively charged when the pH is lowered to 4.8. Such a characteristic of this ligand enables the stationary phase to bind proteins via hydrophobic interactions at low-salt concentrations owing to its high-ligand density. When the pH of the mobile phase is adjusted, desorption of the bound proteins can be readily achieved by electrostatic repulsion via changing the ionizable groups of the proteins to their charged forms.

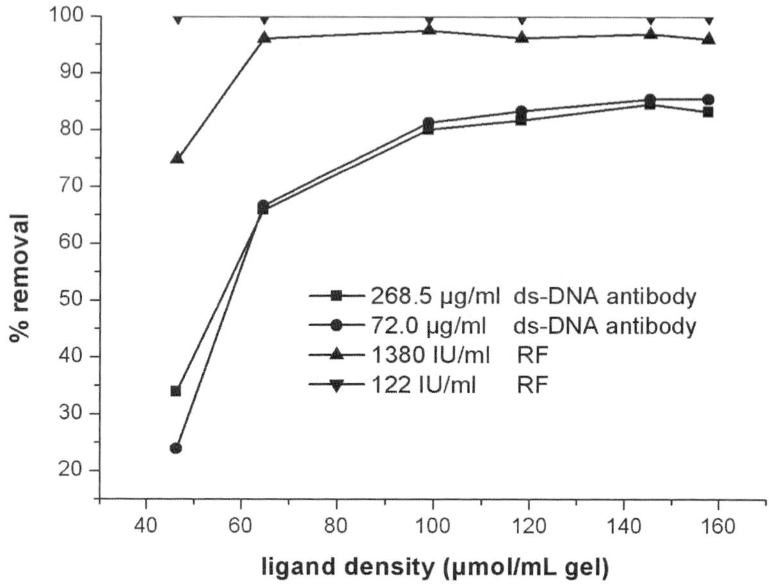

Fig. 22.2. Effect of ligand density on the binding capacity of MEP-Sepharose for RF and anti-ds-DNA antibody in human serum samples (Uzun *et al.*, 2010).

In our study, MEP-based adsorbents were prepared by coupling MEP molecules on Sepharose CL-6B. MEP-based adsorbent has proven effective in the removal of different type of autoantibodies (rheumatoid factor, anti-dsDNA antibodies and antinuclear antibodies) (Ren *et al.*, 2009). In the design of an adsorbent, ligand density is an adjustable parameter for the synthesis of adsorbents aiming at different pathogenic factors according to the class of antibody (Fig. 22.2). MEP-Sepharose with a ligand density of about 100 µmoL/mL gel is capable of removing both IgG and IgM classes of autoantibodies with the performance comparable to protein A immunoadsorbent under the same condition. Moreover, MEP-based adsorbents showed a lower degree of individual differences compared with protein A-Sepharose. About 90% of rheumatoid factors could be removed for all test serum samples. The titers of antinuclear antibody of 11 among 14 serum samples derived from SLE patients were also

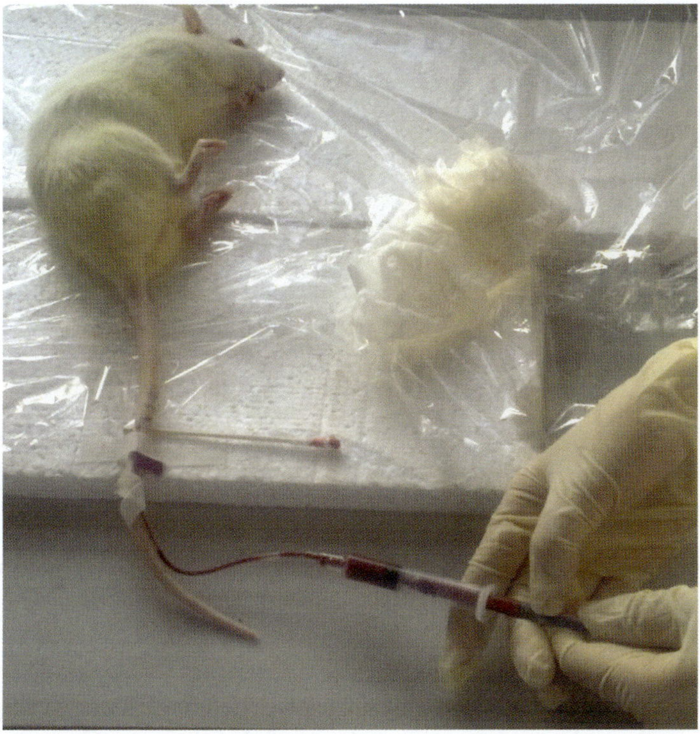

Fig. 22.3. Blood purification treatment of an EAMG rat (Ren *et al.*, 2011).

remarkably reduced. Non-specific adsorption for plasma proteins was limited. Like protein A-based adsorbent, MEP-Sepharose can also be used repeatedly, allowing for a treatment process with two parallel columns repeatedly regenerated. Although the selectivity of MEP-Sepharose is somewhat lower than that of protein A-Sepharose, its higher antibody binding capacity (29.63 mg/mL for Igs from human plasma) can still make it an attractive alternative adsorbent for use in autoantibody removal therapy.

Further investigation of the safety and efficacy of MEP-Sepharose for treating MG was conducted *in vivo* using the animal model of experimental autoimmune MG (EAMG) with female Lewis rats (Fig. 22.3) (Ren *et al.*, 2011). For the EAMG group of rats that

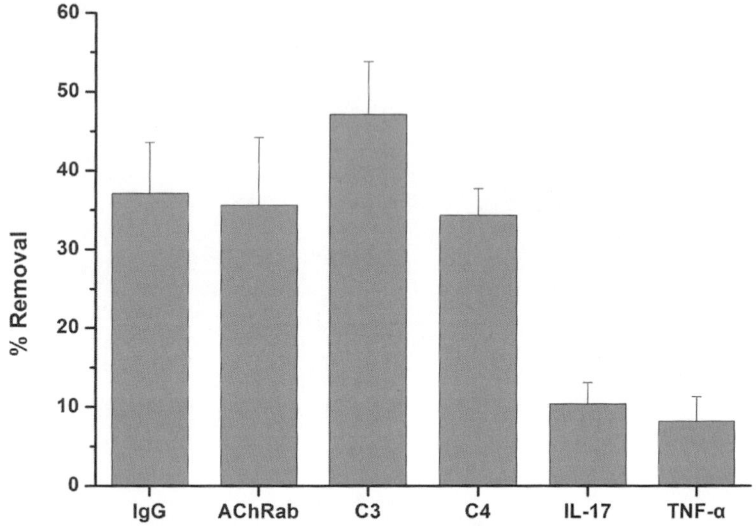

Fig. 22.4. Removal rates of antibodies, complement factors, and proinflammatory cytokines from treated-EAMG rats after a single treatment session (Ren *et al.*, 2011).

were given blood purification treatments, the treatments were carried out when they showed apparent signs of the disease. The treatments provided a significant amelioration of clinical weakness for EAMG rats. After a treatment session, the counts of the blood cell did not significantly change. In contrast, serum levels of total IgG and acetylcholine receptor antibody were significantly reduced. In addition, significant reduction in complement components C3, C4, inflammatory cytokines interleukin-17, and tumor necrosis factor-α were also observed (Fig. 22.4). Remission was observed for rats in the EAMG-treatment group, both in clinical score and in body weight (Fig. 22.5).

Our studies show that MEP-based adsorbent not only removed pathogenic autoantibodies directly from the blood but also modulated cellular immunity through removal of complement components and related proinflammatory cytokines, thereby providing a potentially superior strategy for the treatment of autoimmune diseases, such as MG.

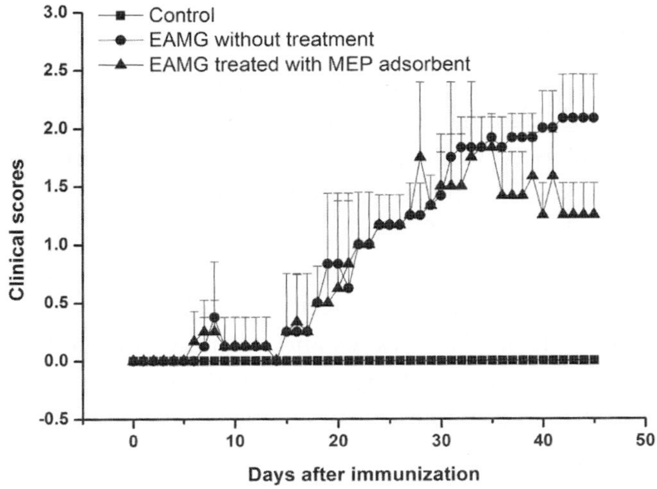

(A)

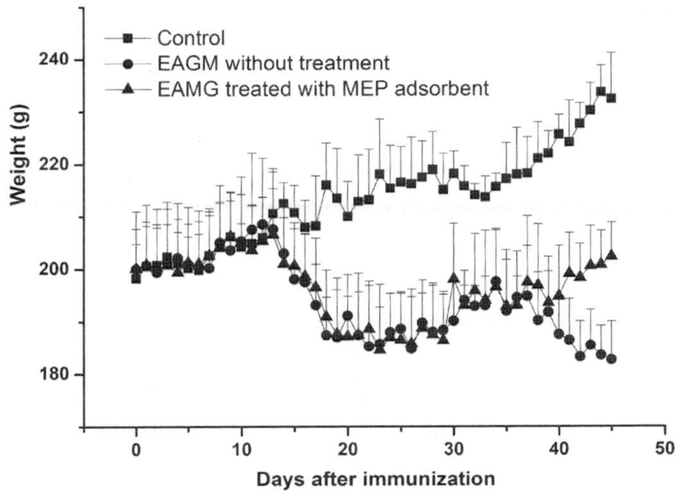

(B)

Fig. 22.5. Clinical score and body weight in treated and untreated EAMG model rats (Ren *et al.*, 2011). Clinical manifestations (A) and body weight (B) were measured in normal rats, EAMG rats, and EAMG rats treated with MEP adsorbent. Hemoperfusion was performed twice for each rat of EAMG-treatment group within five days (day 33 and 37). Data were obtained from six different experiments.

4. Perspectives

Like surgical procedures for tumor treatments, in which pathogenic factors are directly eliminated to cure the disease, immunoadsorption also provides a powerful means to remove autoantibodies and immune complexes that are major pathogenic molecules for autoimmune diseases. It can be foreseen that immunoadsorption will be developed as an important part of immunomodulatory treatment for autoimmune diseases, and more specific adsorbers will be developed not only for specific autoantibodies and immune complexes, but also complement components and related proinflammatory cytokines, to exert more extensive modulation on cellular immunity.

References

Alkan H, Bereli N, Baysal Z, *et al.* (2010) Selective removal of the autoantibodies from rheumatoid arthritis patient plasma using protein A carrying affinity cryogels. *Biochem Eng J* **51:** 153–159.

Baggi F, Ubiali F, Nava S, *et al.* (2008) Effect of IgG immunoadsorption on serum cytokines in MG and LEMS patients. *J Neuroimmunol* **201–202:** 104–110.

Balint JP, Hussein MA, Quagliata F, *et al.* (1996) Modulation of idiotypic and antiidiotypic immunoglobulin G responses in an alloimmune thrombocytopenic patient associated with extracorporeal protein A immunoadsorption. *Artif Organs* **20:** 266–270.

Benny WB, Sutton DMC, Oger J, *et al.* (1999) Clinical evaluation of a staphylococcal protein A immunoadsorption system in the treatment of myasthenia gravis patients. *Transfusion* **39:** 682–687.

Bosch T. (1996) Current status in extracorporeal immunomodulation: Immune disorders. *Artif Organs* **20:** 902–905.

Braun N, Gutenberger S, Erley CM *et al.* (1999) Immunoglobulin and circulating autoimmune complex kinetics during immunoadsorption onto protein A sepharose. *Transfus Sci* **19:** 25–31.

Braun N, Bosch T. (2000) Immunoadsorption, current status and future developments. *Expert Opin Inv Drug* **9:** 2017–2038.

Braun N, Erley C, Klein R, *et al.* (2000) Immunoadsorption onto protein A induces remission in severe systemic lupus erythematosus. *Nephrol Dial Transplant* **15:** 1367–1372.

Carter PH, Zhao Q. (2010) Clinically validated approaches to the treatment of autoimmune diseases. *Expert Opion Inv Drug* **19:** 195–213.

Cooper GS, Stroehla BC. (2003) The epidemiology of autoimmune diseases. *Autoimmun Rev* **2:** 119–125.

Deisenhofer J. (1981) Crystallographic refinement and atomic models of a human Fc fragment and its complex with fragment B of protein A from Staphylococcus aureus at 2.9- and 2.8-A resolution. *Biochemistry-Us* **20:** 2361–2370.

Duncan AR, Winter G. (1988) The binding site for C1q on IgG. *Nature* **332:** 738–740.

Eaton WW, Rose NR, Kalaydjian A, *et al.* (2007) Epidemiology of autoimmune diseases in Denmark. *J Autoimmun* **29:** 1–9.

Ellis JA, Kemp AS, Ponsonby AL. (2014) Gene-environment interaction in autoimmune disease. *Expert Rew Mol Med* **16:** e4.

Felson DT, Lavalley MP, Baldassare AR, *et al.* (1999) The Prosorba column for treatment of refractory rheumatoid arthritis: A randomized, double-blinded, shamcontrolled trial. *Arthritis Rheum* **42:** 2153–2159.

Giles BM, Boackle SA. (2013) Linking complement and anti-dsDNA antibodies in the pathogenesis of systemic lupus erythematosus. *Immunol Res* **55:** 10–21.

Hershko AY, Naparstek Y. (2005) Removal of pathogenic autoantibodies by immunoadsorption, *Ann NY Acad Sci* **1051:** 635–646.

Hirata N, Kuriyama T, Yamawaki N. (2003) Immusorba TR and PH. *Therap Apher Dial* **7:** 85–90.

Jansen M, Schmaldienst S, Banyai S, *et al.* (2001) Treatment of coagulation inhibitors with extracorporeal immunoadsorption (Ig-Therasorb). *Brit J Haematol* **112:** 91–97.

Kao D, Lux A, Schwab I, *et al.* (2014) Targeting B cells and autoantibodies in the therapy of autoimmune diseases. *Semin Immunopathol* **36:** 289–299.

Keller F, Wagner K, Faber U, *et al.* (1983) Elimination kinetics of plasma exchange. *Klin Wochenschr* **61:** 1115–1122.

Knier B, Hemmer B, Korn T. (2014) Novel monoclonal antibodies for therapy of multiple sclerosis. *Expert Opin Biol Th* **14:** 503–513.

Knobel HR, Villinger W, Islinker H. (1975) Chemical analysis and electron microscopy studies of human C1q prepared by different methods. *Eur J Immunol* **5:** 78–81.

Kutsuki H, Takata S, Yamamoto K, *et al.* (1998) Therapeutic selective adsorption of anti-DNA antibody using dextran sulfate cellulose column (Selesorb) for the treatment of systemic lupus erythematosus. *Therap Apher* **2:** 18–24.

Matic G, Bosch T, Ramlow W. (2001) Background and indications for protein A based extracorporeal immunoadsorption. *Therap Apher* **5:** 394–403.

Nakaji S, Hayashi N. (2003) Adsorption column for myasthenia gravis treatment: Medisorba MG-50. *Ther Apher Dial* **7:** 78–84.

Oka K, Nakaji S, Tanihara M, *et al.* (1993) Development of specific immunoadsorbent containing immobilized synthetic peptide of acetylcholine receptor for treatment of myasthenia gravis. *Jpn J Artif Organs* **22:** 194–7.

Reid KBM. (1976) Isolation by partial pepsin digestion of the three collagen-like regions present in subcomponent C1q of the first component of human complement. *Biochem J* **155:** 5–17.

Ren J, Jia L, Xu L, *et al.* (2009) Removal of autoantibodies by 4-mercaptoethylpyridine-based adsorbent. *J Chromatogr B* **877:** 1200–1204.

Ren J, Bai Y, Hao L, *et al.* (2011) Amelioration of experimental autoimmune myasthenia gravis rats by blood purification treatment using 4-mercaptoethylpyridine-based adsorbent. *J Biomed Mater Res A* **98A:** 589–595.

Roque ACA, Silva CSO, Taipa MA. (2007) Affinity-based methodologies and ligands for antibody purification: Advances and perspectives. *J Chromatogr A* **1160:** 44–55.

Samuelsson G. (2001) Extracorporeal immunoadsorption with protein A: Technical aspects and clinical results. *J Clin Apher* **16:** 49–52.

Shibuya N, Sato T, Osame M, *et al.* (1994) Immunoadsorption therapy for myasthenia gravis. *J Neurol Neurosurg Psychiatry* **57:** 578–581.

Stefanutti C, Di Giacomo S, Mareri M, *et al.* (2003) Immunoadsorption apheresis (Selesorb) in the treatment of chronic hepatitis C virus–related type 2 mixed cryoglobulinemia. *Transfus Apheresis Sci* **28:** 207–214.

Stegmayr BG, Almroth G, Berlin G, *et al.* (1999) Plasma exchange or immunoadsorption in patients with rapidly progressive crescentic glomerulonephritis. A Swedish multi-center study. *Int J Artif Organs* **22:** 81–87.

Ullrich H, Kuehnl P. (2004) New trends in specific immunoadsorption. *Transfus Apher Sci* **30:** 223–231.

Uzun L, Yavuz H, Osman B, *et al.* (2010) Poly(hydroxyethyl methacrylate) based affinity membranes for in vitro removal of anti-dsDNA antibodies from SLE plasma. *Int J Biol Macromol* **47:** 44–49.

Vincent A, Palace J, Hilton-Jones D. (2001) Myasthenia gravis. *Lancet* **357:** 2122–2128.

Yang Y, Yu YT, Song JC, *et al.* (1988) A new DNA immune adsorbent for hemoperfusion in SLE therapy: A clinical trial. *Artif Organs* **12:** 444–446.

Yang F, Sundrud MS, Skepner J, *et al.* (2014) Targeting Th17 cells in autoimmune diseases. *Trends Pharmacol Sci* **35:** 493–500.

Chapter 23

Treatment of Dermatonosus by Hemoperfusion

Shi Zhenwei*

1. Psoriasis

Psoriasis, also known as "*Psoriasis vulgaris*", is a long-lasting (chronic), relapsing/remitting, inflammatory skin disease that involves a variety of immune cells. Psoriasis is characterized by skin lesions — including red, scaly patches — in addition to varying degrees of symptoms that are highly detrimental to patients' quality of life. Moreover, it is noteworthy that the incidence of this disease is continuously increasing from year to year. More specifically, in China, the incidence has increased from 0.12% in 1984 to 0.59% more recently. In addition, in European and American countries, the incidence reaches as high as 2~3%, whereas it is relatively low in Japan, falling between 0.2~1%. Clinically, there are four major types of psoriasis: (1) type I, plaque psoriasis (*psoriasis vulgaris*); (2) type II, arthropathic psoriasis; (3) type III, pustular psoriasis; and (4) type IV, erythrodermic psoriasis. The primary pathological features include excessive proliferation of epidermal keratinocytes with parakeratosis; papillary vasodilation; mild vessel wall thickening; peripheral nerve hyperplasia of the skin; and mild to moderate cell infiltration of multinucleated giant cells, lymphocytes, Langerhans cells, macrophages, and mast cells. Furthermore, psoriasis

* Zhenwei (No email address) email of coeditor for this: yaotingyu@nankai.edu.cn.

pathogenesis is highly complex, involving a polygenetic background and related to infections, T cells, dendritic cells and other immune cells, cytokines, angiogenesis-related factors, and neuroimmunomodulation. Recently, because of the satisfactory efficacy of cyclosporine and other calcineurin inhibitors in psoriasis treatment, psoriasis research has seen a shift in direction. Immunological studies on the pathogenesis of psoriasis have been shifted from epidermal keratinocytes (KCs) to the activation of all kinds of immune cells and their secretory products, cytokines, vascular endothelial growth factors (VEGF), etc. Sanchez (2010) pointed out that the immune cells and their secretory products eventually lead to excessive KC proliferation, epidermal hyperplasia, and angiogenesis with significant vasodilation (Sanchez, 2010).

1.1. *Pathogenesis of Psoriasis*

1.1.1. *Genes associated with psoriasis*

Psoriasis is a multi-gene genetic disease, and the pathogenesis is thus related to multiple genes. The first psoriasis-associated genetic factor found that human leukocyte antigen (HLA) is located at human chromosome 6p21.3. HLA-Cw*6, is the most relevant allele, while HLA-Cw*6 and -B*57 are potential susceptibility gene markers for psoriasis. HLA-DRB 1*0701, HLA-DQA1*0201, and DQB 1 *0303 are expressed in type I psoriasis. Gonzalez *et al.* (2000) reported that, compared with HLA-Cw*0602, the β allele of the octamer transcription factor 3B is more significantly associated with plaque psoriasis. An additional gene is the CDSN (corneodesmosin) gene, which is located approximately 160 kb telomeric of HLA-C on 6p21. CDSN is expressed in differentiated keratinocytes and encodes a protein homologous to granule cell layer components. Ameen *et al.* (2005) found that Caucasian patients with psoriasis are closely related to the CDSN allele 5 (+ 619T, + 1240G, and + 1243C) and HLA-Cw6. Another crucial gene is the human endogenous retrovirus K deoxyuridine triphosphatase (HERV.KdUTP) gene that belongs to the tandem-repeat virus HML-2 family and is a susceptibility gene for psoriasis. Foerster *et al.* (2005) found that it is expressed in both the patient's normal and lesional skins, as well as in the peripheral blood. Moreover, the six different proteins that are homologous to the tumor suppressor protein p53 and encoded by p63 genes are crucial for the normal development of ectodermally-derived structures. Gu *et al.* (2006) reported that p63 plays a role in psoriasis. The relationship between the KIR gene, KIR related genes, and the susceptibility genes for psoriasis is clearly complicated, thus further study is needed for elucidation.

1.1.2. *Environmental factors*

Various environmental factors can influence the pathogenesis of psoriasis. One such example is smoking. Smoking stimulates the activation of neutrophils to release peroxidase. Then, these peroxides and peroxidases play important roles in psoriasis, altering the oxidative metabolism of phagocyte, increasing the oxidative metabolism of the inflammatory reaction and enzyme activity and further causing the occurrence or worsening of lesions. In addition, Herron *et al.* (2005) pointed out that the harmful ingredients in tobacco smoke negatively affect the deformability of red blood cells (RBCs), thus reducing the binding of hemoglobin to oxygen and consequently leading to vascular endothelial cell injury.

Another influential environmental factor is alcohol; alcohol can stimulate or aggravate psoriasis. Alcohol can induce vasodilation and increase vascular permeability, and thus facilitate the penetration of neutrophils through the vascular and lead to an infiltration of the skin (lesions). Alcohol can also increase the amount of arachidonic acid in the blood, which is the precursor of such biologically-active substances as prostaglandins and leukotrienes; it can also further inhibit the epidermal adenylate cyclase, decrease cAMP, and increase cGMP, finally leading to the rapid proliferation of epidermal cells.

An improper diet can also aggravate the psoriasis condition. Scarpa *et al.* (2000) reported that in patients with progressive stage plaque psoriasis or arthropathic psoriasis without clinical abdominal symptoms, colonoscopies found that 40% of the patients had congested mucosa in the colon, and 20% of these patients had mucosal edema and granular hyperplasia. Multiple colonic mucosal biopsies revealed that all patients possessed microstructural changes, including a special infiltration of plasma cells and lymphocytes, the formation of lymphoid follicles, active inflammation, and glandular atrophy.

In addition, it is particularly noteworthy that the incidence of psoriasis remains quite low in Africa. Besides genetic factors, the unique diet of the area (with corn as the main component) is certainly a critical factor in maintaining a low incidence in Africa. The finding is reported in Namazi's study. Michaelsson *et al.* (2003) reported that the anti-gluten antibodies IgA and/or IgG exist in the serum of 16% of plaque psoriasis patients. After just three months of adhering to a simple gluten-free diet, the patients' clinical conditions improved remarkably, the previously increased $Ki67^+$ cells had significantly decreased, and the overexpressed transglutaminase was also significantly reduced. Another controlled study conducted by Rucevic *et al.* (2003) showed that a four-week-long low-fat diet resulted in a significantly improved psoriasis condition and reduced the serum lipids.

Furthermore, mental stress can trigger and aggravate psoriasis, implicating nervous-endocrine factors in psoriasis pathogenesis. Usually, psoriasis patients have differing degrees of emotional depression and anxiety. Stress can affect the central nervous system, regulate networking through chemical information factors and receptors shared by the nervous, endocrine, and immune systems, and result in the eventual occurrence and development of the disease. In addition, it has been proven that the α-endorphin level is elevated significantly and correlated with the severity of the psoriasis condition. Neuropeptides may induce changes of immune cell subsets via cytokines and further secrete immunomodulatory cytokines.

1.1.3. *Viral infection*

There is also a potential existing relationship between psoriasis and the persistent infection of certain viruses. For example, the adenovirus has a tropism for keratinocytes and induces quiescent cells into S phase after infection. The E1A and E1B proteins of the adenovirus can inhibit cell apoptosis and keep the infected T cells and keratinocytes in an active state. This results in delayed psoriasis healing.

Another virus is the human parvovirus B19 (PVB19), which is a single-stranded DNA virus that may be the cause of a series of clinical symptoms. Yazici *et al.* (2006) pointed out that the PVB19 DNA is primarily related to antibody IgG, indicating the subclinical activation of PVB19 in psoriasis patients and the important role of the PVB19 infection in the pathophysiology of psoriasis. Moreover, human endogenous retroviruses (HERVs) are the genomic traces from the ancient invasion of active retroviruses and are normal parts of the human genome. HERVs include two open reading frames that encode both a gag protein and a reverse transcriptase and have a role in the pathogenesis of comprehensive human genetic diseases. Moles *et al.* (2005) reported a significant enhancement of HERVs sequencing discovery rate at lesion sites in psoriasis patients.

1.1.4. *Neovascularization*

Abnormal angiogenesis is tightly related to psoriasis occurrences, persistence, and recurrence. Therefore, it is clearly crucial to psoriasis pathogenesis. Under the effect of the endogenous angiogenesis inhibitor, normal skin microvasculature is in a quiescent state, and no proliferation occurs. However, in psoriatic lesions, the original balance is undermined by the increased angiogenic activator, thus microvascular dysplasia occurs. Moreover, VEGFs promote angiogenesis with elevated expression levels in lesional psoriatic plaque. VEGF, $^{+}405$, $^{-}460$ are associated with progressive stage psoriasis.

1.1.5. *Immunological pathogenesis of psoriasis*

1.1.5.1. The role of T cells in psoriatic lesions

Currently, psoriasis is considered to be a persistent, chronic, and inflammatory disease caused by infiltration of T cells. After antigen stimulation, $CD4^{+}$ T lymphocytes initially differentiate into Th0 cells. Th0 cells further differentiate into three cell subsets: Th1, Th2, and Th3. Furthermore, these Th0 cells can also polarize into Th1 or Th2 cells under certain conditions. Understanding this activation is immensely important as T cell activation is the crucial factor in psoriasis pathogenesis. Tonel *et al.* (2009) found that the lymphokines produced by activated T cells stimulate the epidermal proliferation, and the proliferated KCs release other cytokines that further enhance the effect of the activated T cells.

It has also been reported that the activation of T lymphocytes in psoriasis lesions is the mediator for epidermal hyperproliferation. The psoriasis lesions coexist with the epidermal infiltration and T cell activation, and the decrease in (or the disappearance of) infiltrated T cells occurs before lesion recovery. Moreover, there are specific T cell subsets in psoriasis lesions. Yun *et al.* (2010) reported that all human T cell subsets are immunoregulatory cells and are essential for maintaining an immune balance. The inhibition ability of regulatory $CD4^+$ and $CD25^+$ T cells in the peripheral blood and lesions of psoriasis patients is decreased, and this decrease consequently stimulates an enhanced activity of effector T cells.

Additionally, Sweeney *et al.* (2011) demonstrated that psoriasis symptoms are related to the Th1/Th2 cytokines imbalance; more specifically, numerous $CD4^+$ T cells, but only a few $CD8^+$ T cells, accumulate at the papillary dermis around the damaged skin. Hedrick *et al.* (2009) found that Th17 cells — a group of new $CD4^+$ T cell subsets that differ from Th1 or Th2 and primarily secrete IL-17 — as well as the secreted IL-12 and IL-23 are very important in the inflammatory response in psoriasis. The chemokine receptors for the psoriatic-inflamed skin, CCR6 and CCR3, are both expressed by Th17. Initially, it was thought that Th17 cells only produce IL-17a and are associated with autoimmune inflammation processes. However, further studies conducted by Waisman (2011) and Waite *et al.* (2011) demonstrated that Th17 cells also produce other cytokines, such as IL-22 and TNF-α (Waisman, 2011; Waite, 2011).

1.1.5.2. The interaction between T cells and KCs

In vitro experiments conducted by Wrone Smith *et al.* (1996) confirmed that the cloned product of T cells from lesional psoriasis skin can stimulate the proliferation of cultured KCs. In the transferring of uninvolved/symptomless skins from psoriasis patients to SCID mice, only an injection of the activated T cells from psoriasis patients induced a psoriatic plaque, whereas the T cells and skins from healthy individuals did not induce skin lesions. Compared to the KCs from normal skins, the β1 integrins, $^{+}$K1/K10-KC of the uninvolved/symptomless skin from a psoriasis patient have hyperresponsiveness with respect to T cell lymphokines. After T cell-produced cytokines take effect, KCs proliferate and produce several cytokines, such as IL-1, IL-6, and IL-8, which in turn enhance the activation state of the T cells. Furthermore, the activated T cells stimulate keratinocytes via the generated cytokines IFN-γ and TNF-α to produce ICAM-1, and then they mediate the adhesion between T cells and KCs through the LFA-1/ICAM-1 pathway, thus creating a vicious circle between T cells and KCs.

1.1.5.3. Dendritic cells

It has been discovered that psoriasis pathogenesis is closely related to dendritic cells (DC). As the most powerful antigen-presenting cells known, DCs originate from hematopoietic stem cells, existing in the circulatory system and widely distributed in the peripheral tissues in their immature form. Under immune stimulation, the immature DCs in the peripheral blood and tissues migrate to secondary lymphoid tissues and there develop into mature DCs. There are three types of DCs in the skin: Langerhans cell (LC), plasmacytoid DC (pDC), and myeloid DC (mDC). As a special epidermal DC, the LC plays a significant role in the process of delayed hypersensitivity induction and is essential for cellular immunity. The LC-produced cytokines IL-12, IL-20, and TGF can be greatly enhanced, thus activating large numbers of T cells and triggering psoriasis. Furthermore, Fujita *et al.* (2009) discovered that LCs and dermal DCs can induce T cells in peripheral blood and can stimulate the differentiation of $CD4^{+}$T cells into $CD4^{+}$ $CD8^{+}$ T cells, which then produce IL-22. During the inflammatory response in psoriasis, the total amounts of pDCs and mDCs increase. Large amounts of pDCs, mature and immature DCs, and inflammatory dendritic epithelial cells are produced in the psoriatic lesions. Additionally, Zhou *et al.* (2009) found that inflammatory DCs also secrete TNF, IL-12, IL-23, and inducible nitric oxide.

1.1.5.4. Cytokine abnormalities in psoriasis

The IL-1, IL-2, TNF-α, IFN-γ, IL-6, IL-7, IL-8, IL-12 and GM-CSF levels are all enhanced in psoriasis lesions and serum. A summary of the effects that these enhanced cytokine levels have on psoriasis is presented in the following text. IL-1 can activate T cells, promote the proliferation of KCs, stimulate angiogenesis and endothelial cells to express adhesion molecules, and induce other cytokines to amplify the inflammatory cascade. Moreover, IL-6 can promote the proliferation and activation of both KCs and T cells. Paquet *et al.* (1996) demonstrated that the expression of IL-6 and its receptors are most significant in the transition region of skin lesions, indicating a possible relationship between IL-6 and the plaque outward expansion (Paquet and Pierard, 1996). Furthermore, Bonifati *et al.* (1997) have shown that IFN-γ can stimulate KCs to produce IL-7, which in turn stimulates T cells to synthesize IFN-γ and promote the proliferation of T cells. IL-8 has chemotaxis toward neutrophils and T cells and can promote angiogenesis as well as the proliferation of KCs. In addition, IL-4, IL-5 and IL-10 all belong to Th2 cytokines. However, the expression of IL-4 and IL-5 in lesions remains inconsistent across different studies. In the study of Prens *et al.* (1996), IL-4 can inhibit the positive feedback of proinflammatory cytokines in psoriatic lesions and decrease the levels of IL-1, IL-6, IL-8 and TNF-α while increasing the IL-1R antagonist. On the other hand, the role of IL-5 in psoriasis is very difficult to determine because it primarily regulates the growth and differentiation of eosinophils and acts as its chemokine, but eosinophils may not be involved in the pathogenesis of psoriasis. Additionally, Howard *et al.* (1992) found that, IL-10 can inhibit Th1 cytokines and has an anti-inflammatory effect; accordingly, its levels are decreased in psoriasis. In contrast, IL-11 levels are increased in psoriasis. IL-11 is also anti-inflammatory and can inhibit the stimulation of LPS to macrophages from producing TNF-α, IL-6, IL-1β and IL-12. Thus, similar to IL-4, Leng *et al.* (1997) found that increased IL-11 levels might function to control the inflammatory phenomena of psoriasis. Furthermore, the epidermal growth factor (EGF) receptor-ligand system is disordered in psoriatic lesions; the protein expression of EGFR, its ligand TGF-α, and amphiregulin are increased in lesional skins. Cook *et al.* (1997) showed that transgenic mice with the amphiregulin gene had a psoriasis-like skin phenotype, indicating that the aberrant synthesis regulation of the KC amphiregulin might cause psoriasis. Additionally, Mabuchi *et al.* (2012) reported that TNF-α is crucial for mediating psoriatic lesions in the inflammatory cascade in psoriasis. IL-23, on the other hand, stimulates memory T cells to secrete IL-17, which can induce the expression of TNF-α, IL-1β and IL-6 in endothelial and epithelial cells; furthermore, IL-17 can also stimulate the production of chemokines CXCL1, CXCL5, IL-8 and C-C motif ligand 2 (CCL2), which are all related to the neutrophil recruitment in inflammatory reactions. Therefore, the cytokines IL-17, IL-22, IL-23, IL-6 and IL-8 all have particular roles in promoting the pathogenesis and pathology of psoriasis. In addition, IFN-γ is an important cytokine in the pathogenesis of psoriasis. *In vitro* studies conducted by Bonifati *et al.* (1997) have shown that the proliferation of KCs induced by T cell clones is IFN-γ dependent in psoriasis, while the injection of IFN-γ into non-lesional skin can induce dot-like psoriasis lesions around the injection spot. Furthermore, IFN-γ can induce the expression of the mitogen of KCs, TGF-α. It can also stimulate KCs to produce ICAM-1 and HLA-DR and is conducive to inducing T cells into the lesions. However, IFN-γ has a direct anti-proliferative effect on the rapid growth of normal human KCs, and it may inhibit the growth of primary cultured psoriatic epidermal cells as well. In addition, Presta *et al.* (2009) discovered that the fibroblast growth factor basic (FGF2) promotes angiogenesis via the epidermal receptors of all kinds of endothelial cells, including the tyrosine kinase receptor and heparan sulfate proteoglycan. Meanwhile, FGF2 induces chemokines and their receptors, endothelial cell adhesion molecules, and the prostaglandin pathway components to cause an inflammatory responses in lesions. This is critical because angiogenesis and an inflammatory response are the primary pathogenesis of psoriasis. FGF10, which is keratinocyte growth factor 2, can stimulate epithelial cell growth specifically and is an important growth factor in the formation of skin tissues.

1.1.5.5. Vascular endothelial cells, adhesion molecules and complements

The proliferation of endothelial cells, neovascularization and the migration of inflammatory cells into the outside lumen are all important in the pathogenesis of psoriasis. For the lymphocytes within the psoriatic dermal microvascular to extravasate and reach the dermis, they first need to adhere to the endothelial cells and then pass through the vascular basement membrane to reach the dermis via the junction of endothelial cells. The molecular basis of this process lies in the interactions between lymphocytes and the adhesion molecules in vascular endothelial cells, as well as the expression regulation of adhesion molecules by cytokines. The ICAM-1, VCAM-1, and E-selectin in the endothelium and serum in psoriatic lesions are highly expressed and are consistent with disease severity. Furthermore, Sjogren *et al.* (1999) found that the high levels of IL-1α/β, IL-6, IL-8, TNF-α, TGFα/β, and IFN-γ may activate the endothelial cells either directly or indirectly, increase the expression of ICAM-1, VCAM-1, and E-selectin by endothelial cells, and promote the neovascularization of dermal papilla. Moreover, numerous immune cells, such as monocytes, polymorphonuclear cells and T cells, enter the skin by endothelial adhesion through the bloodstream. The interaction between integrin β_2-subunit(CD18), which is expressed in leukocytes, and ICAM-1 plays a critical role in the transendothelial migration and entering the psoriatic lesions of leukocytes. In addition, there are large numbers of immunoglobulin IgG and IgA deposition in psoriatic lesions. High levels of C3a, C4a, C5a, C4b and C5b-9 also exist, and the levels of C3a and C4a are elevated in plasma.

It is important to note that the causes for complement activation remain unclear; nevertheless, infection (including ultra-antigen) may be a potential cause. Complement activation may occur in early psoriatic lesions. The produced complement fragments and complex (C5b-9) slowly enter circulation, and those substances in the epidermis can activate various cells in the skin lesions, including KCs, T cells, macrophages, mast cells, fibroblasts, and endothelial cells. These complement fragments and complexes enter the dermis microvascular from epidermis. As small molecules, C5 and C5des-arg can pass through the vascular wall basement membrane, activate endothelial cells and increase vascular permeability, and the early morphological changes of psoriasis have exhibited abnormal changes of dermal microvascular plexus. The normal endothelium will not allow molecules with a molecular weight greater than 40,000 Da to pass through the vascular wall. However, Fleming *et al.* (1996) pointed out that, with the increased vascular permeability in psoriasis, C5b-9 may enter blood vessels from the epidermis and thus C5b-9 is present in circulation (Fleming *et al.*, 1996).

1.2. *General Treatments of Psoriasis*

While there are currently many available treatments for psoriasis, the disease remains incurable. The following section summarizes current psoriasis treatments.

1.2.1. *Ultraviolet (UV) light therapy*

This therapy refers to a method of irradiation with artificial UV light in the absence of photosensitizers, and it is applicable to the treatment of various skin diseases. The most widely used wavelength is 311 nm narrow-spectrum of UV light, which can directly induce the apoptosis of T cells and inhibit the proliferation of collagen. Studies conducted by Kircik *et al.* (2008) have proven that 63%~80% of patients with psoriasis lesions can recover with just a single course of treatment (one course consists of 10 treatments in total with one treatment every other day).

1.2.2. *Psoralen-Ultraviolet A (PUVA) therapy*

This therapy is commonly known as PUVA. After the patient applies the photosensitizers, such as 8-methoxy psoralen, orally or topically, the skin is exposed to UVA. The therapeutic effect is achieved by repeated, controlled phototoxic reactions. The biggest advantage of PUVA is that severe psoriasis can be controlled without continuous treatment (Khushid *et al.*, 2000; Ashcroff *et al.*, 2000). Additionally, this treatment is undergoing continuous attempts to improve the photochemical treatment, one of which is substituting UVA with shortwave UVB. An efficacy study conducted by Khushid *et al.* (2000) demonstrated that there is no significant difference in efficacy between PUVB and PUVA; however, for the same efficacy, the accumulation of UVB is significantly lower than UVA, and thus, UVB may be safer. Ashcroff *et al.* (2000) have reported that the combination of PUVA with other drugs (calcipotriene, tazarotene, alitretinoin, and methotrexate) may reduce the number of PUVA treatments and side effects as well as enhance efficacy. Most side effects resulting from long-term PUVA therapy are dose-dependent.

1.2.3. *Topical treatments*

The topical treatments for psoriasis include corticosteroids (Ference *et al.*, 2009); topical retinoids (Mehta *et al.*, 2011); vitamin D analogs (Wang *et al.*, 2011); tacrolimus (Harrison *et al.*, 2007); and the composite formula of vitamin D analogues and corticosteroids (Fleming *et al.*, 2010). Thaci *et al.* (2010) analyzed 131 random controlled trials on the topical treatments of chronic plaque psoriasis and found that the topical treatment of psoriasis with calcipotriene in conjunction with potent glucocorticoids exhibits better efficacy and safety over the applications of either of the two drugs alone.

1.2.4. *Methotrexate (MTX)*

There are detrimental side effects associated with MTX; long-term use can cause bone marrow suppression, liver fibrosis, and other adverse reactions. Thus, its usage is strictly administered. Typically, it is used for generalized pustular body and joint disease. However, it can also be used for severe plaque, erythrodermic, and nail psoriasis that are resistant to routine treatments. Lee (2009) reported that the usage of low dose MTX (5mg/week) was beneficial for an 11-year-old female with nail psoriasis, whose every nail was malnourished; after four weeks of treatment, the symptoms began to diminish, and after nine and 13 months of treatment, respectively, all of the patient's fingernails and toenails recovered to normal.

1.2.5. *Cyclosporine*

Cyclosporine may selectively inhibit T lymphocytes at the cellular level; in particular, they suppress helper T lymphocytes from producing IL-2 and other cytokines. At the molecular level, it primarily interferes with the transcription of IL-2 and other cytokine genes. Pereira *et al.* (2006) used cyclosporine to treat six psoriasis patients who were resistant to conventional treatment; after an average of 12-weeks of treatment, the lesions disappeared completely, and there were no obvious side effects.

1.2.6. *Alitretinoin*

Oral alitretinoin or isotretinoin for generalized pustular psoriasis has been reported to have a good efficacy and tolerance. The major adverse reactions for short-term usage include cheilitis, dry skin, scaling, hair loss, abnormal liver function, and elevated blood lipids. It is crucial for long-term users to monitor for any bone damage (Chao *et al.*, 2009).

1.2.7. *Clinical application of combination therapy*

Effective combination therapies as reported by Gustafson *et al.* (2013) and Devaux *et al.* (2012) include: topical vitamin D3 with topical corticosteroids, topical vitamin D3 with systemic therapy, topical tretinoin with light therapy and topical vitamin D3 with light therapy (Devaux *et al.*, 2012; Gustafson *et al.*, 2013).

1.2.8. *Strategies for antibody treatments for psoriasis*

According to the different aspects of the immunological abnormalities found in psoriasis, a variety of strategies for targeted antibody therapy have been developed. Accordingly, many molecules and/or structures are used as the target antigens for antibody therapy.

1.2.8.1. Anti-CD4 antibody

CD4 molecules are mainly distributed around the T cell surface. Through interacting with the MHCH class, CD4 molecules participate in the activation of T cells and antigen-presenting function as well as play a major role in the pathogenesis of psoriasis. Additionally, there have been reports that the treatments of HIV infection, rheumatoid arthritis, multiple sclerosis, and transplant rejection with monoclonal or chimeric antibodies against mouse CD4 have achieved certain effects. Robinet *et al.* (1997) injected monoclonal antibody (mAb) against mouse CD4 into severe psoriasis and atopic dermatitis patients, and their results achieved significant efficacy. A beneficial aspect of this type of treatment is that there are no other serious adverse reactions except for the discomfort associated with mild nausea and vomiting after the initial dose. Another study conducted by Baehelez *et al.* (1998) has also reported satisfactory efficacy by using humanized anti-CD4 mAb to treat plaque psoriasis resistance to other drugs. After four weeks of treatments, the psoriasis area and the severity indexes (PASIs) for patients were decreased by 46% with no significant side effects.

1.2.8.2. Anti-CD25 antibody

By binding to CD25 molecules, anti-CD25 antibodies block the effect of IL-2 and inhibit the activity of T cells; moreover, this therapy has achieved excellent efficacy in the prevention and treatment of transplant rejection. Owen *et al.* (2000) used basiliximba, a chimeric anti-CD25 antibody, in conjunction with cyclosporine A to treat a patient with severe psoriasis who previously demonstrated multi-drug resistance, including cyclosporine A, and obtained satisfactory efficacy. The patient's PASI dropped from 51.0 to 4.0, and no adverse effects occurred. Furthermore, Mrowietz *et al.* (2000) used the same antibody to treat two plaque psoriasis patients. One patient did not respond to the treatment. However, the other patient achieved significant efficacy; after three treatments with 20~40 mg/treatment, the PASI dropped 83%, and the CD25-positive cell count in the peripheral blood decreased significantly.

1.2.8.3. Anti-tumor necrosis factor α (TNF-α) antibody

The anti-TNF-α mAb kills the cells that contain TNF-α on the membrane surface by cytotoxicity, and it can induce the apoptosis of T cells as well. Rioda *et al.* (2006) reported that a treatment of 5 or 10 mg/kg each time may significantly improve patients' quality of life in just two weeks. The disease can be quickly controlled through this treatment for unstable redskins or pustular psoriasis patients. Compared to MTX, there is not any associated renal or liver toxicity; however, it must be injected slowly.

1.2.8.4. CD11a antibody

CD11a is the α chain of the lymphocyte function-associated antigen-1 (LFA-1). It is crucial for the activation and emigration of T cells into the skins as well as for cell toxicity. Treatment of moderate to severe psoriasis with humanized anti-CD11a mAb is currently in phase III clinical trials. In an open-label study conducted by Ogilvie *et al.* (2000), mAb was used to treat 31 severe plaque psoriasis patients. It was found that the mAb can fully or partially block the expression of CD11a in blood cells and skin lesions, and the blocking extent and its duration is dose-dependent. Meanwhile, the treatment process is accompanied by a reduction of CD_3^+ T cells, decreased ICAM-1 as expressed by keratinocytes and endothelial cells, epidermal thinning, and decreased PASI clinically. It is thought that the anti-CD11a mAb improves psoriasis by inhibiting the activation, the emigration towards the skin, and the cytotoxic of the T-cells. The treatment's dose is 0.03–10 mg/kg. Side effects include mild typhoid fever, abdominal discomfort, headache and fever.

1.2.8.5. Anti-IL-8 antibody

In psoriasis pathogenesis, IL-8 may result in a chemotax leading to the emigration of neutrophil and T cells into the epidermis as well as stimulate the activation and proliferation of keratinocytes. *In vitro* experiments conducted by Gottlieg *et al.* (2000) have proven that the anti-IL-8 mAb with neutralizing activity may inhibit the hyperplasia of epidermal keratinocyte in psoriasis and block the chemotaxis of IL-8 toward neutrophil and T cells; thus, it targets psoriasis.

1.2.8.6. Anti-IL-12 antibody

The main biological functions of IL-12 include: promoting the production of cytokines TNF-α and INF-γ, stimulating the activation and proliferation of T cells in collaboration with IL-2, and inducing the formation of Th cell subsets; all of which are important in the pathogenesis of psoriasis. From animal experiments, Tsukahara *et al.* (1999) found that IL-12 may directly induce psoriasis-like skin damage, and the mechanism underlying the anti-IL-12 antibody may down-regulate the expression of TNF-α and INF-γ.

1.2.8.7. Anti-IL-6R antibody

IL-6R is a membrane surface molecule that consists of 468 amino acids. Binding to IL-6, it may stimulate the proliferation of a variety of cells, including KCs. It is also involved in the mediated inflammatory response. Furthermore, Lee *et al.* (1998) have reported that humanized anti-IL-6R mAb has been successfully developed, and its application has preliminarily demonstrated good prospects in the treatment of psoriasis.

1.3. *Hemadsorption Treatment of Psoriasis*

As modern science unveils more and more concerning the etiology and pathogenesis of psoriasis, immune abnormalities are attracting greater amounts of attention. Nowadays, it is thought that the abnormal infiltration of T-lymphocytes into the lesions of psoriasis patients is key to psoriasis pathogenesis. The produced cytokines, such as IL-2, IL-6, IL-8, TNF-α and interferon, play important roles in the development of psoriasis. Furthermore, Ozturk *et al.* (2001) compared the peripheral blood of psoriasis patients with that of normal people and found that the IgG and IgA levels of psoriasis patients are significantly higher than that found in normal people; however, IgM does not exhibit a significant difference. In addition, Nishibu *et al.* (1999) discovered that the peripheral blood mononuclear cells of psoriasis patients secreted high levels of TNF-a, IL-lβ, IL-6 and IL-8 without any stimulation, and the levels were correspondingly decreased with effective treatment. From among these, TNF-α is considered to be the initiating factor for the activation and proliferation of keratinocytes and the inflammatory cascade in psoriasis. Additionally, IL-8 is crucial for neutrophil aggregation, microabscesses and pus formation.

Early clinical studies indicated that most diseases are characterized by immune complex syndrome, characterized by the removal barriers of immune complexes. The study of autoimmune diseases — including psoriasis, lupus erythematosus, and systemic vasculitis — discovered the deposition of plasma immune complexes in the patients' skins. Dmitriev *et al.* (1985) applied hemadsorption techniques to experimental treatments. The results demonstrated that after the first hemadsorption treatment, the immune complexes in the skin transferred from the dermis to the epidermis and horny layer; after three treatments, the immune complexes and dander appeared together as a mixture on the skin surface. The hemadsorption treatment can restore skin secretion and help to remove immune complexes.

As early as 1982, Nasonov *et al.* reported the treatment of psoriasis with hemoperfusion (HP) therapy. In 1987, Kolyadenko *et al.* applied the HP treatment to 89 cases of psoriasis patients with differing degrees of clinical symptoms; the results of their treatment were described with surprising significant therapeutic effects. However, there are negative reports that stand in stark contrast to the aforementioned results. In 1991, Mashkov *et al.* used antiprotease hemosorbent to treat 11 cases of skin disease patients, including two cases of generalized plaque psoriasis, five cases of arthropathic psoriasis, and a single case of erythrodermic psoriasis. The results did not accomplish an ideal therapeutic effect for the generalized and arthropathic psoriasis and produced a significant result only for the erythrodermic psoriasis patient.

It is also noteworthy that this method has good blood compatibility with nearly no complications. However, scientists continue to monitor and study the possible trace of toxins in psoriasis patients' blood. As such, scientists modified the adsorbent in the adsorption column by developing a new hemoperfusion adsorbent and applied it to clinical practice. Soon afterwards, Snezhkova *et al.* (1992) applied the HP method with DNA-coated carbon adsorbents in 14 psoriasis patients; six cases were clinical cured, and six cases demonstrated substantial improvements. In another double-blind study with hemoperfusion therapy, 22 cases received the treatments, and nine cases served as controls. The results showed complete remission rates of 55.5% and 11.2%, respectively, thus demonstrating a significant difference. Since then, the research and application of hemoperfusion for psoriasis patients who have an abnormal increase in certain blood components has become a hot topic in this field. Furthermore, Kanekura *et al.* (2006) used granulocyte and monocyte adsorption apheresis to remove the highly expressed CD11b/CD18 in a patient's blood to achieve therapeutic effects and achieved significant clinical results.

The aforementioned clinical and experimental tests confirmed that hemoperfusion can effectively remove cytokines and inflammatory mediators and adjust immune status. Hemoperfusion purifies blood by adsorbing exogenous or endogenous toxins and their metabolites. Hemodialysis combined with hemoperfusion can not only remove small toxin molecules but also effectively remove toxins that are medium to large molecules and are beyond the range of conventional hemodialysis by adsorption. Moreover, it has been confirmed by Schiffl (2001) and Suzuki *et al.* (2004) that hemoperfusion can effectively remove medium-sized molecules, TNF, IL, etc (Schiffl *et al.*, 2001; Suzuki *et al.*, 2004). The HA resin is a neutral macroporous resin with adsorption features such as a large capacity, high speed, high mechanical strength, excellent biocompatibility and relative selectivity. It primarily adsorbs substances of a molecular weight of 500~5,000 Da and can effectively absorb a variety of protein binding toxins, endotoxin, etc. However, due to the small number of observed cases and the short follow-up times, the long-term prognosis of hemoperfusion needs further clinical validation.

During recent years, many scholars reported the application of blood cell adsorption and separation technology to treat rheumatism and autoimmune diseases. Among the blood leukocyte adsorption separation therapies, the adsorption column Adacolumn (as shown in Fig. 23.1.) can selectively adsorb and remove the activated granulocytes from the peripheral blood. The specific parameters for Adacolumn are shown in Fig. 23.2. The column is filled with specially-designed cellulose acetate beads (CAB). Furthermore, the column is 206 mm long and 60 mm in diameter, and there are 3.5×10^4 beads (with a bead diameter of 2 mm) in each column. Moreover, the column weighs a total of 220 g with a 335 mL capacity, and the volume between the column is 130 mL. Finally, the column meets European standards of quality and safety (JIS T 1001(1992), IEC 601–1).

Adacolumn is an *in vitro* plasma separation and adsorption device. Blood passes through the left or right forearm brachial vein in order to establish vascular access, as shown in Fig. 23.2. According to different treatment goals for various diseases, the number and frequency of the treatment varies. Animal experiments have demonstrated that a 60 minute treatment is sufficient. After the adsorption treatment, the

Fig. 23.1. Adacolumn.

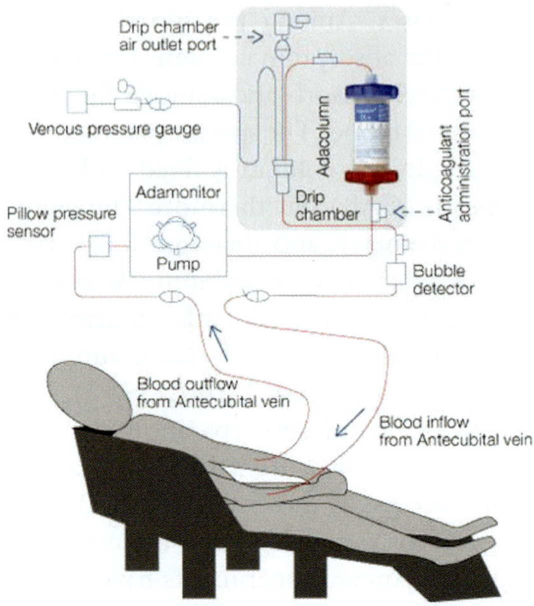

Fig. 23.2. The extracorporeal circulation scheme for Adacolumn.

Adacolumn usually has a thin layer of white blood cells, neutrophils, and monocytes on the column surface. The adsorption saturates after 60 minutes of treatment. For patients with rheumatoid arthritis, treatments should be one to two times/week for at least four continuous weeks and with a blood flow velocity of 30 mL/min during treatment and 60 min/treatment; the total treatment capacity is 1,800 mL. Furthermore, studies show that for patients with active ulcerative colitis, one treatment per week for five continuous weeks would lead to good results. The blood cell counts before and after Adacolumn showed that 65% of neutrophils, 55% of monocytes, and 2% of lymphocytes are absorbed. The cellulose acetate surface of Adacolumn may also adsorb IgG and immune complexes. The principle of leukocytes adsorption is that these leukocytes can produce the Fcγ receptor (FcγR) and complement receptor. The cells adsorbed also include $CD19^{+}$B lymphocytes and $CD56^{+}$ NK cells. Moreover, cellulose acetate adsorbent can adsorb 50% neutrophils. The underlying mechanism is that cellulose acetate can activate and adsorb the complement components iC3b and iC3b; it can also absorb neutrophil adhesion molecules Mac-1 (CD11b/CD18) and thus achieves the purpose of clearing granulocytes. A key aspect of this treatment is that it can remove the effector cells of the inflammatory response, reducing the systemic inflammatory response. Furthermore, it is medication free, dependency and drug-resistance free, and free from any serious side effects. Moreover, the adsorbed leukocytes may release large amounts of active substances, such as cytokines, complement fragment C3a and C5a. These cytokines have a short half-life, cannot reach the bloodstream of the patient in an active form, and is incapable of producing a corresponding inflammatory effect. The study, conducted by Suzuki *et al.* (2004), also demonstrated that after the Adacolumn treatment, the soluble TNF-α receptor I and II levels in the peripheral blood of patients were increased significantly; the soluble TNF-α receptor can neutralize and terminate TNF activity. Another study conducted by Yamamoto *et al.* (2006), showed that the Adacolumn treatment can substantially reduce white blood cells and their active metabolites, such as TNF-α, IL-1β, IL-6, IL-8, etc. At the same time, it can reduce the expression of L-selectin and the chemokine receptor CXCR3. The above results indicate that Adacolumn adsorption has a significantly suppressive effect on leukocyte extravasation.

Saniabadi AR *et al.* (2003) also confirmed the white blood cell counts before and after Adacolumn (as shown in Table 23.1). The column can clear 65% of neutrophils, 55% of monocytes, and 2% of lymphocytes from the blood. Furthermore, the column demonstrates

Table 23.1. Cell Counts before and after Adacolumn Treatment

Cell Type	Adacolumn Inlet	Adacolumn Outlet	*P*-Value
Red blood cell (10^6/μL)	3.6 ± 0.21	3.5 ± 0.24	ns
Lymphocytes (10^3/μL)	1.22 ± 0.14	1.20 ± 0.12	ns
Neutrophils (10^3/μL)	6.8 ± 1.3	2.4 ± 0.4	0.01
Monocytes (10^3/μL)	0.28 ± 0.08	0.12 ± 0.04	0.01
Blood platelet (10^3/μL)	293 ± 22	310 ± 18	ns

a significant cleaning effect toward cytokines produced by leukocytes, such as TNF-α, IL-1β, IL-6, and IL-8. It also down-regulates the expression of L-selectin and the chemokine receptor CXCR3.

In vitro experiments (Hanai *et al.*, 2004) demonstrated that after an hour-long incubation of human whole blood with Adacolumn, significant amounts of the IL-1 receptor antagonist (IL-1ra) and the hepatocyte growth factor (HGF) are detected in the plasma, whereas there is no detectable increase in TNF or IL-1. IL-1ra may antagonize inflammation, and HGF can promote the regeneration of the epidermis, both of which play key roles in the recovery of psoriasis.

The effect of the Adacolumn treatment on cytokines and adhesion molecules primarily relies on neutrophils. It is found that the anti-inflammatory cytokines IL-1ra and IL-10 are increased significantly in the serum at Adacolumn outlet. Furthermore, another study (Hanai *et al.*, 2006) indicated that after the Adacolumn treatment, IFN-γ positive T cells disappeared or were decreased in the living tissue, the IL-1β, IL-6, IL-8 and TNF-α concentrations at the local lesions were decreased significantly, and the ratio of IL-1ra and IL-1β was increased significantly.

Furthermore, Seishima M *et al.* reported the significant effect produced from the application of monocyte adsorption apheresis in the treatment of ulcerative colitis and psoriasis. The authors observed that the histology of pustular psoriasis exhibited neutrophils pustules. Japanese scholars frequently incorporated etretinate and immunosuppressants into treatments; however, some cases exhibited resistance. Seishima M *et al.* (2008) applied monocyte adsorption apheresis in two cases of generalized pustular psoriasis which were cyclosporine A-resistant. One patient was a 44-year-old female diagnosed with obesity and alcoholic cirrhosis, whereas the other was a 66-year-old male with hypertension and cerebral infarction. After two to five treatments, the two patients' lesions were significantly improved, and neither experienced adverse reactions before or after adsorption. Monocyte adsorption apheresis can also be used to effectively treat refractory pustular psoriasis. For example, Kanekura T *et al.* (2006) applied monocyte adsorption apheresis to treat 14 cases of neutrophilic dermatosis and 16 cases of psoriatic arthritis. 77.8% (14/18) of the patients had significantly improved arthritis symptoms, and 75.9% (22/29) of the patients had significantly improved skin lesions, as shown in Fig. 23.3. Compared to the control group, the expression of Mac-1 (CD11b/CD18) dropped significantly after the peripheral mononuclear cells were treated with monocyte adsorption apheresis.

Liumbruno GM *et al.* (2006) applied a weekly lymphocyte adsorption treatment to five cases of psoriasis patients and used the PASI score to evaluate the efficacy of the treatment before and one week after the absorption as shown in Table 23.2.

The results show that all five cases have significantly improved lesions with no side effects. After treatment, the PASI dropped an average of 71.8% with a median of 76%. The average overall blood volume of treatment was 1.03 liters. The average total of lymphocytes cleared are 4.02×10^9/L. Currently, with respect to the specific treatment mechanism, it is thought that the lymphocytes that are removed during each treatment account for only 2% of the total

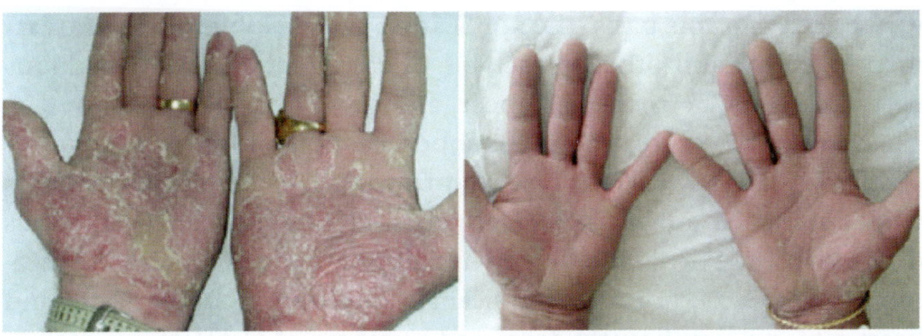

Fig. 23.3. The lesions before and after the adsorption treatment.

Table 23.2. Severity of Lesions (PASI) before and after Absorption Treatment

Patient	Case 1	Case 2	Case 3	Case 4	Case 5
Before	66	33	50	56	29
After	24	8	5	36	2.1

lymphocytes and 20% of the peripheral blood lymphocytes, with an absolute value of 5×10^9. This treatment is primarily based on the non-selective elimination of both activated and non-activated lymphocytes, and the treatment causes further functional changes in the psoriasis immune and inflammatory mechanisms. However, further studies are necessary for distinguishing long-term side effects. Nevertheless, the lymphocyte adsorption method has become an effective therapeutic option and reduces the treatment dose of other traditional methods.

In summary, Adacolumn is a simple extracorporeal treatment. The blood flow rate can be adjusted easily, and the complex arteriovenous shunt can be avoided. Thus, it guarantees an intrinsic safety of the treatment. This technique does not deplete T lymphocytes entirely in terms of immune regulation, which is a beneficial aspect as $CD4^+$ T helper cells and $CD8^+$ cytotoxic T cells are essential components of cellular immunity. Furthermore, over the course of the treatment, there is not a significant change in these cells; however, after the treatment, they follow an increasing trend. This change in the T lymphocytes is called an Adacolumn-induced immunomodulatory effect. On this basis, the application of Adacolumn to the treatment of AIDS and other intractable diseases (such as systemic lupus erythematosus, hepatitis C, cancer, and so on) has been explored.

As an effective treatment, hemoperfusion is associated with many possible risks. However, as reported from various literatures, patients who experience an invasive treatment did not exhibit any obvious discomforts, and liver and kidney functions remained normal; furthermore, there were not any significant changes in blood pressure and heart rate, no significantly decreased numbers of platelets, no gastrointestinal and wound bleeding, nor any other adverse reactions. Moreover, hemoperfusion shortens the treatment duration, relieves patients' suffering, and improves the clinical cure rate. Therefore, the rational use of the hemadsorption treatment in severe psoriasis may be a new treatment in the future.

Zhenwei Shi, Aiying Liu

References

Ameen M, Allen MH, Fisher SA, *et al.* (2005) Corneodesmosin (CDSN) gene association with psoriasis vulgaris in Caucasian but not in Japanese populations. *Clin Exp Dermatol* **30(4):** 414–418.

Ashcroff D, Li W, Williams H, *et al.* (2000) Combination regiments of topical calcipotriene in chronic plaque psoriasis: systematic review of efficacy and tolerability. *Arch Dermatol* **136:** 1536–1543.

Baehelez H, Flsgeul B, Daubertret L, *et al.* (1998) Treatment of recalcitrant plaque psoriasis with a humanized non-depleting antibody to CD4. *J Autoimmunm* **11(1):** 53–62.

Bonifati C, Trento E, Cordiali-Fei P, *et al.* (1997) Increased interleukin-7 concentrations in lesional skin and in the sera of patients with plaque type psoriasis. *Clin Immunol Immunopathol* **83:** 41–48.

Chao PH, Cheng YW, Chung MY. (2009) Generalized pustular psoriasis in a 6-week-old infant. *Pediatr Dermatol* **26(3):** 352–354.

Cook PW, Piepkorn M, Clegg CH, *et al.* (1997) Transgenic expression of the human amphiregulin gene induces a psoriasis like phenotype. *J Clin Invest* **100:** 2286–2291.

Devaux S, Castela A, Archier E, *et al.* (2012) Topical vitamin Danalogues alone or in association with topical steroids for psoriasis: A systematic review. *Journal of the European Academy of Dermatology and Venereology* **26(s3):** 52–60.

Dmitriev AA, Beletskaia LV, Bektimirov RA, *et al.* (1985) Effect of hamosorption on the dynamics of immunopathologic changes in the skin of patients with diseases accompanying the immune complex syndrome. *Ter Arkh* **57(6):** 87–90.

Ference JD, Last AR. (2009) Choosing topical corticosteroids. *American Family Physician* **79(2):** 135–140.

Fleming C, Ganslandt C, Guenther L, *et al.* (2010) Calcipotriol plus betamethaeone dipropionate gel compared with its active components in the same vehicle and the vehicle alone in the treatment of psoriasis vulgaris: A randomized, parallel group, double-blind, exploratory study. *Eur J Dermatol* **20(4):** 465–471.

Fleming CJ, Holmes ER, Mackie RM. (1996) Systemic complement activation in psoriasis vulgaris. *Clin Exp Dermatol* **21:** 415–421.

Foerster J, Nolte I, Junge J, *et al.* (2005) Haplotype sharing analysis identifies a retroviral dUTPase as candidate susceptibility gene for psoriasis. *J Invest Dermatol* **124(1):** 99–102.

Fujita H, Nograles KE, Kikuchi T, *et al.* (2009) Human langerhans cells induce distinct IL-22-producing CD4+ T cells lacking IL-17 production. *Proc Natl Acad Sci USA* **106(51):** 21795–21800.

Gonzalez S, Martinez-Borra J, Del Rio JS, *et al.* (2000) The OTF3 genepoly morphism confers susceptibility to psoriasis independent of the association of HLA-Cw*0602. *J Invest Dermatol* **115(5):** 824–828.

Gottlieg A, Krueger JG, Bright J, *et al.* (2000) Effects of administration of single dose of humanized monoclonal antibody to CD11a on the immunobiology and clinical activity of psoriasis. *J Am Acad Dermatol* **42(3):** 428–435.

Gu X, Lundqvist EN, Coates PJ, *et al.* (2006) Dysregulation of TAp63 mRNA and protein levels in psoriasis. *J Invest Dermatol* **126(1):** 137–141.

Gustafson CJ, Watkins C, Hix E, *et al.* (2013) Combination therapy in psoriasis. *American Journal of Clinical Dermatology* **14(1):** 9–25.

Hanai H, Iida T, Yamada M, *et al.* (2006) Effects of adacolumn selective leukocyte apheresis on plasma cytokines during active disease in patients with active ulcerative colitis. *World J Gastroenterol* 12: 3393–3399.

Hanai H, Watanabe F, Yamada M, *et al.* (2004) Correlation of serum soluble TNF-alpha receptors I and II levels with disease activity in patients with ulcerative colitis. *Am J Gastroenterol* **99:** 1532–1538.

Harrison CA, Bastan R, Peirce MJ, *et al.* (2007) Role of calcineurin in the regulation of human lungmast cell and basophil function by cyclosporine and FK506. *British Journal of Pharmacology* **150(4):** 509–518.

Hedrick MN, Lonsdorf AS, Shirakawa AK, *et al.* (2009) CCR6 is required for IL-23-induced psoriasis-like inflammation in mice. *J Clin Invest* **119(8):** 2317–2329.

Herron MD, Hinckley M, Hoffman MS, *et al.* (2005) Impact of obesity and smoking on psoriasis presentation and management. *Arch Dermatol* **141(12):** 1527–1534.

Howard M, O'Garra A. (1992) Biological properties of interleukin 10. *Immunol Today* **13:** 198–204.

Kanekura T, Hiraishi K, Kawahara K, *et al.* (2006) Granulocyte and monocyte adsorption apheresis for refractory skin disease caused by activated neutrophils and psoriatic arthritis: Evidence that GCAP removes mac-1-expressing neutrophils. *Ther Apher Dial* **10:** 247–256.

Kanekura T, Hiraishi K, Kawahara K, *et al.* (2006) Granulocyte and monocyte adsorption apheresis (GCAP) for refractory skin diseases caused by activated neutrophils and psoriatic arthritis: Evidence that GCAP removes Mac-1-expressing neutrophils. *Ther Apher Dial* **10(3):** 247–256.

Khushid K, Haroon T, Hussain I, *et al.* (2000) Psoralen-ultraviolet A therapy VS, psoralen-ultraviolet B therapy in the treatment of plaque-type psoriasis: Our experience with Fitzpatrick skin type IV. *Int J Dermatol* **39:** 865–867.

Kircik L, Bagel J, Korman N, *et al.* (2008) Utilization of narrow-band ultraviolet light B therapy and etanercept for the treatment of psoriasis (UNITE): Efficacy, safety, and patient-reported outcomes. *Journal of Drugs in Dermatology* 7**(3):** 245.

Kolyadenko VG, Analyan YA, Karpenko VV, *et al.* (1987) Sorption therapy in psoriatic patients. *Biomater Artif Cells Artif Organs* **15:** 281–285.

Lee JY. (2009) Severe 20-nail psoriasis successfully treated by low dose methotrexate. *Dermatol Online J* **15(11):** 8.

Lee KS, Kim SJ, Ryoo YW, *et al.* (1998) All-trans-retinoic acid down-regulates elastin promoter activity elevated by ultraviolet B irradiation in cultured skin fibroblast. *J Dermatol Sci* **17(3):** 182–189.

Leng X, Elias JA. (1997) Interleukin-11 inhibits macrophage interleukin-12 production. *J Immunol* **159:** 2161–2165.

Liumbruno GM, Centoni PE, Molfettini P, *et al.* (2006) Lymphocytapheresis in the treatment of psoriasis vulgaris. *J Clin Apher* **21(3):** 158–164.

Mabuchi T, Chang TW, Quinter S, *et al.* (2012) Chemokine receptors in the pathogenesis and therapy of psoriasis. *J Dermatol Sci* **65(1):** 4–11.

Mashkov OA, Mazer LI, Popov AV, *et al.* (1991) The initial clinical experience of using Ovosorb antiprotease hemosorbent in the combined treatment of dermatoses. *Ter Arkh* **63:** 138–141.

Mehta BH, Amladi ST. (2011) Evaluation of topical 0.1% Tazarotene cream in the treatment of palmoplantar psoriasis: An observer-blinded randomized controlled study. *Indian J Dermatol* **56(1):** 40–43.

Michaelsson G, Ahs S, Hammarstrom I, *et al.* (2003) Gluten-free diet in psoriasis patients with antibodies to gliadin results in decreased expression of tissue transglutaminase and fewer Ki67 + cells in the dermis. *Acta Derm Venereol* **83(6):** 425–429.

Moles JP, Tesniere A, Guilhou JJ. (2005) A new endogenous retroviral sequence is expressed in skin of patients with psoriasis. *Br J Dermatol* **153(1):** 83–89.

Mrowietz U, Zhu KJ, Christpophers E. (2000) Treatment of severe psoriasis with anti-CD25 monoclonal antibodies. *Arch Dermatol* **136(5):** 675–676.

Namazi MR. (2004) Why is psoriasis uncommon in Africans? The influence of dietary factors on the expression of psoriasis. *Int J Dermatol* **43(5):** 391–392.

Nasonor EL, Dnitriev AA, Petrova GN, *et al.* (1982) Effect of benosorption on the level of circulating immune complexes in diseases of the internal organs. *Ter Arkh* **54:** 103–107.

Nishibu A, Han GW, Iwatsuki K, *et al.* (1999) Over expression of monocyte derived cytokines in active psoriasis: A relation to coexistent arthropathy. *J Dermatol Sei* **21(1):** 63–70.

Ogilvie AL, Antoni C, Dechant C, *et al.* (2000) Treatment of psoriatic arthritis with antitumour necrosis with methotrexate. *Br J Dermatol* **144(3):** 587–589.

Owen CM, Harrsion PV. (2000) Successful treatment of severe psoriasis with basiliximab, an interleukin-2 rectpor monoclonal antibody. *Clin Exp Dermatol* **25(3):** 195–197.

Ozturk G, Erbas D, Gelir E, *et al.* (2001) Natural killer cell activity, serum immunoglobulins, complement proteins, and zinc levels in patients with psoriasis vulgaris. *Immunol Invest* **30:** 181–190.

Paquet P, Pierard GE. (1996) Interleukin-6 and the skin. *Int Arch Allergy Immunol* **109:** 308–313.

Pereira TM, Vieira AP, Femandcs JC, *et al.* (2006) Cyclosporin A treatment in severe childhood psoriasis. *J Eur Acad Dermatol Venereol* **20(6):** 651–656.

Prens E, Hegmans J, Chin R, *et al.* (1996) Increased expression of inter leukin-4 receptors on psoriatic epidermal cells. *Am J Pathol* **148:** 1493–1498.

Presta M, Andres G, Leali D, *et al.* (2009) Inflammatory cells and chemokines sustain FGF2-induced angiogenesis. *Eur Cytokine Netw* **20(2):** 39–50.

Rioda WT, Adorni G. (2006) Infliximab treatment in psoriatic arthritis: Our experience. *Acat Biomedica-ateneo Parmense* 77(**2**)**:** 95.

Robinet E, Starmm C, Nieolas JF, *et al.* (1997) CD4 monoclonal antibody administration in atopic dermatitis. *J Am Acad Dermatol* **36(4):** 582–588.

Rucevic I, Perl A, Barisic-Drusko V, *et al.* (2003) The role of the low energy diet in psoriasis vulgaris treatment. *Coll Antropol* 27(**Suppl 1**)**:** 41–48.

Sanchez AP. (2010) Immunopatogenesis of psoriasis. *An Bras Dermatol* **85(5):** 747–749.

Saniabadi AR, Hanai H, Takeuchi K, *et al.* (2003) Adacolumn, an adsorptive carrier based granulocyte and monocyte apheresis device for the treatment of inflammatory and refractory diseases associated with leukocytes. *Ther Apher Dial* 7(**1**)**:** 48–59.

Scarpa R, Manguso F, DArienzo A, *et al.* (2000) Microscopic inflammatory changes in colon of patients with both active psoriasis and psoriatic arthritis without bowel symptoms. *J Rheumatol* **27(5):** 1241–1246.

Schiffl H, Lang SM, Stratakis D, *et al.* (2001) Effects of ultrapure dialysis fluid on nutritional status and inflammatory parameters. *Nephrol Dial Transplant* **16(9):** 1863–1869.

Seishima M, Mizutani Y, Shibuya Y, *et al.* (2008) Efficacy of granulocyte and monocyte adsorption apheresis for pustular psoriasis. *Ther Apher Dial* **12(1):** 13–18.

Sjogren F, Ljunghusen O, Baas A, *et al.* (1999) Expression and function of beta 2 integrin CD11B/CD18 on leukocytes from patients with psoriasis. *Acta Derm Venereol* **79(2):** 105–110.

Snezkkova EA, Kolyadenko VG, Nikolaev VC, *et al.* (1992) DNA-coated carbon adsorbents experimental assessment and results of severe psoriasis treatment. *Biomater Artif Cells Immobilization Biotechnol* **20:** 1201–1221.

Suzuki Y, Yoshimura N, Saniabadi AR, Saito Y. (2004) Selective granulocyte and monocyte adsorptive apheresis as a first line treatment for steroid naive patients with active ulcerative colitis: A prospective uncontrolled study. *Dig Dis Sci* **49:** 565–571.

Sweeney CM, Tobin AM, Kirby B. (2011) Innate immunity in the pathogenesis of psoriasis. *Arch Dermatol Res* **303(10):** 691–705.

Thaci D, Ortonne JP, Chimenti S, *et al.* (2010) A phase IIIb, multicentre, randomized, double-blind, vehicle-controlled study of the efficacy and safety of adalimumab with and without calcipotriol/betamethasone topical treatment in patients with moderate to severe psoriasis: The BELIEVE study. *Br J Dermatol* **163(2):** 402–411.

Tonel G, Conrad C. (2009) Interplay between keratinocytes and immune cell-recent insights into psoriasis pathogenesis. *Int J Biochem Cell Biol* **41(5):** 963–968.

Tsukahara K, Takema Y, Fujimura T, *et al.* (1999) All-trans retinoic acid promotes the repair of tortuosity of elastic fibres in rat skin. *Br J Dermatol* **140(6):** 1048–1053.

Waisman A. (2011) T helper cell populations: As flexible As the skin? *Eur J Immunol* **41(9):** 2539–2543.

Waite JC, Skokos D. (2011) Th17 response and inflammatory auto-immune diseases. *Int J Inflam*, Epub Nov 5, 2011.

Wang RC, Leven B. (2011) Calcipotriol induces autophagy in HeLa cells and keratinocytes. *J Invest Dermatol* **131(4):** 990–993.

Wrone Smith T, Nickoloff BJ. (1996) Dermal injection of immunocytes induces psoriasis. *J Clin Invest* **98:** 1878–1886.

Yamamoto T, Saniabadi AR, Umegae S, Matsumoto K. (2006) Impact of selective leukocyte apheresis on mucosal inflammation and ulcerative colitis: Cytokine profiles and endoscopic findings. *Inflamm Bowel Dis* **12:** 719–726.

Yazici AC, Aslan G, Baz K, *et al.* (2006) A high prevalence of parvovirus B19 DNA in patients with psoriasis. *Arch Dermatol Res* **22(4):** 208–210.

Yun W J, Lee DW, Chang SE, *et al.* (2010) Role of CD4 CD25 FOXP3 regulatory T cells in psoriasis. *Ann Dermatol* **22(4):** 397–403.

Zhou Q, Mrowietz U, Rostami-Yazdi M. (2009) Oxidative stress in the pathogenesis of psoriasis. *Free Radic Biol Med* **47(7):** 891–905.

2. Pemphigus

2.1. *Introduction*

Pemphigus is an autoimmune bullous disease with main manifestations of skin and mucous membrane lesions. Hert (2000) reported that the immunological characteristic of pemphigus is the formation of serum autoantibodies that attack desmoglein, a component of desmosomes, which results in disrupted desmosome functions, the detachment of epidermal cells and the formation of blisters. Pemphigus patients produce IgG antibodies that bind specifically to the desmogleins (Dsg) on epidermal keratinocytes, which cause the release of plasminogen activator; when the protease system is activated, acantholysis occurs, resulting in the formation of mucocutaneous flaccid blisters and bullae. Dsg1, Dsg3 and their autoantibodies are significant to the pathogenesis of pemphigus. Lesions can form in the mouth (56%), skin (12%), or both (32%). Four main types of pemphigus exist, including pemphigus vulgaris, pemphigus vegetans, pemphigus foliaceus and pemphigus erythematosus, as well as several subtypes, such as pemphigus herpetiformis, IgA pemphigus, araneoplastic pemphigus and drug-induced pemphigus.

Pemphigus occurs more commonly in the elderly; it can affect the skin and mucous membranes and severely threaten the lives of patients. Before the advent of corticosteroids, approximately 75% of patients with pemphigus died within one year. After the application of hormone therapy, the mortality rate decreased to approximately 15% to 45%. Since the late 1960s, the combination of immunosuppressant and corticosteroids further decreased the death rates to approximately 5% to 15% (Scully and Challacombe, 2002). Pemphigus treatment remains troublesome due to the variable responses of patients and the progression of medical treatment. Although several treatment protocols are available for pemphigus, no consistent definitions regarding details such as pathogenesis, activity, response to treatment, or end of treatment exist. Therefore, the most effective treatment for pemphigus remains unclear (Stoopler *et al.*, 2003).

2.2. *Pathogenesis of Pemphigus*

2.2.1. *The pemphigus antigen*

The pemphigus antigen is desmoglein in desmosomes. The desmoglein gene is located on 18q12.1 and belongs to the super family of cadherins. The pemphigus vulgaris antigen, Dsg3 (molecular weight

130×10^3 daltons), is primarily expressed in the basal and spinous layers. The pemphigus foliaceus antigen is Dsg1 (molecular weight 160×10^3 daltons) and is primarily distributed from the granular layer to the majority of the epidermis on the basal layer. Dsg1 synthesis gradually increases from the basal layer of the epidermis and reaches the highest expression level in the granular layer (Cowley *et al.*, 1997; Hunt *et al.*, 1999). Dsg3 has five extracellular functional areas (epitope, EC). EC-1 is located at the N-terminus, and EC-5 is located at the C-terminus. EC1-2 and EC3-4 have high affinities to pemphigus vulgaris (pemphigus vulgaris, PV) antibodies, and, thus, are referred to as the "pathogenic epitopes"; the corresponding antibodies are called "pathogenic antibodies". Passive transfer tests have found that PV-related pathogenic epitopes are primarily EC1-2, and the anti-EC1-2 antibody is the most closely related to PV among the anti PV antibodies. Recombinant anti EC1-2 antibody from positive patient serum can cause blisters in neonatal rats and typical histological changes in PV. Therefore, anti EC1-2 antibody detection is significant in the diagnosis, treatment, and prognosis of pemphigus vulgaris (Amagai *et al.*, 1994).

All PV patient sera can react with Dsg3. The three-dimensional conformation of Dsg3 is critical in the activation of autoimmune bullous disease. Recombinant protein gene mutation studies have shown that the presence of Dsg3 extracellular signal peptide is required in the formation of its epitope and is dependent on calcium concentration; that is, low calcium concentration Dsg3 is either reduced in expression or not expressed at all (Denning *et al.*, 1998). Compared with to Dsg3 +/– or +/+, DSG3–/– knockout mice exhibited developmental delay, and skin pathology examinations of those mice revealed obvious pemphigus vulgaris histological changes, such as scabs, release, and forming blisters on the basal layer. In addition, electron microscopy inspections revealed desmosome release and visible hair loss, indicating that the Dsg3 gene is important in the adhesion of deep stratified squamous epithelium and that pemphigus antibodies can interfere directly with intercellular adhesive interaction (Koch *et al.*, 1997).

In recent years, non-Dsg antigenic targets have been revealed (Patel *et al.*, 2010; Kalantari-Dehaghi *et al.*, 2011). However, the pathogenic mechanisms of pemphigus, particularly the genetic factors that produce auto-antibodies, are still largely unknown.

2.2.2. *The pemphigus "clinical drift" phenomenon*

The main antigen of pemphigus vulgaris is Dsg3. The main characteristic is the presence of serum anti-Dsg3 pathogenic autoantibodies (mainly IgG4). However, in some cases, anti-Dsg1 autoantibodies can be detected simultaneously. Previous studies have shown that the pemphigus clinical phenotype is correlated with the Dsg autoantibody phenotypes. For example, patients with the anti-Dsg3 +/anti-Dsg1 - phenotype only exhibit mucosal lesions, but patients with the anti-Dsg3 +/anti-Dsg1 + phenotype may also exhibit visible skin lesions. Severe mucosal-skin lesions are only found in anti-Dsg3 + anti-Dsg1 + patients, suggesting that the presence of anti-Dsg1 antibodies may be associated with the severity of pemphigus. Amagai *et al.* (1999) used ELISA to analyze the Dsg1 and Dsg3 fusion proteins and found that the clinical phenotypes of pemphigus patients are determined by the serum anti-Dsg antibody phenotypes. Thus, the pemphigus "clinical shift" phenomenon was proposed, which states that PV patients with only mucosal lesions have serum antibody phenotype anti-Dsg1(−)Dsg3(+), while patients with both mucosal and skin lesions have serum antibody phenotype anti-Dsg1(+)Dsg3(+), and patients with emerging serum anti-Dsg1(+) have skin lesions throughout the body. Pemphigus foliaceus patients possess serum antibody phenotype anti-Dsg1(+)Dsg3(−). These results suggested that the detection of the serum anti-Dsg antibody phenotype could facilitate the clinical diagnosis, classification and prognosis of pemphigus (Miyagawa *et al.*, 1999).

2.2.3. *Protease and pemphigus*

Kitajima *et al.* (1999) reported that in the pathogenesis of pemphigus, the lysis of spinous cells is related to proteases (e.g. plasminogen activator, PA); the total serum protease activity increases significantly as the total anti-trypsin activity decreases significantly. Plasminogen activator is a serine protease that catalyzes the conversion of plasminogen to plasmin, thereby triggering a series of proteolytic processes that affect the adhesion between cells and causing spinous cell lysis.

Keratinocyte cells can synthesize two types of PA: tPA (tissue-type) and uPA (urokinase-type). tPA protein is mainly distributed in upper epidermal basal cells. Studies have shown that tPA is highly related to the differentiation of keratinocytes (KC). tPA protein and mRNA expression cannot be detected in undamaged skin tissue; however, tPA protein and mRNA level are elevated in damaged skin. This elevated expression is also found in regenerated skin tissue. uPA can hydrolyze the epitope protein on the desmosomal cell surface. The addition of pemphigus autoantibodies and recombinant Dsg3 protein can result in an increased expression of uPA and its receptor in keratinocytes. Pemphigus antibodies alone can cause Dsg3 serine phosphorylation and desmosomal dysfunction resulting from desmosomal separation. uPA expression increases in pemphigus vulgaris. As indicated through skin organ culture models, the high expression level of uPA could be induced by the pemphigus antibody, which further promotes acantholysis. Adding anti-uPA monoclonal antibodies could effectively inhibit pemphigus antibody-induced acantholysis. The finding was reported by Ishikawa *et al.* (2000) and Xue *et al.* (1998). In addition, Mahoney *et al.* (1999) and Asano *et al.* (2001) found that when pemphigus antibodies were passively transferred to uPA knockout and healthy neonatal mice, both the knockout group and the healthy control group had visible bullae and microscopic acantholysis, and the two groups exhibited no significant differences. The same experiment using tPA knockout mice demonstrated similar results, suggesting that PA is not required in the formation of antibody-induced bullae. PA could function through the increased expression level resulting from pemphigus antibody and antigen binding, which promotes proteolysis surrounding keratinocyte cells and increases acantholysis.

2.2.4. *Pemphigus and cell-mediated immunity*

The pathogenicity of pemphigus autoantibodies have been directly or indirectly confirmed through experiments. However, studies have shown that the normal skin of pemphigus patients also exhibits IgG deposition in the intercellular epidermal area. Skin lesions have more mononuclear cell infiltration and increased local cytokine levels. Therefore, cell-mediated immunity has also been speculated to be involved in the pathogenesis of pemphigus. Immunohistochemical studies have found that the infiltration of T cells and monocyte-macrophages and the number of activated T cells in the epidermis and dermis of pemphigus lesions is significantly higher than that of the surrounding skin. The epidermis primarily exhibits monocyte-macrophage infiltration, while the dermis primarily exhibits T-cell infiltration. The infiltrating cells of pemphigus vulgaris mostly culminate at the top of the basal layer, while the infiltrating cells of pemphigus foliceus are scattered throughout the epidermis, as is consistent with the distribution of its antigen (Zillikens *et al.*, 1993).

In vitro lymphocyte proliferation experiments have shown that both recombinant fusion (Dsg1 and Dsg3) proteins can stimulate the proliferation of T lymphocytes in pemphigus patients. Further studies have shown that the ratio of infiltrating T-cells (CD4/CD8) in PV patients with active lesions is 2:1. The $CD4^+$ T lymphocyte in peripheral blood has a proliferative response to Dsg3 and is restricted by human major histocompatibility antigen-II (HLA-II); thus, the T-cells of pemphigus vulgaris patients with HLA-DRB1*0402 and highly homologous HLA-DRB1*11 alleles have proliferative responses to Dsg3. HLA-DRB1 * 0402 and HLA are some of the most relevant loci in pemphigus vulgaris (Lin *et al.*, 1997; Rico *et al.*, 1999). Nishifuji *et al.* (2000) and Lombardi *et al.* (1999) used an enzyme-linked immune spot assay to detect Dsg3-specific self-sensitive B cells in the peripheral blood of pemphigus patients; self-sensitive B cells were found in the peripheral blood of severely afflicted patients, but were not detected in mildly effected or remitted patients. When testing peripheral blood mononuclear cells (PBMC) *in vitro* with pokeweed mitogen and recombinant protein Dsg3 stimulation, B cells that produced anti-Dsg3 autoantibodies were detected in 64% of patients. B cells require the assistance of T cells (mainly $CD4^+$ T cell) in order to produce autoantibodies; when $CD4^+$ T cells are depleted or anti-HLA-DR or anti-HLA-DQ are added, anti-Dsg3 antibodies disappear, proving that HLA-II restrictive $CD4^+$ T cells are important to the production of pemphigus vulgaris autoantibodies.

Monocyte-macrophages and T cells are important sources of cytokines, and monocytemacrophages, T cells, and cytokine secretion are major components of cellular immunity. Previous studies have used immunohistochemistry and *in situ* hybridization to detect Th1 cytokines IL-2 and IFN-γ, Th2 cytokines IL-4, IL-5, IL-13 protein, and mRNA. The results indicated that, in the biopsied skin of pemphigus vulgaris patients, the mixed expression of Th1/Th2 cytokines was the main characteristic, suggesting that cytokines are components of the pathogenesis of pemphigus and participate and mediate inflammatory cell infiltration, complement activation and keratinocyte intercellular adhesion loss (Lopez-Robles *et al.*, 2001). Pemphigus serum IL-6 and TNF-α levels were significantly higher than the control group, and were significantly correlated with the number and area of skin lesions; the IL-6, and TNF-α levels in local blister fluid were significantly higher than those of the serum levels. After hormone therapy, the IL-6, TNF-α and monocyte-macrophage levels significantly decreased, suggesting that IL-6 and TNF-α levels are related to the severity of pemphigus and are involved in the biological conversion of pemphigus clinical phenotype (D'Auria *et al.*, 1997). TNF-α is mainly synthesized and secreted by monocytes and activated T lymphocytes. TNF-α can bind to one of its receptors, TNFRp55 or TNFRp75, and mediate inflammation, such as an autoimmune response. TNF-α is not expressed in normal skin, but is commonly found in the keratinocytes of pemphigus patients, especially in the keratinocytes surrounding the blisters of these patients. Therefore, Lombardi *et al.* (1999) inferred that TNF-α could be a component of the apoptosis of pemphigus keratinocytes and the formation of blisters. Barber *et al.* (1993) reported that TNF-α, IFN-γ, and other cytokines can induce high inducible nitric oxide enzymes (iNOS) expression in macrophages, which produce nitric oxide (iNO), promoting the production of TNF-α in monocytes and thereby exerting a direct cytotoxic effect and aggravating inflammation. Dexamethasone can be used to inhibit TNF-α and IFN-γ induced effects and alleviate inflammation.

IFN-γ is one of the biochemical signals that are used in the communication between immune cells. As an important modulator of biological responses, IFN-γ has a wide range of biological effects. Studies in recent years have found that IFN-γ can induce keratinocyte (KC) expression of major histocompatibility complex (MHC-II) class-II molecules and intercellular adhesion molecules-I (ICAM-1). When KC is stimulated by IFN-γ *in vitro*, its surface clearly expresses the HLA-DR antigen. The IFN-γ induced expression of MHC-II molecules is time-, dose- and self-dependent. IL-2 is primarily produced by activated CD4+ T cells, and can promote T-cell growth, the differentiation and proliferation of B cells, and the inducing of a variety of cytokines and the expression of their receptors in an autocrine manner. Studies have shown that in pemphigus patients, the concentration of blister fluid IL-2 and sIL-2 receptors (T-cell activation markers) is higher than that of serum, suggesting that the presence of locally produced IL-2 and the serum level is proportional to the area of the lesions and the severity of the disease (Smith *et al.*, 1999).

Toto *et al.* (2000) found that in a passive transfer experience targeting mutated mice, $CD28^-$ mice were five times as sensitive to pemphigus as wild-type mice ($CD28^+$). The CD28 molecule is one of the co-stimulatory signals for T-cell activation and plays a central role in the induction of Th2-type cytokine expression. $CD28^{-/-}$ mice skin lesions exhibited low levels of IL-10 mRNA. $IL\text{-}10^{-/-}$ mice experienced more blister formation than the control group. When rIL-10 and the PV antibody were co-administered, the disease activity was significantly inhibited, suggesting that IL-10 is involved in the pathogenesis of pemphigus and could inhibit acantholysis.

2.3. *The Diagnosis of Pemphigus*

Pemphigus is an autoimmune bullous skin and mucosal disease, in which Dsgl and Dsg3 autoantibodies bind to their antigens, thereby activating protease and causing the loss of cell adhesion and acantholysis. Thin-walled blisters and Nikolsky's sign are important bases for clinical diagnosis. However, not all pemphigus patients exhibit positive Nikolsky's signs. Uzun *et al.* (2006) studied the sensitivity and specificity of Nikolsky's sign in the diagnosis of pemphigus. In this study, 123 cases of bullous or erosive lesions were examined. Of the 24 cases positive for Nikolsky's signs, 18 cases were diagnosed as pemphigus, four cases were diagnosed as bullous pemphigoid, one case was diagnosed as linear IgA dermatosis, and one case was diagnosed as staphylococcal scalded skin syndrome. The results demonstrated that, as an indicator of pemphigus, Nikolsky's sign was moderately sensitive, but more specific.

The direct immunofluorescence test has been the gold standard for pemphigus diagnosis, but repetitive histopathological sampling causes stress and pain in patients, making it less acceptable. It is reported by Enno Schmidt *et al.* (2010b) that direct immunofluorescence tests have shown that IgG and/or C3 are deposited between epidermal cells. Schaerer and Trueb (2003) found that pemphigus acantholytic bullae spread from the epidermis to the outer root sheath of hair follicles. This study included examined 50 hairs from each of 15 patients; the experimental methods were the same as those of direct immunofluorescence pathology. Enzyme-linked immunosorbent assay (ELISA) was used to detect serum Dsg1 and Dsg3 antibody levels; in the outer root sheath of early hair, strong positive net-shaped pemphigus antibody fluorescence was apparent. Four of the patients exhibited circulating pemphigus antibodies and obvious blisters, and 11 of the patients were in the stable phase of the disease. The authors concluded that the outer root sheath direct immunofluorescence test reliably diagnoses pemphigus, even when used in circulating antibody-negative patients in the stable phase of the disease.

Sharma *et al.* (2006) used ELISA to detect Dsg1 and Dsg3 antibodies in 27 pemphigus patients, resulting in positive rates of 96.3% and 85.2%; the positive rates of the control group were 7.7% and 0%. These results demonstrated the relatively high sensitivity and reliability of this method. In addition, the test results were correlated with clinical manifestations; patients with mostly mucosal lesions were anti-Dsg3 positive, while patients with both mucosal and skin lesions were anti-Dsg3 and anti-Dsg1 positive. Zagorodniuk *et al.* (2005) investigated 33 cases of pemphigus vulgaris, five cases of pemphigus foliaceus, and 50 cases of healthy patients for the control group. The results indicated that, among the pemphigus vulgaris cases, 26 had positive IIF and ELISA results. All five pemphigus foliaceus cases had positive ELISA results, and three had positive IIF results. Of the 50 control group cases, six had positive ELISA results, and all had negative IIF results, suggesting that, although they are not entirely congruent, both IIF and ELISA can reliably diagnose pemphigus. Since ELISA is slightly more sensitive but less specific than IIF, these two methods could be used complementarily.

Sharma *et al.* (2009) reported that some studies have shown that the anti-Dsg antibody preferentially binds to mature Dsg rather than immature fibrinogen. Therefore, ELISA results can reflect the status of a patient's disease activity. ELISA is easy to operate and accurate, so it is commonly used to detect circulating antibodies and reliably monitor patient conditions. ELISA kits have already been used clinically to detect anti-Dsg1 and Dsg3 antibodies.

NO is an inflammatory mediator, and is related to cancer, psoriasis, urticaria, atopic dermatitis, and other diseases. In pemphigus, inflammatory cells produce NO. Siebra *et al.* (2006) used immunohistochemistry tests to detect NO synthase (iNOS) in skin biopsies. iNOS was detected in all the pemphigus lesions, but none was detected in the healthy skin of the control group. In addition, the serum nitrite levels in the pemphigus patients were significantly different than in the control group ($P < 0.0001$), suggesting that pemphigus patients exhibit elevated serum NO levels, which are associated with increased iNOS content. Therefore, NO and iNOS detection could be used as supplementary examinations for pemphigus diagnosis.

2.4. *General Treatment of Pemphigus*

24.1. *Corticosteroids*

Corticosteroids (corticosteroids, CS) are primarily chosen to treat pemphigus (Femandes *et al.*, 2001). Previous studies have suggested that a combination of oral corticosteroids and alternative medicine is the most effective method of treatment (Chams-Davatchi *et al.*, 2007;

Beissert *et al.*, 2010). The principles of medication are early application, adequate control, reasonable reduction, and maintaining a small amount. When using CS for the systematic immunosuppressive treatment of pemphigus, most patients respond well to small to medium dosage amounts. Studies have shown that dexamethasone pulse therapy does not benefit patients who are using prednisolone or azathioprine for treatment (Mentink *et al.*, 2006). Moderate doses of corticosteroids alone can effectively control pemphigus oral erosion (Chrysomallis *et al.*, 1994).

2.4.2. *Other immunosuppressive drugs*

Several CS adjuvant or alternative agents are available, such as azathioprine (azathioprinum, AZP), mycophenolate mofetil (mycophenolatemofetil, MMF), cyclophosphamide (cyclophosphamide, CTX), chlorambucil (chlorambucil, CB1348), thalidomide (thalidomide), and tripterygium (tripterigium, T-II). Immunosuppressant alone can be applied to patients in a stable condition without severe symptoms. When combined with CS, a more effective treatment and reduced treatment time can be obtained, which can decrease the control and maintenance dose and reduce the side effects of CS.

Azathioprine (azathioprinum, AZP) is the most effective corticosteroids adjuvant, as reported by Chams-Davatchi *et al.* (2007). Azathioprine has a strong inhibitory effect on T lymphocytes and is an antagonist of purine. By inhibiting DNA synthesis, azathioprine can inhibit lymphocyte proliferation, prevent the conversion of lymphocytes into immunoblasts, and thus result in immunosuppression. Azathioprine doses usually range from approximately 50 to 150 mg/d, or 1.0 to 2.0 mg.kg-1.d-1.

Mycophenolate mofetil (mycophenolatemofetil, MMF) is a relatively safe and effective drug (Beissert *et al.*, 2006, 2010; Chams-Davatchi *et al.*, 2007). It primarily functions by interfering with purine metabolism and is highly selective to lymphocytes. It has a strong antiproliferative effect and can significantly reduce the guanylate content in lymphocytes, thereby affecting DNA synthesis and reducing the number of immunoactive lymphocytes. Mycophenolatemofetil can also inhibit the formation of antibodies through the inhibition of the proliferation of B cells. In addition, mycophenolatemofetil is non-toxic to the liver and kidneys, does not suppress bone marrow, and does not cause diabetes, hypertension, or osteoporosis. Other similar drugs do not possess these advantages. The usual dosage of mycophenolatemofetil is two 0.2 mg.kg-1.d-1 doses (Meurer, 2012).

Dapsone, a hormone replacement agent, can be used for pemphigus maintenance treatment (Singh, 2011).

Thalidomide was widely used to treat insomnia, pregnancy reactions and other diseases, but was later banned since it caused fetal malformations. In the past 20 years, it has been used to effectively treat autoimmune diseases, and obtained good results. The usual dosage of thalidomide is 50~100 mg/d orally, separated to 2~3 times, and the medicine should be taken for one week and stopped for one week. Women preparing for childbirth should not use thalidomide. Tripterygium glycosides have strong anti-inflammatory and immunosuppressive effects. They can inhibit the proliferation of healthy T and B cells stimulated by antigens or mitogens. The usual tripterygium glycosides dosage is three doses of approximately 1 to 1.5 mg.kg-1.d-1 or 30 to 60 mg/d. Tripterygium has significant effects on the reproductive system, including menstrual disorders, and significantly reduced sperm counts in males. This drug should be used with caution in children, unmarried women, and young men and women who want to reproduce.

Cyclophosphamide has a similar role to nitrogen mustard in that it is a cell-cycle non-specific drug. It can cause DNA cross-links and inhibit DNA synthesis, as well as interfere with RNA functions and inhibit cellular and humoral immunity. However, studies have shown that the efficacy of dexamethasone and cyclophosphamide pulse therapy is the same as methylprednisolone combined with azathioprine (Rose *et al.*, 2005). Cyclosporine is also used as an immunosuppressant in the treatment of pemphigus; the dosage is 3 to 5 mg/kg per day. Studies have shown that, although cyclosporine is not as effective as azathioprine and mycophenolatemofetil, it is the safest drug for treating pregnant pemphigus patients (Hashimoto, 2008; Schiavo *et al.*, 2010). However, its application in older adults was limited due to its potential side effects, such as high blood pressure and kidney failure.

2.4.3. *Intravenous immune globulin method*

Yumi Aoyama *et al.* (2010) reviewed the initial development of immune globulin treatment for pemphigus and its mechanisms. Specific antibodies in immunoglobulins can efficiently neutralize pathogenic endogenous antibodies, and specifically bind to B cell receptors, thereby reducing receptor functions and interfering with the production of immune complexes. In addition, immune globulin regulates complement-mediated cell cytotoxicity. This method requires the use of a combination of CS and immunosuppressive drugs, and it is safe and effective for steroid-tolerant patients and patients with apparent side effects. It is also suitable for patients unable to use immunosuppressants, e.g. patients young in age or with a family history of cancer.

In 2009, Japanese researchers Amagai *et al.* conducted a clinical experiment summary. This summary was the first multi-center, randomized, double-blind, placebo-controlled experiment. The experiments confirmed that a dose of intravenous immunoglobulin 400 mg/kg once every five days is an effective method of adjuvant hormonal and/or immunosuppressive therapy.

Studies prior to this were not double-blind; the number of patients was relatively small, and the dose of immunoglobulin was varied from three days of 2 g/kg/D doses to five days of 400 mg/kg/D doses (Baum *et al.*, 2006; Ahmed *et al.*, 2006). Furthermore, changes in the disease were not quantified.

2.4.4. *Photochemical therapy and hematopoietic stem cell transplantation*

Photodynamic therapy is a new technology that uses photochemical reactions to diagnose and treat diseases. The host immune response can be activated by changing the immune activity of light-sensitive cells. Thus, photodynamic therapy can be used as an adjuvant therapy, especially for the treatment of patients with drug resistance.

Chinese scholars were the first to use autologous peripheral blood stem cell transplantation to treat a severe pemphigus patient who had received long-term traditional hormonal and immunosuppressive therapy without success. The new therapy was successful and became a new method of pemphigus treatment.

2.5. *Circulative Clearing Therapy for Pemphigus*

2.5.1. *Plasmapheresis*

In plasmapheresis, patient's blood is drawn, and plasma and cellular components are separated with blood separation technology. The plasma is discarded, and the cellular components are returned to the patient with the same amount of healthy human plasma or alternative plasma. This method allows for the removal of pathological substances, such as autoantibodies, immune complexes, high viscosity proteins, cytokines, and protein binding poisons.

Pemphigus pathogenesis functions through the presence of serum autoantibodies, which can bind to epidermal cell desmosomes and activate the fibrin protease system, thereby causing epidermal acantholysis and epidermal blisters. An antibody titer is usually positively correlated with the severity of the disease. Plasma exchange and immune adsorption, one of the fastest methods of relieving clinical symptoms, can remove pemphigus antibodies and various cytokines from pemphigus patients. Pemphigus patients possess a specific serum anti-epidermal basement membrane antibody. Plasma exchange therapy for pemphigus patients in 1978, but few studies (conducted by Tan-Lim *et al.* (1990), Sondergaard *et al.* (1995), Turner *et al.* (2000)) have confirmed the therapeutic effects of plasmapheresis (Tan-Lim *et al.*, 1990; Sondergaard *et al.*, 1995; Turner *et al.*, 2000).

However, several research centers have found that, of 34 pemphigus cases, those treated with hormone therapy (0.5 mg/kg/D) combined with 10 plasma exchanges in four weeks exhibited no significant differences to patients treated with hormone therapy alone in terms of clinical effect and antibody levels (Guillaume *et al.*, 1988).

Sagi *et al.* (2011) used hormone therapy combined with monthly plasma exchange therapy to treat seven patients. After three cycles (three times a week) of plasma exchange, 86% of patients with the disease were alleviated with no side effects. This result was congruent with those in the studies conducted by Saraceno *et al.* (2008) and Ranugha *et al.* (2012).

Matthew *et al.* (2000) have found that although plasma exchange reduces circulating antibody levels, the negative feedback mechanism can simultaneously activate B cells to secrete more pathogenic autoantibodies, producing a rebound phenomenon. Therefore, blood purification should be combined with corticosteroids and immunosuppressive therapy. In Enler's study, it is reported that plasma exchange therapy is suitable for patients that are not sensitive to corticosteroids and immunosuppressants but should be conducted continuously and multiple times in order to eradicate autoantibodies.

Shang Yingbin *et al.* (2004) used glucocorticoids and/or immunosuppressive therapy to treat eight pemphigus patients on plasma exchange therapy. All of the patients quickly experienced complete or partial remission. In this study, a total of eight patients (six males and two females) aged 39 to 74 years with disease durations of three months to 10 years were confirmed as having pemphigus with histopathological and immunohistochemical pathology tests. The patients were all treated with prednisone (50~100 mg/d) and/or immunosuppressants supplemented with one to three plasma exchange therapies performed every other week. In this method, a constant pump was used to withdraw blood and separate plasma and cells with a membrane plasma separator. The plasma was discarded, and the cellular components, combined with an equal amount of healthy human plasma or albumin solution, were re-injected into the patients. Two weeks after the plasma exchange, eight of the patients exhibited skin rash relief and no new blisters; one of the two patients with mucosal damage exhibited symptom alleviation; two of the patients, who had previously demonstrated positive Nikolsky's signs, were negative for Nikolsky's sign within 10 days; and five of the patients experienced improved systemic symptoms. Of four patients tested for the pemphigus antibody one to two weeks after plasmapheresis, three cases had reduced antibody titers (by 40, 60 and 70%), and the remaining patient exhibited no change. After the first administration, the prednisone dosage was reduced by 10 mg/d for three of the cases and reduced by 20 mg/d for the remaining case. One typical patient, a 53-year-old male, was admitted for "recurrent oral ulcers for ten months, whole body blisters for two months, aggravating within two weeks." The patient was diagnosed with pemphigus vulgaris through histopathological and immunohistochemical dermatology examinations eight months ago. After treatment with a 60 mg/d dose of prednisone, the patient had improved his condition, and the prednisone dosage was decreased to 25 mg/d. However, three months ago, new blisters formed and aggravated two weeks ago, when blisters formed across whole body along with erosion and oozing. The patient was admitted to the hospital on January 20th, 1997. The patient had developed diabetes two years prior. Physical examinations indicated no abnormalities. The dermatological examination revealed blisters of various sizes on the top of the head, forehead, torso, limbs, and thigh, as well as a positive Nikolsky's sign with erosion, oozing, and crusting on the back and inner thigh. In addition, blisters the size of green beans were found in the sublingual area. The laboratory test results indicated blood levels of WBC and Hb of 16.0 × 109/L and 178 g/L, respectively, and urine levels of Glu and fasting blood glucose of 4+ and 3.87 mmol/L, respectively. The biochemical and immunological test results indicated a two-hour postprandial blood glucose of 7.56 mmol/L; normal liver function, renal function, blood potassium, and calcium level; and positive pemphigus antibody titers of 1:128. The patient was diagnosed with pemphigus vulgaris and was treated with 65 mg/d of prednisone, two 0.6 g doses of intravenous cyclophosphamide every other week, and support for symptoms. However, his condition did not improve. On February 27th, the prednisone dosage was increased to 80 mg/d, but his condition still did not improve and was coupled with elevated blood sugar, a fasting plasma glucose of 8.8 mmol/L, and a slight decrease in blood potassium levels. On March 17th, 150 mg of cyclosporine A was administered twice orally each day. The patient still exhibited erosion, oozing, and redness on the original skin lesions, as well as pain at the lesions and a positive Nikolsky's sign. Therefore, on March 21st, plasma exchange therapy was implemented. Ten days after the treatment, the patient's entire body rash was significantly reduced, the scabs sloughed off, and the Nikolsky's sign was negative. Furthermore, the pemphigus antibody titers were negative. The prednisone dosage was reduced to 70 mg/d. On April 4th, the lesions were almost eradicated, and the patient was discharged from the hospital.

Xiao Ruiqing *et al.* (1993) studied an 18-year-old female patient who reported the appearance of back skin erosion with exudates and pain one morning two months prior to the admission. After a few days, the erosion surface gradually expanded, and thumb blisters were apparent where the erosion surface was visible after ulceration. The patient was first treated at a local hospital, but there was no obvious improvement; in fact, new blisters formed. Afterwards, she was admitted to dermatology department in the hospital. Approximately 90% of the skin on the body had manifestations; the distal limbs were less severely afflicted. The skin on the entire body exhibited erythema erosion, and most of the erosion surface had a thick layer of brown scabs. The limbs had visible thumb or goose egg size blisters, and most of the scab wall was broken. The patient was also Nikolsky's sign positive, direct immunofluorescence positive, epidermal cell indirect fluorescent positive, and was diagnosed with pemphigus vulgaris. She was treated with 60 mg/d of prednisone, aminomethylbenzoic acid, and potassium chloride. On May 10th, the prednisone was increased to 80 mg/d, but no significant improvement was achieved; new blisters still formed, with broad bean to egg sizes. The pemphigus antibody titer was 1:1280. On May 5th, 19th, and 23rd, plasma exchanges were conducted. Considering the patient's lighter weight (40 kg), each therapy was separated into three phases. In each phase, 300 mL of blood was drawn and centrifuged in order to remove plasma; then, the cells were returned to the body in order to perform the next phase. By the end of the third phase, red blood cells and healthy plasma were subjected to UV irradiation and oxygenation and then returned to the patient. After three plasma exchange therapies, the titer of the pemphigus antibody decreased to 1:80; the scabs on the face, body, and limbs sloughed off, and no new blisters were formed. The patient was discharged from the hospital on June 17th.

Guo Yingjun *et al.* (2007) conducted a retrospective analysis of 10 cases of pemphigus, seven of which were bullous pemphigoid patients who received plasma exchange therapy. The clinical, histopathological, and direct immunofluorescence tests were consistent with the diagnosis of pemphigus, bullous pemphigoid. These patients received glucocorticoids and/or immunosuppressive therapy, as well as one to three plasma exchanges using the double filtration plasmapheresis (DFFP) method. This method essentially uses two filters with different diameters to remove pathogenic substances, such as antibodies and immune complexes; the remaining blood is returned to the patient. In each therapy, 2 L of plasma was exchanged, and one or two therapies were administered each week. The results indicated that all 10 pemphigus patients were given 40 to 100 mg/d of prednisone. After the plasma exchanges, all of the patients exhibited alleviation in skin rash, two of the four patients with mucosal lesions saw improvement, four patients had negative Nikolsky's indexes within several days, five cases experienced improvement on the entire body, and three of the four patients tested with antibody titers had reduced titers. In three of the cases, the glucocorticoid dosage was reduced, and the other was reduced by 20 mg/d compared to the first dosage after the third plasma exchange therapy. Of the seven patients with bullous pemphigoid, four were given only prednisone (60–105 mg/d), one was given cyclosporine A (200 mg) twice per day orally, and two were given glucocorticoids with cyclophosphamide (0.6 mg) every other day. After the plasma exchange, all skin rashes were alleviated, one patient with mucosal lesions exhibited significant improvement, and the four patients with manifestations on the entire body exhibited alleviated symptoms. In addition, the four patients that used glucocorticoids reduced their dosages to approximately 45 to 80 mg/d, and, of the other two patients taking glucocorticoids combined with an immunosuppressant, one stopped taking the immunosuppressant.

Lu Haigang *et al.* (2000) examined the treatment of two bullous pemphigoid cases using plasmapheresis. One patient, a 58-year-old male was admitted to a hospital after 45 days of tension blister development on the entire body and redness with itching. A systematic examination showed no abnormalities. The head, face, torso, and limbs all had skin erythema; soy bean to broad bean size blisters were present on the erythema, some with ulcerations; and the Nikolsky's test was negative. Laboratory tests: WBC 14.7 × 109/L, N 0.83, L0.15, E 0.02, immune complexes (+), IgG 18.12g/L, IgA 1.21 g/L, IgM 18.12 g/L, liver function and chest radiograph were normal. The tissue biopsy was consistent with bullous pemphigoid. Immune pathological examinations showed C3 linear deposition at the basement membrane, indicating bullous pemphigoid. The patient was given 80 mg/d of prednisone for three days, but it had no effect. The prednisone was then increased to 105 mg/d, but it still had no effect after 15 days. Then, the patient received plasma exchange therapy. One week after plasmapheresis treatment, the original blisters dried and formed scabs; the prednisone dosage was gradually reduced. At a two-month follow-up, the patients had no recurrence of blisters. Another patient, an 80-year-old male, experienced intermittent tension blisters on the entire body for 20 years; the patient had a relapse for 20 days before hospitalization. A systematic examination revealed no abnormalities. The limbs, torso, and palmoplantar all had erythema, and nail-sized blisters were present, most of which exhibited ulceration. The Nikolsky's test was negative. The immunoglobulin levels, liver and kidney function, and chest radiograph were all normal. The pathological diagnosis was bullous pemphigoid. Immune pathological examinations revealed IgG and C3 linear deposition at the basement membrane, indicating bullous pemphigoid. The patient was treated with 60 mg/d of prednisone; on days six and 12, plasma exchange therapy was conducted. Three days after the therapy, the blisters dried up and the prednisone dosage was gradually reduced. Two months of follow-up examinations determined no recurrence.

Yang Shuang *et al.* (2009) observed 34 cases of severe pemphigus or bullous pemphigoid patients for 22.8 ± 2.0 days. Based on the method of treatment, patients were separated into the plasma exchange group (PE group) or the conventional treatment group. The PE group included 15 cases, 10 of which were male, and five of which were female, with an age range of 23 to 80. Seven of these cases were pemphigus vulgaris, four of the cases were pemphigus erythematosus, and four of the cases were bullous pemphigoid. The conventional treatment group included 19 cases, 11 of which were male and eight of which were female. The ages of the patients in this group ranged from approximately 34 to 77, with an average age of 60.6. Eleven of the cases were pemphigus vulgaris, three of the cases were pemphigus erythematosus, one case was pemphigus vegetans, and four of the cases were bullous pemphigoid. The plasma exchange therapy was conducted with Plasauta-iQ plasma switches (Asahi Kasei Corporation, Japan), double filtration plasmapheresis (DFPP), an op-08 series plasma separator, and an AC/EC series plasma component separator. The replacement fluid was 5.7% albumin solution, the blood flow rate was approximately 80 to 100 mL/min, and the volume of each replacement was approximately 2–3 L. Each plasmapheresis lasted approximately two to three hours and was performed one or two times per week. The first plasma exchange for the 15 patients in the PE group was conducted after glucocorticoids were taken for 12.93 ± 2.18 days; while glucocorticoids and an immunosuppressant were taken simultaneously, a total of 27 plasma exchanges were performed for an average of 1.8 plasma exchanges per person while glucocorticoids and immunosuppressant were used simultaneously. The conventional treatment group used only glucocorticoids and an immunosuppressant (cyclophosphamide). Statistically, before treatment, the two groups of patients were not significantly different in age, sex, weight, disease duration, disease severity score, antibody titer, type, or use of immunosuppressive agents.

A comparison of the hormone dosages and times are shown in Table 23.3. The initial glucocorticoid dosage in the PE group and the cumulative dosage up to the reduction in dosage prior to the maximum control dosage was significantly lower than that of the conventional treatment group ($P < 0.05$); however, the difference in dosage reduction time and the duration of no new blister formation between the two groups was not statistically significant ($P > 0.05$).

Table 23.3. Comparison of Patient Hormone Dosage and Treatment Time

Group	Initial Glucocorticoid Dosage (mg/kg.d)	Maximum Control Dosage (mg/kg.d)	Cumulative Dosage (mg/kg)	Dosage Reduction Time (d)	Duration of No New Blister Formation (d)
Plasma Exchange group	0.81 ± 0.09*	0.96 ± 0.08*	18.67 ± 1.99*	22.9 ± 2.1	14.4 ± 2.4
conventional treatment group	1.15 ± 0.08	1.22 ± 0.09	28.29 ± 2.24	23.3 ± 1.9	14.4 ± 2.1

*P < 0.05.

The patients in both groups exhibited changes in autoantibody titers after the treatment. In the PE group, 11 pemphigus vulgaris patients had significantly decreased antibody titers ($P < 0.05$), from 1: (105.6 ± 33.3) to 1: (27.8 ± 13.8) after 12.57 ± 1.29 days. Four of the bullous pemphigoid patients' anti-basement membrane antibody titers significantly decreased ($P < 0.05$) from 1: (108.0 ± 55.2) to 1: (16.5 ± 8.9) prior to plasmapheresis. After plasmapheresis, the antibody titers decreased by 76.6 ± 7.8%. The conventional treatment group had antibody titers of 1: (68.6 ± 21.0) prior to the treatment. After the conventional treatment, the average antibody titer was 1: (30.1 ± 8.5), which decreased by 39.0 ± 9.0%. The antibody titer reduction rates of the two groups were significantly different ($P < 0.01$).

Approximately 10–80% of bullous pemphigoid patients have anti-basement membrane zone antibodies. Plasmapheresis is believed to be able to rapidly remove autoantibodies, immune complexes and pro-inflammatory factors, lower inflammatory mediator blood concentrations, regulate the immune system and restore complements and coagulation factors and condition their functions. Plasmapheresis can also enhance the cleanup function of the reticuloendothelial system, supplement substances required in the body, and thereby reduce immune and inflammatory responses and rapidly assist in controlling disease. Since plasmapheresis removes large amounts of autoantibodies, the body has a negative feedback regulation mechanism that can activate and increase pathogenic B lymphocyte proliferation and produce more autoantibodies; this increase in production can even exceed the pre-plasmapheresis antibody levels. Therefore, immunosuppressant should be used before and after plasmapheresis in order to reduce the activity of B lymphocytes and avoid the "rebound phenomenon." Due to the high cost of plasmapheresis, a comprehensive assessment of the patient, including age, severity of illness, consolidated basis of disease, hormone contraindications, response to corticosteroids and immunosuppressive treatment, and other factors, as well as understanding of the indication, is necessary. Plasmapheresis is especially suitable for patients with severe illness, high circulating antibody titers, and glucocorticoids contraindications who are intolerant to long-term high-dose hormone therapy. It is also suitable for use in patients that are insensitive to conventional corticosteroid and immunosuppressive combination therapy.

In summary, plasma exchange therapy can significantly reduce serum autoantibodies and complications. For severely afflicted patients who are intolerant to large doses of hormones or immunosuppressants, plasmapheresis can be used for a better therapeutic effect.

2.5.2. *Immunoadsorption therapy*

Due to its strict indication requirement and the possibility of removing some normal components, plasmapheresis development primarily depends on the advancement of technology, especially the development of immune adsorption techniques (Ruocco *et al.*, 2005; Pagoux, 2007).

Since plasmapheresis and plasma separation technology nonspecifically remove plasma proteins, such as coagulation factors, hormones and albumin, and immunoadsorption can achieve the specific adsorption of immunoglobulin and circulating immune complexes with high selectivity, immunoadsorption has been used to treat various autoimmune diseases, such as systemic lupus erythematosus and Guillain–Barré syndrome. Recent studies have used the immunoadsorption method to treat pemphigus. Studies have shown that immunoadsorption can significantly decrease serum Dsg3 autoantibodies and alleviate clinical symptoms. In recent years, immunoadsorption has gradually replaced plasmapheresis due to its clear advantages. Braun *et al.* (2000) summarized the main advantages of immunoadsorption are the selective removal of immunoglobulin — the replacement of plasma components, such as human albumin or fresh plasma, is not necessary; each treatment cycle allows for the simultaneous processing of two to three times the volume of the plasma; and the risk of side effects and allergic reactions is low.

Previous studies have found that, after three days of continuous immunoadsorption treatment, antibody levels can be decreased by 95%. Typically, within 24 to 48 hours after immunoadsorption treatment, autoantibodies rebound to 40% of the initial level, which might be related to tissue infiltration. Nevertheless, as an adjuvant therapy, immunoadsorption assay can maintain low autoantibody levels. After one month of immunoadsorption treatment, antibody levels can be reduced by 80%; after six months antibody levels can be reduced by 90%; after 12 months antibodies can be maintained at that reduced level. As proven in clinical trials, immunoadsorption therapy can rapidly control this disease within several weeks (Schmidt *et al.*, 2003).

Guidance to immunoadsorption therapy has been further refined in studies (Zillikens *et al.*, 1993). High affinity reusable immunoadsorption systems can be better used clinically, such as Immunosorba (Germany), Globaffin (Germany), and Therasorb (Germany). In general, when taking corticosteroids (1–2 mg/kg/d) and an immunosuppressant, patients can be treated with two to four cycles of immunoadsorption for three to four weeks until clinical improvement is achieved.

Staphylococcal protein A (SPA) is a surface protein in bacterial cell walls. It can bind with the Fc portion of human plasma lgG and has a high specificity and high sensitivity. Protein A immunoadsorption (PAIA) is a method that uses agar to carry SPA as the adsorption column to remove pathogenic antibodies in the plasma of patients with immune injury. PAIA can not only eliminate pathogenic autoantibodies, but also regulate immune functions with little damage to inherent blood components.

Shimanovich *et al.* (2006) treated nine pemphigus patients using PAIA combined with high-dose glucocorticoid therapy; azathioprine or mycophenolate mofetil was added during the hormone reduction phase. After treatment, the circulating antibody level was significantly reduced, and the disease was controlled within four weeks. Afterwards, the PAIA remission stage was extended to 26 months. In a recent study, five patients with severe pemphigus were treated with an immunoadsorption therapy, which used the Staphylococcus aureus protein A adsorption column, for three consecutive days and then once every one to four weeks. Five patients responded well to treatment, indicating that protein A adsorption treatment can effectively reduce anti-Dsg1, Dsg3, and IgG levels, with an average reduction rate of 76% (Schmidt *et al.*, 2003).

Frost *et al.* (2005) reported a typical case involving a 56-year-old female patient with a 14-year history of pemphigus. The patient had bullous skin and severe mucous membrane lesions and experienced severe side effects when using glucocorticoid. She was clinically diagnosed with refractory pemphigus. The author used the protein A immunoadsorption treatment along with an immunosuppressant. The immunoadsorption treatment was conducted once per week and 19 times total. After treatment, the pemphigus symptoms were significantly relieved; the skin lesions gradually disappeared, and the serum pathogenic autoantibody Dsg1 and Dsg3 levels were significantly reduced. However, during two adsorption treatments, the patient's antibody levels rebounded significantly; therefore, standard immunosuppressive therapy was added to the treatment plan, and the serum autoantibodies were monitored. Protein A immunoadsorption provides an effective method of treatment for severely afflicted patients experiencing poor treatment efficacy and significantly extends the remission stage.

The Globaffin immunoadsorption treatment system includes two Globaffin adsorbers and, therefore, is a dual adsorption column system. The adsorption column can be reused, and is produced by the German company Fresenius. Each adsorption column contains 60 mL of matrix; the matrix is based on CL-4B Sepharose, with 10 mmol/L of sodium citrate solution (pH = 4.0) as a buffer. At least one of the adsorbers contains 250 mg of PGAM146 peptide covalently bound to Sepharose. Prior to treatment, each adsorber must be washed and shocked with 2,000 mL of sterile, non-pyrogenic saline in order to remove tiny residual particles. During treatment, the adsorption column needs to be regenerated with an acid solution. After treatment, the Globaffin adsorber is stored in a toxic antibacterial NaN3 buffer. In order to facilitate preservation, each adsorber should be washed with 350 mL of preservation buffer (PBS 0.1% sodium azide). After rinsing and perfusion, the adsorber needs to be maintained at 2–10°C during the adsorption treatment. The Globaffin immunoadsorption treatment system uses centrifugation plasma separation technology. The operating parameters, adsorption capacity of the column, filling volume (250 mL per cycle), desorption volume, regeneration capacity, adsorption processing speed, and treatment cycles are controlled by the adsorption system device (Medicap, Ulrichstein). Before connecting a patient's vessel to begin the treatment, the adsorber and piping should be rinsed with 2,000 mL of 0.9% saline. In Globaffin immunoadsorption treatment, the plasma alternately passes through the adsorbers; one adsorber adsorbs and removes immunoglobulins, and the other adsorber simultaneously regenerates with the glycine buffer. Each plasma therapeutic dose is calculated before treatment. The vascular access point should be located in the forearm cubital vein or central vein. For anticoagulation purposes, the patient should be injected intravenously with 1250 IU of unfractionated heparin before immunoadsorption; for *in vitro* anticoagulation, sodium citrate [(ACD-A, 2.2%; 1: 15-1: 20 v/v to blood)] is used. Plasma is driven through the blood pump in order to pass the Globaffin adsorption system. The separated plasma is perfused through the adsorption column at a perfusion rate of 11 to 28 mL/min, with a common speed of 24 mL/min. The volume of the plasma in each adsorption cycle is 250 mL. After immunoadsorption, the plasma and other blood components are refluxed into the patients' blood circulation. For the first day of treatment, the amount of adsorbed plasma should be 1.5 times that of the calculated theoretical plasma volume for the adsorption treatment.

Eming *et al.* (2006) used the Globaffin immunoadsorption system to treat four cases of pemphigus vulgaris and two cases of chronic, severe pemphigus foliaceus. The treatment lasted four consecutive days and was conducted once every four weeks. Throughout the immunoadsorption treatment, the patients received continuous immunosuppressive therapy, including glucocorticoids and non-steroidal drugs according to the severity of illness. At the end of each of the immunoadsorption treatments, the patients were injected intravenously with 10 g of immunoglobulin. According to each patient's condition, a high dose (750 mg/m^2 IV) of cyclophosphamide was given after the first day of immunoadsorption treatment. In the four weeks of the intermittent period, according to the patient's condition, six impact therapies were administered. Patients could take several plasma immunoadsorption therapies based on the severity of their conditions. The results suggested that immunoadsorption therapy (IA) yields excellent clinical effects. Immunoadsorption can effectively reduce the levels of total plasma IgG. After four days of treatment, the IgG level decreased from 880 ± 271 mg/dl to 160 ± 100 mg/dl, and the average reduction rate was 82%. In the second course of immunoadsorption treatment, the plasma total IgG level decreased from 704 ± 117 mg/dl to 220 ± 29 mg/dl, and the average reduction rate was 69%. The plasma concentrations of the other plasma proteins, such as albumin and immunoglobulins IgA, IgM, and IgE, were not affected. Thus, the Globaffin immunoadsorption system can specifically adsorb immunoglobulin IgG. Immunoadsorption therapy reduced the levels of the Dsg1/Dsg3-reactive IgG antibody. The results indicated that, after the first course of immunoadsorption, the Dsg1 antibody levels declined by 68 ± 20.8%, and the Dsg3 antibody declined by 59.5 ± 15.5%. The Globaffin adsorption system, on average, can lower the level of circulating antibodies by 50 to 70% while relieving the patients' clinical symptoms. Figure 23.2 shows that immunoadsorption treatment can effectively restore pemphigus lesions. Therefore, severe and treatment-resistant pemphigus patients could benefit significantly from plasma adsorption treatment. In addition, the use of clearing serum Dsg-reactive autoantibodies is a promising treatment method for pemphigus.

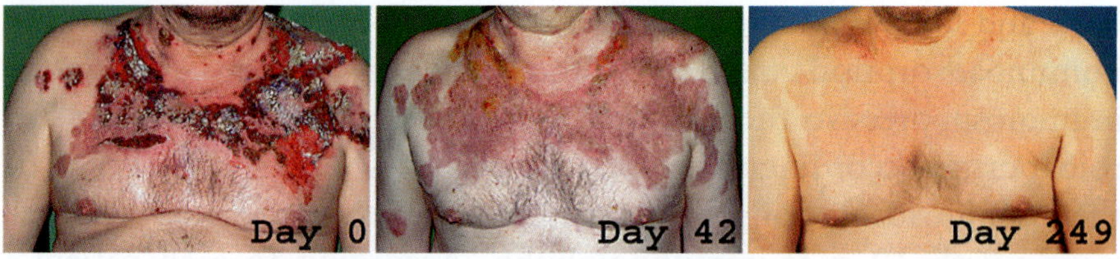

Fig. 23.2. Skin lesion improvement after immunoadsorption therapy.

Luftl *et al.* (2003) used the tryptophan-coated polyvinyl adsorption system to treat nine pemphigus patients, seven of which with pemphigus vulgaris and two of which with pemphigus foliaceus. The nine patients exhibited acantholysis and bullous formation, IgG and C3 deposition in epidermal cells, were positive for immunofluorescence positive, and were positive for indirect immunofluorescence of circulating autoantibodies.

The clinical manifestations were skin lesions and the presence of anti-Dsg1 antibodies. While receiving glucocorticoid treatment, each patient received two immunoadsorption therapies with 48-hour intervals. Each time, 1.5 to 2.3 liters of plasma were separated and passed through the adsorber for immunoadsorption treatment. Temporary extracorporeal circulation was established through a peripheral vein or temporary deep venous catheter. A polyethylene plasma separator OP05 (Asahi Medical, Tokyo, Japan and Diamed, Cologne, Germany) was connected to a tryptophan-coated polyvinyl adsorber, TR350 (Asahi Medical and Diamed), for immunoadsorption. The blood was pumped into the plasma separator at a rate of 40 to 60 mL/min using a high-precision peristaltic pump. Unfractionated heparin was used for anticoagulation; the initial dose was 5,000 IU, and the maintenance dose was 1,000 IU per hour. In order to prevent the rebound of the pemphigus antibody after each treatment, a 1.5 mg/kg dose of prednisone was used after each immunoadsorption. The results indicated that all of the patients exhibited significantly decreased serum anti-Dsg1 and anti-Dsg3 IgG levels. After the first adsorption treatment, the anti-Dsg1 antibody optical density value (OD value) decreased from 0.53 ± 0.29 to 0.40 ± 0.29. After the second adsorption treatment, the anti-Dsg1 antibody OD values decreased from 0.52 ± 0.3 to 0.42 ± 0.31. Similarly, after the first adsorption treatment, the anti-Dsg3 antibody OD value decreased from 0.46 ± 0.13 to 0.36 ± 0.16; after the second treatment, the OD value of the anti-Dsg3 antibodies decreased from 0.42 ± 0.1 to 0.22 ± 0.05. All of these results were statistically significant. The tryptophan-coated polyvinyl adsorption system decreased the Dsg serum autoantibody reactivity levels and significantly improved the clinical symptoms (see Fig. 23.3).

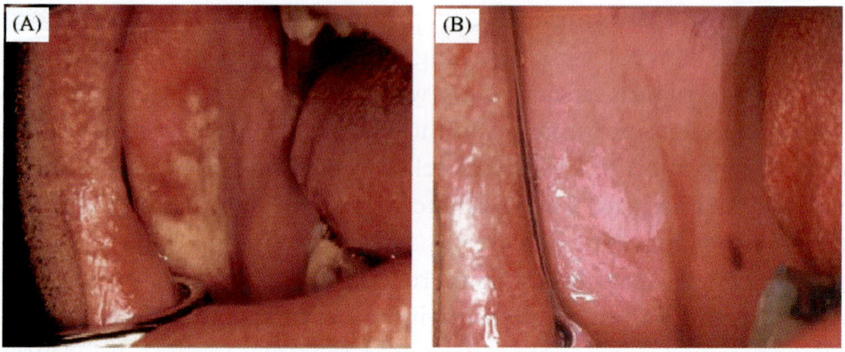

Fig. 23.3. Changes in mucosal lesions before and after immunoadsorption treatment.

Common side effects of immunoadsorption therapy include deep vein thrombosis, transient paresthesia, hypotension, bradycardia, allergic reactions, and sepsis (Schmidt *et al.*, 2003; Shimanovich *et al.*, 2006).

The most important advantages of immunoadsorption therapy are high selectivity in clearing pathogens, reduced loss of useful plasma ingredients, and no common risks in plasma exchange therapy. Various experimental and clinical studies have demonstrated that immunoadsorption and immunosuppressant combination therapy could significantly improve the clinical manifestations of severe pemphigus patients. Among the different immunoadsorption techniques, the tryptophan-coated adsorption column is expensive; although the protein A adsorption column can effectively reduce plasma IgG levels, it is much more expensive than the tryptophan-coated adsorption column, and its effectiveness in pemphigus treatment has not been studied in controlled clinical experiments. Therefore, immunoadsorption combined with immunosuppressive therapy is primarily used to treat severe and refractory pemphigus patients and, initially, cases of high antibody levels. Recent case reports have indicated that immunoadsorption therapy is a safe and effective method of treatment.

Zhenwei Shi, Shaoxia Zi

References

Ahmed AR, Spigelman Z, Cavacini LA, *et al.* (2006) Treatment of pemphigus vulgaris with rituximab and intravenous immune globulin. *N Engl J Med* **355:** 1772–1779.

Amagai M, Hashimoto T, Shimizu N, *et al.* (1994) Absorption of pathogenic autoantibodies by the extracelluar domamin of pemphigus vulgaris antigen produced by baculovirus. *J Invest* **94(1):** 59–63.

Amagai M, Ikeda S, Shimizu H, *et al.* (2009) A randomized double-blind trial of intravenous immunoglobulin for pemphigus. *J Am Acad Dermatol* **60:** 595–603.

Amagai M, Tsunoda K, Zillikens D, *et al.* (1999) The clinical phenotype of pemphigus is defined by the anti-desmogelin autoantibody profile. *J Am Acad Dermatol* **40(2 pt 1):** 167–170.

Asano S, Seishima M, Kitajima Y. (2001) Phosphatidylinositol specific phospholipase C cleaves urokinase plasminogen activator receptor from the cell surface and leads to inhibition of pemphigus-IgG-induced acantholysis in DJM1 cells, a squamous cell carcinoma line. *Clin Exp Dermatol* **26(3):** 289–95.

Barber AE, Coyle SM, Maano MA, *et al.* (1993) Glucocorticoid therapy alters hormonal and cytokine responses to endotoxin in man. *J Immunol* **150(3):** 1999–2002.

Baum S, Scope A, Barzilai A, *et al.* (2006) The role of IVIG treatment in severe pemphigus vulgaris. *J Eur Acad Dermatol Venereol* **20:** 548–552.

Beissert S, Mimouni D, Kanwar AJ, *et al.* (2010) Treating pemphigus vulgaris with prednisone and mycophenolate mofetil: A multicenter, randomized, placebo-controlled trial. *J Invest Dermatol* **130:** 2041–8.

Beissert S, Werfel T, Frieling U, *et al.* (2006) A comparison of oral methyprednisolone plus azathioprine or mycophenolate mofetil for the treatment of pemphigus. *Arch Dermatol* **142:** 1447–54.

Braun N, Bosch T. (2000) Immunoadsorption, current status and future developments. *Expert Opin Investig Drugs* **9:** 2017–38.

Chams-Davatchi C, Esmaili N, Daneshpazhooh M, *et al.* (2007) Randomized controlled open-label trial of four treatment regimens for pemphigus vulgaris. *J Am Acad Dermatol* **57:** 622–8.

Chrysomallis F, Ioannides D, Teknetzis A, *et al.* (1994) Treatment of oral pemphigus vulgaris. *Int J Dermatol* **33:** 803–7.

Cotterill JA, Barker DJ, Millard LG. (1978) Plasma exchange in the treatment of pemphigus vulgaris. *Br J Dermatol* **98:** 243.

Cowley CM, Simrak D, Marsden MD, *et al.* (1997) A YAC contig joining the desmocollin and desmoglein loci on human chromosome 18 and ordering of the desmocollin genes. *Genomics* **42(2):** 208–216.

D'Auria L, Bonifati C, Mussi A, *et al.* (1997) Cytokines in the sera of patients with pemphigus vulgaris: Interleukin-6 and tumour necrosis factor-alpha levels are significantly increased as compared to healthy subjects and correlate with disease activity. *Eur Cytokine Netw* **8(4):** 383–387.

Denning MF, Guy SG, Ellerbroek SM, *et al.* (1998) The expression of desmoglein isoforms in cultured human keratinocytes is regulated by calcium, serum, and protein kinase C. *Exp Cell Res* **239(1):** 50–59.

Eming R, Rech J, Barth S, *et al.* (2006) Prolonged clinical remission of patients with severe pemphigus upon rapid removal of desmoglein reactive autoantibodies by immunoadsorption. *Dermatology* **212(2):** 177–187.

Enler HH, Loffer H, Christophers E. (1987) Synchronization of plasmapheresis and pulse cyclophosphamide therapy in pemphigus vulgaris. *Arch Dermatol* **123:** 1205–1210.

Enno Schmidt, Detlef Zillikens. (2010) Modern diagnosis of autoimmune blistering skin diseases. *Autoimmunity Reviews* **10:** 84–89.

Femandes NC, Perez M. (2001) Treatment of pemphigus vulgaris and pemphigus foliacues experience with 71 patients over a 20 year period. *Rev Inst Med Trop Sao Paulo* **43:** 45–50.

Frost N, Messer G, Fierlbeck G, *et al.* (2005) Treatment of pemphigus vulgaris with protein A immunoadsorption: Case report of long-term history showing favorable outcome. *Ann N Y Acad Sci* **1051:** 591–596.

Guillaume JC, Roujeau JC, Morel P, *et al.* (1988) Controlled study of plasma exchange in pemphigus. *Arch Dermatol* **124:** 1659–63.

Guo Y, Shang Y, Zhao Y, *et al.* (2007) Efficacy analysis of plasma exchange therapy for pemphigus and bullous pemphigoid. *China Journal of Leprosy and Skin Diseases* **23(2):** 140–141.

Hashimoto T. (2008) Treatment strategies for pemphigus vulgaris in Japan. *Expert Opin Pharmacother* **9:** 1519–30.

Hert M. (2000) Humoral and cellular autoimmunitin autoimmune bullous skin disorders. *Arch Allergy Immunol* **122:** 91–100.

Hunt DM, Sahota VK, Taylor K, *et al.* (1999) Clustered cadherin genes: A sequence-ready contig for the desmosomal cadherin locus on human chromosome 18. *Genomics* **62(3):** 445–455.

Ishikawa H, Li K, Tamai K, *et al.* (2000) Cloning of the mouse desmoglein 3 gene(Dsg3): Interspecies conservation within the cadherin super family. *Exp Dermatol* **9(4):** 229–239.

Kalantari-Dehaghi M, Molina DM, Farhadieh M, *et al.* (2011) New targets of pemphigus vulgaris antibodies identified by protein array technology. *Exp Dermatol* **20:** 154–6.

Kitajima Y, Aoyama Y, Seishima M. (1999) Transmembrane signaling for adhesive regulation of desmosomes and hemidesmosomes, and for cell-cell detachment induced by pemphigus IgG in cultured keratinocytes: Involvement of protein kinase C. *J Invest Dermatol Symp Proc* **4(2):** 137–144.

Koch PJ, Mahoney MG, Ishikawa H, *et al.* (1997) Targeted disruption of the pemphigus vulgaris antigen (desmoglein 3) gene in mice causes loss of keratinocyte cell adhesion with a phenotype similar to pemphigus vulgaris. *J Cell Biol* **137(5):** 1091–1092.

Lin MS, Swartz SJ, Lopez A, *et al.* (1997) Tlymphocytes from a subset of patients with pemphigus vulgaris respond to both Dsg3 and Dsg1. *J Invest Dermatol* **109(6):** 734–737.

Lombardi ML, Mercuro O, Ruocco V, *et al.* (1999) Common human leukocyte antigen alleles in pemphigus vulgaris and pemphigus foliaceus Italian patients. *J Invest Dermatol* **113(1):** 107–110.

Lopez-Robles E, Avalos-Diaz E, Vega-Memie E, *et al.* (2001) TNFalpha and IL-6 are mediators in the blistering process of pemphigus. *Int J Dermatol* **40(3):** 185–188.

Lu H, Ning X, He W. (2000) Two cases of plasma replacement therapy for bullous pemphigoid. *Journal of Clinical Dermatology* **29(4):** 218.

Luft M, Stauber A, Mainka A, *et al.* (2003) Successful removal of pathogenic autoantibodies in pemphigus by immunoadsorption with a tryptophan linked polyvinylalcohol adsorber. *British Journal of Dermatology* **149:** 598–605.

Mahoney MG, Wang ZH, Stanley JR. (1999) Pemphigus vulgaris and pemphigus foliaceus antibodies are pathogenic in plasminogen activator knockout mice. *J Invest Dermatol* **113(1):** 22–25.

Matthew ST, David S, Daniel NS. (2000) The use of plasmapheresis and immune suppression in the treatment of pemphigus vulgaris. *J Am Acad Dermatol* **43(6):** 1059–1064.

Mentink LF, Mackenzie MW, Toth GG, *et al.* (2006) Randomized controlled trial of adjuvant oral dexamethasone pulse therapy in pemphigus vulgaris. PEMPULS trial. *Arch Dermatol* **142:** 570–6.

Meurer M. (2012) Immunosuppressive therapy for autoimmune bullous diseases. *Clinics in Dermatol* **30:** 78–83.

Miyagawa S, Amagai M, Iida T, *et al.* (1999) Late development of antidesmoglein 1 antibodies in pemphigus vulgaris: Correlation with disease progression. *Br J Dermatol* **141(6):** 1084–1087.

Nishifuji K, Amagai M, Kuwana M, *et al.* (2000) Detection of antigen-specific B cells in patients with pemphigus vulgaris by enzyme-linked immunospot assay: Requirement of T cell collaboration for auto antibody production. *J Invest Dermatol* **114(1):** 88–94.

Pagoux C. (2007) Plasma exchange for systemic lupus erythematosus. *Transfus Apher Sci* **36:** 187–193.

Patel M, Furstenberg G, Hazelton J, *et al.* (2010) Development of protein microarrays to investigate autoantibody profiles in pemphigus vulgaris. *J Invest Dermatol* **130:** S19 [abstract: 549].

Ranugha PSS, Kumari R, *et al.* (2012) Therapeutic plasma exchange as a crisis option in severe pemphigus vulgaris. *Indian Journal of Dermatology, Venereology, and Leprology* **78(4):** 508–510.

Rico MJ, Benning C, Weingart ES, *et al.* (1999) Characterization of skin cytokines in bullous pemphigoid and pemphigus vulgaris. *Br J Dermatol* **140(6):** 1079–1086.

Rose E, Wever S, Zillikens D, *et al.* (2005) Intravenous dexamethasone-cyclophosphamide pulse therapy in comparison with oral methylprednisolone-azathioprine therapy in patients with pemphigus vulgaris: Results of a multicenter prospectively randomized study. *J Dtsch Dermatol Ges* **3:** 200–6.

Ruocco E, Baroni A, Wolf R, *et al.* (2005) Life-threatening bullous dermatoses: Pemphigus vulgaris. *Clin Dermatol* **23:** 223–226.

Sagi L, Baum S, Gendelman V, *et al.* (2011) The role of therapeutic plasma exchange in pemphigus vulgaris. *J Eur Acad Dermatol Venereol* **25:** 82–6.

Saraceno R, Ruzzetti M, Lanti A, *et al.* (2008) Therapeutic options in an immunocompromised patient with pemphigus vulgaris: Potential interest of plasmapheresis and extracorporeal photochemotherapy. *Eur J Dermatol* **18:** 354–6.

Schaerer L, Trueb RM. (2003) Direct immunofluorescence of plucked hair in pemphigus. *Arch Dermatol* **139(2):** 228–229.

Schiavo AL, Puca RV, Ruocco V, *et al.* (2010) Adjuvant drugs in autoimmune bullous diseases, efficacy versus safety: Facts and controversies. *Clin Dermatol* **28:** 337–43.

Schmidt E, Klinker E, Opitz A, *et al.* (2003) Protein A immunoadsorption: A novel and effective adjuvant treatment of severe pemphigus. *Br J Dermatol* **148:** 1222–1229.

Schmidt E, Zillikens D. (2010a) Immunoadsorption in dermatology. *Arch Dermatol Res* **302:** 241–53.

Scully C, Challacombe ST. (2002) Pemphigus vulgaris: Update on etiopathogenesis, oral manifestations, and management. *Crit Rev Oral Biol Med* **13:** 397–408.

Shang Y, Han S, Guo Y, *et al.* (2004) Analysis of therapeutic effect of plasma exchange in pemphigus. *Journal of Dalian Medical University* **26(4):** 288–289.

Sharma PM, Choi EJ, Kuroda K, *et al.* (2009) Pathogenic antidesmoglein MAbs show variable ELISA activity because of preferential binding of mature versus protein isoforms of desmoglein 3. *J Invest Dermatol* **129:** 2309–12.

Sharma VK, Prasad HR, Khandpur S, *et al.* (2006) Evaluation of desmoglein enzyme-linked immunosorbent assay (ELISA) in Indian patients with pemphigus vulgaris. *Int J Dermatol* **45(5):** 518–522.

Shimanovich I, Herzog S, Schmidt E, *et al.* (2006) Improved protocol for treatment of pemphigus vulgaris with protein A immunoadsorption. *Clin Exp Dermatol* **31:** 768–74.

Shimanovich I, Herzog S, Shmidt E, *et al.* (2006) Improved protocol for treatment of pemphigus vulgaris with protein A immunoadsorption. *Clin Exp Dermatol* **31(6):** 768–774.

Siebra MX, Santos MA, Almeida TLP, *et al.* (2006) Evidence for the participation of nitric oxide in pemphigus. *Brazil J Med Biol Res* 39(5): 671–675.

Singh S. (2011) Evidence-based treatments for pemphigus vulgaris, pemphigus foliaceus, and bullous pemphigoid: A systematic review. *Indian Journal of Dermatology, Venereology, and Leprology* 77**(4):** 456–69.

Smith DA, Germolec DR. (1999) Introduction to immunology and autoimmunity. *Environ Health Perspect* **107(5):** 661–665.

Sondergaard K, Carstend J, Jorgensen J, *et al.* (1995) The steroid-sparing effect of long-term plasmapheresis in pemphigus. *Acta Derm Venereol* **75:** 150–2.

Stoopler ET, Pinto A, DeRossi SS. (2003) Pemphigus: Update for the general practitioner. *NY State Dent J* **69:** 30–32.

Tan-Lim R, Bystryn JC. (1990) Effect of plasmapheresis therapy on circulating levels of pemphigus antibodies. *J Am Acad Dermatol* **22:** 35–40.

Toto P, Feliciani C, Amerio P, *et al.* (2000) Immune modulation in pemphigus vulgaris: Role of CD28 and IL-10. *J Immunol* **164(1):** 522–529.

Turner MS, Sutton D, Sauder DN. (2000) The use of plasmapheresis and immunosuppression in the treatment of pemphigus vulgaris. *J Am Acad Dermatol* **43:** 1058–64.

Uzun S, Durdu M. (2006) The specificity and sensitivity of Nikolskiy sign in the diagnosis of pemphigus. *J Am Acad Dermatol* **54(3):** 411–415.

Xiao R, Liang X, Ling W, *et al.* (1993) One case of pemphigus vulgaris treatment using manual plasma replacement therapy and blood quantum method. *Third Military Medical University* **15(1):** 33.

Xue W, Hashimoto K, Toi Y. (1998) Functional involvement of urokinase-type plasminogen activator receptor in pemphigus acantholysis. *J Cutan Pathol* **25(9):** 469–474.

Yang S, Jiang S, Wang L, *et al.* (2009) Clinical efficacy of plasma exchange therapy for autoimmune bullous skin diseases. *Chinese Journal of Blood Purification* **8(3):** 134–136.

Yumi A. (2010) What's new in i.v. immunoglobulin therapy and pemphigus: High-dose i.v. immunoglobulin therapy and its mode of action for treatment of pemphigus. *J Dermatol* **37:** 239–245.

Zagorodniuk I, Weltfriend S, Shtruminger L. (2005) A comparison of anti desmoglein antibodies and indirect immunofluorescence in the sero diagnosis of pemphigus vulgaris. *Int J Dermatol* **44(7):** 541–544.

Zillikens D, Derfler K, Eming R, *et al.* (2007) Recommendations for the use of immunoapheresis in the treatment of autoimmune bullous diseases. *J Dtsch Dermatol Ges* **5:** 881–7.

Zillikens D, Rady PL, Goldblum RM, *et al.* (1993) Evidence for cell-mediated immune mechanisms in the pathology of pemphigus. *Br J Dermatol* **128(6):** 636–643.

Chapter 24

Use of Sorption-based Methods in Treatment of Asthma and Allergic Diseases

Vladimir Viktorovich Komov*, Sergey Andreevich Krynskiy*, Elisaveta Aleksandrovna Snezhkova†

Abstract

In this chapter a 30-year experience of clinical use of sorption-based methods in treatment of allergic diseases is analyzed. Clinical and laboratory results of therapy with selective (anti-IgE) and non-selective hemosorbents, with enterosorption and with fractioned plasmapheresis in patients with asthma, urticaria, recurrent Quincke's edema, atopic dermatitis and other allergic diseases who underwent treatment in the laboratory of clinical immunology and in the Republican Center for sorption-based methods of treatment based in the Scientific Researcher Institute (SRI) of physical-chemical medicine are provided. Sorption-based methods were used both as

---

*Vladimir Viktorovich Komov — Republican Center for Extracorporeal Methods of Treatment of SRI of Physical-Chemical Medicine, Federal Medical-Biological Agency, Moscow, Russia. vladokom@mail.ru.

Sergey Andreevich Krynskiy — Laboratory of Clinical Immunology of SRI of Physical-Chemical Medicine, Federal Medical-Biological Agency, Moscow, Russia.

†Elisaveta Aleksandrovna Snezhkova — Institute of Experimental Pathology, Oncology and Radiobiology named after R.E. Kavetsky NAS, Kiev, Ukraine.

monotherapy and along with other methods of treatment. The total number of patients is more than 2,000; the maximal follow-up period in patients receiving program efferent therapy was 28 years. A high therapeutic and preventive effect of sorption-based methods in treatment of allergic diseases is demonstrated. The mean duration of remission in patients with atopic asthma was 16.7 ± 4.2 months, in patients with infection-mediated and mixed asthma it was 13.5 ± 3.6 months, in patients with aspirin-mediated asthma it was 11.4 ± 4.8 months. The probability of progression of pre-asthmatic state into clinical asthma was reduced four times in comparison to that in patients receiving only the drug therapy.

1. Introduction: Types and Clinical Impact of Allergic Reactions

Nowadays there is a notable increase in the frequency of allergic diseases that started several decades ago. The prevalence in the general population is estimated by different authors as 15–30%.[2] Despite all the advances in allergology and pharmacology, it seems that medical science cannot find a solution for this problem yet.

1.1. *Types of Allergic Reactions*

Allergic reactions are pathologic forms of immune response characterized by sensitization of the immune system to recurrent encounters with allergens. Their pathogenesis can be divided into three stages:

(I) Pathoimmune stage, when hypersensitivity to the allergen develops;
(II) Pathochemical stage, when a recurrent encounter with the allergen causes a release of biologically active substances (BAS);
(III) Pathophysiological stage, when the released BASs cause pathology in the organs and systems of the body.

Allergic reactions are classified into four types, after Gell P.G.H. and Coombs P.R.F (Gell and Coombs, 1964; Khaitov, 2013). If we describe in a simple way the processes taking place during their development, their pathogenesis can be reviewed as follows. Allergic reactions of the I type are mediated by antibodies of the IgE isotype (sometimes of the IgG4 isotype). The allergen is bound by these IgE antibodies to the Fc-receptors of cells. If the allergen is encountered again, an immediate allergic reaction takes place. Allergic reactions of the II type (cytotoxic reactions) are mediated by antibodies (most commonly by IgG or IgM) that bind allergens on cytoplasmic membranes. These reactions involve the activation of complement system and other pathological mechanisms carried out by NK cells, CD8+ cells, lysosomal enzymes and other BASs. This results in the death of target cells and causes certain syndromes and diseases.

Allergic reactions of the III type are immune complex-mediated and involve the deposition of damaging immune complexes in organs of the body (Arthus reaction). If a large number of antigen molecules is inoculated into the blood, specific IgG and IgM antibodies bind to the antigen leading to the formation of immune complexes (ICs). As ICs are deposited and penetrability of the vascular wall increases, ICs leave the blood flow and precipitate in tissues, imparting these with antigenic capacities. This triggers cascades of immune reactions that lead to autoimmune diseases (rheumatoid arthritis, dermatomyositis, etc.).

Allergic reactions of the IV type are also known as delayed hypersensitivity reactions. They are cell-mediated and can be triggered by various types of antigens. These reactions require sensibilization of antigen-specific T-helper cells ($CD4^+$). The sensibilized lymphocytes secrete mediator molecules that trigger cascades involving cell-mediated and humoral immune mechanisms, action of cytokines and free radicals. Delayed hypersensitivity reactions are important for counteracting various infectious agents, however, in certain conditions their regulation is disrupted leading to fibrosis, deposition of collagen and other forms of pathology. This can result in diseases such as asthma, eczema, dermatitis, transplant rejection and collagenous diseases.

In 2003 HRs of the V type were described by Rajan TV. These are due to direct receptor-mediated stimulation (thyrotoxicosis) or suppression (diabetes mellitus, myasthenia gravis) of cell function by antibodies (Rajan, 2003).

Therefore even a simple description of allergic reactions shows that their pathogenesis involves various systems of the body and complex multi-step cascades of biochemical and pathophysiological reactions that are regulated by neurohormonal mechanisms. This causes substantial difficulties in the search for effective prophylactics and treatment of allergic, autoimmune and immune diseases.

1.2. *Allergic Diseases and Basic Principles of Their Treatment*

There are several dozens of diseases that involve allergic reactions. However, the term "allergic diseases" is only used for the diseases that are caused by type I hypersensitivity reactions.

Type I allergic reactions are mediated by IgE and (less commonly) by IgG4 antibodies. The allergen is taken up by antigen-presenting cells, fragmented into short peptides and presented to Th2 lymphocytes. These cells produce a variety of cytokines and express the ligand for CD40 (CD40L/CD154), thus causing B-lymphocytes to produce specific IgE antibodies. IgE binds to type 1 and type 2 Fc-receptors (CD23) expressed by cells of the mucus membranes and connective tissues, mast cells, basophils, B-lymphocytes, monocytes, eosinophils and, probably, T-lymphocytes. CD23 secreted into the blood flow induces synthesis of IgE by B-lymphocytes. IgE antibodies become fixed on the surface of mast cells and basophils. After a repeated contact with the antigen they bind to it and induce a cascade of chemical reactions (pathochemical stage) that lead to degranulation of the cell and secretion of histamine, serotonin, acetylcholine and other molecular mediators of allergy. These molecules cause effects in target organs and tissues of the body such as contraction of smooth muscles, stimulation of secretion in exocrine glands, rise in vascular permeability, edema, and hypersensitivity of pain receptors (pathophysiological stage). These reactions become clinically detectable in minutes after the contact with allergen. Vascular permeability is affected and adhesion molecules become expressed on cells. Secretion of chemokines causes infiltration of tissues by lymphocytes, eosinophils and mast cells. After activation these cells secrete more pro-inflammatory mediators, causing the delayed phase of AR. They can also induce hypersensitivity to unspecific agents, like cold, emotional stress and physical exercise. These pathogenic mechanisms may cause conditions such as urticaria, allergic rhinitis and conjunctivitis, pollinosis, food allergy, atopic asthma, Quincke's edema and anaphylaxis (Khaitov, 2013).

Therefore, allergic reactions are mediated by interaction of allergen with membrane-bound IgE antibodies that causes secretion of histamine and other mediators. There are more than three types of histamine receptors in the body. H2 and H3 receptors affect vascular tonus and consequently blood pressure level, cardiac rhythm, vascular permeability, gastric secretion and intestinal peristalsis. Activation of H1-receptors in the nasopharynx, trachea and bronchus causes edema and bronchospasm. Clinically this can manifest as bronchial asthma (BA), vasomotor rhinitis and conjunctivitis. Activation of H1-receptors of the skin causes urticaria that is accompanied by edema, rash and itching. Histamine-mediated reactions are important in the pathogenesis of Quincke's edema, Lyell's syndrome and other urgent conditions.

1.2.1. *Pharmacotherapy of allergic diseases*

Pharmacotherapy is the most important method in treatment of allergic diseases. In particular, H1-receptor blockers (antihistamine drugs) are the basis of treatment. These drugs act by lowering histamine levels in blood and suppressing histamine-mediated reactions. There are several generations of H1-blockers. They belong to different chemical classes and have different selectivity, pharmacokinetics and pharmacodynamics. However, all H1-blockers act by reducing symptoms and do not affect the pathogenesis of ARs. They cannot completely block the action of histamine and other mediators on the organs and systems of the body. Besides, they have a wide spectrum of side effects.

Other classes of drugs that are used for treatment of ARs are compounds that increase the histamine-binding capacity of blood plasma ("Histaglobin," Biobasal AG, Sveden). These drugs also inhibit the release of histamine by mast cells, suppressing IgE-mediated reactivity of basophils. However, they also have only a symptomatic effect. Besides, these drugs contain human proteins that might provoke allergic and autoimmune reactions.

The pharmacologic industry produces a variety of drugs (i.e. cromoglicic acid derivates, such as sodium cromoglicate, nedocromil, etc.) that reduce the release of allergic mediators by stabilizing the membranes of mast cells, thrombocytes, macrophages and eosinophils. Cromoglicates also inhibit Ca-dependent reactions and the effects of phosphodiesterase, diminishing IgE-dependent secretion of allergic mediators. Therefore, they affect all stages of allergic reactions. But their therapeutic effect is not especially strong and is achieved only after long-term courses of treatment.

Another class of drugs widely used in AD is glucocorticoid hormones. These synthetic analogs of cortisol have profound effects on physiologic processes, including allergic reactions. They also have strong anti-inflammatory and immunosuppressive capacities, regulate the blood pressure, vascular permeability, etc. Glucocorticoids are the most potent agents for treatment of allergic diseases. However, they cause a variety of side effects if used for long periods of time and, not uncommonly, these side effects affect the well-being of a person as profoundly as the principal medical condition. Topical use of glucocorticoids is only a partial solution for the problem. Therefore, GC use is limited and these drugs cannot solve the problem.

Symptomatic drugs are widely used for reducing the clinical manifestations of ARs. These include the broncholytic drugs (methylxanthins, β2-adrenomymetic drugs, etc.) that are used for treatment of exacerbations of asthma. Sympathomimetic drugs are used for treatment of rhinitis and conjunctivitis. However, these drugs also have a variety of side effects that limit their clinical use.

New insights into the pathogenesis of allergic diseases allowed the creation of drugs that act by blocking leukotriene receptors (zafirlukast, montelukast). They block the basic pathogenic steps of ARs, eliminating or preventing bronchospasm, increased vascular permeability, increased secretion of bronchial glands, disruption of mucociliary clearance, eosinophil infiltration and increased bronchial reactivity. However there is not yet much experience on the clinical use of these drugs, as well as of drugs based on cytokines. Pharmacological interventions that affect such profound mechanisms of immune regulation require thorough investigation and long-term follow-up.

Therefore, the problem of pharmacological treatment of allergic diseases is far from being solved as of yet. Most of the drugs currently available either have symptomatic effects or affect only certain pathogenic steps of ARs, causing side effects that have clinical impact comparable to that of the primary medical condition.

1.2.2. *Allergen-specific immunotherapy*

Allergen-specific immunotherapy (AST) is a pathogenesis-targeted method used in treatment of allergic diseases. It is based on the idea of making the patient's cells tolerant to a certain antigen and thus reducing the consequences of contracting this antigen in the future. AST has a stable positive effect in 70–90% of cases. The method is comparatively safe if used correctly (the frequency of complications is lower than 0.01%). However, its use also has certain limitations. Most importantly, it is not suitable for use in patients with severe allergy. It also should not be used if the patient has comorbid conditions such as severe asthma, immunodeficiency, infectious and oncologic diseases, and in patients receiving β-adrenoblocking drugs. During treatment there is always a risk of anaphylactic reactions. These considerations limit the clinical use of AST.

2. Physical-Chemical Basis and Clinical Effects of Efferent Methods of Therapy

2.1. *Historical Perspective on Efferent Methods of Therapy in Russia*

In the '70s and '80s, extracorporeal methods other than hemodialysis (plasmapheresis (PA), hemosorption (GS), etc.) were introduced into the clinical practice. These methods are based on using physical and chemical processes to remove certain regulatory factors from the bloodstream. Y.M. Lopuhin used the name "efferent methods of treatment" (lat. efferens — to excrete) for this group of medical interventions. In the Soviet Union, the methods based on sorption were introduced preferentially. Effective non-specific sorbents were developed, i.e. SKN, KAU, SKT-6A, Carbocol, VNIITU, etc. Carbon matrixes were usually used for the hemosorbents, hence the term "hemocarboperfusion" appeared. Several samples of specific sorbents were introduced, including immunosorbents for household dust. Effective technical maintenance and methodological support soon allowed the introduction of hamosorption into clinical practice, including emergency brigades and outpatient primary care. In 1987 the All-Union and Republican Centers for sorption-based methods of treatment were created, and plasmapheresis and hamosorption got their place in treatment of a wide spectrum of diseases. In particular, scientific reports and publications confirmed their effectiveness in the treatment of allergic diseases: asthma, pollinosis, urticarial, atopic dermatitis, etc. Our comparative study that assessed the effectiveness of HS and PA for treatment of early stages of asthma demonstrated comparable results for these methods that significantly exceeded the effect of pharmacotherapy (Komov, 1991a, 1991b). In the late '80s as many as 130,000–150,000 HS sessions per year were performed in the USSR. Unfortunately in 1991–95, Russian medicine underwent a substantial decline as a result of political and social changes in the former USSR.

Due to various objective and subjective factors, development of efferent medicine methods and most of all sorption-based therapy in Russia suffered severe damage. Effective sorbents were now produced abroad, and the Russian medical industry was unable to satisfy the need for sorbents and consumable materials. The use of ineffective sorbents brought discredit to the method. Nevertheless, for the last few years the interest in sorption-based therapy has been steadily increasing. Therapy based on the use of specific sorbents is attracting the most attention, but according to our experience the therapeutic potential of non-specific HS is high and far from being completely explored (Komov, 2012).

2.2. *Physical and Chemical Basis for Therapeutic Efficiency of Sorption-Based Methods of Therapy*

Elimination of chemical compounds that takes place if a fluid is perfused through a sorbent is based on three physical–chemical processes: adsorption (uptake through surface interaction), absorption (uptake by the special structure of the sorbent) and chemisorption (uptake through chemical reactions). Specific surface of up-to-date hemosorbents is up to 2,000 m^2/g and more, and the pore volume is up to 90% of the total volume of adsorbents. The specificity of sorption is determined by natural or introduced chemical centers and by the proportion of micro-, meso- and macropores.

Most non-specific sorbents can bind molecules with molecular mass from several decades to several thousand Da (Nikolaev, 1984; Lopukhin and Molodenkov, 1985; Kostutchenko, 2003). Blood perfusion leads to elimination of cells, in particular neutrophils and monocytes. Primarily old functionally inactive cells are adsorbed, activating the outflow of respective blood components into the circulation. This induces a renewal of circulating blood cells and therefore activates the functional systems of homeostasis (Komov *et al.*, 2010a). Some sorbents can remove circulating immune complexes (CIC) as effectively as plasmapheresis (Fig. 24.1).

However, the capacity of sorbents or homeostasis components differs and is usually quite limited. Table 24.1 contains data for sorption of total plasma protein by some sorbents (Komov *et al.*, 2010b). Taking into account that the volume of sorption columns is not more than 300 cm^3, the protein loss during HS is insignificantly low. This makes this method (unlike plasmapheresis) suitable for patients with hypoproteinemia.

The common view is that the effect of efferent therapy is determined by the spectrum and quantity of removed compounds. This is true for their use in toxicology. However the experience of their use in other settings, in particular for treatment of allergy and autoimmune disorders, shows that the therapeutic effect is not totally determined by elimination of substances.

First, most efferent interventions (hamosorption, hemapheresis, plasma sorption) involve the contact of blood with large volumes of external surface, through the use of extracorporeal circuit systems. Despite the effort put into making all the components of the circuits minimally damaging for the blood, this contact is a significant stress and induces stress reactions that involve coagulation cascade, cytokines and other systems of neurohumoral regulation.

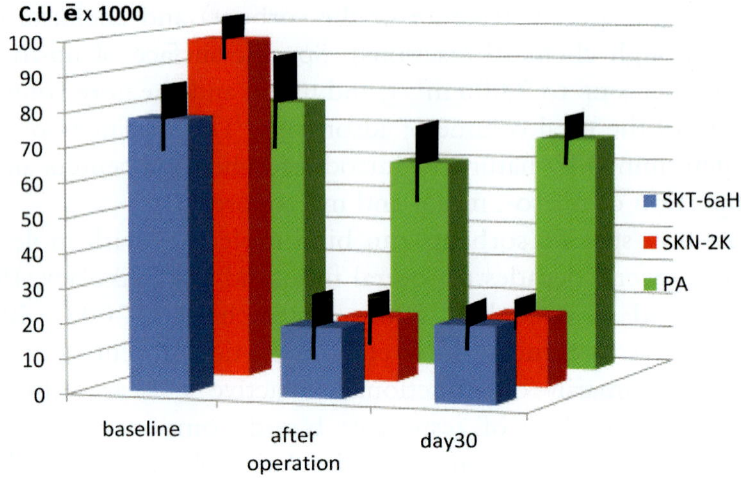

Fig. 24.1. The effect of hemosorption with sorbents "SKT-6aVCh" and "SCN-2K" and plasmapheresis on the blood level of circulating immune complexes.

Vertical axis — the level of CICs (conventional units e x 1000); horizontal axis — control points. Hemosorption is more effective than plasmapheresis for CIC elimination and the effect is more stable.

Table 24.1. Sorption Capacity of Some Sorbents Determined for Total Plasma Protein ($M \pm m$)

Sorbent	SKT-6a-VT (Russia)	ARI (Chemviron Carbon, UK)	«HABO» (Russia)
Sorption capacity (mg/cm^3)	21.66 ± 0.9	24.74 ± 0.3	6.93 ± 0.2

Second, the effect of treatment is not always completely dose-determined. The comparison of different HS and PA regimens in patients with asthma has shown that increase of perfusion to values more than 1 circulating blood volume (CBV) or effusion of more than 1.5 circulating plasma volume in PA does not lead to any significant increase in clinical and laboratory effect (Komov, 1991a).

Third, efferent methods express effects that are not due to elimination. In particular, they affect physical and chemical behavior of plasma proteins, functional activity of blood cells and cells of other organs (liver, adrenal glands, endocrine and nervous system) and influence primarily the parameters that are affected by the disease, returning them back to normal (Komov, 1991a, 1991b; Komov *et al.*, 2010b). We researched the results of affine chromatography of plasma proteins before (A) and after (B) an incubation with hemosorbent "SKT-6aVC." Figure 24.2 illustrates the result of a typical experiment. The hemosorbent caused more than a two-fold increase of sorption–desorption potentials for both abundant and minor proteins. This is a sign of a significant increase in their functional activity (Komov *et al.*, 2010b).

There are data that hemosorption causes re-activation of cell receptors and changes in pharmacokinetics and pharmacodynamics of drugs such as cortisol and theophylline (Komov *et al.*, 2010a). There are significant rebound fluctuations of biologically active substances such as cytokines, interferons and neurohumoral mediators in the early period after HS and PA. Sometimes their levels exceed the baseline ones. This means there might be other, non-sorption-based mechanisms of elimination (Voinov, 2010; Kirkovskij, 2012). Numerous reports at the XXII conference of the Moscow Hemapheresis Society stated that non-specific efferent therapy typically has a normalizing effect and targets mostly the parameters that were altered in the patient. Moreover, the use of ET in patients with inflammatory-based diseases leads to a temporary "cytokine crisis" after which the levels of pro-inflammatory cytokines drop in about a day, while the levels of anti-inflammatory interleukins and interferons stay elevated until the inflammatory reaction subsides (Komov *et al.*, 2010a; Kirkovskij, 2012).

3. Our Clinical Experience in The Use of Sorption-based Methods in Treatment of Allergic Diseases

Since the establishment of our laboratory (1985), we have been using hemosorption in treatment of asthma, pollinosis and other allergic

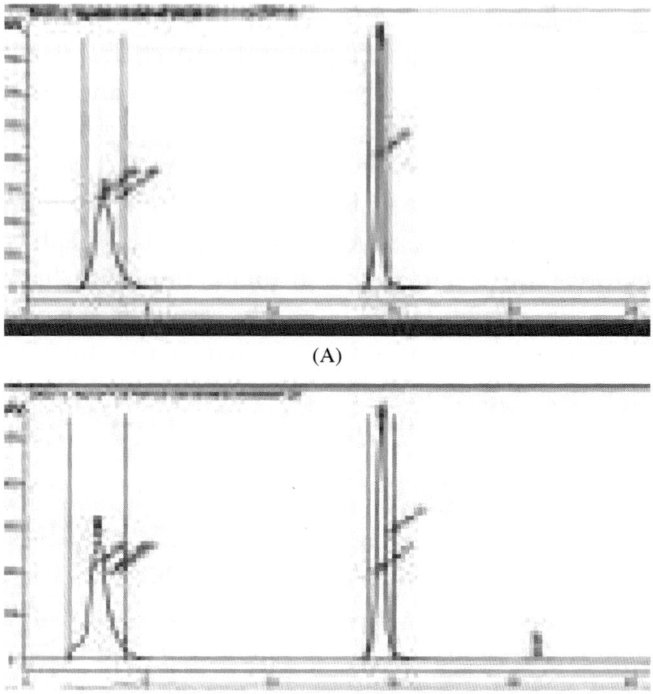

(A)

(B)

Fig. 24.2. Affine chromatography of the blood plasma proteome before (A) and after (B) incubation with a sorbent (a typical result from a series of experiments). Vertical axis — electric potential of reactions (mAV); horizontal axis — time (min). A significant increase of electric potential of sorption–desorption reactions was observed after incubation with a sorbent (B). This is a sign of a significant increase in the functional activity of major and minor plasma proteins. (Reproduced from Komov VV, *et al.* (2010) The effect of sorption on plasma proteins (Kostutchenko, 2003).

and autoimmune diseases. Hemosorption was used in complex with other methods and in some cases as a monotherapy. In 1991 we established the clinical approach based on the use of ET at early stages of diseases (Komov, 1991a). As of now we have conducted more than 3,000 sorption procedures in these patients. Selected groups of patients were followed up in average for 4.3 ± 1.7 years. In

67 patients the follow-up period was more than three years; in 32 patients it was more than 10 years; in seven patients it was more than 20 years. The most common diseases in the patients were vasomotor rhinitis and conjunctivitis, allergic urticaria and dermatitis. These diseases are not life-threatening; however they significantly impact the quality of life and in the absence of appropriate treatment, as the diseases advance, they might give rise to secondary bacterial infections, endogenous intoxication and other life-threatening complications. A.D. Ado and P.K. Bulatov regard these diseases as "pre-asthmatic state," labeling them as "risk factors for asthma development."[14] Our research on these diseases has also lead us to view that they are "a pre-clinical stage of asthma characterized by latent hyperactivity of the bronchial receptor system (BRS) underlain by an allergic, acute, chronic respiratory disease or by a respiratory disease of complex origin." To objectify the differential diagnostics of this asthma stage we performed spirometry with loading tests that detect latent bronchial hyperactivity in patients that belonged to risk groups (Komov, 1991a). For the load tests, bronchoconstrictor drugs were used per *inhalationem* or, in patients with supposed exercise-induced asthma, physical exercise was used. To detect aspirin-induced asthma, a prostaglandinF2α (PgF2α) drug was used. Sensitivity index (IS) was calculated as follows: **SI = lg (ΔFEV1/threshold loading dose)**, where ΔFEV1 is the difference between forced expiratory volume (FEV) at 1 second before and after load, and "threshold loading dose" is the level of drug or physical exercise that affected the FEV (Didkovskij *et al.*, 1988). Our data shows that pre-asthmatic state progresses into asthma in more than 50% of patients that receive pharmacologic therapy indicated by guidelines.

3.1. *Non-selective Hemosorption*

Pharmacologic therapy had low efficiency in treatment of pre-asthmatic state, as in most patients treated this way the disease progressed into asthma. Therefore, we included efferent therapy into the treatment. Later we conducted a retrospective study where we assessed the effectiveness of HS and PA in this group of patients. The results of three years of follow-up were compared in two groups of patients. In the first group ($n = 22$) efferent therapy was included into the treatment (14 patients were treated with hemosorption and eight patients were treated by plasmapheresis); in the second group ($n = 24$) efferent therapy was not used. Patients in two groups were matchable for sex, age and baseline clinical findings. After three years of follow-up, pre-asthmatic state progressed into asthma (episodes of shortness of breath were detected) in four cases in the first group (18.2%) and in 20 cases in the second group (83.8%) (Komov *et al.*, 2010a).

3.1.1. *Clinical case no. 1*

Patient K.B., born 1955, presented in May 1988 with severe rhinitis and lacrimation, throat tickling and episodes of coughing. These symptoms had been present for five years, appearing at the end of March–start of October and continuing until September. H-1 blockers and adrenomymetics (Suprastin, Galazolin, Naftisin, etc.) brought only partial relief and their effect lasted fever than 60 min. Eosinophilia (10%) and increase in IgE levels (153 ME/ml) were notable in laboratory findings. Bronchial hyperreactivity was detected in spirometric loading tests with Obsidan, Acetylcholine and physical exercise (SI was 2.4, 1.6, 1.9, respectively). Hypersensitivity to house dust (4 series) and other household antigens, as well as to pollen of hazel, alder, birch and cereals was detected in scarification tests (to 22 allergens in total, 64 "+" in total). Episodes of shortness of breath were not detected. Lung auscultation did not reveal wheezing. Based on the amnestic data, physical examination, eosinophilia, bronchial hyperreactivity and signs of polyvalent allergy the patient was diagnosed with "pre-asthmatic state, atopic form; pollinosis; sensibilization to household, food and pollen allergens. Pre-asthmatic state for exercise-induced asthma." After hemosorption eosinophil blood level decreased to 5% IgE levels decreased to 86 ME/mL, SI to Obsidan, Acetylcholine and physical exercise decreased from 2.4 to 0.5, from 1.6 to 0.4 and from 1.9 to 0.1, respectively (N = 0). Recurrent scarification tests detected hypersensitivity to 15 allergens in total and there were 37 "+" in total. Spring and summer allergic episodes became less frequent and less severe. In March 1989 the patient underwent hemosorption again. In 1989 allergic reactions were not frequent and were easily controlled by episodic intake of H1-blockers and deconjestants. The patient was receiving HS procedures on the 1st–15th March for 26 years. Twice we tried to stop using HS which didn't affect the severity of symptoms in the according year. However, in the spring of the following year the patient experienced a worsening of vasomotor symptoms of pollinosis and HS use was resumed. A plasmapheresis procedure conducted at the patient's request was less effective, as the H1-blockers intake has increased after the procedure. No worsening of symptoms, progression of allergic reactions, of skin and bronchial hypersensitivity was detected during the follow-up. Scarification tests in 2007 detected hypersensitivity "+ +" to 11 food and household allergens. No shortness of breath episodes were detected during the follow-up.

3.1.2. *Clinical case no. 2*

Patient Ts.V., born 1955, presented in November 1986 with frequent episodes of shortness of breath and episodes of severe asthma fewer than four times a week. The patient had experienced these episodes at home, after physical exercise, in forests. In spring and summer severe rhinitis and lacrimation were present. The patient reported that he lived in an old rural house with an oven. On examination, a reversible obstruction of terminal and medium bronchial tree, sensibilisation to household allergens (dust, mold) and pollen (hazel, birch, cereals), bronchial obstruction induced by physical exercise were detected. IgE level was 156 ME/ml. The patient was taking H1-blockers intranasally and intraconjunctivally, as well as sympathomimetic drugs in inhalations. The patient was diagnosed with "Atopic bronchial asthma of moderate severity: sensibilisation to household and pollen allergens. Pollinosis. Exercise-induced asthma." In December 1986 the patient underwent HS: 3600 ml of blood (80% of CBV) were perfused through carbon-based sorbent SKN-1K (300 cm^3*). For the next year, episodes of shortness of breath were less frequent and less severe and there were no severe asthma episodes. In 1987 HS was conducted again, as a prophylactic measure. For the next 10 years there were almost no signs of shortness of breath, and other signs of allergy were significantly less severe. The patient was engaging in sport* *activities (jogging, swimming, football). He was still living in the old house and periodically engaged in carpet trade. No professional medical care was necessary. In 1997 the patient caught a cold and suffered a severe pneumonia. He started experiencing shortness of breath and severe asthma episodes once again. In addition to pharmacological therapy, the HS procedure was conducted in the patient for the third time. In six days, asthma episodes ceased. In April FEV was decreased by 27%. Bronchodilator drugs increased the FEV to 13% below normal. From that time the patient started experiencing shortness of breath in dusty places and during respiratory infections, indicating that his asthma became partially infection-mediated. Severe asthma episodes were very rare and were easily managed with sympathomimetic inhalations. In 2006 the patient suffered influenza and severe asthma episodes became as frequent as 2–3 times a week. Three weeks after reconvalescence HS was conducted for the fourth time, after which asthma episodes ceased, shortness of breath and other symptoms of allergy became mild and infrequent. There was only a limited need for supportive pharmacologic therapy. Two more HS procedures were conducted before August 2014 (in 2008 and 2012) as a prophylaxis for asthma exacerbations, as the patient was experiencing exacerbations of pollinosis at that time. The patient is employed and active. There are no asthma episodes, other symptoms of allergy are mild and infrequent and do not require systematic pharmacologic therapy.*

These cases demonstrate that sorption therapy has a high clinical and prophylactic effect in patients with asthma.

One of the most important mechanisms mediating therapeutic effects of efferent therapy in asthma is the decrease in skin and bronchial hyperreactivity. Figure 24.3 illustrates the effect of HS (A) and PA (B) on SI to various provoking factors: acetylcholine, nonselective adrenoblockers (Obzidan), PgF2α, physical exercise in patients with pre-asthmatic state. In all cases a decrease of SE is observed. Also there is a significant decrease of total "+" index detected by scarification tests.

In Fig. 24.4, the dynamics of IgE levels after HS and PA are plotted. There is a notable increase of IgE level in the one day after the procedure, although increased clinical manifestations of atopic reactions are observed only in 3% of patients after HS and in 22.7% after PA.

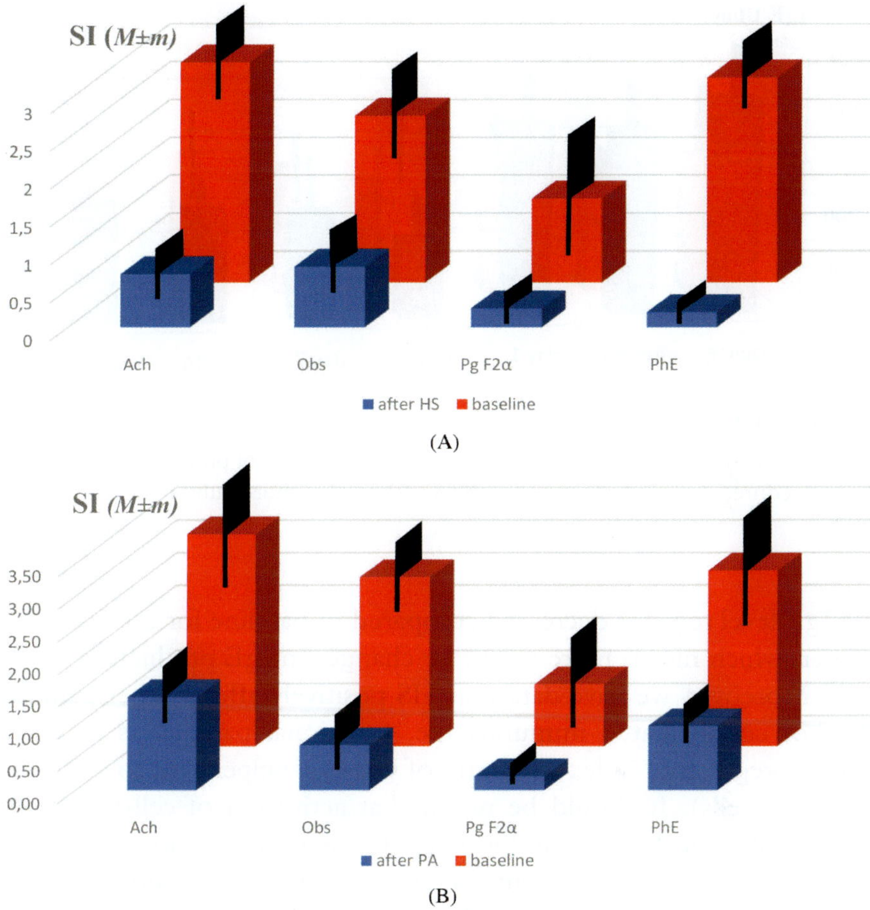

Fig. 24.3. The impact of HS (A) and PA (B) on sensitivity index (IS) of bronchial receptors to the mediators of bronchial constriction.

Vertical axis — sensitivity index. Horizontal axis — mediators of bronchial constriction: Ach — acetylcholine; Obs — obsidian; Pg F2α — prostaglandin F2α; PhE — physical exercise. Both methods significantly decreased the sensitivity of bronchial receptors to all mediators of constriction that were researched.

In many cases the allergic reactions are quite pronounced, sometimes reactions to new allergens are observed. After two days following HS, the IgE level significantly ($P < 0.05$) decreases.

Therefore, if the contact of patient with allergens after the procedure cannot be avoided, HS should be preferred over PA. Interestingly,

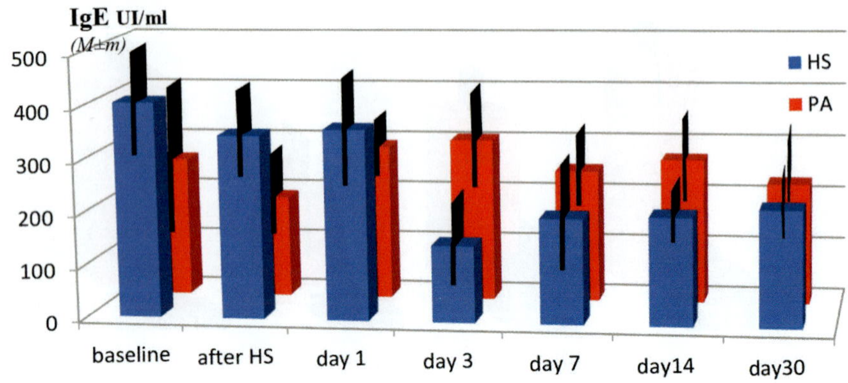

Fig. 24.4. The impact of hamosorption and plasmapheresis on IgE level. Vertical axis — IgE level (IU/ml). Horizontal axis — control points: baseline, after the procedure, day 1, dvels is more delayed but also more substantial and more stable.

the IgE level remains decreased compared to baseline for 30 days and longer, which means there might be changes in IgE production by B cells. Moreover, we have data that HS positively affects other parameters of cell-mediated immunity, i.e. functional parameters such as immunoregulatory index (the ratio of $CD4^+$ T helper cells to $CD8^+$ T cytotoxic cells). It should be noted that activation of cell-mediated homeostatic mechanisms induced by HS is more pronounced than that induced by other efferent therapy methods. Other authors confirm these data (Kostutchenko, 2003; Kirkovskij, 2012). Figure 24.5 illustrates changes in general fagocytic activity of neutrophils (GFAn) in a group of patients with asthma ($n = 57$) after HS. The fluctuations of mean parameters are not significant. However, in patients with increased GFAn at baseline ($n = 14$) and with decreased GFAn at baseline ($n = 21$) (compared to normal levels: $2.08 < n < 2.3$) there is a tendency of normalization of GFAn after HS that remains evident for 30 days. Endogenous cortisol, sex hormones, thyroid hormones, tetrazolium nitroblue test results and some other humoral and cell parameters have a similar dynamics.

One should bear in mind that HS might induce a short-term decrease in cell-mediated immunity that lasts up to 10–12 hr

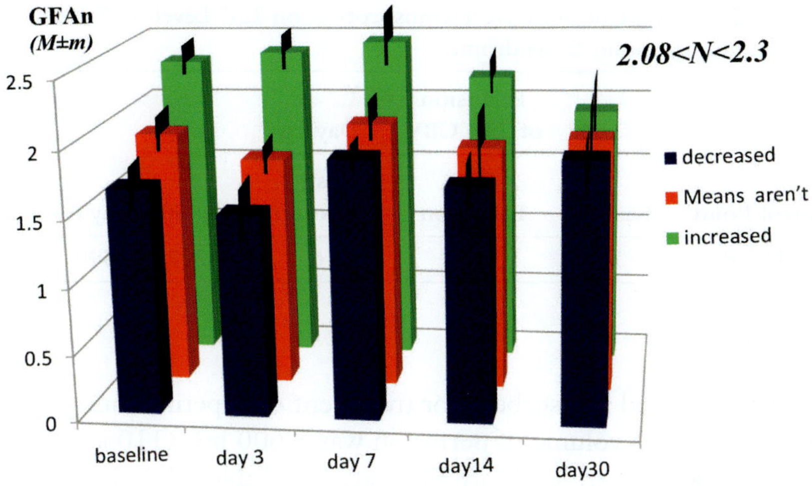

Fig. 24.5. The effect of hemosorption on general fagocytic activity of neutrophils (GFAn) in patients who had increased (green row) and decreased (blue row) GFAn at baseline. Means are not significantly affected (red row).

Vertical axis — GFAn (M ± m); horizontal axis — control points (days).

(Komov *et al.*, 2010a). Therefore, theoretically during the first hours after HS patients have increased risk of infections and/or exacerbations of chronic infections. These complications can be avoided by irradiation of 10–15% of perfused blood with low-intensity "red" laser ($W = 2$mW; $\lambda = 635$ ηm).

3.2. *Selective Hamosorption*

Modern medical science and pharmaceutic industry focuses on increasing the specificity of treatment to effectively target pathogenic mechanisms of diseases. In accordance with this tendency specific sorbents that allow for selective elimination of pathogenically important factors are created. As described earlier, the pathogenesis of allergic diseases is based on IgE-dependent mechanisms. A biologically specific "Anti-IgEsorbent" has been developed for treatment of these diseases by "Farmavit" scientific-industrial association, Belorussia. For the first

Table 24.2. Impact of Selective Hemosorption on IgE Level in Patient with Hyperimmunoglobulin E Syndrome

		Perfusion of 75% CBV	Day 14			
Control Point	Baseline	Pre-Perfusion	Post-Perfusion	Day 30	Day 60	
IgE, IU/ml	716	545	516	367	390	580

time we have used this sorbent for treatment of hyperimmunoglobulin E syndrome. The volume of perfusion was 5,600 mL (110% of CBV). The dynamics of IgE levels are illustrated in Table 24.2. Just as after non-specific HS, the effect fully develops only at the definite time after hemoperfusion and is maintained for significant periods of time.

On the second day after HS there was a significant decrease in itching of the skin and in hyperemia of the oral cavity and of the rectum. By the end of the first week, the patient reported significant subjective amelioration, improvement of appetite and of functional activity that was previously impaired because of pain syndrome. However after that a gradual increase in IgE levels accompanied by clinical impairment was observed, and the follow-up was terminated on the request of the patient.

Perfusion of 1 CBV through a specific sorbent leads to a 43.7 ± 14.4% decrease of IgE levels in patients with allergic symptoms and does not decrease levels of other immunoglobulins. A prompt interruption of allergic symptoms during the procedure is reached, and this is an important advantage compared to standard hemocarboperfusion. However, the duration of remission is 1–4 months, which is substantially less than that after non-specific HS. We compared the results of treatment in two clinically comparable groups of patients with allergic diseases. Patients of the first group ($n = 5$) were receiving antiIgE-sorption and for patients of the second group ($n = 17$), the blood was perfused first through non-specific carbon-based hemosorbent (VNIITU-I) and then through an antiIgE-sorbent. The results are plotted in Fig. 24.6.

For the first 14 days after the procedure, the elimination efficiency and the IgE dynamics did not differ significantly. However by day 30 a trend of increasing IgE levels was detected in Group 1, while in Group 2 they remained decreased. There was no effect on other immune parameters in Group 1, demonstrating a high specificity of treatment. In Group 2 there was a significant normalizing effect on circulating immune complexes level, on IgG and IgM levels, on some parameters of cell immunity and on cytokines. There were also other positive effects characteristic for hemocarboperfusion. The duration of remission in Group 1 was 2.2 ± 0.4 months. In Group 2 it was 16.6 ± 4.4 months. Moreover, in two patients in Group 2 the remission lasted more than 20 months. Therefore, antiIgE-sorption leads to a prompt but comparatively short-term effect. Still, in some cases, particularly in patients with polyetiological diseases (i.e. asthma, atopic dermatitis), a combination of non-specific and selective sorption is preferable, allowing for therapeutic effect on multiple pathogenetic mechanisms and for increased duration of remission.

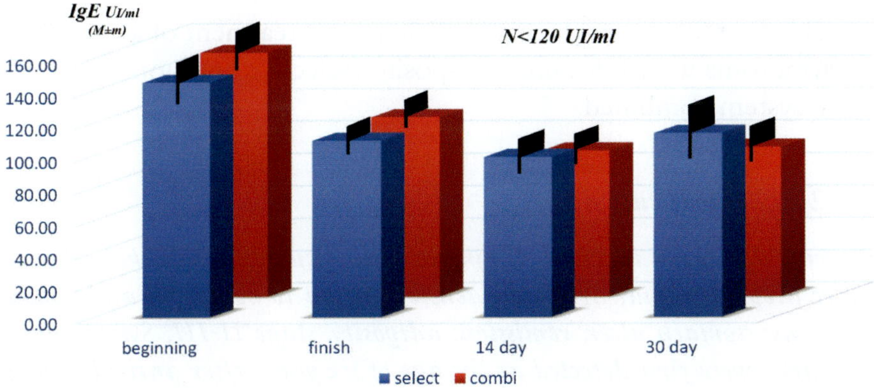

Fig. 24.6. The impact of sorption with biospecific (blue row) and combined (red row) sorbents on IgE level. By day 30 after HS with a selective sorbent a trend towards Ig accumulation could be observed. If a combined sorbent was used, the effect was more stable.

3.2.1. *Clinical case no. 3*

Patient H.L., 43 years old, presented January 15, 2012, with symptoms of urticaria that manifested a day before. The patient had experienced recurrent symptoms of urticaria for 3.5 years, and the symptoms became as frequent as 1–2 times a month. Total IgE level was 92–130 ME/mL; antigen-specific tests detected "++" reaction on equine epidermis, goat milk, shrimps. However the intake of shrimps never caused urticaria in the patient and she denied consuming goat milk or having contacts with horses. Specific HS was conducted, with perfusion of 3,600 mL of blood (100% CBV) through an antiIgE-sorbent. In 40 minutes after the start of perfusion (1,800 mL) the skin lesions vanished and the itching stopped. IgE level immediately before and after HS was 128 ME/mL, and on day 16 it was 78 ME/mL. Despite prompt clinical and laboratory effect, there was a relapse of urticaria on March 24th. A second HS procedure was conducted, with successive perfusion through a carbon sorbent (VNIITU-I) and a selective sorbent. The skin lesions vanished after the perfusion of 2,000 mL of blood. The total volume of perfusion was 4,100 mL, the IgE level decreased from 103 to 87 ME/mL and consequently remained normal. There were no allergic reactions by August 2014.

This case illustrates the high efficiency of treatment of acute allergic symptoms with HS using composite (selective and non-selective) DHP system combined.

3.2.2. *Clinical case no. 4*

Patient U.L., born in 1957; diagnosed with recurrent Quincke's edema; severe atopic asthma; urticaria; sensibilisation to household and pollen allergens; stomach ulcer, remission; adiposity, stage II-III. Symptoms of urticaria were first detected at the age of six years after antirabic vaccination. From that time on there was a recurrent skin rash and paroxysms of coughing. The patient never underwent medical examination for her symptoms and did not receive systematic treatment. In 1981 after vaccination she suffered an anaphylactic reaction, Quincke's edema and urticaria. A high IgE level was detected. Recurrent urticarial reactions *and Quincke's edema were managed by intravenous and oral steroids that caused an increase in body mass and an exacerbation of gastritis. Exacerbations of allergic symptoms were sometimes provoked by psychological stress and sometimes there were no detectable provoking factors. In 1997 when the patient was receiving a course of steroid therapy there was a first manifestation of gastric ulcer that required hospitalization. After hospital treatment the patient contacted the laboratory of clinical immunology to receive treatment for allergic symptoms. There was a notable increase in CICs (129 conventional units* $*10^3$*, 35 < N < 85) and normal IgE level (57 ME/mL, N < 120 ME/mL).*

Taking into account that the patient was in remission for gastric ulcer, extracorporeal therapy was prescribed, including two procedures of discrete PA with total exfusion of 1,900 mL of plasma and one procedure of HS (hemosorbent Carbocol 200 ml, volume of perfusion 4,150 mL of blood). Following the treatment, there was a remission that lasted more than three years. Consequently the patient experienced exacerbations once in 8–12 months, usually manifesting as urticaria and/or edema of lips and face. The exacerbations were managed with H1-blockers and 1–2 intravenous infusions of steroids (equivalent to 60–90 mg of prednisolone) or by extracorporeal therapy. One of the episodes is described more extensively. 22.04.2009 the patient experienced an edema of lips and face while being on duty. She received 12 mg of dexamethasone intravenously. On examination 27.04.2009 there was a significant increase of eosinophil cationic protein (ECP) to 146 ng/mL (N < 24 ng/mL), IL-8 and TNF, while IgE level was normal. Oral metypred 8 mL a day was prescribed, however on the next visit, 13.05.2009, the patient reported skin hyperemia and itching, an increase in IgE, IL-8, TNF levels. Therefore, 18.05.2009 a HS procedure was conducted. The dynamics of some parameters is shown in Table 24.3. The data shows that immunoglobulin changes during HS procedure.

On follow-up, 20.05.2009 there were no skin lesions and no edema, and there was a significant positive dynamics of IgE level, ECP, IL-8. There was a stable clinical and laboratory remission as for 20.11.2009.

Consequently procedures of extracorporeal therapy were conducted in connection with exacerbations. Oral metypred in doses up to 12 mg a day generally caused gastrointestinal adverse reactions and increases in body

Table 24.3. Laboratory Findings during Exacerbation in Patient U.L.

				18.05.2009 Hemosorption (VNIITU, Perfusion Volume 6 L)	20.05.2009		
Name	Normal Range	27.04.2009	13.05.2009	Baseline	After HS	28.05.2009	
Protein	66–87 g/L		69	74	67	75	70
Glucose	3.9–6.4 mmol/L	4.1	—	3.9	2.7	4.2	4.5
Ig A	0.8–2.5 mg/mL	1.19	1.23	1.49	1.24	1.20	1.21
Ig M	0.4–2.3 mg/mL	2.74	2.34	2.73	2.45	2.82	2.57
Ig G	7.0–16 mg/mL	11.21	10.40	12.48	11.52	10.50	9.79
Ig E	< 87 IU/mL	67.8	78.0	113.9	101.3	63.8	69.3
экБ	< 24 ng/mL	146.0	85.3	112	176	47.2	19.5
Il 6	< 4.1 pg/mL	< 2	< 2	2.43	2.37	< 2	< 2
Il 8	< 62 pg/mL	284.0	383.0	857	382	163.0	93.0
TNF	< 8.1 pg/mL	11.5	9.4	6.5	6.3	11.7	8.0

mass, and the dose had to be diminished to 0. However the treatment was not systematic, and there was a gradual decrease of remission time and an augmentation of severity of exacerbations. Now the dose has to be maintained at 5 mg a day to avoid relapses of Quincke's edema. Mean duration of period between courses of treatment has declined from 11.6 ± 1.3 months in 1997–2007 years to 5.4 ± 1.7 months in 2007–2014 years. Increased level of ECP (up to 176 ng/mL, N < 24 ng/mL) and recurrent increases in C-reactive protein and IgE levels (up to 563 ME/ml) are detected. It should be noted that relapses of Quincke's edema are not always accompanied by increases in IgE levels. This confirms that the disease has multifactorial origin in this case.

In total the patient received 26 courses of efferent therapy including combined and separate procedures of HS and PA. There is a statistically insignificant increase of intervals between treatment courses after combined therapy.

As of August 2014 the patient is employed and active. No severe life-threatening exacerbations have been detected.

This case illustrates that combination of extracorporeal and pharmacologic therapy allows to control severe allergic diseases for many years when standard therapy is not applicable or not effective.

3.3. *Plasma Sorption*

Plasma sorption has been considered in the literature (Kostutchenko, 2003; Voinov, 2010) as an alternative to HS for treatment of allergic diseases. We also have investigated this issue. This method is based on sorption of plasma instead of whole blood. The most important argument for plasma sorption is that this method is less damaging for blood cells than HS. However sorbents used in HS nowadays have high compatibility with blood elements and the increase in free hemoglobin during the procedure is no more than 3–5%. Furthermore our data does not confirm that plasmasorption (PS) has advantages over HS in treatment of allergic diseases. This is due to lower efficiency of cell-dependent mechanisms in PS. At the same time PS is more expensive and has higher technical requirements. Based on our clinical experience, we stopped using PS in treatment of allergic and autoimmune diseases.

3.4. *Enterosorption*

In some patients performing hemoperfusion is not recommended. In these cases sorbents can be given orally for treatment of chronic diseases (enterosorption, ES). Our experience shows that ES courses can be quite effective if the treatment is systematic, and such courses allow controlling asthma, pollinosis, recurrent urticaria and other allergic diseases. Frequency and duration of ES courses is determined by severity of the disease. Typically course duration is 14–30 days and the frequency of courses is 2–4 times a year. For prophylaxis sorbents are given in daily doses once a day in the evening. For treatment of acute conditions sorbents are given as often as three and more times a day. Daily dose is determined depending on the sorbent used and on the body mass of the patient. Nowadays there are effective oral sorbents available in Russia: VNIITU-2, Polyfepan, Enterosgel, Entegnin. Informed use of these sorbents allows to significantly increase the quality of life of patients with allergic diseases while the adverse reactions are minimal.

3.5. *Hemosorption Use in Treatment of Other Allergic Diseases*

The above material mainly deals with HS use in asthma. However there is a considerable experience for HS use in other forms of allergy: atopic dermatitis, urticaria, Lyell's syndrome, etc. Early use of HS in complex therapy allows increasing the effect of treatment, lowering the drug dependence of patients, lowering intoxication and decreasing duration of acute forms of allergic diseases. The tactics for use of sorption-based methods is similar for asthma and other allergic diseases.

3.6. *Adverse Effects of Hemosorption*

The literature, especially that printed in the '70s to early '80s, describes severe adverse effects of HS like embolism with fragments of the sorbent, air embolism, anaphylactic reactions during perfusion and in the early post-sorption period, septic complications. Our long-term experience of HS use allows us to declare with responsibility that all these reactions are iatrogenic. The technology used in modern sorption columns and the methodology for the procedure completely prevents air and sorbent embolism. The volume of sorption circuit is 75–150 mL depending on the size of sorption column; therefore critical decreases of CBV are impossible in patients with body mass more than 40 kg. Furthermore, blood exfusion is typically performed in an isovolumic manner: saline is infused in the volume similar to that of exfused blood. In 10–15% of cases after perfusion of 10–12% of CBV there is tachycardia and, less commonly, dizziness. We consider this as adaptation of vascular system to the connection of the "3rd circulation circuit" (extracorporeal circuit). These symptoms last for 15–30 sec and vanish spontaneously or are managed by temporally stopping the perfusion. They can also be prevented by lowering the speed of perfusion at the start of the procedure. Sometimes at 2–5 min after the start of perfusion there is hyperemia and fever of the face which are thought to be due to hemodilution and start of anticoagulant effect and therefore are not threatening and vanish spontaneously. In some cases, usually in asthenic and elderly people, there are orthostatic reactions immediately after the procedure. To prevent these reactions the patient should stay in a lying/sitting position for some time after the procedure and stand up with care. To prevent hypoglycemia due to carbohydrate sorption, the patient should drink a glass of sweetened warm tea or coffee after the procedure. A common reaction is fever up to 38.0°C and higher, starting in 1.5–2 hr after the end of HS. This corresponds to a release of cytokines into the blood, including pyrogenic ones. Typically the fewer is not subjectively distressing and vanishes spontaneously in 3–9 hr or, less commonly, requires an intake of antipyretic drugs.

Rarely pyrogenic reactions might develop during the procedure. In such cases the reactions are managed by intravenous infusion of 10 mL of 20% calcium gluconate solution or by 20–40 mg of prednisolone.

3.7. *General Considerations on the Use of Efferent Therapy in Treatment of Allergic Diseases*

Surveys of the literature allow making out four main trends of using efferent therapy in treatment of allergic diseases:

(1) Highly potent elimination of pathochemical factors from circulating blood that aims on depleting their reserves in the body and on preventing the replenishment of their reserves.
(2) Other efferent methods in severe disease, when the methods used are not effective.
(3) Periodic use for prophylaxis of disease progression.
(4) Use of efferent methods to manage exacerbations of chronic diseases.

Historically the efferent methods (hemodialysis, plasmapheresis, hemosorption) were used to treat intoxications of various origins; therefore the credo of extracorporeal detoxification was "More perfusion volume for more effect." Initially, the same approach was followed when use of HS started to be established in other branches of medicine, like allergology and immunology. Most recommendations for HS use in allergology and immunology that were written in the '70s–'80s prescribe courses of 3–5 procedures conducted at intervals of 48–71 hr, with perfusion volume of 1.5–2 CBV per procedure. In some patients the treatment caused definite disturbances of blood coagulation and of the endocrine system, as well as other adverse effects. Sometimes the severity of adverse effects matched that of the primary disease. Moreover, the effect of treatment was determined based on the degree of decrease of chosen laboratory markers. However the immune factors form multiple connections, therefore most immune functions can be compensated by alternative mechanisms of their activation. It is a well-known fact in allergology that the severity of atopic reactions does not always directly correspond to IgE levels and the functional state in asthma does not always correspond to the results of spirometry tests. The impossibility of explaining the effect of HS just by a decrease of certain laboratory markers is one of the main arguments used against this method nowadays.

The second trend — the use of efferent methods only in severe disease, when other methods are not effective, was also established historically when the efferent medicine was being formed. The consequences of efferent methods use in conditions of poor methodological and technical support suggested that HS is a high risk procedure. Blood compatibility of the first clinically available sorbents was critically low, therefore high doses of anticoagulants had to be used and in many patients massive hemolysis developed due to blood cell trauma. As a result, complications like pyrogenic reactions, bleedings and anemia after the procedure were common. Because of technical defects of sorption columns there was always the danger of sorbent fragments getting into the bloodstream. Due to these factors, HS started to be regarded as a "last-resort method" and the trend of "using HS when no other methods are effective" was established.

From the moment HS and PA found their place in treatment of asthma the programs for course treatment were constantly changing. We modified parameters such as the number of procedures in a course and the volume of perfusion (the volume of plasma exfusion in PA) per procedure and per course of treatment. We came to the conclusion that in most cases a single perfusion of 50–75% of CBV is sufficient for a notable clinical effect. A stable clinical and laboratory effect develops by the 10th–14th day after the procedure. A lot of patients note a temporary worsening of condition in the first 24–48 hr that is unspecific and is caused by factors such as improved rheological properties of the blood due to massive anticoagulant therapy and hemodilution, activation of stress-responsive mechanisms and compensatory release of catecholamines and other mediators into the bloodstream. In 15% of cases, mostly in patients with severe allergy, repeats of the procedure were required in 7–10 days. In case of PA, the optimal program was 2–3 procedures per course with total exfusion of 50–75% of CPV. Further increase of procedure impact did not lead to significant increases in clinical and laboratory effect but increased the risk of complications (Komov, 1991a). Our further clinical practice has proven these conclusions.

One of the most important questions in determining the program of treatment is "When?" After researching the dynamics of clinical and laboratory effects, we attempted to use HS and PA for prevention of seasonal exacerbations. We have found that conducting efferent therapy 20–25 days before the supposed start of seasonal exacerbations either prevents or significantly reduces clinical manifestations, i.e. rhinitis, conjunctivitis and urticaria. In this way it is possible to diminish drug use in the patients by 50–100%. The clinical effect of HS procedure or one PA course conducted as described above was retained for the whole season including spring, summer and autumn.

In treating allergic diseases that are not characterized by seasonal exacerbation the optimal time for conducting efferent therapy is, in our opinion, the onset of exacerbation. The efferent therapy should be prescribed if standard therapy for 1–2 days does not control the clinical manifestations. In such cases HS is recommended, as PA can sometimes even aggravate the allergic symptoms, probably because of the rebound effect of IgE.

If the patient presents at the peak of exacerbation efferent therapy is recommended in cases when pharmacologic therapy is not effective. In such cases fractional PA should be preferred as it has a significant effect on the rheological properties of the blood. The alleviation of load on the small circuit of circulation is beneficial as well as moderate detoxification mediated by elimination of pathochemical agents including high-molecular substances. HS should be conducted only when the condition of the patient is stabilized, at the onset of reconvalescence period. This approach allows reaching a fuller and more stable remission.

According to our lengthy experience, treatment courses including both HS and PA procedures are often quite effective. Presumably this is because such treatment allows taking advantage of the effects of both methods. In severe cases when there is a marked tolerance to treatment or a history of such tolerance we reinforce the drug therapy by one HS procedure with perfusion of 60–100% CBV and one to two PA procedures with exfusion of 20–25% CPV per procedure. The sequence of procedures is determined by clinical symptoms. If allergic symptoms prevail, there are signs of intoxication and a history of high-dose drug therapy then HS with selective and non-selective sorbents is conducted first. After that one to two PA procedures are conducted with an interval of two to three days. The number of PA procedures is determined depending on the condition of the patient. If there are active infections, a common situation in severe dermatoses and asthma, than it is more beneficial to conduct PA first. The risk of aggravation of allergy is minimized by the drug therapy. After an interval of 48–72 hr a combined or non-selective HS procedure is conducted and then, depending on the clinical situation, PA is conducted once again. If the clinical course is severe, blood irradiation courses ($W = 2$mW; $\lambda = 635\eta$m; $t = 20$ min) are conducted between the procedures. Blood irradiation in these regimens has a significant anti-inflammatory and immunomodulatory effect and positively affects the rheological properties of blood and sputum.

In severe cases a method of combined therapy is recommended — extracorporeal immunopharmacotherapy. When preparing the sorption extracorporeal circuit, the sorbent is saturated with steroid hormones (Metypred 1,000 mg, Dexamethasone 100 mg) in saline. The drug molecules are bound by the sorbent and when blood is perfused through the sorption column, the drug interacts with blood cells. In blood probes taken from the deferent part of the circuit only traces of steroids are found. Perfusion of blood through sorbents modified as described allows managing the most severe forms of allergic reactions including Quincke's edema, Lyell's syndrome etc., not including adverse reactions of high-dose hormonal treatment.

When used as described above, efferent methods are highly effective in treatment of allergic diseases. In our practice the mean duration of remission in atopic asthma patients is 16.7 ± 4.2 months (four months to 10 years), in infection-mediated and mixed asthma it is 13.5 ± 3.6 months and in aspirin-induced asthma it is 11.4 ± 4.8 months. Use of HS in this group of patients significantly reduces the duration of hospital treatment and manifestations of bronchial and skin hypersensitivity. Furthermore, outpatient HS use substantially diminishes the need for hospital treatment. This program of treatment sustains full or partial remission for many years.

Conclusions

(1) Non-specific efferent therapy, especially hemosorption, is a highly effective method that has beneficial effects on basic pathogenic mechanisms of allergic reactions.
(2) The program use of HS controls the clinical course of allergic diseases, preventing severe forms of diseases, substantially reducing the need for drug therapy, in particular the need for steroid use and, moreover, diminishing the need for hospital treatment.
(3) Selective sorption is a highly effective method of intensive symptomatic therapy of IgE-dependent diseases. In treatment of allergic diseases, combinational use of selective and biospecific antiIgE-sorbents causes prompt and stable clinical effect.
(4) In mild cases of allergic diseases, sorption-based methods including enterosorption can be effectively used as a monotherapy. Treatment of severe cases requires also the use of other methods: plasmapheresis, rational drug therapy, laser irradiation of the blood, etc.
(5) In contrast to plasmapheresis, hemosorption is a protein-conserving method of therapy and can be used in patients with hypoproteinemia. Combined use of specific and non-specific hemosorption allows managing all forms of allergic diseases without resorting to protein-based plasma substitutes. With HS, a clinical effect comparable to that of fractional PA can be attained with smaller numbers of procedures.

References

Didkovskij NA, Treskunov VK, Chemeleva TK, *et al.* (1988) Haemosorption in treatment of the asthma beginning and nonsevere bronchial asthma. *Clinical Medicine* **12:** 53–57.

Fedoseev GB, Chlopotova GP. (2001) *Bronchial Asthma.* Meditcina, Moscow.

Gell PGH, Coombs PRA. (1964) *Clinical Aspects of Immunology.* Philadelephia, FA Davis Co.

Khaitov RM. (2013) *Immunology Manual for Physicians*, 2nd ed. Medpress-info, Moscow.

Kirkovskij VV. (2012) *Physico-chemical methods of homeostasis correction* (monograph). Russian Physician, Moscow.

Komov VV. (1991a) Comparative assessment of haemosorption and plasmapheresis efficacy in the treatment of bronchial asthma on early stage. PhD thesis, Moscow.

Komov VV. (1991b) Haemosorptional methods of treatment. (Haemosorption) In: *Blood's Correction with Transfusion (Manual for Physicians)*, Ragimov AA (ed.) Practical Medicine, Moscow, pp. 198–221.

Kostutchenko AL (ed.) (2003) *Efferent Therapy* (monograph). Edition "Foliant," Saint Petersburg.

Komov VV, Kablachova NA, Kompanejetz IA, Jelkin DG. (2010). Nonspesific haemoperfusion: Realities and perspectives. *Vestnik Rossiiskoi akademii meditsinskikh nauk* **1:** 12–17.

Komov VV, Mochekovskij SA, Fomtchenkova EA. (2010) Impact of haemosorption on plasma proteins. *Efferent, Physical and Chemical Medicine* **2:** 32–35.

Komov VV. (2012) Clinical possibilities of nonspecific haemosorption. *Efferent, Physical and Chemical Medicine* **2:** 20–24.

Lopukhin JM, Molodenkov MN. (1985) *Haemosorption* (monograph), 2nd ed. Medecine, Moscow.

Nikolaev VG. (1984) *Method of Haemocarboperfusion in Experiment and Clinic* (monograph). Kiev, 359 pp.

Rajan TV. (2003) The Gell-Coombs classification of hypersensitivity reactions: A re-interpretation. *Trends Immunol* **24(7):** 376–9.

Voinov VA. (2010). *Efferent therapy. Membrane's Plasmapheresis* (monograph), 5th ed. Treckpore Technology, Moscow, 365 pp.

Chapter 25

Immunoadsorption for Collagen and Rheumatic Diseases

Ken Yamaji* and Hiroshi Tsuda†

1. Introduction

The field of therapeutics has seen remarkable progress in recent years, which has made mainstream drug treatment possible for collagen and rheumatic diseases. However, treatment of intractable cases where drug effectiveness is poor is a challenge. Furthermore, organ damage, concurrent illnesses or allergic reactions make adequate drug therapy impossible. For such cases, therapeutic apheresis is very significant, and it is important how this should be valued related to drug therapies. Therapeutic apheresis for collagen and rheumatic diseases involves the removal of factors that cause and exacerbate the disease; the aim of immunoadsorption, in particular, is to improve the clinical condition of patients with autoimmune disease by selectively removing pathogenic immune-complexes and autoantibodies from their plasma.

In this paper, our discussion includes the indications for immunoadsorption, such as collagen and rheumatic diseases, the relevant conditions and types, as well as the latest understanding related to methods and clinical efficacy.

---

*Juntendo University, Department of Internal Medicine and Rheumatology, Japan.
†Juntendo Tokyo Koto Geriatric Medical Center, Department of General Medicine, Japan.

2. Indications and Target Conditions

Under the umbrella of collagen and rheumatic diseases, the main indications are considered to be rheumatoid arthritis (RA) and systemic lupus erythematosus (SLE). In Japan, immune-adsorption is often indicated for intractable conditions such as RA with vasculitis, and in the SLE group, lupus nephritis and central nervous system lupus. In addition, therapeutic apheresis is also indicated for intractable conditions that coexist with collagen and rheumatic diseases, including thrombotic thrombocytopenic purpura (TTP)/hemolytic uremic syndrome (HUS), macroglobulinemia, antiphospholipid syndrome (APS), macrophage activation syndrome (MAS), and hemophagocytic syndrome (HPS). However, for these conditions, therapeutic apheresis other than immunoadsorption (e.g., plasma exchange [PE]) is generally indicated.

3. Types and Methods

During therapeutic apheresis for collagen and rheumatic diseases, various methods can be selected depending on the condition, including PE, double-filtration plasmapheresis (DFPP), cryofiltration, immunoadsorption plasmapheresis (IAPP), leukocytapheresis (LCAP), granulocyte monocyte apheresis (GMA), endotoxin adsorption therapy, and polymyxin B-immobilized fiber (PMX). Here, however, we will discuss immunoadsorption.

Immunoadsorption is a method of plasma adsorption (PA) or plasma perfusion (PP). An anticoagulant is added to the blood extracted from the body, which is then separated into blood cells and plasma by centrifugal separation or membrane separation in a plasma separator. A plasma pump then takes the separated plasma to an adsorber unit, wherein specific pathogenic factors are removed by adsorption using material with high affinity to the specific plasma component. The "purified" plasma is then returned to the body, together with the separated blood cells (Fig. 25.1).

The following adsorption columns are used in Japan: Immusorba PH-350 and Immusorba TR-350 (both from Asahi Kasei Medical

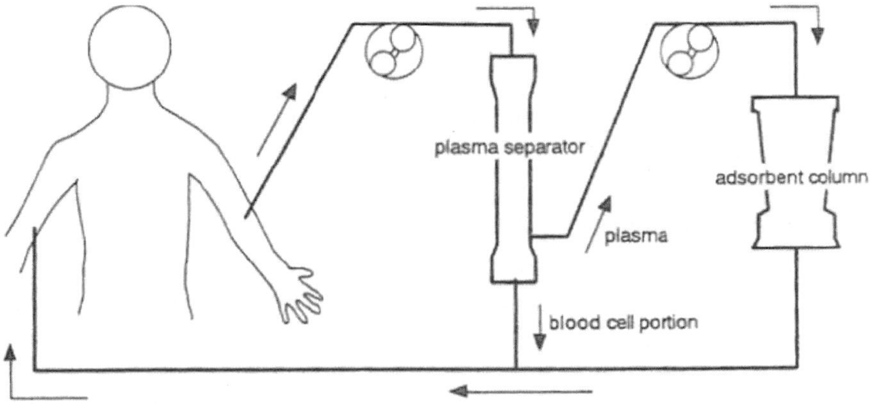

Fig. 25.1. Schematic diagram of extracorporeal immunoadsorption using membrane separation.

Co., Ltd., Tokyo, Japan) as well as Selesorb and Liposorber (both from Kaneka Medix Corporation, Tokyo, Japan). Immusorba PH-350 utilizes multiple interaction with hydrophobicity and ionicity of immobilized ligand; phenylalanine is immobilized as a ligand onto polyvinyl alcohol gel and rheumatoid factors such as immune complexes and anti-DNA antibodies are adsorbed by the column. In Immusorba TR-350, tryptophan is immobilized and acetylcholine receptor (AChR) antibodies are adsorbed. Selesorb and Liposorber utilize the electrostatic attraction between positively and negatively charged molecules; negatively charged polyanion dextran sulfate, used as a ligand, adsorbs immune complexes, anti-DNA antibodies, anticardiolipin (anti-CL) antibodies, and low-density lipopolysaccharides (LDL). Immusorba PH-350 and Selesorb are used in collagen and rheumatic diseases. In Europe and the US, the following adsorption columns are used: Prosorba (Cypress Bioscience, Inc., CA, USA), containing protein A immobilized on silica, which binds specifically to IgG and immune complexes. Agarose-immobilized antibodies are used in Immunosorba (Fresenius Hemocare, Bad Homburg, Germany), Therasorb (containing polyclonal sheep IgG), and Globaffin (containing synthetic peptide).

3.1. *Phenylalanine Columns*

Immusorba PH-350 (Fig. 25.2) is an adsorption column in which phenylalanine, a hydrophobic amino acid, is immobilized as a ligand on a porous, polyvinyl alcohol column (Fig. 25.3). Adsorption is mainly by hydrophobic bonding, but also includes some amount of electrostatic bonding (Fig. 25.4) (Hirata *et al.*, 2003).

This system effectively adsorbs rheumatoid factors, immune complexes, and anti-DNA antibodies, and is therefore used in the treatment of autoimmune disorders such as RA, SLE, and other collagen and rheumatic diseases. The Immusorba PH-350 column was shown to have an adsorption capacity equal to or greater than protein A sepharose CL4B (Pharmacia Fine Chemical AB, Uppsala, Sweden), which is known to strongly adsorb immune complexes, specifically IgG, in comparison to Immusorba PH-350 (Yamazaki *et al.*, 1986).

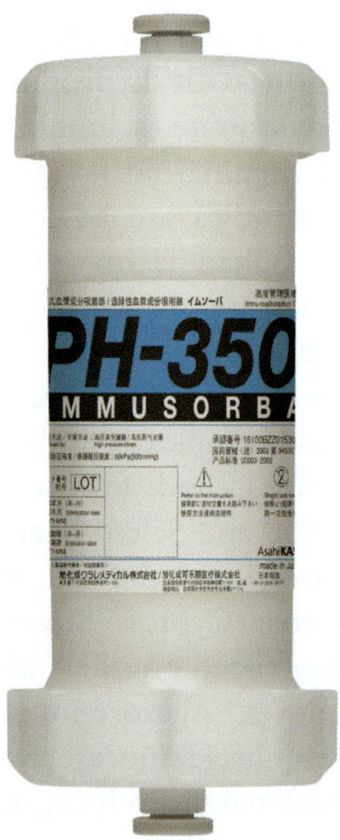

Fig. 25.2. Immusorba column.

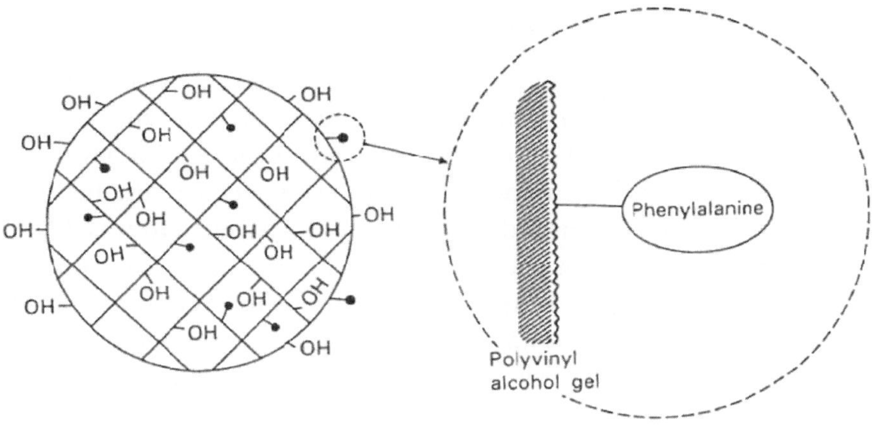

Fig. 25.3. Diagrammatic illustration of chemical structure of Immusorba PH-350. (Reproduced from literature with some alterations.)

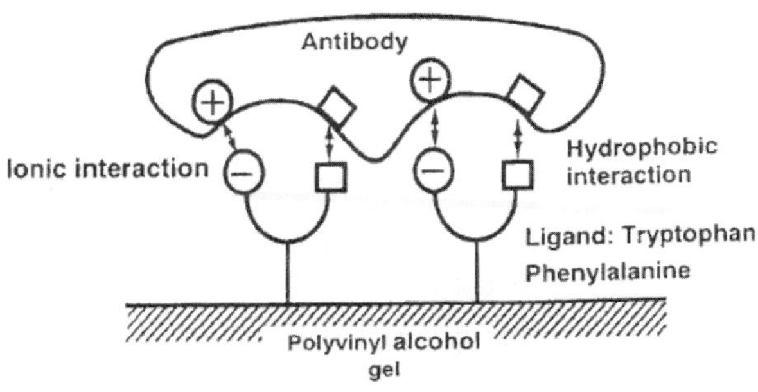

Fig. 25.4. Schematic illustrations of interaction between antibody and Immusorba column. (Reproduced from literature with some alterations.)

The volume of plasma processed and the intervals between treatments differ depending on the clinical condition. However, the targeted plasma processing volume is approximated from the patient's body mass and tends to be between 2 and 3 L. The treatment is usually administered once a week, but the frequency may increase to two or three times a week during acute clinical exacerbation, with the intervals between treatments gradually prolonged as per clinical outcome, after which the patient is transitioned to maintenance treatment at a frequency of two to four times a month.

3.2. *Dextran Sulfate Columns*

Selesorb (Fig. 25.5), manufactured by Kaneka Medix Corporation, Tokyo, is an adsorption column that utilizes dextran sulfate as a ligand. It adsorbs positively charged anticardiolipin antibodies, anti-DNA antibodies, and immune complexes using electrostatic binding (Fig. 25.6). Liposorber uses dextran sulfate as a ligand and was developed to adsorb LDL; it has a smaller pore size and higher adsorption of autoantibodies. The adsorption selectivity of the Selesorb column to plasma components is as follows: adsorption efficiency for anti-DNA antibodies at 38%, anti-cardiolipin antibodies at 53%, and circulating

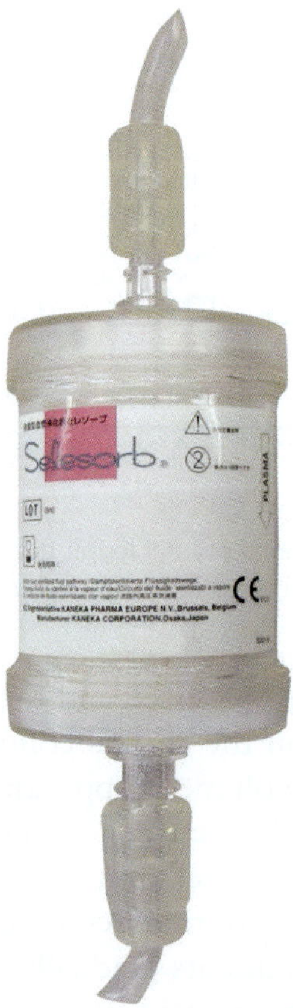

Fig. 25.5. Selesorb column.

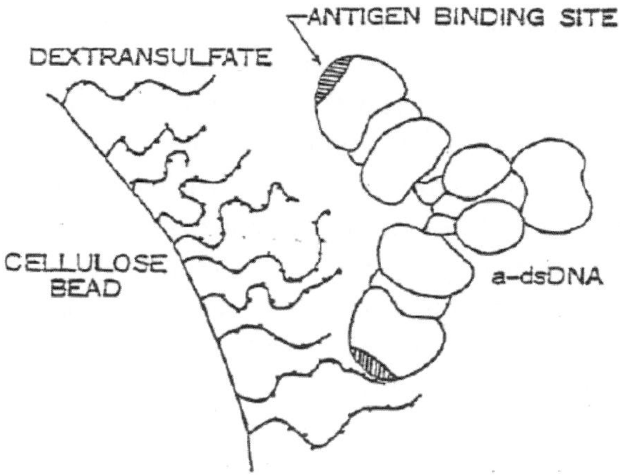

Fig. 25.6. Schematic illustrations of the binding between a dextran sulfate cellulose bead and anti-DNA antibody. (Reproduced from literature with some alterations.)

immune complexes at 74%, while total protein, albumin, IgG, and C3 are only slightly adsorbed (Kutsuki *et al.*, 1998).

The volume of plasma processed and the intervals between treatments differ depending on the clinical condition. However, the targeted plasma processing volume is approximated from the patient's body mass and tends to be between 2,000 and 3,000 mL. The treatment is usually administered once a week, but it is sometimes administered two or three times a week during periods of acute clinical exacerbation, with the intervals between treatments being gradually lengthened in line with evaluation of the clinical impact, and the patient transitioning to maintenance treatment of two to four times a month.

3.3. *Protein A Columns*

The two extracorporeal protein A columns, the Immunosorba and Prosorba columns, utilize highly purified protein A covalently coupled to a relatively inert solid phase matrix. Although these devices differ in design and utilization, both have demonstrated clinical effectiveness in several autoimmune illnesses, with manageable adverse effects. Immunosorba by Frescenius Hemocare, Bad Homburg, Germany uses staphylococcal Protein A as a ligand, which binds to an FcY site on IgG and, via alternative binding sites, also to IgM and IgA. The columns contain purified Staphylococcus protein A covalently attached to a Sepharose® bead support matrix. The column has been licensed in the US for the treatment of immune-mediated haemophilia. Prosorba by Cypress Bioscience, San Diego, CA, USA contains an adsorbent that consists of silica beads as the base material and highly purified staphylococcal protein A as the ligand. In these columns, protein A has a high affinity to IgG, rheumatoid factors, circulating immune complexes which may play a role in pathogenesis in rheumatic diseases. Although promising clinical responses were observed, there was also substantial toxicity. We now understand this toxicity to be due to leakage of protein A and enterotoxins from the formalin-fixed *Staphylococcus aureus*. As a result, more stable protein A matrices were developed.

3.4. *Ig-Columns*

Ig-Therasorb by Miltenyi Biotec, Bergisch Gladbach, Germany uses polyclonal sheep IgG which are directed against the human Ig κ and λ chain and against the IgG heavy chain. The sterile and pyrogen-free Ig-Therasorb glass column contains polyclonal sheep anti-human IgG antibodies (heavy chain- and light chain-specific) conjugated to cyanogen bromide-activated Sepharose beads. The anti-Ig column with a total volume of 300 ml removes specifically IgG (subclasses 1–4), IgM, IgA, circulating immune complexes and fragments of immunoglobulins with an average Ig reduction of 60–70% in every cycle.

3.5. *Gam-Columns*

Globaffin by Fresenius Medical Care, Bad Homburg, Germany employs the synthetic peptide Gam 146 as the ligand, which shares the binding sites, at least partly, with protein A. Peptides as ligands for immunoadsorption exhibit several potential advantages over native proteins. Globaffin is capable of binding immunoglobulins independent from their antigen specificity and thus, applicable in transplant recipients and several antibody mediated autoimmune diseases. Among others, the most important disorders suitable for the treatment with Globaffin are rheumatoid arthritis, systemic lupus erythematosus, and acute renal transplant rejection.

4. Immunoadsorption Plasmapheresis to Treat Collagen and Rheumatic Diseases

4.1. *Rheumatoid Arthritis (RA)*

In Japan, the therapeutic apheresis method routinely used to treat rheumatoid arthritis is LCAP, and its clinical efficacy has been reported by a number of institutions (Kempe *et al.*, 2004; Onuma *et al.*, 2006; Eguchi *et al.*, 2007). Additionally, research has been carried out to clarify the mechanism, and the following have been implicated to contribute to clinical efficacy: (i) mobilization of juvenile cells from bone marrow, (ii) mobilization of activated leukocytes from the inflammation site, (iii) control of cell infiltration to the inflammation site, (iv) correction of cytokine imbalance, and (v) the removal of activated platelets. The therapeutic effect is believed to be a combination of these effects (Hidaka *et al.*, 1999, 2001; Yamaji *et al.*, 2002).

When it was first developed, studies on the effectiveness of immunoadsorption on disease activity were centered on RA joint symptoms and joint findings. Although clinical efficacy on par with that of PE or DFPP and cyrofiltration was confirmed, no evidence could be found indicating a significant reduction in disease activity. Generally, this technique has been used for RA with vasculitis symptoms such as mononeuritis multiplex, skin sores, subcutaneous nodules, episcleritis, pleurisy/pericarditis, myocarditis, interstitial pneumonia, and organ infarction, and has been associated with clinical efficacy (Kobayashi *et al.*, 1984; Kumagae and Inoue, 1985; Takahashi *et al.*, 1986).

In Europe and the US, the effectiveness of immunoadsorption on RA symptoms, especially using Prosorba®, has been reported.

Furst *et al.* (2000) carried out a double-blind, randomized, placebo-controlled study using immunoadsorption with Prosorba® in 99 rheumatoid arthritis patients who were refractory, with high disease activity despite treatment with at least two disease modifying antirheumatic drugs (DMARDs) including methotrexate. Significant improvement in accordance with the American College of Rheumatology (ACR) score, was observed, where 41.7% of the group that underwent immunoadsorption as opposed to only 15.6% of the control group, achieved 20% improvement (ACR 20%). Felson *et al.* (1999) carried out a double-blind, randomized, sham-controlled, multicenter trial in 91 patients similar to those in Furst's study, treated at the frequency of once a week for 12 weeks (total length of the trial 19 or 20 weeks). Upon evaluation at seven or eight weeks after treatment, significant improvement was shown, with 31.9% of the group that received immunoadsorption as opposed to only 11.4% of the sham-control group achieving ACR 20% improvement.

In recent years, due to the remarkable progress in drug therapy involving biological preparations, the prognosis for RA has improved dramatically. Current RA treatment targets clinical remission, thus curbing the progress of joint destruction and functional deficit is now possible. On the other hand, it is also true that some patients are unable to receive adequate drug therapy perhaps due to age, side effects, or concurrent illnesses. What is now required from immunoadsorption in RA treatment is superior long-term efficacy and safety for these patients.

4.2. *Systemic Lupus Erythematosus (SLE)*

A variety of autoantibodies are often found in patients with SLE, but central to the clinical condition is tissue damage due to anti-double-stranded DNA antibodies and their immune complexes. The site of tissue damage differs between patients, but as the name suggests, damage occurs in various organs including kidneys, skin, lungs, and the brain. In short, autoantibodies and immune complexes play an important role in the pathogenesis of SLE. The main significance of therapeutic apheresis in SLE lies in the removal of autoantibodies and immune complexes that are integral to the disease and in impeding the progress of tissue damage. Therefore, the conditions for which therapeutic apheresis is indicated are intractable conditions where establishing good prognosis is important, such as lupus nephritis, central nervous system (CNS) lupus, and blood abnormalities. For these intractable diseases, therapeutic apheresis is administered along with drugs such as adrenal cortical steroids or immunosuppressants. Autoantibodies such as immune complexes and anti-DNA antibodies are removed from the plasma as early as possible, curbing the progress of tissue damage, which is thought to influence long-term prognosis.

We have previously reported our findings regarding immunoadsorption in SLE using the Immusorba PH-350 column, which uses phenylalanine as a ligand (Hidaka *et al.*, 2001; Sugimoto *et al.*, 2006). A total of 29 patients with SLE accompanied by lupus nephritis were divided into four groups: one treated with steroid only; the second, with steroid-pulse therapy; the third, with immunoadsorption; and the fourth, with intravenous cyclophosphamide pulse therapy (IVCY). The immunoadsorption group showed a significantly higher rate of improvement in anti-DNA antibodies than the steroid-only group at two weeks post treatment, and this improvement was superior to that in the steroid pulse-therapy and steroid-only groups at four weeks. At this time point, it was also recognized that creatinine clearance was significantly better in the immunoadsorption group than the other treatment groups. Consequently, the immunoadsorption group achieved not only a reduction in their steroid dosage earlier that the other groups, but at four weeks, a significantly higher rate of dose reduction was demonstrated compared with the IVCY group.

Clinical efficacy has also been reported for immunoadsorption via a Selesorb column, which uses dextran sulfate as a ligand. Suzuki *et al.* (1991) administered immunoadsorption at an average of 3.7 times to 19 SLE patients with various organ pathologies such as lupus nephritis, CNS lupus, and blood abnormalities. The mean SLE disease activity index (SLEDAI), an indicator of SLE disease activity, fell significantly, from 10.2 to 4.2, and the removal rate for anti-DNA antibodies was reported at 34% after a single treatment administration.

In Europe and the US, clinical efficacy has been reported for immunoadsorption using a Therasorb column that contains polyclonal sheep IgG. Graninger *et al.* (2002) reported the efficacy of immunoadsorption using the Therasorb column in SLE 5 patients with intractable comorbidities: one with gastrointestinal hemorrhage due to vasculitis, one with pneumonitis and carditis, one with hemolytic anemia and severe leucopenia, one with CNS involvement and thrombopenia, and one with pancytopenia. Graninger's group confirmed remission in three patients and an improvement in blood abnormality in the case with severe leucopenia and hemolytic anemia. For the patient with gastrointestinal hemorrhage due to vasculitis, bleeding stopped at one week after the commencement of IVCY and immunoadsorption. Side effects prompted an end to IVCY after two cycles, but even after terminating therapy, the situation remained stable. However, gastrointestinal hemorrhage reoccurred a year later, and the patient died from septicemia caused by a new cycle of IVCY. Further, Biesenbach *et al.* (2009) studied eight lupus nephritis patients treated with immunoadsorption using a Therasorb column. Their condition improved from low to moderate disease activity. In the course of the study, six of the eight patients switched to the Immunosorba column and the remaining two to Globaffin column. Even after the switch, quantities of IgG, IgM, and anti-dsDNA antibodies continued to fall and the level of disease activity remained low in all patients. Efficacy was reported in all cases, even in those with high disease activity whose first-line extracorporeal treatment was Immunosorba column (six cases) or Globaffin column (one case).

While previous reports showed clinical efficacy during acute disease, we (Yamaji *et al.*, 2008) reported clinical efficacy over the long-term using therapeutic apheresis during acute disease. A total of 38 lupus nephritis patients with high disease activity treated with large doses of steroids were divided into three groups: the first received therapeutic apheresis (PP group), the second received IVCY (IVCY group), and the third received both, therapeutic apheresis and IVCY (synchronized-therapy group). Long-term outcomes over at least five years were assessed, and the rate of complete remission was higher in the synchronized-therapy group than in the PP or IVCY group, and the rate of recurrence was also lower. We believe that this demonstrates that early multi-faceted treatment has a positive impact on long-term outcomes.

The effect of therapeutic apheresis in SLE is attributed to early removal of factors linked to pathogenesis, such as anti-ds-DNA antibodies and immune complexes to improve clinical condition and prevent the progression of tissue damage. Appropriate treatment during acute clinical phases not only makes improvement in the long term possible, but dosage of drugs such as steroids and immunosuppressants is also likely to be reduced. Thus, therapeutic apheresis can be considered to have strong clinical significance.

5. Summary

Drug therapy for collagen and rheumatic diseases has shown remarkable progress in the recent years, such as development of immunosuppressants, targeted drugs, and biological agents, thus assuring improvement in prognosis. Nonetheless, several exceptions exist, such as (i) intractable cases where drugs have little effect; (ii) hard-to-treat cases, where adequate drug therapy is not possible, perhaps due to organ pathology, side effects, allergies, infections, age, or pregnancy; and (iii) cases associated with intractable clinical conditions. As pathogens and factors related to pathogenesis are gradually clarified, clinical efficacy can be expected from therapeutic apheresis that specifically removes these factors, combined with drug therapy. In addition to the therapeutic efficacy of the process itself, the impact of drug therapy can be increased, and a reduction in dosage of drugs that cause significant tissue damage can be effected. Therefore, the process has great clinical significance.

Immunoadsorption, in particular, unlike plasma exchange and DFPP, utilizes a high-affinity column that selectively removes autoantibodies and immune complexes, leaving other plasma components intact. There is no need to replenish fresh frozen plasma or blood products such as albumin and gamma globulin preparations. Immunoadsorption is thus superior in terms of safety, as the risk of infection or allergic reaction relating to these preparations can be avoided.

We anticipate future investigations of application of synchronized therapy using drugs and therapeutic apheresis, most notably immunoadsorption, in combination to treat intractable clinical conditions such as connective tissue and rheumatic diseases.

References

Biesenbach P, Schmaldienst S, Smolen JS, *et al.* (2009) Immunoadsorption in SLE: Three different high affinity columns are adequately effective in removing autoantibodies and controlling disease activity. *Atheroscler Suppl* **10(5):** 114–121.

Eguchi K, Saito K, Tanaka Y, *et al.* (2007) Enhanced effect of high-dose leukocytapheresis using a large filter in rheumatoid arthritis. *Mod Rheumatol* **17:** 481–85.

Felson DT, Lavalley MP, Baldassare AR, *et al.* (1999) The prosorba column for the treatment of refractory rheumatoid arthritis: A randomized, double-blind, sham-controlled trial. *Arth Rheum* **42(10):** 2163–2159.

Furst D, Felson D, Thoren G, *et al.* (2000) Immunoadsorption for the treatment of rheumatoid arthritis: Final results of a randomized trial. *Ther Apher* **4(5):** 363–373.

Graninger M, Schmaldienst S, Derfler K, Grainger WB. (2002) Immunoadsorption therapy (Therasorb) in patients with severe lupus erythematosus. *Acta Med Austriaca* **29:** 26–29.

Hidaka T, Suzuki K, Matsuki Y, *et al.* (1999) Changes in CD4+ T lymphocyte subsets in circulating blood and synovial fluid following filtration leukocytapheresis therapy in patients with rheumatoid arthritis. *Ther Apher* **3:** 178–185.

Hidaka T, Suzuki K, Kawakami M, *et al.* (2001) Dynamic changes in cytokine levels in serum and synovial fluid following filtration leukocytapheresis therapy in patients with rheumatoid arthritis. *J Clin Apher* **16:** 74–81.

Hirata N, Kuriyama T, Yamawaki N. (2003) Immusorba TR and PH. *Ther Apher Dial* **7(1):** 85–90.

Kempe K, Tsuda H, Yamaji K, *et al.* (2004) Filtration leukocytapheresis therapy in the treatment of rheumatoid arthritis patients resistant to or failed with methotrexate. *Ther Apher Dial* **8:** 197–205.

Kobayashi N, Okudaira K, Takahashi K, *et al.* (1984) Plasma cleaning using immunoadsorbent IM-P for patients with rheumatoid arthritis. *Ther Plasmapheresis* **(IV):** 153–157.

Kumagae M, Inoue N. (1985) Rheumatoid arthritis — clinical trial with immunoadsorbent IM-P 350. *Immunohaematology* **7(4):** 448–451.

Kutsuki H, Takata S, Yamamoto K, Tani N. (1998) Therapeutic selective adsorption of anti-DNA antibody using dextran sulfate cellulose column (Selesorb) for the treatment of systemic lupus erythematosus. *Ther Apher* **2:** 18–24.

Onuma S, Yamaji K, Tsuda H, *et al.* (2006) Investigation of the clinical effect of large volume leukocytapheresis on methotrexate-resistant rheumatoid arthritis. *Ther Apher Dial* **10:** 404–11.

Sugimoto K, Yamaji K, Yang KS, *et al.* (2006) Immunoadsorption plasmapheresis using a phenylalanine column as an effective treatment for lupus nephritis. *Ther Apher Dial* **10(2):** 187–192.

Suzuki K, Kawagoe M, Horiuchi A, *et al.* (1991) Clinical evaluation of immunoadsorption in patients with systemic lupus erythematosus using two dextran sulfate cellulose columns (SL-01, Selesorb®) in an automated column regeneration unit. *Prog Med* **11:** 3007–3021.

Takahashi K, Yoshinoya S, Miyamoto T, Yoshizawa H. (1986) Comparison of therapeutic plasma exchange and hydrophobic amino acid adsorbent therapy in management of rheumatoid arthritis. *Nihon Naika Gakkai Zasshi* **75:** 362–374.

Yamaji K, Yang KS, Tsuda H *et al.* (2002) Fluctuations in the peripheral blood leukocyte and platelet counts in leukocytapheresis in healthy volunteers. *Ther Apher* **6:** 402–12.

Yamaji K, Kim YJ, Tsuda H, Takasaki Y. (2008) Long-term clinical outcomes of synchronized therapy with plasmapheresis and intravenous cyclophosphamide pulse therapy in the treatment of steroid-resistant lupus nephritis. *Ther Apher Dial* **12(4):** 298–305.

Yamazaki Z, Idezuki Y, Tsuda N, *et al.* (1986) Extracorporeal immunoadsorption. *Ther Apher* **(VI):** 113–121.

Chapter 26

Treatment of Rheumatoid Arthritis by Hemoperfusion

Hequn Zou, Xinyu Liu*

1. The Characteristics of Rheumatoid Arthritis (RA)

Rheumatoid arthritis (RA) is a chronic, systemic inflammatory disease characterized as symmetric synovitis (Harris, 1990). People of any age can be affected, but middle age onset cases are more often seen. Women are twice as likely to develop the disease as men. RA may cause joint destruction, deformity, disability, and even death if not properly treated (Hochberg, 1981).

RA is the most frequently encountered inflammatory rheumatic disease, with a prevalence of 0.5% to 1% in the adult population (Scott *et al.*, 2010). The outcome of this disease substantially impacts the patient and his/her family with large economic loss, poor quality of

---

*Institute of Nephrology and Urology, The Third Affiliated Hospital of Southern Medical University, Guangzhou 510630, Guangdong, China.

Correspondence: Hequn Zou, E-mail: hequnzou@hotmail.com, Prof. Hequn Zou, MD, Post member, International Society of Nephrology (ISN) Research Committee; Committee Member, Chinese Internal Medicine Association; Vice-president, Guangdong Internal Medicine Association; Director, Institute of Nephrology and Urology, The Third Affiliated Hospital of Southern Medical University.

life and frequent medical visits and hospitalizations (Filipovic *et al.*, 2011).

1.1. *Etiology and Pathogenesis*

The exact etiology of RA is not fully understood yet. Even though many factors including MHC-II antigens, inflammatory mediators, rheumatoid factors, cytokines, chemokines, proteolytic enzymes and superoxide have been demonstrated to be associated with the development of the disease. There is no consensus currently achieved on the pathogenesis of RA. It is speculated that the complex risk factors, including environmental factors (bacteria, mycoplasma or virus infections may, in a certain way, induce the development and progression of RA), hereditary susceptibility (DRB1 genes encoding the human leukocyte antigen (HLA)-DR4 and HLA-DR1 molecules are related to RA), immune system disorders (such as pathological activation of $CD4^+$ T cells) and so on play key roles in the development and progression of RA.

1.2. *Clinical Manifestation*

The duration and severity of RA vary among individuals. Some patients have long periods of remission with inactive arthritis and few or no symptoms. Other patients might present constant symptoms for months, even with acute exacerbation. Early RA affects small joints from fingers to hands and from toes to feet. The specific symptoms usually include morning stiffness, warm, painful and swollen joints. Along with the disease progression and constant active inflammation, tendon tethering and joint erosion, destruction will occur and deteriorate, eventually lead to joint deformity and disability. Besides these effects, some patients may experience multiple organ damage involving skin, lungs, heart, eyes, etc., due to the diffuse inflammation.

Rheumatoid factors (RF) are a variety of antibodies that are detectable in 70% to 90% of RA. However, RF can also present in individuals without RA, for example patients with chronic infection, e.g. hepatitis, other rheumatic or auto-immune diseases, or even normal elderly people. Generally, RA patients with negative serum rheumatoid factor have a less severe course. Mild to moderate anemia with normal RBC size often presents in most patients in the active stage of the disease, while white cell count is usually in the normal range. Eosinophils and platelets might increase in several cases. Serum IgG, IgM and IgA levels may increase, and complement levels may be slightly elevated or normal in RA patients. The indexes reflecting general inflammation, such as erythrocyte sedimentation rate (ESR) and C reactive protein (CRP), usually show increased level, but they are non-specific. Anti-cyclic citrullinated peptide antibody (anti-CCP) is an index which is of higher specificity to the diagnosis of RA.

1.3. *Diagnosis*

The 2010 ACR/EULAR Rheumatoid Arthritis Classification Criteria were released in 2010, which could replace the 1987 ACR criteria; it is suitable for early RA diagnosis. The new criteria established a point value between 0 and 10, and a patient with a point total of 6 or higher could be unequivocally classified as an RA case provided he/she had synovitis in at least one joint and no other diagnosis could explain the synovitis. A Disease Activity Score of 28 joints (DAS28) is widely used as an indicator of RA disease activity and response to treatment.

1.4. *Treatment*

General treatment of RA includes rest, joint immobilization during the acute stage, joint function training during the recovery stage, physical therapy, and so on. The target of treatment for RA is disease activity suppression to achieve clinical remission.

The classical medications for treating RA are NSAIDs, disease-modifying antirheumatic drugs (DMARDs), steroids, plant medicine, and so on. As reported in a Finnish study, remission was achieved in 35% of patients treated with DMARDs which started within four months of diagnosis, compared with only 10% of those who received treatment later (Mottonen *et al.*, 2002).

NSAIDs could relieve the symptoms of arthritics but could not stop disease progression, and DMARDs are simultaneously needed. The gastrointestinal side effects should be considered in every patient taking NSAIDs. Low-dose of glucocorticoids have disease-modifying potential and joint damage may increase in patients with glucocorticoids discontinuation (Kirwan, 1995; Hickling *et al.*, 1998). However, doctors should balance the benefits and adverse effects of glucocorticoids. Long-term glucocorticoid usage may induce osteoporosis, hyperglycemia, hypertension, weight gain, fluid retention, etc. Surgery is considered only for patients having unacceptable pain, limited function or loss of motion induced by severe structural joint damage.

Although many cases of RA could be successfully treated with the above methods and the disability rate and mortality rate were decreasing, refractory cases persist. Continuously applying large doses of the above drugs can bring significant side effects, as mentioned above. In recent years, hemoperfusion techniques have been used for the treatment of RA and other rheumatic and autoimmune diseases. Some pathogenic factors in the plasma of RA patients could be effectively removed by hemoperfusion, which may have promising prospects.

2. Terms and Technical Features of Hemoperfusion and General Efficacy on Rheumatic and Autoimmune Diseases

2.1. *The Term "Hemoperfusion"*

Technically, hemoperfusion includes both whole blood perfusion and plasma perfusion. Plasma perfusion is more often called plasma adsorption (PA). According to the mechanism of adsorption, as well as the features of adsorption columns, plasma perfusion can be divided into immunoadsorption and non-immune adsorption.

From another point of view, plasma perfusion belongs to plasmapheresis therapies which also include centrifugal plasmapheresis, plasma exchange (PE), double filtration plasmapheresis (DFPP), etc. Plasmapheresis represents a new series of blood purification techniques which was based on plasma separation (separating plasma from whole blood and driving back blood cells to the patient body) followed by removing pathogenic substances from plasma with various absorbers. Two modalities of plasma separating procedures are currently available. Centrifugal plasmapheresis could separate blood into different fractions with their specific gravity by using a normal centrifuge. Membrane filtration plasmapheresis could split cellular constitutes and plasma by applying a filter with a semipermeable membrane with appropriate pore size, and the next step is the infusion of equal quantity of fresh healthy plasma (PE), plasma component separation (DFPP) or plasma adsorption (PA).

Currently, various kinds of adsorbers for specific targets are applicable in clinic. Generally, the principle of plasma adsorption includes molecular sieve adsorption and immune adsorption. For the former, the adsorbents fixed in the adsorber carry a certain amount of electric charge and have definite apertures, for adsorbing plasma substances with appropriate electric charge and size based on the principle of the molecular sieve. While for the latter, pathogenic materials are adsorbed by means of specific antigen–antibody binding reaction.

Both centrifugal and membrane filtration plasmapheresis approaches are compatible with secondary plasma processing methods. Compared to the traditional plasmapheresis technique, plasma exchange (PE), plasma perfusion (plasma adsorption) could more specifically remove pathogenic substances, meanwhile minimizing the amount of plasma product usage.

2.2. *SPA Immunoadsorption*

SPA was found over 50 years ago as a component of the cell wall of *Staphylococcus aureus*. It was revealed that SPA consists of a cell wall binding structure with five domains named C, B, A, D and E from the cell wall side. Such a distinct structure feature could explain the multiple immunoglobulin-binding characteristics of SPA (Forsgren and Sjoquist, 1967).

Two kinds of SPA-based extracorporeal immunoadsorption system, namely SPA-silica (Prosorba) and SPA-Sepharose (Immunosorba), have been widely used for the main indications including rheumatoid arthritis, transplant rejection, etc. Both of the systems were applied after initial plasma separation, the first step of plasmapheresis therapies, therefore the adsorption systems need to be combined with the primary separator. Immunoglobulin and circulating immune complexes (CIC) could be adsorbed by the SPA column. Then purified plasma was infused into the bloodstream of the patients after converging with blood cells (Matic *et al.*, 2001). A major difference between Immunosorba and Prosorba is that the former has almost unlimited adsorption capacity. During treatment with Immunosorba, two columns work alternately, as one column is perfused with plasma when another one is on-line regenerated (Samuelsson, 1999).

In addition to the well-known ability of binding immunoglobulin, SPA could also stimulate immunocytes and induce cytokine release (Alonso and Daggett, 2000). The immunomodulatory effects of SPA immnoadsorption therapy have also been reported. It was shown that Immunosorba not only has good ability to remove IgG and CIC, but it also has the ability to affect the number of immune cells, as well as the Th1/Th2 cytokine expression. The plasma processing capacity of Prosorba during one time of plasma adsorption therapy is two liters and the treating time for each procedure is usually less than two hours, therefore IgG and CIC of a body could not be removed by adsorption itself as much as we expected. It is in accordance the IgG and CIC levels not showing a constant decrease in patients after SPA immnoadsorption therapy. Sometimes, there was no difference of either IgG or CIC level before and after the treatment, or the level could be even higher after therapy in some cases. Thus, it is speculated that the therapeutic mechanism for SPA immnoadsorption might be mainly its immunomodulation effect. The structure of IgG and CIC might be changed when passing the SPA adsorbent column, which results in the enhanced immunogenicity of Fab fragment. The ratio of antigen to antibody changed when plasma passed through the SPA column.

3. Application of Hemoperfusion Therapies in Treating RA

Plasma perfusion (plasma adsorption) can immediately be used to relieve severe diseases without considering the shortage of plasma resource, and there is less possibility of allergic reaction due to the infusion of heterogeneous plasma.

Hemoperfusion, either whole blood perfusion or plasma perfusion (plasma adsorption), has been widely applied in rheumatic and autoimmune diseases. A number of studies have shown that hemoperfusion, when combined with pharmacological treatment, is effective in treating severe or refractory rheumatic and autoimmune diseases and inducing remission (Gong *et al.*, 2013).

In our experience, the efficacy of hemoperfusion is significant on a variety of rheumatic and autoimmune diseases including RA (Liu *et al.*, 2012). In recent years, our blood purification center has applied hemoperfusion, including whole blood perfusion and the plasma perfusion (plasma adsorption), in the treatment of more than 100 cases of severe or refractory rheumatic and autoimmune diseases. One of the important reasons for completion of the therapies was the great support of the medical insurance system of our city (Guangzhou, China). In the process, we deeply recognized the great future of hemoperfusion. Meanwhile we preliminarily accumulated a series of experiences in preventing and treating hemoperfusion-related complications, which occurred more frequently than in hemodialysis. The high prevalence of complications during hemoperfusion is mainly due to the poor biological and chemical stability of plasma separators and adsorbents compared to dialyzers used in hemodialysis.

3.1. *Staphylococcal Protein A (SPA) Immnoadsorption in Treating RA*

By means of chemical cross-linking technique, Kolb *et al.* synthesized the IgG polymer in 1990. It showed a high RF clearance rate (reaching 94%) in an *in vitro* experiment. Nevertheless, this polymer is too tiny to be used in clinic. Joseph and his colleagues combined protein A with silica for removing IgG-RF and immune complex from RA patients (Balint, 1996). After several rounds of plasma adsorption therapy, the IgG-RF level was significantly decreased. Quanwen *et al.* compared five different kinds of adsorbents for removal of RF. Final results revealed that the adsorbent with phenylalanine directly linked to the cellulose beads via epichlordrin had the highest adsorption capability, even though the other adsorbents could also reduce the RF concentration in patient blood, which indicated that adsorbent with phenylalanine might be a promising one used specifically for RA patients (Matic *et al.*, 2000). The currently available immunoadsorbent columns clinically used in the treatment of RA are presented in Table 26.1.

For past years, the most commonly used adsorbent in clinic for RA is staphylococcal protein A (SPA), thus hereby introduced firstly is SPA immnoadsorption therapy.

The Prosorba system was first approved by the Food and Drug Administration (FDA) in the United States for treating TTP in 1987. It was then approved, in 1999, for treating refractory rheumatoid arthritis patients resistant to conventional medication therapy including DMARDs. The decision was mainly based on several prior clinical studies. In 1994, Wiesenbutter *et al.* (1994) reported

Table 26.1. Currently Available Immunoadsorbent Columns for Clinical Treatment

Column	Manufacturer	Ligand	Matrix	Target Substance
Prosorba	Fresenius HemoCare, Inc.	Protein A	Silica	IgG, IgA, IgM, IgM RF
Immunosorba	Fresenius HemoCare, Inc.	Protein A	Sepharose	IgG, IgA, IgM
Immusorba PH	Asahi Medical Co.	Phenylalanine	Polyvinyl alcohol	Immunocomplex, Rheumatoid factor
Immusorba TR	Asahi Medical Co.	Tryptophan	Polyvinyl alcohol	Anti-acetylcholine receptor antibody, Immunocomplex, Anti-DNA antibody

an open prospective trial about the evaluation of safety and efficacy of immunoadsorption therapy with Prosorba in the treatment of 11 refractory RA patients. After scheduled 12–15 times treatment over 9–15 weeks, significant improvement was observed in most patients, with two cases achieving long-term clinical remission according to the American College of Rheumatology criteria. A randomized, double blind, sham-controlled trial was performed by Felson DT and his colleagues (Felson *et al.*, 1999) in 91 refractory RA patients. After the completion of 12 weekly treatments, 31.9% of patients receiving Prosorba immunoadsorption therapy experienced improvement according to ACR response criteria, while only 11.4% of the patients in the sham-treatment group achieved improvement. Caldwell *et al.* (1999) reported another six months of open label pilot study which enrolled 15 refractory RA patients. The results showed that, with Paulus 50% criteria, nine patients improved at the fourth month after starting weekly immunoadsorption therapy, with one patient reaching improvement of >20% Paulus criteria.

3.2. *Other Immunoadsorbent Columns in Treating RA*

In 1982, Yamazaki *et al.* developed two types of physicochemical affinity immunoadsorbent which then became commercial products widely used in clinic for treating autoimmune diseases including RA. They first reported effective treatment of four cases of RA by using IM-TR and IM-PH. Their successful trails also open a new window for the application of plasma perfusion therapies in treating rheumatic diseases and even auto-immune diseases. The products were named Immunsorba PH-350 and Immunsorba TR-350, and are manufactured by Asahi Medical Co. Ltd (Tokyo, Japan). The specific hydrophobic amino acids such as phenylalanine (PH-350) or tryptophan (TR-350) are fixed on resin microspheres as ligands. The target substances were removed by the adsorbents through a hydrophobic bonding with the ligands. In theory, they are safer than other adsorbents since the ligands also exist in the normal human body. Thus there is no concern about the ligands leaching from the column into the circulation. Moreover, high efficiency in adsorbing rheumatoid factors and immune complex has been observed, comparatively more than protein A adsorbent.

We have applied the plasma perfusion therapy on 25 RA patients in recent years. After all sessions, the ESR, CRP, RF, IgG, IgA and IgM levels of the patients were significantly reduced, the numbers of joints with pain or swelling went down, and morning stiffness times were shortened. VAS score of rest pain, overall status evaluated by patients or doctors and HAQ score were all decreased. Two cases acquired remarkable improvement according to ACR 70 standard, 12 cases achieve improvement according to ACR 20 standard, and 11 cases showed a non-significant effect according to ACR 20 standard. The total effective rate was 56.00%.

We have also investigated the effectiveness of plasma adsorption on 21 SLE patients. The IgG, IgA and IgM levels of the patients were significantly decreased, while the decreased C4 level elevated. Anti-dsDNA antibody was positive in 11 patients before plasma adsorption and turned negative after the therapy. In addition, SLEDAI score, 24-hour urine protein amount, blood urea nitrogen, and serum creatinine levels were decreased compared to pre-treatment levels. Four cases (19.05%) showed excellent improvement with a decreased SLEDAI score of more than 5, nine cases achieved efficiency (42.86%) with a decreased SLEDAI score of 3–5, and eight cases showed non-significant effect (38.10%) with a decreased SLEDAI score of less than 3. The total effective rate was 61.90%. Adverse reaction occurred in two of the 21 cases, and the adverse reaction rate was 9.52%. The adverse reaction was respiratory tract infection in one case and hypertension in another one. The adverse reactions were slight. They were released after treatment and it did not affect the therapies.

We also treated 12 patients with vasculitis causing multiple organ injury by means of combination therapy with both plasma adsorption (weekly for three weeks) and drugs (with methylprednisolone 500 mg intravenously for three days followed by oral prednisone at a dose of 1 mg/kg daily, and meanwhile with oral cyclophosphamide of 2 mg/kg/day). The change in the 24-hour urine protein, Scr, and CRP levels during the period from the start of the treatment to one month after the treatment was significant. One month after the initiation of the treatment, six patients with acute kidney injury (AKI) recovered with serum creatinine in the normal range after one month, seven patients with acute heart failure saw complete recovery, and the patient with acute respiratory failure recovered after one month.

Fifteen patients with ankylosing spondylitis were also treated with plasma perfusion therapy in our center. After the treatment, morning stiffness time was shortened, Schober's test scores were increased, numbers of joints with swelling were decreased, and the ESR, CRP, IgG, IgA and IgM levels were significantly decreased with a significant difference, compared to pre-treatment levels ($P < 0.05$). Five cases got excellent efficacy (33.33%) according to ASAS 70 standard, seven cases got efficacy (46.67%) according to ASAS 20 standard, and three cases got non-significant efficacy (20.00%) according to ASAS 20 standard. The total effective rate was 80.00%. During the treatment, adverse reaction happened only in two cases, who got hypersensitive reaction, with an adverse reaction rate of 13.33%.

3.3. *Other Plasmapheresis Techniques in Treating RA*

A relatively new technique, double filtration plasmapheresis (DFPP) has recently been used in the treatment of RA. Plasma component separators with customized pore size could selectively remove large molecular weight substances, such as rheumatoid factor and IgM. Yu *et al.* (2007) conducted a controlled study to investigate the efficacy of DFPP in the treatment of RA patients in active phase. Eighty-two patients were assigned to two groups, the DFPP group (DFPP was performed once a week for two to three sessions) and the non-DFPP group, and all patients of the two groups were administered with sulfasalazine plus methotrexate for over three months. After the last treatment, 100%, 92.9% and 81.0% patients in the DFPP group got ACR20, ACR50 and ACR70 improvements. Comparatively, 17.5%, 0% and 0% of patients in non-DFPP group acquired such improvement — the difference was significant. More significant changes in Health Assessment Questionnaire scores and disease activity scores were also observed in the DFPP group compared to the non-DFPP group. After seven to 22 months follow-up after DFPP treatment, the improvement could be maintained. Liu *et al.* (1997) reported a study of 21 cases of RA patients treated by DFPP. It was shown that, after the DFPP therapy (an average of 3.7 ± 1.1 sessions per patient within 1–2 weeks), five out of eight patients with positive rheumatoid factor turned negative. Improvement of function capacities was observed in all patients.

To further investigate the effect of DFPP in patients with early stage RA, Yu *et al.* (2012) lately performed another comparative study which found that combination treatment with leflunomide, methotrexate (MTX) and DFPP therapy in patients with early stage highly active RA achieved greater improvement in synovitis and bone edema, compared to those who received only leflunomide and MTX. In additon, ACR20, ACR50, ACR70 and ACR90 responses and DAS28 remission rates in the DFPP group were also significantly greater than in the non-DFPP group. The results suggest that combination therapy with DFPP and DMARDs is superior to DMARDs alone for treating patients with early stage RA with high disease activity.

4. Complications of Hemoperfusion

In the last 10 years, more and more centers used plasma adsorption (the plasma perfusion) instead of whole blood perfusion to treat rheumatic and autoimmune diseases, as it has a higher clearance rate of target substances. However, it has been shown by our more than years of experience that plasma perfusion therapy may not be a relatively safe procedure if the medical staff perform the therapy without knowing well how to prevent the potential complications. Patients receiving plasma perfusion could suffer severe adverse effects including hypersensitivity, hypotension and shock. The shock could result from hypersensitivity, membrane rupture, hemolysis, etc. By now the complication rate of the plasma perfusion has become less than 10%, as our staff knows so well the prevention of adverse effects related to plasma perfusion (Liu *et al.*, 2012).

The most commonly reported adverse effects of blood perfusion therapy for RA patients were short term joint pain and swelling. Such arthritic flare could recover spontaneously without sequelae.

Apart from those, other complications are not specifically for RA patients, but could occur in all patients during the therapeutic plasmapheresis procedure. Most of them are related to the procedure itself, or to the replacement solutions. The main notable complications are listed below:

(1) Allergy: The main source inducing allergic reaction during the hemoperfusion therapy might include membrane incompatibility of plasma separator or plasma component separators, and ligands leaching from the column into the circulation. All this does not exist in traditional plasma exchange therapy. Therefore some steroids might be recommended, especially in patients having allergic history.

(2) Hypotension often results from inadequate volume resuscitation and has been reported in association with the development of myocardial ischemia and infarction during hemoperfusion. The use of angiotensin-converting enzyme inhibitors may increase the risk of hypotension during the procedure, because these drugs inhibit the metabolism of bradykinin, which is formed during column perfusion. Therefore, withholding the medicine prior to the treatment is necessary.

(3) Hypocalcemia sometimes happens during plasmapheresis, which may result in the prolongation of the QT interval and predispose the development of ventricular dysrhythmias. This effect may be potentiated by hypomagnesemia in patients with abnormal hepatic function. Calcium also plays an important role in myocardial and smooth muscle contractility. Any replacement fluid has possibility to cause ionized hypocalcemia because of citrate (FFP), direct sequestration (albumin), or dilution (crystalloid). Hepatic metabolism of citrate becomes bicarbonate ion which is subsequently excreted in urine. Patients with abnormal renal function may develop metabolic alkalosis after citrate administration, causing increased calcium sequestration by plasma proteins in exchange for hydrogen ions. Prophylactic infusing of a 10% calcium chloride has been demonstrated to decrease citrate-related complications to 1%.

(4) Plasmapheresis-associated coagulopathy is a significant perioperative concern that is often multifactorial in nature.
(5) Hemolysis happening during plasmapheresis is usually the result of membrane incompatibility or stability either physical or biochemical of plasma separator or plasma component, absorbents or ligands leaching into the circulation, high transmembrane pressure due to high concentration of some plasma components, or even fresh plasma.

5. Conclusions

Rheumatoid arthritis is one kind of autoimmune diseases with high prevalence worldwide, and patients with high activity and/or who are resistant to conventional treatment represent an obviously increased risk and economic burden. Although a number of new drugs, such as tumor necrosis factor alpha inhibitors, have emerged recently, there is currently also a demand for effective treatment strategies with reasonable side effects.

The efficacy and side effects of hemoperfusion therapies have been comprehensively evaluated by a number of studies. Consensus has been reached about the application of hemoperfusion therapies for highly active and/or refractory RA patients. With significant advances in the field, hemoperfusion technique has higher blood/plasma processing capacity, increased specificity, and fewer side effects. There is a trend recently that early application of proper hemoperfusion therapies for RA patients, even for those not in the high active stage, may also bring benefits. However, the optimal timing and dose of hemoperfusion treatment remain to be determined.

References

Alonso DO, Daggett V. (2000) Staphylococcal protein A: Unfolding pathways, unfolded states, and differences between the B and E domains. *Proc Natl Acad Sci USA* **97(1):** 133–138.

Balint JP. (1996) Immune modulation associated with extracorporeal immunoadsorption treatments utilizing protein A/silica columns. *Artif Organs* **20(8):** 906–913.

Caldwell J, Gendreau RM, Furst D, *et al.* (1999) A pilot study using a staph protein A column (Prosorba) to treat refractory rheumatoid arthritis. *J Rheumatol* **26(8):** 1657–1662.

Felson DT, LaValley MP, Baldassare AR, *et al.* (1999) The Prosorba column for treatment of refractory rheumatoid arthritis: A randomized, double-blind, sham-controlled trial. *Arthritis and Rheumatism* **42(10):** 2153–2159.

Filipovic I, Walker D, Forster F, Curry AS. (2011) Quantifying the economic burden of productivity loss in rheumatoid arthritis. *Rheumatology* **50(6):** 1083–1090.

Forsgren A, Sjoquist J. (1967) "Protein A" from Staphylococcus aureus. 3. Reaction with rabbit gamma-globulin. *J Immunology* **99(1):** 19–24.

Gong D, Ji D, Xu B, Liu Z. (2013) More selective removal of myeloperoxidase-anti-neutrophil cytoplasmic antibody from the circulation of patients with vasculitides using a novel double-filtration plasmapheresis therapy. *Therapeutic Apheresis and Dialysis* **17(1):** 93–98.

Harris ED. (1990) Rheumatoid arthritis. Pathophysiology and implications for therapy. *New Engl J Med* **322(18):** 1277–1289.

Hickling P, Jacoby RK, Kirwan JR. (1998) Joint destruction after glucocorticoids are withdrawn in early rheumatoid arthritis. Arthritis and Rheumatism Council Low Dose Glucocorticoid Study Group. *Brit J Rheumatol* **37(9):** 930–936.

Hochberg MC. (1981) Adult and juvenile rheumatoid arthritis: Current epidemiologic concepts. *Epidemiol Rev* **3:** 27–44.

Kirwan JR. (1995) The effect of glucocorticoids on joint destruction in rheumatoid arthritis. The Arthritis and Rheumatism Council Low-Dose Glucocorticoid Study Group. *New Engl J Med* **333(3):** 142–146.

Liu JD, Zhang C, Li WS, Lun LD. (1997) Double filtration plasmapheresis for the treatment of rheumatoid arthritis: A study of 21 cases. *Artif Organs* **21(2):** 96–98.

Liu X, Wang X, Li B, *et al.* (2012) Application of plasmapheresis in rheumatoid and auto-immune diseases. *Chin J Blood Purification* **11(12):** 646–649.

Matic G, Schutt W, Winkler RE, *et al.* (2000) Extracorporeal removal of circulating immune complexes: From non-selective to patient-specific. *Blood Purification* **18(2):** 156–160.

Matic G, Bosch T, Ramlow W. (2001) Background and indications for protein A-based extracorporeal immunoadsorption. *Therapeutic Apheresis* **5(5):** 394–403.

Mottonen T, Hannonen P, Korpela M, *et al.* (2002) Therapy, F. I. R. T. G. F. R. A. C., Delay to institution of therapy and induction of remission using single-drug or combination-disease-modifying antirheumatic drug therapy in early rheumatoid arthritis. *Arthritis Rheum* **46(4):** 894–898.

Samuelsson G. (1999) What's happening? Protein A columns: Current concepts and recent advances. *Transfus Sci* **21(3):** 215–217.

Scott DL, Wolfe F, Huizinga TW. (2010) Rheumatoid arthritis. *Lancet* **376(9746):** 1094–1108.

Wiesenhutter CW, Irish BL, Bertram JH. (1994) Treatment of patients with refractory rheumatoid arthritis with extracorporeal protein A immunoadsorption columns: A pilot trial. *J Rheumatol* **21(5):** 804–812.

Yamazaki Z, Fujimori Y, Takahama T, *et al.* (1982) Efficiency and biocompatibility of a new immunosorbent. *Transactions — American Society for Artificial Internal Organs* **28:** 318–323.

Yu X, Ma J, Tian J, *et al.* (2007) A controlled study of double filtration plasmapheresis in the treatment of active rheumatoid arthritis. *Journal of Clinical Rheumatology: Practical Reports on Rheumatic & Musculoskeletal Diseases* **13(4):** 193–198.

Yu X, Wang L, Xu P, *et al.* (2012) Effects of double filtration plasmapheresis, leflunomide, and methotrexate on inflammatory changes found through magnetic resonance imaging in early rheumatoid arthritis. *J Rheumatol* **39(6):** 1171–1178.

Chapter 27

Immunomodulator of Nervous System Disease

Hidenori Matsuo*, MD

1. Introduction

Immunomodulation with adsorbent to treat neurological disorders first appeared in the 1980s. Immunoadsorption plasmapheresis (IA) has been applied for treatment of neuroimmunological disorders, such as myasthenia gravis (MG), autoimmune neuropathy and multiple sclerosis (MS). Plasma exchange (PE) is effective in temporarily relieving the symptoms of the diseases. However, because PE removes plasma protein unselectively, when repeated many times, the lack of selectivity and need for replacement fluids are important drawbacks. Since the pathogenic substances in these disorders exist in mainly IgG fraction, more selective reduction of IgG should become more efficient treatment. For this purpose, IA has been developed. IA can be the treatment of choice when intensive apheretic protocols or long-term treatments must be performed, in patients needing frequent apheresis to keep a stable clinical condition, in case of unresponsiveness to corticosteroids and immunosuppressive treatments, or failure with PE or intravenous immunoglobulins, and in patients with severe contraindications to long-term corticosteroids.

---

*National Hospital Organization, Nagasaki Kawatana Medical Center, Japan.

Immusorba TR (IM-TR) and PH (IM-PH) were devised with non-biological materials as affinity ligands to remove pathogenic autoantibodies. The adsorbents of IM-TR and IM-PH are polyvinyl-alcohol gel immobilized with tryptophan and phenylalanine as ligands, respectively. IM-TR is clinically applied for treatment of autoimmune neurological diseases including MG, Guillain–Barré syndrome (GBS), chronic inflammatory demyelinating polyneuropathy (CIDP), and MS. IM-PH is also used for GBS and MS.

Staphylococcal protein A, a potent ligand for immunoglobulins that interacts negligibly with other plasma proteins, has been also used as a ligand for IA in neuroimmunological disorders.

2. Immunomodulation by Immunoadsorption Plasmapheresis

IA can remove immunoglobulins including IgG autoantibodies and bring clinical effects in antibody-mediated autoimmune diseases. However, its other effects on the immune system have not been fully investigated. Therapeutic effects of IA in autoantibody mediated diseases can be attributed to three major mechanisms: immediate intravascular reduction of auto-antibody concentration, pulsed induction of antibody redistribution, and subsequent immunomodulatory changes.

It was reported that protein A IA induced a greater increase in activated T and cytolytic T cells which correlated with a greater decrease of B cells (Schneidewind *et al.*, 2002). Baggi *et al.* (2008) observed upregulation of interleukin (IL)-10, an anti-inflammatory and B cells growth factor, and reduction of pro-inflammatory factors such as IL-18 and IL-17, in both MG and LEMS patients after IA. They speculated that the massive removal of antibodies might induce modifications of the cytokine balance linked to T and B cell mediated autoimmunity. However, these observations were based on a limited sample and may not be specific for IA. Further immunological investigation is required to verify these findings.

3. Immunoadsorption in Myasthenia Gravis

3.1. *Immunoadsorption with IM-TR*

MG is an autoimmune disease in which autoantibodies to the acetylcholine receptor (anti-AChR) interfere with neuromuscular transmission. These anti-AChR antibodies belong to IgG, mainly IgG3. The first reports of IA in MG was in 1983, using IM-TR (Sato *et al.*, 1988). Thereafter, several investigators reported that the efficacy of IA was comparative to PE in MG (Hosokawa and Oyamaguchi, 1990; Ichikawa *et al.*, 1993; Splendiani *et al.*, 1992; Yamazaki *et al.*, 1989). In 1994, Shibuya *et al.* (1994) first reported the results of a multicenter trial to evaluate the efficacy of IA for severe generalized MG. Twenty patients with MG who were concurrently receiving high dose prednisolone and azathioprine therapy were treated with IM-TR. Among the 20 patients who received five IA treatments within a period of 10 days, 11 showed pronounced improvement of myasthenic weakness. Circulating anti-AChR antibodies were reduced by about 60% by treating one plasma volume. No serious complications occurred during 100 procedures. They concluded that IA with IM-TR was useful in controlling symptoms in patients with severe MG who were otherwise unresponsive. Grob *et al.* (1995) treated 16 patients with moderately severe to severe generalized MG by IA of 2,500 ml plasma on each of four alternate days. Fourteen patients who completed treatment all had significant improvement in strength (six excellent, six good and two fair), which began a mean of 42 hours after the first IA, reached a maximum four days after the fourth IA (mean, 250% of baseline strength), and returned to baseline over a mean of two months. Thirty-seven grams of plasma proteins were removed during each IA, which required no replacement, compared with 175 grams during plasma exchange, which requires replacement with albumin. Anti-AChR antibody fell to a mean of 23% of original level, fibrinogen to 26%, C4 to 29%, IgM to 33%, IgG to 35%, CH50 to 41%, C3 to 42%, IgA to 54% and albumin to 76%. All proteins, including anti-AChR antibodies, returned to their original levels within one to three weeks after the last IA, while improvement in strength lasted a mean of six weeks longer. One seronegative patient had excellent improvement lasting more than a month. Activated complement C5a and white blood cell count rose during each IA, while activated complement C3a fell, and each returned to its original level within hours.

While IA had been demonstrated as effective treatment especially in acute situations of myasthenic crisis, controlled data comparing clinical safety and efficacy of PE and IA were not sufficient. In 2011, Köhler *et al.* (2011) reported the results of a prospective randomized controlled clinical trial. They investigated 19 patients with myasthenic crisis, who were randomized to receive either PE ($n = 10$) or IA ($n = 9$) in addition to combined drug treatment. Patients received 3–5 (mean 3.5 for PE, and 3.4 for IA) treatments over a period of seven days with a predefined treatment volume of 1.5 l plasma. Clinical courses were monitored using disease-specific clinical scores. After initiation of IA or PE the mean value of myasthenia scores decreased equally until Day 14 of the post-treatment phase. Patients from both treatment groups improved to a stable clinical status. Substantial reduction of anti-AChR antibodies was documented after each session of PE or IA. In the treatment period 16 adverse effects (seven serious adverse events, SAE) in the PE and 10 (1 SAE) in the IA group were observed. Thus, IA proved to be equally effective compared with PE treatment in patients with myasthenic crisis. Three to five treatment sessions using low plasma volume dosage of 20–25 mL/kg were adequate to improve clinically relevant symptoms significantly in most patients.

Yeh and Chiu (2000) conducted a controlled study evaluating the differences between IA and DFPP in clinical effects and serological changes. Five patients with generalized MG (clinical states IIb and III) were enrolled; each patient received IA and DFPP on separate occasions. IM-TR was used for IA, while Evaflux 4A was used as the plasma fractionator for DFPP. Each course of treatment consisted of five sessions of apheresis. Both treatments effectively ameliorated symptoms of MG. There were no significant changes in MG score between the two groups (IA vs. DFPP: 2.2 vs. 2.6, $P > 0.5$). IA had a higher clearance rate of anti-AChR antibodies than DFPP (66 % vs. 54%, $P < 0.05$), while DFPP removed more IgA (72% vs. 21%, $P < 0.05$) and IgM (89% vs. 57%, $P < 0.01$) than did IA. Although IA removed anti-AChR antibodies more effectively than DFPP, the clinical effects between these two treatments were similar.

MG patients with autoantibodies to muscle-specific tyrosine kinase (MuSK) represent a distinct subset of those with this disease. Most of anti-MuSK antibodies belongs to IgG4. As mentioned above, IA with IM-TR can adsorb mainly IgG3, next IgG1, then IgG2, but not efficiently IgG4. Therefore, IA with IM-TR is not recommended in treatment of this particular subgroup.

3.2. *Immunoadsorption with Protein A*

Protein A-Sepharose has been evaluated as a possible agent for the selective, extracorporeal removal of anti-AChR antibodies from the sera of patients with MG. Protein A-Sepharose effectively adsorbed anti-AChR IgG antibodies of subclasses 1 and 2 to the same extent (i.e. from 50 to 60% decrease) as the total IgG, but anti-AChR antibody of subclass IgG3 was less effectively adsorbed (Somnier and Langvad, 1989). A full-scale *in vitro* experiment using one plasma volume and an automatic continuous monitor demonstrated the removal of 43% of the circulating anti-AChR antibodies and the IgG within 70 min, which is comparable to the amount removed by the unselective, membrane PE technique. Berta *et al.* (1994) treated three patients with severe immunosuppression-resistant MG with protein A IA. Neurological impairment significantly improved in all patients. After IA of 1.5–2 plasma volumes per session, the mean percentage reductions for serum IgG and specific autoantibodies were 71 and 82%, respectively. No major side effects occurred. Protein A IA appears to be a safe, efficient treatment for selected MG patients requiring prolonged apheresis. The Canadian Apheresis Group undertook a study at two Canadian apheresis centers to clinically evaluate a staphylococcal protein A IA system (EXCORIM) in MG patients and concluded the IA system was safe and well tolerated (Benny *et al.*, 1999). Ten of 12 patients had improvement in their neurologic status.

3.3. *Approaches to Specific Removal of Anti-Acetylcholine Receptor Antibodies in Myasthenia Gravis*

Several approaches to remove anti-AChR antibodies from blood circulation in MG patients has been made using the immunoadsorbent with an AChR peptide or polypeptide as ligands.

Takamori *et al.* (1996) used a synthetic peptide of AChR. The amino acid sequence of the peptide is modified from the segment of α183–200 of the *Torpedo* AChR protein, and the segment is the ACh binding site on AChR and the target site of anti-AChR antibody. The adsorbent showed specific adsorption characteristics to the anti-AChR antibody (blocking antibody) *in vitro*, and became an adsorption column Medisorba MG-50 (Kuraray Medical Inc.) for the treatment of MG (Nakaji and Hayashi, 2003). The adsorbent in this column is composed of cellulose beads as carrier material and covalent-bound synthetic peptide as a ligand that has a specific affinity to a pathogenic anti-AChR antibody of MG. Clinically, MG-50 is used in plasma-perfusion therapy, and it is recognized that MG-50 specifically reduces blocking antibody titer and improves MG symptoms. MG-50 was approved in Japan. However, because the target of this adsorbent is too limited to remove "binding antibodies" which are major pathogenic antibodies in MG, this has not been used widely in clinical setting.

Several other groups have also developed new adsorbents which can selectively remove anti-AChR experimentally (Guo *et al.*, 2005; Psaridi-Linardaki *et al.*, 2005). Psaridi-Linardaki *et al.* (2005) used the Sepharose-immobilized N-terminal extracellular domain of human muscle AChRα-subunit. This immunoadsorbents removed 60–94% of the anti-AChR antibodies in 10 sera and a mean of 35% from all samples combined. In addition, their group immobilized the β-, γ- and ε-AChR extracellular domains (ECDs) on Sepharose and tested them as immunoadsorbents using sera from 50 MG patients (Kostelidou *et al.*, 2007). A given ECD removed a different percentage of autoantibodies from different sera and different ECDs removed different percentages from the same serum; on average, the β-, γ- and ε-AChR — ECDs removed 22, 20 and 15.5% of the autoantibodies, respectively. The combined use of two (α + γ) or four (α + β + γ + ε) ECDs in a single immunoadsorbent resulted in much higher (often additive) IA. These results suggest that the combined use of all AChR ECDs could provide the basis for a novel, antigen-specific therapy for MG. By the same method, they also developed two different immunoadsorbents which very efficiently and selectively depleted the MuSK antibodies (Zisimopoulou *et al.*, 2014).

4. Immunoadsorption in Guillain-Barré Syndrome and Fisher's Syndrome

4.1. *Immunoadsorption with IM-TR*

Acute and chronic autoimmune neuropathies, including GBS are often characterized by the presence of autoantibodies that react with neural gangliosides. Evidence from human and animal studies indicates that anti-ganglioside antibodies play a primary neuropathogenic role, and their rapid elimination from the circulation through specific IA therapy thus has the potential to ameliorate the course of the disease. Patients with GBS and FS often receive IA as well as PE and DFPP.

There have been no RCTs, but several clinical studies demonstrated the efficacy of IA with IM-TR in GBS[21-28] (Table 27.1). Haupt *et al.* (1996) treated 11 patients by PE, 13 patients by IA using IM-TR, and 21 patients by IA followed by intravenous IgG (IVIg), showing IA and PE were equally effective; sequential treatment with IA followed by IVIg may be superior to plasma treatment alone. Diener *et al.* (2001) performed a randomized, multicenter study comparing IVIg (23 patients), PE (26 patients), and IA (18 patients) in GBS. The primary endpoint, at least a one grade improvement in the functional score by day 28, was achieved by 16/20 patients after

Table 27.1. Clinical Studies of Immunoadsorption Therapy in Guillain–Barré Syndrome

IA (*n*)	PE (*n*)	Other Treatment Group (*n*)	Efficacy of IA	Authors Year
13	11	IA + IVIg 21	Effective No significant difference between IA and PE	Haupt *et al.* (1996)
18	26	IVIg 23	Effective No significant difference between 3 groups	Diener *et al.* (2001)
21	—	—	No controls Effective comparing RCTs in the past	Takei *et al.* (2002)
20	3	DFPP 11	Effective No significant difference between 3 groups	Okamiya *et al.* (2004)
39	14	IVIg 10	Effective No significant difference between 3 groups	Seta *et al.* (2005)
9	5	IA + PE 5	Effective as well as PE (severe GBS)	Marn *et al.* (2009)
10	—	IA + IVIg 20	Effective as well as IVIg (severe GBS)	Galldkis *et al.* (2009)

IA: immunoadsorption, PE: plasma exchange, IVIg: intravenous immunoglobulin, DFPP: double filtration plasmapheresis.

IVIg, 15/21 patients with PE, and 7/14 patients with IA (difference not significant).

Sera from patients with Fisher's syndrome in the acute phase contain IgG anti-GQ1b ganglioside antibody. A batchwise adsorption test showed that an IM-TR gel adsorbed a larger amount of the IgG anti-GQ1b antibody than did an IM-PH column. An *ex vivo* plasma perfusion study done with the IM-TR column confirmed that it effectively adsorbs the IgG anti-GQ1b antibody. Results of adsorption tests done with various amino acid-immobilized gels suggest that both the hydrophobic force of the side chain and the anionic charge of the carboxylic acid in tryptophan are important in the adsorption of the autoantibody by the IM-TR gel (Yuki, 1996). IA using the IM-TR column offers an alternative type of plasmapheresis for Fisher's syndrome.

4.2. *Immunoadsorption with Disialylgalactose Glycoconjugate*

In 2004, Willson *et al.* (2004) synthesized the terminal trisaccharide, NeuAc(α2-8)NeuAc(α2-3)Gal common to GQ1b and GD3, and conjugated it to bovine serum albumin (BSA). This disialylgalactose glycoconjugate (DSG-BSA) binds anti-GQ1b antibodies in 32/58 (55%) human sera containing IgG or IgM anti-GQ1b antibodies at titres up to 1/130,000; it also binds a wide range of mouse monoclonal anti-GQ1b and -GD3 antibodies. When conjugated to Sepharose as mock therapeutic immmunoaffinity columns, the immobilized trisaccharide (DSG-Sepharose) eliminates anti-GQ1b antibodies from positive sera in proportion to their level of binding to DSG-BSA. Although this is still experimental, oligosaccharide-specific IA therapy may provide a new therapeutic approach to anti-GQ1b antibody-associated syndromes that could be applied to clinical practice. Furthermore, modification of the immobilized oligosaccharide epitopes to incorporate other glycan structures may allow this approach to be adapted to other forms of autoimmune neuropathy associated with uniform anti-glycolipid antibody profiles.

5. Immunoadsorption in Chronic Inflammatory Demyelinating Polyradiculoneuropathy

CIDP is rather characterized by a relapsing or slowly progressing disease course, whereas GBS is usually acute and monophasic in nature. PE is proven of benefit in CIDP (Hahn *et al.*, 1996), but there have been no randomized controlled trials for IA. IA removes IgG more effectively than plasma exchange, and should have fewer adverse effects because it does not require the administration of albumin or donor plasma. Therefore, IA has theoretical advantages in treatment of CIDP. First-line treatment options for CIDP are corticosteroids, IVIg, and PE. However, first-line therapy fails in a significant number of patients, and long-term maintenance treatment still remains a therapeutic challenge. Previous reports of IA in CIDP using various types of column have described beneficial responses, although none was randomized or controlled. For example, Galldiks *et al.* (2011) treated 14 patients with CIDP using tryptophan-IA with significant improvement, and concluded IA could be an effective and safe option for CIDP patients with unsatisfactory response to first-line treatment options and for long-term maintenance treatment. Zinman *et al.* (2005) also reported the efficacy and safety of selective immunoglobulin removal by EXCORIM(protein A) IA in a pilot trial in CIDP patients. However, as CIDP is a heterogeneous condition, some have a different pathogenesis, and a small proportion of CIDP sera has IgG or IgM antibodies to various proteins and glycolipids in peripheral myelin. Hadden *et al.* (2002) reported the superior efficacy of PE to IA as the result of removal of an IgM antibody in a single case.

6. Immunoadsorption in Multiple Sclerosis

IA for treatment of MS was first introduced in Japan in the 1980s, when IM-PH and IM-TR immnoadsorbent column were developed. Yamazaki *et al.* (1989) reported that IM-PH plasma perfusion has been successfully used in patients with rheumatoid arthritis, systemic lupus erythematosus, and MS in 1989. Hosokawa *et al.* (1989) also reported their experience of treatment of 16 MS patients by IA using a column containing polyvinyl alcohol gel. They found that most patients showed clinical improvements with no adverse effects. In addition, activity of natural killer (NK) cells and helper T cells was significantly increased after IA. Schmitte *et al.* (1990) performed a controlled pilot study, comparing the efficacy of IA using IM-PH and PE in RR-MS. The found IA with IM-PH was almost equally effective to PE, and had a superior safety profile. Thereafter, several investigators have reported successful results of IA using IM-TR, not IM-PH, for treatment of MS. However, there have been no RCTs to demonstrate the efficacy to date.

Therapeutic effects of PE and IA in autoantibody-mediated diseases can be attributed to three major mechanisms: immediate intravascular reduction of (auto-)antibody concentration, pulsed induction of antibody redistribution, and subsequent immunomodulatory changes. In MS, the exact mechanism which leads to clinical improvement has not been fully elucidated. Koziolek *et al.* (2012) prospectively treated 11 patients with MS who had optic neuritis by IA with IM-TR. In addition to clinical assessments, they also performed proteomic analyses to analyze column-bound proteins as well as corresponding changes in patients' sera. They found soluble interleukin-2 receptor decreased in responders, whereas in non-responders it did not. Proteomic analyses of proteins adsorbed to IA columns revealed that several significant immunological proteins as well as central nervous system (CNS) protein fragments, including myelin basic protein, had been removed by IA. Although the pathogenic role of antibodies in MS is still controversial, Pedotti *et al.* (2013) reported an interesting study. They transferred to mice with experimental autoimmune encephalomyelitis (EAE), animal model of MS, IgG antibodies purified from a MS patient presenting a dramatic clinical improvement during relapse after selective IgG removal with IA. Passive transfer of the patient's IgG exacerbated motor paralysis and increased mouse CNS inflammation and demyelination. Binding of the patient's IgG was demonstrated in mouse CNS, with a diffuse staining of white matter oligodendrocytes.

7. Immunoadsorption in Neuromyelitis Optica

Neuromyelitis optica (NMO), Devic's disease, is another devastating neurological disease characterized by severe optic neuritis and transverse, longitudinally extensive myelitis. It was for a long time regarded as a variant of MS, although with major distinctive pathological features. A major breakthrough in our understanding of the disease was the discovery of a pathogenic antibody against the astrocyte water channel protein aquaporin-4 (AQP4) with high diagnostic sensitivity and specificity for NMO, which has led to general acceptance that this condition is distinct from MS (Wingerchuk and Weinshenker, 2014).

In the acute phase of NMO, PE should be considered as the second choice treatment when corticosteroid pulse therapy is unsuccessful. It is believed that the beneficial effects of PE occur through the elimination of pathogenic humoral and plasma factors, including autoantibodies, complement components, and cytokines. IA has also been used in treatment of NMO due to the theoretical rationale of elimination of pathogenic antibodies, such as anti-AQP4. In our experience, IA is as effective as PE in treatment of NMO, although there have been no clinical trials demonstrating the efficacy.

Table 27.2. Other Neurologic Disorders Treated by Immunoadsorption and the Target Antigen(s)

Neurologic Disorders	Target Antigen(s)
Autoimuune encephalitis	NMDA receptor
Limbic encephalitis	AMPA receptor
	$GABA_B$ receptor
	VGKC Complex (Lgl-1)
	VGCC
	Glycine receptor
	Metabotropic GLuR
Rasmussenn encephalitis	GLuR ε2
Acquired neuromyotonia (Isaacs' syndrome)	VGKC complex (Caspar 2)
Autoimmune autonomic ganglionopathies	gAChR
Stiff-person syndrome	GAD 65
	amphiphysin
Paraneoplastic syndrome	Hu, Yo, Ri, Ma2, amphiphysin, etc.
	Tr, ANNA-3, PCA-2, Zic4, etc.
	VGKC, NMDA, etc.

NMDA: N-methyl-D-aspartic acid.
AMAR: alpha-amino-3-hydroxy-5-methyl-4-isoxazolepropionic acid.
$GABA_B$: gamma-aminobutyric acid (GABA) B, VGKC: voltage-gated potassium channels.
Lgl-1: leucine-rich glioma –inactivated 1, VGCC: voltage-gated calcium channels.
GLuR: glutamate receptor, Caspar-2: contactin associated protein 2.
gAChR: ganglionic acetylcholine receptor.
GAD: glutamic acid decarboxylase.

8. Immunoadsorption in other Neurological Disorders

IA has been applied for treatment of several other neurologic disorders, including autoimmune encephalitis, Rasmussen's encephalitis (Antozzi *et al.*, 1998), Isaacs' syndrome, stiff-person syndrome and paraneoplastic neurologic syndrome (Table 27.2).

8.1. *Immunoadsorption in Acquired Neuromyotonia*

Acquired neuromyotonia (Isaacs' syndrome) is a rare, immune-mediated disorder of the neuromuscular endplate characterized by cramping, myokymia, fasciculations and excessive sweating. The frequent presence of antibodies against voltage-gated potassium channels (VGKC) in the serum provides a rationale for the successful application of plasmapheresis. The successful application of IA in neuromyotonia has so far only been described in case reports (Nakatsuji *et al.*, 2000; Antozzi *et al.*, 2005).

Autoimmune ion channel disorders of the peripheral nervous system include myasthenia gravis, Lambert–Eaton myasthenic syndrome, acquired neuromyotonia and autoimmune autonomic ganglionopathies. These disorders are characterized by the common feature of being mediated by IgG autoantibodies against identified target antigens. Moreover, experimental animal models have been identified for these diseases that respond to immunotherapy and are improved by PE. On this basis, autoimmune ion channel disorders represent the ideal candidate for therapeutic apheresis.

8.2. *Immunoadsorption in Autoimmune Encephalitis*

Recently, several types of encephalitis turned out to be associated with autoimmune pathogenesis. A new category of treatment-responsive encephalitis has been proposed in association with antibodies to neuronal cell membrane antigens, including VGKC, N-methyl-D-aspartic acid receptor (NMDAR), an amino-3-hydroxy-5-methyl-4-isoxazolepropionic acid receptor (AMPAR), gamma-aminobutyric acid (GABA) B receptor and other antigens that have not yet been characterized. Among them, anti-NMDAR encephalitis is a distinct disorder characterized by the predictable sequential development of symptoms; prodromal symptoms are initially noted, followed by prominent psychiatric symptoms, seizures, an unresponsive/catatonic state, hypoventilation and involuntary orofacial-limb movements (Dalmau *et al.*, 2007). This type of encephalitis is often associated with ovarian teratoma in young women. Anti-NMDAR encephalitis is a neuroimmune syndrome in patients with autoantibodies recognizing extracellular epitopes of NMDAR, and the autoantibodies attenuate NMDAR function through the internalization of NMDAR. In such autoimmune encephalitis, early start of immunotherapy correlates with good outcome and is strongly recommended. First-line therapy consists of corticosteroids, IVIg or PE, alone or combined, followed by immunosupressant. Theoretically, selective extracorporeal elimination of autoantibodies and immune complexes with IA is expected to be as effective as PE, although the use of IA for patients with autoimmune encephalitis has not yet been investigated separately.

Köhler *et al.* (2015) performed a retrospective analysis of patients with autoimmune encephalitis who were treated with tryptophan IA. Twelve patients were positive for specific autoantibodies (NMDAR, GABA, GAD, Lgl-1). Patients received a series of a median of six IA treatments. Eleven patients improved by at least one point in the modified Rankin Scale after IA. For autoimmune encephalitis rapid elimination of autoantibodies with selective IA seems to be an effective therapeutic option as part of multimodal immune therapy.

8.3. *Immunoadsorption in Paraneoplastic Syndromes*

Although controlled data are limited, PE and IA might be effective in classic paraneoplastic syndromes of the CNS, such as limbic encephalitis or cerebellar degeneration (Cher *et al.*, 1995). As another autoimmune and, sometimes, even paraneoplastic disease, the stiff-person syndrome is characterized by central hyperexcitability and cramping or muscle spasms. In some, but not all cases, antibodies against amphiphysin or glutamate decarboxylase (GAD) can be found. The relevance of such humoral factors in stiff-person syndrome has been confirmed by passive transfer of serum from patients to animal models. Some patients suffering from stiff-person syndrome can be well managed with corticosteroids. In other cases, a benefit from PE or IA was described in case reports (Hayashi *et al.*, 1999; Hagiwara *et al.*, 2001).

References

Antozzi C, Granata T, Aurisano N, *et al.* (1998) Long-term selective IgG immuno-adsorption improves Rasmussen's encephalitis. *Neurology* **51:** 302–5.

Antozzi C, Frassoni C, Vincent A, *et al.* (2005) Sequential antibodies to potassium channels and glutamic acid decarboxylase in neuromyotonia. *Neurology* **64:** 1290–3.

Baggi F, Ubiali F, Nava S, *et al.* (2008) Effect of IgG immunoadsorption on serum cytokines in MG and LEMS patients. *J Neuroimmunol* **201–202:** 104–10.

Benny WB, Sutton DM, Oger J, *et al.* (1999) Clinical evaluation of a staphylococcal protein A immunoadsorption system in the treatment of myasthenia gravis patients. *Transfusion* **39:** 682–7.

Berta E, Confalonieri P, Simoncini O, *et al.* (1994) Removal of antiacetylcholine receptor antibodies by protein-A immunoadsorption in myasthenia gravis. *Int J Artif Organs* **17:** 603–8.

Cher LM, Hochberg FH, Teruya J, *et al.* (1995) Therapy for paraneoplastic neurologic syndromes in six patients with protein A column immunoadsorption. *Cancer* **75:** 1678–83.

Dalmau J, Tüzün E, Wu HY, *et al.* (2007) Paraneoplastic anti-N-methyl-D-aspartate receptor encephalitis associated with ovarian teratoma. *Ann Neurol* **61:** 25–36.

Diener HC, Haupt WF, Kloss TM, *et al.*; Study Group. (2001) A preliminary, randomized, multicenter study comparing intravenous immunoglobulin, plasma exchange, and immune adsorption in Guillain-Barré syndrome. *Eur Neurol* **46:** 107–9.

Galldiks N, Dohmen C, Neveling M, *et al.* (2009) Selective immune adsorption treatment of severe Guillain Barré syndrome in the intensive care unit. *Neurocrit Care* **11:** 317–21.

Galldiks N, Burghaus L, Dohmen C, *et al.* (2011) Immunoadsorption in patients with chronic inflammatory demyelinating polyradiculoneuropathy with unsatisfactory response to first-line treatment. *Eur Neurol* **66:** 183–9.

Grob D, Simpson D, Mitsumoto H, *et al.* (1995) Treatment of myasthenia gravis by immunoadsorption of plasma. *Neurology* **45:** 338–44.

Guo CY, Li ZY, Xu MQ, Yuan JM. (2005) Preparation of an immunoadsorbent coupled with a recombinant antigen to remove anti-acetylcholine receptor antibodies in abnormal serum. *J Immunol Methods* **303:** 142–7.

Hadden RD, Bensa S, Lunn MP, Hughes RA. (2002) Immunoadsorption inferior to plasma exchange in a patient with chronic inflammatory demyelinating polyradiculoneuropathy. *J Neurol Neurosurg Psychiatry* **72:** 644–6.

Hagiwara H, Enomoto-Nakatani S, Sakai K, *et al.* (2001) Stiff-person syndrome associated with invasive thymoma: A case report. *J Neurol Sci* **193:** 59–62.

Hayashi A, Nakamagoe K, Ohkoshi N, *et al.* (1999) Double filtration plasma exchange and immunoadsorption therapy in a case of stiff-man syndrome with negative anti-GAD antibody. *J Med* **30:** 321–7.

Hahn AF, Bolton CF, Pillay N, *et al.* (1996) Plasma exchange therapy in chronic inflammatory demyelinating polyneuropathy: A double-blind, sham-controlled, cross-over study. *Brain* **119:** 1055–66.

Haupt WF, Rosenow F, van der Ven C, *et al.* (1996) Sequential treatment of Guillain-Barré syndrome with extracoporeal elimination and intravenous immunoglobulin. *J Neurol Sci* **137:** 145–149.

Hosokawa S, Oyamaguchi A, Yoshida O. (1989) Successful immunoadsorption with membrane plasmapheresis for multiple sclerosis. *ASAIO Trans* **35:** 576–7.

Hosokawa S, Oyamaguchi A. (1990) Safety, stability, and effectiveness of immunoadsorption under membrane plasmapheresis treatment for myasthenia gravis. *ASAIO Trans* **36(3):** M207–8.

Ichikawa M, Koh CS, Hata Y, *et al.* (1993) Immunoadsorption plasmapheresis for severe generalised myasthenia gravis. *Arch Dis Child* **69:** 236–8.

Köhler W, Bucka C, Klingel R. (2011) A randomized and controlled study comparing immunoadsorption and plasma exchange in myasthenic crisis. *J Clin Apher* **26:** 347–55.

Kostelidou K, Trakas N, Tzartos SJ. (2007) Extracellular domains of the beta, gamma and epsilon subunits of the human acetylcholine receptor as immunoadsorbents for myasthenic autoantibodies: A combination of immunoadsorbents results in increased efficiency. *J Neuroimmunol* **190:** 44–52.

Koziolek MJ, Tampe D, Bähr M, *et al.* (2012) Immunoadsorption therapy in patients with multiple sclerosis with steroid-refractory optical neuritis. *J Neuroinflammation* **9:** 80.

Köhler W, Ehrlich S, Dohmen C, *et al.* (2015) Tryptophan immunoadsorption for the treatment of autoimmune encephalitis. *Eur J Neurol* **22:** 203–6.

Marn Pernat A, Buturović-Ponikvar J, Svigelj V, Ponikvar R. (2009) Guillain-Barré syndrome treated by membrane plasma exchange and/or immunoadsorption. *Ther Apher Dial* **13:** 310–3.

Nakatsuji Y, Kaido M, Sugai F, *et al.* (2000) Isaacs' syndrome successfully treated by immunoadsorption plasmapheresis. *Acta Neurol Scand* **102:** 271–3.

Nakaji S, Hayashi N. (2003) Adsorption column for myasthenia gravis treatment: MedisorbaMG-50 *Ther Aphe Dial* **7:** 78–84.

Okamiya S1, Ogino M, Ogino Y, *et al.* (2004) Tryptophan-immobilized column-based immunoadsorption as the choice method for plasmapheresis in Guillain-Barré syndrome. *Ther Apher Dial* **8:** 248–53.

Pedotti R, Musio S, Scabeni S, *et al.* (2013) Exacerbation of experimental autoimmune encephalomyelitis by passive transfer of IgG antibodies from a multiple sclerosis patient responsive to immunoadsorption. *J Neuroimmunol* **262:** 19–26.

Psaridi-Linardaki L, Trakas N, Mamalaki A, Tzartos SJ. (2005) Specific immunoadsorption of the autoantibodies from myasthenic patients using the extracellular domain of the human muscle acetylcholine receptor alpha-subunit. Development of an antigen-specific therapeutic strategy. *J Neuroimmunol* **159:** 183–91.

Sato T, Ishigaki Y, Komiya T, Tsuda H. (1988) Therapeutic immunoadsorption of acetylcholine receptor antibodies in myasthenia gravis. *Ann N Y Acad Sci* **540:** 554–6.

Schmitt E, Behm E, Buddenhagen F, *et al.* (1990) Immunoadsorption (IA) versus plasma exchange (PE) in multiple sclerosis — first results of a double blind controlled trial. *Prog Clin Biol Res* **337:** 289–92.

Schneidewind-Müller JM, Winkler RE, Tiess M, *et al.* (2002) Changes in lymphocytic cluster distribution during extracorporeal immunoadsorption. *Artif Organs* **26:** 140–4.

Seta T, Nagayama H, Katsura K, *et al.* (2005). Factors influencing outcome in Guillain-Barré syndrome: Comparison of plasma adsorption against other treatments. *Clin Neurol Neurosurg* **107:** 491–6.

Shibuya N, Sato T, Osame M, *et al.* (1994) Immunoadsorption therapy for myasthenia gravis. *J Neurol Neurosurg Psychiatry* **57:** 578–81.

Somnier FE, Langvad E. (1989) Plasma exchange with selective immunoadsorption of anti-acetylcholine receptor antibodies. *J Neuroimmunol* **22:** 123–7.

Splendiani G, Passalacqua S, Barbera G, *et al.* (1992) Semi-selective immunoadsorption treatment in myasthenia gravis. *Biomater Artif Cell Im* **20:** 1145–51.

Takamori M, Ide Y. (1996) Specific removal of anti-acetylcholine receptor antibodies in patients with myasthenia gravis. *Transfus Sci* **17:** 445–53.

Takei H, Komaba Y, Araki T, *et al.* (2002) Plasma immunoadsorption therapy for Guillain-Barré syndrome: Critical day for initiation. *J Nippon Med Sch* **69:** 557–63.

Willison HJ, Townson K, Veitch J, *et al.* (2004) Synthetic disialylgalactose immunoadsorbents deplete anti-GQ1b antibodies from autoimmune neuropathy sera. *Brain* **127:** 680–91.

Wingerchuk DM, Weinshenker BG. (2014) Neuromyelitis optica (Devic's syndrome). *Handb Clin Neurol* **122:** 581–99.

Yamazaki Z, Idezuki Y, Inoue N, *et al.* (1989) Extracorporeal immunoadsorption with IM-PH or IM-TR column. *Biomater Artif Cell* **17(2):** 117–24.

Yeh JH, Chiu HC. (2000) Comparison between double-filtration plasmapheresis and immunoadsorption plasmapheresis in the treatment of patients with myasthenia gravis. *J Neurol* **247:** 510–3.

Yuki N. (1996) Tryptophan-immobilized column adsorbs immunoglobulin G anti-GQ1b antibody from Fisher's syndrome: A new approach to treatment. *Neurology* **46:** 1644–51.

Zinman LH, Sutton D, Ng E, *et al.* (2005) A pilot study to compare the use of the Excorim staphylococcal protein immunoadsorption system and

IVIG in chronic inflammatory demyelinating polyneuropathy. *Transfus Apher Sci* **33:** 317–24.

Zisimopoulou P, Lagoumintzis G, Poulas K, Tzartos SJ. (2014) Expression of extracellular domains of muscle specific kinase (MuSK) and use as immunoadsorbents for the development of an antigen-specific therapy. *J Neuroimmunol* **276(1–2):** 150–8.

V

Selective Cell Removal

Chapter 28

Leukocyte Apheresis Using Adsorbent

Yoshihiro Endo*

Cytapheresis is a medical technique to remove or separate blood components from the patient's peripheral blood. This section provides a description of leukocyte apheresis using adsorbent, which is the most popular procedure in therapeutic apheresis.

It is well known that leukocytes adhere to foreign bodies when they come into contact with them. This phenomenon was applied to leukocyte reduction filter for blood transfusion. Leukocyte removal or reduction was applied to clinical therapy in the 1970s. Since then, its efficacy of treatment in several diseases has been reported. In 1973, Dr. Pauls *et al.* reported the efficacy of thoracic duct lymphocyte drainage in rheumatoid arthritis.

There have been many attempts to reduce white blood cells in patients, particularly those with rheumatoid arthritis. Clinical studies have reported efficacy of lymphocyte depletion by continuous flow cell centrifugation (Karsh *et al.*, 1979), total lymphoid irradiation (Kotzin *et al.*, 1981) and photopheresis technique (Malawista *et al.*, 1991). However, the above treatments have not been established as a common therapeutic procedure because of the complicated procedure, the high medical cost and/or the adverse effect. In 1985, leukocyte removal therapy using leukocyte reduction filter for blood transfusion was applied for rheumatoid arthritis. Apart from lymphocyte depletion

---

* Professor of Clinical Nursing Shiga University of Medical Science, Japan.

by continuous flow cell centrifugation, leukocyte apheresis was classified into two methods, removal method by filter or beads, and centrifugal method. This section covers both basic and clinical aspects of adsorbents, filter type adsorbent CELLSORBA® and bead type adsorbent Adacolumn® and Immunopure®.

Dr. Nishimura and Dr. Sukegawa, researchers from Asahi Kasei Medical Co. Ltd, describe basic aspects of filter type adsorbent CELLSORBA®. The clinical aspect of CELLSORBA® is described by Dr. Sawada, who applied the first leukocyte apheresis therapy with leukocyte removal filter for inflammatory bowel disease (Sawada *et al.*, 1995).

Regarding bead adsorbents, Dr. Saniabadi and Dr. Shima, researchers at JIMRO Co. Ltd, which performs as a manufacturer, provide detailed information on the basic aspects of Adacolumn®.

As for the clinical aspect of Adacolumn®, Dr. Yokoyama and Dr. Fukunaga, who are famous doctors researching a mechanism of the leukocyte apheresis for inflammatory bowel disease, described the efficacy of Adacolumn® in the treatment of the inflammatory bowel disease (Yokoyama *et al.*, 2007).

A multi-center clinical trial on the efficacy of Adacolumn® in patients with generalized pustular psoriasis, which is a chronic autoimmune disease characterized by fever, erythema, and sterile pustules over large areas of the skin, was conducted. After that, it was decided that pustular psoriasis was to be covered by public health insurance in October 2012. Prof. Kanekura, who early on applied Adacolumn® to patients with pustular psoriasis (Kanekura *et al.*, 2003) and was one of the main researchers in that trial, proposes a detailed explanation on the therapy.

Another bead type adsorbent, Immunopure, is the latest clinically available adsorbent in Europe. Dr. Waltz and Dr. Ramlow, who are the contributors in developing Immunopure and first applied it in patients with active ulcerative colitis, explain the basic and the clinical aspects of the absorbent (Ramlow *et al.*, 2013).

These leukocyte adsorbents would be expected to treat various intractable diseases and elucidate their mechanism of pathophysiology.

References

Kanekura T, Yoshii N, Yonezawa T, *et al.* (2003) Treatment of pustular psoriasis with granulocyte and monocyte adsorption apheresis. *J Am Acad Dermatol* **49:** 329–332.

Karsh J, Wright DG, Klippel JH, *et al.* (1979) Lymphocyte depletion by continuous flow cell centrifugation in rheumatoid arthritis: Clinical effects, *Arthritis Rheum* **22:** 1055–1059.

Kotzin BL, Strober S, Engleman EG, *et al.* (1981) Treatment of intractable rheumatoid arthritis with total lymphoid irradiation. *N Engl J Med* **305:** 969–976.

Malawista SE, Trock D, Edelson RL. (1991) Photopheresis for rheumatoid arthritis. *Ann NY Acad Sci* **636:** 217–26.

Paulus HE, Machleder H, Bangert R, *et al.* (1973) A case report: Thoracic duct lymphocyte drainage in rheumatoid arthritis. *Clin Immunol Immunopathol* **1:** 173–181.

Ramlow W, Waitz G, Sparmann G, *et al.* (2013) First human application of a novel adsorptive-type cytapheresis module in patients with active ulcerative colitis: a pilot study. *Ther Apher Dial* **17:** 339–47.

Sawada K, Ohnishi K, Fukui S, *et al.* (1995) Leukocytapheresis therapy, performed with leukocyte removal filter, for inflammatory bowel disease. *J Gastroenterol* **30:** 322–329.

Yokoyama Y, Fukunaga K, Fukuda Y, *et al.* (2007) Demonstration of low-regulatory CD25High+CD4+ and high-pro-inflammatory CD28-CD4+ T-Cell subsets in patients with ulcerative colitis: Modified by selective granulocyte and monocyte adsorption apheresis. *Dig Dis Sci* **52:** 2725–31.

Chapter 29

White Blood Cell Adsorption: LCAP (Basic Aspect)

Takao Nishimura, Takeshi Sukegawa

1. Introduction

In leukocyte apheresis, the patient's blood is treated outside the body and then returned to the circulatory system which is connected with a patient. The key to this clinical application is how to remove target cells which relate to clinical disorders.

The principles governing the technique of separation of cells, including leukocytes, can be classified into (i) difference in the density or volume of cells, (ii) difference in adherence property between cells and substrate, (iii) difference in the behavior of cells in an electrical field, (iv) difference in the behavior of cells in a magnetic field, and (v) difference in the immunological behavior of cells. The most commonly applied cell separation technique is the centrifugation method, which is based on (i), but the filter method, based on (ii), is increasingly commonly used for the rapid separation of leukocytes with immune function.

As shown in Table 29.1, different methods of separation or removal of leukocytic-series cells have been performed in medical scenarios. Leukocyte apheresis is one of the major applications associated with cell separation technology. Depending on the principle of apheresis technology, current leukocyte apheresis is technically classified

Table 29.1.

Field	Application (example)	Target of Separation	Purpose
Blood transfusion	Production of leukoreduced blood components	Leukocyte in blood components for transfusion	Prevention of side effects due to blood transfusion
Extracorporeal blood circulation treatment	Leukocyte apheresis	Abnormal leukocytes in patient's blood	Treatment for immuno-associated diseases
Regenerative medicine	Separation/ concentration of hematopoietic stem cell	Hematopoietic stem cells in umbilical cord blood	Suppression of storage cost in order to volume reduction, suppression of side effects
Cellular immunotherapy	Separation/ activation of immunologic cell	Immunologic cells in the patient's blood	Cancer treatments

into three types: (a) removal of non-specific leukocyte population, (b) removal of most granulocytes and monocytes, and c) removal of most lymphocytes.

In this chapter, we describe a leukocyte apheresis system for removing non-specific leukocyte population (a) using the CELLSORBA® system, which uses the cyto-adsorption technology with a microfiber non-woven cloth manufactured by Asahi Kasei Medical Co., Ltd. We explain the basic principles of leukocyte apheresis using microfibers, the basic structure and product design of CELLSORBA®, and the techniques required to conduct safe and efficient leukocyte apheresis therapy.

2. Basic Principle of Leukocyte Apheresis with Microfiber Non-Woven Cloth[1]

Initial experiments were focused on the diameter and dimensional structure of the fibers comprising the filter media to examine its leukocyte-capturing properties. As shown in Fig. 29.1, the leukocyte apheresis rate is higher at smaller mean fiber diameters, particularly at less than 3 μm. As can be seen in Fig. 29.2, moreover, non-woven

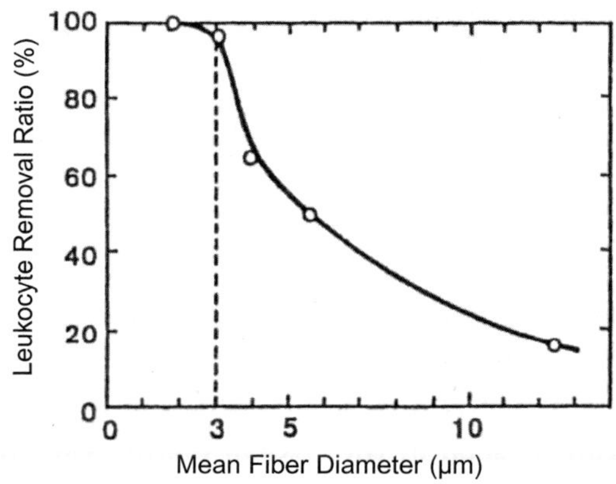

Fig. 29.1. Leucocyte removal ratio based on fiber diameter of nonwoven cloth.

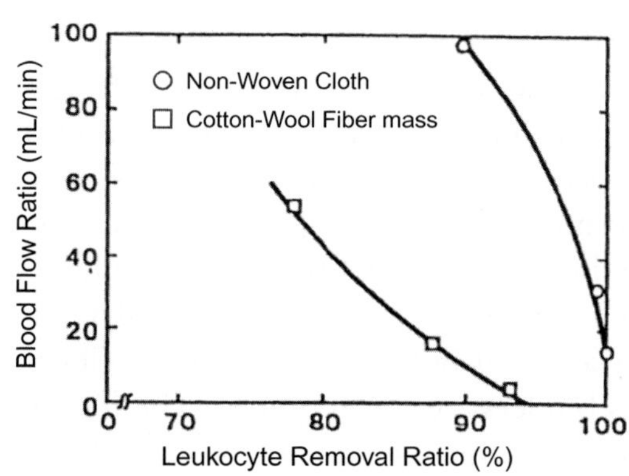

Fig. 29.2. Comparison of performance of non-woven cloth with that of cotton-wool fiber mass.

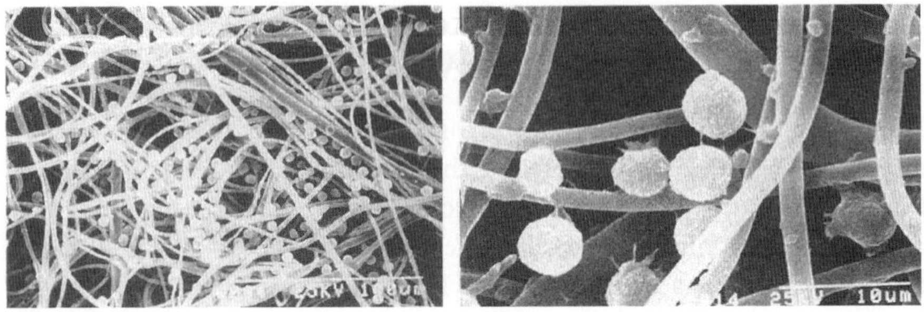

Fig. 29.3. Leukocytes trapped with microfiber non-woven cloth (left: low magnification, right: high magnification).

cloth with a mean fiber diameter of 1.8 μm achieves a high blood processing rate while maintaining a higher leukocyte removal rate than seen with cotton-wool fiber mass.

As shown on the right-hand side of Fig. 29.3, leukocytes that are 6–20 μm in size are sterically trapped near the fiber tangles in the microfiber non-woven cloth with a mean fiber diameter of 1.8 μm. When the fiber diameter was large, such steric trapping was rarely observed. As can be seen in the overall picture on the left-hand side of Fig. 3, moreover, non-woven cloth appears to provide a uniform network of fibers, unlike cotton-wool fiber mass, thus lowering the risk of "short-circuiting" of blood. These mechanisms allow a high leukocyte removal rate to be combined with a high blood flow rate.

This polyester microfiber non-woven cloth is prepared by melt blowing, as shown in Fig. 29.4. Polyethylene terephthalate (PET) or a similar thermoplastic polymer is heated to liquefy it, then extruded through a die comprising a line of spinnerets, then stretched by a gas at subsonic speed to make microfibers. It is finally deposited as non-woven cloth on a net and reeled off. Since its shape is maintained by the interlocking and self-bonding of the microfibers, possibly toxic additives such as binders are not required. The filter media using this melt blowing method achieves high porosity, so blood flow is smooth.

This microfiber non-woven cloth can be produced industrially in large quantities in a clean environment, allowing it to be commercialized as a filter for medical use.

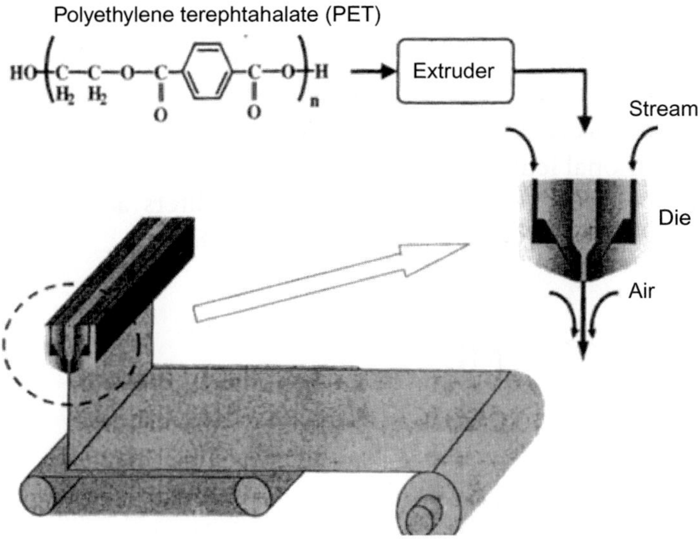

Fig. 29.4. Manufacturing process of microfiber non-woven cloth by melt blowing.

3. Application of Leukocyte Removal Filters to Blood Transfusion

To adopt the leukocyte-capturing technique using microfiber non-woven cloth in clinical practice, we attempted to create its application to the removal of leucocytes during blood transfusion, which was widely recognized in the 1980s.

Blood transfusion is an effective therapeutic method of replenishing components essential for life support, but since leukocytes that are present in various blood preparations can cause adverse reactions, it is desirable to remove them as effectively as possible.

Asahi Kasei Medical Co., Ltd. developed their leukocyte removal filter for erythrocyte components by collaborative research with Japanese and overseas researchers in blood transfusion field (Sepacell R Type: 1985).

Sepacell R-500 can be used to treat erythrocyte components equivalent to 400–1,000 mL of whole blood within 20 minutes. It achieves a leukocyte removal rate of 99% or more, a platelet removal rate of 90% or more, and an erythrocyte recovery rate of 90% or more. It has been confirmed to have no adverse effects on erythrocyte function. As shown in Fig. 29.5, it has been clinically confirmed that it excels in the prevention of adverse reactions to blood transfusions over conventional leukocyte removal methods, such as centrifugation. The non-woven type of leukocyte removal filters are disseminated globally and widely used for erythrocyte transfusion.

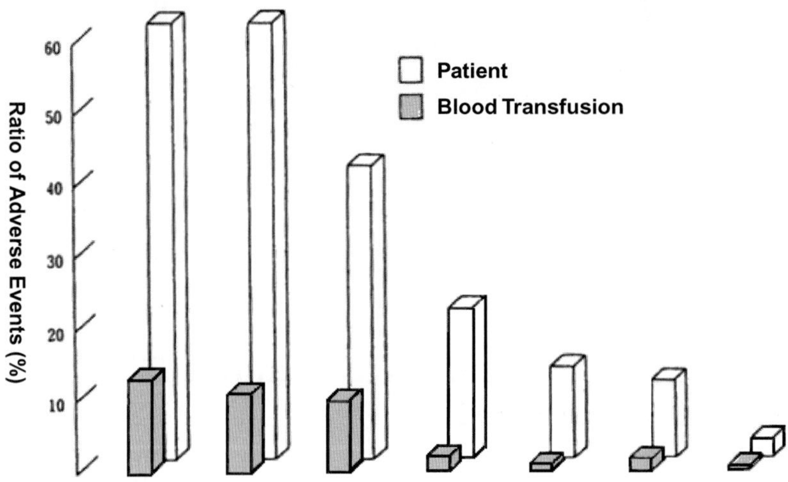

Fig. 29.5. Manufacturing method and incidence of adverse reaction.

Term	JAN 1980 DEC 1980	JAN 1981 DEC 1981	JAN 1982 OCT 1982	NOV 1982 DEC 1983	JAN 1984 APR 1985	MAY 1985 OCT 1985	NOV 1985 MAY 1986
Number of Patients	75	79	80	81	81	81	80
Unit of Blood Transfusion	1122	1179	847	1455	1630	654	611
Manufacturing Process	Centriguge			Filtration by cotton fiber		Cooling Centrifuge+ Filtration	Filtration by Sepacell

Following its popularization, further technical benefits were demanded by the market. Covering the fiber surface with a polymer bearing hydrophilic and positively-charged groups significantly improved the wettability by blood of PET non-woven cloth. The result was increased bioavailability of filter elements and improved leukocyte-trapping capacity due to the positively-charged groups, which led to ultra-high performance of 99.999% (1995). It became the *de facto* standard for high-performance filters and represents a major contributor to the improved safety of blood transfusions worldwide.

Figure 29.6 shows the residual rates of granulocytes, lymphocytes and platelets in relation to the thickness of non-woven cloth when the

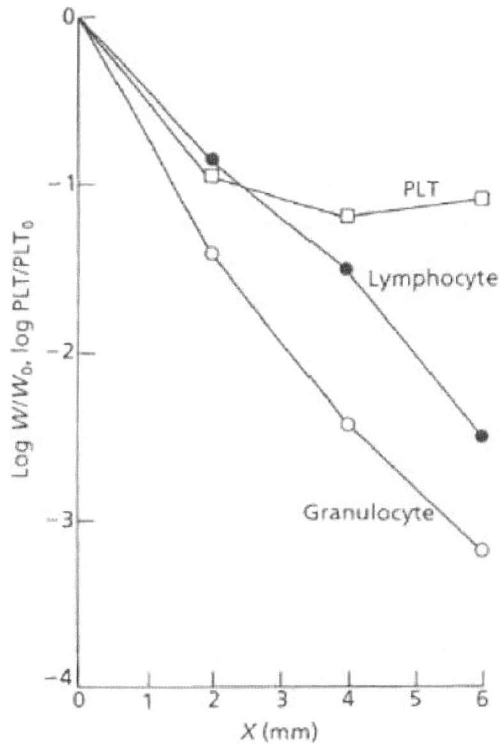

Fig. 29.6. Residual ratios of granulocytes, lymphocytes and platelets in relation to the thickness of non-woven cloth.

erythrocyte preparation was stored at 4°C for three days (Nishimura *et al.*, 1994). The residual rates of granulocytes and lymphocytes decreased logarithmically as a function of the thickness of the non-woven cloth. It can be seen from the downward slope of the line that granulocytes and lymphocytes consist of cell populations with uniform adsorptivity and that granulocytes are more likely to be adsorbed than lymphocytes. On the other hand, about 10% of platelets passed through, irrespective of the thickness of the non-woven cloth. It appears that about 10% of platelets lose their adsorptivity during storage at 4°C for three days.

For blood transfusions, leukocytes are removed from the blood of healthy subjects to prevent the occurrence of adverse reactions in patients during the process.

Asahi Kasei Medical Co., Ltd. attempted to develop a leukocyte apheresis system, CELLSORBA®, which removes the abnormal leukocytes of patients in extracorporeally circulated blood, by utilizing the numerous findings accumulated through the development of a leukocyte apheresis filter for the blood transfusion process (Nishimura, 1993).

4. CELLSORBA® System for Leukocyte Apheresis

CELLSORBA is a therapeutic device designed to reduce inflammation caused by active disease by applying leukocyte apheresis (LCAP) to remove leukocytes (granulocytes, monocytes and lymphocytes) in the peripheral blood by extracorporeal circulation. The CELLSORBA filter consists of non-woven cloth made of polyester microfibers, which is layered in cylindrical form into the column. The leukocytes in the blood that have passed through the non-woven cloth filter consisting of these microfibers are removed by filtration and adsorption.

LACP is a therapeutic method developed in Japan, in which blood is withdrawn from a vein in one arm of the patient at a blood flow rate of about 30 to 50 mL/min, mixed with anticoagulant in the circuit immediately after withdrawal so as to prevent coagulation of

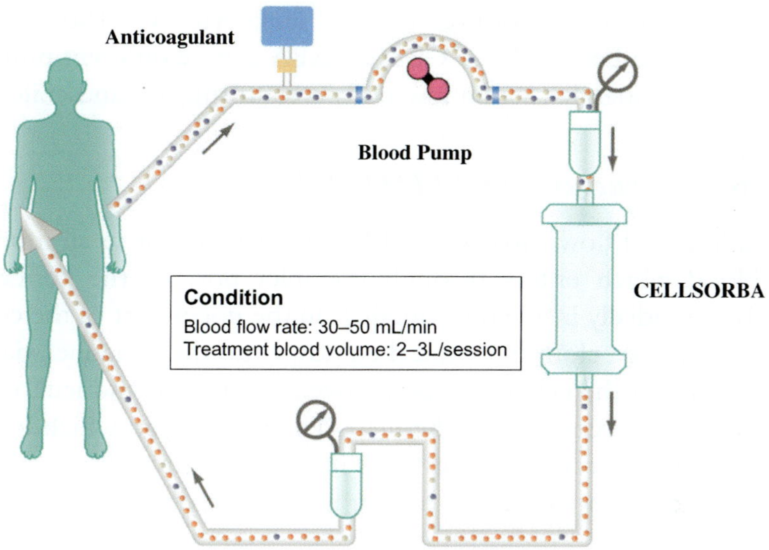

Fig. 29.7. Scheme of leukocyte apheresis treatment.

the blood, and delivered to CELLSORBA. The blood from which leukocytes have been removed with CELLSORBA is returned to a vein in the other arm of the patient at the same blood flow rate. This extracorporeal procedure is performed five times, at a frequency of once a week, processing a blood volume of about 2–3 L each time. If the inflammation caused by the disease is serious, the treatment may be performed more frequently, such as twice a week.

4.1. *Types of CELLSORBA*

There are two CELLSORBA product types: CELLSORBA and CELLSORBA-E, depending on the target disease. There are two models of each product type. CELLSORBA used for rheumatoid arthritis is called CS-100 and CS-180S, according to the volume of blood processed, and CELLSORBA-E, for inflammatory bowel disease, has two models: EX and EI for different patient body weights. CS-100 and CS-180S have differently structured leukocyte apheresis filters. When a larger volume of blood is to be processed and the filter is likely to clog because of high levels of leukocytes, CS-180S is normally used. The EX and EI of CELLSORBA-E have different priming volumes. EI is mainly used in low-body-weight subjects and children.

4.2. *Basic Structure of CELLSORBA*

The basic blood flow through CELLSORBA is similar in all models. The blood which enters through the inlet port of the container spreads out radially inside the container in the upper part of the cylindrical non-woven cloth filter and then flows into the container via the outer surface of the cylindrical non-woven cloth filter, and leukocytes and platelets are removed by the filter. The blood from which leukocytes, etc., have been removed flows into the central hollow core of the cylindrical non-woven cloth filter and exits from the outlet port

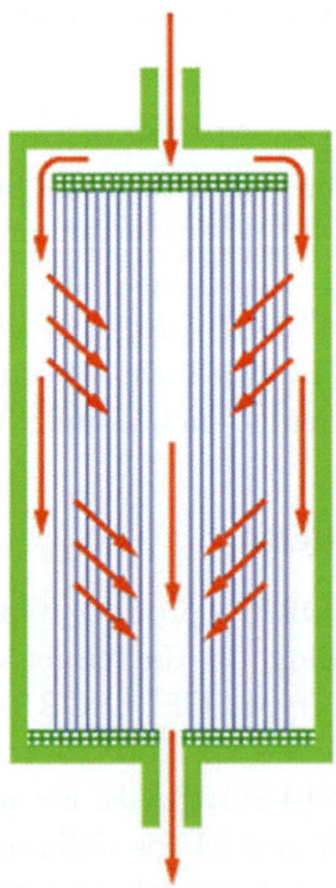

Fig. 29.8. Basic blood flow through CELLSORBA.

of the filter that communicates with the outlet port of the container. In CELLSORBA, since the filter may clog with leukocytes or platelets, it is recommended to make the direction of blood flow initially through the outer surface, which has a larger filtration area than the circumferential surface.

4.3. *CELLSORBA CS-100 and CS-180S*

CELLSORBA is used to improve clinical symptoms in patients with rheumatoid arthritis that is resistant to drug therapy. Rheumatoid arthritis is a disease in which inflammation occurs in all the joints: this causes pain, swelling or deformation and results in damage to joints and dysfunction if the disease progresses. Activated leukocytes are involved in this inflammation of the joints. If they enter the joint, substances that induce inflammation (mediators) are released that cause inflammation inside the joint. Since the synovial thickens and joint fluid increases due to this inflammation, the joint swells. If the activated leukocytes stay in the joint and prolong inflammation, they may cause breakage of cartilage or bones in the joint. CELLSORBA removes the activated leukocytes in the blood and thus reduces inflammation.

The difference between the CELLSORBA CS-100 and CS-180S is, as shown in Table 29.2, the method of winding of the non-woven

Table 29.2.

Trade Name		Cellsorba		CELLSORBA-E	
Model		CS-100	CS-180S	EX	EI
Main country of sale		Japan	Japan	Japan and Europe	Japan
Target disease		Rheumatoid arthritis		Inflammatory bowel disease*	
Priming volume		170 mL	270 mL	170 mL	90 mL
Filter structure		Cascade type	Spiral Type	Cascade type	Cascade type
Material	Filter	Polyethylene terephthalate (PET)			
	Housing	Polycarbonate			

*Approved in Japan for ulcerative colitis.

cloth filter. CELLSORBA CS-100 has a "cascade" filter structure. The main filter, consisting of microfibers, is placed in the inner circumferential surface of the cylindrical filter structure and the prefilter, consisting of fibers thicker than those of the main filter, is positioned on the outer side. When the blood flows from the outside to the inside of the cylindrical non-woven filter, more active leukocytes and platelets are trapped by the prefilter in the outer layer, thus reducing the burden on the main filter; and by trapping leukocytes stepwise, the removal efficiency of the main filter is increased.

On the other hand, CS-180S has a spiral filter structure, which is made by winding a spacer together within the non-woven cloth (main filter) consisting of microfibers that are thinner than the non-woven cloth used in CS-100. This is intended to improve the volume of leukocyte apheresis by increasing the area of the main filter and preventing clogging of the main filter. Normally, the blood flows from the outer surface to the inner surface of the cylindrical non-woven cloth filter, but if the outer part becomes clogged with activated leukocytes or platelets, the blood flows into the space made by the spacer and then to the inner circumference, thus preventing clogging. This increases the surface area of main filter, so even if a large volume of blood is treated, clogging is unlikely to occur, and a high degree of leukocyte removal can be acquired.

The leukocyte removal rate of CS-100 and CS-180S in 15 patients with rheumatoid arthritis was, as shown in [Figure], 91.8% (CS-100) and 97.3% (CS-180S) during the treatment of 3 L of blood in each case. A high leukocyte removal rate was achieved with both devices. Comparing the removal rate using each differential leukocyte count showed the granulocyte removal rate to be the same, at 99.5%, between CS-100 and CS-180S, but the lymphocyte removal rate was significantly different, at 34.8% for CS-100 and 73.4% for CS-180S (Iwamoto and Ueki, 2000).[4]

As a result of clinical research to investigate the clinical effects of CELLSORBA in 57 patients with rheumatoid arthritis that was resistant to drug therapy, a high therapeutic effect of 64–79% was achieved, far exceeding the 20% improvement (ACR20) basis set by the American College of Rheumatology (ACR) (Felson, *et al.*, 1993).

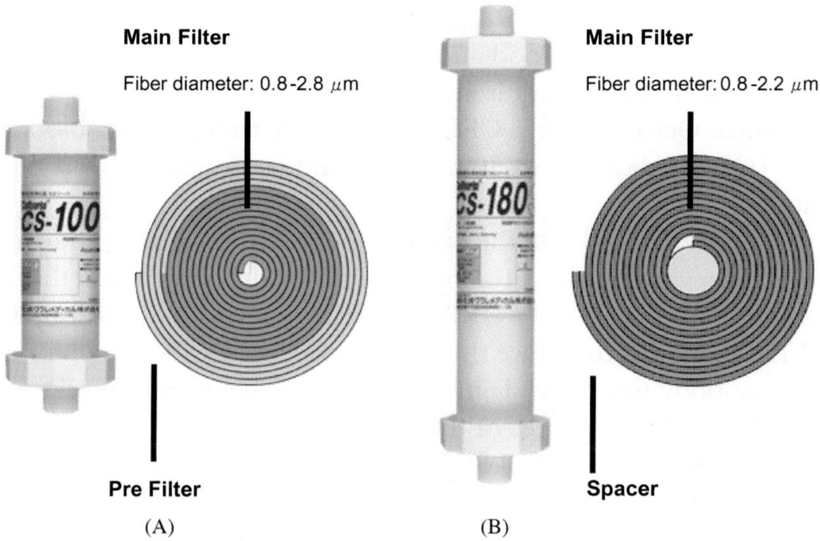

Fig. 29.9. Comparison of the structure between CELLSORBA CS-100 and CS-180S. (A) CS-100: Cascade-type filter structure, which consists of main filter and pre-filter. (B) CS-180S: Spiral-type filter structure, which consists of main filter and spacer.

Moreover, the number of swollen joints and the number of joints showing tenderness decreased from 12 to four, and from 13 to five on average. CRP also decreased (Hidaka *et al.*, 1999; Ueki *et al.*, 2000).

4.4. *CELLSORBA-E Type (EX and EI)*

CELLSORBA-E Type is used to improve clinical condition and for induction of remission in ulcerative colitis that is in the active phase. Ulcerative colitis is a disease in which inflammation occurs in the large-intestinal mucosa that results in erosions (sores) or ulcers. It appears that leukocytes are implicated in inflammation in the large intestine. Leukocytes are stimulated for certain reasons and activated. Various substances (mediators) involved in inflammation are released from the activated leukocytes and inflammation is induced in the intestinal mucosa. CELLSORBA-E removes the activated leukocytes and platelets from the blood and relieves inflammation.

CELLSORBA-EX acquired approval in Japan and has been marketed since October 2001. CELLSORBA-EX circulates the blood at a flow (Qb) of about 30 to 50 mL/min and treats about 2 or 3 L per treatment.

CELLSORBA-EI has the same filter structure, but as shown in Table 2 previously, the priming volume is different from that of CELLSORBA-EX. CELLSORBA-EI was obtained by miniaturizing CELLSORBA-EX to reduce the priming volume by about half for use with low-body-weight and infant patients. Therefore, the external circumferential surface area of the cylindrical non-woven cloth filter is also about half that of CELLSORBA-EX. To make the linear speed of the blood passing through the surface of non-woven cloth filter similar to that of CELLSORBA-EX, the extracorporeal flow rate of blood was reduced by half (Qb = about 15–25 mL/min). Since the volume held in the filter is also half that in CELLSORBA-EX, the volume of blood treated is also reduced to half that treated by CELLSORBA-EX (about 1.0–1.5 L).

The leukocyte removal rate for CELLSORBA-EX in 14 patients with ulcerative colitis can reach about 94% at 15 minutes after the start of treatment, but decreases to about 65% an hour after the start. This is due to the fact that the removal rate for granulocytes and monocytes, etc., is maintained at about 100% from start to the end of treatment, but the lymphocyte removal rate decreases with elapsing of treatment time. Moreover, it has been found that the lymphocyte removal rate correlates with the platelet removal rate. The principles of leukocyte apheresis with CELLSORBA-E are considered to be the following two methods: filtration with non-woven cloth filter and adsorption because of the phagocytosis rate of leukocytes. Most of the granulocytes and monocytes adsorb to the filter fiber themselves due to their strong phagocytic capacity. On the other hand, lymphocytes have weak phagocytic capacity, and are thus removed mainly by the effect of filtration, but it also appears that lymphocytes are removed by another process: platelets attach to the fiber surfaces and form fibrin, to which the lymphocytes attach.

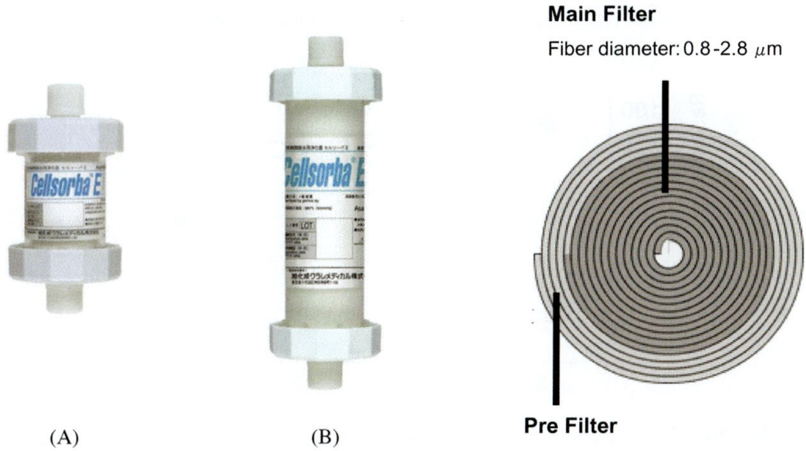

Fig. 29.10. Comparison of the structure between CELLSORBA-EX and EI. (A) CELLSORBA-EX: Cascade-type filter structure. (B) CELLSORBA-EI: Cascade-type filter structure, which is smaller than EX.

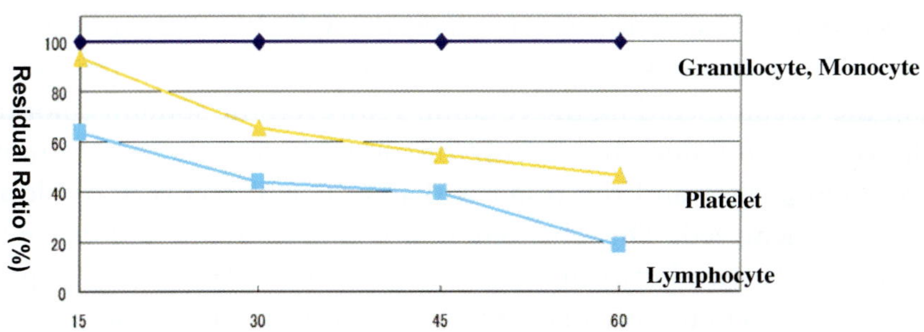

Fig. 29.11. Residual ratios of leukocytes depending on treatment time.

4.5. *Efficacy and Safety of CELLSORBA*

4.5.1 *Prevention of clogging by blood coagulation*

In CELLSORBA, if too low a dosage of anticoagulant is used or insufficient mixing occurs, blood coagulation may be accelerated and the non-woven cloth filter will clog, resulting in increased pressure in

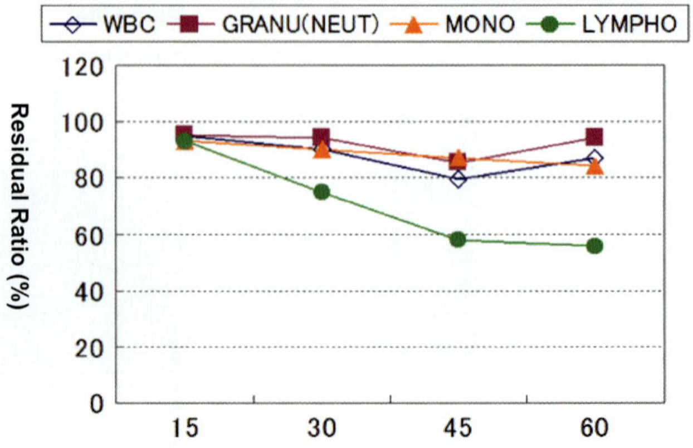

Fig. 29.12. Residual ratios of leukocytes depending on treatment time.

CELLSORBA during extracorporeal circulation. This pressure may vary according to the condition of the blood being treated. After extracorporeal circulation is complete, the CELLSORBA column is rinsed with isotonic sodium chloride solution to recover the blood from the circuit including the column. However, the sodium chloride solution is sometimes not evenly distributed over the whole non-woven cloth filter and the result is that sufficient recovery of blood cannot be achieved. This phenomenon is caused by the difference in the flow resistance between the inlet flow path of blood, which present in outer circumferential surface of the non-woven cloth filter, and the output flow path of the blood, which passes via the inner circumferential path. Computer simulation was therefore employed to arrive at an optimum shape. It was found that if the resistant body was gradually made thinner, from about 1/2 to 3/4 of the length of the non-woven cloth filter, and the resistant body's diameter falls to zero near the outlet, the speed of the blood flowing in the radial direction through the non-woven cloth filter in the direction of the length of non-woven cloth filter is made almost uniform. Adopting this technique allows clogging during treatment to be prevented, and restoration of blood with isotonic sodium chloride solution after the end of treatment can reduce blood loss.

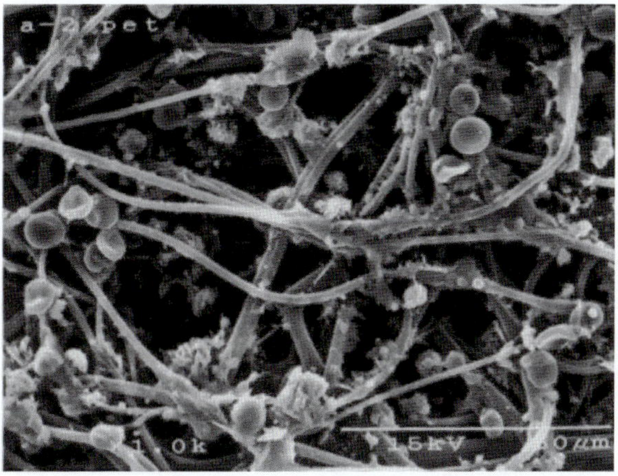

Fig. 29.13. Leukocytes trapped with microfiber non-woven cloth after CELLSORBA treatment.

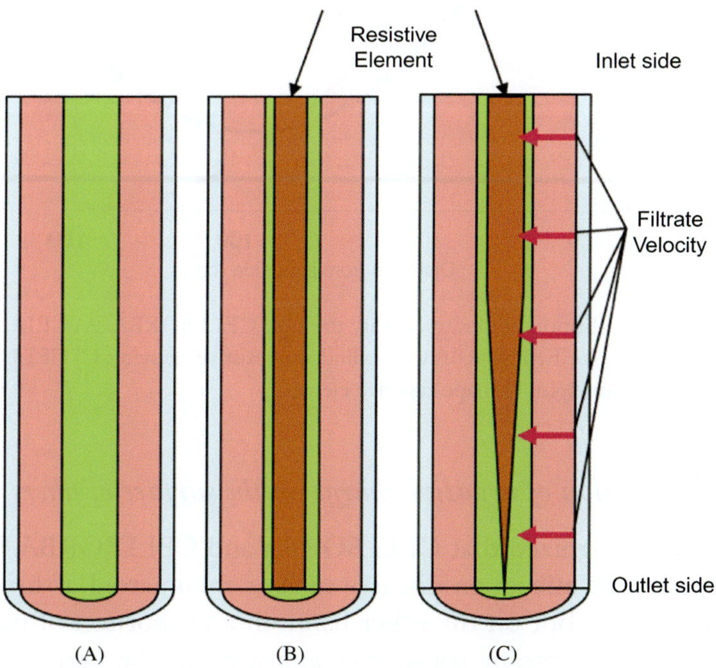

Fig. 29.14. Images of blood flow through CELLSORBA. A) CELLSORBA before prevention; B) CELLSORBA with cylindrical resistive element; C) CELLSORBA after prevention with spindle shape resistive element.

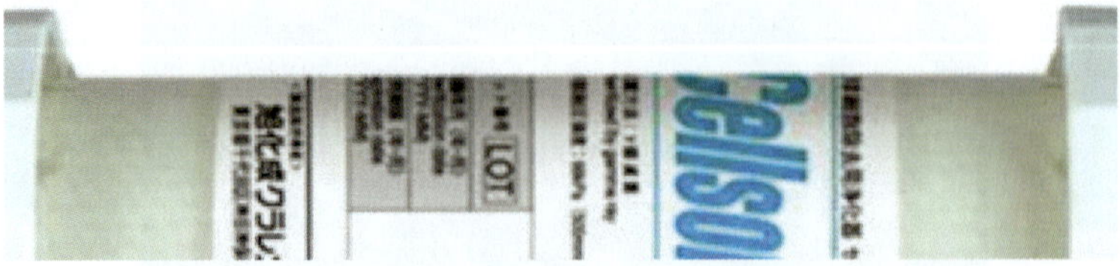

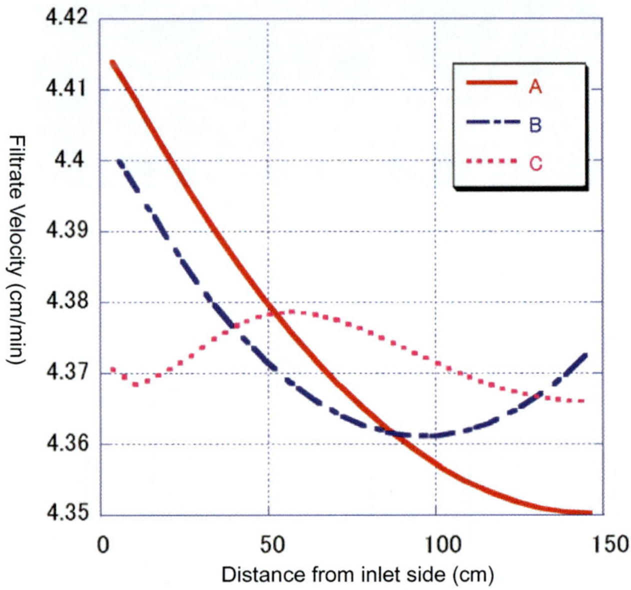

Fig. 29.15. Distribution of filtrate velocity through CELLSORBA. A) CELLSORBA before prevention; B) CELLSORBA with cylindrical resistive element; C) CELLSORBA after prevention with spindle shape resistive element.

4.5.2. *Inhibition of negative charge on the surface of fibers*

It has been recognized that CELLSORBA and CELLSORBA-E have high therapeutic effects on diseases such as rheumatoid arthritis and ulcerative colitis, but on the other hand, in a proportion of patients, anaphylactoid symptoms sometimes appeared in the early phase of treatment. These included hot flashes and a fall in blood pressure. It is difficult to predict their occurrence. The main causes of these anaphylactoid symptoms are because the non-woven cloth filter is made

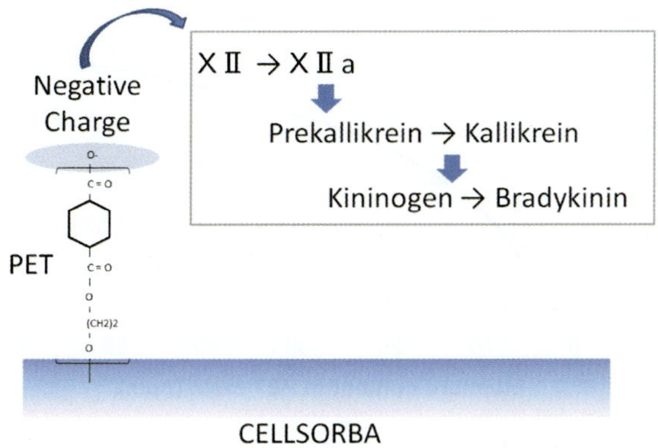

Fig. 29.16. Generation mechanism of Bradykinin.

of polyester fiber, and bradykinins are produced in the blood due to the negative charge on the surface of this fiber. When the blood comes into contact with the negative charge on the surfaces of materials such as glass, blood coagulation factor XII is activated, resulting in bradykinin production from a polymer kininogen via prekallikrein and kallikrein. In the polyethylene terephthalate used in the non-woven cloth, negatively-charged carboxyl groups are present as polymer terminals. Since, however, there are very few negative charges on the surface of polyethylene terephthalate non-woven cloth, and its surface is not flat and smooth, it proved difficult to quantify the degree of electric charge, even by acid–alkali titration and various surface analysis methods. Onodera *et al.* developed a method of quantifying the negative charge on the surface of non-woven cloth utilizing a reaction that produces complex triiodide ions from iodine and iodide ions. They discovered a technique for minimizing the amount of negative charge on the surface of non-woven cloth by controlling the water content of polyethylene terephthalate resin while spinning the non-woven cloth (Onodera and Yoshida).

4.5.3. *Adsorptivity of drug*

CELLSORBA and CELLSORBA-E are often used in combination with the drugs used to treat rheumatoid arthritis and ulcerative colitis.

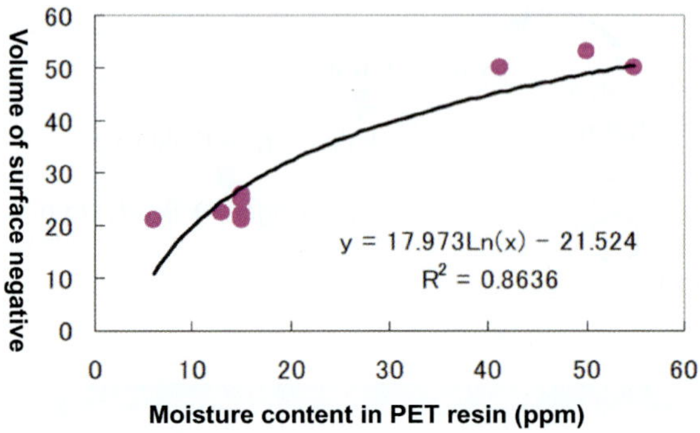

Fig. 29.17. Relationship between moisture content and surface negative charge volume.

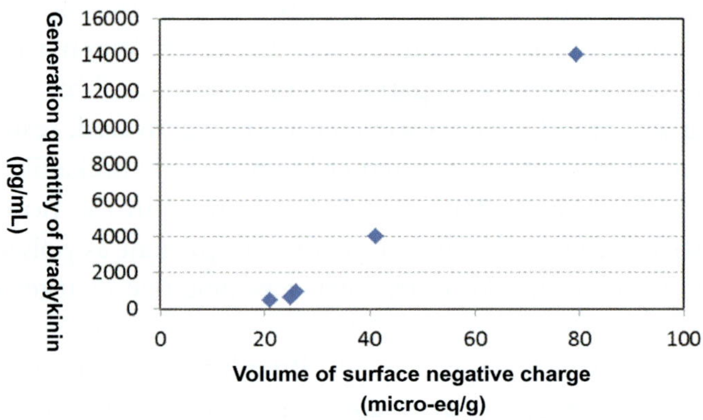

Fig. 29.18. Quantity of bradykinin generated by surface negative charge.

However, there is as yet little information on the adsorptivity of these drugs to CELLSORBA. A mini-scale column, which is approximately 1/9-scale of CELLSORBA-E (model EI), was therefore prepared to evaluate the adsorption characteristics of two drugs, prednisolone (AbuRuz *et al.*, 2003) and mesalazine (5-ASA) (Bystrowska *et al.*, 2000), commonly used to treat ulcerative colitis in Japan. The mini-scale column was obtained by molding and packing about 1/9 volume of non-woven cloth cylindrically in a polycarbonate container 20 mm in diameter and 36 mm in height that was then radiation-sterilized. Through this mini-scale column, about 174 mL of blood spiked with a drug was passed at a flow rate of about 3 mL/min. The drug concentration in the blood was determined after 10, 20 and 40 minutes, at the end, and 10 minutes after returning the blood. The results showed that prednisolone tended to be adsorbed at about 6% at 10, 20 and 40 minutes, and mesalazine (5-ASA) tended to be adsorbed at about 11% at 10 minutes. On returning the blood, the concentration again decreased due to the influence of dilution with isotonic sodium chloride solution. The drug that had adsorbed to the non-woven cloth filter desorbed, but no pattern of concentration by elution was observed. The margin of error using this determination system is about 15%, but since a significant difference was observed at the above points of determination, there is a distinct possibility that adsorption is taking place. These adsorptions may be attributable to hydrophobic interaction between various materials and drugs. It should be possible to provide safer treatment by investigating the adsorptivity of specific drugs.

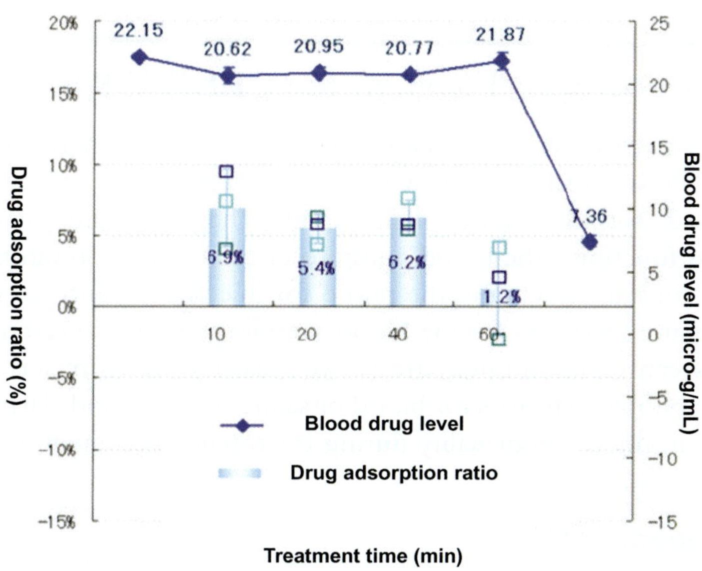

Fig. 29.19. Adsorption characteristics of mesalazine (5-ASA). Before treatment, the level was 22.15 μg. Blood was returned after 60 min, and 10 min after this the level was 7.36 μg.

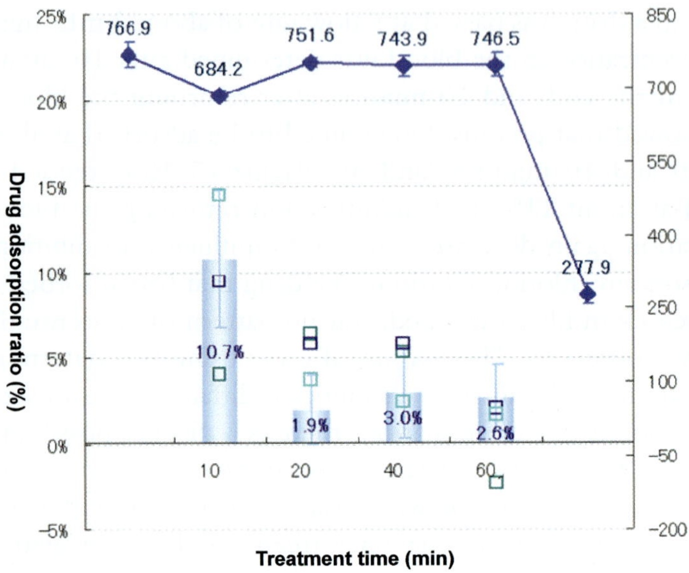

Fig. 29.20. Adsorption characteristics of prednisolone. Before treatment, the level was 766.9 ng. Blood was returned after 60 min, and 10 min after this the level was 277.9 ng.

5. Extracorporeal System for CELLSORBA

For LCAP treatment, a suitable extracorporeal system is required, such as Plasauto-LC for direct hemoperfusion (DHP) procedure or Plasuato-Σ for multi-functional system. These machines have the required functions, the blood pump to circulate the blood in the circuit by constant flow rate, the pump for injection of anti-coagulant, the pressure monitors for (a) blood sampling point, (b) arterial side which means column inlet, and (c) venous side which means return side, air-bubble detectors for blood or saline solution, and alarm functions to notify of abnormality during treatment to co-medicals.

5.1 *Plasauto-LC*

Plasauto-LC was developed for exclusive use for direct hemoperfusion (DHP) treatment and has been supplied in Japan since 2001. Features of this machine are: (1) miniaturization of its size by focusing on required functions, (2) adopting new tubing set exclusively designed for this machine, and (3) display of step-by-step instruction in the color panel. Due to these features, Plasauto-LC is well adopted by doctors and co-medicals who are not familiar with the extracorporeal treatment procedure.

The blood pump of Plasauto-LC has a wide range of flow rate capability of 10–150 mL/min, and is therefore well adopted for various types of patient condition, such as body weight or physical conditions. A transfuse pump for injection of anticoagulant maintains the mixing ratio of blood and anticoagulant at the operator's setting in accordance with automatic calculation by machine. Loading of tubing set is automated by the pump's function. Such user-friendly features of the exclusive system drive diffusing of LCAP treatment in Japan.

5.2 *Plasauto-Σ*

Plasauto-Σ is a multi-functional type of extracorporeal system, which also adopts plasma exchange (PE), acute renal replacement therapy (CRRT) and other types, including LCAP. It has three major features: (1) Speed: The cassette type of tubing was exclusively designed for this machine, and makes setting procedure much faster and easier, achieving 75% reduction of setting time compared to other therapeutic apheresis systems. (2) Stability: An air-free chamber is applied for this system to minimize the contact of blood with air, and circuit volume is also minimized. The accuracy of dialysis volume especially for CRRT procedure is significantly improved to four times higher than that of similar systems. This provides flexibility of treatment condition for higher flow rate, for longer extracorporeal process and for pediatric patients. (3) Safety: There are several automatic control

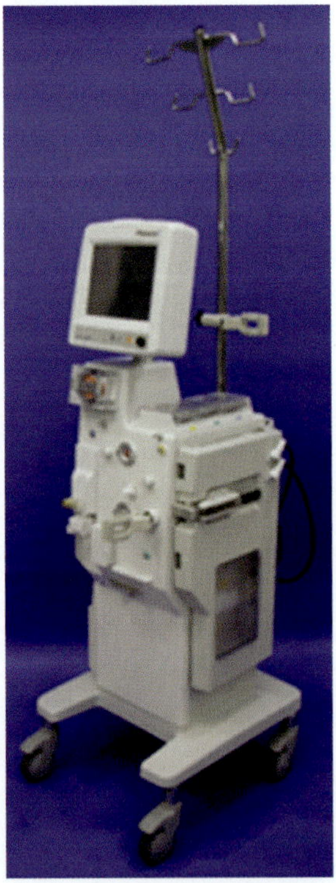

functions to extend its safety use, such as auto-priming and auto-flow control. There is visual guidance for treatment steps and also for trouble shooting effectively, supporting safet operation.

Plasauto-Σ (a brand name in Japan: ACH-Σ) is well accepted not only in Japan but also European countries, China and other areas.

6. Extension of the Technique to Separation of Targeted Materials by Antibody Fixation (Onodera *et al.*, 1998)

If the separation of specific immune cells can be readily performed, the impact will be significant and should open new paths for development of novel therapeutic methods. By fixing monoclonal antibodies to the non-woven cloth of polystyrene microfibers and introducing active groups using the reaction shown in Fig. 29.21, it has become possible to specifically remove CD4-positive cells (helper T lymphocytes), as shown in Fig. 29.22.

Targeted separation and removal of all cells differentiable with monoclonal antibodies is possible in principle, so widespread application can be expected.

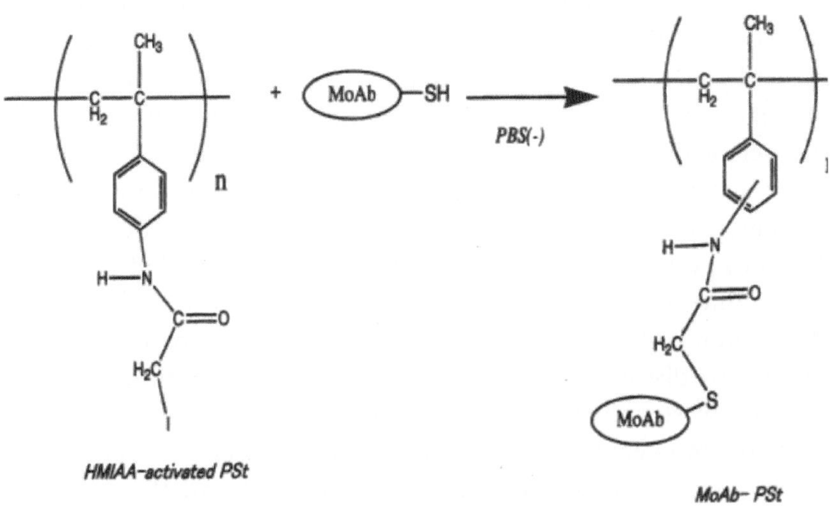

Fig. 29.21. Antibody fixation reaction.

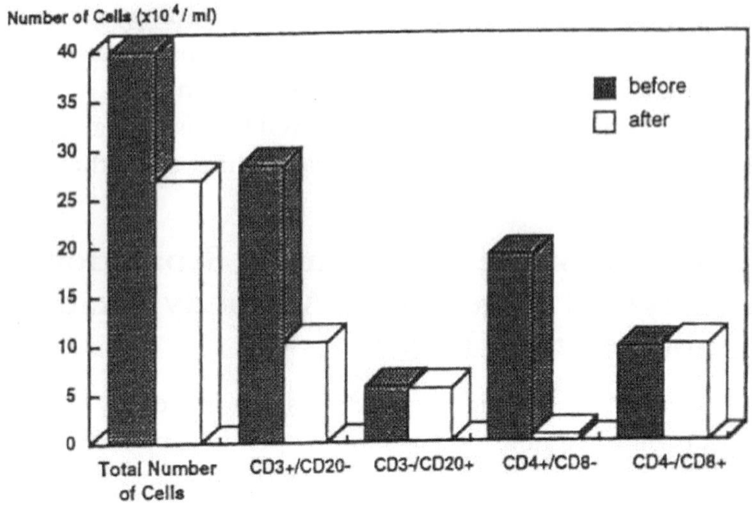

Fig. 29.22. Specific separation with antibody-fixing non-woven cloth.

7. Conclusion

CELLSORBA® was originally developed based on the design concept that the leukocyte-trapping technique by microfiber non-woven cloth can be applied for use in extracorporeal circulation therapy. The processing of blood using non-woven cloth is characterized by the fact that leukocyte apheresis of granulocytes with high adhesiveness to the surfaces of the material as well as lymphocytes can be comprehensively performed.

Extracorporeal circulation therapy requires the capacity to process several liters of blood. Special attention was paid to the design of this column to permit the device to be employed safely and effectively for the continuous processing of extracorporeally circulated blood and to make it amenable to various methods and conditions of processing suited to the disease in question and the patient's condition, resulting in its present structure.

There are likely to be numerous conditions under which immune cells are involved in the mechanism of disease, and thus there is a wide potential range of applications of leukocyte apheresis therapy.

In inflammatory bowel disease and rheumatism, therapeutic methods have already been established in combination with drug treatment. Leukocyte apheresis therapy is a therapeutic method with considerable future potential.

References

AbuRuz S1, Millership J, Heaney L, McElnay J. (2003) Simple liquid chromatography method for the rapid simultaneous determination of prednisolone and cortisol in plasma and urine using hydrophilic lipophilic balanced solid phase extraction cartridges. *Journal of Chromatography* **798:** 193–201.

Bystrowska B1, Nowak J, Brandys J. (2000) Validation of a LC method for the determination of 5-aminosalicylic acid and its metabolite in plasma and urine. *Journal of Pharmaceutical and Biomedical Analysis* **22:** 341–347.

Felson DT, *et al.* (1993) The american college of rheumatology preliminary core set of disease activity measures for rheumatoid arthritis clinical trials. *Arthritis Rheum* **36:** 729–740.

Hidaka T, *et al.* (1999) Filtration leukocytapheresis therapy in rheumatoid arthritis. *Arthritis Rheum* **42:** 431–437.

Iwamoto and Ueki. (2000) Development of leukocyte apheresis system for extracorporeal circulation therapy. *Japanese Journal of Apheresis* **19:** 38–41.

Nishimura T, Kuroda T, Mizoguchi Y, *et al.* (1989) Advanced methods for leucocyte removal by blood filtration. In: Brozobic B (ed.), *The Role of Leucocyte Depletion in Blood Transfusion Practice.* Blackwell Scientific Publications, Oxford, pp. 35–40.

Nishimura T. (1993) Polymer materials for blood purification. In: Tsuruta T, Hayashi T, Kataoka K, *et al.* (eds.), *Biomedical Applications of Polymeric Materials.* CRC Press, pp. 191–218.

Nishimura T, Oka S, Yamawaki N. (1994) Technical aspects of leucocyte depletion. In: Hogman CF (ed.), *Leucocyte Depletion of Blood Components.* VU University Press, Amsterdam, pp. 51–58.

Onodera H, Abe Y, Yoshida M, *et al.* (1998) A new device for selective removal of CD4+ T Cells. *Therapeutic Apheresis* **2(1):** 37–42.

Onodera H, Yoshida M. EP 0561379B Filter medium having a limited surface negative charge for treating a blood material.

Onodera H, Yoshida M. Patent No. 3176752: Blood filtering material.

Ueki Y, *et al.* (2000) Evaluation of filtration leukocytapheresis use in the treatment of patients with rheumatoid arthritis. ***Rheumatology*** **39:** 165–171.

Chapter 30

Leukocytapheresis (LCAP) with Cellsorba Leukocyte Removal Filter Column as One Therapeutic Option for Inflammatory Bowel Disease

Koji Sawada*

1. Introduction

Inflammatory bowel disease (IBD) is a chronic disease with repeated relapses and remissions which is, caused by severe intestinal inflammation. Although the etiology of IBD is not fully revealed, leukocytes (including neutrophils, monocytes, and lymphocytes) are elevated with activation behavior in the lesion and leukocyte-produced inflammatory cytokines are considered to be important factors in the pathogenesis of IBD. Accordingly, leukocytes should be appropriate targets for the therapy for IBD. In clinical studies, striking clinical efficacy and safety has been reported with leukocytapheresis, which might be an effective adjunct to therapy for IBD to promote remission

*Chionkai, Dojima General and Gastroenterology Clinic (Inflammatory bowel disease & liver disease), B1F Shin-Fujita Bild. Dojima 2-4-27, Kita-ku, Osaka, 530-0003, TEL 06-6343-1000/FAX 06-6343-1005, E-mail: k-sawada@dojima-gi-clinic.jp.

and taper conventional drug dosage, especially leading to a marked reduction in the dose of corticosteroids. The results should further understanding of the pathophysiology of IBD.

In the early 19th century, centrifugation (Fig. 30.1) became a common practice for separating blood components for therapeutic purposes (Abel *et al.*, 1924). Leukapheresis was performed for the treatment of leukemia to reduce leukemic white blood cells (Lowenthal, 1977; Goldfinger *et al.*, 1980; Lanc, 1980). Lymphocytapheresis (to reduce lymphocytes) has been a frequently used therapeutic term applied to diseases that have a suspected cellular immune mechanism, such as rheumatoid arthritis (Wallance *et al.*, 1980), multiple sclerosis (Hauser *et al.*, 1984) myasthenia gravis (Valbonesi *et al.*, 1982), and renal allograft rejection (Kurland *et al.*, 1980).

Several reports have described the application of apheresis to the treatment of inflammatory bowel disease (IBD), especially Crohn's disease (CD). Plasmapheresis (Holdstock *et al.*, 1979) and T-lymphocyte apheresis by centrifugation (Bicks *et al.*, 1995) have been reported to exert anti-inflammatory effects in the active stage of CD and induce long-term remission. It is difficult to remove

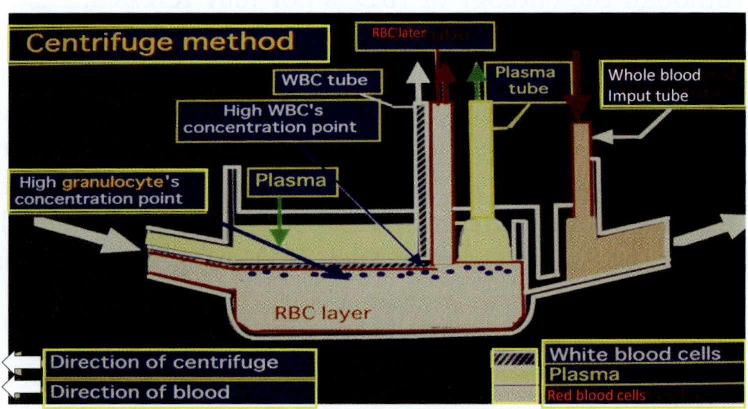

Fig. 30.1. Even though activated granulocytes (Neurophils) are target removed, they are difficult to remove via the centrifugal method because their gravity is greater than that of monocytes and lymphocytes so that they are buried and exist in the red blood cell layer.

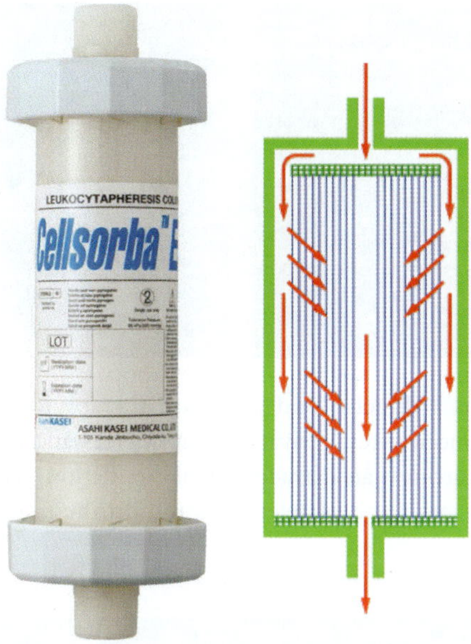

Fig. 30.2A. Cellsorba® E column (Asahi Kasei Medical Co., Ltd., Tokyo, Japan) consists of nonwoven polyester fiber in a polycarbonate container sterilized by γ-ray irradiation and has a priming volume of 200 mL. When anti-coagulated whole blood passes through the column from outside to inside, white blood cells adhere to the filter of the column.

targeting cells, activated granulocytes (neutrophils) because of those gravity. To remove the activated granulocytes, the other new techniques have been considered.

Cytapheresis (CAP) with filter could involve removing white blood cells, especially activated granulocytes, from whole blood cells that are inappropriately increased in the activated and promoting diseases. Leukocytapheresis (LCAP) was administred with Cellsorba® E leukocyte removal filter column (Asahi Kasei Medical Co., Ltd., Tokyo, Japan) (Fig. 30.2A). Leukocyte removal in LCAP is effected by adherence to fibers in the filter (Fig. 30.2B). LCAP is performed with the Plasauto® LC apheresis unit (Fig. 30.3A; Asahi Kasei Medical Co., Ltd.) equipped with a Cellsorba® E. This therapy is a direct hemoperfusion system (Fig. 30.3B). Moderate regular LCAP is

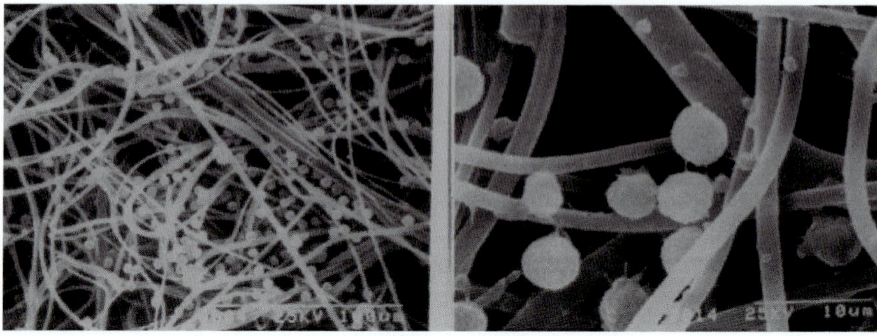

Fig. 30.2B. LCAP is a direct blood perfusion therapy for removing any desired level of granulocytes, monocytes, activated lymphocytes and activated platelets from the peripheral blood through an extracorporeal circulation. This microscopic figure shows that many leukocytes and platelets adhered on the nonwoven fibers less than 3 μm of their diameter.

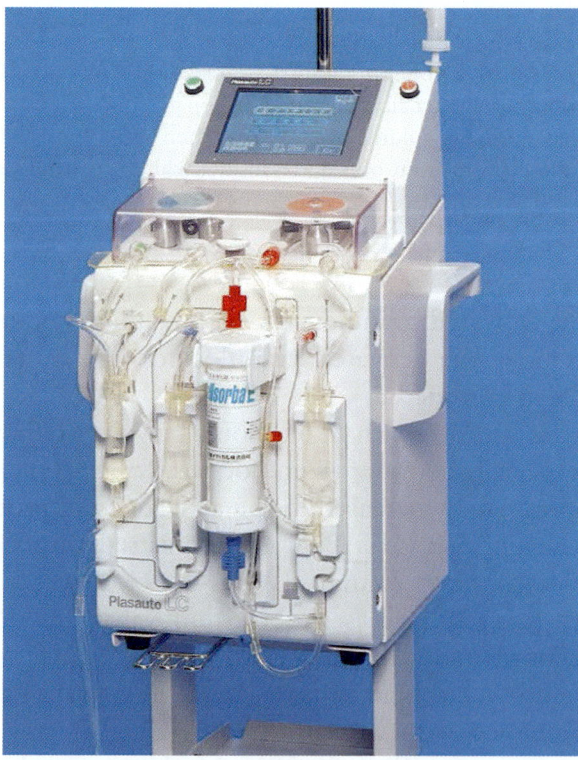

Fig. 30.3A. Plasauto® LC apheresis unit (Asahi Kasei Medical Co., Ltd.) for LCAP treatment consists of two pumps and three pressure gages.

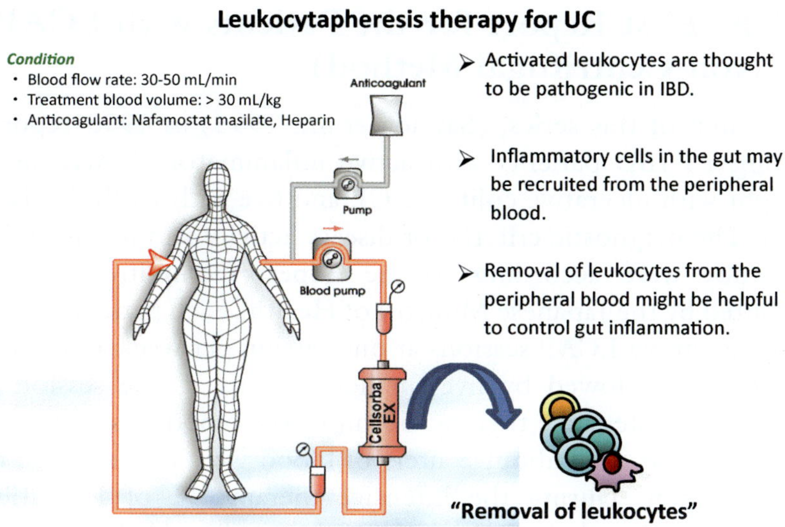

Fig. 30.3B. 3,000 ml of whole blood are processed at a blood flow rate of 50 ml/min. Access and return lines are connected via cubital veins and anti-coagulated by Nafamostat mesilate or heparin. Then the anti-coagulated whole blood is passed through the leukocyte-removing column and retuned to the cubital vein in the other arm.

performed once weekly for 10 weeks or intensive LCAP is performed twice weekly for five weeks. It is well known that intensive CAP is more effective than moderate regular CAP (Sakuraba *et al.*, 2009; Sawada *et al.*, 2005). So, the frequency of CAP treatment is related to its efficiency.

LCAP has been reported associated with a significant decrease in the number of several pro-inflammatory cytokines produced by peripheral blood leukocytes and might be key factors in its pathogenesis (Hendrickson *et al.*, 2001; Grimm *et al.*, 1995). Several review papers (Sawada, 2003; Mitsuyama *et al.*, 2009; Fridrik and Ingvar, 2008) suggests a larger scale clinical study would be needed for further evaluation for LCAP efficacy and adverse effects. According to this suggestion, a larger clinical study has been planned and performed.

2. The First Report for the Patients with LCAP (Non-Centrifugal Method)

In the first of this series, (Sawada *et al.*, 1995) in 1995 reported treating 13 IBD patients with active inflammatory bowel disease — eight with ulcerative colitis (UC), and five with Crohn's disease (CD). The diagnostic criteria for disease activity and inclusion into the study were according to the Expert Committee on IBD appointed by the Japanese Ministry of Health. Each patient received five consecutive LCAP sessions at one session per week as an intensive therapy followed by five further sessions at one session per month as maintenance therapy. 50 ml whole blood was treated per minute so that total about 3 liters of blood were processed in one LCAP session. During the intensive therapy, 11 of 13 patients (84.6%) significantly improved; six of these achieved clinical remission. At the end of the maintenance therapy, eight patients (61.5%) had maintained clinical response without any additional medication. Additionally, flow cytometry revealed that patients who improved had a higher percentage of $HLADR^+$, $HLADR^+CD3^+$ and $HLADR^+CD8^+$ cells at baseline (entry). The levels of afore-mentioned cells, C-reactive protein (CRP) and erythrocyte sedimentation rate (ESR) had decreased to normal range by the end of the therapy. In contrast, in patients who showed poor response to LCAP, all the afore-mentioned parameters were normal or near normal at baseline. Therefore, the results of this preliminary study suggested that patients with high levels of $HLADR^+$, $HLADR^+CD3^+$ and $HLADR^+CD8^+$ cells, CRP and ESR at baseline may be responders to LCAP.

3. Overshoot Phenomenon for the Leukocyte Count after LCAP (Fig. 30.4)

In their second study in 1997, Sawada *et al.*, 1997 reported treating 45 UC patients (21 severe, 15 moderately severe and nine mild); treatment regimens were similar to those described for the afore-mentioned

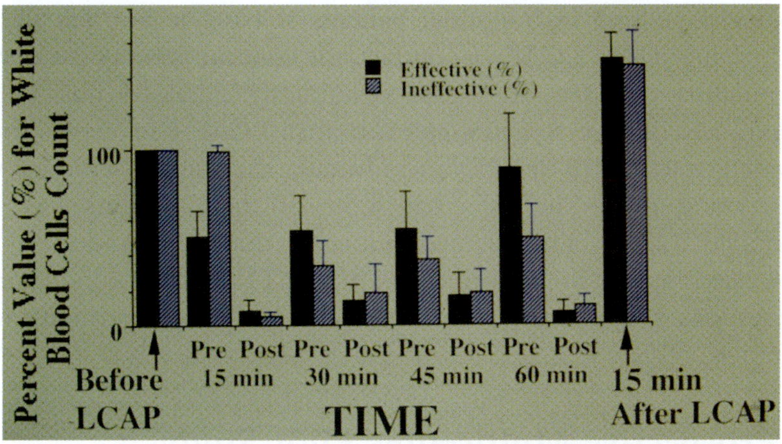

Fig. 30.4. During LCAP, WBC decreased to less than 60% of the initial value on the part of pre-column. However, it started to increase at 60 min passed on the part of pre-column even though Cellsorba® E column has capacity to remove WBC effectively for more than 60 min. Then WBC was increased to about 170% at 15 min after LCAP procedure was completed. This is the "overshoot phenomenon of homeostasis."

report (Sawada *et al.*, 1995). During the time course of LCAP, ongoing conventional drugs were continued or tapered if the patient improved. They monitored changes in differential leukocyte counts by flow cytometry and cytokine production by peripheral blood mononuclear leukocytes (PBML). Clinical improvement was observed in 35 of 45 patients (77.8%) during intensive therapy. In 32 of 35 patients, the improvements continued throughout the maintenance therapy. The most striking observations were a decrease in the patients' leukocyte counts of up to 60 percent during the first 30 minutes of cytapheresis treatment followed by an increase of up to 170 percent of baseline level 15 minutes after the completion of LCAP. The results indicated that an initial fall in the level of peripheral blood leukocytes triggers an influx from the marginal pools into the circulation. Furthermore, the cytapheresis had a marked effect on the amount of tumor necrosis factor-alpha (TNFα), interleukin-1beta (IL-1β), IL-2, IL-8 and interferon gamma produced by PBML. The level of these pro-inflammatory cytokines was high in responder patients at baseline but was significantly reduced after LCAP. In contrast, the concentration of the immunoregulatory cytokine, IL-4, was significantly increased after LCAP. Additionally, flow cytometry revealed that LCAP had depleted the activated fraction of leukocytes. Overall, the clinical efficacy was encouraging given the fact that most of these patients had not responded to conventional medications at that time Andoh *et al.*, 2004 also reported in 2004 that IL-1β- and TNFα-induced inflammatory responses suppressed by leukocytapheresis therapy in patients with ulcerative colitis.

4. Prospective Multicenter Pilot Study for Crohn's Disease using LCAP

In a prospective multicenter pilot study for Crohn's disease (CD), we studied the effect and safety of an additional LCAP into ongoing drug therapy in refractory CD patients who were not responding for the treatment with salicylic acids and elemental diet or parental nutrition for more than two weeks but without steroid and immune suppressants (Kosaka *et al.*, 1999) treated 18 patients with active CD using this LCAP procedure in 1999 and nutritional indices, inflammatory reactions, flow cytometry profiles, and cytokine production by PBML were monitored at baseline and after the intensive and maintenance LCAP session and compared between baseline and post each session. Results showed that nine of 18 patients (50%) attained remission at the end of the intensive therapy with a significant fall in CRP and ESR. The nine non-responders had longer periods of disease duration together with more severely affected sites before the therapy. In 14 of 18 patients (77.8%), the nutritional indices, International Organization of Inflammatory Bowel Disease (IOIBD) score, and Crohn's Disease Activity Index (CDAI) significantly improved. Compared with non-responders, the nine responders had significantly higher baseline $CD4^{+}CD45^{+}$ cells ratio and IL2 produced by PBML, however they had significantly lower activated leukocytes. There were no severe adverse effects.

5. Multicenter Randomized Controlled Trial for the Treatment of Ulcerative Colitis with a Leukocytapheresis Column

To evaluate LCAP effects and adverse events, a controlled multicenter trial (Sawada *et al.*, 2003) with randomized assignment for 76 active-stage UC patients was planned and run in 2003. In the LCAP group (39 patients), LCAP weekly for five weeks as an intensive therapy was added to the ongoing drug therapy, while steroids were maintained but not increased, and then LCAP was gradually reduced to once every four weeks as a maintenance therapy totally in an approximately one year. In the high dose prednisolone (h-PSL) treatment group (37 patients), PSL was added or increased 30–40 mg/day (0.5 mg/kg/day) for moderately severe and 60–80 mg/day (1.0 mg/kg/day) for severe patients and gradually tapered. The results showed that the LCAP with low-dose corticosteroids (26.9 mg/day on average) treatment group had significantly higher efficacy (74% (29/39) vs. 38% (14/37), $p = 0.005$) than h-PSL (47.9 mg/day on average) treatment group and had significantly lower incidence (24% vs. 68%, $p < 0.001$) of adverse events (Fig. 30.5). The patients were able to continue the trial for a longer period in the LCAP treatment group than the h-PSL group ($p = 0.012$). Clinical activity and endoscopic indexes showed the LCAP group had better improvements than the h-PSL group.

6. Leukocytapheresis in Ulcerative Colitis: Results of a Multicenter Double-Blind Prospective Case-Control Study with Sham Apheresis as Placebo Treatment

In order to elucidate the placebo effect of extracorporeal treatment including anticoagulants, a multicenter study using active and sham apheresis devices was conducted in a double-blind, prospective case-control study by (Sawada *et al.*, 2005) in 2005. Twenty-five patients with active of severe and moderately severe grade were enrolled and assigned to the active group or the sham group. Six patients were

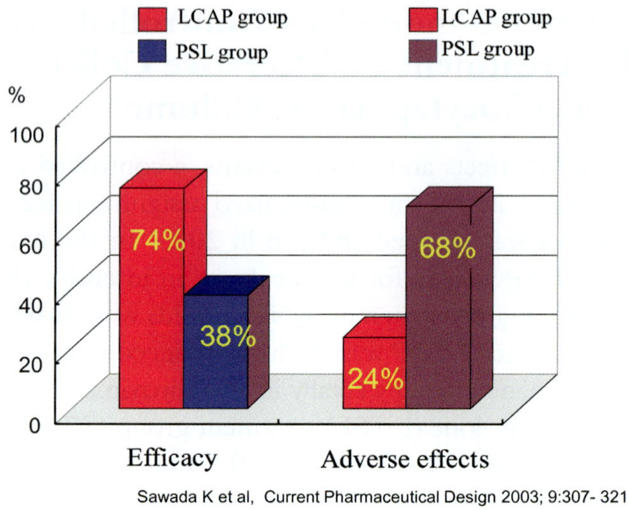

Fig. 30.5. Efficacy and adverse events for "Multicenter Randomized Controlled Trial for the Treatment of Ulcerative Colitis with a Leukocytapheresis Column."

excluded from the study and 19 (10 in the active group and nine in the sham group) were evaluated. LCAP (treatment using an active device or a sham device) was performed once a week for five weeks, followed by two additional sessions during the next four weeks at two-week intervals. Steroids and other medications were continued at the same dosage for next four weeks, which included a two-week pre-observation period and the first two weeks after the start of the LCAP treatment. New medications or increases in the dosage of previous ongoing medication were prohibited until evaluation was conducted. The results showed that the clinical activity index (CAI) value of UC in active group had significantly greater improvement (80%, 8/10) than for the sham group (33%, 3/9; $p < 0.05$) (Fig. 30.6), and adverse effects were observed in five patients (one in the active group and four in the sham). None of these effects was severe and none of the sessions was terminated as a consequence of the adverse effects.

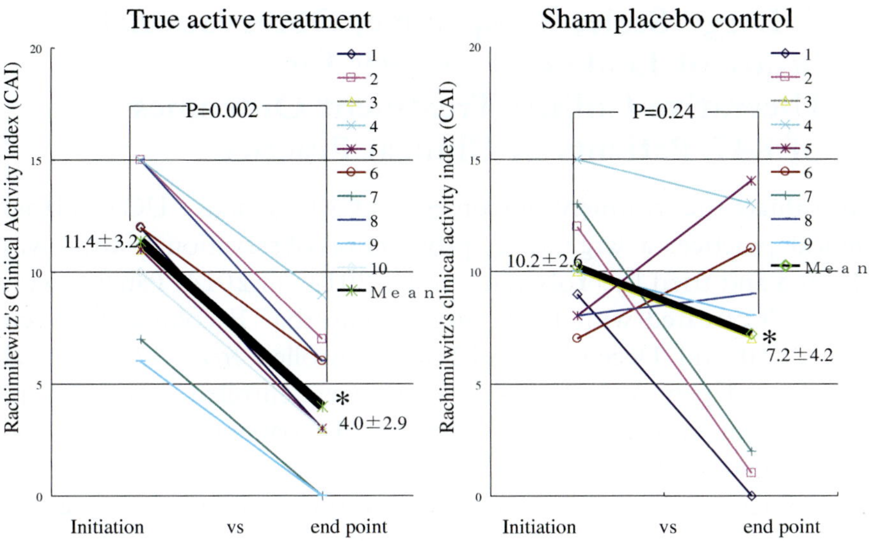

Fig. 30.6. Clinical activity index (CAI) for active group and sham group.

7. Leukocytapheresis for Management of Fulminant Ulcerative Colitis with Toxic Megacolon

The results of a pilot trial on the application of LCAP for fulminant UC patients with toxic megacolon (TM) were reported by (Sawada *et al.*, 2005) The effectiveness and safety of LCAP were evaluated in six fulminant UC patients with toxic megacolon (TM). LCAP sessions were performed three times per week for two weeks, followed by four further times in the next four weeks. After completion of therapy, four patients improved in TM and went into the remission stage of UC. The remaining two patients had to undergo total colectomy, however, the symptoms had been mitigated by LCAP and the operations were completed without any problems. These results suggest that frequent LCAP procedure is an additional effective and safe option for TM management in preventing colectomy or for bridging to a safer operation.

8. A Large-Scale, Prospective, Observational Study of Leukocytapheresis for Ulcerative Colitis: Treatment Outcomes of 847 Patients in Clinical Practice

To evaluate the treatment outcomes of LCAP for active UC in a large clinical practice, a large-scale, prospective observational study was planned and tried by (Yokoyama *et al.*, 2014) in 2014. Patients with active UC treated with LCAP using Cellsorba® E column between May 2010 and December 2012 were enrolled from 116 medical facilities in Japan. A total of 847 patients were enrolled, and 623 were available for efficacy analysis. Out of 847 patients, 80.3% of the patients had moderate to severe disease activity, and 67.6% were steroid refractory. As concomitant medication, 5-aminosalicylic acids, corticosteroids, and thiopurines were administered to 94.8, 63.8, and 32.8% of the patients, respectively. In addition, infliximab and tacrolimus were concomitantly used in 5.8 and 12.3%, respectively. Intensive LCAP (4 LCAP sessions within the first two weeks) was used in > 70% of the patients. Adverse events were seen in 10.3% (87/847), which were severe in only five patients (0.6%). Any concomitant medications did not increase the incidence of adverse events. Intensive LCAP was as safe as the conventional weekly procedure. The overall clinical remission rate was 68.9% (429/623), and the mucosal healing rate was 62.5% (145/232) (Fig. 30.7). Clinical remission was achieved more frequently and rapidly in the intensive group than in the weekly group. This large-scale study indicates that LCAP, including intensive procedure, is a safe and effective therapeutic option for active UC patients.

9. Safety of Leukocytapheresis for Patients with UC ((Yokoyama *et al.*, 2014; Nagase *et al.*, 1998; Ashizuka *et al.*, 2006; Muratov *et al.*, 2010)

The overall incidence of adverse effects was around 10%. The main adverse effects observed during and after cytapheresis were headache,

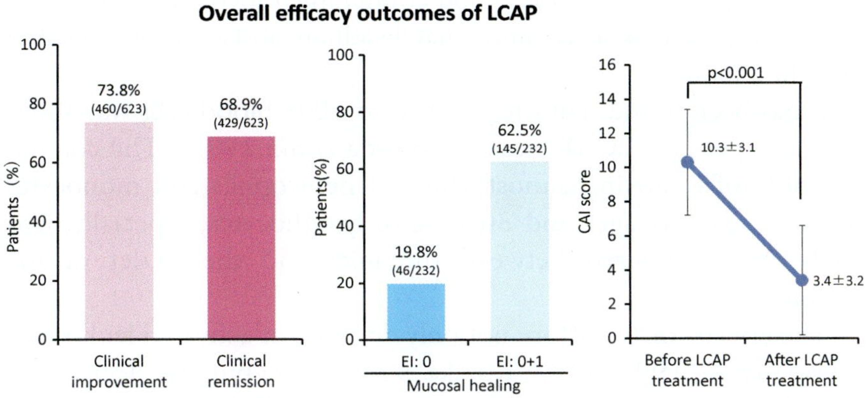

(A) Clinical improvement and remission rates

(B) Mucosal healing rates

(C) CAI scores between before and after the LCAP treatments

Y. Yokoyama, K. Matsuoka, T. Kobayashi et al., Journal of Crohn's and Colitis (2014) 8, 981-991.

Fig. 30.7. Overall efficacy outcomes of LCAP in the 623 patients. (A) Clinical improvement and remission rates at two weeks after the last LCAP session. (B) Mucosal healing rates in the patients whose endoscopic findings were available (n = 232). (C) Comparison of the CAI scores between before and after the LCAP treatments. The data shown are mean ± standard deviation values.

nausea and slight fever, and these events are commonly associated with extracorporeal circulation like dialysis. Almost all the adverse events were mild to moderate, and all the patients either recovered from the events or showed significant improvement.

10. Mechanism of LCAP Action

Initially, this filter was made for the removal of leukocytes from transfusion blood in order to inhibit symptoms of graft-versus-host disease. From this there emerged a hint that massive leukocyte removal with this filter can use to treat patients with activated leukocyte-associated diseases.

The development of the leukocytapheresis column was based on the inherent tendency of leukocytes to adhere to foreign bodies, and discovery of their particularly strong adhesion to ultrafine polyester fibers (Takenaka, 1996). Upon contact with these fibers, during the passage of whole blood containing anticoagulant through the column, the leukocytes immediately adhere and are thus separated from the blood.

It has been found that the removal of white blood cells was about 1.6×10^{10} per one LCAP session (Sawada *et al.*, 1997). The column is capable of removing almost 100% of neutrophils and monocytes including macrophages and 30–60% of lymphocytes, especially activated when measured between the inlet and the outlet of the column.

The effects for LCAP are not only decrease of activated leukocytes (Sawada *et al.*, 1995, 1997; Kosaka *et al.*, 1999) but also decrease of pro-inflammatory cytokines. (Sawada *et al.*, 1997; Andoh *et al.* 2004; Kosaka *et al.*, 1999)

Recently, biological drugs such as infliximab (IFX) (Sands *et al.*, 2008; Hanauer *et al.*, 2002; Sands *et al.*, 2004; Dretzke *et al.*, 2011) and adalimmab (Dretzke *et al.*, 2011), and immunomodulating drugs such as tacrolimus (Tac) (Thin *et al.*, 2013; Gerich *et al.*, 2013) have been used for treating IBD patients. These drugs so strongly depressed the congenital immunological function that we have to take care of infectious disease, especially tuberculosis.

11. Japanese Food "WASHOKU," Traditional Dietary Cultures of the Japanese May Well Match Cytapheresis Treatment

On the other hand, fat from fatty foods was reported to affect leukocyte activity and permeability from vessels to the lesion increase. (Smith, 2007; Teixeira *et al.*, 2011; Ji *et al.*, 2011; Sell and Eckel, 2010; Lee *et al.*, 2010; de Silva and Korzenik, 2014) However, traditional Japanese food does not affect leukocyte activity so much because of its low fat content. In 2013, "WASHOKU," traditional dietary cultures of the Japanese, was inscribed Intangible Cultural Heritage (ICH) of Unesco. WASHOKU is a social practice based on an essential spirit of the Japanese, "respect for nature." It contributes to not only strengthening familial and community ties, healthy life, long life expectancy and prevention of obesity, but may also decrease fat-associated activity of diseases as one of aggravated factors because it requires well-balanced and low-fat diets.

Cytapheresis may be effective in the patients who eat traditional Japanese food because this does not activate leukocytes which would attack intestinal tissue again. So that the activity of a disease settles after activated leukocytes were removed. However new activated leukocytes appear after IBD patients eat fatty food. Thus much stronger treatment would be needed to stop leukocyte re-activation. If IBD patients want to eat fatty food, they should select the treatment of biological drugs or immunomodulating drugs. But the patients must be always taking notice of infectious diseases such as viral and bacterial infections, especially tuberculosis. There are no severe adverse effects in the cytapheresis.

Key words: inflammatory bowel disease, leukocytapheresis, leukocyte removal filter, low fat diet, WASHOKU.

References

Abel JJ, Rowntree LG, Turner BB. (1924) Plasma removal with return of corpuscles (plasmapheresis). *J Pharmacol Exp Ther* 5: 625–31.

Andoh *et al.* (2004) Suppression of interleukin-1beta- and tumor necrosis factor-alpha-induced inflammatory responses by leukocytapheresis therapy in patients with ulcerative colitis. *J Gastroenterol* 39: 1150–7.

Ashizuka *et al.* (2006) Leukocytapheresis for ulcerative colitis: A comparative study of anticoagulant (nafamostat mesilate vs. dalteparin sodium) for reducing clinical complications. *Ther Apher Dial* 10: 54–8.

Bicks RO, Groshart KD. (1989) The current status of T-lymphocyte apheresis (TLA) treatment of Crohn's disease. *J Clin Gastroenterol* 11: 136–8.

Bicks RO, Groshart KD. (1995) Tempest about T-lymphocyte apheresis in Crohn's disease. *Gastroenterology* 108: 952–953.

de Silva PS, Korzenik JR. (2014) Inflammatory bowel disease: A focus on the involvement of dietary fats. *International Journal of Interferon, Cytokine and Mediator Research* 6: 19–25.

Dretzke J, Edlin R, Round J, *et al.* (2011) A systematic review and economic evaluation of the use of tumour necrosis factor-alpha (TNF-α) inhibitors, adalimumab and infliximab, for Crohn's disease. *Health Technol Assess* 15(6): 1–244.

Fridrik TS, Ingvar B. (2008) Leukocytapheresis for the treatment of IBD. *Nature Reviews Gastroenterology and Hepatology* 5: 509–16.

Gerich ME, Pardi DS, Bruining DH, *et al.* (2013) Tacrolimus salvage in anti-tumor necrosis factor antibody treatment-refractory Crohn's disease. *Inflamm Bowel Dis* 19(6): 1107–11.

Goldfinger D, Capostagno V, Lowe C, *et al.* (1980) Use of long-term leukapheresis in the treatment of chronic lymphocytic leukemia. *Transfusion* 20: 450–4.

Grimm MC, Pullman WE, Bennett GM, *et al.* (1995) Direct evidence of monocyte recruitment to inflammatory bowel disease mucosa. *J Gastrenterol Hepatol* 10: 387–95.

Hanauer SB, Feagan BG, Lichtenstein GR, *et al.* (2002) ACCENT I Study Group. Maintenance infliximab for Crohn's disease: The ACCENT I randomised trial. *Lancet* 359: 1541–1549.

Hauser SL, Fosburg M, Kevy SV, Weiner HL. (1984) Lymphocytapheresis in chronic progressive multiple sclerosis: Immunologic and clinical effect. *Neurology* 34: 922–6.

Hendrickson BA, Gokhale R, Cho JH. (2001) Clinical aspects and pathophysiology of inflammatory bowel disease. *Clin Microbiol Rev* 15: 79–94.

Holdstock GE, Fisher JA, Hamblin TJ, Loehry C. (1979) Plasmapheresis in Crohn's disease. *Digestion* 19: 197–201.

Ji Y, Sakata Y, Tso P. (2011) Nutrient-induced inflammation in the intestine. *Curr Opin Clin Nutr Metab Care* 14: 315–21.

Kosaka T, Sawada K, Ohnishi K, *et al.* (1999) Effect of leukocytapheresis therapy using a leukocyte removal filter in Crohn's disease. *Intern Med* 38: 102–11.

Kurland J, Franklin S, Goldfinger D. (1980) Treatment of renal allograft rejection by exchange plasmalymphocytapheresis. *Transfusion* 20: 455–7.

Lanc TA. (1980) Continuous-flow leukapheresis for rapid cytoreduction in leukemia. *Transfusion* 20: 455–7.

Lee IS, Shin G, Choue R. (2010) Shifts in diet from high fat to high carbohydrate improved levels of adipokines and pro-inflammatory cytokines in mice fed a high-fat diet. *Endocr J* 57: 39–5.

Lowenthal CM. (1977) Chronic leukemias: Treatment by leukapheresis. *Exo Hematol* 5(Suppl): 73–84.

Mitsuyama K, Yamasaki H, Kuwaki K, *et al.* (2009) Recent understanding of leukocytapheresis (LCAP) for the treatment of inflammatory bowel disease. *Curr Pharm Des* 15(18): 2110–9.

Muratov *et al.* (2010) Safety and tolerability of a modified filter-type device for leukocytapheresis using ACD-A as anticoagulant in patients with mild to moderately active ulcerative colitis. Results of a pilot study. *J Clin Apher* 25: 287–93.

Nagase *et al.* (1998) Complications of leukocytapheresis. *Ther Apher* 2: 120–4.

Sakuraba A, Motoya S, Watanabe K, *et al.* (2009) An open-label prospective randomized multicenter study shows very rapid remission of ulcerative colitis by intensive granulocyte and monocyte adsorptive apheresis as compered with routine weekly treatment. *Am J Gastroenterol* 104: 2990–5.

Sands BE, Blank MA, Patel K, van Deventer SJ; ACCENT II Study. (2004) Long-term treatment of rectovaginal fistulas in Crohn's disease: Response to infliximab in the ACCENT II Study. *Clin Gastroenterol Hepatol* 2(10): 912–20.

Sands BE, Sandborn WJ, Feagan B, *et al.* (2008) A randomaized, double-blind, sham-controlled study of granulocyte/monocyte apheresis for active ulcerative colitis. *Gastroenterology* 135: 400–9.

Sawada K, Egashira A, Ohnishi K, *et al.* (2005) Leukocytapheresis for management of fuluminant ulcerative colitis with toxic megacolon. *Dig Dis Sci* 50(4): 767–73.

Sawada K, Kusugami K, Suzuki Y, *et al.* (2005) Leukocytapheresis in ulcerative colitis: Results of a multicenter double-blind prospective case-control study with sham apheresis as placebo treatment. *Am J Gastroenterol* 100: 1362–9.

Sawada K, Muto T, Shimoyama T, *et al.* (2003) Multicenter randomized controlled trail for the treatment of ulcerative colitis with a leukocytapheresis column. *Curr Pharm Des* 9: 307–21.

Sawada K, Ohnishi K, Fukui S, *et al.* (1995) Leukocytapheresis therapy, performed with leukocyte removal filter, for inflammatory bowel disease. *J Gastroenterol* 30: 322–9.

Sawada K, Ohnishi K, Kosaka T, Shimoyama T (1997) Leukocytapheresis with leukocyte removal filter as new therapy for ulcerative colitis. *Ther Apher* 1: 207–11.

Sawada K. (2003) Leukocytapheresis as an adjunct to conventional medication for inflammatory bowel disease. *Dis Colon Rectum* 46(Suppl): 566–77.

Sell H, Eckel J. (2010) Adipose tissue inflammation: Novel insight into the role of macrophages and lymphocytes. *Curr Opin Clin Nutr Metab Care* 13: 366–370.

Smith CW. (2007) Diet and leukocytes. *Am J Clin Nutr* 5: 1257–8.

Takenaka Y. (1996) Lymphocytapheresis. *Artif Organs* 20(8): 914–6.

Teixeira LG, Leonel AJ, Aguilar EC, *et al.* (2011) The combination of high-fat diet-induced obesity and chronic ulcerative colitis reciprocally exacerbates adipose tissue and colon inflammation. *Lipids in Health and Disease* 10: 204–11.

Thin LW, Murray K, Lawrance IC. (2013) Oral tacrolimus for the treatment of refractory inflammatory bowel disease in the biologic era. *Inflamm Bowel Dis* 19(7): 1490–8.

Valbonesi M, Garelli S, Zerbi D, *et al.* (1982) Plasma exchange combined with cytotoxic drugs and lymphocytapheresis for myasthenia gravis. *Vox Sang* 43: 142–6.

Wallance D, Goldfinger D, Lowe C. (1980) A double-blind, controlled study of lymphoplasmapheresis versus sham apheresis in rheumatoid arthritis. *N Eng J Med* 306: 1406–10.

Yokoyama Y, Matsuoka K, Kobayashi T, *et al.* (2014) A large-scale, prospective, observational study of leukocytapheresis for ulcerative colitis: Treatment outcomes of 847 patients in clinical practice. *J Crohn's Colitis* 8: 981–91.

Chapter 31

Adacolumn for Hemoperfusion to Deplete Inflammatory Leucocytes as an Alternative or Complementary to Drug Therapy in Patients with Immune Disorders: Basic Mechanisms and Concepts for Therapeutic Efficacy

Abbi R. Saniabadi*,†, Chikako Shima*,
Tomotaka Tanaka‡, Koji Sawada§

Abstract

Chronic diseases like rheumatoid arthritis (RA) and inflammatory bowel diseases (IBD), which include ulcerative colitis (UC) and Crohn's disease (CD), are debilitating life-long morbid conditions caused by a dysregulated immune profile in the affected individuals.

* JIMRO Co. Ltd., Gunma, Japan.
† Institute of Pharmacology, Hamamatsu University School of Medicine.
‡ Department of Gastroenterology, Akitsu Prefectural Hospital, Hiroshima.
§ Chionkai Dojima General & Gastroenterology Clinic, Osaka, Japan.
Correspondence address: Dr. A R Saniabadi,
JIMRO, 351-1 Nishiyokote Machi, Takasaki, Gunma 370-0021, Japan
Tel: 81 273531411; Fax: 81 273 531770; E-mail: Saniabadi@jimro.co.jp.

These conditions afflict millions of individuals throughout the world with symptoms that impair performance and quality of life. Considering IBD per se, the chronic nature of the immune disorders means that patients may require life-long medications often at high doses, but the currently available pharmacological preparations are aimed at reducing the symptoms or suppressing exacerbations, without a lasting effect on the underlying cause. Further, long-term drug therapy often leads to drug dependency and loss of response together with adverse side effects. Indeed, drug side effects become an additional morbidity factor in many patients on long-term medications.

However, the efficacy of anti-tumor necrosis factor (TNF)-α biologics has validated the role of inflammatory cytokines, notably TNF-α in the exacerbation of IBD. Given that inflammatory cytokines are released by patients' own cellular elements including myeloid lineage leucocytes, which in patients with active IBD are elevated with activation behavior and prolonged survival, selective depletion of these leucocytes should promote disease remission or at least enhance drug efficacy with the possibility of reducing drug dosage. Based on this thinking, hemoperfusion to selectively deplete excess and activated myeloid leucocytes with the Adacolumn has been applied to treat patients with dysregulated immune profile. Basically, the Adacolumn medical device is filled with specially designed cellulose acetate beads of 2mm diameter bathed in sterile saline, which serve as the column leucocytapheresis carriers. During hemoperfusion, the carriers selectively adsorb from the blood in the column FcγR (fragment crystallizable gamma receptor) and complement receptor expressing leucocytes. Hemoperfusion with the Adacolumn is popularly known as adsorptive granulocyte and monocyte apheresis (GMA). Pre-and post-column blood cell counts have shown that the carriers selectively adsorb about 65% of granulocytes and 55% of monocytes together with a significant fraction of platelets, while lymphocytes are spared; less than 2% adsorb, and red cell count is unaffected. GMA has been applied to treat patients with IBD in Japan and in the European Union countries. Efficacy outcomes have been impressive as well as disappointing. In fact the clinical response to GMA seems to define the patients' disease course and severity of the mucosal damage at entry. The best responders have been first episode cases (up to 100%) followed by steroid naïve and patients with a short duration of active IBD. Patients with deep ulcers and extensive loss of mucosal tissue do not respond well. Likewise, patients with a long duration of IBD and exposure to multiple pharmacologicals including corticosteroids are not likely to benefit from GMA if they have badly damaged mucosal tissue. It is clinically relevant to stress that patients who respond well to GMA have a good long-term disease course by avoiding drugs including corticosteroids at an early stage of their IBD. Additionally, GMA is very much favored by patients for its good safety profile, as well as for being a non-pharmacologic treatment strategy. GMA reminds us of phlebotomy as a major medical practice at the time of Hippocrates, but in patients with IBD, there is a scope for removing from the body the sources of pro-inflammatory cytokines, potentially to achieve disease remission, and this should fulfill a desire to treat without drugs. The bottom line is that by introducing GMA at an early stage following the onset of IBD or before patients develop extensive mucosal damage and become refractory to medications, many patients should respond well to GMA and be spared from pharmacologicals.

Keywords: Hemoperfusion; Inflammatory bowel disease; Ulcerative colitis; Suppressing UC relapse; Rheumatoid arthritis; Hepatitis C virus, Behcet's disease; Pyoderma gangrenosum; Myeloid lineage leucocytes; Adsorptive granulocytes/monocytes apheresis; Complement activation fragments; Treating disease without drugs.

1. Introduction

1.1. *The Evolution of Apheresis as a Therapeutic Practice*

The term hemoperfusion means passage of a patient's blood through a column containing absorptive material, which can remove from the blood specific toxic substances. Hemoperfusion found clinical application in the 1940s to remove small and medium size molecules for which conventional haemodialysis was considered to be inadequate. Historically, the major application of hemoperfusion includes removing waste products from the blood in patients with renal or liver insufficiency, supporting life before and after transplantation. In broader terms, hemoperfusion is an "apheresis" procedure. Indeed, apheresis means to take away or to remove disease causing elements; it is an extracorporeal therapy (Abel *et al.*, 1924; Goldfinger *et al.*, 1980; Lazarus, 1991; Eshleman and Silberstein, 1991). Likewise, in classical sense, the most basic practice in apheresis involves passing of a patient's blood through a medical device where removal of components takes place; it is a kind of bloodletting. In the early 20th century, centrifugation was introduced for selective depletion of blood cells to achieve a therapeutic effect (Morse *et al.*, 1966). Even in today's biologic-based therapy targeting a pathologic cytokine, chemokine or an adhesion molecule, apheresis is considered as a non-drug intervention to manage diseases for which there is no effective drug-based intervention, like removing elevated or activated cells, which threaten well-being (Morse *et al.*, 1966; Buckner *et al.*, 1969; Curtis *et al.*, 1972; Saniabadi *et al.*, 2014).

Strategies that selectively remove plasma components or blood cells as therapeutic interventions have generally been safe and not associated with serious side effects in a significant number of patients. Recently, apheresis has been considered as an effective alternative to drug therapy or as an adjunct to conventional medications to enhance drug efficacy with potential to reduce standard medication dosage like the granulocyte and monocyte apheresis (GMA) trials in patients with inflammatory bowel disease (IBD) where corticosteroids were avoided or were tapered to a minimum (Lerebours *et al.*, 1994; Shimoyama *et al.*, 2001; Hanai *et al.*, 2002, 2004a; Suzuki *et al.*, 2004; Cohen, 2005). However, today's selective removal of constituents of circulating blood to achieve a therapeutic effect by apheresis is reminiscent of the rather crude technique of bloodletting (phlebotomy) and its therapeutic application at the time of Hippocrates (460–377 BC) in Ancient Greece. Bloodletting was widely practiced as a major therapeutic intervention in the ancient civilization of Egypt from where it expanded to Ancient Greece and other parts of the world. The perception then was that disease reflects presence of disease-causing elements in the blood and bloodletting was to expel the pathologic agents. Bloodletting was routinely and extensively practiced for diseases like inflammation, fever, hypertension, and other undiagnosed diseases; its most high profile patient was the first US president, George Washington. Bloodletting as a major medical practice was subsequently popularized by Claudius Galen (129–203 AC), a Greek physician who practiced in Rome and became a well respected authority in medicine for over 1500 years. In the late 19th century, bloodletting under a more modern name, phlebotomy was popularly used to treat conditions like haemochromatosis, polycythaemia vera and others (Clutterbuck, 1840). It may just be the folly of the past, but it is hard to imagine this procedure would have been so widely practiced for such a long time had it not been associated with efficacy.

Even in this era of modern medicine and if we consider IBD, and innovations in biologic therapy, it is clinically relevant to bear in mind that drug therapy by its very nature involves adding a foreign substance or substances to the body system and although initially effective, can lead to the disease becoming drug dependent or drug refractory. Hence, a therapeutic strategy based on a non-drug intervention, a correction or support of body's natural processes like GMA, if effective, should have an unrivalled advantage over pharmacologic therapy (Shimoyama *et al.*, 2001; Hanai *et al.*, 2002, 2003, 2004; Suzuki *et al.*, 2004; Cohen, 2005; Hanai, 2006; Suzuki *et al.*, 2006; Aoki *et al.*, 2007; Tanaka *et al.*, 2008; Saniabadi and Hanai, 2010; Tanaka *et al.*, 2010; Yoshimura *et al.*, 2012).

2. The Work Featured in this Chapter

In this chapter, we have reviewed the fundamental concepts that led to the development of the Adacolumn for therapeutic depletion of myeloid lineage leucocytes by adsorptive granulocyte and monocyte apheresis, widely known as GMA, following publication of clinical reports by Hanai and colleagues in 2004 (Hanai *et al.*, 2004b; Saniabadi and Hanai, 2010). However, in this chapter, the mode of actions of the Adacolumn GMA in terms of changes in cytokine or chemokines (Hanai *et al.*, 2011), which potentially contribute to clinical efficacy outcomes, together with relevant concepts for achieving maximum therapeutic efficacy from this non-pharmacologic treatment strategy are reviewed. Finally, we have highlighted representative efficacy outcomes and how these are intimately related to patients' disease profile at entry. This is particularly relevant in relation to patients with IBD given that the expressions of IBD as well as the causes are diverse. Accordingly, IBD response to medical intervention is defying the basic rules (Mulder *et al.*, 2014). We believe that applying clinically relevant concepts should help to select patients who may respond well to GMA and avoid drug therapy at an early stage, or otherwise stop futile GMA therapy in patients who may not benefit (Saniabadi *et al.*, 2014). This thinking should spare medical resources and shorten morbidity time in many patients because while GMA responders have a chance of being spared from unnecessary drug therapy, GMA non-responders may opt for an alternative treatment. Clinical efficacy outcomes are covered in more detail in other chapters in this issue by other authors.

3. The Adacolumn Design Features

The Adacolumn (Fig. 31.1) was developed in April 1988 by the then-called Japan Immunoresearch Laboratories (Gunma, Japan), and was called the G-1 column (see below). As yet, there has been no fundamental alteration in the prototype Adacolumns. The currently produced columns are virtually identical to the original design, except perhaps the size; one group reports using a 22,000-bead column. The currently produced Adacolumns are about 206 mm in length, 60 mm in diameter, and are made of polycarbonate (Saniabadi *et al.*, 2003b). The column volume is about 335 mL, filled with 215–220 g cellulose acetate beads (approximately 35,000 pieces) of 2 mm in diameter, bathed in sterile physiologic saline. The beads serve as the column leucocytapheresis carriers. The volume of the bathing saline is 130 mL, approximately equal to the column void volume (less residual air in the column). It is essential to bear in mind that the basic scientific phenomena observed during clinical use of the Adacolumn or when an investigation is carried out on the carriers *in vitro* are almost entirely attributable to the carriers and this in turn depends on the selectivity of the carriers for granulocytes (neutrophils, eosinophils) and monocytes/macrophages, referred to

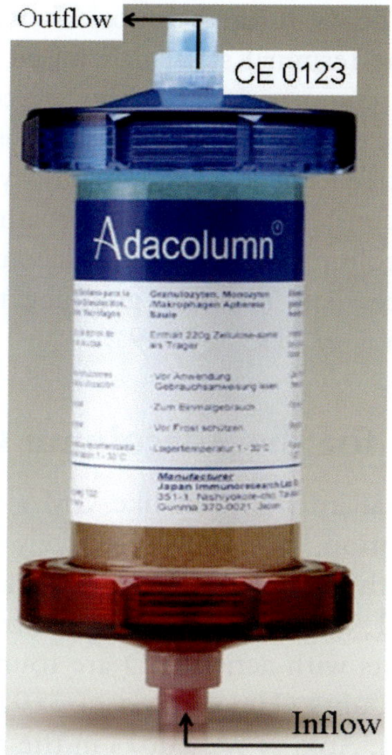

Fig. 31.1. Photograph of a fully assembled Adacolumn. The single use polycarbonate column has a capacity of 335 mL, filled with about 220 g cellulose acetate beads (about 35,000 pieces) of 2mm in diameter as the column adsorptive leucocytapheresis carriers. The carriers are bathed in sterile physiological saline; the volume of the saline in the column is about 130 mL, approximately equal to the column void volume. The Adacolumn is the first and the only adsorptive type leucocytapheresis medical device in clinical use.

as myeloid lineage leucocytes (Saniabadi *et al.*, 2014; Yoshimura *et al.*, 2012).

4. The Perceptions that Inspired Development of the Adacolumn

The letter "G" is related to a cell growth and differentiation phase in cancer progression called G1; the G1 arrest is a key to treatment efficacy. It is believed that cytotoxic T-lymphocytes (CTLs) can efficiently eliminate newly formed tumors. It was noted that some cancer patients in the terminal stage had raised granulocyte counts (granulocytosis), but compromised lymphocyte counts, hence a high granulocyte÷lymphocyte (G÷L) ratio. It was thought that lymphocytes were being destroyed by activated granulocytes. In other words, granulocytosis was considered as an exacerbating factor in cancer progression. A G-1 column, later named Adacolumn, was developed to take away the excess and activated granulocytes to normalise the G ÷ L ratio (Tabuchi *et al.*, 1995, 1999).

5. Target Cells for the Adacolumn GMA

For an extracorporeal intervention like GMA to be a novel non-drug therapeutic option, it should selectively deplete leucocytes, which in patients with inflammatory diseases like IBD are elevated, as seen in Table 31.1, and are thought to contribute to the disease pathogenesis. Patients with active IBD are found to have compromised lymphocyte counts (Heimann *et al.*, 1985; Saniabadi *et al.*, 2003a, 2003b, 2014; Aoki *et al.*, 2007) in the presence of elevated and activated granulocytes and monocytes/macrophages (Meuret *et al.*, 1978; Cassatella, 1995; Brannigan *et al.*, 2000; Saniabadi *et al.*, 2003a, 2003b, 2005, 2014). Additionally, certain sub-populations of lymphocytes like the CD4(+)CD25(+) phenotype known as the regulatory T-cells (Treg) have essential immunoregulatory function and therefore are indispensable to the host (Liu *et al.*, 2003; Huber *et al.*, 2004; Maul *et al.*, 2005; Uhlig and Powrie, 2005; Kanai *et al.*, 2005; Yokoyama *et al.*, 2007; Cuadrado *et al.*, 2008). Based on these understandings, GMA is designed to spare lymphocytes, seen in Table 31.2 and Fig. 31.2, but boost Treg phenotype (Saniabadi *et al.*, 2003b; Yokoyama *et al.*, 2007; Cuadrado *et al.*, 2008). As shown in Table 31.2, the carriers remove from the blood in the column most of the granulocytes, monocytes/macrophages together with a significant fraction of platelets (Saniabadi *et al.*, 2003b; Yokoyama *et al.*, 2007; Cuadrado *et al.*, 2008). Surprisingly, the procedure has been associated with a sustained increase in absolute lymphocyte counts following a course of GMA (Saniabadi *et al.*,

Table 31.1. Neutrophil (Granulocyte) Counts in Healthy Controls, Inflammatory and Non-Inflammatory Conditions

Condition	Neutrophils (×10³/μL)	Author
Healthy controls	3.4 ± 0.4	Fujimori, *et al.* (1996)
Ulcerative Colitis (UC)	6.4 ± 0.4	Shimoyama, *et al.* (2001)
Crohn's disease	7.8 ± 1.3	Meuret, *et al.* (1978)
Severe UC	9.3 ± 0.5	Hanai, *et al.* (2002)
Advanced cancer	5.2 ± 0.7	Tabuchi, *et al.* (1999)
Behcet's disease	6.13 ± 2.94	Namba *et al.* (2006)
Rheumatoid arthritis	5.5 ± 0.3	Fujimori *et al.* (1996)
An arthritic joint contains	5 × 10⁹ neutrophils	Edwards and Hallett (1997)
Osteoarthritis	3.8 ± 0.1	Fujimori *et al.* (1996)

Pro-inflammatory cytokines including tumour necrosis factor (TNF)-α, and interleukin (IL)-1β as well as corticosteroids are known to promote neutrophil survival. Osteoarthritis is not an inflammatory condition, and therefore, patients have neutrophil counts in the normal range. As patients with dysregulated immune function may receive corticosteroid as an immunosuppressant, the medication together with a high pro-inflammatory cytokine (TNF-α, and IL-1β) profile potentially set a vicious cycle, which worsens the disease and impairs medication efficacy. The authors believe that reducing the extra load of neutrophils in these patients should help the immune system to set itself to a more normal function, or at least enhance medication efficacy. Modified from Saniabadi A, *et al.*, *Ther Apher Dial* 2003a; 7: 48–59.

Table 31.2. Blood Cell Counts at the Adacolumn Inflow and Outflow Points During Adsorptive Granulocytes/Monocyte Apheresis (GMA) Therapy of Patients with Rheumatoid Arthritis. Test Samples were Taken for Blood Counts at Least 30 Minutes after the Start of GMA to Ensure that the Column Outflow Count was not Compromised by Dilution within the Column (Prior to Hemoperfusion, The Adacolumn and its Flow Lines are Primed with Sterile Saline)

Cell Population	Inflow	Outflow	Significance Level
Red Cells (10⁶/μL)	3.6 ± 0.22	3.5 ± 0.25	ns
Granulocytes (10³/μL)	6.8 ± 1.4	2.4 ± 0.5	0.01
Monocytes (10³/μL)	0.28 ± 0.07	0.12 ± 0.03	0.01
Lymphocytes (10³/μL)	1.22 ± 0.15	1.20 ± 0.14	ns
Platelets (10³/μL)	306 ± 8	256 ± 9	0.001

Modified from Saniabadi A, *et al.*, *Ther Apher Dial* 2003a; 7: 48–59.

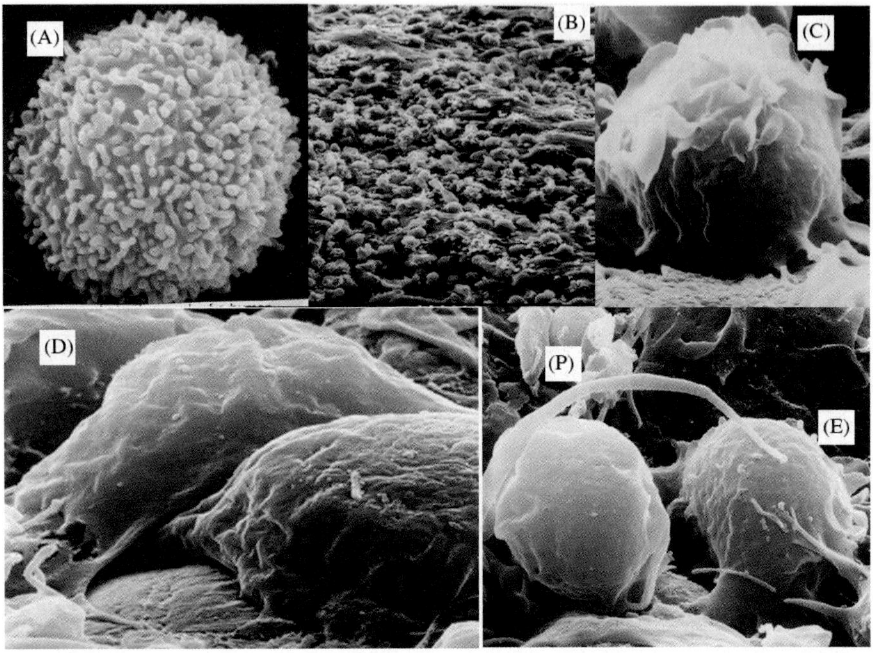

Fig. 31.2. Scanning photomicrographs of leucocytes adsorbed onto an Adacolumn carrier (cellulose acetate bead) representing basic events on the surface of the Adacolumn carriers during adsorptive granulocyte and monocyte apheresis (GMA) in clinical setting (see Fig. 31.3 as well). Normal (non-activated and non adherent) neutrophils and monocyte/macrophages express extensive surface microvilli (A to E). The villi bear various receptors like tumor necrosis factor (TNF) receptors, and L selectin, which are shed upon adsorption of the cells to the carriers. Further as seen, adsorbed leucocytes undergo extensive release reaction. Up to now, interleukin (IL)-1receptor antagonist (IL-1ra), hepatocyte growth factor (HGF), IL-10 and soluble TNF receptors have been measured. Accordingly, during GMA, the blood, which returns to patients from the column-outflow may be likened to a biologic cocktail containing a vast number of soluble substances released by the adsorbed leucocytes in the column (see text). The adsorptive nature of Adacolumn leucocytapheresis is thought to be the most intriguing feature of the Adacolumn system. In (B), low power view of adsorbed cells onto the carrier are seen; (C) shows an activated monocyte/macrophage; (D) and (E) show neutrophils; (P) shows adsorbed platelets.

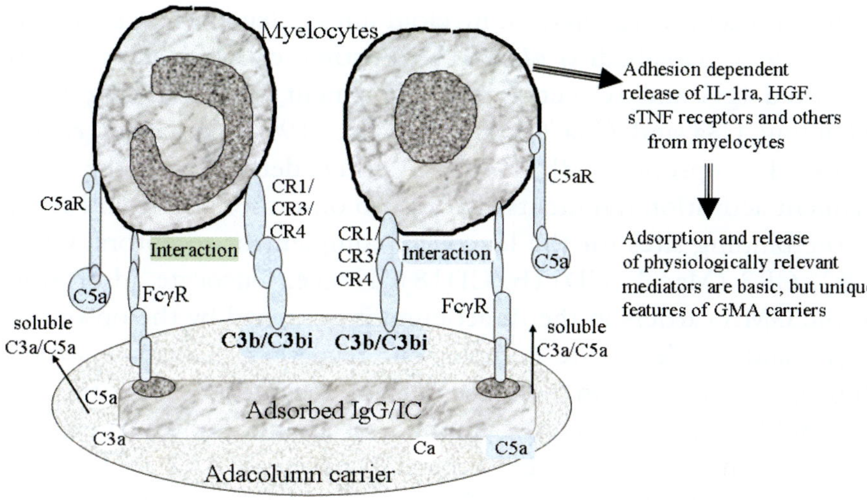

Fig. 31.3. A tentative illustration of the mechanisms which mediate the selective adsorption of myeloid lineage leucocytes to the Adacolumn granulocyte and monocyte apheresis cellulose acetate carriers. The first event is adhesion of plasma immunoglobulin (IgG) and immune complexes (IC) to the carriers, which serve as the binding sites for the fragment crystallizable gamma receptor (FcγR) on myeloid leucocytes. Then complement activation (helped by IgG and IC) generates C3a, C5a, C3b/C3bi fragments. Of these, C3b/C3bi (opsonins) adsorb onto the carriers and serve as the binding sites for the complement receptors (CR) on myeloid leucocytes. Adhesion of leucocytes results in the release of interleukin-1 receptor antagonist, hepatocyte growth factor, soluble tumor necrosis factor receptors, all with therapeutic effects. From Saniabadi A *et al.* (2014) *World J Gastroenterol.*

2003a; Suzuki *et al.*, 2004; Aoki *et al.*, 2007). The increase in lymphocyte counts includes the CD4(+)CD25(+) Treg phenotype (Yokoyama *et al.*, 2007; Cuadrado *et al.*, 2008). The mechanisms for sparing lymphocytes are briefly described here and illustrated in Fig. 31.3. Patients with IBD may have immune complexes (IC) in their plasma (D'Arrigo *et al.*, 1994; Hiraishi *et al.*, 2003; Saniabadi *et al.*, 2005). Cellulose acetate adsorbs immunoglobulin G (IgG) and IC from the plasma (D'Arrigo *et al.*, 1994; Hiraishi *et al.*, 2003; Saniabadi *et al.*, 2005). Upon adsorption, the binding sites on IgG and IC become available for the Fcγ (fragment crystallizable gamma) receptors (FcγRs) on myeloid lineage leucocytes (D'Arrigo *et al.*, 1994; Saniabadi *et al.*, 2014). Further, cellulose acetate with adsorbed IgG and IC generates complement activation fragments including C3a and C5a (D'Arrigo *et al.*, 1994; Saniabadi *et al.*, 2014). The opsonins C3b/C3bi and others derived from the complement activation fragments also adsorb onto the carriers and serve as the binding sites for the leucocyte complement receptors, CR1, CR2, CR3 (Mac-1, CD11b/CD18). Hence, leucocyte adsorption to the GMA carriers in the Adacolumn is governed by the opsonins, FcγRs and the leucocyte complement receptors (Saniabadi *et al.*, 2005). The expression of these sets of receptors is a feature of myeloid lineage leucocytes. Lymphocytes are not known to express complement receptors except on small subsets of B, T and natural killer (NK) cells. Similarly, FcγRs are not widely expressed on lymphocytes except on small populations of CD19+B cells and CD56+NK cells (Hiraishi *et al.*, 2003). These basic phenomena proceed well on the carriers and lend GMA selectivity (Fig. 31.3).

To our knowledge, there is no published data showing elevated circulating lymphocytes in patients with active IBD, and in one of the best controlled studies on lymphocytapheresis in IBD, Lerebours and colleagues (Lerebours *et al.*, 1994) selectively depleted circulating lymphocytes in patients with Crohn's disease (CD). At the end of an 18-month follow-up, the clinical outcome in the lymphocytapheresis group was 21% inferior to that of the control, non-lymphocytapheresis group. This well planned and controlled clinical trial, together with the afore reviewed reports, rule out any therapeutic benefit from removing lymphocytes in patients with IBD (Lerebours *et al.*, 1994).

6. The Adacolumn Hemoperfusion System

As shown in Fig. 31.4, the Adacolumn medical device is for direct hemoperfusion, meaning that patient's blood flows directly into the column via an inflow line and from the column outflow line, blood returns to the patient. The extracorporeal circulation through the Adacolumn is by using a standard blood circuit line and pump/monitor

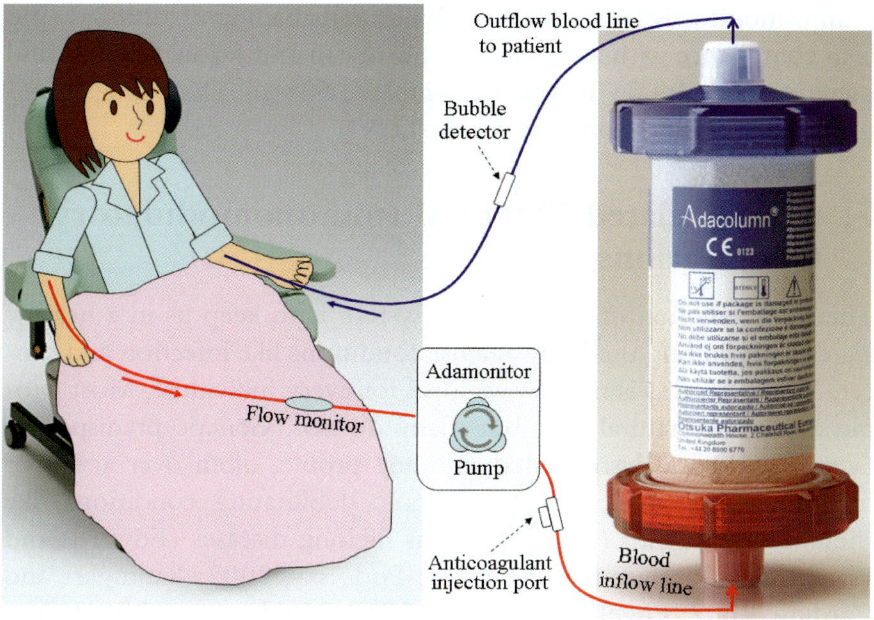

Fig. 31.4. Hemoperfusion outline of the JIMRO Adacolumn adsorptive granulocyte and monocyte apheresis (GMA) system. As seen in this figure, the Adacolumn is for direct hemoperfusion. Patient's blood usually from the antecubital vein in one arm flows into the column and from the column-outflow, blood returns to the patient via the antecubital vein in the contralateral arm. A standard blood circuit line, a pump/Adamonitor are also seen. The capacity of the circuit line is 80 mL. The Adamonitor may be considered as the nerve center of the Adacolumn hemoperfusion system. In addition to running the pump, it has other functional units including flow rate setting, and time setting panels, pressure monitoring and fault detecting alarm systems. If the pressure in the blood line deviates from the preset value, an alarm sound will warn the operator and the system blood flow is automatically switched off. Similarly, in the event of a fault in the flow system, such as an air bubble in the blood line, automatically the blood flow is switched off and an alarm sound will warn the operator.

(seen in Fig. 31.4). Currently, the blood circuit line used for the Adacolumn apheresis is manufactured under the Japan Ministry of Health Certification (Saniabadi, *et al.*, 2003a). The material of the circuit lines is PVC (polyvinylchloride), with an 80 mL capacity. The pump/monitor is ISO 13485 certified (Saniabadi *et al.*, 2003a), and is in compliance with all applicable European and Japanese safety and quality standards (JIS T 1001 (1992), IEC 601-1). The pump/monitor has EC certification (CE-Mark, TUV).

7. GMA Related Potential Immunomodulatory Observations

The basic function of the immune system is to keep us in a healthy state by defending the body against conditions like infection and cancer. However, this is not always ideal for every individual. Most people enjoy this natural gift, but many do not. Genetic predisposing factors lead to a dysregulated immune profile, often over active or inappropriately active causing chronic debilitating conditions like rheumatoid arthritis (RA) or IBD as lifelong diseases (Edwards and Hallett, 1997; Allison *et al.*, 1998; Podolsky, 2002; Baumgart and Carding, 2007; Engel and Neurath, 2010; Mulder *et al.*, 2014). As a non-pharmacologic treatment intervention, GMA is expected to normalize the dysregulated immune system either by removing key contributing factors like elevated/activated myeloid lineage leucocytes seen in Table 31.2 and Fig. 31.2, or otherwise facilitating release of disease-modifying substances from the blood within the column, which returns to the patients via the blood return line seen in Fig. 31.4. In line with this thinking, Fig. 31.2 shows scanning photomicrographs of leucocytes adsorbed onto an Adacolumn carrier (cellulose acetate bead) representing basic events on the surface of the Adacolumn carriers during GMA in clinical setting (see Figs. 31.3 and 31.4 as well). Normal, non-activated and non-adherent neutrophils and monocytes/macrophages express extensive surface microvilli. The villi bear various receptors like TNF receptors and L selectin, which are shed upon adsorption of the cells to the carriers. Further as seen, adsorbed leucocytes undergo extensive release reaction. Accordingly, during GMA, the blood, which returns to patients from the column outflow, may be likened to a biologic cocktail containing a vast number of soluble substances released by adherent leucocytes in the column (see below). The adsorptive nature of the Adacolumn leucocytapheresis is thought to be the most intriguing feature of this hemoperfusion system (Saniabadi *et al.*, 2005, 2014).

Regarding the impact of GMA on inflammatory cytokine producing leucocytes, Hanai *et al.* (2008) reported a very marked effect by GMA on inflammatory $CD14^+CD16^+$ monocytes; this monocyte phenotype was depleted to the levels seen in healthy persons. The $CD14^+CD16^+$ phenotype is known to be a major source of TNF (Belge *et al.*, 2002) and shows dramatic expansion in inflammatory conditions (Scherberich, 2003). This action of GMA should significantly alleviate inflammation in patients with immune dysfunction. Additionally, as reviewed above, GMA has been associated with a sustained increase in circulating lymphocytes including the CD4(+) CD25(+) Treg phenotype (Yokoyama *et al.*, 2007; Cuadrado *et al.*, 2008). In spite of these reports, the clinical efficacy associated with a course of GMA is unlikely to reflect its effects on peripheral leucocytes per se. It is inferred that additional mechanisms of actions might follow a course of GMA. As seen in Figs. 31.2 and 31.3, leucocytes that bear the FcγR and complement receptors adhere to the GMA carriers (Saniabadi *et al.*, 2003b, 2005, 2014). Within the column, an array of substances both toxic and non-toxic, but anti-inflammatory as well are generated, which reach the patients' circulation via the column outflow that returns to patients. Among these, cytokines, complement activation fragments like C3a and C5a are of short half-life and may not reach patients' circulation in fully active form. Hanai *et al.* (2004, 2006) and Suzuki *et al.* (2004) independently reported a significant increase in the column outflow blood levels of soluble TNF receptors I, and II, known to neutralize TNF-α without invoking TNF-like actions (Mohler *et al.*, 1993). Further, *ex vivo* studies reported GMA-related suppression of inflammatory cytokines including TNF-α, IL-1β, IL-6 and IL-8 released by leucocytes (Kashiwagi *et al.*, 2002; Hanai *et al.*, 2011). GMA appears to have a similar effect on leucocyte trafficking receptors (Saniabadi, 2005).

Further, *in vitro* studies by Takeda *et al.* (2004) showed that exposure of human blood to the GMA carriers releases significant amounts of IL-1 receptor antagonist (IL-1ra) and hepatocyte growth factor (HGF), as seen in Fig. 31.5. IL-1ra has an important role in

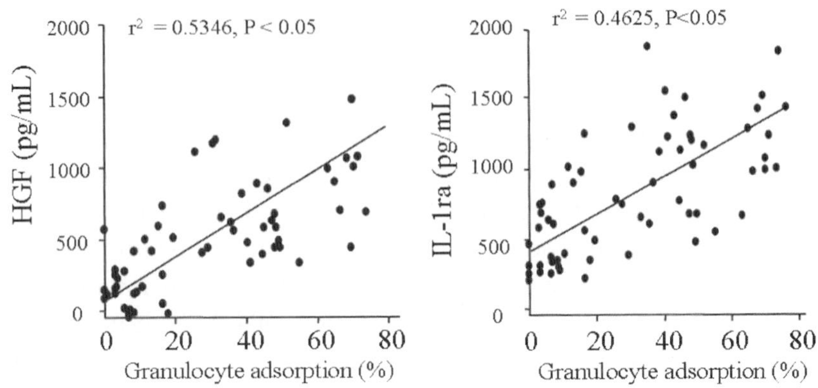

Fig. 31.5. Adsorption dependent release of interleukin-1 receptor antagonist (IL-1ra), and hepatocyte growth factor (HGF) from myeloid lineage leucocytes (granulocytes and monocytes) in *in vitro* setting. HGF is known to promote ulcer healing and cell regeneration, while IL-1ra is strongly anti-inflammatory. *Source*: Takeda Y *et al.* (2004) *Inflamm Res* **53:** 277–283.

the control of inflammation in the mucosa (Dinarello and Thompson, 1991), while HGF is known to promote mucosal epithelial cell regeneration, an essential step in ulcer healing (Tahara *et al.*, 2003). The release of these substances should proceed inside the Adacolumn during GMA in clinical settings. Further, examinations of mucosal biopsy specimens suggest that GMA suppresses cytokine profiles within the mucosa (Muratov *et al.*, 2006; Yamamoto *et al.*, 2006). A wary reader may wonder why this is possible with a system that impacts peripheral blood leucocytes. One answer might be that GMA reduces the number of leucocytes that are destined for the mucosa, partially by depletion (Tanaka *et al.*, 2008; Saniabadi *et al.*, 2014) and partially by down-modulating the adhesion receptors on the remaining leucocytes (Saniabadi *et al.*, 2005). As a result, the density of cells which are known to produce inflammatory cytokines or otherwise damage the mucosal tissue by granule proteases or reactive oxygen species is reduced. Thus, Muratov *et al.* (2006) found a marked decrease in tissue interferon (IFN)-γ and TNF-α positive T-cells in clinical responders after GMA. In parallel, significantly lower levels of IFN-γ producing lymphocytes were detected in the peripheral blood. IFN-γ positive T-cells in pretreatment biopsies completely disappeared or decreased in post-treatment biopsies sampled two weeks after the last GMA session in responders and appeared to predict the maintenance of long-term remission, up to 12 months. However, our view is that the impact of GMA on the CD4(+)CD25(+) regulatory T-cells (Yokoyama *et al.*, 2007; Cuadrado *et al.*, 2008) reviewed above is potentially very interesting in patients with immune dysfunction in whom the immune system is in a state of exuberant activity. Additionally, the column outflow, which contains a vast number of released substances warrants investigation for the presence of new anti-inflammatory mediators. Therefore, with the aforementioned effects, GMA should be a non-drug treatment intervention in patients with disordered immune profile.

8. Down-Modulation of the Chemokine Receptor CXCR3

Figure 31.6 shows down-modulation of the chemokine receptor CXCR3 on leucocytes following a course of six Adacolumn GMA sessions over six weeks (Biswas *et al.*, 2003). The last GMA session was on day 35, but maximum down-modulation of CXCR3 was seen on observation day 70, well beyond the last GMA session. The patient had refractory psoriasis that had not responded to drug therapy over several years. CXCR3 is a homing receptor for T-cells. Chemokines and their receptors regulate the migration of leucocytes to inflammatory lesions during immune response (Gillitzer, 2001; Rottman *et al.*, 2001). In Fig. 31.6, the level of CXCR3 seen on day 70 is within the normal range for all the three leucocyte populations shown.

9. Effect of GMA on Tissue Level of Myeloid Leucocytes

As reviewed above, one pathologic feature of active IBD is presence of a large number of granulocytes and monocytes/macrophages in the colonic mucosa (Hanai *et al.*, 2006; Tanaka *et al.*, 2008; Saniabadi

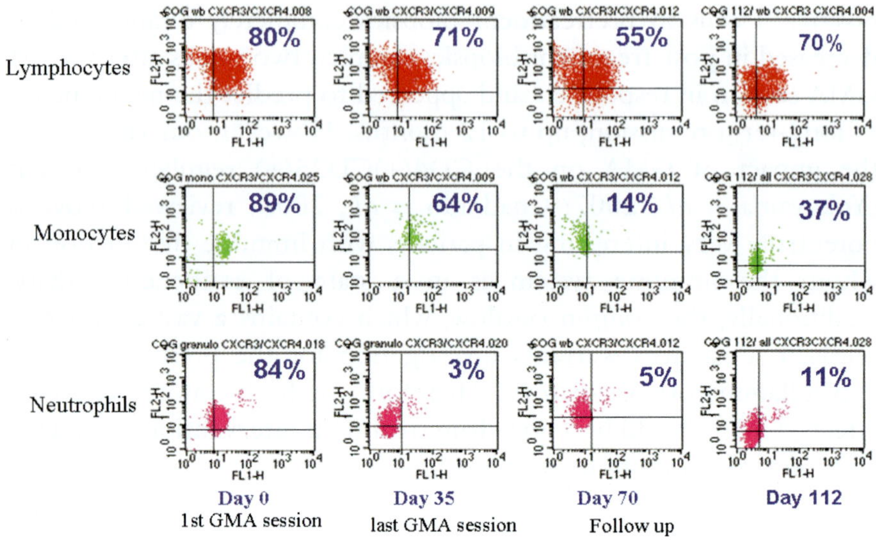

Fig. 31.6. Reduction of CXCR3 positive lymphocytes, monocytes and granulocytes in a patient with immune disorder treated with the Adacolumn GMA. This figure shows down-modulation of the chemokine receptor CXCR3 on circulating blood leucocytes following a course of six GMA sessions over six weeks (see Biswas *et al.*, 2003). The last GMA session was on day 35 (Biswas *et al.*, 2003). The figure shows maximum depletion of CXCR3-expressing leucocytes on day 70 well into the observation time. The level seen at day 70 is within the normal range for all these leucocyte populations. From Saniabadi A *et al.* (2005) *J Clin Apher* **20:** 171–184.

et al., 2014). In clinical settings, GMA is applied as an extracorporeal procedure to reduce the excess and activated myeloid lineage leucocytes from patients' systemic circulation. One could logically argue that removal of the leucocytes from the circulating blood should lead to depletion of these cells in the colonic mucosa as well where they are believed to exacerbate and perpetuate IBD (Tibble *et al.*, 2000; Limburg *et al.*, 2000). Figure 31.7 shows typical immunohistochemical images taken from colonic biopsy specimens in a patient with total colitis in active stage and following GMA induced UC remission. The figure shows that the colonic mucosal tissue is densely infiltrated by myeloid lineage leucocytes with the formation of crypt abscesses and GMA has reduced the concentration of these leucocytes in the mucosa. It is appropriate to mention here that the images seen

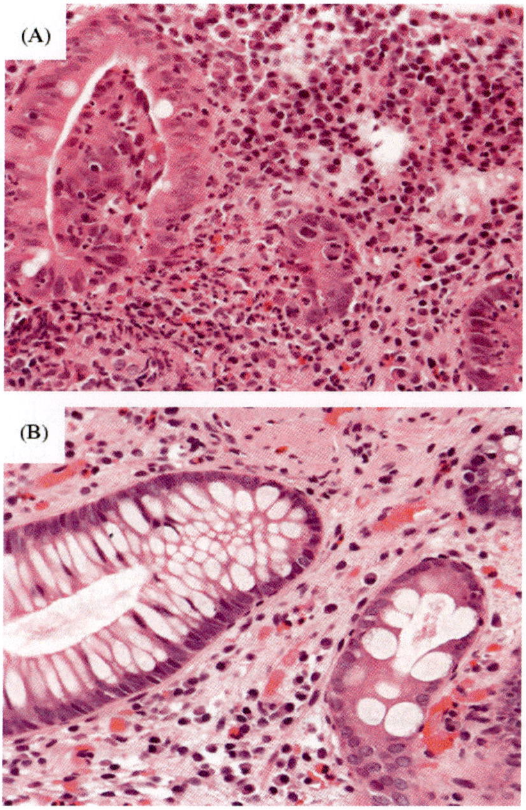

Fig. 31.7. Typical immunohistochemical images by processing colonic biopsy specimens in a patient with total colitis in active stage (A) and following Adacolumn GMA induced remission. (B) This figure shows that the mucosal tissue is densely infiltrated by myeloid lineage leucocytes and GMA has reduced the concentration of these leucocytes in the mucosa. The specimens seen in this figure are from a male patient with moderately severe UC and corticosteroid naïve (a GMA responder feature). GMA = adsorptive granulocyte/monocyte apheresis. Modified from Tanaka T *et al.* (2010) *Gastroenterol Clin Biol* **34:** 687–695.

in Fig. 31.7 are typical for IBD lesions in the large intestine because similar images are uncommon in biopsies from small intestinal IBD lesions found in most patients with CD. It might be equally true to say that CD patients with the IBD lesions confined only to the small intestine (the large intestine being unaffected) may not be the right candidates for GMA therapy (Matsui *et al.*, 2003).

10. The Incurable Nature of a Dysfunctioning Immune System

As stated above inflammatory conditions like RA and IBD by their very nature are incurable as far as the fundamental causes of the diseases are the targets of medications (Allison *et al.*, 1998; Podolsky, 2002; Mulder *et al.*, 2014). Nonetheless, geno/phenotypically, these conditions have a number of common denominators including abnormal cytokine expression, aberrant antigen–antibody complexes, T-cell anomalies, and an increased numbers of granulocytes and monocytes/macrophages, which show activation behavior and prolonged survival (McCarthy *et al.*, 1991; Lefkowitz *et al.*, 1995; Schreiber *et al.*, 1999; Brannigan *et al.*, 2000; Papadakis and Targan, 2000). Thus, the interaction between granulocytes and macrophages is thought to induce a state of chronic inflammation, which contributes to the persistent inflammatory nature of these diseases (Lefkowitz *et al.*, 1995). The known factors are in part responsible for the perpetuation of these diseases. Additionally, the chronic nature of these disorders means that afflicted individuals need long-term drug therapy, and this can lead to refractoriness and drug-related adverse side effects, which often add to the disease complexity (Taffet and Das, 1983; Present, 2000; Brown *et al.*, 2002; Cullen *et al.*, 2011). Therefore, while we do not understand the genetic predisposing factors, the known exacerbating factors can be targeted. For example, both RA and IBD respond to anti-cytokine antibodies, notably, the anti-TNF-α antibodies (van Dullemen *et al.*, 1995; Rutgeerts *et al.*, 2005; Lovell *et al.*, 2008). This validates the roles of certain cytokines as the exacerbating factors in inflammatory immune dysfunction. Further, cytokines like TNF-α, IL-1β and others are produced by the patients' cellular elements, notably myeloid lineage leucocytes (Cassatella, 1995; Nikolaus *et al.*, 1998). Hence, elevated circulating granulocytes and monocytes/macrophages appear logical targets of therapy by selective leucocytapheresis in patients with immune dysfunction.

11. The Logics of Therapeutic GMA in IBD Patients

CD and UC are the two most common phenotypes of IBD, which is marked by inflammation in the lining of the gastrointestinal (GI) organs during active disease (Allison *et al.*, 1998; Podolsky, 2002). However, whereas UC is confined to the colon and the rectum (large intestine), CD can impact any part of the digestive tract, predominantly the ileum, and common symptoms may include abdominal discomfort, diarrhoea, rectal bleeding, weight loss and fever. The most common symptoms of UC include abdominal discomfort and blood or pus in diarrhoea. There is no known single cause for CD or UC, although many researchers believe that the interaction between genes, the body's immune system, and environmental factors may play a role. The aim of CD and UC treatments is to induce remission and achieve extended periods of time when patients do not experience symptoms (Allison *et al.*, 1998; Mulder *et al.*, 2014). Historically, patients with active IBD have conventionally been treated with an aminosalicylate like 5-aminosalicylic acid or sulphasalazine often in combination with a corticosteroid with consideration of nutritional support if necessary (Truelove and Jewell, 1974; O'Keefe, 1996; Allison *et al.*, 1998; Hanauer, 2004). However, as seen in Table 31.1 and reviewed above, patients with inflammatory immune dysfunction may harbor elevated peripheral granulocytes (McCarthy *et al.*, 1991; Saniabadi *et al.*, 2003b, 2005, 2014) in the presence of compromised lymphocytes (Heimann and Aufses, 1985; Saniabadi *et al.*, 2003b, 2014; Suzuki *et al.*, 2004; Aoki *et al.*, 2007). Factors suspected to support granulocytes in IBD include inflammatory cytokines (Lee *et al.*, 1993) and paradoxically corticosteroids (Meagher *et al.*, 1996), which are given to most patients with IBD. Indeed, neutrophil activation and prolonged survival is a feature of persistent intestinal inflammation and histological examinations of the mucosal tissue in biopsy specimens from patients with active IBD reveals a spectrum of pathologic manifestations among which presence of an abundance of neutrophils (Fig. 31.7) accounts not only for the morphologic lesions in IBD but also for the prevailing patterns of mucosal inflammation (McCarthy *et al.*, 1991; Allison *et al.*, 1998; Brannigan *et al.*, 2000; Tibble *et al.*, 2000; Limburg *et al.*, 2000). When activated, neutrophils together with monocytes/macrophages produce an array of pleiotropic inflammatory cytokines (Nikolaus *et al.*, 1998; Belge *et al.*, 2002) that can perpetuate the inflammatory condition. This indicates that in patients with active IBD, a vicious circle prevails in which activated leucocytes, inflammatory cytokines and even corticosteroids are known to be major players. Hence, if we can selectively deplete the excess and activated myelocytes, one parameter of the inflammatory profile is taken out of the cycle. This should cool down the inflammatory profile so to speak, and allow healing to take place. Then, the immune system should be able to adjust itself to a healthier functional state. It is like saying depleting the extra load of activated leucocytes might be likened to an effective and safe biologic therapy in IBD. It is an attractive therapeutic principle and should be achievable in skillful hands — removing from the body one major disease factor instead of adding to the body a foreign substance(s). Currently, this thinking has given rise to the inference that elevated and activated inflammatory cytokine-releasing leucocytes are attractive targets of therapy in IBD.

12. Effective Dosage of GMA in UC Patients

The evolution of modern medicine has relied on the outcomes of clinical trials to determine the dosage of drugs with maximum efficacy margin and minimum adverse side effect. Fortunately for GMA, which is a non-drug treatment strategy, reliance on clinical trial outcomes has been less demanding or at least lack of them has not been associated with serious side effects. Nonetheless, it is a basic requirement to know the most effective frequency and the number of GMA sessions for patients with mild, moderate or severe IBD as this can help to save time and cost. The reality is that up to now GMA treatment has been an empirical practice. Some centers (Hanai *et al.*, 2003; Tanaka *et al.*, 2013; Saniabadi *et al.*, 2014) administer two GMA sessions/week in the first two to three weeks and then one session/week up to 10 or 11 sessions. Hanai *et al.* (2003) described that although patients with steroid naïve UC responded well to five GMA sessions, steroid refractory patients with severe UC responded better to 10 sessions. In contrast, Suzuki *et al.* (2004, 2006), Aoki *et al.* (2007) administer two GMA sessions/week and cease when clinical activity index (CAI) decreases to 4 or less (clinical remission level); patients who do not improve after several sessions are classified as non-responders.

Regarding duration of one GMA session, Kanke *et al.* (2004) found that 90 minutes was significantly better than 60 minutes per GMA session. Likewise, Yoshimura, *et al.* (2012) increased the processed blood volume from the conventional 1,800 mL per GMA session to over 3,000 mL per session by prolonging the GMA session time. In this study, the efficacy rate in the higher processed blood volume group was significantly better than in the 1,800 mL per session group. Further, there is evidence to suggest that intensive GMA is more effective and induces remission in a shorter time when compared with slow weekly GMA (Sakuraba *et al.*, 2009; Kato *et al.*, 2012). Thus Sakuraba *et al.* (2009) assigned 112 patients with moderately active UC to two groups. Group 1 patients received one GMA session/week, while group 2 patients received two sessions/week, up to 10 sessions in both groups. The remission rate in group 1 was 46.7%, while that in group 2 was 73.1%. Further, the mean time to remission was 28.1 days in group 1 and 16.3 days in group 2. However, in spite of this outcome, the full clinical response to GMA may not be immediately evident. For example, in patients with RA, there was a sustained increase in CD4+ T-lymphocytes up to 12 weeks following the last GMA session (Saniabadi *et al.*, 2003a). Similarly, there was a striking down-modulation of the pro-inflammatory chemokine receptor CXCR3 on leucocytes several weeks after the last GMA session (Saniabadi *et al.*, 2005), seen in Figure 31.6. Clearly further investigations are warranted for establishing the optimum frequency and duration of GMA session.

13. GMA Responder and Non-Responder Features of IBD Patients

As reviewed above, published data suggest that first episode and steroid naive patients respond particularly well to GMA (Hanai *et al.*, 2003; Cohen, 2005; Suzuki *et al.*, 2006, 2011; Tanaka *et al.*, 2008, 2011; Saniabadi et al., 2014). Characteristically they respond faster with fewer GMA sessions and have a high cumulative rate of remission. Thus, the remission rate in steroid naïve patients reported by Suzuki *et al.* (2004) was 85%. Similarly, Tanaka *et al.* (2008) treated a cohort of 45 patients, 26 steroid naive and 19 steroid dependent. Each patient could receive up to a maximum of 11 GMA sessions (or until CAI decreased to 4 or less). At week 12, the response rate (CAI ≤4) in steroid naïve subgroup was 22 of 26 patients (84.6%) and that in steroid dependent sub-group was 11 of 19 (57.9%) patients. Colonoscopy revealed that most non-responders in both groups had deep colonic ulcers and extensive loss of the mucosal tissue ((Tanaka *et al.*, 2010; Sacco *et al.*, 2014). Further, Tanaka *et al.* were the only group who looked at the impact of GMA on leucocyte level in the colonic mucosa. Colonoscopic biopsies revealed massive infiltration of the colonic mucosa by neutrophils (Fig. 31.7), and GMA was associated with a striking reduction of neutrophils in the mucosa (Suzuki *et al.*, 2006; Tanaka *et al.*, 2010, 2011). Tanaka's colonoscopic observations echo those of Suzuki *et al.* (2004) a few years earlier, who also reported that the only three non-responders in their cohort of 20 steroid naïve patients had deep colonic ulcers. In a very thorough study by Suzuki *et al.* (2006) the authors aimed at determining the responders (or non-responders) to GMA. Their major findings were as follows. Seven days after the last GMA session, 20 of 28 patients (71.4%) achieved clinical remission including all eight patients who had their first UC episode. The mean duration of UC in the eight first episode cases was 3.4 months compared with 40.2 months for all 28 patients and 65.4 months for the eight patients who did not respond at all. The response to GMA seemed to be independent of baseline CAI in that study. The eight non-responders were given additional conventional medications (CM) or cyclosporin A (CsA) if the former failed. Two patients responded to CM, three to CsA and three underwent colectomy. The authors concluded that first UC episode and short disease duration appear to be relevant predictors of response to GMA in the UC clinical setting. Very recently, Yokoyama and colleagues (Yokoyama *et al.*, 2013) also reported that a short duration of active UC was a good predictor of response to GMA. In other words, the best time to apply GMA was found to be immediately after a clinical relapse (Yokoyama *et al.*, 2013). The authors have stated that GMA might be an effective first line medication for steroid naïve patients (Suzuki *et al.*, 2004; Cohen, 2005). It would appear that the clinical response in patients with chronic continuous UC (Hanai *et al.*, 2003) and patients with deep colonic lesions (Saniabadi *et al.*, 2004, Suzuki *et al.*, 2004; Tanaka *et al.*, 2010, 2011, Sacco *et al.*, 2014) might be somewhat less satisfactory than in those with steroid naïve UC (Suzuki *et al.*, 2004; Cohen, 2005; Tanaka *et al.*, 2008, 2010, 2013) or those experiencing their first UC episode (Suzuki *et al.*, 2006).

14. Colonoscopic Features of GMA Responders and Non-Responders

As reviewed above, several studies have reported that any UC patient with a fair level of intact colonic mucosa is a potential responder to GMA (Suzuki *et al.*, 2006; Tanaka, 2008, 2010; Saniabadi *et al.*, 2014; Sacco *et al.*, 2014). With this in mind, there are physicians who undertake colonoscopic evaluations in all patients with an active flare of IBD, and the evaluation is used to decide to treat or not to treat the patient by GMA (Suzuki *et al.*, 2006; Tanaka *et al.*, 2008, 2014; Saniabadi *et al.*, 2014; Sacco *et al.*, 2014). However, as stated above, by applying GMA at an early stage (Yokoyama *et al.*, 2013), first episode cases respond well (Suzuki *et al.*, 2006), and the centers who apply this strategy may not have many drug refractory patients with badly damaged mucosal tissue. Even patients with a near equal CAI score may have very different mucosal damage status, indicating that

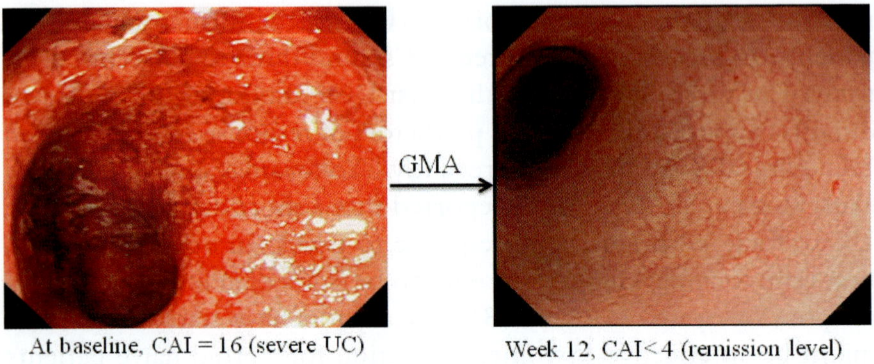

Fig. 31.8. Typical endoscopic features of patients who may respond well to GMA and avoid drug based medications. This case was from a subgroup of patients who were identified as good responders by colonoscopy. This patient was steroid naïve with severe UC, clinical activity index (CAI >16), yet good responder to GMA, firstly because the mucosal tissue was preserved and secondly, the patient was corticosteroid naïve at the time of relapse. The photographs show complete restoration of the mucosal vascular patterns at post GMA. *Source*: Tanaka T *et al.* (2011) *J Gastroint Dig Syst* **1:** 104.

CAI per se does not reflect the full extent of mucosal damage in patients with IBD (Suzuki *et al.*, 2006; Tanaka *et al.*, 2010, 2011; Sacco *et al.*, 2014). Figure 31.8 shows typical colonoscopy features of patients who may respond well to GMA and be spared from drug-based medications. This case was among a subgroup of patients who were identified as responders by colonoscopy. The patient was steroid naïve with severe UC based on clinical evaluation (CAI >16), but without colonoscopy. Nonetheless this patient was a good responder to GMA. First, the mucosal tissue was preserved and second, the patient was not exposed to multiple drugs prior to GMA. The photographs show complete restoration of the mucosal vascular patterns post GMA. Almost all such cases readily respond to GMA. Without colonoscopy, clinical evaluation would have identified such patients as having severe UC, and a potential candidate for steroid therapy. However, such responders to GMA without corticosteroids should have a good long-term clinical course (Yamamoto *et al.*, 2006, 2007, 2012) by avoiding drugs during their initial active UC phase.

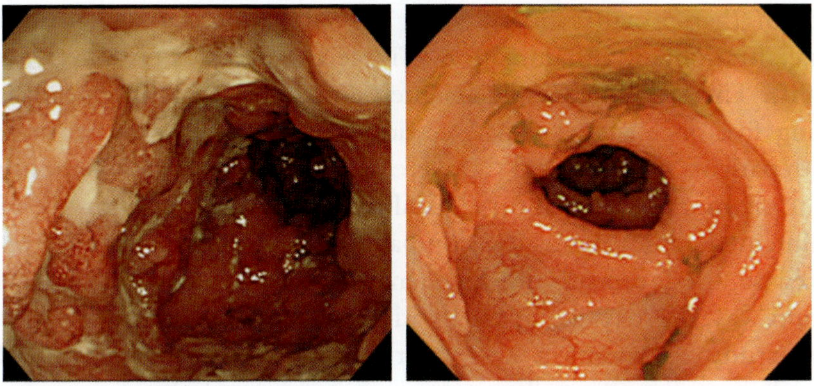

Fig. 31.9. Colonoscopy images showing deep and extensive colonic lesions together with inflammatory polyps and contact bleeding. However, most patients with the entry mucosal damage seen in this figure are medication refractory and unlikely to respond to GMA, some opt for colectomy. *Source*: Tanaka T *et al.* (2011) *J Gastroint Dig Syst* **1:** 104.

In contrast, Fig. 31.9 shows colonoscopic features of GMA non-responder patients. Here, colonoscopy photographs from a typical GMA non-responder patient at entry and at week 12 are presented. The image at entry reveals deep and extensive colonic lesions with virtually no mucosal tissue left at the lesion sites. However, no patient with the entry mucosal damage seen in this figure shows any marked fall in CAI; a few may have fulminant UC, and be candidates for surgical intervention.

15. GMA to Suppress Clinical Relapse in IBD Patients

To prevent an IBD in remission from flaring up has obvious advantages. First, morbidity is intercepted and second, medical resources are spared. Based on this thinking, Maiden *el al.* (2008) in London identified patients at a high risk of clinical relapse on the basis of a very high faecal calprotectin level. Calprotectin is a neutrophil specific protein that provides quantitative measure of intestinal neutrophil concentration and inflammation (Tibble *et al.*, 2000; Limburg *et al.*, 2000). This study randomly assigned patients to Adacolumn, undergoing five once-a-week GMA sessions, or to unchanged treatment. Both patients with UC and CD were included. Follow up was monthly up to six months to monitor clinical relapse. In the GMA group, 62% maintained clinical remission compared to 24% in the control group. Likewise, the mean survival time (maintained remission) in the GMA group was 181 days compared with 104 days in the control group. This study represents a new approach to the treatment of IBD, namely targeting the inflammatory component of the disease at an asymptomatic stage.

16. GMA in other Diseases Involving Leucocytes

16.1. *Rheumatoid Arthritis*

It is noteworthy that the GMA studies in patients with RA were undertaken and the outcomes were published (Fujimori *et al.*, 1996; Kyogoku and Kasukawa, 1998; Kashiwagi *et al.*, 1998; Nagashima *et al.*, 1998; Sanmartí *et al.*, 2005) before the studies in patients with IBD. In one multicenter clinical trial (Kashiwagi *et al.*, 1998; Saniabadi *et al.*, 2003a), a total of 164 patients were randomly assigned to groups I and II. Of these, 143 were eligible, 79 in group I and 64 in group II. Patients in group I received one GMA session per week for four consecutive weeks; group II patients received two sessions per week for four consecutive weeks plus 12 weeks follow-up. The selection criteria with respect to disease and drug therapy were according to the American College of Rheumatology guidelines (Kashiwagi *et al.*, 1998; Nagashima *et al.*, 1998; Sanmartí *et al.*, 2005). If we consider week 6 (two weeks after the last GMA session), there were significant improvements in the following parameters of RA symptoms: active joint score, tender joint score, swollen joint score, duration of morning stiffness, grip strength and activity of daily life. However, there were no marked difference in these parameters between the group with one session per week and the group with two sessions per week (Kashiwagi *et al.*, 1998). There was also GMA-related down-modulation of LCAM-1 (L-selectin), up-regulation of Mac-1, suppression of pro-inflammatory cytokines produced by leucocytes and an unexpected increase in $CD4^+$ T cells (Saniabadi *et al.*, 2003a, 2005). The outcomes of the studies in RA are summarized in Table 31.3.

16.2. *Patients with HIV Infection*

In patients with human immunodeficiency virus (HIV) infection, monocytes/macrophages are known to serve as reservoirs for HIV replication and dissemination. Additionally, these leucocytes produce large amounts of cytokines like TNF-α, which is a major pathologic factor in patients with HIV (for reviews and clinical outcomes, see Refs. 101–105). Further, as stated above, an earlier study in patients with RA (Saniabadi *et al.*, 2003a) reported that GMA was associated with a significant increase in $CD4^+$ T-cells (the cells which are depleted by HIV). Therefore, in patients with HIV infection, GMA was expected to selectively deplete monocytes/macrophages, which are major cellular reservoirs of HIV and to increase $CD4^+$ T-cells as the most desirable effect (Beretta *et al.*, 2002; Hasson *et al.*, 2002, 2007). In the first HIV clinical trial, patients received eight GMA sessions over an eight-week time course then were followed for a further 12 weeks. Initial assessment did not show a marked difference in $CD4^+$ T-cell increase between GMA patients and controls. However, follow-up data showed a marked increase in $CD4^+$ T-cells, but only in patients who had undetectable HIV RNA and a very low $CD4^+$ T-cell counts at baseline. Additionally, TNF-α was strongly suppressed in patients who had received six GMA sessions (Hasson *et al.*, 2002; Saniabadi *et al.*, 2003a). The results indicated that for GMA to have clinical efficacy in patients with HIV infection, suppression of HIV replication by anti-HIV drugs was essential (drug therapy should be effective).

Table 31.3. Summary of Several Measurements of Rheumatoid Arthritis (RA) Symptoms and Markers of Leucocyte Inflammatory Activities after Adsorptive Granulocyte/Monocyte Apheresis (GMA) with the Adacolumn Inpatients with RA or Inflammatory Bowel Disease (IBD)

Measurement	Outcome after GMA	Author
Tender joint count (RA)	Markedly improved and sustained up to week 16 (final observation time point)	Fujimori, *et al.* 1996; Nagashima, *et al.* 1998; Sanmati, *et al.* 2005
Grip strength (RA)	Markedly improved	Fujimori, *et al.* 1996; Kashiwagi, *et al.* 1998.
Swollen joint count (RA)	Markedly improved	Fujimori, *et al.* 1996; Nagashima, *et al.* 1998; Sanmati, *et al.* 2005
Duration of morning stiffness (RA)	Markedly improved	Fujimori; *et al.* 1996., Kashiwagi *et al.* 1998
Leucocyte produced TNF-α, interleukin (IL)-β, and IL-6 (IBD)	Release of these inflammatory cytokines was markedly decreased when stimulated *ex-vivo.*	Kashiwagi *et al-* 2002; Saniabadi *et al.* 2003
Inflammatory cytokine producing myeloid lineage leucocytes (IBD)	Depletion of TNF-producing CD14+CD16+DR++ monocyte phenotype to normal levels	Hanai, *et al.* 2008
L-selectin, Mac-1 on leucocytes	L-selectin was down-regulated, while Mac-1 was up-regulated at post column	Saniabadi *et al.* 2005
Bone marrow CD 15+ CD 16- (RA)	The proliferative response of these immature neutrophils was decreased in *in-vitro* setting	Nagashima *et al.* 1998
Peripheral lymphocytes (RA or lBD)	Sustained increase in the absolute count up to week 16; subsequently found to include CD4(+) CD25(+) Treg phenotype	Nagashima *et al.* 1998; Saniabadi *et al.* 2003. Yokoyama *et al.* 2007

It should be noted that up-regulation of Mac-1 (CD IIb/CD 18, CR3) in the presence of L selectin (CD62L) being down-regulated is not sufficient to promote leucocyte infiltration of tissues, because the first step in leucocyte interaction with the vessel wall is mediated by L selectin. As shown, laboratory analyses revealed a significant increase in the regulatory T lymphocytes (Treg) and depletion of tumour necrosis factor (TNF)-α producing CD14+CD16+DR++ monocyte phenotype associated with GMA. For full text see original references. From Saniabadi A, *et al. Ther Apher Dial* 2003 (updated with new data).

16.3. *Treatment of Pyoderma Gangrenosum and Psoriasis-Like Skin Lesions*

In Fig. 31.10, typical efficacy outcome for the Adacolumn GMA in patients with pyoderma gangrenosum (PG) lesions is seen. It is interesting to know that PG and psoriasis-like skin lesions are reported to be the best responders to the Adacolumn GMA (Kanekura and Maruyama, 2002; Ohmori *et al.*, 2003; Kanekura *et al.*, 2003, 2005; Mori *et al.*, 2004; Saniabadi *et al.*, 2005; Seishima *et al.*, 2008; Ikeda *et al.*, 2013). To our knowledge, every treated case of PG has responded well, and yet it all happened by chance (Ohmori *et al.*, 2003). A patient with IBD associated pyoderma gangrenosum was treated for CD, but the response of pyoderma lesions was more striking than anything expected from GMA (Ohmori *et al.*, 2003). The young patient had very stressful deep refractory lesions over a long period of time in addition to CD. The lesions fully remitted after 10 GMA sessions (Ohmori *et al.*, 2003; Saniabadi *et al.*, 2005).

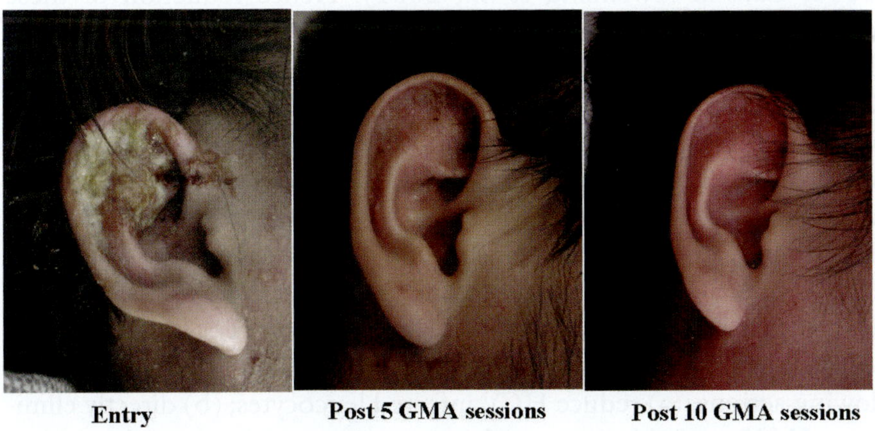

Fig. 31.10. Adacolumn GMA induced remission of pyoderma gangrenosum (PG) lesions on the earlobe of a 19-year-old male patient with Crohn's disease (CD), which was refractory to conventional medications (see Ohmori *et al.*, Ref. 106). Up to 2% of patients with inflammatory bowel disease (IBD) may develop PG at some time point during their IBD (see Ref. 48). *Source*: Saniabadi A *et al.* (2005) *J Clin Apher* **20:** 171–84.

Subsequently, other authors (Kanekura and Maruyama, 2002; Kanekura *et al.*, 2003, 2005a; Seishima *et al.*, 2008; Mori *et al.*, 2004; Ikeda *et al.*, 2013) independently reported cases of pyoderma or psoriatic skin lesions responding well to GMA. The mechanism associated with the efficacy of GMA in these conditions is not very clear. However, the chemokine receptor CXCR3 is known to have a major role in the initiation and perpetuation of inflammatory skin lesions (Gillitzer, 2001; Rottman *et al.*, 2001; Biswas *et al.*, 2003) and is strongly down-modulated by GMA (Fig. 31.6). This action, if duplicated by follow-up studies, can serve as a major breakthrough in the efforts to understand the mechanisms of clinical efficacy of Adacolumn GMA in these clinical settings.

16.4. *Hepatitis C Infection*

In patients with hepatitis C (HCV) infection, granulocytes and monocytes/macrophages are thought to be extra-hepatic sites of HCV replication and dissemination (Bouffard *et al.*, 1992; Cravatto *et al*, 2000; Caussin-Scwemling *et al.*, 2001). Hence reduction of these cells should minimize inflammation and contribute to the clearance of HCV. Based on these understandings, Sawada and colleagues (Sawada *et al.*, 2003) reported successful treatment of a patient with HCV-UC co-morbidity by applying GMA, one session/day for five consecutive days and then one session/week for 24 consecutive weeks during active anti-HCV drug therapy. This patient could not be treated with IFN-α alone, which is the most common medication for HCV, but also is a pro-inflammatory cytokine and can provoke IBD relapse (given that the patient had UC as well). In their report, Sawada and colleagues suggest that in patients with HCV, GMA may have the following actions, (a) reduce HCV infected leucocytes; (b) directly eliminate HCV probably by complement generated in the column (see Fig. 31.3); (c) serve as an adjunct to INF-α to suppress relapse in patients with HCV-UC co-morbidity and (d) to enhance the efficacy of INF-α. Following this unexpected response, several other larger studies were undertaken and all of those studies reported efficacy. In spite of these early outcomes (Bouffard *et al.*, 1992; Cravatto *et al.*,

2000; Caussin-Scwemling *et al.*, 2001; Sawada *et al.*, 2003a, 2003b, 2005a, 2005b; Diepolder *et al.*, 2005), use of GMA in patients with HCV is not continuing. It is likely that GMA sessions are required for the entire duration of IFN-α therapy (several months), which means a very large number of Adacolumns must be used, making the treatment too expensive.

16.5. *Behcet's Disease*

Behcet's disease (BD) is an immune disorder of unknown aetiology, characterized by recurrent ocular, vascular, central nervous system, articular, mucocutaneous and gastrointestinal manifestations. Most manifestations of BD are self-limiting, but repeated attacks of uveitis are a major cause of blindness (Sakane *et al.*, 1999). BD is prevalent and in the past has been a significant cause of morbidity in countries along the ancient Silk Road and the Mediterranean basin. Biopsies from active lesions of BD show presence of large number of neutrophils in the absence of infection, and neutrophils from patients with BD show increased superoxide anion production, enhanced chemotaxis and excessive release of granular enzymes, indicating neutrophil hyperactivity (Sakane *et al.*, 1999). Similarly, levels of circulating TNF-α, IL-1β and IL-8 are high, and it is believed that these cytokines are involved in neutrophil activation (Sahin *et al.*, 1996; Ehrlich, 1997; Sakane *et al.*, 1999). Therefore, it was thought that patients with BD should respond favorably to therapeutic reduction of neutrophils. Accordingly, up to now several groups have applied GMA to patients with BD refractory to conventional medications (Ehrlich, 1997; Garcia *et al.*, 2004; Kanekura *et al.*, 2005b; Sonoda *et al.*, 2005; Namba *et al.*, 2006). All of these studies were uncontrolled, but were designed to allow each patient to serve as his or her own control. For example, Sonoda *et al.* (2005) and Namba *et al.* (2006) monitored the frequencies of ocular attacks over six months in the presence of optimum conventional drug therapy. GMA was then added, and after five sessions over five consecutive weeks, monitoring continued for a further six months. The number of ocular episodes per patient for the first six months was 4.21 ± 1.6 and for the six months post GMA was 2.93 ± 1.39; 64% improved and 36% did not change or worsened. The authors concluded that patients with long duration of disease are better responders as compared with patients with a short duration (Sonoda *et al.*, 2005; Namba *et al.*, 2006), contrary to the outcome in patients with UC reported by Suzuki *et al.* (2006) Sonoda *et al.* (2005) also reported a marked alleviation of retinal inflammation (edema), while Kanekura *et al.* (2005b) reported a dramatic improvement of orogenital lesions.

17. Conclusions & Future Perspectives

When inflammatory diseases are viewed from a fundamental standing, most are the outcomes of mechanisms that reflect elevated or activated natural components of the body's own immune system. Additionally, in many cases drugs are not very effective and may complicate the condition as an additional morbidity factor. Therefore, even in this age of modern medicine, the concept of treating disease by removing from the blood circulating causative agents is very intriguing and can be expanded to minimize the need for drug therapy or potentially as an adjunct to maximize drug efficacy and minimize side effects. In this endeavor, the therapeutic value of GMA would be related to the success of identifying the causative factors as well as to the selectivity of the GMA actions to eliminate the target agents or otherwise generate the missing factor. Conditions like IBD and RA are debilitating chronic diseases reflecting a dysregulated exuberant immune behavior, which afflict millions of individuals throughout the world and have symptoms that impair quality of life and ability to function. Until now, the focus of drug therapy has been to reduce symptoms with no impact by drugs on the fundamental cause of the disease. However, it is now widely understood that patients with IBD, RA, BD or pyoderma bear as yet uncharacterized genetic predisposing factors, and because of this lack of understanding, hitherto, drug therapy of these conditions has been dictated by an empirical thinking rather than the drug being targeted at the mechanism of the immune disorder. This is like saying that at least until now, genetic diseases by their very nature have defied treatment and therefore are incurable. Likewise, the chronic nature of these conditions means that affected individuals require lifelong medications; this together with the empirical approach to drug therapy have been a major factor for refractoriness and adverse drug effects, which further complicate the treatment. However, in recent years, certain cytokines like TNF, IL-1, IL-6, IL-8, IL-23 and others have been recognized to be major factors in the exacerbation and the perpetuation of inflammatory conditions. In fact, the efficacy of anti-TNF antibodies in IBD has validated this new perception. With this background in mind, major sources of inflammatory cytokines include circulating granulocytes and monocytes/macrophages, which in patients with inflammatory diseases are elevated with activation behavior. In line with this thinking, the Adacolumn GMA that can selectively deplete elevated circulating leucocytes should be a therapeutically relevant non-pharmacologic intervention in patients with immune disorders. In particular, depletion of elevated TNF producing $CD14^{+}CD16^{+}$ phenotype should alleviate the inflammatory intensity, and this should allow healing to proceed. Additionally, an increase in the regulatory CD4(+)CD25(+) T-cell phenotype associated with GMA is potentially very interesting. Additional anti-inflammatory agents could be present among the vast numbers of substances that are released from the adherent cells in the Adacolumn during GMA in clinical setting.

On the clinical efficacy side, patients with IBD who are steroid naïve or have a short duration of relapse phase are good responders while patients with deep colonic lesions and extensive loss of the mucosal tissue have shown to be unlikely responders. Hitherto clinical reports indicate that GMA in patients with IBD has spared corticosteroids and other unsafe pharmacologic preparations. Because, GMA, instead of adding to the body systems as drug therapy does, removes from the body, and hence it is not expected to induce dependency or refractoriness and has an unrivalled safety profile. However, the authors wish to reflect comments from some physicians that in addition to patients' baseline demographic features, operational skills and favorable conditions during GMA are important factors for optimal clinical efficacy. These include maintenance of the recommended blood flow rate for the duration of GMA (non-intermittent flow) and an adequate ambient temperature to avoid cooling of blood during GMA. Additionally, the impact of flow rate (>30 mL/minute) and duration of one GMA session (>60 minutes) and the frequency of GMA need further investigation. The bottom line is that in IBD patients, leucocytes are elevated with activation behavior and are known sources of inflammatory cytokines. Therefore removing the extra load of leucocytes should produce efficacy in many patients, while in those who do not respond to GMA monotherapy, removing the activated myeloid leucocytes should at least enhance drug efficacy. Finally, it is very important to bear in mind that patients with IBD who respond to GMA at an early stage of their disease and avoid pharmacologicals have a very favorable clinical course by avoiding drugs including corticosteroids at the initial stages of their IBD.

18. Acknowledgements

Dr. Saniabadi and Ms Shima are scientists affiliated to JIMRO. Dr. Saniabadi also holds an academic position at the Institute of Pharmacology, Hamamatsu University School of Medicine.

All the authors acknowledge having no conflict of interest in connection with the publication of the article in this chapter.

List of Abbreviations used in this Chapter

AC	after Christ
BC	before Christ
BD	Behcet's disease
CAI	clinical activity index
CD	Crohn's disease
CE	conforming to European standards
CR	complement receptor
CTL	cytotoxic T-lymphocyte
EC	European Commission
FcγR	fragment crystallizable gamma receptor

GI	gastrointestinal
GMA	granulocyte and monocyte apheresis
HCV	hepatitis C virus
HGF	hepatocyte growth factor
HIV	Human Immunodeficiency Virus
IBD	inflammatory bowel disease
IC	immune complex
IEC	International Electrotechnical Commission
IFN	interferon
IgG	immunoglobulin G
IL	interleukin
IL-1ra	interleukin 1 receptor antagonist
ISO	international organization for standardization
JIS	Japanese industrial standards
LECAM-1	intercellular adhesion molecule 1
NK	natural killer
PG	pyoderma gangrenosum
PVC	polyvinylchloride
RA	rheumatoid arthritis
TNF	tumor necrosis factor
Treg	regulatory T-cell
TUV	TÜV SÜD
UC	ulcerative colitis
US	United States

References

Abel JJ, Rowntree LG, Turner BB. (1924) Plasma removal with return of corpuscles (plasmapheresis). *J Pharmacol Exp Ther* **5:** 625–631.

Allison MC, Dhillon AP, Lewis WG, Pounder RE. (1998) *Inflammatory Bowel Disease.* Mosby, London, pp. 9–95.

Aoki H, Nakamura K, Yoshimatsu Y, *et al.* (2007) Adacolumn selective leukocyte adsorption apheresis in patients with active ulcerative colitis: Clinical efficacy, effects on plasma IL-8, and expression of Toll-like receptor 2 on granulocytes. *Dig Dis Sci* **52:** 1427–1433.

Baumgart DC, Carding SR. (2007) Inflammatory bowel disease: Cause and immunobiology. *Lancet* **369:** 1627–1640.

Belge KU, Dayyani F, Horelt A, Ziegler-Heibrock L. (2002) The proinflammatory CD14+CD16+DR++ monocytes are a major source of TNF. *J Immunol* **168:** 3536–3542.

Beretta A, Hasson H, Saniabadi A, *et al.* (2000) Ex-vivo purging of circulating monocytes results in immunovirologic improvement in partially HAART responder HIV infected patients. *J Biol Regul Homeost Agents* **14:** 27–31.

Beretta A, Hasson H, Saniabadi A, *et al.* (2002) Selective granulocyte/monocyte apheresis in the treatment of HIV-infected patients: Short-term and long-term effects on immunological and virological parameters. *Perfusion* **17:** 47–51.

Biswas P, Mantelli B, Hasson H, *et al.* (2003) *In vivo* modulation of leukocyte trafficking receptor following therapeutic purging of myeloid cells: Implications for treatment of HIV infection and other immune disorders. *Clin Immunol* **109:** 355–358.

Bouffard P, Hayashi PH, Acevedo R, *et al.* (1992) Hepatitis C is detected in a monocyte macrophage subpopulation of peripheral blood mononuclear cells of infected patients. *J Infect Dis* **166:** 1276–1280.

Brannigan AE, O'Connell PR, Hurley H. (2000) Neutrophil apoptosis is delayed in patients with inflammatory bowel disease. *Shock* **13:** 361–366.

Brown SL, Greene MH, Gershon SK, *et al.* (2002) Tumor necrosis factor antagonist therapy and lymphoma development: Twenty-six cases reported to the Food and Drug Administration. *Arthritis Rheum* **46:** 3151–3158.

Buckner D, Graw RG Jr, Eisel RJ, *et al.* (1969) Leukapheresis by continous flow centrifugation (CFC) in patients with chronic myelocytic leukemia (CML). *Blood* **33:** 353–369.

Cassatella MA. (1995) The production of cytokines by polymorphonuclear neutrophils. *Immunol Today* **16:** 21–26.

Caussin-Scwemling C, Scmitt C, Stoll-Keller F. (2001) Study of the infection of human blood derived monocyte/macrophage with hepatitis C virus *in vitro*. *J Med Virol* **65:** 14–22.

Clutterbuck H. (1840) *On the Proper Administration of Blood-Letting for the Prevention and Cure of Disease.* S. Highley, London.

Cohen RD. (2005) Treating ulcerative colitis without medications — look mom, no drugs. *Gastroenterology* **128:** 235–236.

Cohen RD. (2005) Treating ulcerative colitis without medications — "Look Mom, No Drugs!" *Gastroenterology* **128:** 235–236.

Cravatto M, Pozzato G, *et al.* (2000) Peripheral blood neutrophils from hepatitis C virus-infected patients are replication sites for the virus. *Haematologica* **85:** 356–61.

Cuadrado E, Alonso M, de Juan MD, *et al.* (2008) Regulatory T cells in patients with inflammatory bowel diseases treated with adacolumn granulocytapheresis. *World J Gastroenterol* **14:** 1521–1527.

Cullen G, Kroshinsky D, Cheifetz S, Korzenik JR. (2011) Psoriasis associated with anti-tumour necrosis factor therapy in inflammatory bowel disease: A new series and a review of 120 cases from the literature. *Aliment Pharmacol Ther* **34:** 1318–1327.

Curtis JE, Hersh EM, Freireich EJ. (1972) Leukoapheresis therapy of chronic lymphocytic leukemia. *Blood* **39:** 163–175.

D'Arrigo C, Candal-Couto JJ, Greer M, *et al.* (1994) Human neutrophil Fc receptor mediated adhesion under flow: A hallow fiber model of intravascular arrest. *Clin Exp Immunol* **100:** 173–179.

Diepolder HM, Kashiwagi N, Teuber G, *et al.* (2005) Leucocytapheresis with Adacolumn enhances HCV-specific proliferative responses in patients infected with hepatitis C virus genotype 1. *J Med Virol* **77:** 209–215.

Dinarello CA, Thompson RC. (1991) Blocking IL-1: Interleukin 1 receptor antagonist *in vivo* and *in vitro. Immunol Today* **12:** 404–410.

Edwards SW, Hallett MB. (1997) Seeing the wood for the trees: The forgotten role of neutrophils in rheumatoid arthritis. *Immunol Today* **18:** 320–324.

Ehrlich GE. (1997) Vaculitis in Behcet's disease. *Int Rev Immunol* **14:** 81–88.

Engel MA, Neurath MF. (2010) New pathophysiological insights and modern treatment of IBD. *J Gastroenterol* **45:** 571–583.

Eshleman JR, Silberstein LE. (1991) The development of selective plasmapheresis as a means for specific therapeutic intervention (editorial). *Transfusion* **31:** 96–98.

Fujimori J, Yoshino S, Koiwa M, *et al.* (1996) Improvement in rheumatoid arthritis following application of an extracorporeal granulotrap column, G-1. *Rheumatology Int* **15:** 175–180.

Garcia O, Vidaller Palacin A, Munoz Sanchez J, Roca Linares G. (2004) Granulocyte-Apheresis in ocular Behcet's disease resistant to medical treatment (Engl Abstract). *Arch Soc Esp Oftalmol* **9:** 335–340.

Gillitzer R. (2001) Inflammation in human skin: A model to study chemokine-mediated leukocyte migration *in vivo. J Pathol* **194:** 398–405.

Goldfinger D, Capostagno V, Lowe C, *et al.* (1980) Use of long-term leukapheresis in the treatment of chronic lymphocytic leukemia. *Transfusion* **20:** 450–454.

Hanai H, Iida T, Takeuchi K, *et al.* (2008) Adsorptive depletion of elevated proinflammatory CD14+CD16+DR++ monocytes in patients with inflammatory bowel disease. *Am J Gastroenterol* **103:** 1210–1216.

Hanai H, Iida T, Watanabe F, *et al.* (2006) Effects of Adacolumn selective leukocytapheresis on plasma cytokines during active disease in patients with ulcerative colitis. *World J Gastroenterol* **12:** 3393–3399.

Hanai H, Takeda Y, Eberhardson M, *et al.* (2011) The mode of actions of the Adacolumn therapeutic leucocytapheresis in patients with inflammatory bowel disease: A concise review. *Clin Exp Immunol* **163:** 50–58.

Hanai H, Watanabe F, Saniabadi AR, *et al.* (2002) Therapeutic efficacy of granulocyte and monocyte adsorption apheresis in severe active ulcerative colitis. *Dig Dis Sci* **47:** 2349–2353.

Hanai H, Watanabe F, Takeuchi K, *et al.* (2003) Leukocyte adsorptive apheresis for the treatment of active ulcerative colitis: A prospective, uncontrolled, pilot study. *Clin Gastroenterol Hepatol* **1:** 28–35.

Hanai H, Watanabe F, Yamada M, *et al.* (2004a) Adsorptive granulocyte and monocyte apheresis versus prednisolone in patients with corticosteroid dependent moderately severe ulcerative colitis. *Digestion* **70:** 36–44.

Hanai H, Watanabe F, Yamada M, *et al.* (2004b) Correlation of serum soluble TNF-alpha receptors I and II levels with disease activity in patients with ulcerative colitis. *Am J Gastroenterol* **99:** 1532–1538.

Hanai H. (2006) Positions of selective leukocytapheresis in the medical therapy of ulcerative colitis. *World J Gastroenterol* **12:** 7568–7577.

Hanauer SB. (2004) Medical therapy of ulcerative colitis. *Gastroenterology* **126:** 1582–1592.

Hasson H, Mantelli B, Biswas P, *et al.* (2007) Favorable outcome of *ex-vivo* purging of monocytes after the reintroduction of treatment after interruption in patients infected with multidrug resistant HIV-1. *J Med Virol* **79:** 1640–1649.

Hasson H, Saniabadi A, Alfano M, *et al.* (2002) Granulocyte/monocyte apheresis induces sustained increases in CD4 T cells in HIV-infected patients: With poor CD4 restoration after suppression of viral replication by HAART. *J Biol Regul Homeost Agents* **16:** 58–63.

Heimann TM, Aufses AH Jr. (1985) The role of peripheral lymphocytes in the prediction of recurrence in Crohn's disease. *Surg Gynecol Obstet* **160:** 295–298.

Hiraishi K, Takeda Y, Shiobara N, *et al.* (2003) Studies on the mechanisms of leukocyte adhesion to cellulose acetate beads: An in vitro model to assess the efficacy of cellulose acetate carrier-based granulocyte and monocyte adsorptive apheresis. *Ther Apher Dial* **7:** 334–340.

Huber S, Schramm C, Lehr HA, *et al.* (2004) Cutting edge: TGF-beta signaling is required for the *in vivo* expansion and immunosuppressive capacity of regulatory CD4+CD25+ T cells. *J Immunol* **173:** 6526–6531.

Ikeda S, Takahashi H, Suga Y, *et al.* (2013) Therapeutic depletion of myeloid lineage leukocytes in patients with generalized pustular psoriasis indicates a major role for neutrophils in the immunopathogenesis of psoriasis. *J Am Acad Dermatol* **68:** 609–617.

Kanai T, Watanabe M. (2005) Clinical application of human CD4+ CD25+ regulatory T cells for the treatment of inflammatory bowel diseases (Review). *Expert Opin Biol Ther* **5:** 451–462.

Kanekura T, Gushi A, Iwata M, *et al.* (2005a) Treatment of Behçet's disease with granulocyte and monocyte adsorption apheresis. *J Am Acad Dermatol* **51 (Suppl 2):** S83–S87.

Kanekura T, Kawahara K, Maruyama I, Kanzaki T. (2005b) Treatment of pyoderma gangrenosum with granulocyte and monocyte adsorption apheresis. *Ther Apher Dial* **9:** 292–296.

Kanekura T, Maruyama I. (2002) Granulocyte and monocyte adsorption apheresis for pyoderma gangrenosum. *J Am Acad Dermatol* **47:** 320–321.

Kanekura T, Yoshii N, Yonezawa T, *et al.* (2003) Treatment of pustular psoriasis with granulocyte and monocyte adsorption apheresis. *J Am Acad Dermatol* **49:** 329–332.

Kanke K, Nakano, M, Hiraishi H. (2004) Clinical evaluation of granulocyte/monocyte apheresis therapy for active ulcerative colitis. *Dig Liver Dis* **36:** 811–817.

Kashiwagi N, Hirata I, Kasukawa R. (1998) A role for granulocyte and monocyte apheresis in the treatment of rheumatoid arthritis. *Ther Apher* **2:** 134–141.

Kashiwagi N, Saniabadi A, Shimoyama T. (2002) Immunomodulatory effects of granulocyte and monocyte adsorption apheresis as a treatment for patients with ulcerative colitis. *Dig Dis Sci* **47:** 1334–1341.

Kato S, Hosomi E, Amano F, *et al.* (2012) The efficacy of intensive granulocyte and monocyte adsorption apheresis in a patient with Crohn's disease complicated by extensive subcutaneous aseptic neutrophilic abscesses. *J Crohns Colitis* **6:** 787–791.

Kyogoku M, Kasukawa R. (1998) Clinical and basic studies on the G-1 column, a new extracorporeal therapeutic device in controlling rheumatoid arthritis. *Inflamm Res* **47 (suppl 3):** S166–S176.

Lazarus HM. (1991) Selective in vivo removal of rheumatoid factor by an extracorporeal treatment device in rheumatoid arthritis patients. *Transfusion* **31:** 122–128.

Lazzarin A, Hasson H, Saniabadi A, *et al.* (2001) Immunovirological improvement in partially HAART responder HIV-infected patients by monocyte adsorption apheresis. *J Clin Apher* **16:** 33–4.

Lee A, Whyte M, Haslett C. (1993) Inhibition of apoptosis and prolongation of neutrophil functional longevity by inflammatory mediators. *J Leukoc Biol* **54:** 283–288.

Lefkowitz DL, Mills K, Lefkowitz SS, *et al.* (1995) Neutrophil-macrophage interaction: A paradigm for chronic inflammation. *Med Hypotheses* **44:** 58–62.

Lerebours E, Bussel A, Modigliani R, *et al.* (1994) Treatment of Crohn's disease by lymphocyte apheresis: A randomized controlled trial. Groupe d'Etudes Thérapeutiques des Affections Inflammatoires Digestives. *Gastroenterology* **107:** 357–361.

Limburg P, David M, Ahlquist A, Sandborn WJ. (2000) Faecal calprotectin levels predict colorectal inflammation among patients with chronic diarrhoea referred for colonoscopy. *Am J Gastroenterol* **95:** 2831–2837.

Liu H, Hu B, Xu D, Liew FY. (2003) CD4+CD25+ regulatory T cells cure murine colitis: The role of IL-10, TGF-beta, and CTLA4. *J Immunol* **171:** 5012–5017.

Lovell DJ, Ruperto N, Goodman S, *et al.*; Pediatric Rheumatology Collaborative Study Group; Pediatric Rheumatology International Trials Organisation. (2008) Adalimumab with or without methotrexate in juvenile rheumatoid arthritis. *N Engl J Med* **359:** 810–20.

Maiden L, Takeuchi K, Baur R, *et al.* (2008) Selective white cell apheresis reduces relapse rates in patients with IBD at significant risk of clinical relapse. *Inflamm Bowel Dis* **14:** 1413–1418.

Matsui T, Nishimura T, Matake H, *et al.* (2003) Granulocytapheresis for Crohn's disease: A report on seven refractory patients. *Am J Gastroenterol* **98:** 511–512.

Maul J, Loddenkemper C, Mundt P, *et al.* (2005) Peripheral and intestinal regulatory CD4+ CD25(high) T cells in inflammatory bowel disease. *Gastroenterology* **128:** 1868–1878.

McCarthy DA, Rampton DS, Liu Y-C. (1991) Peripheral blood neutrophils in inflammatory bowel disease: Morphological evidence of *in vivo* activation in active disease. *Clin Exp Immunol* **86:** 489–493.

Meagher LC, Cousin JM, Seckl JR, Haslett C. (1996) Opposing effects of glucocorticoids on the rate of appoptosis in neutrophilic and eosinophilic granulocytes. *J Immunol* **156:** 4422–4428.

Meuret G, Bitzi A, Hammer B. (1978) Macrophage turnover in Crohn's disease and ulcerative colitis. *Gastroenterology* **74:** 501–503.

Mohler KM, Torrance DS, Smith GA, Widmer MB. (1993) Soluble tumour necrosis factor (TNF) receptors are effective therapeutic agents in lethal endotoximia and function simultaneously as both TNF carriers and TNF antagonists. *J Immunol* **151:** 1548–1561.

Mori S, Nagashima M, Yoshida K, *et al.* (2004) Granulocyte adsorptive apheresis for leg ulcers complicated by rheumatoid arthritis: A report on three successfully treated cases. *Int J Dermatol* **43:** 732–735.

Morse EE, Carbone PP, Freireich EJ, *et al.* (1966) Repeated leukapheresis of patients with chronic myelocytic leukemia. *Transfusion* **6:** 175–182.

Mulder DJ, Noble AJ, Justinich CJ, Duffin JM. (2014) A tale of two diseases: The history of inflammatory bowel disease. *J Crohns Colitis* **8:** 341–348.

Muratov V, Lundahl J, Ulfgren AK, *et al.* (2006) Down-regulation of interferon-gamma parallels clinical response to selective leukocyte apheresis in patients with inflammatory bowel disease: A 12-month follow-up study. *Int J Colorectal Dis* **21:** 493–504.

Nagashima M, Yoshino S, Tanaka H, *et al.* (1998) Granulocyte and monocyte apheresis suppresses symptoms of rheumatoid arthritis: A pilot study. *Rheumatology Int* **18:** 113–118.

Namba K, Sonoda KH, Kitamei H, *et al.* (2006) Granulocytapheresis in patients with refractory ocular Behcet's disease. *J Clin Apher* **21:** 121–128.

Nikolaus S, Bauditz J, Gionchetti P. (1998) Increased secretion of pro-inflammatory cytokines by circulating polymorphonuclear neutrophils and regulation by interleukin-10 during intestinal inflammation. *Gut* **42: 4**70–476.

O'Keefe SJ. (1996) Nutrition and gastrointestinal disease. *Scand J Gastroenterol (Suppl)* **220:** 52–59.

Ohmori T, Yamagiwa A, Nakamura I, *et al.* (2003) Treatment of pyoderma gangrenosum associated with Crohn's disease. *Am J Gastroenterol* **98:** 2101–2102.

Papadakis KA, Targan SR. (2000) Role of cytokines in the pathogenesis of inflammatory bowel disease. *Annu Rev Med* **51:** 289–298.

Podolsky DK. (2002) Inflammatory bowel disease. *N Engl J Med* **347:** 417–429.

Powrie F, Read S, Mottet C, *et al.* (2003) Control of immune pathology by regulatory T cells. *Novartis Found Symp* **252:** 92–98.

Present DH. (2000) How to do without steroids in inflammatory bowel disease. *Inflamm Bowel Dis* **6:** 48–57.

Rottman JB, Smith TL, Ganley KG, *et al.* (2001) Potential role of the chemokine receptors CXCR3, CXCR4, and the integrin alphaEbeta 7 in the pathogenesis of psoriasis vulgaris. *Lab Invest* **81:** 335–347.

Rutgeerts P, Sandborn WJ, Feagan B, Hanauer S. (2005) Infliximab for induction and maintenance therapy for ulcerative colitis. *N Engl J Med* **353:** 2462–2476.

Sacco R, Tanaka T, Yamamoto T, *et al.* (2014) Adacolumn leucocytapheresis for ulcerative colitis: Clinical and endoscopic features of responders and unresponders. *Expert Rev Gastroenterol Hepatol* **27:** 1–7.

Sahin S, Lawrence R, Direskeneli H, *et al.* (1996) Monocyte activity in Behcet's disease. *Br J Rheumatol* **35:** 424–429.

Sakane T, Takeno M, Suzuki N, Inaba G. (1999) Current Concepts: Behcet's disease. *N Engl J Med* **341:** 1284–1291.

Sakuraba A, Motoya S, Watanabe K, *et al.* (2009) An open-label prospective randomized multicenter study shows very rapid remission of ulcerative colitis by intensive granulocyte and monocyte adsorptive apheresis as compared with routine weekly treatment. *Am J Gastroenterol* **104:** 2990–2995.

Saniabadi AR, Hanai H, Fukunaga K, *et al.* (2003a) Therapeutic leukocytapheresis for inflammatory bowel disease. *Transfus Apher Sci* **37:** 191–200.

Saniabadi AR, Hanai H, Takeuchi K, *et al.* (2003b) Adacolumn, an adsorptive carrier based granulocyte and monocyte apheresis device for the treatment of inflammatory and refractory diseases associated with leukocytes. *Ther Apher Dial* **7:** 48–59.

Saniabadi AR, Hanai H, Suzuki Y, *et al.* (2005) Adacolumn for selective leukocytapheresis as a non-pharmacological treatment for patients with disorders of the immune system: An adjunct or an alternative to drug therapy? *J Clin Apher* **20:** 171–84.

Saniabadi AR, Hanai H. (2010) Therapeutic apheresis from the early civilizations to the twenty-first century. *Gastroenterol Clin Biol* **34:** 645–648.

Saniabadi AR, Tanaka T, Ohmori T, *et al.* (2014) Treating inflammatory bowel disease by adsorptive leucocytapheresis: A desire to treat without drugs. *World J Gastroenterol* **20:** 9699–9715.

Sanmartí R, Marsal S, Valverde J, *et al.* (2005) Adsorptive granulocyte/monocyte apheresis for the treatment of refractory rheumatoid arthritis: An open pilot multicentre trial. *Rheumatology (Oxford)* **44:** 1140–1144.

Sawada K, Kashiwamura S, Okamura H, *et al.* (2005) Selective granulocyte and monocyte apheresis as a new adjunct to enhance the efficacy of interferon-alpha + ribavirin in patients with high plasma hepatitis C virus. *Dig Liver Dis* **37:** 515–521.

Sawada K, Masaki N, Hayashi S, *et al.* (2005a) Immunomodulatory effects of selective leucocytapheresis as a new adjunct to interferon-alpha2b plus ribavirin combination therapy: A prospective study in patients with high plasma HCV viraemia. *J Viral Hepat* **12:** 274–282.

Sawada K, Ohnishi K, Fukunaga K, *et al.* (2003) Granulocyte and monocyte adsorptive apheresis for patients with chronic hepatitis C virus infection: A report on six cases with high plasma viremia. *Ther Apher Dial* **7:** 547–53.

Sawada K, Ohnishi K, Fukunaga K, Shimoyama T. (2003a) A new treatment for HCV-ulcerative colitis co-morbidity intolerant to INF-alpha. *Am J Gastroenterol* **98:** 228–229.

Scherberich JE. (2003) Proinflammatory blood monocytes: Main effector and target cells in systemic and renal disease; background and therapeutic implications. *Int J Pharmaco Ther* **41:** 459–464.

Schreiber S, Nikolaus S, Hampe J. (1999) Tumour necrosis factor alpha and interleukin 1beta in relapse of Crohn's disease. *Lancet* **353:** 459–461.

Seishima M, Mizutani Y, Shibuya Y, *et al.* (2008) Efficacy of granulocyte and monocyte adsorption apheresis for pustular psoriasis. *Ther Apher Dial* **12:** 13–18.

Shimoyama T, Sawada K, Hiwatashi N, *et al.* (2001) Safety and efficacy of granulocyte and monocyte adsorption apheresis in patients with active ulcerative colitis: A multicenter study. *J Clin Apher* **16:** 1–9.

Sonoda KH, Inaba S, Ariyama A, *et al.* (2005) Therapeutic neutrophil apheresis in patients with ocular Behçet disease. *Arch Ophthalmol* **123:** 267–269.

Suzuki Y, Yoshimura N, Fukuda K, *et al.* (2006) A retrospective search for predictors of clinical response to selective granulocyte and monocyte apheresis in patients with ulcerative colitis. *Dig Dis Sci* **51:** 2031–2038.

Suzuki Y, Yoshimura N, Saniabadi AR, Saito Y. (2004) Selective granulocyte and monocyte adsorptive apheresis as a first-line treatment for steroid naïve patients with active ulcerative colitis: A prospective uncontrolled study. *Dig Dis Sci* **49:** 565–571.

Tabuchi T, Ubukata H, Saniabadi AR, Soma T. (1999) Granulocyte apheresis as a possible new approach in cancer therapy: A pilot study involving two cases. *Cancer Detect Prev* **23:** 417–421.

Tabuchi T, Ubukata H, Sato S, *et al.* (1995) Granulocytapheresis as a possible cancer treatment. *Anticancer Res* **15:** 985–990.

Taffet SL, Das KM. (1983) Sulphasalazine-adverse effects and desensitization. *Dig Dis Sci* **28:** 833–842.

Tahara Y, Ido A, Yamamoto S, *et al.* (2003) Hepatocyte growth factor facilitates colonic mucosal repair in experimental ulcerative colitis in rats. *J Pharmacol Exp Ther* **307:** 146–151.

Takeda Y, Shiobara N, Saniabadi AR, *et al.* (2004) Adhesion dependent release of hepatocyte growth factor and interleukin-1 receptor antagonist from human blood granulocytes and monocytes: Evidence for the involvement of plasma IgG, complement C3 and β2 integrin. *Inflamm Res* **53:** 277–283.

Tanaka T, Hanai H, Bjarnason I. (2011) Adacolumn therapeutic leucocytapheresis for ulcerative colitis: Clinical and endoscopic features of responders and unresponders to this non-pharmacologic intervention. *J Gastroint Dig Syst* **1:** 104.

Tanaka T, Okanobu H, Kuga Y, *et al.* (2010) Clinical and endoscopic features of responders and non-responders to adsorptive leucocytapheresis: A report based on 120 patients with active ulcerative colitis. *Gastroenterol Clin Biol* **34:** 687–695.

Tanaka T, Okanobu H, Yoshimi S, *et al.* (2008) In patients with ulcerative colitis, adsorptive depletion of granulocytes and monocytes impacts mucosal level of neutrophils and clinically is most effective in steroid naïve patients. *Dig Liver Dis* **40:** 731–736.

Tanaka T, Sugiyama S, Goishi H, *et al.* (2013) Treatment of children and adolescents with ulcerative colitis by adsorptive depletion of myeloid lineage leucocytes as monotherapy or in combination with low dose prednisolone after failure of first-line medications. *BMC Gastroenterology* **13:** 130.

Tibble JA, Sigthorsson G, Bridger D, *et al.* (2000) Surrogate markers of intestinal inflammation are predictive of relapse in patients with inflammatory bowel disease. *Gastroenterology* **119:** 15–22.

Truelove SC, Jewell DP. (1974) Intensive intravenous regimens for severe attacks of ulcerative colitis. *Lancet* **1:** 1067–1070.

Uhlig HH, Powrie F. (2005) The role of mucosal T lymphocytes in regulating intestinal inflammation. *Springer Semin Immunopathol* **27:** 167–180.

van Dullemen HM, van Deventer SJ, Hommes DW, *et al.* (1995) Treatment of Crohn's disease with anti-tumour necrosis factor chimeric monoclonal antibody (cA2). *Gastroenterology* **109:** 129–135.

Yamamoto T, Saniabadi AR, Maruyama Y, *et al.* (2007) Factors affecting clinical and endoscopic efficacies of selective leucocytapheresis for ulcerative colitis. *Dig Liver Dis* **39:** 626–633.

Yamamoto T, Saniabadi AR, Umegae S, Matsumoto K. (2006) Impact of selective leukocytapheresis on mucosal inflammation and ulcerative colitis: Cytokine profiles and endoscopic findings. *Inflamm Bowel Dis* **12:** 719–726.

Yamamoto T, Umegae S, Matsumoto K. (2012) Long-term clinical impact of early introduction of granulocyte and monocyte adsorptive apheresis in new onset, moderately active, extensive ulcerative colitis. *J Crohns Colitis* **6:** 750–755.

Yokoyama Y, Fukunaga K, Fukuda Y, *et al.* (2007) Demonstration of low-regulatory CD25High+CD4+ and high-pro-inflammatory CD28-CD4+ T-Cell subsets in patients with ulcerative colitis: Modified by selective granulocyte and monocyte adsorption apheresis. *Dig Dis Sci* **52:** 2725–2731.

Yokoyama Y, Kawai M, Fukunaga K, *et al.* (2013) Looking for predictive factors of clinical response to adsorptive granulocyte and monocyte apheresis in patients with ulcerative colitis: Markers of response to GMA. *BMC Gastroenterol* **13:** 27.

Yoshimura N, Tadami T, Kawaguchi T, *et al.* (2012) Processed blood volume impacts clinical efficacy in patients with ulcerative colitis undergoing adsorptive depletion of myeloid lineage leucocytes. *J Gastroenterol* **47:** 49–55.

Chapter 32

Efficacy Outcomes for Adacolumn Adsorptive Granulocyte and Monocyte Apheresis Applied as a Non-Pharmacologic Treatment Option in Patients with Inflammatory Bowel Disease

Yoko Yokoyama* and Ken Fukunaga

Abstract

Adacolumn is a single use medical device for direct hemoperfusion in therapeutic settings. The currently produced Adacolumns are about 206 mm in length and 60 mm in diameter, filled with acetate beads as the column leucocytapheresis carriers. The carriers selectively adsorb FcγR (fragment crystallizable gamma receptor) and complement receptor bearing myeloid lineage leucocytes. Adsorptive apheresis with the Adacolumn is popularly known as GMA, which stands for granulocyte and monocyte apheresis. Pre- and post-column cell

* Division of Internal Medicine, Department of Inflammatory Bowel Disease, Hyogo College of Medicine, Hyogo, Japan.
Correspondence address: Dr. Yoko Yokoyama MD, PhD,
Division of Internal Medicine, Department of Inflammatory Bowel Disease, Hyogo College of Medicine, 1-1 Mukogawa, Nishinomiya, Hyogo 663-8501, Japan.
Tel: +81 798 456662; Fax: +81 798 456661; E-mail: yoko0502@hyo-med.ac.jp.

counts have shown that GMA depletes about 65% of granulocytes and 55% of monocytes together with a significant fraction of platelets from the blood, which passes through the column. Red cell count is unaffected, while less than 2% of lymphocytes are removed. In fact, GMA has been associated with a sustained increase in absolute lymphocyte count up to several months post last GMA session. The increase in lymphocyte count includes the $CD4^{+}CD25^{+High}$ T-cell phenotype known as the regulatory T-cells (Treg), which have essential immunoregulatory function in the intestinal mucosa.

Inflammatory bowel diseases (IBD), which include ulcerative colitis (UC) and Crohn's disease (CD), are chronic relapsing-remitting inflammatory disorders that afflict millions of individuals throughout the world with debilitating symptoms which impair quality of life and performance. Although the etiology of IBD is not fully understood at present, elevated and activated myeloid lineage leucocytes potentially are significant factors in the exacerbation and the perpetuation of IBD by releasing inflammatory cytokines. Accordingly, selective depletion of these cells by GMA with the Adacolumn has been introduced as a non-pharmacologic treatment option for patients with active IBD. GMA is expected to be effective as monotherapy in many patients with the first UC episode and who are steroid naïve. In other UC patients, GMA, at least should enhance the efficacy of pharmacologics. Patients with intractable UC require life-long medications with conventional drugs including novel anti-tumor necrosis factor (TNF) biologics, and this may lead to drug dependency, loss of response, together with adverse side effects, which add complexity to the disease. It is very important to bear in mind that GMA as a non-pharmacologic treatment intervention, if effective has multiple advantages over pharmacologics including lack of refractoriness or loss of response and freedom from serious adverse side effects often seen with pharmacologics. Therefore, it is true to say that the most unrivalled feature of GMA is its safety as well as it being a non-drug treatment strategy. For these features, GMA with the Adacolumn is very much favored by patients as well as by physicians. Efficacy outcomes in first episode cases, steroid naïve cases as well as in patients with a very short duration of active IBD prior to GMA have been over 70%. In contrast, patients with extensive deep IBD lesions and those with a long duration of active IBD refractory to multiple pharmacologics including new biologics are not very likely to benefit from GMA therapy. Further, patients who receive GMA therapy during an early phase of their active IBD attain a favorable clinical course by avoiding pharmacologics from the start.

1. Introduction

Ulcerative Colitis (UC) and Crohn's disease (CD) are the two major phenotypes of the idiopathic inflammatory bowel disease (IBD). These immune disorders afflict millions of individuals with symptoms that impair wellbeing and quality of life (Mulder *et al.*, 2014). Whereas UC is confined to the colon and the rectum (large intestine), the expression of CD can be from the mouth to the perianal region. CD. Currently, a plethora of pharmacologicals are available for the treatment of IBD flare ups, but as yet none offers a permanent cure, at best suppressing a clinical relapse or prolonging quiescent periods of IBD. However drug therapy of IBD has been associated with adverse side effects, which add complexity to the disease (Akerkar *et al.*, 1997; Present, 2000). Therefore, a non-drug intervention to normalize the dysregulated immune profile is highly desirable and should be favored by the patients as well as by the physicians. With this in mind, an Adacolumn has been developed for selective depletion of myeloid lineage leucocytes by adsorptive apheresis in patients with active IBD (Saniabadi *et al.*, 2003; Saniabadi *et al.*, 2007; Yokoyama *et al.*, 2007; Sakuraba *et al.*, 2009; Hanai *et al.*, 2011; Cabriada *et al.*, 2012; Tanaka *et al.*, 2013; Fukuchi *et al.*, 2013). Adsorptive apheresis with the Adacolumn is popularly referred to as GMA, which stands for granulocyte and monocyte apheresis. Pre- and post-column cell counts have shown that GMA depletes about 65% of granulocytes and 55% of monocytes together with a significant fraction of platelets from the blood, which passes through the column (Saniabadi *et al.*, 2003, 2007). Red cell count is unaffected, while less than 2% of lymphocytes are removed (Saniabadi *et al.*, 2003). In fact, GMA has been associated with a sustained increase in absolute lymphocyte count up to several months post last GMA session (Saniabadi *et al.*, 2003).

In Japan, GMA with the Adacolumn has been officially approved as one treatment option for patients with IBD, while in the European Union countries, the Adacolumn has CE Mark (conforming to the European standards), which allows its therapeutic application for IBD patients (Saniabadi *et al.*, 2003). However, in Japan, GMA was first approved as one treatment option for patients with UC in 2000, and then for patients with CD in 2009. Additionally, intensive GMA with two sessions per week instead of one session per week (Sakuraba *et al.*, 2009) was approved for UC patients in 2010. Further, in UC patients, retrospective studies have reported an inverse correlation between disease duration and GMA efficacy (Suzuki *et al.*, 2006; Yokoyama *et al.*, 2013). In this chapter, we have endeavored to highlight the clinical efficacy of GMA in patients with IBD. More specific coverage of the Adacolumn GMA design features with emphasis on basic mode of action (Hanai *et al.*, 2011) are described in another chapter of this book.

2. GMA, Mechanisms of Action in Brief

GMA leucocytapheresis carriers are made of cellulose acetate beads (JIMRO, Takasaki, Japan). Briefly, the Adacolumn is filled with specially designed cellulose beads of 2 mm diameter bathed in sterile saline. The carriers serve as the column adsorptive leucocytapheresis carriers for FcγR (fragment crystalizable gamma receptor) and complement receptor bearing cells (Hanai *et al.*, 2011). Hanai *et al.* reported that there was a strong correlation between clinical activity and peripheral blood expression of the pro-inflammatory $CD14^{+}CD16^{+}$ monocyte phenotype in IBD patients Ref. 14. Further, GMA significantly depleted the $CD14^{+}CD16^{+}$ monocytes, which are known to be major sources of tumor necrosis factor (TNF)-α. Other actions of GMA include down-modulation of L-selectin expression on granulocytes and reduction of leucocyte chemotaxis (Kashiwagi *et al.*, 2002). In a previous study (Yokoyama *et al.*, 2007), we found that patients with active UC had compromised expression of the $CD4^{+}CD25^{+High}$ T-cell phenotype known as regulatory T-cells (Treg), while the expression of the pro-inflammatory $CD28^-CD4^+$ T-cell phenotype was elevated. GMA was associated with a significant increase in the Treg phenotype (Yokoyama *et al.*, 2007).

3. GMA Procedures

Adacolumn is a single use medical device for direct hemoperfusion (Saniabadi *et al.*, 2003). This means the patient's blood flows into the column and within the column selective adsorption of myeloid leucocytes together with some platelets proceeds and from the column outflow, blood returns to the patient. The currently produced Adacolumns are about 206 mm in length, 60 mm in diameter, and are made of polycarbonate (Saniabadi *et al.*, 2003). The column volume is about 335 mL. The volume of the bathing saline is 130 mL. Blood access is via a venous puncture in the antecubital vein in one arm and from the column outflow, blood returns to the patient via a venous puncture in the contralateral arm. The duration of one GMA session is 60 minutes, at 30 mL/minute. An optimum dose of sodium heparin (up to 2,000 U/session) is administered during the GMA therapy as an anti-coagulant, mainly for the blood flow lines including the column.

4. GMA in the Treatment of Patients with UC

4.1. *The Existing Concept*

Although the etiology of UC is still not fully understood, elevated and activated myeloid lineage leucocytes (granulocytes and monocytes) potentially are significant factors in the exacerbation and the perpetuation of IBD by releasing inflammatory cytokines (Yokoyama *et al.*, 2007; Saniabadi *et al.*, 2007; Xavier and Podolsky, 2007; Mizoguchi and Mizoguchi, 2008; Hanai *et al.*, 2008; Yokoyama *et al.*, 2013). Accordingly, selective depletion of these cells by GMA with an Adacolumn has been applied as a non-pharmacologic treatment option for patients with active UC (Saniabadi *et al.*, 2003, 2007; Cabriada *et al.*, 2012). GMA is expected to be effective as monotherapy in many patients with the first UC episode and who are steroid naïve (Suzuki *et al.*, 2006; 2004). In other UC patients, GMA should at least enhance the efficacy of pharmacologics by depleting the inflammatory leucocytes.

Patients with intractable UC require lifelong medications with conventional pharmacologics including new anti-TNF biologics, and this may lead to drug dependency, and loss of response, together with adverse side effects (Thomas, 1984; Faubion *et al.*, 2001; Ali *et al.*, 2009; Lichtenstein *et al.*, 2012; Kawakami *et al.*, 2013). Corticosteroids are potentially effective, but are associated with high risk of adverse events like osteoporosis, increased susceptibility to infections, exacerbation of hypertension and diabetes mellitus (Thomas, 1984; Faubion *et al.*, 2001; Ali *et al.*, 2009; Lichtenstein *et al.*, 2012). Additionally, corticosteroids concomitant with other immunosuppressive drugs like thiopurines and biologics has increased the incidence of severe adverse events including opportunistic infection (Faubion *et al.*, 2001; Ali *et al.*, 2009; Lichtenstein *et al.*, 2012). It is clinically relevant to say that patients should benefit most from GMA if this non-pharmacologic treatment option is applied when an individual experiences an episode of IBD for the first time, and not after the patient has been exposed to multiple pharmacologics over a long period of time and the patient's IBD has become refractory to the available pharmacologics (Suzuki *et al.*, 2004). This notion is further reviewed below. Further, it is very important to bear in mind that GMA as a non-pharmacologic treatment intervention if effective has multiple advantages over pharmacologics including lack of refractoriness, free from serious adverse side effects often seen with pharmacologics. Therefore, it is true to say that the most unrivalled feature of GMA is its safety as well as its being a non-drug treatment strategy. For these features, GMA with the Adacolumn is very much favored by patients as well as by physicians.

4.2. *The Efficacy of GMA in UC Patients*

According to the national guidelines in Japan, GMA may be included in the treatment of UC patients with moderate to severe active disease, with total or left-sided colitis, steroid-dependent or steroid refractory UC. Currently, we induce remission with GMA in patients with UC refractory to 5-amininosalyicylic acid (5-ASA) preparations who wish to avoid corticosteroids. The publication of the first clinical experience with GMA in patients with UC by Shimoyama *et al.* (2001) generated significant interest in this non-pharmacological treatment intervention for patients with IBD, which is based on depleting self-activated myeloid lineage leucocytes (Shimoyama *et al.*, 2001; Matsui *et al.*, 2002; Hanai *et al.*, 2003; Domenech *et al.*, 2004; Naganuma *et al.*, 2004; Fukuda *et al.*, 2004; Cohen, 2005; Muratov *et al.*, 2006; Bresci *et al.*, 2006; Saniabadi *et al.*, 2007; Caprilli and D'Ovidio, 2007; Aoki *et al.*, 2007; Sands *et al.*, 2008; Tanaka *et al.*, 2008; Hibi *et al.*, 2009; Ruuska *et al.*, 2009; Tanaka *et al.*, 2010; Saniabadi and Hanai, 2010; Fukuchi *et al.*, 2011; Yoshimura *et al.*, 2012; Takayama *et al.*, 2012; Yamamoto *et al.*, 2012; Kato *et al.*, 2012; Sands *et al.*, 2012; Sacco *et al.*, 2013, 2014). In 2004, Suzuki *et al.* reported an 85% efficacy rate, or 17 of 20 patients they treated with GMA as the first-line medication. Then controlled studies by Sands *et al.* (2008, 2012) reported no statistically significant difference between the sham arm and the GMA arm. The patients included in the latter studies had a very different demography, had a long history of IBD, exposure to multiple pharmacologics, and their IBD had become refractory to the available pharmacologics, including new anti-TNF biologics. The severity and the extent of mucosal damage are also key factors. In contrast, all of the patients in Suzuki's study were steroid naïve, most with a short UC duration, and only needed first line medication (Suzuki *et al.*, 2004). Indeed, Suzuki *et al.* reported that the only three GMA non-responders in that study had defective mucosal tissue and deep ulcers (Suzuki *et al.*, 2004). In a subsequent study, Suzuki *et al.* reported that all patients with a short duration of UC (< 3 months) and first episode cases readily responded to GMA, while all of the eight GMA non-responders in that study had a very long duration of UC together with a long history of exposure to multiple conventional medications (Suzuki *et al.*, 2006). The observations by Suzuki *et al.* (2004, 2006) were independently confirmed by Tanaka and colleagues in much larger cohorts of UC patients (Tanaka *et al.*, 2008, 2010, 2013). The bottom line is that patients' demography, in particular mucosal tissue damage level at baseline, can help to select patients who are most likely to benefit from GMA and avoid pharmacologics.

With respect to the GMA treatment course, in routine treatment practice, the frequency of GMA has been one session per week, at 30 mL/min for 60 min. This is equivalent to 1,800 mL of processed blood volume per GMA session. This treatment protocol might be inadequate for optimum efficacy. In line with this thinking, Sakuraba *et al.* (2009) found that two GMA sessions per week produced significantly higher efficacy rate in a shorter time as compared with one session per week (equal number of sessions in each arm). Similarly, Yoshimura *et al.* (2012) reported that ≥ 3,000 mL of processed blood volume per GMA session was significantly more effective than 1,800 mL per session.

4.3. *The Efficacy Outcomes for GMA Evaluated in a Retrospective Setting*

To know the predictive factors of clinical response to GMA or to any other treatment intervention in patients with IBD is of paramount importance. Predictive factors could be a biomarker like C-reactive protein (CRP), a cytokine, one or more among the patients' demographic variables at baseline. A predictive of response to GMA can aid the physicians to select patients for the start of therapy, while patients who do not bear an appropriate predictive factor may opt for an alternative therapy. The bottom line is that knowing a predictive factor of clinical response to any therapeutic intervention should aid the physicians in stopping futile use of medical resources.

With the aforementioned background in mind, we undertook a retrospective investigation to determine predictive factors of clinical response to GMA (Yokoyama *et al.*, 2013). The outcomes were reported for UC patients who had been treated with GMA between April 2005 and March 2009 at our IBD center (Yokoyama *et al.*, 2013). In the original study, 43 outpatients with active UC had been given 10 consecutive weekly GMA sessions, at one session per week as seen in Fig. 32.1. Among the 43 patients, 15 were male, and 28 were

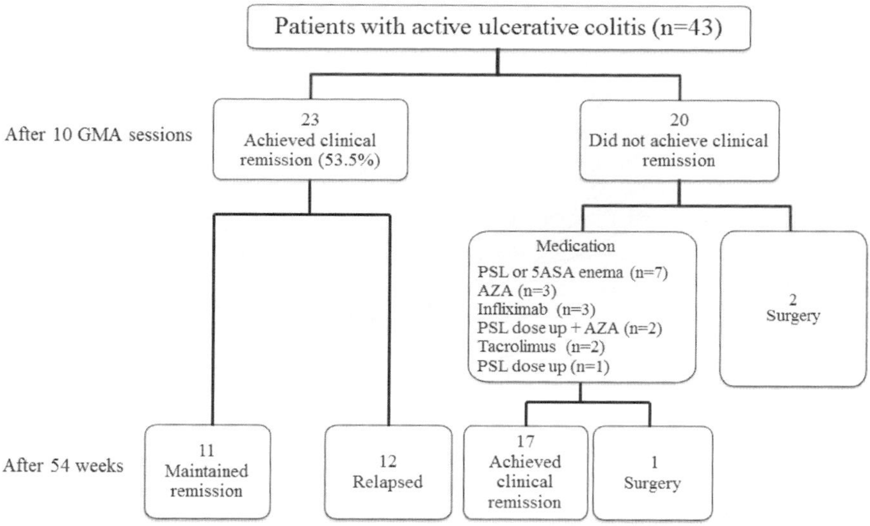

Fig. 32.1. Treatment of patients and summary of the clinical outcomes. GMA, adsorptive granulocyte and monocyte apheresis; PSL, prednisolone; AZA, azathioprine; 5-ASA, 5-aminosalicylic acid.
(*Source*: Yokoyama Y *et al.* (2013) *BMC Gastroenterol* **13:** 27.)

female, average age 41.5 years, range 18–71 years, all with a definitive diagnosis of UC. Figure 32.1 shows that after 10 once-a-week GMA sessions, the overall remission rate was 53.5%, or 23 of 43 patients. Among the 23 responders, nine patients (39.1%) achieved steroid-free remission. Three of the 43 patients received ≤ 5 GMA sessions and one patient opted for an alternative medication. Therefore, 40 of the 43 initially planned patients received 10 GMA sessions. The change in the clinical activity index (CAI) score for all 40 patients who received at least five GMA sessions is presented in Fig. 32.2. In the 23 responders, the average CAI score improved from 8.7 ± 1.8 (baseline) to 3.8 ± 0.7 ($P < 0.001$). Similarly, the average endoscopic index (EI) score ($n = 23$) was significantly better, 8.5 ± 2.4 vs 1.8 ± 1.9, ($P < 0.05$), and 56.5% of those who achieved clinical remission had mucosal healing as well. Further, the daily prednisolone (PSL, a corticosteroid) dose was significantly tapered from 18.7 ± 14.4 mg/day at baseline to 5.2 ± 6.1 mg/day ($P < 0.05$) in responders after 10 GMA sessions.

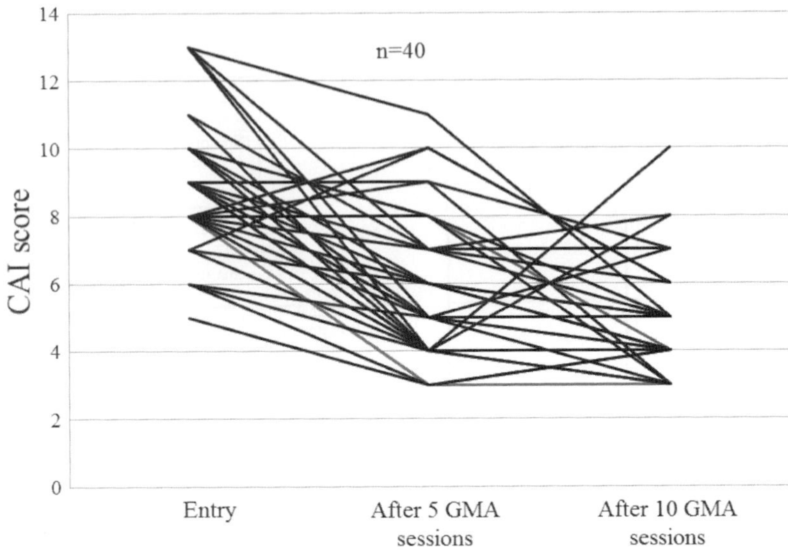

Fig. 32.2. The clinical efficacy of GMA based on changes in Lichiger's clinical activity index (CAI) evaluated at entry, after five and 10 GMA sessions.
(*Source*: Yokoyama Y *et al.* (2013) *BMC Gastroenterol* **13:** 27.)

Colonoscopic findings in three typical responder patients who achieved mucosal healing are shown in Fig. 32.3. Univariate analysis showed that there was no significant difference in age, gender, extent of colitis, CAI score, daily PSL dose and steroid response (steroid-refractory or steroid-dependent) at baseline between the remission and non-remission groups (Table 32.1). However, the cumulative PSL dose before the first GMA session in the non-remission group was significantly higher than in the remission group, 939.2 ± 1125.9 vs 4578.5 ± 4752.6 ($P = 0.006$). Likewise, the duration of UC was significantly shorter in the remission group as compared with the non-remission group, 6.6 ± 7.3 years vs 8.6 ± 6.8 years ($P = 0.036$). More importantly, the mean interval between relapse and the first GMA session was significantly shorter in the remission group as compared with the non-remission group, 27.6 ± 31.0 days vs 57.7 ± 54.4 days ($P = 0.016$). Another variable which tended to be significantly different between the remission and the non-remission groups at baseline

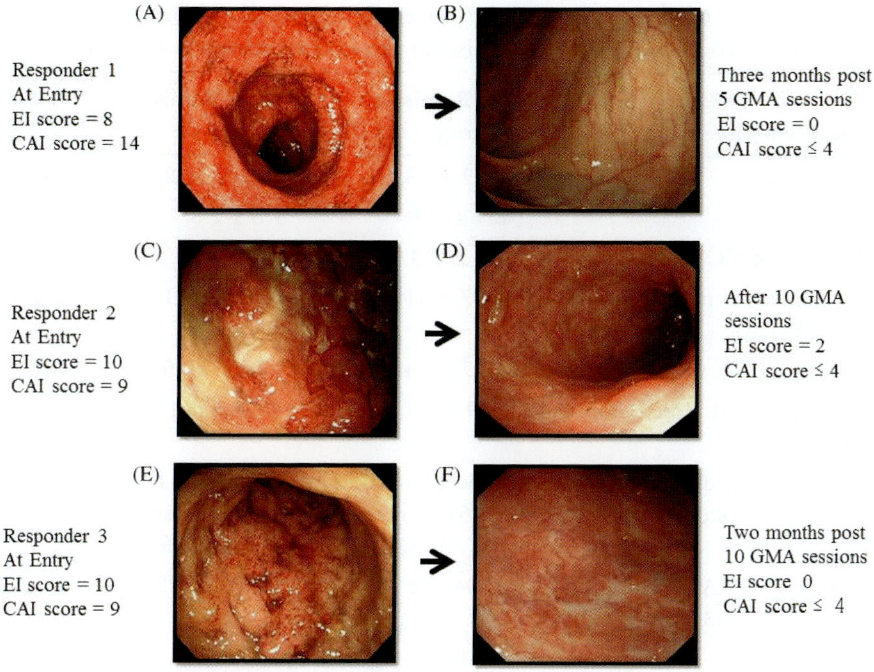

Fig. 32.3. Endoscopic findings in three typical responders to GMA. All patients achieved complete remission after weekly GMA sessions. One of these patients (Fig. 32.3A) with the first UC episode had absence of vascular patterns, spontaneous bleeding, erythema, oedema and friability at entry. This patient showed an early response to GMA after receiving just five GMA sessions and achieved mucosal healing (MH) a few months later (Fig. 32.3B). The second patient with steroid-dependent UC had extensive ulcers, oedematous mucosa at entry (Fig. 32.3C). This patient achieved clinical remission and partial MH after 10 GMA sessions (Fig. 32.3D). The third patient had a short duration of UC and absence of vascular patterns, spontaneous bleeding, erythema, ulcers and friability at entry (Fig. 32.3E). This patient achieved MH a few months later after 10 GMA sessions (Fig. 32.3F). These three patients had maintained remission at week 54 following the last GMA session.
Source: Yokoyama Y *et al.* (2013) *BMC Gastroenterol* **13:** 27.

was white blood cell (WBC) count, 8,153.0 ± 2480.9/μL vs 9,617.4 ± 2235.4/μL, respectively ($P = 0.032$). Multiple logistic regression analysis showed that the interval between relapse and the first GMA session was a significant and independent predictive factor of response

Table 32.1. Comparison Patinets' Demographic Background with the Remission Group and the Non-Remission Group in Univariate Analysis

Measurements	Remission Group ($n = 23$)	Non-Remission Group ($n = 20$)	P Value
Age (years)	39.2 ± 13.6	44.1 ± 16.2	ns
CAI (score)	8.9 ± 2.1	9.3 ± 2.3	ns
Duration of UC (years)	6.1 ± 7.3	8.6 ± 6.8	0.036
Interval between relapse and 1st GMA (days)	27.7 ± 31.0	76.2 ± 88.1	0.016
Daily PSL dose at 1st GMA (mg)	18.7 ± 14.4	17.8 ± 10.5	ns
Cumulative PSL dose before 1st GMA (mg)	939.2 ± 1125.9	4578.5 ± 4752.6	0.006
Steroid response (steroid-dependent/steroid-resistant) (n (%))	dependent 16/23 (69.6) resistant 7/23 (30.4)	dependent 17/20 (85) resistant 3/20 (15)	ns
WBC at 1st GMA (counts/mm3)	8304.1 ± 2432.3	9572.9 ± 2283.4	0.032
Granulocyte at 1st GMA (%)	72.4 ± 11.7	77.7 ± 9.1	ns
Hb at 1st GMA (g/dl)	12.0 ± 1.7	12.7 ± 2.0	ns
CRP at 1st GMA (mg/dl)	0.7 ± 1.0	0.7 ± 0.8	ns

WBC, white blood cell; GMA, granulocyte and monocyte apheresis; Hb, haemoglobin; CRP, C-reactive protein, ns : not significant ($p > 0.05$).

to GMA ($P = 0.016$) (Table 32.2). Therefore, GMA at an early stage following a clinical relapse is more likely to induce remission of an active UC than if applied late. Additionally, baseline WBC count was a significant and independent predictive factor for response to GMA ($P = 0.025$) (Table 32.2). Based on these findings, currently, select patients for GMA therapy with these three factors in mind, namely the interval between relapse and the first GMA session, cumulative PSL dose and WBC count at the first GMA session. Potentially, such data should spare medical cost and reduce morbidity time for many patients, relevant for decision making in clinical settings.

Table 32.2. The Outcomes of Multiple Logistic Regression Analysis for Determining Predictive Factors of Clinical Response to GMA in Patients with Active Ulcerative Colitis

Measurements	Beta	*P* Value	Odds Ratio	95% Confidence Internal	
				Lower	Upper
Interval between relapse and 1st GMA	−0.025	0.016	0.976	0.956	0.995
WBC at 1st GMA	0.000	0.025	1.000	0.999	1.000

WBC, white blood cell; GMA, granulocyte and monocyte apheresis.

We often have the opportunity to treat UC patients who are found to be intolerant to salicylate based first-line medications. Among the 5-aminosalicylic acid preparations, adverse events are most common with salazosulfapyridine (sulphasalazine). Sulphapiridine of salazosulfapyridine may cause adverse events (Saniabadi *et al.*, 2014). Previously, one study reported 21 cases with mesalazine-induced pneumonia (Sullivan, 1987). Onset of mesalazine-induced pneumonia was not related to mesalazine doses or intervals, and interstitial pneumonia was most common (Sullivan, 1987). In such settings, GMA should be a safe option. In line with this assertion, Suzuki *et al.* (2004) and Fukuchi *et al.* (2011) reported a high efficacy rate for GMA in UC patients with corticosteroid-naïve background at an early-stage of active disease. The efficacy of GMA may increase further by applying an intensive treatment protocol (Sakuraba *et al.*, 2009; Fukuchi *et al.*, 2011).

4.4. *The Relationship between GMA Processed Blood Volume and Bodyweight*

Just like predictive factors for clinical response to GMA, determination of the processed blood volume per one GMA session in relation to patients' bodyweight (BW) should guide to better efficacy outcomes. With this in mind, Kikuyama *et al.* (2011) undertook an investigation aimed at determining the relationship between the processed blood volume (PV) per kg BW and the efficacy of GMA in clinical settings. Thirty-three patients were included for remission induction therapy with five weekly GMA sessions at a standard PV of 1,800 mL per GMA session, regardless of patients' BW. Then, in a retrospective setting, the patients were divided into three groups: High (H) BW ($\geq$ 65 kg, $n = 11$); medium (M) BW < 65 kg ($n = 12$), and low (L) BW ($\leq$ 50 kg, $n = 10$). CAI and other efficacy parameters were evaluated at baseline, one week post third GMA (Week 4) and one week post fifth GMA (Week 6). The average BW was 70.9 ± 6.2 kg in the HBW group, 55.8 ± 4.5 kg in the MBW group, and 46.8 ± 1.2 kg in the LBW group. Accordingly, the mean PV/BW values (mL/kg) in the three groups were 25.6 ± 2.12, 32.5 ± 2.50 and 38.7 ± 1.0, respectively with the routinely applied 1,800 mL PV. The LBW group had only female patients (lighter BW gender among the people of Japan). Relative to baseline, significant improvements of CAI scores were seen at both Week 4 and Week 6 in all three groups ($P <$ 0.05). Further, at week 6, a significantly higher remission rate was achieved by the LBW group (80.0%) vs the MBW (33.3%) or the HBW (27.3%) group ($P < 0.03$). According to this GMA evaluation study, the lower-limit of optimum PV/kg should be higher than 38.7 mL/kg for clinical efficacy to be significantly greater than the routine GMA method. Further, as already mentioned above, Yoshimura *et al.* (2012) reported that more than 3,000 mL of processed blood volume per GMA session was significantly better than 1,800 mL per session. These findings together with the reports by Sakuraba *et al.*, who reported that intensive GMA as two sessions per week had better efficacy outcomes as compared with weekly GMA, even when an equal number of GMA sessions are factored into the analyses (Sakuraba *et al.*, 2009). Based on these findings, the frequency and optimum number of GMA sessions for maximum efficacy remain unknown because none of the aforementioned studies reached a level (GMA frequency or PV) above which no further increase in efficacy was observed.

4.5. *GMA as Maintenance Therapy in UC Patients*

In therapeutic settings, patients with active IBD who achieve clinical remission need to receive remission maintenance therapy to sustain the quiescent state of their IBD achieved by remission induction therapy. In this setting, experience with efficacy outcomes to justify the clinical relevance of GMA as maintenance therapy has been reported (Fukunaga *et al.*, 2012). For some time, we were interested in evaluating the efficacy of GMA as maintenance therapy in a multi-center, prospective and randomized-control setting. Fukunaga *et al.* (2012) investigated the efficacy of monthly GMA as maintenance therapy to suppress UC relapse in a prospective setting. They included 33 patients with UC refractory to routine corticosteroid therapy. These 33 patients had already achieved clinical remission by weekly GMA (Fukunaga *et al.*, 2012). Within 14 days following the last weekly GMA session, the eligible patients were randomized to one of the three study arms: True, Sham or Control for maintenance therapy (Fig. 32.4). Patients in the True arm received one GMA session per week, while patients in the Sham arm received an equal volume of processed blood without the column (column circulation lines were used instead). In the Control arm, patients continued with their ongoing conventional medications and did not receive any blood processing. The primary endpoint was a non-relapse rate at week 48. Relapse was assumed if a patient had to receive new medication or to increase the dose of an ongoing medication (including more frequent GMA session) to suppress an active UC. At weeks 48, relapse avoidance rates in True, Sham and Control arms were 40.0, 9.1 and 18.2%, respectively, clearly in favor of GMA, but without the difference reaching statistical significance level (Fig. 32.5). However, in patients who had tapered their corticosteroid dose to < 20 mg/day during the GMA remission induction therapy, relapse avoidance rate was better in the True arm than in the other two arms ($P < 0.05$) (Fig. 32.6). The authors concluded that the efficacy of maintenance GMA therapy was closely related to the daily corticosteroids dose during remission induction therapy. In brief, the patients who could

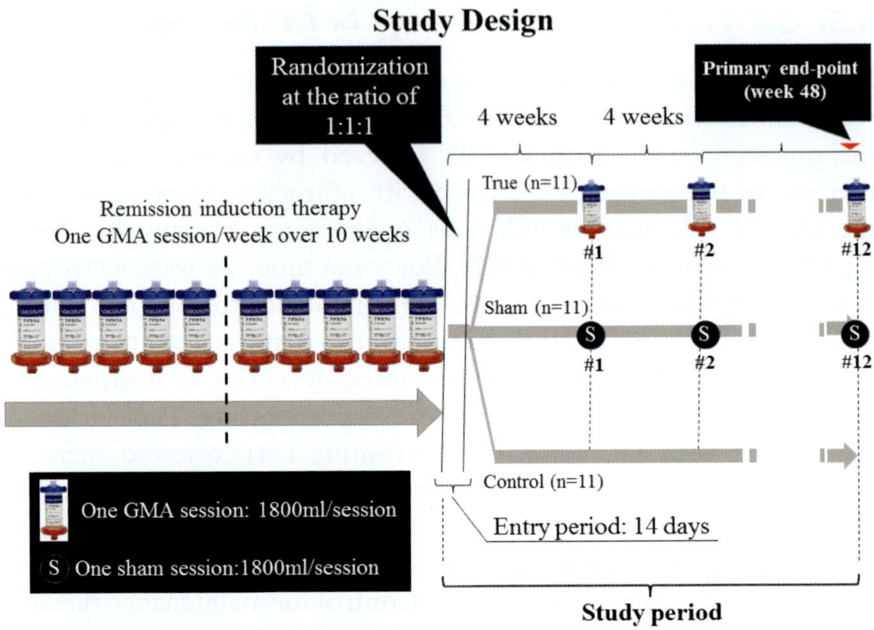

Fig. 32.4. Thirty-three patients had already achieved clinical remission by weekly GMA. Within 14 days following the last weekly GMA session, the eligible patients were randomized to one of the three study arms: True, Sham or Control for maintenance therapy. Patients in the True arm received one GMA session per week, while patients in the Sham arm received an equal volume of processed blood, without the column. In the Control arm, patients continued with their ongoing conventional medications and did not receive any blood processing. The primary endpoint was a non-relapse rate at week 48.
Source: Fukunaga K *et al.* (2012) *Gut Liver* **6**: 427–433.

reduce their daily corticosteroid doses during the GMA remission induction therapy had good prognosis. This study was the first attempt to assess the efficacy of GMA as maintenance therapy in a prospective randomized, double blind, sham-control setting.

4.6. *Factors Associated with Long-term Prognosis of UC patients Undergoing GMA*

Patients with UC have elevated/activated myeloid lineage leucocytes and may respond favorably to GMA. However, there are patients who

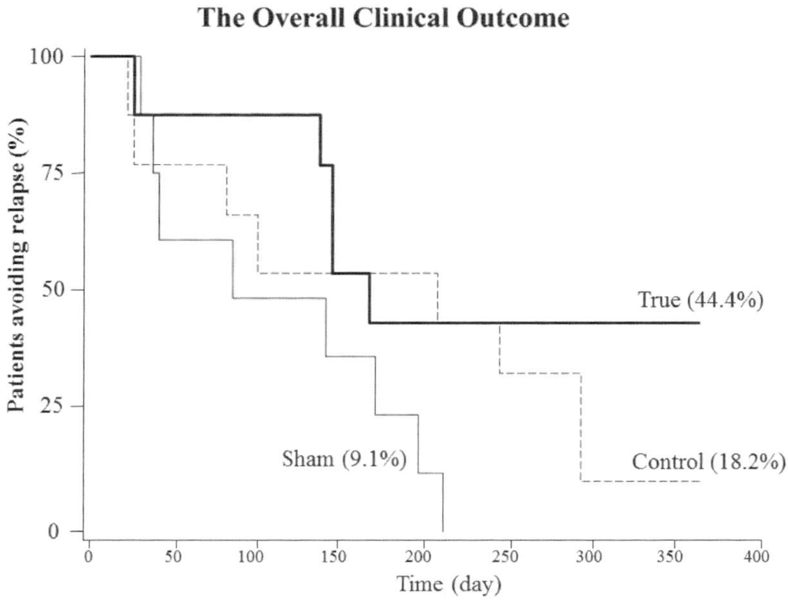

Fig. 32.5. At week 48, relapse avoidance rates in True, Sham, and Control were 44.4, 9.1, and 18.2%, respectively, showing better efficacy for active GMA. *Source*: Fukunaga K *et al.* (2012) *Gut Liver* **6:** 427–433.

respond well to GMA and patients who do not benefit. Therefore, predictive factors of GMA efficacy need to be defined.

In a prospective multi-center setting, 200 UC patients at 32 institutes received one GMA session/week over 10 weeks (Yokoyama *et al.*, 2015). Patients who achieved remission were followed for 12 months. Clinical activity index CAI $\leq$ 3 meant remission, while response meant CAI decreased by $\geq$ 3. Life quality was evaluated by inflammatory bowel disease questionnaire (IBDQ).

Table 32.3 shows demographic variables of the 200 patients of this study (94 female and 106 male). The average CAI score at entry was 10.1 ± 0.3 and the average IBDQ score was 135.9 ± 2.6. One hundred and seventy-five of the 200 patients (87.8%) had relapsed and were recruited for remission induction. One hundred and twenty patients (60%) were steroid refractory or steroid dependent, and 97 of these (48.5%) were receiving corticosteroid (prednisolone) at entry. In addition, 31 patients (15.5%) were receiving azathioprine

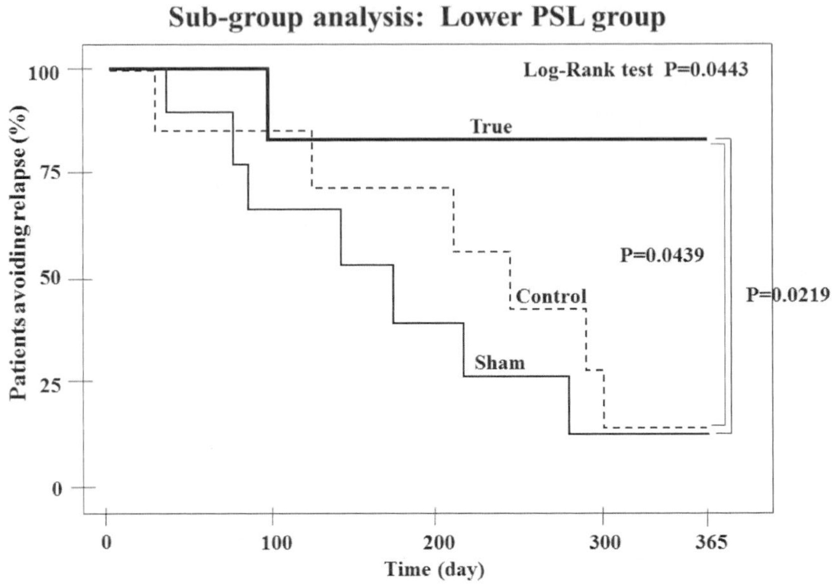

Fig. 32.6. In patients who could taper their PSL dose to < 20 mg/day during remission induction therapy, relapse avoidance rate in True was better vs Sham ($P < 0.03$) or Control ($P < 0.05$). Similarly, the Kaplan–Meier survival analysis indicated better prognosis for True vs Sham or Control ($P < 0.05$).
Source: Fukunaga K *et al.* (2012) *Gut Liver* **6**: 427–433.

(AZA) as an immunomodulator. No patient was receiving other immune suppressive drugs, like tacrolimus, cyclosporine or an anti-TNF-α biologic.

Post final GMA, remission, response, and no response rates were 67.0, 15.0 and 18%, respectively (Fig. 32.7). The remission group had a significant fall in myeloid leucocytes and platelets. Corticosteroid dose decreased ($P < 0.001$); 49 of 97 patients on corticosteroids became steroid-free. Baseline CAI was lower in the remission group vs non-remission ($P < 0.01$), while IBDQ was higher in the remission group vs non-remission ($P < 0.05$) (Table 32.4). After 12 months, 52 of 134 patients had maintained remission (Fig. 32.7). Disease duration was longer in the relapsed group vs maintained remission

Table 32.3. Patients' Demographic Variables at Entry ($n = 200$)

Demography		**Subgroup (%)**	**Number of patients**
Gender	Male	106 (53.0%)	200
	Female	94 (47.0%)	
Clinical course	First attack	25 (12.2%)	200
	Relapse	175 (87.8%)	
Disease location	Proctitis	5 (2.5%)	
	Total colitis	125 (62.5%)	200
	Left-sided colitis	70 (35.0%)	
Severity	Fulminant	2 (1.0%)	200
	Severe	48 (24.0%)	
	Moderate	150 (75.0%)	
Response to steroid	Steroid refractory	51 (25.5%)	
	Steroid dependent	69 (34.5%)	200
	Steroid responder	80 (40.0%)	
Age (year)		41.2 ± 0.9	200
Disease duration (month)		87.5 ± 7.3	158
Clinical activity index (CAI)		10.1 ± 0.3	170
Quality of life (IBDQ)		135.9 ± 2.6	158
C-reactive protein (CRP)		3.2 ± 0.7	54
Matts' endoscopic grade		3.1 ± 0.7	131
Concomitant corticosteroid use		97 (48.5%)	200
Concomitant immunomodulator use		31 (15.5%)	200

group ($P = 0.041$) (Table 32.6). Male gender, first UC episode, and corticosteroid responder were significant factors for maintaining remission, whereas corticosteroid dependent UC was associating with relapse (Table 32.6, Fig. 32.8).

In conclusion, there are UC patients who respond well to GMA, but also subgroups who seem not to benefit from GMA, reflecting different entry disease features. With this in mind, this study focused on factors which potentially identify a patient as a responder or otherwise as a non-responder to GMA. For the first time, this study

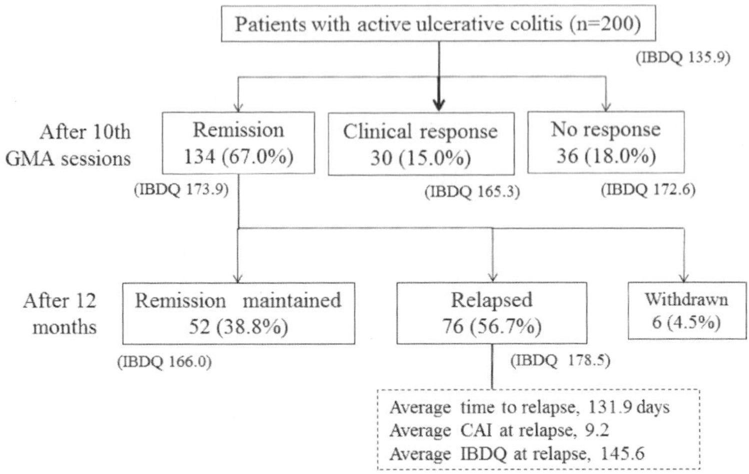

Fig. 32.7. Study design and the overall treatment outcomes. At appropriate points, the average IBDQ score is shown (below each box).
Source: Yokoyama Y *et al.* (2015) *Cytotherapy* (in press).

Table 32.4. Comparison of Background Factors Between GMA Responders and Non-Responders

Demography		**Remission (134)**	**Non-Remission (66)**	***P* Value**
Gender	Male	73 (54.5%)	33 (50.0%)	0.558
	Female	61 (45.5%)	33 (50.0%)	
Clinical course	First attack	17 (12.1%)	8 (87.7%)	0.970
	Relapse	87.9% (117)	12.3% (58)	
Disease location	Rectal	2 (1.5%)	3 (37.9%)	0.322
	Total colitis	87 (64.9%)	38 (57.6%)	
	Left-sided	45 (33.6%)	25 (4.5%)	
Severity	Fulminant	1 (0.8%)	1 (1.5%)	0.845
	Severe	33 (24.6%)	15 (22.7%)	
	Moderate	100 (74.6%)	50 (75.8%)	
Response to steroid	Steroid refractory	32 (23.9%)	19 (28.8%)	0.688
	Steroid dependent	46 (34.3%)	23 (34.8%)	
	Steroid responder	56 (41.8%)	24 (36.4%)	

(*Continued*)

Table 32.4. (*Continued*)

Demography	**Remission (134)**	**Non-Remission (66)**	***P* Value**
Age (year)	40.4 ± 1.0 (134)	43.0 ± 1.8 (66)	0.193
Disease duration (month)	87.0 ± 8.8 (109)	88.6 ± 13.0 (49)	0.917
Clinical activity index (CAI)	9.47 ± 0.4 (107)	11.1 ± 0.4 (63)	0.005**
Quality of life (IBDQ)	139.6 ± 3.2 (106)	128.3 ± 4.2 (52)	0.038*
C-reactive protein (CRP)	3.0 ± 0.7 (50)	5.2 ± 4.3 (4)	0.416
Matts' endoscopic grading	3.2 ± 0.7 (89)	3.1 ± 0.7 (42)	0.433
Concomitant corticosteroid use	70 (52.2%)	27 (16.3%)	0.190
Concomitant immunomodulator use	21 (15.7%)	10 (15.2%)	1.000

Table 32.5. Comparison of Blood Cell Numbers before and after the 10th GMA Session

		Pre-GMA	Post-GMA	*P* Value
Total white blood cell (/μL)	Remission	9309.0 ± 3506.6	5981.4 ± 2757.5	0.001**
	Non-remission	8582.2 ± 2535.9	7943.0 ± 2691.3	0.614
Lymphocytes (/μL)	Remission	1653.0 ± 739.1	1635.1 ± 690.6	0.928
	Non-remission	1514.6 ± 732.4	1767.5 ± 587.5	0.428
Granulocytes (/μL)	Remission	6855.4 ± 3412.1	5073.1 ± 2813.7	0.008**
	Non-remission	6292.7 ± 2554.7	5636.8 ± 2885.9	0.694
Monocytes (/μL)	Remission	606.2 ± 381.2	403.2 ± 163.1	<0.001**
	Non-remission	474.3 ± 239.4	447.5 ± 188.9	0.302
Platelets 104 /μL	Remission	35.4 ± 12.0	29.4 ± 8.0	<0.001**
	Non-remission	32.2 ± 9.4	33.1 ± 9.3	0.256

WBC, white blood cells, Non-remission includes no response and clinical response.

Table 32.6. Comparison of Background Factors between Remission-maintained and Relapsed Patients

Demography		Maintain Remission (52)	Relapsed (76)	*P* Value
Gender	Male	34 (65.4%)	36 (47.4%)	0.044*
	Female	18 (34.6%)	40 (52.6%)	
Clinical course	First attack	11 (19.6%)	5 (6.6%)	0.025*
	Relapse	41 (80.4%)	71 (93.4%)	
Disease location	Rectum	0 (0.0%)	2 (2.6%)	0.445
	Total colitis	33 (63.5%)	50 (65.8%)	
	Left-sided	19 (36.5%)	24 (31.6%)	
Severity	Fulminant	1 (2.0%)	0 (0.0%)	0.103
	Severe	9 (17.3%)	24 (31.6%)	
	Moderate	42 (80.7%)	52 (68.4%)	
Response to steroid	Steroid refractory	15 (28.8%)	16 (21.0%)	0.001**
	Steroid dependent	8 (15.4%)	36 (47.4%)	
	Steroid responders	29 (55.8%)	24 (31.6%)	
Age (year)		38.7 ± 1.6 (52)	41.6 ± 1.5 (76)	0.196
Disease duration (month)		63.5 ± 11.5 (40)	101.6 ± 12.6 (64)	0.041*
Clinical activity index (CAI)		10.1 ± 0.5 (43)	9.2 ± 0.5 (59)	0.203
Quality of life (IBDQ)		130.7 ± 5.2 (42)	145.6 ± 4.0 (59)	0.025*
C-reactive protein (CRP)		2.7 ± 5.0 (18)	3.1 ± 4.9 (29)	0.381
Matts' endoscopic grading		3.3 ± 0.7 (32)	3.1 ± 0.7 (52)	0.196
Concomitant corticosteroid use		20 (38.5%)	45 (59.2%)	0.077
Concomitant immunomodulator use		11 (21.2%)	9 (11.8%)	0.215

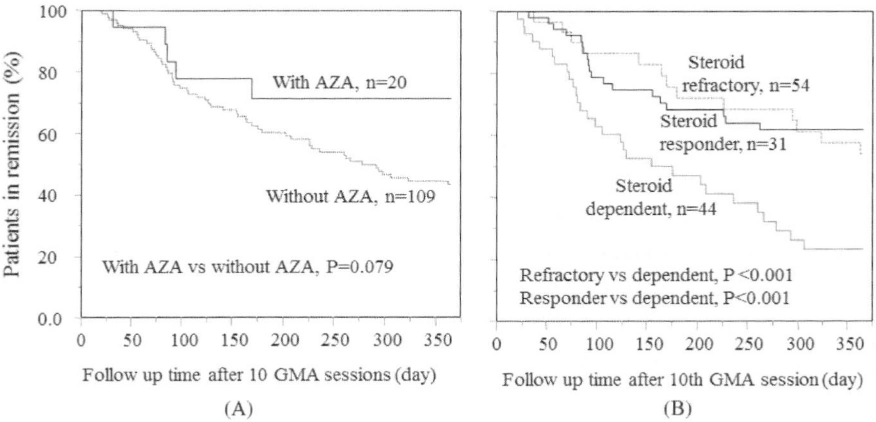

Fig. 32.8. The Kaplan–Meier survival graphs showing fraction of patients in remission after induction therapy by GMA. (A) Remission rate in patients with or without azathioprine (AZA). (B) Remission rate in patients with steroid-refractory, steroid-dependent or steroid-responder ulcerative colitis (UC).
Source: Yokoyama Y *et al.* (2015) *Cytotherapy* (in press).

found that patients who achieved remission by GMA showed a significant fall in myeloid lineage leucocytes and platelets post GMA, which was not seen in non-responders (Table 32.5). Patients with a low (mild to moderate) baseline CAI are more likely to respond well to GMA as compared with patients with a high baseline CAI. Likewise, patients with a high baseline IBDQ are more likely to respond favorably to GMA as compared with patients with a low baseline IBDQ. Regarding maintenance of GMA-induced remission, first UC episode and corticosteroid responder features were significant factors for remission maintenance, whereas corticosteroid dependency was associated with a short maintenance time. These finding should generate discussion in clinical settings and help avoid futile use of an expensive GMA therapy. Additionally, we like to state here that GMA is very much favored by patients for its excellent safety profile as well as for being a non-pharmacologic treatment intervention.

5. GMA in the Treatment of Crohn's Disease

5.1. *Background*

As stated above, GMA was introduced as one therapeutic option for patients with active UC in 2000, while for patients with CD, GMA received official recognition in 2009. Therefore, there are many more publications on GMA in UC setting than in CD condition. Additionally, whereas UC is confined to the large intestine (colon and rectum), the majority of patients with CD show small intestinal involvement as well (Allison *et al.*, 1998). Further, although biopsy analyses have shown presence of vast number of myeloid lineage leucocytes in the colonic mucosa of UC patients (Allison *et al.*, 1998; Suzuki *et al.*, 2004; Lichtenstein *et al.*, 2012), such data are not available for small intestinal mucosa. This means that the role of myeloid lineage leucocytes in small intestinal CD lesions is yet to be established. Nonetheless the majority of CD patients have colonic lesions, and therefore GMA appears to have produced significant efficacy in patients with active CD (Domenech *et al.*, 2004; Matsui *et al.*, 2002; Fukuda *et al.*, 2004; Bresci *et al.*, 2006), very impressive outcomes (Bresci *et al.*, 2006) as well as disappointing outcome (Sands *et al.*, 2012). This is reminiscent of what we described for UC patients. Patients with severe CD refractory to currently available pharmacologicals may not respond well to GMA (Sands *et al.*, 2012). However, CD lesions in the large intestine which may respond well to GMA are more likely to cause bloody diarrhea and abdominal discomfort. Accordingly, GMA received favor as a treatment option for patients with colonic CD. Hitherto, GMA has been applied mainly to patients with CD refractory to anti-TNF biologics. Therefore, the best responder CD candidate patients for GMA are yet to be identified. In this section, we have presented preliminary data from a few studies.

5.2. *The Efficacy of GMA in CD Patients*

The first clinical trial in patients with CD refractory to conventional medications was undertaken by Fukuda and colleagues in Japan (Fukuda *et al.*, 2004) and the outcome led to the reimbursement approval for patients with CD. However, the first study in CD was reported by Matsui *et al.* (2002) on seven patients who were refractory to standard medications. Each patient received five GMA sessions over five consecutive weeks. Five of seven patients achieved remission. It is clinically relevant to mention here that the only two GMA non-responders in Matsui's study had primarily small intestinal lesions. Subsequently, Fukuda *et al.* (2004) reported an efficacy rate of 52% by applying five GMA sessions to each of 21 patients with severe CD. However, it is imperative to elaborate that the patients Fukuda *et al.* included had received conventional medications including two weeks of optimum nutritional therapy (a routine treatment for CD in Japan), and only patients who remained with a high CD clinical activity index (CDAI) score received GMA. Therefore, 52% remission rate in a cohort of patients with severe and medication refractory CD was very encouraging. Further, Domenech *et al.* (2004) treated 12 steroid dependent patients with CD. The remission rate was 70%. Similarly, Muratov *et al.* (2006) treated seven patients with CD who had relapsed while on optimum conventional medication and a few on biologics. The median CDAI decreased from 290 at week 1 to 184 at week 7 and to 128.5 (remission level) at 12 months. Perhaps the best efficacy outcome for GMA in CD patients is reported by Bresci *et al.* (2006) in a cohort of corticosteroid refractory CD patients. These authors reported an efficacy rate of 83.3%; all of these patients became steroid free (Bresci *et al.*, 2006).

5.3. *The Efficacy of GMA in a CD Case Complicated by Hepatitis B Virus*

As stated above, a unique feature of GMA is that it is a non-pharmacologic treatment intervention designed to remove from the body cellular elements which are inappropriately elevated, show activation behavior and contribute to disease progression (Saniabadi *et al.*, 2007). With this action, GMA is not likely to cause refractoriness or loss of response often seen with pharmacologics, and has shown an excellent safety profile. Further, it is clinically relevant to bear in mind that patients with IBD receive immunosuppressive medications to keep their IBD in remission. Immunosuppressive therapy by its very

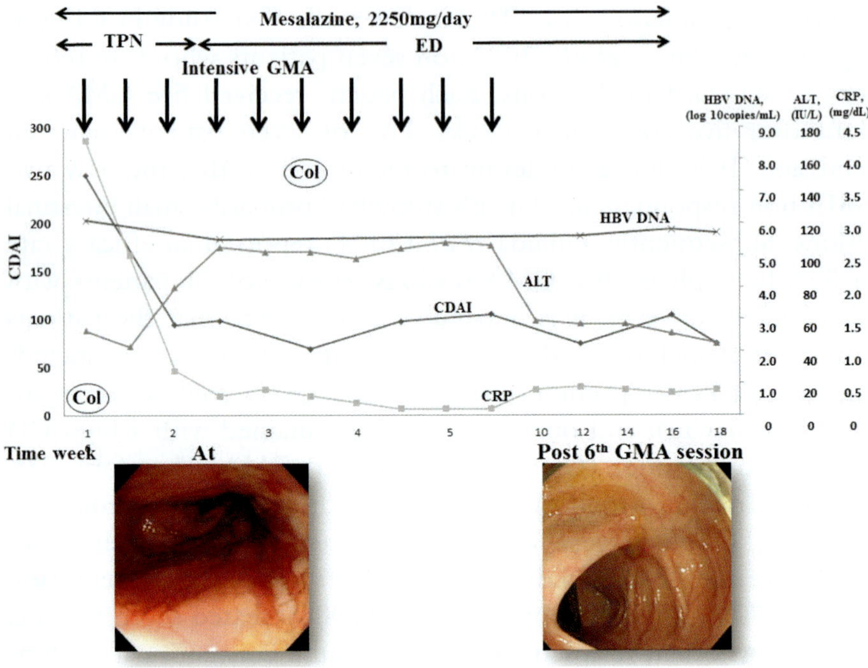

Fig. 32.9. Changes in Crohn's disease (CD) activity index (CDAI), alanine aminotransferase (ALT), and C-reactive protein (CRP) levels from baseline to the time when patient achieved clinical remission, up to week 20. This case received 10 GMA sessions, and endoscopic findings at admission and post six GMA sessions are presented and show loss of mucosal vascular patterns at baseline and restoration of the mucosal microvasculature post-GMA. This patient achieved clinical remission and improvement of mucosal lesions without HBV reactivation during or beyond the course of intensive GMA therapy; background HBV-DNA remained stable. ED = enteral diet, TPN = total parenteral nutrition.

Source: Yokoyama Y *et al.* (2014) *Cytotherapy* **16:** 821–825.

nature can promote reactivation of an otherwise inactive HBV in asymptomatic carriers (Loras *et al.*, 2009).

However, HBV infection is one of the most frequent viral infections of the liver (Lok *et al.*, 2007; Ojiro *et al.*,2008; Rahier *et al.*, 2009; Jason *et al.*, 2010; Yokoyama *et al.*, 2014). Studies in Europe found that the prevalence of HBV and hepatitis C virus (HCV) in patients with IBD was similar to the level in the general population,[56-59] while in Japan, the prevalence of HBV and HCV in IBD

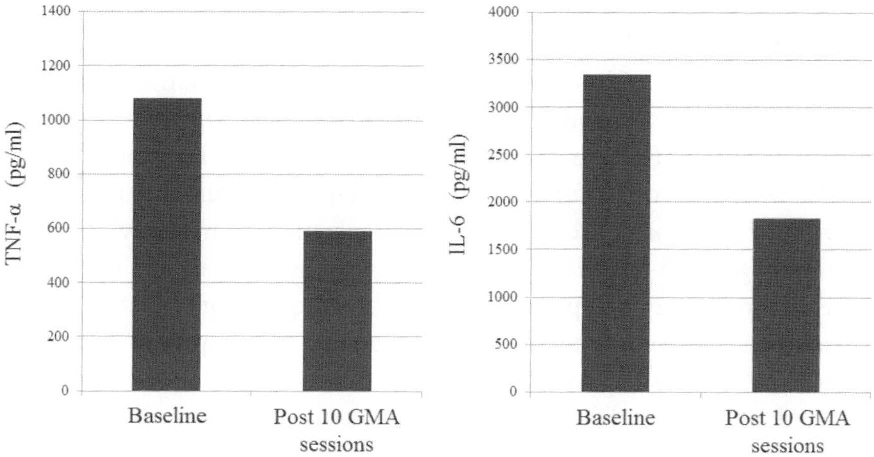

Fig. 32.10. Comparision of TNF-α and IL-6 levels between before and after GMA therapy. The serum these cytokines levels were significantly decreased after 10 GMA sessions ($P < 0.05$).
Source: Yokoyama Y *et al.* (2014) *Cytotherapy* **16:** 821–825.

patients is unknown at present. Loras *et al.* (2009) reported that combination therapy with more than two immunosuppressive agents was a significant factor of HBV reactivation in IBD patients with chronic HBV infection. Further, recently there have been several case reports of HBV reactivation after anti-TNF biologic therapy in combination with corticosteroids and/or azathioprine (Ojiro *et al.*, 2008). Therefore ECCO and the Japanese guidelines recommend antiviral prophylactic measures starting two weeks prior to the initiation of aggressive immunosuppressive therapy in HBsAg-positive IBD patients. With this background in mind, we applied GMA as a non-pharmacologic treatment option in a CD case who was HBsAg positive and with a high HBV-DNA load (Yokoyama *et al.*, 2014). Further, from this patient, blood samples were taken before every GMA sessions to measure serum levels of TNF-α and interleukin IL-6. GMA induced stable remission (CDAI 105) and endoscopic improvement without HBV reactivation or safety concerns (Fig. 32.9). Further, CD remission was paralleled by suppression of TNF-α and IL-6, as seen in Fig. 32.10. It should be logical to say that in this CD case, GMA has been a rescue therapy.

6. Concluding Remarks

The current state of knowledge assumes that more than one predisposing factor is responsible for the expression of inflammatory bowel disease (IBD) in affected individuals. Likewise, the mechanisms, which account for the chronic relapsing-remitting nature of IBD (exacerbations) are diverse. This lack of a clear knowledge on the etiology and the exacerbating factors has led to an empirical approach to drug therapy rather than the drug being targeted at the cause of the morbid conditions. Additionally, the empirical approach to medication has led to refractoriness and loss of response together with adverse side effects as additional complications. Further, in recent years, the efficacy of certain anti-TNF biologics, notably infliximab, indicates that pleiotropic cytokines like TNF-α, IL-1β and others are major parts of the exacerbation and the perpetuation factors in IBD patients. However, major sources of inflammatory cytokines include activated myeloid lineage leucocytes, which in IBD patients are elevated with activation behaviour. Therefore, in the setting of IBD therapy, GMA with the Adacolumn offers a unique opportunity; GMA can conveniently and safely deplete the extra load of leucocytes from the circulation and this per se should alleviate inflammation and promote IBD remission, or at least enhance drug efficacy.

Acknowledgements

The authors hereby acknowledge having no conflict of interest in connection with the publication of this book chapter.

List of Abbreviations used in this Chapter

5-ASA, 5-amininosalyicylic acid
AZA, azathioprine
BW, bodyweight
CAI, clinical activity index
CD, Crohn's disease

CDAI, Crohn's disease clinical activity index
CRP, C-reactive protein
EI, endoscopic index
FcγR, fragment crystallizable gamma receptor
GMA, Adsorptive Granulocyte and Monocyte Apheresis
HBW, high bodyweight
HBV, Hepatitis B virus
HCV, Hepatitis C virus
IBD, inflammatory bowel disease
IBDQ, inflammatory bowel disease questionnaire
IL-6, interleukin-6
LBW, low bodyweight
MBW, medium bodyweight
PSL, prednisolone
PV, processed blood volume
TNF, tumor necrosis factor
Treg, regulatory T-cells
UC, Ulcerative Colitis
WBC, white blood cell

References

Akerkar GA, Peppercorn MA, Hamel MB, Parker RA. (1997) Corticosteroid-associated complications in elderly Crohn's disease patients. *Am J Gastroenterol* **92:** 461–4.

Ali T, Lam D, Bronze MS, Humphrey MB. (2009) Osteoporosis in inflammatory bowel disease. *Am J Med* **122:** 599–604.

Allison MC, Dhillon AP, Lewis WG, Pounder RE. (1998) *Inflammatory Bowel Disease.* Mosby, London, pp. 9–95.

Aoki H, Nakamura K, Yoshimatsu Y, *et al.* (2007) Adacolumn Selective Leukocyte Adsorption Apheresis in Patients with Active Ulcerative Colitis:Clinical Efficacy, Effects on Plasma IL-8, and Expression of Toll-like Recepter 2 on Granulocytes. **52:** 1427–1433.

Bresci G, Parisi G, Bertoni M. (2006) Treatment of active steroid-refractory inflammatory bowel disease with granulocytapheresis: Our experience with a prospective study. *World J Gastroenterol* **12(14):** 2201–2204.

Cabriada JL, Domenech E, Ibargoyen N, *et al.* (2012) Leukocytapheresis for steroid-dependent ulcerative colitis in clinical practice: Results of a nationwide Spanish registry. *J Gastroenterol* **47:** 359–65.

Caprilli R, D'Ovidio V. (2007) Leukocytapheresis as promising therapy for inflammatory bowel disease. *Dig Liver Dis* **39:** 435–7.

Cohen RD. (2005) Treating ulcerative colitis without medications — "Look Mom, No Drugs!" *Gastroenterology* **128:** 235–6.

Domenech E, Hinojosa J, Esteve-Comas M. (2004) Spanish Group for the Study of Crohn's Disease and Ulcerative Colitis (GETECCU). Granulocyteaphaeresis in steroid-dependent inflammatory bowel disease: A prospective, open, pilot study. *Aliment Pharmacol Ther* **20:** 1347–52.

Faubion WA Jr, Loftus EV Jr, Harmsen WS, *et al.* (2001) The natural history of corticosteroid therapy for inflammatory bowel disease: A population-based study. *Gastroenterology* **121:** 255–60.

Fukuchi T, Nakase H, Ito D, *et al.* (2011) Rapid induction of mucosal healing by intensive granulocyte and monocyte adsorptive aphaeresis in active ulcerative colitis patients without concomitant corticosteroid therapy. *Aliment Pharmacol Ther* **34:** 576–86.

Fukuchi T, Nakase H, Matsuura M, *et al.* (2013) Effect of intensive granulocyte and monocyte adsorptive apheresis in patients with ulcerative colitis positive for cytomegalovirus. *J Crohns Colitis* **7:** 803–11.

Fukuda Y, Matsui T, Suzuki Y, *et al.* (2004) Adsorptive granulocyte and monocyte apheresis for refractory Crohn's disease: An open multicenter prospective study. *J Gastroenterol* **39:** 1158–1164.

Fukunaga K, Yokoyama Y, Kamokozuru K, *et al.* (2012) Adsorptive granulocyte/monocyte apheresis for the maintenance of remission in patients with ulcerative colitis: A prospective randomized, double blind, sham-controlled clinical trial. *Gut Liver* **6:** 427–433.

Hanai H, Iida T, Takeuchi K. (1008) Adsorptive depletion of elevated pro-inflammatory CD14+CD16+DR++ monocytes in patients with inflammatory bowel disease. *Am J Gastroenterol* **103:** 1210–1216.

Hanai H, Takeda Y, Eberhardson M, *et al.* (2011) The mode of actions of the Adacolumn therapeutic leucocytapheresis in patients with inflammatory bowel disease: A concise review. *Clin Exp Immunol* **163:** 50–8.

Hanai H, Watanabe F, Takeuchi K, *et al.* (2003) Leukocyte adsorptive apheresis for the treatment of active ulcerative colitis: A prospective, uncontrolled, pilot study. *Clin Gastroenterol Hepatol* **1:** 28–35.

Hibi T, Sameshima Y, Sekiguchi Y, *et al.* (2009) Treating ulcerative colitis by Adacolumn therapeutic leucocytapheresis: Clinical efficacy and safety based on surveillance of 656 patients in 53 centres in Japan. *Dig Liver Dis* **41:** 570–7.

Jason KH, Fernade V, Norah T, Uma M. (2010) Viral hepatitis and inflammatory bowel disease. *Inflamm Bowel Dis* **16:** 925–32.

Kashiwagi N, Sugimura K, Koiwai H, Saniabadi A (2002). Immunomodulatory effects of granulocyte and monocyte adsorption apheresis as a treatment for patinets with ulcerative colitis. Dig Dis Sci; **47**: 1334-41.

Kato S, Hosomi E, Amano F, *et al.* (2012) The efficacy of intensive granulocyte and monocyte adsorption apheresis in a patient with Crohn's disease complicated by extensive subcutaneous aseptic neutrophilic abscesses. *J Crohn's Colitis* **6:** 787–91.

Kawakami A, Tanaka M, Nishigami M, *et al.* (2013) Relationship between non-adherence to aminosalicylate medication and the risk of clinical relapse among Japanese patients with ulcerative colitis in clinical remission: A prospective cohort study. *J Gastroenterol* **48:** 1006–15.

Kikuyama R, Fukunaga K, Kawai M, *et al.* (2011) Relevance of the processed blood volume per granulocyte and monocyte apheresis session to its clinical efficacy in patients with ulcerative colitis. *Ther Apher Dial* **15:** 360–366.

Lichtenstein GR, Feagan BG, Cohen RD, *et al.* (2012) Serious infection and mortality in patients with Crohn's disease: More than 5 years of follow-up in the TREAT registry. *Am J Gastroenterol* **107:** 1409–22.

Lok AS, McMahon BJ. (2007) Chronic hepatitis B. *Hepatology* **45:** 507–39.

Loras C, Saro C, Gonzalez-Huix F, *et al.* (2009) Prevalence and factors related to hepatitis B and C in inflammatory bowel disease patients in Spain: A nationwide, multicenter study. *Am J Gastroenterol* **104:** 57–63.

Matsui T, Nishimura T, Matake H, *et al.* (2002) Granulocytetapheresis for Crohn's disease: A report on seven refractory patinets. *Am J Gastroenterol* **98(2):** 511–512.

Mizoguchi A, Mizoguchi E. (2008) Inflammatory bowel disease, past, present and the future: Lessons from animal models. *Gastroenterology* **43:** 1–17.

Mulder DJ, Noble AJ, Justinich CJ, Duffin JM. (2014) A tale of two diseases: The history of inflammatory bowel disease. *J Crohns Colitis* **8:** 341–348.

Muratov V, Lundahl J, Ulfgren AK, *et al.* (2006) Down-regulation of interferon-gamma parallels clinical response to selective leukocyte apheresis in patients with inflammatory bowel disease: A 12-month follow-up study. *Int J Colorectal Dis* **21:** 493–504.

Muzzi A, Ciani F, Bianchini D. (1995) Adverse pulmonary effects of mesalamine. *Chest* **108:** 1181.

Naganuma M, Funakoshi S, Sakuraba A, *et al.* (2004) Granulocytetapheresis is useful as an alternative therapy in patients with steroid-refractory or -dependent ulcerative colitis. *Inflamm Bowel Dis* **10:** 251–7.

Ojiro K, Naganuma M, Ebinuma H, *et al.* (2008) Reactivation of hepatitis B in a patient with Crohn's disease treated using infliximab. *J Gastoroenterol* **43:** 397–401.

Present DH. (2000) How to do without steroids in inflammatory bowel disease. *Inflamm Bowel Dis* **6:** 48–57.

Rahier JF, Ben-Horin S, Chowers Y, *et al.* (2009) European evidence-based consensus on the prevention, diagnosis and management of opportunistic infection in inflammatory bowel disease. *J Crohn's Colitis* **3:** 47–91.

Ruuska T, Wewer V, Lindgren F, *et al.* (2009) Granulocyte-monocyte adsorptive apheresis in pediatric inflammatory bowel disease; results, practical issue, safety, and future perspectives. *Inflamm Bowel Dis* **15:** 1049–54.

Sacco R, Romano A, Mazzoni A, *et al.* (2013) Granulocytapheresis in steroid-dependent and steroid-resistant patients with inflammatory bowel disease: A prospective observational study. *J Crohns Colitis* **7:** e692–697.

Sacco R, Tanaka T, Yamamoto T, *et al.* (2014) Adacolumn leucocytapheresis for ulcerative colitis: Clinical and endoscopic features of responders and unresponders. *Expert Rev Gastroenterol Hepatol* **27:** 1–7.

Sakuraba A, Motoya S, Watanabe K, *et al.* (2009) An open-label prospective randomized multicenter study shows very rapid remission of ulcerative colitis by intensive granulocyte and monocyte adsorptive apheresis as compared with routine weekly treatment. *Am J Gastroenterol* **104:** 2990–9.

Sands BE, Katz S, Wolf DC, *et al.* (2012) A randomized, double-blind, sham-controlled study of granulocyte/monocyte apheresis for moderate to severe Crohn's disease. *Gut* **64:** 1288–1294.

Sands BE, Sandborn WJ, Feagan B, *et al.* (2008) A randomized, double-blind, sham-controlled study of granulocyte/monocyte apheresis for active ulcerative colitis. *Gastroenterology* **135:** 400–9.

Saniabadi AR, Hanai H, Fukunaga K, *et al.* (2007) Therapeutic leukocytapheresis for inflammatory bowel disease. *Transfus Apher Sci* **37:** 191–200.

Saniabadi AR, Hanai H, Takeuchi K, *et al.* (2003) Adacolumn, an adsorptive carrier based granulocyte and monocyte apheresis device for the treatment of inflammatory and refractory diseases associated with leukocytes. *Ther Apher Dial* **7:** 48–59.

Saniabadi AR, Hanai H. (2010) Therapeutic apheresis from the early civilizations to the twenty-first century. *Gastroenterol Clin Biol* **34:** 645–8.

Saniabadi AR, Tanaka T, Ohmori T, *et al.* (2014) Treating inflammatory bowel disease by adsorptive leucocytapheresis: A desire to treat without drugs. *World J Gastroenterol* **20:** 9699–9715.

Shimoyama T, Sawada K, Hiwatashi H, *et al.* (2001). Safety and efficacy of granulocytes and monocyte adsorption apheresis in patients with active ulcerative colitis: A multiple study. *J Clin Apher* **16:** 1–9.

Sullivan SN. (1987) Sulfasalazine lung. Desensitization to sulfasalazine and treatment with acrylic coated 5-ASA and azodisalicylate. *J Clin Gastroenterol* **9:** 461–3.

Suzuki Y, Yoshimura N, Fukuda K, *et al.* (2006) A retrospective search for predictors of clinical response to selective granulocyte and monocyte apheresis in patients with ulcerative colitis. *Dig Dis Sci* **51:** 2031–2038.

Suzuki Y, Yoshimura N, Saniabadi AR, Saito Y. (2004) Selective granulocyte and monocyte adsorptive apheresis as a first-line treatment for steroid naïve patients with active ulcerative colitis: A prospective uncontrolled study. *Dig Dis Sci* **49:** 565–71.

Takayama T, Kanai T, Matsuoka K, *et al.* (2012) Long-term prognosis of patients with ulcerative colitis treated with cytapheresis therapy. *J Crohn's Colits* **7:** 49–54.

Tanaka T, Okanobu H, Kuga Y, *et al.* (2010) Clinical and endoscopic features of responders and non-responders to adsorptive leucocytapheresis: A report based on 120 patients with active ulcerative colitis. *Gastroenterol Clin Biol* **34:** 687–95.

Tanaka T, Okanobu H, Yoshimi S, *et al.* (2008) Adsorptive depletion of granulocytes and monocytes impacts mucosal level of neutrophils and clinically is most effective in steroid naïve patients. *Dig Liver Dis* **40:** 731–6.

Tanaka T, Sugiyama S, Goishi H, *et al.* (2013) Treatment of children and adolescents with ulcerative colitis by adsorptive depletion of myeloid lineage leucocytes as monotherapy or in combination with low dose prednisolone after failure of first-line medications. *BMC Gastroenterol* **13:** 130.

Thomas TP. (1984) The complications of systemic corticosteroid therapy in the elderly. A retrospective study. *Gerontology* **30:** 60–5.

Xavier RJ, Podolsky DK. (2007) Unravelling the pathogenesis of inflammatory bowel disease. *Nature* **448:** 427–34.

Yamamoto T, Umegae S, Matsumoto K. (2012) Long-term clinical impact of early introduction of granulocyte and monocyte adsorptive apheresis in new onset, moderately active, extensive ulcerative colitis. *J Crohn's Colitis* **6:** 750–5.

Yokoyama Y, Fukunaga K, Fukuda Y, *et al.* (2007) Demonstration of low-regulatory CD25High+CD4+ and high-pro-inflammatory CD28-CD4+ T-Cell subsets in patients with ulcerative colitis: Modified by selective granulocyte and monocyte adsorption apheresis. *Dig Dis Sci* **52:** 2725–31.

Yokoyama Y, Fukunaga K, Kamikozuru K, *et al.* (2014) Crohn's disease complicated by hepatitis B virus successfully treated with the use of adsorptive depletion of myeloid lineage leucocytes to suppress inflammatory cytokine profile. *Cytotherapy* **16:** 821–825.

Yokoyama Y, Kawai M, Fukunaga K, *et al.* (2013) Looking for predictive factors of clinical response to adsorptive granulocyte and monocyte apheresis in patients with ulcerative colitis: Markers of response to GMA. *BMC Gastroenterol* **13:** 27.

Yokoyama Y, Watanabe K, Ito H, *et al.* (2015) Factors associated with treatment outcome, and long-term prognosis of patients with ulcerative colitis undergoing selective depletion of myeloid lineage leucocytes: A prospective multicenter study. *Cytotherapy* (in press).

Yoshimura N, Tadami T, Kawaguchi T, *et al.* (2012) Processed blood volume impacts clinical efficacy in patients with ulcerative colitis undergoing adsorptive depletion of myeloid lineage leucocytes. *J Gastroenterol* **47:** 49–55.

Chapter 33

Granulocyte and Monocyte Adsorption Apheresis (GMA) in Dermatology

Takuro Kanekura*

1. Introduction

1.1. *Neutrophils*

Neutrophilic leucocyte is a major hematopoietic cellular component in biophylaxis such as defense against bacteria or fungi. In case of inflammation, neutrophils infiltrate into the site of infection and ingest or kill invasive organisms with their bactericidal and phagocytic properties, which are mediated by various enzymes and reactive oxygen species. Important enzymes produced by neutrophils include acidic protease (cathepsin), collagenase, elastase, phosphatase, ribonuclease, deoxyribonuclease, lysozyme, and phospholipase. Oxygen uptake by activated neutrophils is increased and much of oxygen is converted to H_2O_2, a bactericidal substance. Superoxide anion (O_2^-), a potent bactericidal and cytotoxic molecule, is derived from oxidation of hypoxanthine or by reduction of oxygen. In activation of neutrophils, complement cascade plays a pivotal role (Steinhoff *et al.*, 2004).

Pathological conditions in which complement system is activated independent of infection are known. Ulcerative colitis (UC) and

---

*Department of Dermatology, Kagoshima University Graduate School of Medical and Dental Sciences, 8-35-1 Sakuragaoka, Kagoshima 890-8520, Japan, Tel.: 099-275-5388 FAX: 099-275-1134, E-mail: takurok@m2.kufm.kagoshima-u.ac.jp.

rheumatoid arthritis (RA) are representative diseases of this sort, and neutrophils in these conditions express the cell surface activation marker Mac-1 (CD11b/CD18). Non-infectious inflammatory disorders of the skin attributed to activated neutrophils include pyoderma gangrenosum (PG), Behçet's disease (BD), Sweet's syndrome (SS), adult-onset Still's disease (ADSD), erythema nodosum (EN), erythema elevatum diutinum, pustular psoriasis (PP), and vascular disorders characterized histopathologically by leucocytoclastic vasculitis such as cutaneous allergic vasculitis (CAV) and polyarteritis nodosa.

1.2. *Pyoderma Gangrenosum*

Various skin diseases are known to have systemic associations. When examining skin lesions, dermatologists consider the possibility of underlying systemic disorders. PG, an intractable skin disease characterized by sterile skin ulcerations, is one such skin disorder (Callen, 1998). The lesions of PG begin as vesiculopustules that resemble acne or folliculitis. They rapidly become turgid and ulcerated, and their size often increases to 10 cm or more. As the ulcer enlarges peripherally, the skin at the margin becomes purplish red and undermined. The histological appearance is not pathognomonic but manifests features of a large and fairly well-circumscribed sterile abscess that, at onset, is composed largely of neutrophils. In a fully developed ulcer, epithelioid and giant cells as well as numerous neutrophils are present in the infiltrates. A variety of systemic diseases such as UC, RA, aortitis syndrome, monoclonal hypergammaglobulinemia and myelogenous leukemia are found in association with PG. About a half to two-thirds of patients with PG have or develop systemic diseases. Various systemic treatments, including the administration of corticosteroids, salazopyrine, and immunosuppressants such as cyclosporin or tacrolimus are only partially successful (Barham *et al.*, 2004).

1.3. *Arthropathy Associated with Neutrophilic Dermatosis*

Neutrophilic dermatoses are often associated with arthropathy. Joint pain is a symptom of BB, SS, and AOSD. Psoriatic arthritis (PsA), pustulotic arthro-ostitis (PAO), and SAPHO (synovitis, acne, pustulosis, hyperostosis, osteiris) syndrome are well-known complications of skin lesions and arthropathy. Skin lesions are often seen in patients with RA.

1.4. *Granulocyte and Monocyte Adsorption Apheresis*

Granulocyte and monocyte adsorption apheresis (GMA), developed and provided by the Japan Immunoresearch Laboratories (Takasaki, Japan), is an extracorporeal apheresis instrument designed to remove pathogenic granulocytes. The GMA column is filled with specially designed cellulose acetate (CA) beads, which serve as the column adsorptive leucocytapheresis carriers (Saniabadi *et al.*, 2003; Hiraishi *et al.*, 2003). The carriers adsorb about 65% of granulocytes, 55% of monocytes/macrophages and a significant fraction of platelets from the blood in the column. These cells bear the Fcγ receptor (FcγR) and complement receptors; lymphocytes are spared and subsequently increase, except minor sub-populations (Saniabadi *et al.*, 2003, 2007; Hanai *et al.*, 2011). GMA was initially approved by the Japan Ministry of Health, Labour and Welfare for the treatment of UC (Saniabadi *et al.*, 2003). Because UC is often associated with PG, as mentioned above, and both UC and PG are characterized by neutrophilic infiltration histologically, we anticipated that GMA might be beneficial in treating PG.

1.5. *Assessment of the Therapeutic Efficacy of GMA in Pyoderma Gangrenosum*

The therapeutic efficacy of GMA on skin disease was first assessed in a 38-year-old man with PG (Kanekura *et al.*, 2002). He was referred to our department with skin lesions that had first appeared three years earlier. Physical examination revealed irregularly shaped palm-sized ulcerative lesions on his bilateral axillae, back, and buttocks. Many pustules and small ulcers were present on the margins of the lesions (Fig. 33.1A). The lesions undermined the skin, and there were fistulae with pyoid discharge. A biopsy specimen obtained from a lesion on his back revealed dense cellular infiltration consisting of dominant

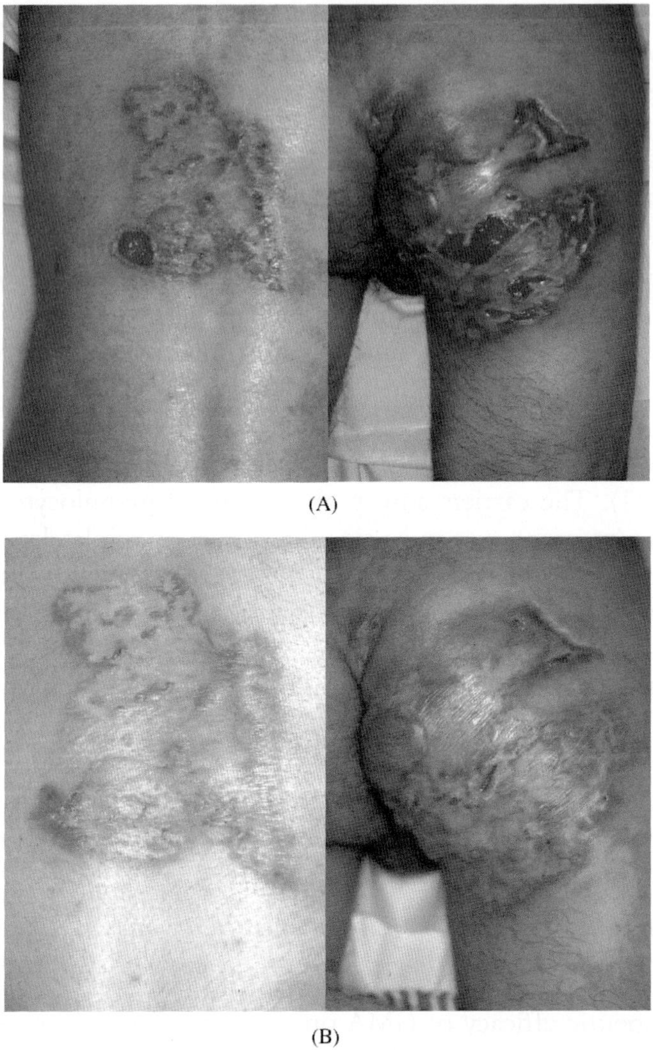

Fig. 33.1. (A) Skin lesions of PG before GMA treatment. Irregularly shaped ulcerative lesions on the back and buttock. (B) The lesions cleared after the fourth treatment session.

neutrophils and plasma cells throughout the entire dermis. A diagnosis of PG was made based on the clinical and pathological features. Laboratory findings were as follows: erythrocyte sedimentation rate (ESR), 102 mm/hr; C-reactive protein (CRP), 6.9 mg/dL (normal range < 0.30 mg/mL); white blood cell (WBC) count, 12,400/mm^3 (75% neutrophils). General examinations revealed no systemic disorders. The lesions had shown little response to salicylazosulfapyridine at 2 g/day for four weeks and prednisolone at 30 mg/day for one week. GMA consisting of five sessions weekly was started without any other systemic or topical remedy after obtaining the prior written informed consent. After the second session, dramatic improvement was observed. The ulcer beds became vividly fresh in color, the discharge disappeared, the margins dried up, and the lesion sizes decreased remarkably. The lesions cleared up completely after the fourth treatment session (Fig. 33.1B). ESR, CRP, and WBC count decreased to 75 mm/hr, 0.17 mg/dL, and 5,500/mm^3 (51% neutrophils), respectively. Thus, our first study of the efficacy of GMA for a skin disease returned excellent results.

1.6. *Activated Neutrophils and Mac-1*

Activated granulocytes express Mac-1, a cell-surface adhesive molecule that belongs to the integrin family (Diamond and Springer, 1993). CD11b (integrin αM subunit) combines with CD18 (integrin β2 subunit) to form a heterodimer Mac-1 (Corbi *et al.*, 1988; Arnaout *et al.*, 1988) known as complement receptor 3 (CR3) (Beller *et al.*, 1982; Cai and Wright, 1996). The CA beads in the GMA column activate and adsorb complement component iC3b, a ligand for Mac-1 (Cheung *et al.*, 1991; Hiraishi *et al.*, 2003). Based on these facts, it was demonstrated that the column traps activated granulocytes by binding Mac-1 expressed on the granulocytes to iC3b adsorbed on the beads (Fig. 33.2) (Kanekura *et al.*, 2006a).

2. Clinical Study of GMA for Skin Disorders and Related Arthropathy

2.1. *Patients*

Based on the remarkable treatment outcome in the first patient with PG, a clinical study of the effectiveness of GMA for treating various skin diseases and related arthropathy was conducted. A total of 73

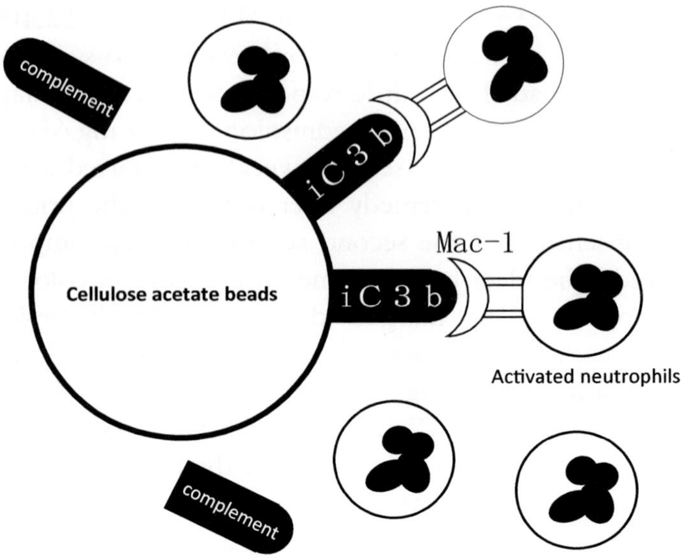

Fig. 33.2. Schematic presentation of the possible mechanism of GMA to remove activated granulocytes. Cellulose acetate beads in the GMA column trap activated granulocytes by binding Mac-1 expressed on the granulocytes to iC3b adsorbed on the beads.

patients were enrolled in our study for the evaluation of skin lesions. Of these patients, 36 had PP, 13 had BD, 10 had PG, and three had systemic lupus erythematosus (SLE) skin rash. EN, CAV, acrodermatitis continua of Hallopeau and SS were seen in two patients, respectively. The other three patients had Reiter's disease (RD), AOSD, and leg ulcers associated with RA. Arthropathy was associated in 26 patients. Of 26 patients with arthropathy, 21 had PsA and the other five patients had BD, SS, RD, AOSD and RA (Refs.).

2.2. *GMA*

GMA is an extracorporeal-type apheresis unit that uses a column containing 3×10^4 pieces of 2 mm-diameter cellulose acetate beads. In the current series, GMA was performed five or 10 times at five-day intervals. If the lesion showed amelioration but persisted at the completion of the fifth treatment session, five more sessions were added. In each session, 1,800 mL of blood is drained from the cubital vein of one arm, passed through the GMA column at a flow rate of 30 mL/min, and returned to the cubital vein of the other arm. GMA was performed without adding any other systemic remedy; in patients who already received medication, the dose was left unchanged. GMA was carried out after obtaining the prior written informed consent of all participants. All GMA therapy procedures were approved by the Human Investigation Committee of Kagoshima University Graduate School of Medical and Dental Sciences.

2.3. *Clinical and Laboratory Evaluation*

The severity of the patients' skin lesions was evaluated clinically based on overall lesional assessment (OLA)[27] and recorded on a 5-point scale where 0 = none, 1 = slight, 2 = mild, 3 = moderate, and 4 = severe. Symptoms considered for OLA were plaque elevation, erythema, and scaling for PsA; pustules, erythema, and scaling for PP, RD, and acrodermatitis continua; area and depth of ulcerations, granulation, and pain for PG, BD, and CAV; erythema, induration, and tenderness for SS and EN. The response to GMA was rated as excellent in patients who improved by four grades (i.e. grade 4 to grade 0), as good when they improved by two or three grades, and as moderate when improvement was by a single grade; no change and exacerbation were also recorded. All clinical evaluations were made by three independent physicians. If their assessments differed, the median value of their judgment was recorded as the treatment outcome.

The degree of arthropathy, assessed by the patients based on joint pain, the number of affected joints and range of joint motion, was represented on a 10-cm visual analog scale (VAS). Treatment responses were recorded as excellent, good, and moderate when the VAS score was 20% or less, 50% or less, and 90% or less, respectively, of the baseline score, or as no change or exacerbation.

The WBC count, neutrophil count, CRP, and the expression level of Mac-1 on peripheral neutrophils were measured before starting

GMA therapy and one day after the last treatment session. Mac-1 expression was analyzed by flow cytometry using whole blood and mouse anti-human CD11b antibody that is not a blocking antibody.

2.4. *Results of the Clinical Study*

2.4.1. *Response of skin lesions to GMA*

Skin lesions of 73 patients were evaluated. Excellent response was observed in 14 patients; six with PP, three with BD and PG, and one with EN and CAV. Good and moderate responses were obtained in 34 and 15 patients, respectively. The lesions in 10 patients; five with PP, one with BD, two with SLE, one with acrodermatitis continua of Hallopeau and SS were unchanged. In 63 of the 73 evaluated patients (86.3%), the skin lesions showed improvement by GMA (Table 33.1).

Table 33.1. Response of Skin Lesions to GMA

		Excellent	Good	Moderate	Unchanged
Pustular psoriasis	($n = 36$)	6	16	9	5
Behçet's disease	($n = 13$)	3	6	3	1
Pyoderma gangrenosum	($n = 10$)	3	4	3	
SLE	($n = 3$)		1		2
Erythema nodosum	($n = 2$)	1	1		
Cutaneous allergic vasculitis	($n = 2$)	1	1		
Acrodermatitis continua	($n = 2$)		1		1
Sweet's syndrome	($n = 2$)		1		1
Reiter's disease	($n = 1$)		1		
Adelt-onset Still's disease	($n = 1$)		1		
Leg ulcer associated with RA	($n = 1$)		1		
Total	($n = 73$)	14	34	15	10

RA: Rheumatoid arthritis.

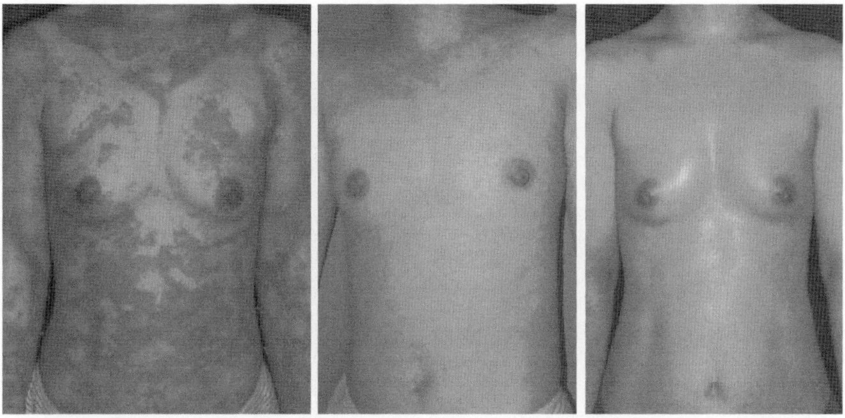

Fig. 33.3. Skin lesions of GPP before and after GMA treatment. Irregularly-shaped erythematous lesions with numerous pustules on the trunk and extremities showed dramatic improvement after the first treatment and completely disappeared after the fifth session.

Representative patients are shown in Figs. 33.3 and 33.4. A patient with generalized PP had irregularly-shaped erythematous lesions with numerous pustules on her trunk and extremities. They had appeared two months prior to her visit to our department. The lesions showed dramatic improvement after the first treatment and completely disappeared after the fifth session (Fig. 33.3). PG that manifested itself as a large and painful ulcer on the patient's lower leg had appeared two years before and persisted until GMA therapy and subsided after the 10th treatment session (Fig. 33.4).

2.4.2. *Response of arthropathy to GMA*

Arthropathy was evaluated in 26 patients. An excellent response was obtained in five patients; three with PsA, one with BD, and one with AOSD. Good and moderate responses were observed in 10 and six patients, respectively. The arthropathy of five patients remained unchanged. Overall, arthropathy improved in 21 of the 26 patients (80.8%) (Table 33.2).

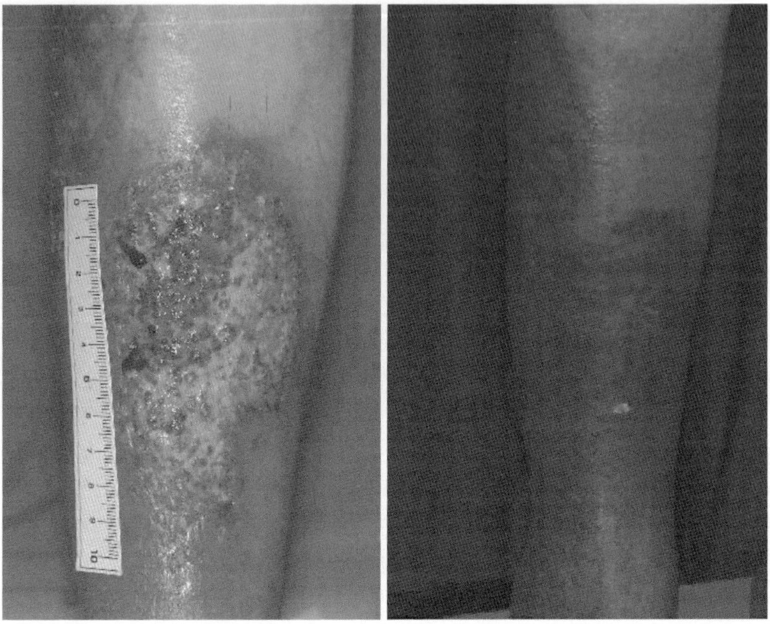

Fig. 33.4. Skin lesions of PG before and after GMA treatment. A large and painful ulcer on the patient's lower leg subsided after the 10th treatment session of GMA.

Table 33.2. Response of Arthropathy to GMA

		Excellent	Good	Moderate	Unchanged
Psoriatic arthropathy	($n = 21$)	3	8	6	4
Behçet's disease	($n = 1$)	1			
Sweet's syndrome	($n = 1$)				1
Reite's disease	($n = 1$)		1		
Adult-onset Still's disease	($n = 1$)	1			
RA associated with leg ulcer	($n = 1$)		1		
Total	($n = 26$)	5	10	6	5

RA: Rheumatoid arthritis.

2.4.3. *Laboratory examinations*

The mean values of the WBC- and neutrophil count, CRP, and Mac-1 expression level before starting GMA therapy were 8,048/μL, 5,625/μL, 2.58 mg/dL, and 34.5 MFI (mean fluorescence intensity), respectively. They decreased to 6,983/μL, 4,555/μL, 1.48 mg/dL, and 20.2 MFI with statistical significance.

2.4.4. *Adverse events*

No adverse events such as flushing, dizziness, nausea, thrombosis, and embolism were observed in the study.

2.5. *Conclusion of the Clinical Study*

This clinical study revealed that GMA is effective and safe for the treatment of various skin diseases and related arthropathy attributable to activated granulocytes. More than 80% of the patients benefited from GMA and no adverse effects were encountered. Inflammatory indices such as the WBC- and neutrophil count and CRP decreased in response to GMA. The expression level of Mac-1 was inhibited by GMA, suggesting a possible mechanism by which GMA exerts its therapeutic effects. Based on the favorable outcome of our clinical study, a prospective multi-center study of GMA for PP was designed.

3. Prospective and Multi-Center Study of GMA for Pustular Psoriasis

3.1. *Pustular Psoriasis*

Generalized pustular psoriasis (GPP) is a chronic autoimmune disease characterized by fever, erythema, and sterile pustules over large areas of the skin. The etiology and factors to perpetuate the disease are not understood well, and GPP does not respond well to pharmacological interventions currently available (Griffiths *et al.*, 2004; Seishima *et al.*, 2008; Shukuya *et al.*, 2011). Biopsy specimens from skin lesions of GPP, notably from pustules, reveal prominent neutrophil infiltration into the epidermis that forms the spongiform pustules of Kogoj and unilocular intraepidermal pustule. Since a dysregulated immune profile involving elevated myeloid lineage leucocytes, neutrophils and monocytes/macrophages is a major factor in the pathogenesis of GPP (Viguier *et al.*, 2004; Michalak-Stoma *et al.*, 2011; Fujisawa *et al.*, 2011; Botti *et al.*, 2012; Johnson-Huang *et al.*, 2012), elevated and activated neutrophils are appropriate targets for therapeutic intervention, and patients with GPP should respond favorably to selective depletion of increased neutrophils by GMA that has been developed for selective depletion of myeloid lineage leucocytes (Saniabadi *et al.*, 2003). In light of these backgrounds, a multi-center study on the efficacy of GMA in patients with GPP was conducted.

3.2. *Patients*

A total of 15 patients with moderate to severe GPP from 11 participating institutions were enrolled. Patients who had pustules covering more than 10% of the skin were considered to be eligible. These patients received conventional medications; oral etretinate, cyclosporin, methotrexate, prednisolone, or topical corticosteroid and vitamin D3 (Table 33.3). These drugs were allowed to continue if they had started well in advance of entry of GMA but without any change in the dosage. Patients who participated in other clinical investigations, received a biological agent within the past eight months, received phototherapy within the past two weeks, or underwent extracorporeal therapy within the past six months were excluded.

3.3. *Determination of GPP Severity and Treatment Efficacy*

Disease severity and efficacy outcomes were determined according to the Guidelines for the Management of GPP edited by the Japan

Table 33.3. Baseline Demography of 15 Patients with GPP Included in the Study

Demography	Measurement	
Male/Female	11/4	
Average age (years)	50.3 ± 12.9, range: 31–71	
Average weight (kg)	69.27 ± 11.10, range: 41.5–88.0	
Past or Concomitant Medication	**Number of Refractory Patients**	**Number of Patients Intolerant or Unwilling to Take**
Etretinate	5	5
Cyclosporin	4	3
Prednisolone	5	2
Methotrexate	3	1
Biologics	2	2

Table 33.4. Severity Score Introduced by Guidelines for the Management of GPP Proposed by Japan Ministry of Health Research Group for Rare and Intractable Dermatological Diseases Dermatological Diseases

Skin Symptoms (Score 0–9)	3 Points	2 Points	1 Point	0 Points
Area of erythema	75%	< 75%–25%	< 25%	0%
Area of erythema with pustules	50%	< 50%–10%	< 10%	0%
Area of edema	50%	< 50%–10%	< 10%	0%

Laboratory Variables (Scores 0–8)	2 Points	1 Point	0 Points
Body temperature, °C	38.5	< 38.5–37.0	< 37.0
WBC, /μL	15,000	< 15,000–10,000	< 10,000
CRP, mg/dL	7.0	< 7.0–0.3	< 0.3
ALB, g/dL	< 3.0	< 3.8–3.0	3.8

Ministry of Health Research Group for Rare and Intractable Dermatological Diseases (published in Japanese) (Table 33.4). The total GPP severity score ranges from 0 to 17; points ≤ 6 = mild, 7 to 10 = moderate, and 11 to 17 = severe. These scores were translated from sub-scores; the extent (% of total body surface area) of erythema, pustules on erythematous lesions, and edema is given a score of 0 to 9 points with 3 points (maximum) reflecting ≥ 75% erythema, ≥ 50% erythema with pustules or ≥ 50% edema, while 0 indicates normal skin. Other relevant variables including body temperature (BT), WBC counts, serum albumin (ALB), and the inflammation marker, CRP, are scored from 0 to 8. A score of 2 points (maximum) is given as follows; BT ≥ 38.5°C = 2, WBC ≥ 15,000/μL = 2, CRP ≥ 7 mg/dL = 2 and ALB < 3g/dL = 2. The score for BT < 37.0°C, WBC < 10,000/μL, CRP < 0.3 mg/dL, and ALB ≥ 3.8 g/dL is 0. Intermediate scores are listed in Table 33.5. These scores were determined before the first GMA session and then after the fifth sessions, usually two weeks following the last session. The color improvement score was added to the overall severity score of the evaluation data and at least a 40% reduction in the overall score was judged as response to the therapy. The results were evaluated using appropriate statistical analyses.

3.4. *GMA Procedures*

Each patient received one GMA session weekly over five consecutive weeks. The duration of one session was 60 min at 30 mL/min. A small dose of sodium heparin (2,000 units/session) was administered as an anti-coagulant.

3.5. *Ethical Considerations*

The study protocol was reviewed and approved by the Japan Ministry of Health, Labour and Welfare, as well as by the Institutional Review Board at each investigation site. Written informed consent was obtained from all included patients. The study was conducted within Japanese pharmaceutical laws and in accordance with Good Clinical Practice Guidelines for investigations involving medical devices. Strict adherence to the Declaration of Helsinki was made through the study.

3.6. *Results of the Multi-Center Study*

3.6.1. *Patients' baseline demographic variables*

Of the 15 eligible patients, 11 were male and four were female. Their age ranged from 31 to 71 years with an average age of 50.3 ± 12.9. Patients' average weight was 69.27 ± 11.10 kg, ranging from 41.5 to 88.0 (Table 33.3). Most patients were refractory or intolerant to conventional medications and a few were unwilling to take the medications.

3.6.2. *Study safety and patient compliance*

One patient was withdrawn from the study. This patient did not complete the first GMA session because of allergic reaction to nafamostat mesilate, an anti-coagulant used during extracorporeal circulation. In all other patients, heparin was used as the anti-coagulant and all 14 patients completed their five GMA sessions according to the protocol. Safety evaluation was based on all 15 patients in whom GMA therapy was initiated. GMA was not associated with any abnormal laboratory finding in any patient. During the study, 16 adverse events were observed in three patients but none was serious; these three patients completed all GMA sessions. Of 16 adverse events, 14 were reported by one patient. They included transient headache, dizziness and light headedness on standing, chills, and feeling of weakness after a GMA session. These symptoms are often reported during extracorporeal circulation in a minority of patients (Shimoyama *et al.*, 2001). Bullous pemphigoid (BP), an autoimmune blistering disease, was complicated and slightly worsened during GMA therapy in one patient. The causality was judged as "unknown" because BP exacerbated when the patient's GPP was improved. CT scan showed shadows in the lungs in one patient during the follow-up period. The effect of the device was not ruled out but also suspected of concomitantly administered methotrexate, previously administered infliximab, or the primary disease. The shadows in this patient subsequently disappeared without medical intervention.

3.6.3. *Efficacy outcomes*

The overall severity score significantly improved relative to entry ($n = 14$, $P = 0.0027$), and 12 of the 14 patients (85.7%) who completed the five GMA sessions were judged as responders. The extent of erythema ($n = 14$, $P = 0.0042$), pustules ($n = 14$, $P = 0.0031$) and edema ($n = 14$, $P = 0.0014$) were significantly improved. The extent of erytherma, pustules on erythematous lesions, and edema on the skin were all markedly reduced (Fig. 33.5); pustules and edema disappeared in seven and 10 patients, respectively. In Figs. 33.6 and 33.7, typical images of GPP lesions from the affected areas of skin before and after the course of GMA are presented. The figures show that the effect of GMA was striking in most responder patients. Additionally, the quality of life parameter for patients with skin diseases called Dermatology Life Quality Index (DLQI)

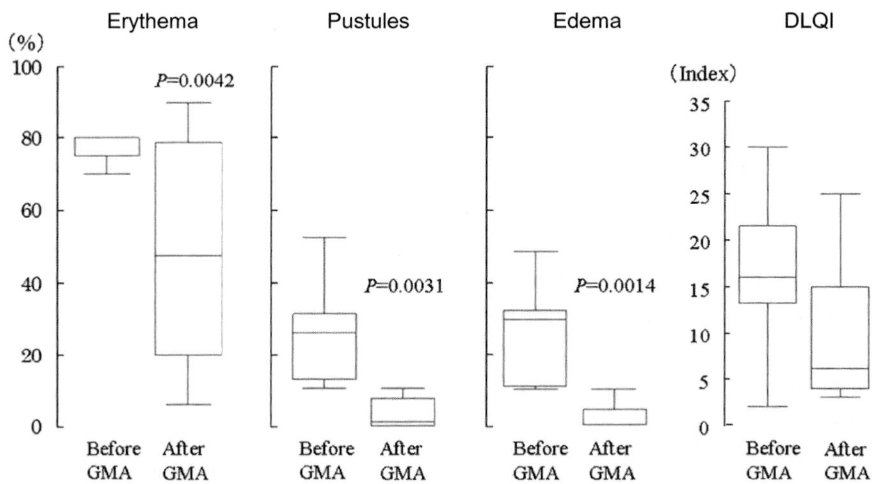

Fig. 33.5. GPP treatment outcomes by box plots showing changes in the extent (%) of psoriatic lesions, severity scores listed in Table 33.4 and quality of life/well-being in 14 patients who completed the GMA treatment course. The figure shows a significant fall in the extent of erythroderma, pustules, edema, and a significant improvement in the Dermatology Life Quality Index (DLQI).

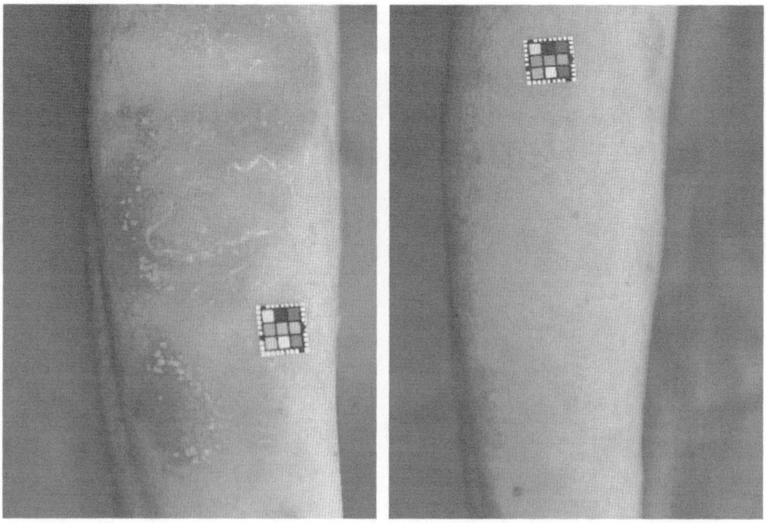

(A) (B)

Fig. 33.6. GPP lesions before (A) and after GMA (B). (A) At entry, over an 80% of the patient's skin was covered by erythroderma, 49% by pustules and 9% by edema. (B) After GMA treatment, the area of erythroderma decreased to 20% with complete disappearance of pustules and edema. The most severely affected skin areas before and after GMA are shown.

(Finlay and Khan, 1994) showed significant improvement (P = 0.0016), reflecting better daily function and quality of life after a course of GMA (Fig. 33.5). Further, 10 of the 12 responders maintained their clinical response for more than 10 weeks after the last GMA session without any change in their medications. Accordingly, the efficacy of this non-pharmacologic intervention targeting myeloid linage leukocytes is potentially very interesting in this clinical setting.

Low ALB (< 3.8 g/dL) in four patients at baseline improved to normal range after the course of GMA. Likewise, seven of 10 patients with a mild elevation of CRP (> 0.30 mg/dL) at baseline showed reduced CRP level. But laboratory findings did not show statistically significant changes (Ikeda *et al.*, 2013).

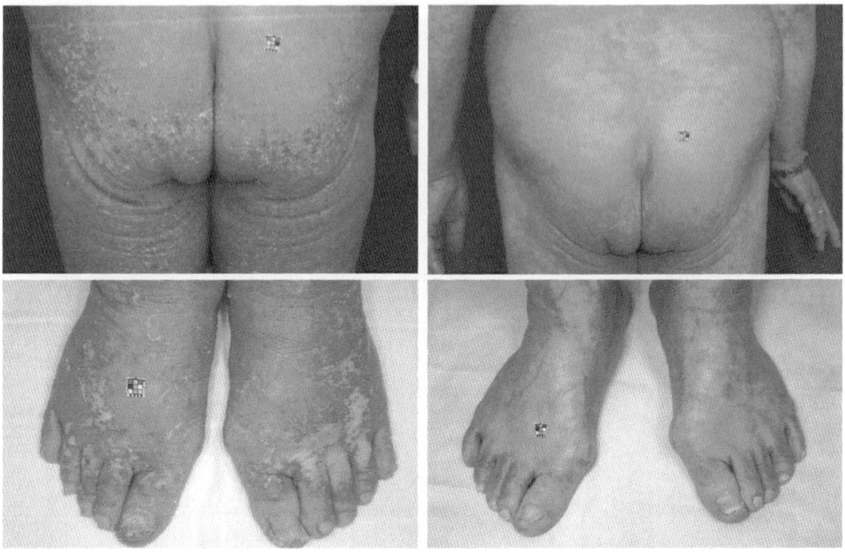

(A) (B)

Fig. 33.7. GPP lesions before (A) and after GMA (B). This figure shows near complete disappearance of psoriatic lesions, erythematous plaques, pustules, and edema after a course of GMA in a case whose skin was largely affected by GPP.

3.7. *Conclusion of the Multi-Center Study*

This study demonstrated the safety and efficacy of GMA for patients with active GPP who received conventional medications. Based on the GPP severity scores, the majority of this cohort of 15 patients with medication refractory GPP responded to GMA with a significant improvement of their disease and achieved an improved quality of life. The overall scores for erythema, pustules, and edema were significantly decreased. 85.7% of the patients were judged as responders to GMA, potentially a very significant outcome given that GMA is a non-pharmacologic treatment intervention. Further, the fall in GPP scores were paralleled by the disappearance of GPP lesions at most severely affected skin areas. The majority of patients who responded to GMA maintained their clinical response for over 10 weeks. In addition, the GMA was not associated with any severe adverse side effects.

With this excellent outcome of the prospective and multi-centered study, GMA was approved for the treatment of PP by the Japan Ministry of Health, Labour and Welfare in 2012.

4. Conclusion and the Future of GMA for Skin Diseases

In a series of the clinical and multi-center studies, GMA has turned out to be a safe and effective treatment for refractory skin diseases attributable to activated neutrophils. GMA is a non-pharmacological intervention designed to remove activated neutrophils and macrophages. On the basis of the nature of this apheresis, GMA is a promising treatment modality for various diseases including inflammatory bowel disease and rheumatoid arthritis and so forth as well as neutrophilic dermatoses. Currently, biologic agents such as anti-TNF biologics have validated efficacy in these diseases. A combination of selective depletion of activated myeloid lineage leucocytes by GMA and newly developed biologics is expected to produce better efficacy and reduce the side effects seen with the use of biologics. Clinical trials in various settings are underway at present and warranted in the future.

References

Arnaout MA, Gupta SK, Pierce MW, Tenen DG. (1988) Amino acid sequence of the alpha subunit of human leukocyte adhesion receptor Mo1 (complement receptor type 3). *J Cell Biol* **106:** 2153–2158.

Barham KL, Jorizzo JL, Grattan B, Cox NH. (2004) Vasculitis and neutrophilic vascular reactions. In: Burns T, Breathnach S, Cox N, Griffith C (eds.), *Rook's Textbook of Dermatology*, Blackwell Science, Malden, MA, pp. 49.1–49.46.

Beller DI, Springer TA, Schreiber RD. (1982) Anti-Mac-1 selectively inhibits the mouse and human type three complement receptor. *J Exp Med* **156:** 1000–1009.

Botti E, Spallone G, Caruso R, *et al.* (2012) Psoriasis, from pathogenesis to therapeutic strategies: IL-21 as a novel potential therapeutic target. *Curr Pharm Biotechnol* **13:** 1861–1867.

Cai TQ, Wright SD. (1996) Human leukocyte elastase is an endogenous ligand for the integrin CR3 (CD11b/CD18, Mac-1, alpha M beta 2) and modulates polymorphonuclear leukocyte adhesion. *J Exp Med* **184:** 1213–1223.

Callen JP. (1998) Pyoderma gangrenosum. *Lancet* **351:** 581–585.

Cheung AK, Hohnholt M, Gilson J. (1991) Adherence of neutrophils to hemodialysis membranes: Role of complement receptors. *Kidney Int* **40:** 1123–1133.

Corbi AL, Kishimoto TK, Miller LJ, Springer TA. (1988) The human leukocyte adhesion glycoprotein Mac-1 (complement receptor type 3, CD11b) alpha subunit. Cloning, primary structure, and relation to the integrins, von Willebrand factor and factor B. *J Biol Chem* **263:** 12403–12411.

Diamond MS, Springer TA. (1993) A subpopulation of Mac-1 (CD11b/CD18) molecules mediates neutrophil adhesion to ICAM-1 and fibrinogen. *J Cell Biol* **120:** 545–556.

Finlay AY, Khan GK. (1994) Dermatology Life Quality Index (DLQI) — A simple practical measure for routine clinical use. *Clin Expt Dermatol* **19:** 210–216.

Fujisawa T, Murase K, Okumura Y, *et al.* (2011) Generalized pustular psoriasis successfully treated with granulocyte and monocyte adsorption apheresis. *Ther Apher Dial* **15:** 374–378.

Griffiths CEM, Camp RDR, Barker J. (2004) Psoriasis. In: Burns T, Breathnach S, Cox N, Griffith C (eds.), *Rook's Textbook of Dermatology.* Blackwell Science, Malden, MA, pp. 33.51–35.662.

Hanai H, Takeda Y, Gruber R, *et al.* (2011) The mode of actions of the Adacolumn therapeutic leucocytapheresis in patients with inflammatory bowel disease: A concise review. *Clin Exp Immunol* **163:** 50–58.

Hiraishi K, Takeda Y, Shiobara N, *et al.* (2003) Studies on the mechanisms of leukocyte adhesion to cellulose acetate beads: An *in vitro* model to assess the efficacy of cellulose acetate carrier-based granulocyte and monocyte adsorption apheresis. *Ther Apher Dial* **7:** 334–340.

Ikeda S, Takahashi H, Suga Y, *et al.* (2013) Therapeutic depletion of myeloid lineage leukocytes in patients with generalized pustular psoriasis indicates a major role for neutrophils in the immunopathogenesis of psoriasis. *J Am Acad Dermatol* **68:** 609–617.

Johnson-Huang LM, Pensabene CA, Shah KR, *et al.* (2012) Post-therapeutic relapse of psoriasis after CD11a blockade is associated with T cells and inflammatory myeloid DCs. *PLoS One* **7:** e30308.

Kanekura T, Gushi A, Iwata M, *et al.* (2004) Treatment of Behcet's disease with granulocyte and monocyte adsorption apheresis. *J Am Acad Dermatol* **51:** S83–87.

Kanekura T, Hashiguchi T, Mera Y, *et al.* (2004a) Improvement of SLE skin rash with granulocyte and monocyte adsorption apheresis. *Dermatology* **208:** 79–80.

Kanekura T, Hiraishi K, Kawahara K, *et al.* (2006) Granulocyte and monocyte adsorption apheresis (GCAP) for refractory skin diseases caused by activated neutrophils and psoriatic arthritis: Evidence that GCAP removes Mac-1-expressing neutrophils. *Ther Apher Dial* **10:** 247–256.

Kanekura T, Kawahara K, Maruyama I, Kanzaki T. (2005) Treatment of pyoderma gangrenosum with granulocyte and monocyte adsorption apheresis. *Ther Apher Dial* **9:** 292–296.

Kanekura T, Maruyama I, Kanzaki T. (2002) Granulocyte and monocyte adsorption apheresis for pyoderma gangrenosum. *J Am Acad Dermatol* **47:** 320–321.

Kanekura T, Mochitomi Y, Fujimoto S, *et al.* (2005) Granulocyte and monocyte adsorption apheresis for leg ulcers in a patient with rheumatoid arthritis. *J Am Acad Dermatol* **52:** 1108–1109.

Kanekura T, Terasaki K, Higashi Y, *et al.* (2004) Improvement of adult Still's disease with granulocyte and monocyte adsorption apheresis. *Clin Exp Dermatol* **29:** 410–2.

Kanekura T, Yoshii N, Kawahara K, *et al.* (2006) Granulocyte and monocyte adsorption apheresis for cutaneous allergic vasculitis. *Ther Apher Dial* **10:** 287–290.

Kanekura T, Yoshii N, Yonezawa T, *et al.* (2003) Treatment of pustular psoriasis with granulocyte and monocyte adsorption apheresis. *J Am Acad Dermatol* **49:** 329–332.

Kanekura T, Yoshii N, Yonezawa T, *et al.* (2003) Treatment of pustular psoriasis with granulocyte and monocyte adsorption apheresis. *J Am Acad Dermatol* **49:** 329–32.

Kanekura T. (2005) Granulocyte and monocyte adsorption apheresis for refractory skin diseases. *Jpn J Apher* **24:** 179–189.

Michalak-Stoma A, Pietrzak A, Szepietowski JC, *et al.* (2011) Cytokine network in psoriasis revisited. *Eur Cytokine Netw* **22:** 160–168.

Sakanoue M, Takeda K, Kawai K, Kanekura T. (2013) Granulocyte and monocyte adsorption apheresis for refractory skin diseases due to activated neutrophils, psoriasis, and associated arthropathy. *Ther Apher Dial* **17:** 287–290.

Saniabadi AR, Hanai H, Fukunaga K. (2007) Therapeutic leucocytapheresis for inflammatory bowel disease. *Transfus Apher Sci* **37:** 191–200.

Saniabadi AR, Hanai H, Takeuchi K, *et al.* (2003) Adacolumn, an adsorptive carrier based granulocyte and monocyte apheresis device for the treatment of inflammatory and refractory diseases associated with leukocytes. *Ther Apher Dial* **7:** 48–59.

Seishima M, Mizutani Y, Shibuya Y, *et al.* (1008) Efficacy of granulocyte and monocyte adsorption apheresis for pustular psoriasis. *Ther Apher Dial* **12:** 13–18.

Shimoyama T, Sawada K, Hiwatashi N, Saniabadi AR. (2001) Safety and efficacy of granulocytes and monocyte adsorption apheresis in patients with active ulcerative colitis: A multicenter study. *J Clin Apher* **16:** 1–9.

Shukuya R, Hasegawa T, Niwa Y, *et al.* (2011) Granulocyte and monocyte adsorption apheresis for generalized pustular psoriasis. *J Dermatol* **38:** 1130–1134.

Steinhoff M, Griffiths MK, Church MK, Luger TA. (2004) Inflammation. In: Burns T, Breathnach S, Cox N, Griffith C (eds.), *Rook's Textbook of Dermatology*. Blackwell Science, Malden, MA, pp. 9.1–9.67.

Viguier M, Matthieu A, Zagdanski A, *et al.* (2004) High frequency of cholestasis in generalized pustular psoriasis: Evidence for neutrophilic involvement of the biliary tract. *Hepatology* **40:** 452–458.

Weinstein GD, Koo JYM, Krueger GG, *et al.*, Tazaroten Cream Clinical Study Group. (2003) Tazarotene cream in the treatment of psoriasis: Two multicenter, double-blind, randomized, vehicle-controlled studies of the safety and efficacy of tazarotene creams 0.05% and 0.1% applied once daily for 12 weeks. *J Am Acad Dermatol* **48:** 760–767.

Yoshifuku A, Oyama K, Ibusuki A, *et al.* (2012) Granulocyte and monocyte adsorption apheresis as an effective treatment for Reiter disease. *Clin Exp Dermatol* **37:** 241–244.

Chapter 34

New White Blood Cell Adsorbent: Immunopure

Grit Waitz*,‡, Heinrich Prophet†, Wolfgang Ramlow†

Abstract

Immunopure is a novel granulocyte/monocyte adsorption apheresis device (GMCAP) based on polyarylate resin (PAR) beads with a diameter of 1 mm for the treatment of patients with ulcerative colitis (UC). The device has been CE certified since 2011. In addition to other adsorbers for the treatment of UC, Immunopure offers a stronger platelet adsorption capacity while inducing only a low complement activation, making it attractive for UC patients with high platelet numbers and coagulation disorders. *In vitro*, after 60 min of circulation the module selectively reduced platelets by 76%, neutrophils by 42% and monocytes by 63%, while lymphocytes were minorly affected (only 3% reduction rate). Furthermore, the module showed a strong adsorption of inflammatory $CD14^{+}CD16^{+}$ monocytes and platelet leukocyte aggregates (PLAs). In animal studies with healthy premature pigs, the Immunopure device has been shown to be safe

*,‡Corresponding author: Grit Waitz, Apheresis Center Rostock (ACR), NephroCare Rostock GmbH, Nobelstr. 53, 18059 Rostock, Phone: +49 381 4050 287, Fax: +49 381 4050 221, Email: g.waitz@praxisverbund-rostock.de.

†Apheresis Center Rostock (ACR), NephroCare Rostock GmbH, Nobelstr. 53, 18059 Rostock, Phone: +49 381 4050 219, Fax: +49 381 4050 234, Email: ramlow@apherese.de, prophet@apherese.de.

and tolerable. Two uncontrolled studies in active UC patients demonstrated that the Immunopure device was safe, well tolerated and clinically efficient with clinical remission rates of 80% and 67%, respectively. Laboratory data indicated a strong adsorption of platelets, PLAs, monocytes including $CD14^{+}CD16^{+}$ monocytes and granulocytes in these studies. However, like other adsorbers for the treatment of UC, cytapheresis using the Immunopure adsorber did not show any long-term influence on cell populations and seems to act by regulating cell populations and subsequently immune reactions. Additional studies are necessary to further elucidate the mode of action and the clinical benefit of Immunopure cytapheresis in controlled outcome studies.

1. Introduction

1.1. *Platelets as Therapeutic Targets in Inflammatory Bowel Disease*

Chronic inflammatory conditions such as inflammatory bowel disease (IBD) are associated with both inflammation and coagulation. Various cell populations are elevated in IBD, including circulating and/or tissue levels of macrophages, dendritic cells, neutrophils, platelets and plasma cells (Harries *et al.*, 1983; Sartor, 2006; Roberts-Thomson *et al.*, 2011), releasing a variety of inflammatory cytokines (Fig. 34.1). IBD is currently incurable. Standard medication like 5-aminosalicylic acid (5-ASA), corticosteroids, azathioprine and monoclonal antibodies to tumor necrosis factor alpha (TNFα) aims to suppress the symptoms of IBD by inducing anti-inflammatory effects (Burger and Travis, 2011). In spite of a good proportion of patients that improve or remit as a result of established therapies, a considerable number of cases are refractory to steroids and/or immunosuppressive drugs or steroid-dependent and develop complications due to drug toxicity as well as increased risk of colorectal cancer.

With the rationale to remove inflammatory mediators like granulocytes, monocytes and lymphocytes from the blood of patients with

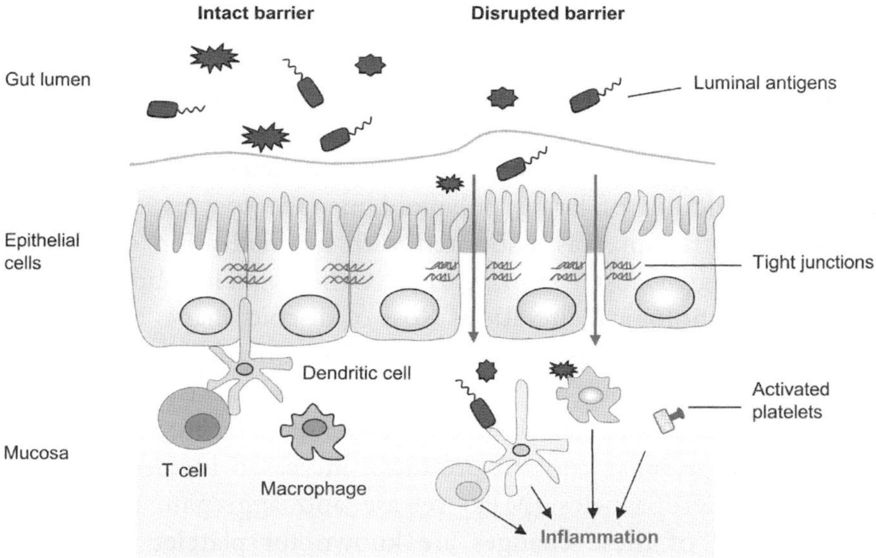

Fig. 34.1. Luminal antigens in genetically susceptible hosts are able to disrupt the epithelial barrier, triggering the activation of innate and adaptive immune responses. Multiple cell types are elevated in patients with IBD including macrophages, dendritic cells, neutrophils, platelets and plasma cells. Modified according to Xavier and Podolsky (2007).

IBD, adsorptive cytapheresis has been clinically applied for more than a decade (Shimoyama, 1999, 2001; Shibata *et al.*, 2003; Hanai *et al.*, 2008; Sigurbjörnsson and Bjarnason, 2008). However, aside from non-immune cells like endothelial, mesenchymal, nerve cells as well as platelets have become a focus of attention as key players in the IBD inflammatory cascade (Danese, 2008). Besides their traditional role in maintaining hemostasis, an increasing number of studies has focused on platelets as amplifiers of inflammation. Microvascular thrombosis and thromboembolism is frequent in patients with IBD (Irving *et al.*, 2005; Danese *et al.*, 2007; Yoshida and Granger, 2009). There is an increased incidence of thrombosis accompanied with abnormalities in platelets and platelet function (Voudoukis *et al.*, 2014). The platelet number is increased during active disease and correlates with disease severity. Therefore, thrombocytosis has been proposed as a biomarker of disease activity in IBD (Harries *et al.*, 1983; Nielsen *et al.*, 2000). However, platelet count increase is not due only to inflammation, but iron deficiency may also lead to the production of larger polyploid megakaryocytes, while iron replacement therapy was associated with normalization of platelet count (Kulnigg-Dabsch *et al.*, 2012). Other platelet abnormalities include morphological alterations (loss of discoid shape, acquisition of pseudopodia, increase in size, granular content augmentation, mean platelet volume (MPV) decrease, platelet distribution width (PDW) increase, plateletcrit (PCT) increase, overproduction and excretion of granular content products, increased incorporation of receptors, microparticles (MPs) release and aggregate formation. However, all of these changes are known for platelet activation, which is high in IBD patients (Voudoukis *et al.*, 2014).

The primary role of platelets, small nucleus-free fragments with 1–6 μm diameter and a lifespan of 5–9 d derived from bone marrow megakaryocytes, is to stop bleeding and to keep hemostasis via binding of von Willebrand factor to collagen from subendothelial layers at the site of injury. After adhesion, platelets become activated and clot the injured tissue. However, there is evidence that inflammation and coagulation are linked in a way that enables both to activate and propagate each other. Today, coagulation is regarded as an integral part of the innate immune response and inflammation as a potent prothrombotic stimulus that activates the coagulation cascade (Esmon *et al.*, 2011). Platelets are able to express toll-like receptors, internalize pathogens, and stimulate the formation of extracellular DNA nets by neutrophils (Youssefian *et al.*, 2002; Clark *et al.*, 2007; Cognasse *et al.*, 2007; Ma and Kubes, 2007). However, they may also act as mediators between innate and adaptive immune systems. In IBD, granulocytes and platelets are the first cells mobilized to the sites of inflammation. There they interact with lymphocytes to orchestrate the inflammatory response. Platelets can produce cytokines like TNFα (Limb *et al.*, 1999) and activate various cell types via contact with CD40L and secretion of soluble CD40L (Danese, 2007).

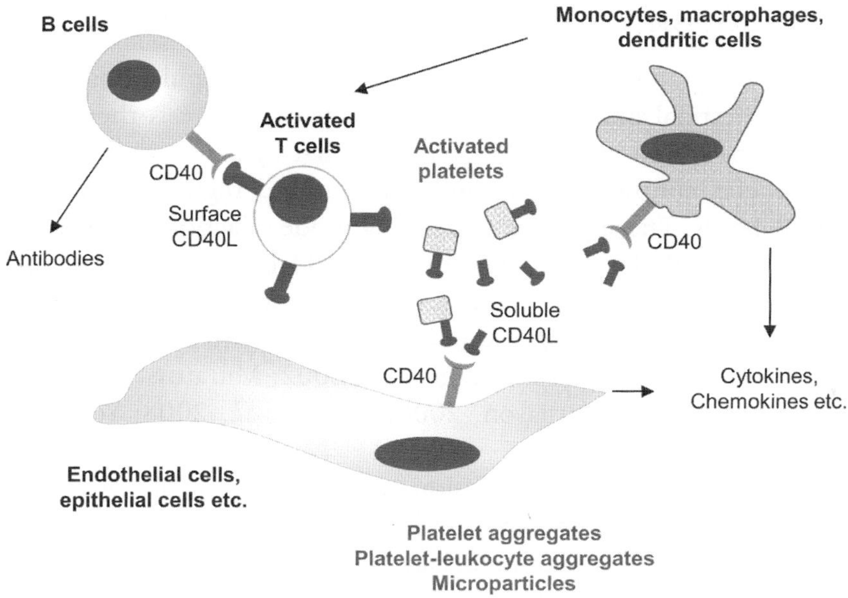

Fig. 34.2. Platelets and CD40/CD40L system. Pleiotropic effects result from the interaction of activated platelets expressing surface CD40L or soluble CD40L with immune and non-immune cell including immunoglobulin, cytokine and chemokine production. Modified according to Danese *et al.* (2004).

Engagement of CD40 on B cells leads to antibody production. Monocytes, macrophages, dendritic cells and non-immune cells activated through the CD40 pathway may produce a variety of cytokines, chemokines, reactive oxygen species, adhesion molecules etc. Furthermore, platelets do also adhere to other platelets, form complexes with leukocytes (PLA) and secrete microparticles (MPs), thereby aggravating the inflammatory process and the risk for thromboembolic events (Fig. 34.2).

Increased PLA and MP formation is observed in many chronic inflammatory disorders and may have an important role in IBD (Michelson *et al.*, 2001; Andoh *et al.*, 2005). Taking into account the high procoagulant and pro-inflammatory characteristics of platelets, PLAs and platelet-derived MPs, these particles could be ambitious targets for IBD therapies.

1.2. *Adsorptive Cytapheresis — Current Situation*

Adsorptive cytapheresis selectively removes cell populations taking part in intestinal inflammation. The effects of apheresis therapy are probably triggered by a dual mechanism, first by direct adsorption of cells to the column and second via immune regulation of non-adsorbed cells. The supposed underlying mode of action of adsorptive cytapheresis includes a reduction of activated leukocytes, downregulation of pro-inflammatory cytokines, modification of adhesion molecule expression, effects on the innate immune system and induction of regulatory processes (Kashiwagi *et al.*, 2002; Muratov *et al.*, 2006; Sigurbjörnsson and Bjarnason, 2008; Waitz *et al.*, 2008; Yokoyama *et al.*, 2007).

Currently, two blood perfusion systems are generally applied in Japan (as a standard therapy) and Europe (approved but without reimbursement). The commercially available cytapheresis systems preferentially adsorb granulocytes (46%), monocytes/macrophages (54%) and only a small number of lymphocytes (18.5%) and platelets (19%) (GMCAP) (Ramlow *et al.*, 2005) or lymphocytes (30–60%), platelets (35%) in addition to granulocytes and monocytes (almost 100% of cells in the blood line of the device) (leukocytapheresis, LCAP) (Shibata *et al.*, 2003). Both methods have been reported to be effective, however, success varied (Shimoyama, 1999, 2001; Shibata *et al.*, 2003; Hanai *et al.*, 2008; Sigurbjörnsson and Bjarnason, 2008). Remission rates deviated from nearly 100% (Suzuki *et al.*, 2006) to 17% being not superior compared to sham apheresis (Sands *et al.*, 2008). These results reflect the fact that it has to be considered that not every patient is suitable for apheresis, and that critical patient selection especially in large clinical trials is crucial. Best responders for adsorptive cytapheresis are patients early after initial diagnosis (Suzuki *et al.*, 2006), steroid-naïve patients (Hanai *et al.*, 2003; Suzuki *et al.*, 2004; Tanaka *et al.*, 2008, 2010), patients with markedly increased disease activity (Matsumoto *et al.*, 2008) shortly after clinical relapse (Yokoyama *et al.*, 2013). In contrast, patients with low disease activity, deep colonic lesions, a long medical history and refractory to existing drugs may possibly not benefit from apheresis (Saniabadi *et al.*, 2014). However, the big opportunity of cytapheresis treatment may be a slowdown of disease progression if the therapy is applied at an early stage of the disease, possibly preventing exacerbations and colectomy.

Predictive factors for the efficacy of adsorptive cytapheresis have rarely been reported. However, several reports have indicated that a decrease in platelets may be an early marker for a beneficial response in patients with severe UC (Fukunaga *et al.*, 2006; Takemoto *et al.*, 2009). This is in accordance with the observation that platelet parameters seem to display good predictive value regarding disease activity (Voudoukis *et al.*, 2014). Therefore, attempts have been made to develop a novel cytapheresis device with a stronger capacity to adsorb platelets.

1.3. *Immunopure Adsorber*

The basic idea was to develop an efficient adsorber using a material that is well proven for a long time in patients on maintenance hemodialysis with regard to both biocompatibility and suggested adsorptive properties. Synthetic polymer membranes are widely used as dialysis membranes. Polyester-polymer alloy (PEPA) membrane was developed and met the performance requirements for a hollow fiber in 1990 by Nikkiso Co., Japan. It is a synthetic polymer high-flux membrane with a good biocompatibility and has been shown to offer features like endotoxin and free light chain retentive capability (Igoshi *et al.*, 2011). PEPA consists of two polymer components: polyethersulfone (PES) and polyarylate (PAR). PAR is an aromatic polyester polymer (Fig. 34.3) which is hydrophobic with superior heat resistance, recovery performance and mechanical properties.

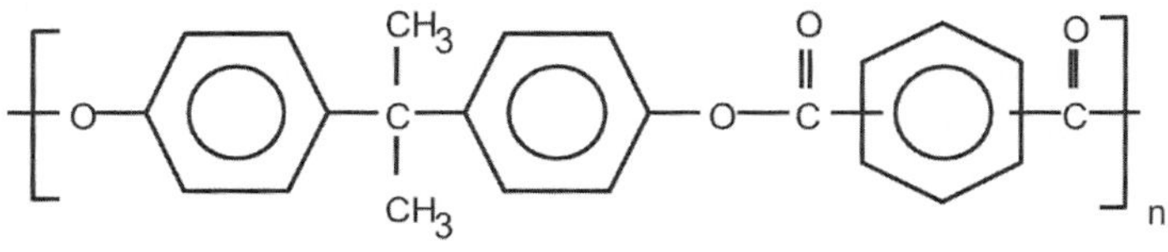

Fig. 34.3. Molecular structure of PAR (modified according to Igoshi *et al.*, 2011).

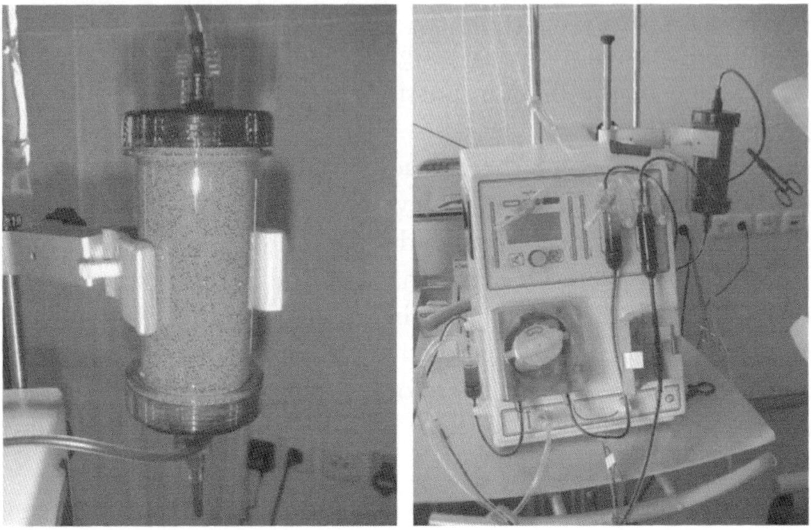

Fig. 34.4. Immunopure device (Nikkiso Co., Japan) (left) and blood monitor (LPM-01, Nikkiso Co., Japan) (right) during apheresis treatment.

Immunopure is a novel medical device based on PAR beads with a diameter of 1 mm to be used in a simple hemoperfusion setting provided by Nikkiso Co., Japan. The apheresis system comprises the Immunopure device, blood lines (ABT-002PX, Nikkiso Co., Japan) and the blood monitor (LPM-01, Nikkiso Co., Japan). The adsorption device has a length of 185 mm, a diameter of 59 mm, a total weight of 537 g, a total volume of 350 mL, and a priming volume of 139 mL (Fig. 34.4). The Immunopure device has been CE certified since 2011. In clinical application it is recommended to use standard heparin anticoagulation and the standard treatment protocol with a treatment time of 60 min once a week over five weeks. *In vitro* experiments as well as *in vivo* investigations in pigs (see Section 2) and humans (see Section 3) revealed that the module adsorbs granulocytes, monocytes/macrophages and platelets selectively. The results of the pilot study demonstrated a good safety profile and promising clinical remission rates (80% at week 10) (Ramlow *et al.*, 2013).

2. Basic Aspects

2.1. *In Vitro Circulation Experiments — Setup*

To analyze the adsorption behavior of the Immunopure adsorber beads, *in vitro* circulation experiments using mini-modules in downscaled circulation experiments and blood of healthy volunteers were performed. Investigations were accomplished with a special focus on the influence of the adsorber on platelets, PLAs and platelet-derived MPs. An adsorber mini-module consisting of PAR beads (diameter: 1 mm) placed in a syringe with a volume of 40 mL was designed (Fig. 34.5).

The *in vitro* experiments were performed utilizing a downscaled experimental setting with mini-module dimensions calculated from commercially available cytapheresis modules, which have a volume of about 300 mL, a cross section area of $A = 22.9\ cm^2$, and are operated with a blood flow of $Q = 30$ mL/min. The contact time of blood to the adsorber beads depends on the blood speed v, which yields from the blood flow Q and the cross section area A ($v = Q/A = 1.3$ cm/min). The cross-section area of the 40 mL mini-module is

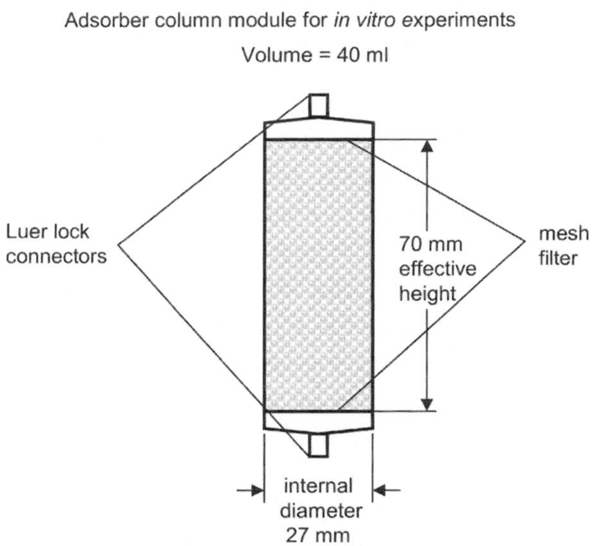

Fig. 34.5. Dimensions of the adsorber minimodule.

5.7 cm^2, therefore a blood flow of Q = 7.4 mL/min was necessary to achieve the same blood speed as in a clinical setting. As tubing system regular sets for infusion pumps were used (pump tube diameter about 1.5 mm).

Adsorber mini-modules filled with PAR beads and empty sham modules of same void volume were tested in parallel. After pre-rinsing with physiological saline, donor blood (500 mL divided into two parts of 250 mL each, 1.0 IU/mL Heparin, ratiopharm, Ulm, Germany) passed from the blood bag through the primed mini-modules and displaced the physiological saline. The circulation to the blood bag was closed when almost all of the physiological saline within the tubes was displaced by blood. Experiments were performed at a steady flow rate (7.4 mL/min) for a test period of 60 min. Collection of the baseline (0 min) blood samples was started after the column had been completely filled with blood. Collection of the influx and efflux samples was performed at the column entry and exit 15 min and 60 min after starting recirculation. Temperature was maintained at 37°C throughout the experiment by incubating the blood bag in a thermostatic water bath under gentle rotation (Fig. 34.6).

2.2. *In Vitro Circulation Experiments — Results*

The results of blood count analyses showed a significant reduction of platelets to 23.7%, leukocytes to 70.0%, neutrophils to 58.3%, monocytes to 36.8% and lymphocytes to 96.9% after 60 min of blood circulation through the mini-module containing beads from the Immunopure device compared to the 0-min-values (100%) (Fig. 34.7).

Clearly, platelets showed the strongest reduction, followed by monocytes and neutrophils with a maximum adsorption rate at 15 min. In contrast, lymphocytes showed minimal changes. The strong platelet adsorption capacity also became apparent in electron microscope photographs (Fig. 34.8).

The hemoglobin and hematocrit values remained constant throughout the observation period during all circulation experiments showing that there were no dilution or sedimentation effects. However, data were hematocrit corrected. There was a significant difference

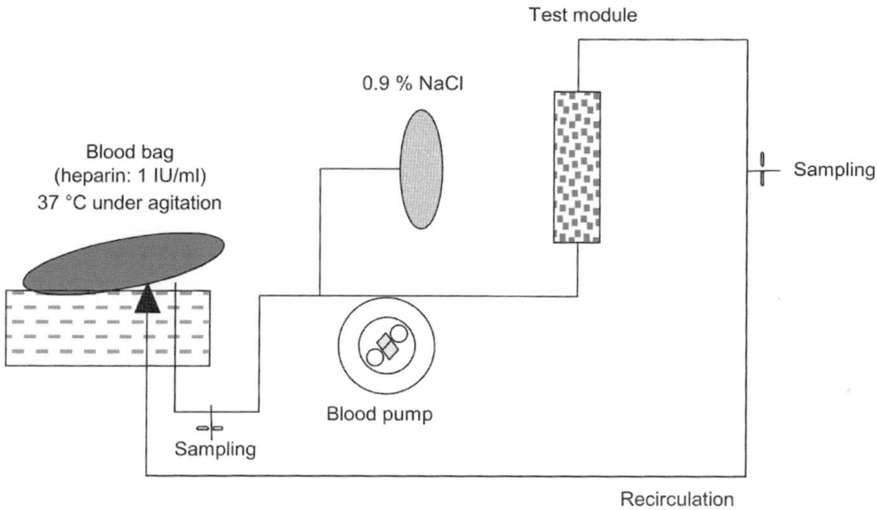

Fig. 34.6. A miniaturized module consisting of polyarylate resin beads (1 mm diameter) from Immunopure with a volume of 40 ml was designed. 250 ml of heparinized (1.0 IE/mL) blood from a healthy volunteer passed through the primed mini-module (7.5 mL/min, 37°C) for a test period of 60 min.

between filled modules and empty test modules with regard to platelet, leukocyte, monocyte, neutrophil and eosinophil adsorption. In this circulation experiments, adsorption decreased after 15 min of recirculation suggesting a saturation phenomenon of the mini-modules from that point on.

Furthermore, analyses with regard to adsorption of different cell populations were performed using flow cytometry. In detail, $CD3^+CD4^+$ T helper cells as well as $CD3^+CD8^+$ cytotoxic T cells did not show any adsorption to the Immunopure beads (not shown). On the contrary, there was a strong decrease of $CD10^+$ granulocytes to 63%, leukocytes expressing the cell adhesion proteins CD11b and L-selectin to 57%, $CD14^+$ monocytes/macrophages to 39% and inflammatory $CD14^+CD16^+$ monocytes to 41% after 15 min in the filled modules (Fig. 34.9). $CD14^+CD16^+$ monocytes exhibit a pro-inflammatory phenotype and have been shown to be reduced after GMCAP *in vivo* (Hanai *et al.*, 2008).

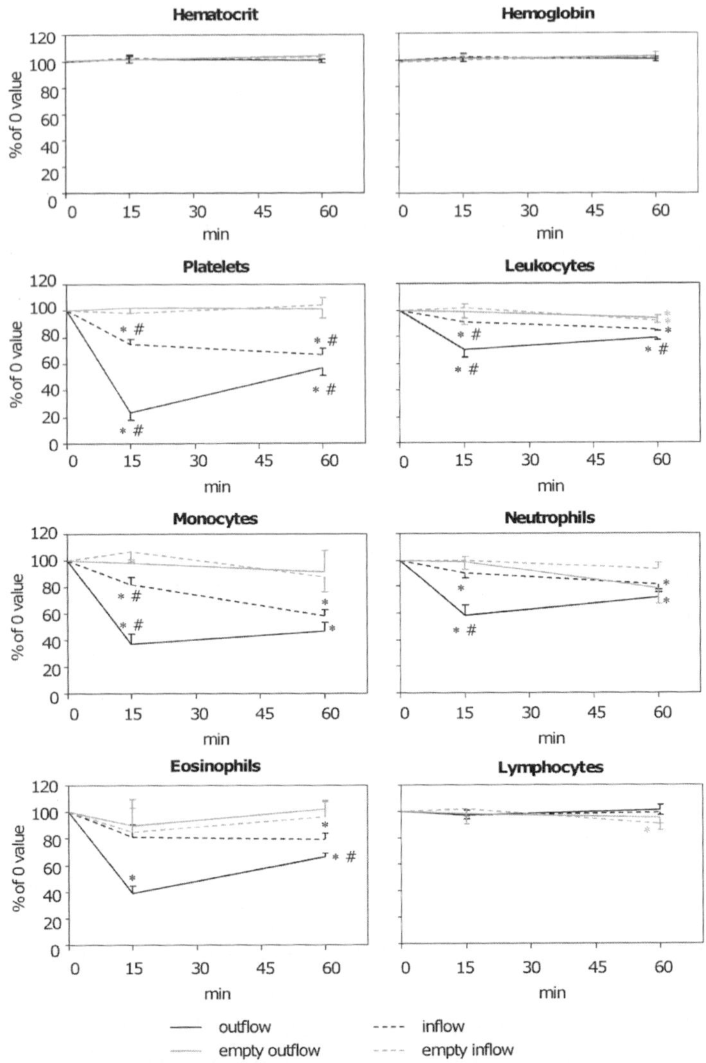

Fig. 34.7. Peripheral blood counts of platelets, leukocytes, monocytes, neutrophils, and eosinophils were significantly reduced by the Immunopure beads. Results (mean/SEM) were compared with baseline values at time point 0 min (100%) at the beginning of each experiment using the Wilcoxon test ($n = 6$) (*$P < 0.05$) and with values of the empty modules ($n = 6$) using the U test (#$P < 0.05$).

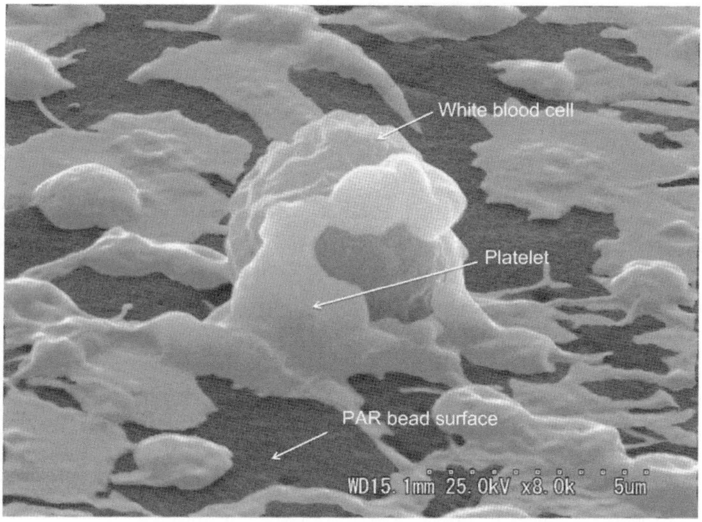

Fig. 34.8. Electron microscope photograph of the PAR bead surface with an attached white blood cell and numerous platelets (kindly provided by Nikkiso Co., Japan).

PLAs with $CD14^+$ monocytes, $CD11b^+$ cells, and $CD3^+$ T cells were strongly reduced by the Immunopure beads *in vitro,* too (Fig. 34.10). PLAs are formed during inflammation via binding of P-selectin on thrombin-stimulated platelets to its specific ligand on the leukocyte surface, P-selectin glycoprotein ligand-1. Then, PLAs are stabilized via binding of the leukocyte surface Mac-1 to platelet surface glycoprotein Ib (GPIb), integrin $\alpha IIb\beta 3$, and/or junctional adhesion molecule 3. Leukocyte tethering by platelet P-selectin also triggers the expression of transcription factors such as nuclear factor kappa B (NF-kB), which activates gene transcription of pro-inflammatory molecules leading to an inflammatory status of the leukocytes (Cerletti *et al.*, 2012).

PLAs are also formed in healthy individuals, but an increased number has been described in several inflammatory and thrombotic conditions, including IBD (Michelson *et al.*, 2001; Pamuk *et al.*, 2006). Our own assessments were in accordance with these observations. Patients with UC had significant higher platelet numbers, $CD42b^+CD14^+$ and $CD42b^+CD11b^+$ PLAs (Fig. 34.11).

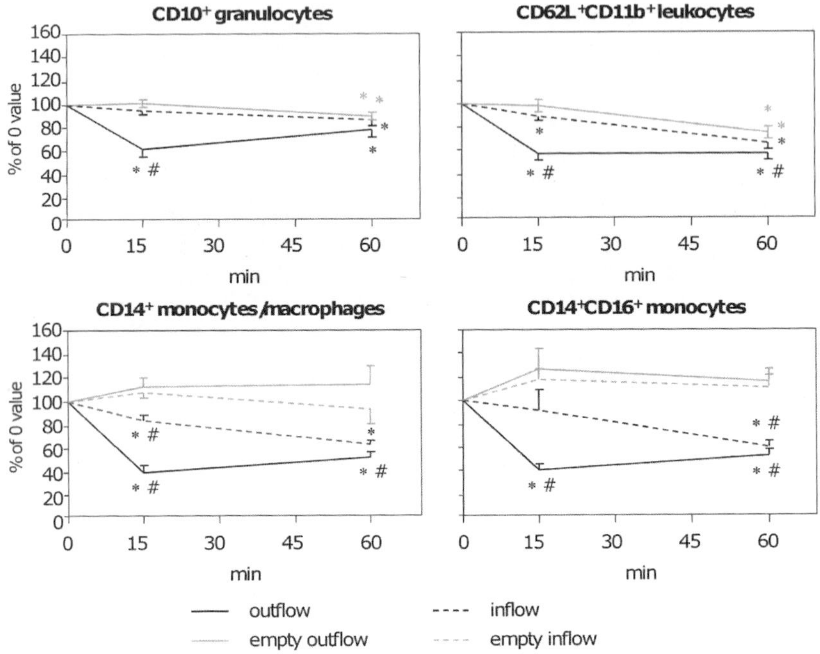

Fig. 34.9. Flow cytometry analyses of $CD10^+$ granulocytes, $CD62L^+CD11b^+$ leukocytes, $CD14^+$ monocytes and $CD14^+CD16^+$ monocytes after circulation of whole blood through Immunopure and sham mini-modules. Results (mean/SEM) were compared with baseline values at time point 0 min (100%) at the beginning of each experiment using the Wilcoxon test ($n = 6$) ($*P < 0.05$) and with values of the empty modules ($n = 6$) using the U test ($\#P < 0.05$).

Elevated PLA numbers in IBD may result from enhanced platelet and neutrophil activation and could contribute to the exacerbation of the inflammatory process. Increased numbers of PLAs were found in the mesenteric vasculature in patients with UC, but not in controls, which supports the hypothesis that the activated vascular endothelium stimulates PLA formation in UC. PLA formation seems to play a role in the pathogenesis of the mucosal lesion and contributes to the risk of systemic thrombosis in patients with UC (Irving *et al.*, 2008). Therefore, therapies that prevent or diminish PLA formation may be useful in patients with IBD. It has been shown that PLAs were reduced by antiplatelet agents like clopidogrel, prasugrel (reduction of P-selectin expression) and abciximab (blocking of glycoprotein

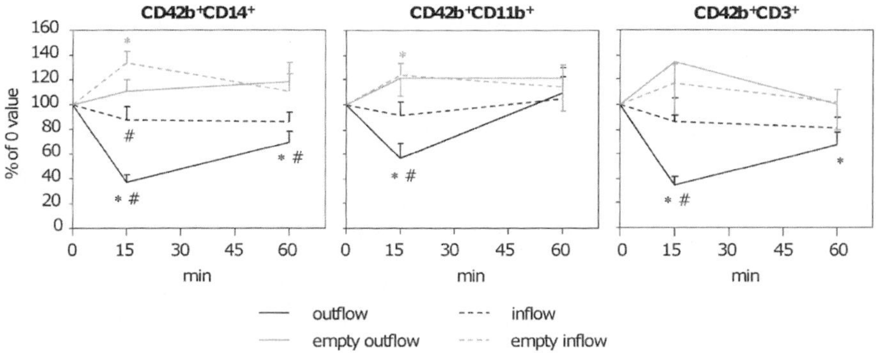

Fig. 34.10. Flow cytometry analyses of PLAs stained with the CD42b platelet marker combined with CD14, CD11b and CD3. Results (mean/SEM) were compared with baseline values at time point 0 min (100%) at the beginning of each experiment ($n = 6$) (*$P < 0.05$, Wilcoxon test) and with values of the empty modules ($n = 6$) (#$P < 0.05$, Mann–Whitney U test).

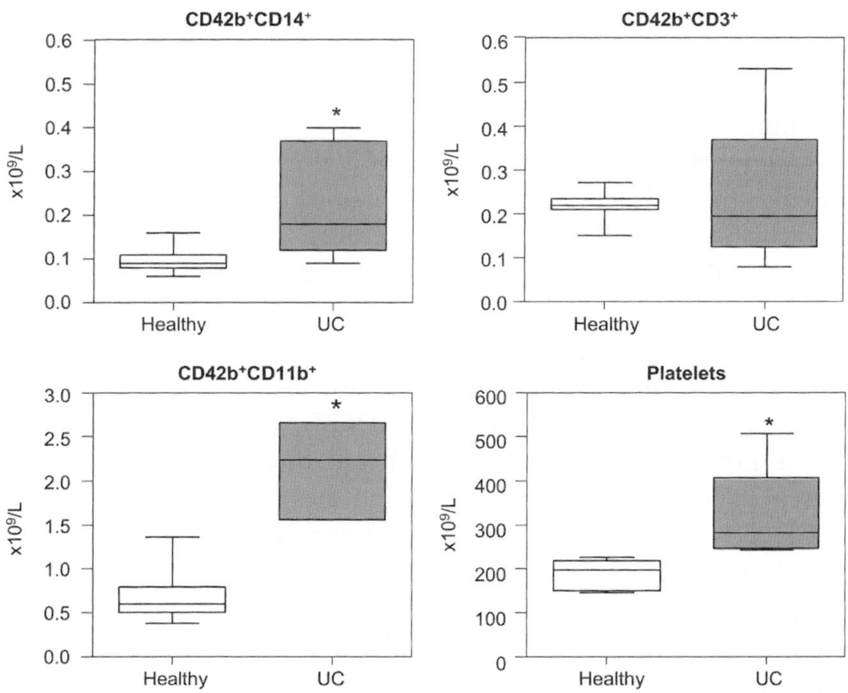

Fig. 34.11. Flow cytometry analyses of PLAs stained with the CD42b platelet marker combined with CD14, CD11b and CD3 and platelets (blood count): Median, healthy controls ($n = 8$) vs. patients with UC ($n = 6$) *$P < 0.05$ Mann–Whitney U test.

IIb/IIIa receptor) *in vivo*, and by epoprostenol (inhibition of intracellular signaling pathways) and low molecular weight heparins *in vitro* (Cerletti *et al.*, 2012). However, effective adsorption of PLAs by apheresis may be an interesting approach during active flare ups.

Platelet-derived MPs are very small vesicles (< 1 mm) generated from the plasma membrane upon platelet activation. To date, there are few studies about platelet-derived MPs in IBD, however, along with activated platelets they are thought to be involved in the pathogenesis of venous thromboembolism in IBD patients (Andoh *et al.*, 2005). For the analysis of platelet-derived MPs after *ex vivo* circulation of blood from healthy volunteers through the Immunopure and sham mini-modules, platelet-free plasma was obtained. Analyses were performed in a modified form according to Robert *et al.* (2009). Platelet-derived microparticles were defined according to their size as particles smaller than 0.9 μm, Annexin V^+/$CD42b^+$ and counted using Truecount tubes (BD, Heidelberg, Germany). Flow cytometry analyses revealed no influence of the Immunopure beads on platelet-derived MPs. Likewise, there were no significant differences utilizing filled mini-modules and empty sham modules.

Additionally, there were a number of flow cytometry parameters, in particular those which are expressed or down-regulated upon activation, which increased/decreased in the filled as well as in the empty sham modules (Table 34.1). The expression of CD62L, which is reflected by the mean channel fluorescence of these cells in flow cytometry, is known to be reduced on leukocytes after contact with artificial surfaces due to their activation. In contrast, the expression of CD11b is enhanced after leukocyte activation. CD62L expression was decreased and CD11b expression was increased, but there were no significant differences between the test modules and the sham modules.

CD25, the alpha chain of the IL-2 receptor, is present on activated T cells. A slightly enhanced expression of this T cell activation marker was observed after circulation through the filled Immunopure as well as through the empty sham mini-modules without any differences between the two module types. The same was true for the platelet activation markers CD63 and CD62, both of them were increased by each module type. These results clearly demonstrate the

Table 34.1. **Flow Cytometry Analyses of CD62L (Mean Channel Fluorescence), CD11b (Mean Channel Fluorescence), $CD4^+CD25^+$ Cells, $CD42b^+CD62P^+$ and $CD42b^+CD62P^+$ Platelets after Circulation of Whole Blood through Immunopure and Sham Mini-modules. Results (Mean ± SD) were Compared with Baseline Values at Time Point 0 min (100%) at the Beginning of Each Experiment (* $P < 0.05$)**

		Inflow	Outflow	Inflow	Outflow
Parameter	Module	15 min	15 min	60 min	60 min
Mean CD62L	Filled	82.6 ± 12.9*	77.6 ± 9.5*	54.5 ± 12.3*	57.2 ± 16.3*
	Sham	93.2 ± 13.8	89.7 ± 17.7	62.9 ± 15.0*	65.3 ± 13.9*
Mean CD11b	Filled	132.9 ± 53.6	153.5 ± 47.1*	185.7 ± 67.5*	181.7 ± 64.3*
	Sham	142.4 ± 39.8*	131.5 ± 30.8*	176.1 ± 36.2*	159.0 ± 38.7*
$CD4^+CD25^+$	Filled	135.2 ± 75.9	137.0 ± 73.0	125.0 ± 47.1	127.6 ± 41.6
	Sham	159.3 ± 146.6	128.0 ± 76.8	126.1 ± 65.5	97.2 ± 30.7
$CD42b^+CD63^+$	Filled	207.4 ± 79.1*	225.4 ± 117.2*	310.5 ± 249.8	276.9 ± 198.6
	Sham	158.4 ± 78.8	187.2 ± 80.6*	313.5 ± 248.4*	231.0 ± 106.6*
$CD42b^+CD62P^+$	Filled	584.4 ± 519.5*	625.7 ± 461.4*	535.9 ± 429.9*	509.9 ± 342.3*
	Sham	304.8 ± 305.3	308.7 ± 228.8*	481.6 ± 479.7	377.1 ± 237.5*

"sham" effect, which becomes apparent when blood gets into contact with artificial surfaces like tubing systems, housing material, daylight, or is influenced by shear stress due to the blood pump. Therefore, sham apheresis cannot be directly compared with placebos in drug trials. *Per se*, sham apheresis should be considered as a kind of treatment having an ever stronger influence than placebos. This should be taken into account when performing sham-controlled apheresis studies.

Despite the similar increase of $CD63^+$ and $CD62P^+$ platelets using both module types, plasma soluble CD40L concentrations were two to threefold higher in circulation experiments using the PAR bead filled mini-modules indicating enhanced platelet activation by the adsorber material. However, soluble P-selectin was not increased in the plasma by the Immunopure mini-modules compared to the sham modules (Fig. 34.12).

After contact of platelets with negatively charged surfaces like glass or adsorber materials, the intrinsic coagulation cascade (contact activation pathway) is activated. The intrinsic pathway starts with the formation of the primary complex by high-molecular-weight kininogen (HMWK), prekallikrein, and factor XII (Hageman factor).

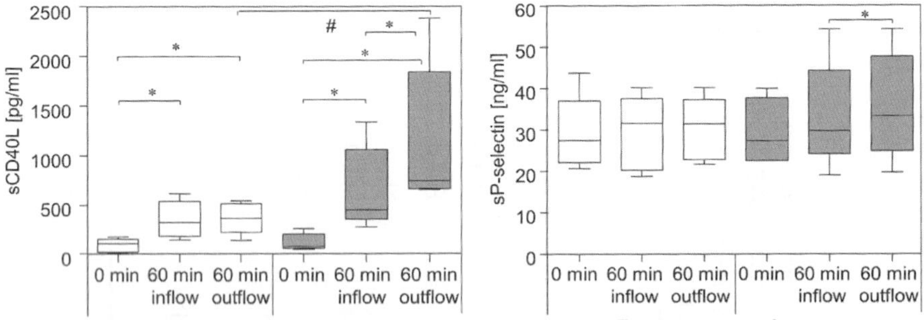

Fig. 34.12. Median plasma concentrations of soluble CD40L (left) and soluble P-selectin (right) as determined by enzyme-linked immunosorbent assay for each circulation experiment (empty modules $n = 6$, filled Immunopure modules $n = 6$). Results were compared with baseline values at time point 0 min (*$P < 0.05$, Wilcoxon test) and with values of the empty modules (#$P < 0.05$, Mann–Whitney U test).

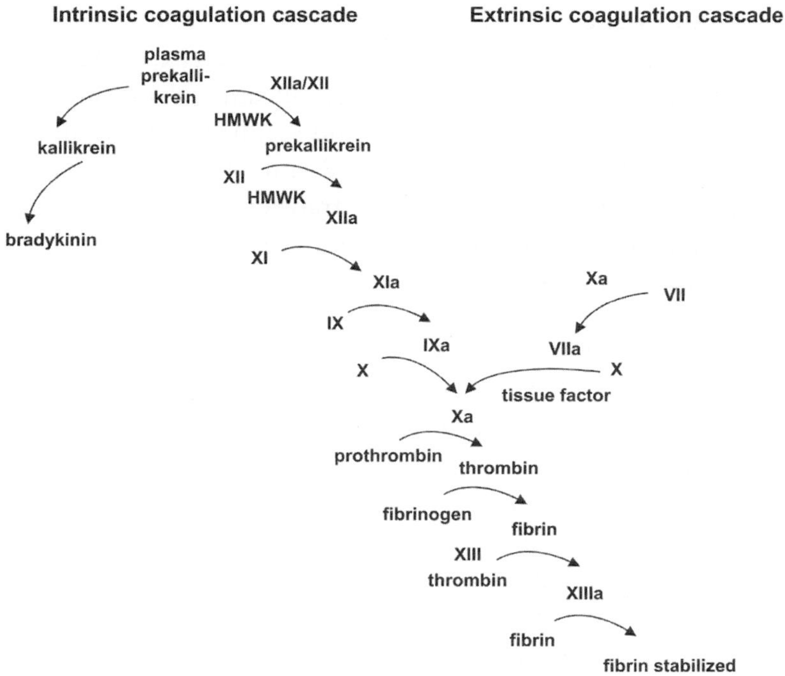

Fig. 34.13. Coagulation cascade modified according to Davie and Mac Farlane (High-molecular-weight kininogen (HMWK)) (Earl, 2003).

Prekallikrein is then converted to kallikrein. Activation of factor XII is followed by activation of factor XI and factor IX. The intrinsic and extrinsic pathways converge at the level of factor X activation. Factor Xa activates prothrombin to thrombin in the presence of the cofactor factor Va, and thrombin subsequently triggers conversion from fibrinogen to fibrin (Fig. 34.13).

Activated platelets express glycoprotein IIb/IIIa on their surface, which can bind fibrinogen. Subsequently, PLAs may be formed directly on the Immunopure adsorber surface, or preformed PLAs may be catched from the circulation. To prove whether leukocytes, platelets and fibrinogen are attached to the bead surface, several beads were removed from the Immunopure mini-modules after the circulation experiments and fixed in 3% paraformaldehyde.

Immunofluorescence was performed using the following primary antibodies: CD42b, fibrinogen gamma chain, CD14, CD11b, CD3 (BD Biosciences, San Diego, USA; abcam, Cambridge, GB; Immunotools, Friesoythe, Germany) and secondary antibodies: anti-mouse Alexa 546, anti-rabbit Alexa 488 (Life Technologies GmbH, Darmstadt, Germany). Beads were transferred on a coverglass-bottom dish and analyzed using a confocal laser microscope (Fluoview FV10i, Olympus America Inc., Center Valley, USA). Samples without primary antibody as well as beads without contact to blood served as controls (Fig. 34.14).

Indeed, confocal laser microscopy pictures showed an adsorption of fibrinogen, CD42b+ platelets, $CD11b^+$, $CD14^+$, and $CD3^+$ leukocytes to the bead surface. In summary, the suggested mechanisms of granulocyte, monocyte and platelet adhesion to the Immunopure device may be postulated as follows: Naïve platelets from the bloodstream get into contact with the artificial surface, are activated and release different mediators (factor XII, factor XI) which induce the conversion of fibrinogen to fibrin. Platelets adhered to fibrin form complexes with leukocytes. Moreover, the adsorber surface is enveloped with different proteins (complement, immune complexes) similar to processes as described for Adacolumn beads (Saniabadi, 2014) which may be recognized by Fc-gamma receptors of monocytes/macrophages and neutrophils (Fig. 34.15). However, this has to be further elucidated.

In addition to the described *in vitro* results, complement activation induced by the Immunopure mini-module was very low (plasma C5a mean ± standard deviation: from 0.55 ± 0.53 at 0 min to 2.40 ± 1.51 ng/mL at 30 min of circulation, $n = 5$), possibly due to the fast envelopment of the beads with fibrin. This suggests a very good biocompatibility which is a favorable characteristic for clinical application.

2.3. *Adsorptive-Type Cytapheresis in a Pig Model*

After *in vitro* analyses, the safety and technical feasibility of the Immunopure device was tested in large healthy pigs. The advantage

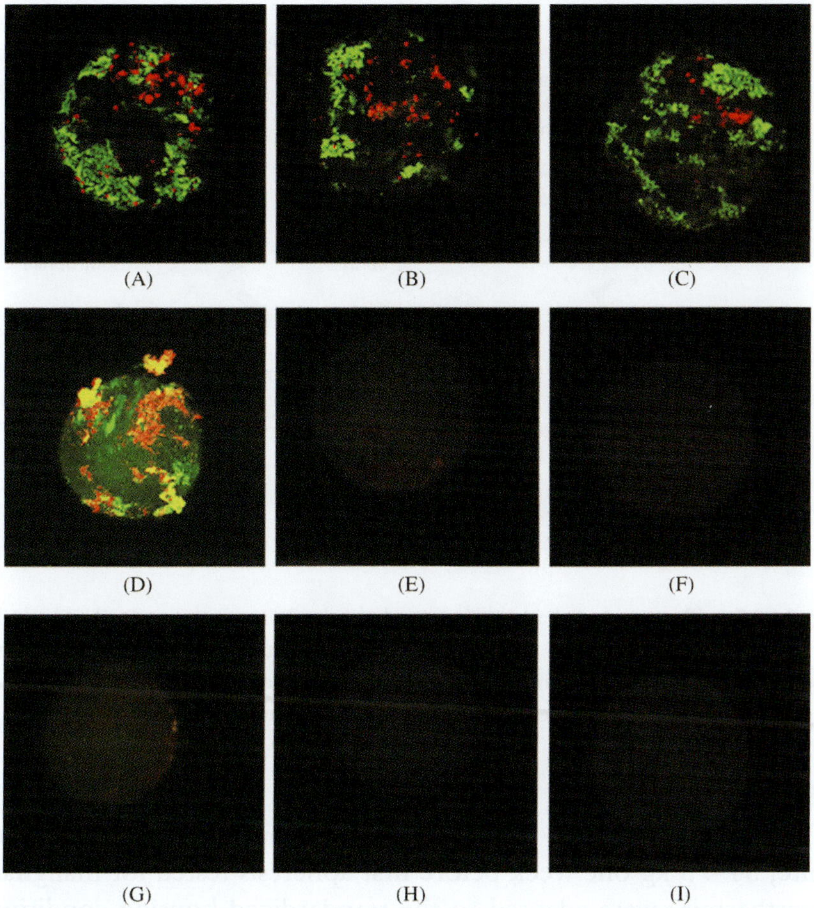

Fig. 34.14. Confocal laser microscopy pictures of Immunopure beads after circulation experiments. Staining: (A) CD42b (green)/CD11b (red); (B) CD42b (green)/CD14 (red); (C) CD42b (green)/CD3 (red); (D) fibrinogen (green)/CD42b (red); E-H: controls without blood; (E) CD42b (green)/CD11b (red); (F) CD42b (green)/CD14 (red); (G) CD42b (green)/CD3 (red); (H) fibrinogen (green)/CD42b (red), (I) control without primary antibody, after blood circulation.

of this pig model over other animal models is that it resembles humans by size and blood volume. This allowed the testing of the whole extracorporeal system setting in its original size as suggested for later human application.

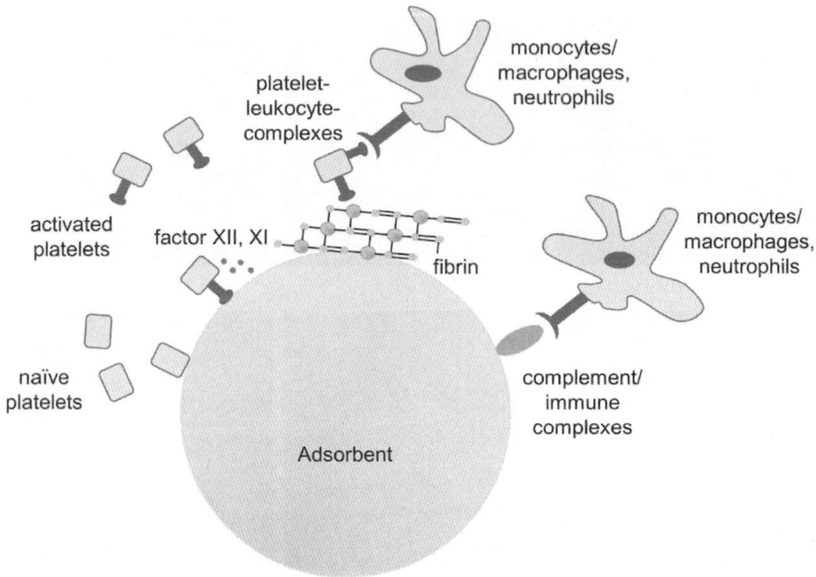

Fig. 34.15. The suggested adsorption mode of the Immunopure adsorber includes both fibrin formation on the bead surface with subsequent formation of PLAs and adsorption of proteins like complement/immune complexes which are recognized by Fc-gamma receptors on monocytes/macrophages and neutrophils.

A total of five large sized premature female pigs (German large white, 54–73 kg one week before first apheresis tested for malignant hyperthermia) were adapted to the standardized housing conditions and handling procedures of the institute (Core Unit for Biomedical Research, Medical University of Vienna) several weeks before the treatments. The animals had free access to standardized food and drinking water.

This study was carried out in strict accordance with the Austrian Federal Act on the protection of animals. The protocol was approved by the Committee on the Ethics of Animal Experiments of the Medical University Vienna and the Austrian Ministry of Science and Research (permit number: 66.009/191-II/10b/2008). All interventions including cytapheresis treatments were performed under anesthesia (induced by i.v. thiopental, followed by ventilation with

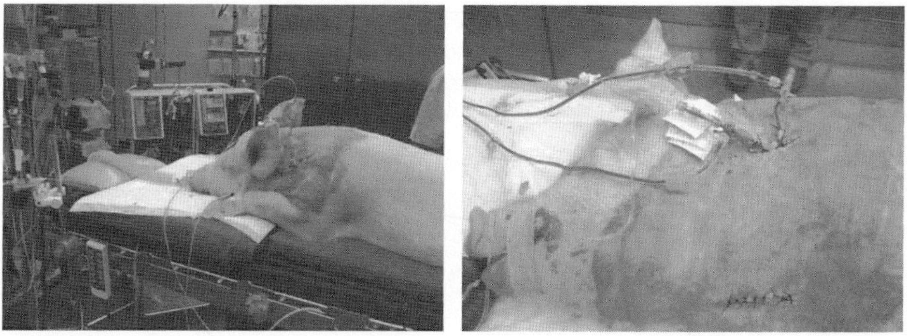

Fig. 34.16. Following arterial and venous catheter insertion, the pigs received a total of five apheresis treatments with the investigational device under anesthesia. The treatment time was 120 min per session with a blood flow of approximately 30 ml/min.

oxygen-room air mixture and 1.5–2.5 vol% isoflurane, piritramide 0.1 mg/kg/hr, i. v.) and all efforts were made to minimize suffering. Analgetic and antibiotic therapy were started during the surgical catheter insertion (fentanyl plasters 50 μg/hr, cefazoline for three days, 15 mg/kg i. v. every eight hours). The animals first received an arterio-venous catheter implantation followed by a total of five apheresis treatments with the novel PAR beads containing module under anesthesia (Fig. 34.16).

The first apheresis session was performed immediately after the catheter insertion. During the first two weeks the pigs received two apheresis treatments per week with a treatment free interval of three to four days. The last treatment session was performed in the third week. Blood samples were taken one week prior to surgical catheter insertion (week 0, W0), at final evaluation (seven days after the fifth treatment session, final) and at the end of the experiment (9–10 days after the fifth treatment session, follow up), as well as before (0 min), during and at the end of each apheresis treatment (15, 60 and 120 min of treatment time).

In spite of the catheter implantation as well as frequent anesthetic interventions during the apheresis treatments, the pigs developed normally and showed a steady increase in body weight over time (Fig. 34.17). After recovery from anesthesia the animals showed

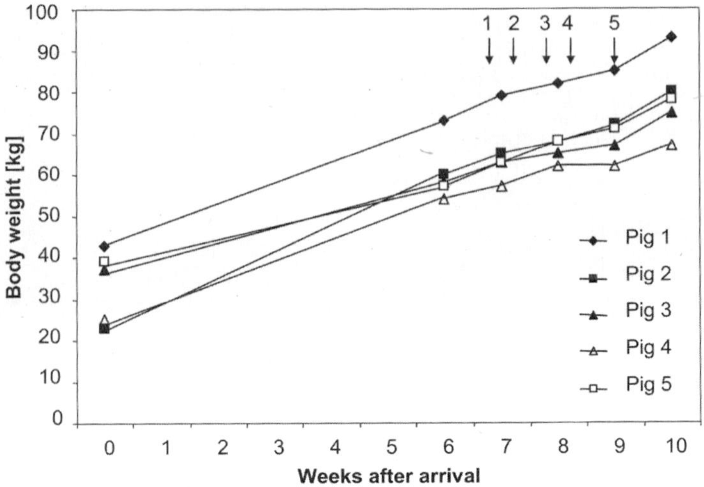

Fig. 34.17. The body weight of the growing pigs was not markedly influenced by the apheresis treatments. Apheresis treatments were performed during the seventh (first and second treatment), eighth (third and fourth treatment) and nineth (fifth treatment) week after arrival at the animal unit (marked with arrows).

a normal behavior, interest in companion animals, and normal food and water uptake during the whole observation period. Using the neurologic deficient score the animals showed a physiologic neural status during the whole observation period. All apheresis treatments with the investigational device were completed regularly according to the study protocol.

Concerning safety, no adverse events were observed during this animal trial. There were no serious long-term impacts of the apheresis treatments with the novel device on the chemical blood parameters apart from increased ACT and aPTT values before the first treatment session caused by earlier heparinization after catheter insertion. A special aspect in pigs is to achieve an optimal coagulation state to avoid thrombi formation and embolism over the time period of the apheresis procedure. Therefore, a relatively high dose of heparin was necessary, however, this was well tolerated by the animals without the occurrence of bleeding disorders. Altogether, the apheresis treatments with the novel device demonstrated a good safety profile. All measured safety parameters remained unchanged, or in case of minor

Table 34.2. Changes in Chemical Parameters during Apheresis Treatments of Pigs (Mean ± SD, n = 5, 25 Treatments)

Parameter	Start Values	15 Min Inflow	60 Min Inflow	120 Min Inflow
aPTT [s]	70.4 ± 123.5	400.0 ± 0.0*	400.0 ± 0.0*	400.0 ± 0.0*
ACT [s]	305.2 ± 378.4	812.0 ± 541.3*	1369.1 ± 364.5*	1276.0 ± 412.8*
Fibrinogen [mg/dL]	168.8 ± 68.0	148.7 ± 56.0*	146.1 ± 56.4*	145.8 ± 56.8*
Total protein [g/dL]	5.8 ± 0.4	5.2 ± 0.4*	5.1 ± 0.4*	5.0 ± 0.4*
Albumin [g/dL]	3.7 ± 0.5*	3.3 ± 0.5*	3.2 ± 0.5*	3.2 ± 0.5*
Bilirubin [mg/dL]	0.09 ± 0.06	0.12 ± 0.06*	0.12 ± 0.06*	0.11 ± 0.05
free Hb [g/dL]	3.3 ± 1.7	2.5 ± 1.2*	2.9 ± 1.6	2.6 ± 0.9
Glucose [mg/dL]	88.4 ± 12.3	78.9 ± 18.0*	89.4 ± 18.8	102.8 ± 9.0*
AST [U/L]	40.0 ± 37.6	35.8 ± 32.0*	33.0 ± 28.9*	31.4 ± 26.7*
ALT [U/L]	95.8 ± 67.5	84.9 ± 59.6*	83.4 ± 57.4*	81.8 ± 55.9*
LDH [U/L]	628.9 ± 281.4	543.8 ± 240.0*	524.6 ± 214.9*	504.4 ± 190.8*
AP [U/L]	84.6 ± 18.5	78.4 ± 19.5*	81.0 ± 20.2*	85.2 ± 19.1
γ-GT [U/L]	29.4 ± 4.3	25.9 ± 3.7*	25.8 ± 3.5*	25.7 ± 3.5*
CK [U/L]	3769.7 ± 6089.4	3232.6 ± 5086.4*	3070.4 ± 4924.8*	2833.6 ± 4484.3*
Crea [mg/dL]	1.1 ± 0.1	1.0 ± 0.2	1.1 ± 0.2*	1.2 ± 0.2*
BUN [mg/dL]	6.1 ± 1.0	6.2 ± 1.0	6.7 ± 1.3*	7.5 ± 1.5*
K^+ [mmol/L]	3.8 ± 0.2	3.8 ± 0.3	4.2 ± 0.3*	4.5 ± 0.4*
Na^+ [mmol/L]	139.0 ± 1.2	139.0 ± 1.1	138.7 ± 1.3	138.0 ± 1.7*
$Ca2^+$ [mmol/L]	1.2 ± 0.0	1.2 ± 0.0	1.2 ± 0.0	1.2 ± 0.0
Cl^- [mmol/L]	100.7 ± 2.0	101.7 ± 2.0*	101.2 ± 1.9	100.6 ± 2.2
Lactate [mmol/L]	0.8 ± 0.2	0.9 ± 0.3	1.1 ± 0.4*	0.9 ± 0.3

*$p < 0.05$ vs. start value, Wilcoxon-test.

changes, clinically irrelevant, especially in terms of inter-treatment comparisons (Tables 34.2 and 34.3).

Catheter-related problems occurred prior to six treatment sessions and were solved by switching to veno-venous apheresis (in four cases) or a new preparation of the arterial line (in two cases). However, this did not influence the technical feasibility of the novel device (defined as the ability to perform apheresis treatments with the module for 120 min at a blood flow rate of approximately 30 mL/min without replacement) which was 100%.

Table 34.3. Comparisons of Values One Week before Catheter Insertion (W0) with Pre-Treatment Data of First to Fifth Apheresis Treatment Session in the Pigs as well as with Final Values of the Follow-Up Examinations (Mean, $n = 5$). Tn: treatment number 1 to 5, final: one week after T5, follow-up: 9–10 days after T5

Parameter	W0	T1	T2	T3	T4	T5	Final	Follow up
aPTT [s]	13.2	284.1	18.7	15.6	17.4	16.0	23.8	16.8
ACT [s]	110.0	993.6*	137.0	136.0	160.0	99.2	128.8	113.8
Fibrinogen [mg/dL]	143.5	132.4	284.8	150.4	127.4	149.2	149.0	193.3
Total protein [g/dL]	5.7	5.3	6.2	5.8	5.6	6.0	6.4	6.3
Albumin [g/dL]	3.4	3.3	3.6	4.0	3.8	3.8	4.1	4.2
Bilirubin [mg/dL]	0.07	0.17	0.04	0.09	0.10	0.06	0.06	0.09
free Hb [g/dL]	3.8	2.1	2.6	4.2	5.1	2.4	5.0	3.2
Glucose [mg/dL]	103.5	93.6	91.6	79.2	89.0	88.6	98.2	104.8
AST [U/L]	18.5	20.6	68.6	53.0	38.0	19.8	20.8	18.8
ALT [U/L]	43.3	41.6	93.6	134.8	123.8	85.4	62.4	57.4
LDH [U/L]	583.0	507.2	787.6	674.0	648.4	527.4	521.6	471.4
AP [U/L]	110.5	111.4	84.6	70.6	71.8	84.8	97.4	96.2
γ-GT [U/L]	30.5	27.8	28.4	27.4	30.2	33.0	34.6	33.4
CK [U/L]	2162.3	2189.2	7461.4	6945.0	1773.4	479.4	813.0	773.2
Crea [mg/dL]	1.17	1.12	1.00	0.98	1.08	1.10	1.20	1.18
BUN [mg/dL]	6.8	6.6	5.8	5.4	6.4	6.2	6.8	5.6
K^+ [mmol/L]	3.46	4.06*	3.72	3.68	3.70*	3.66	3.74*	3.48
Na^+ [mmol/L]	140.0	140.2	138.6	139.4	138.4*	138.6	138.4*	140.0
$Ca2^+$ [mmol/L]	1.27	1.21	1.26	1.24	1.22	1.24*	1.31	1.26
Cl^- [mmol/L]	103.0	103.0	100.2	101.0	99.8	99.6	100.8	102.8
Lactate [mmol/L]	1.6	1.0	0.9	0.8*	0.8*	0.7*	0.7*	1.4
ESR [mm/h]	7.6	—	—	—	—	—	7.0	6.3

*$p < 0.05$ vs. W0, Wilcoxon-test.

Concerning the performance of the Immunopure device, a strong removal of platelets, monocytes and eosinophils to a maximum of 53, 69, and 44%, respectively, in the outflow, could be demonstrated. Neutrophil granulocytes and basophils were moderately reduced in

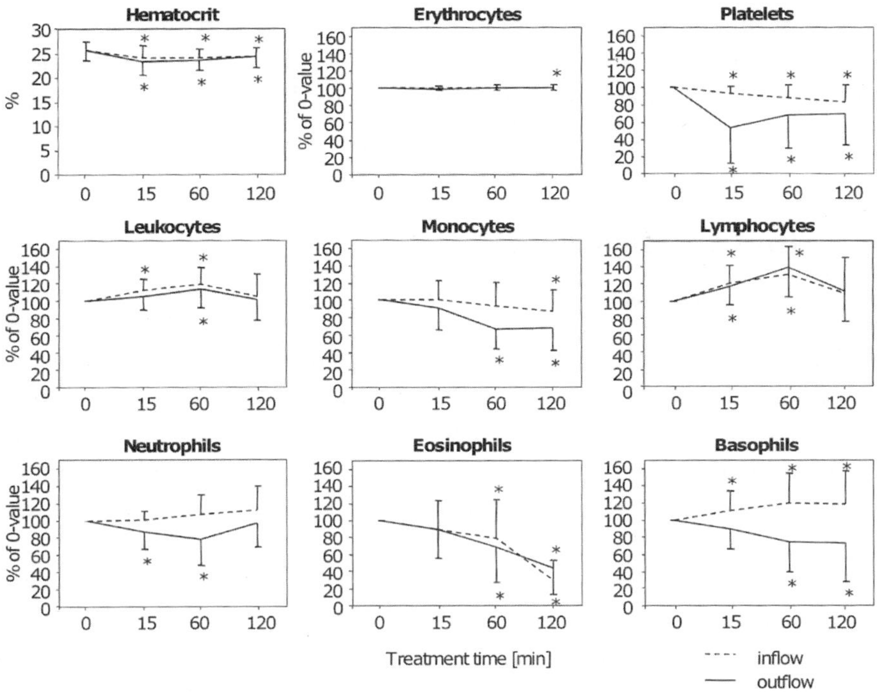

Fig. 34.18. Blood count graphs — intra-treatment impacts during and after the apheresis treatments of pigs. Comparison of baseline values (0 min) vs. 15, 60 and 120 min values (five animals, 25 treatments, mean ± SD, absolute counts are expressed as % of 0-min-value, $*P < 0.05$ vs. 0-min-value, Wilcoxon test).

the outflow sample, but increased after 120 min of treatment time (Fig. 34.18).

The reason for this may be an overshoot phenomenon, which has been observed previously leading to the delivery of cells from marginal or other pools (Yamaji *et al.*, 2002). The increase in leukocytes over time may also be ascribed to this phenomenon. There was also an increase of lymphocytes with a maximum at 60 min of treatment. These effects were not observed in humans (see Section 3) and seem to be characteristic for pigs or may be due to the longer treatment time. There was no long-term impact of the apheresis treatments on the blood cell counts.

In summary, this study demonstrated that a large animal model using the German large white pig is suitable for the safety evaluation of extracorporeal devices. The Immunopure device was shown to be safe and tolerable in this animal model.

3. Clinical Aspects

3.1. *Introduction*

Therapeutic strategies in chronic UC depend on localization, intensity of the illness, and response to steroids. Treatment goals include treatment of symptoms, induction of clinical remission, maintenance of steroid-free remission, prevention of hospital admission and surgery, mucosal healing, improvement of life quality, and avoidance of disability (Ordás *et al.*, 2012).

In general, medication for UC mainly includes 5-aminosalicylic acid (5-ASA), corticosteroids, immunosuppressive drugs, and monoclonal antibodies to TNF-α. 5-ASA is the first line treatment for mildly to moderately active UC. Proctitis is treated with topical 5-ASA. Combining topical 5-ASA with oral 5-ASA or topical steroid is considered for escalation of treatment. Systemic corticosteroids are appropriate if symptoms of active colitis do not respond to 5-ASA (Dignass *et al.*, 2010).

Patients with bloody diarrhea ≥ 6 per day and any signs of systemic toxicity suffer from severe colitis. They should be admitted to hospital for intensive treatment. Then, intravenous corticosteroids are the mainstay of conventional therapy. The response to intravenous steroids should be assessed around the third day. Second line therapy consists of either ciclosporin, infliximab or tacrolimus. If there is no improvement within 4–7 days of salvage therapy, colectomy is recommended. Third line medical therapy may be considered at a specialist center (Dignass *et al.*, 2010).

In clinical routine, about 70% of patients respond to the first course of corticosteroids, but 22% develop steroid dependency during the first year of treatment, and only half of them maintain steroid-free remission (Faubion *et al.*, 2001). Steroid-dependent patients are treated with azathioprine/mercaptopurine. However, some patients exhibit UC that is refractory to corticosteroids or immunosuppressive drugs. Outpatients with steroid refractory disease should be treated with anti-TNF therapy or tacrolimus. Surgery used to be the ultimate therapeutic measure to cure, however, imposing other restrictions and complications. Thus, a treatment alternative, if possible with no or minor toxicity, is needed to handle refractory or steroid-dependent individuals, to save steroids and to treat high risk or other specific patient groups (pregnant women, children, toxic megacolon).

Cytapheresis has the potential to avoid drug-mediated side effects and to treat steroid-refractory patients while being not associated with safety concerns or serious side effects. However, the treatment is expensive, and its therapeutic success varied, also due to the fact that the patients' background and treatments are often very heterogeneous. Therefore, careful patient selection may improve efficacy and cost-benefit ratio of this treatment option. In addition, new developments like the Immunopure adsorber may be of special interest in UC patients with high platelet numbers and/or coagulation disorders.

3.2. *Uncontrolled Studies — Results*

In a pilot study including 10 patients with UC, the Immunopure device was shown to be safe, well tolerated and clinically efficient (80% remission rate). It offered very good biocompatibility and platelet elimination capacity (Ramlow *et al.*, 2013). In addition to the animal study (see Section 2.3), this uncontrolled human pilot study provided convincing data to achieve approval status of the Immunopure device in Europe (CE marking) in 2011. On average, the Clinical Activity Index [CAI according to Rachmilewitz (1989)] was reduced from 8.5 ± 1.5 (screening value) to 4.5 ± 4.1 at W6 and 2.2 ± 2.7 at W10 ($P < 0.05$). Clinical remission in UC is defined as a CAI score of ≤ 4. Clinical response is defined as CAI drop ≥ 3 or CAI ≤ 4. Endoscopic remission according to Rachmilewitz (1989) (EI < 4) was achieved in four of the nine patients (44%) who consented to endoscopic examination. The average EI of the nine eligible patients decreased from 7.4 ± 1.6 (screening value) to 5.2 ± 3.2 at W10

($p < 0.05$). There was a strong reduction of monocytes (outflow to 37.3% after 30 min), neutrophil granulocytes (outflow to 53.2% after 30 min), eosinophils (outflow to 51.5 % after 15 min) and platelets (outflow to 20.3% after 15 min) during the treatment.

The aim of a subsequent post-marketing surveillance study was to further evaluate the influence of the Immunopure module on platelets, platelet activation and PLAs in UC patients similar to the *in vitro* experiments (see Section 2.2). This non-interventional study according to the German law on medical devices §23b was conducted in accordance with the principle of ISO 14155 good clinical practice and the Declaration of Helsinki. Informed consent was received by all patients after the ethics committee of the University of Rostock had approved the study protocol. A total of six patients (three males, three females, mean age: 40.5 years, minimum age: 27 years, maximum age: 52 years) with moderately to severely active UC of 6–13, defined by the CAI according to Rachmilewitz, who had failed to achieve long-term remission with steroids and/or immunosuppessive agents or who had contraindications or were intolerant to high doses of steroids and/or immunosuppressive agents, were recruited. The mean disease duration was 7.8 years. Fifty percent of the patients suffering from active UC exhibited a pancolitis, 50% a left sided colitis (distal colitis). In general, the patients' medications were comprised of 5-ASA, tacrolimus and steroids (Table 34.4). Disease activity was

Table 34.4. Patient Data of Post-Marketing Surveillance Study

Patient No.	Sex	Age	Diagnosis†	Treatment	CAI Before	CAI W10
1	f	26	E3	Tacrolimus, 5-ASA	13	4
2‡	m	45	E2	5-ASA, Glucocorticoids, Tacrolimus	10	10
3	f	50	E2	5-ASA, Glucocorticoids	8	2
4‡	m	51	E3	5-ASA, Glucocorticoids, Tacrolimus	6	1
5	m	29	E2	5-ASA, Glucocorticoids	7	3
6	f	33	E3	5-ASA	13	5

†Montreal classification [54]: E2: left sided UC (distal UC), E3: extensive UC (pancolitis).
‡non-responder.

evaluated by assessing the CAI at baseline and 10 weeks after the first apheresis treatment.

The average CAI was significantly reduced from 9.5 ± 3.0 (screening value) to 4.2 ± 3.2 at W10 ($p < 0.05$). Remission of the disease was achieved in four of the six patients (67%), response in five out of the six patients (83%) at W10 (Table 34.4).

In accordance with the pilot study (Ramlow *et al.*, 2013), cytapheresis treatment with the Immunopure adsorber strongly decreased platelets, monocytes and granulocytes to a maximum of 33, 33 and 58%, respectively (Fig. 34.19). The hemoglobin and hematocrit values were slightly decreased during the treatment; values were hematocrit corrected. In accordance with the *in vitro* results (Fig. 34.7), there was a maximum of platelet adsorption after 15 min. After 15 min of treatment platelets increased in the outflow suggesting a saturation of the column with these cells. This behavior excludes safety problems with low platelet numbers; values had never fallen below 180×10^9/L in these patients. However, there was a different adsorption behavior for leukocytes, monocytes, neutrophils, and eosinophils which showed a continuous decrease over time. At the end of the treatment platelet and eosinophil numbers had a slightly, but significantly lowered level than before apheresis. Nevertheless, there were no significant differences of the single parameters between the different apheresis sessions.

Flow cytometry analyses confirmed these results showing a maximum decrease of $CD10^+$ granulocytes to 52%, leukocytes expressing the cell adhesion proteins CD11b and L-selectin to 50%, $CD14^+$ monocytes/macrophages to 31% and inflammatory $CD14^+CD16^+$ monocytes to 53% (Fig. 34.20).

In vivo, PLAs with $CD14^+$ monocytes, $CD3^+$ T cells, and $CD11b^+$ cells were strongly reduced by Immunopure apheresis too (Fig. 34.21). Values were reduced to a maximum of 29% ($CD42b^+CD14^+$, $CD42b^+CD11b^+$) and 30% ($CD42b^+CD3^+$). Similar to the adsorption of platelets, values increased again after 15 minutes towards the end of the treatment. After 60 minutes, inflow values were also significantly reduced to 84% ($CD42b^+CD14^+$) and 82% ($CD42b^+CD3^+$). However, there were no long-term effects on PLAs.

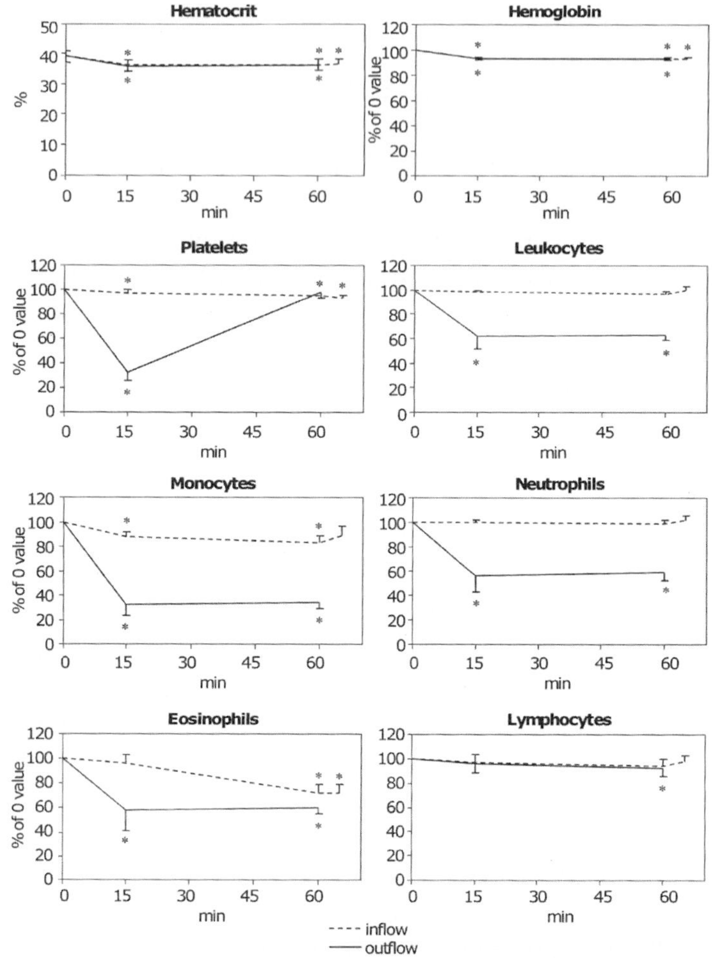

Fig. 34.19. Peripheral blood counts during and after treatment of patients with the Immunopure adsorber. Results were compared with baseline values at time point 0 min (100%) at the beginning of each treatment using the Wilcoxon test (six patients, five treatments each, mean/SEM out of 30 treatments) (*$P < 0.05$).

There was a minimal increase in platelet-derived MPs during the apheresis treatment, possibly due to an enhanced platelet activation. There was also a significant decrease in the inflow/outflow values after 15 minutes of treatment, indicating an MP adsorption at this

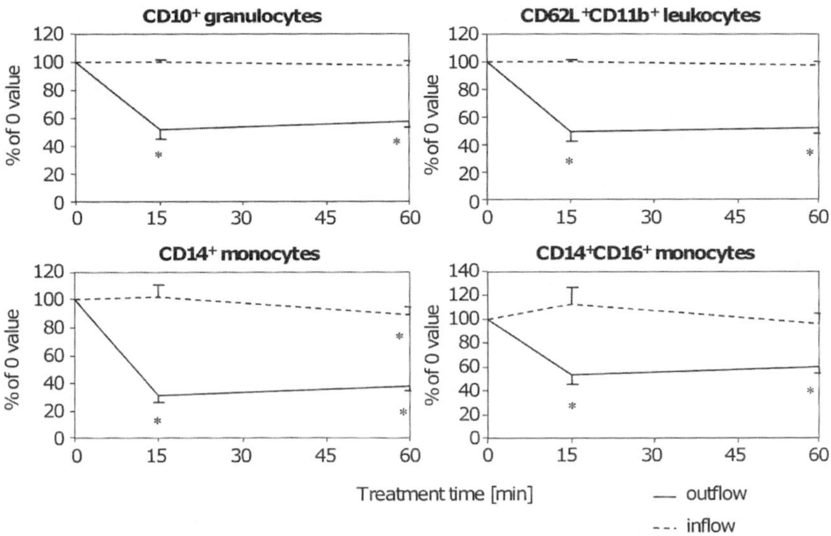

Fig. 34.20. Flow cytometry analyses of CD10$^+$ granulocytes, CD62L$^+$CD11b$^+$ leukocytes, CD14$^+$ monocytes and CD14$^+$CD16$^+$ monocytes during and after treatment of patients with the Immunopure adsorber. Results were compared with baseline values at time point 0 min (100%) at the beginning of the apheresis treatment using the Wilcoxon test (six patients, mean/SEM of first, third and fifth treatment session) ($*P < 0.05$).

time point (Fig. 34.22). However, differences were relatively small indicating that they were of minor relevance.

In accordance with the *in vitro* results (see Table 34.1), several activation parameters were elevated or down regulated due to activation of leukocytes or platelets. CD62L expression was slightly decreased and CD11b expression was moderately increased, suggesting a relatively low leukocyte activation. This also applied to T cell activation (CD4$^+$CD25$^+$). In contrast, platelet activation was relatively strong with large deviations (Table 34.5). However, as seen in the *in vitro* experiments, most of these markers were also changed using sham modules.

The high expression of the platelet activation markers CD63 and CD62P during Immunopure apheresis was accompanied by increased plasma soluble CD40L concentrations. Levels were six to sevenfold

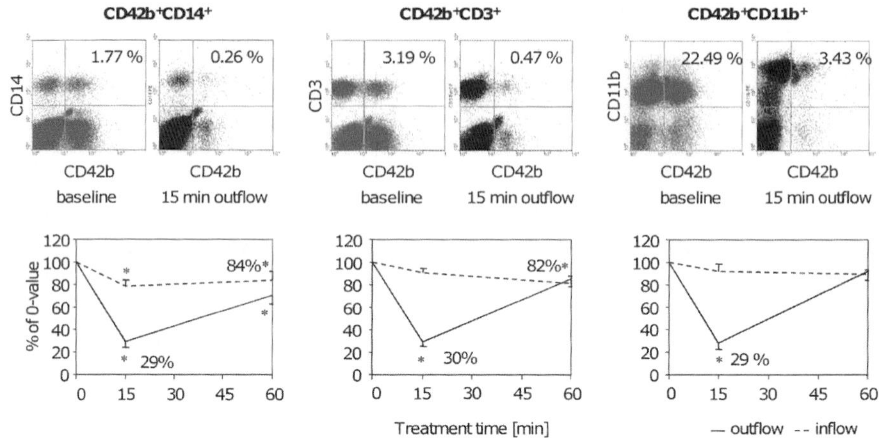

Fig. 34.21. Flow cytometry analyses of PLAs stained with the CD42b platelet marker combined with CD14, CD11b and CD3 during and after treatment of patients with the Immunopure adsorber. Results (mean/SEM) were compared with baseline values at time point 0 min (100%) at the beginning of the apheresis treatment (six patients, mean/SEM of first, third and fifth treatment session) (*$P < 0.05$, Wilcoxon test).

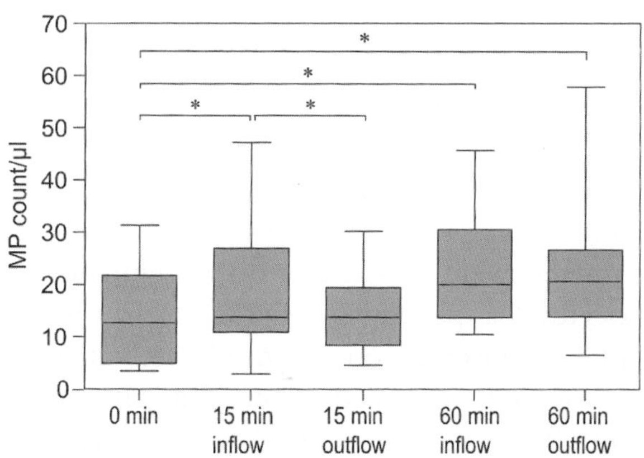

Fig. 34.22. Flow cytometry analyses of platelet-derived MPs (Annexin V^+/$CD42b^+$, <0.9 µm) during and after treatment of patients with the Immunopure apheresis (six patients, median of first, third and fifth treatment session) (*$P < 0.05$, Wilcoxon test).

Table 34.5. Flow Cytometry Analyses of CD62L (Mean Channel Fluorescence), CD11b (Mean Channel Fluorescence), $CD4^+CD25^+$ Cells, $CD42b^+CD62P^+$ and $CD42b^+CD62P^+$ Platelets after Treatment of Patients with Immunopure Apheresis. Results (Six Patients, mean ± SD of First, Third and Fifth Treatment Session) were Compared with Baseline Values at Time Point 0 min (100%) at the Beginning of Each Experiment (*$P < 0.05$)

	Inflow	Outflow	Inflow	Outflow
Parameter	15 min	15 min	60 min	60 min
Mean CD62L	100.1 ± 24.5	92.7 ± 26.5	97.1 ± 23.3	91.0 ± 23.1*
Mean CD11b	103.0 ± 28.2	145.7 ± 46.8*	93.3 ± 16.7	175.2 ± 56.0*
$CD4^+CD25^+$	112.2 ± 51.3	116.4 ± 56.9	111.5 ± 53.4	110.0 ± 51.1
$CD42b^+CD63^+$	102.5 ± 51.0	207.5 ± 112.9*	102.5 ± 43.0	285.2 ± 228.7*
$CD42b^+CD62P^+$	180.6 ± 205.8	901.7 ± 1159.1*	170.3 ± 150.0	1753.1 ± 2379.6*

higher after 60 min of treatment compared to baseline levels before treatment. Whether these increased soluble CD40L concentrations may have any consequences needs to be further elucidated. However, it seems to be only a temporary effect of the adsorber column as values return to basic levels before the next apheresis session starts. Soluble P-selectin concentrations were not significantly altered during Immunopure apheresis treatments. There was only a 1.3-fold increase during the third treatment session (Fig. 34.23).

In conclusion, these uncontrolled studies showed that Immunopure treatments were well tolerated and remarkably reduced disease activity. In addition to monocytes including proinflammatory $CD14^+CD16^+$ monocytes and granulocytes, Immunopure is able to adsorb platelets and platelet aggregates. Since UC is associated with abnormalities in platelet and leukocyte number and function, their temporary reduction by the Immunopure device seems to be related to clinical improvement. However, there was also a kind of inflammatory stimulus at the column, e.g. leading to leukocyte and platelet activation. Usually, the immune system has the capability to regulate immune responses, allowing robust responses against pathogens while avoiding autoimmunity. In UC, this capability is constrained, leading to an overshooting immune response. Possibly, adsorptive cytapheresis

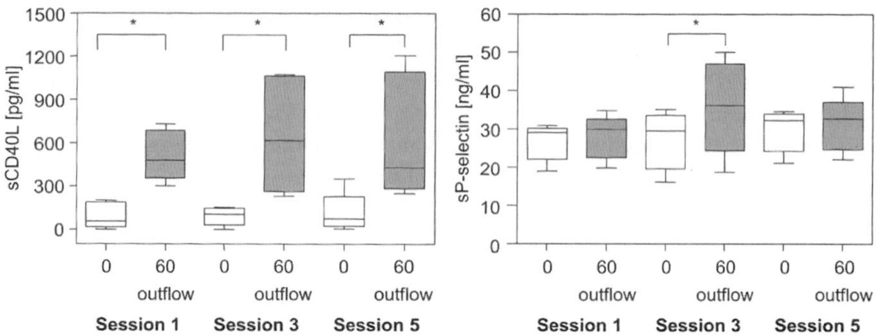

Fig. 34.23. Plasma concentrations of soluble CD40L (left) and soluble P-selectin (right) as determined by enzyme-linked immunosorbent assay before and after treatment of patients with the Immunopure adsorber (six patients, median of first, third and fifth treatment session) (*$P < 0.05$, Wilcoxon test).

is like a second stimulus which activates the immune system outside the body, finally leading to induction of regulatory processes that might also affect the course of UC. Generally, the following mechanisms are conceivable and have already partly been discussed for the mode of action of other cytapheresis adsorbers: a temporary removal of reactive cells including platelets and PLAs, a parallel activation of leukocytes and platelets that return to the patients' circulation and transfer signals to other cells, leading to mobilization of naïve and regulatory cells.

3.3. *Controlled Studies*

The clinical benefit of Immunopure on efficacy parameters merits further evaluation in controlled trials and studies with larger patient numbers. To date, there are no results of controlled studies investigating the Immunopure device. One multicenter trial is ongoing comparing Immunopure versus oral prednisolone in a small number of patients. However, further sufficiently powered, controlled studies are necessary, especially with regard to patients with active disease that are really supposed to respond to apheresis treatments. In particular, these are patients early after initial diagnosis with moderate to severe disease activity as well as steroid-naive patients only on 5-aminosalicylic acid immediately after a clinical relapse. This should be the target group for initial cytapheresis trials, especially when apheresis treatment should be compared with sham-apheresis. In a second step, enhanced adsorptive cytapheresis protocols should be tested which possibly offer advantages for patients with a longer disease history and combined medication for UC. Moreover, international approved end points and disease indices should be applied.

Acknowledgements

This work was supported by NIKKISO Co. Ltd., Tokyo, Japan. We thank Udo Losert, Helga Bergmeister and Roberto Plasenzotti for performing the animal trial; Peter Ahrenholz for planning the *in vitro* experiments; and Matthias Schneider, Grynet Quintern, Anja Ramlow, Annett Bieber, Manuela Seidl and Katja Vedder for excellent technical support.

Abbreviations

azathioprine (Aza)
5-aminosalicylic acid (5-ASA)
clinical activity index (CAI)
endoscopic index (EI)
granulocyte/monocyte adsorption apheresis (GMCAP)
inflammatory bowel diseases (IBD)
leukocytapheresis (LCAP)
microparticle (MP)
mean platelet volume (MPV)
polyarylate resin (PAR)
plateletcrit (PCT)
platelet distribution width (PDW)
Polyester-polymer alloy (PEPA)
polyethersulfone (PES)
platelet leukocyte aggregate (PLA)
tumour necrosis factor alpha (TNFα)
ulcerative colitis (UC)

References

Andoh A, Tsujikawa T, Hata K, *et al.* (2005) Elevated circulating platelet-derived microparticles in patients with active inflammatory bowel disease. *Am J Gastroenterol* **100:** 2042–8.

Burger D, Travis S. (2011) Conventional medical management of inflammatory bowel disease. *Gastroenterology* **140:** 1827–37.

Cerletti C, Tamburrelli C, Izzi B, *et al.* (2012) Platelet-leukocyte interactions in thrombosis. *Thromb Res* **129:** 263–6.

Clark SR, Ma AC, Tavener SA, *et al.* (2007) Platelet TLR4 activates neutrophil extracellular traps to ensnare bacteria in septic blood. *Nat Med* **13:** 463–9.

Cognasse F, Lafarge S, Chavarin P, *et al.* (2007) Lipopolysaccharide induces sCD40L release through human platelets TLR4, but not TLR2 and TLR9. *Intensive Care Med* **33:** 382–4.

Danese S, Papa A, Saibeni S, *et al.* (2007) Inflammation and coagulation in inflammatory bowel disease: The clot thickens. *Am J Gastroenterol* **102:** 174–86.

Danese S, Sans M, Fiocchi C. (2004) The CD40/CD40L costimulatory pathway in inflammatory bowel disease. *Gut* **53:** 1035–43.

Danese S. (2007) Inflammation and the mucosal microcirculation in inflammatory bowel disease: The ebb and flow. *Curr Opin Gastroenterol* **23:** 384–9.

Danese S. (2008) Nonimmune cells in inflammatory bowel disease: From victim to villain. *Trends Immunol* **29:** 555–64.

Dignass A, Van Assche G, Lindsay JO, *et al.*; European Crohn's and Colitis Organisation (ECCO). (2010) The second European evidence-based consensus on the diagnosis and management of Crohn's disease: Current management. *J Crohns Colitis* **4:** 28–62.

Earl W. (2003) A brief historical review of the waterfall/cascade of blood coagulation. *J Biol Chem* **278:** 50819–32.

Esmon CT, Xu J, Lupu F. (2011) Innate immunity and coagulation. *J Thromb Haemost* **9 (Suppl 1):** 182–8.

Faubion WA Jr, Loftus EV Jr, Harmsen WS, *et al.* (2001) The natural history of corticosteroid therapy for inflammatory bowel disease: A population-based study. *Gastroenterology* **121:** 255–60.

Fukunaga K, Fukuda Y, Yokoyama Y, *et al.* (2006) Activated platelets as a possible early marker to predict clinical efficacy of leukocytapheresis in severe ulcerative colitis patients. *J Gastroenterol* **41:** 524–32.

Hanai H, Iida T, Takeuchi K, *et al.* (2008) Adsorptive depletion of elevated proinflammatory CD14+CD16+DR++ monocytes in patients with inflammatory bowel disease. *Am J Gastroenterol* **103:** 1210–6.

Hanai H, Iida T, Takeuchi K, *et al.* (2008) Intensive granulocyte and monocyte adsorption versus intravenous prednisolone in patients with severe ulcerative colitis: An unblinded randomised multi-centre controlled study. *Dig Liver Dis* **40:** 433–40.

Hanai H, Watanabe F, Takeuchi K, *et al.* (2003) Leukocyte adsorptive apheresis for the treatment of active ulcerative colitis: A prospective, uncontrolled, pilot study. *Clin Gastroenterol Hepatol* **1:** 28–35.

Harries AD, Fitzsimons E, Fifield R, *et al.* (1983) Platelet count: A simple measure of activity in Crohn's disease. *Br Med J* **286:** 1476.

Igoshi T, Tomisawa N, Hori Y, Jinbo Y. (2011) Polyester polymer alloy as a high-performance membrane. *Contrib Nephrol* **173:** 148–55.

Irving PM, Macey MG, Feakins RM, *et al.* (2008) Platelet-leucocyte aggregates form in the mesenteric vasculature in patients with ulcerative colitis. *Eur J Gastroenterol Hepatol* **20:** 283–9.

Irving PM, Pasi KJ, Rampton DS. (2005) Thrombosis and inflammatory bowel disease. *Clin Gastroenterol Hepatol* **3:** 617–28.

Kashiwagi N, Sugimura K, Koiwai H, *et al.* (2002) Immunomodulatory effects of granulocyte and monocyte adsorption apheresis as a treatment for patients with ulcerative colitis. *Dig Dis Sci* **47:** 1334–41.

Kulnigg-Dabsch S, Evstatiev R, Dejaco C, Gasche C. (2012) Effect of iron therapy on platelet counts in patients with inflammatory bowel disease-associated anemia. *PLoS One* **7:** e34520.

Limb GA, Webster L, Soomro H, *et al.* (1999) Platelet expression of tumour necrosis factor-alpha (TNF-α), TNF receptors and intercellular adhesion molecule-1 (ICAM-1) in patients with proliferative diabetic retinopathy. *Clin Exp Immunol* **118:** 213–8.

Ma AC, Kubes P. (2008) Platelets, neutrophils, and neutrophil extracellular traps (NETs) in sepsis. *J Thromb Haemost* **6:** 415–20.

Matsumoto T, Andoh A, Okawa K, *et al.* (2008) Multivariate analysis for factors predicting rapid response of leukocytapheresis in patients with steroid-resistant ulcerative colitis: A multicenter prospective open-label study. *Ther Apher Dial* **12:** 484–90.

Michelson AD, Barnard MR, Krueger LA, *et al.* (2001) Circulating monocyte-platelet aggregates are a more sensitive marker of in vivo platelet activation than platelet surface P-selectin: Studies in baboons, human coronary intervention, and human acute myocardial infarction. *Circulation* **25(104):** 1533–7.

Muratov V, Lundahl J, Ulfgren AK, *et al.* (2006) Down-regulation of interferon-gamma parallels clinical response to selective leukocyte apheresis in patients with inflammatory bowel disease: A 12-month follow-up study. *Int J Colorectal Dis* **21:** 493–504.

Nielsen OH, Vainer B, Madsen SM, *et al.* (2000) Established and emerging biological activity markers of inflammatory bowel disease. *Am J Gastroenterol* **95:** 359–67.

Ordás I, Eckmann L, Talamini M, *et al.* (2012) Ulcerative colitis. *Lancet* **380:** 1606–19.

Pamuk GE, Vural O, Turgut B, *et al.* (2006) Increased circulating platelet-neutrophil, platelet-monocyte complexes, and platelet activation in patients with ulcerative colitis: A comparative study. *Am J Hematol* **81:** 753–9.

Rachmilewitz D (on behalf of an international study group). (1989) Coated mesalazine (5-aminosalicylic acid) versus sulphasalazine in the treatment of active ulcerative colitis. *Br Med J* **298:** 82–6.

Ramlow W, Emmrich J, Ahrenholz P, *et al.* (2005) In vitro and in vivo evaluation of Adacolumn cytapheresis in healthy subjects. *J Clin Apher* **20:** 72–80.

Ramlow W, Waitz G, Sparmann G, *et al.* (2013) First human application of a novel adsorptive-type cytapheresis module in patients with active ulcerative colitis: A pilot study. *Ther Apher Dial* **17:** 339–47.

Robert S, Poncelet P, Lacroix R, *et al.* (2009) Standardization of platelet–derived microparticle counting using calibrated beads and a Cytomics FC500 routine flow cytometer: First step towards multicenter studies? *J Thromb Haemost* **7:** 190–7.

Roberts-Thomson IC, Fon J, Uylaki W, *et al.* (2011) Cells, cytokines and inflammatory bowel disease: A clinical perspective. *Expert Rev Gastroenterol Hepatol* **5:** 703–16.

Sands BE, Sandborn WJ, Feagan B, *et al.*; Adacolumn Study Group. (2008) A randomized, double-blind, sham-controlled study of granulocyte/monocyte apheresis for active ulcerative colitis. *Gastroenterology* **135:** 400–9.

Saniabadi AR, Tanaka T, Ohmori T, *et al.* (2014) Treating inflammatory bowel disease by adsorptive leucocytapheresis: A desire to treat without drugs. *World J Gastroenterol* **20:** 9699–715.

Sartor RB. (2006) Mechanisms of disease: Pathogenesis of Crohn's disease and ulcerative colitis. *Nat Clin Pract Gastroenterol Hepatol* **3:** 390–407.

Satsangi J, Silverberg MS, Vermeire S, Colombel JF. (2006) The Montreal classification of inflammatory bowel disease: Controversies, consensus, and implications. *Gut* **55:** 749–53.

Shibata H, Kuriyama T, Yamawaki N. (2003) Cellsorba. *Ther Apher Dial* **7:** 44–7.

Shimoyama T, Sawada K, Hiwatashi N, *et al.* (2001) Safety and efficacy of granulocyte and monocyte adsorption apheresis in patients with active ulcerative colitis: A multicenter study. *J Clin Apher* **16:** 1–9.

Shimoyama T, Sawada K, Kiwatashi N, *et al.* (1999) Granulocyte and monocyte apheresis with the G-1 column in the treatment of patients with active ulcerative colitis. *Jpn J Apher* **18:** 117–31.

Sigurbjörnsson FT, Bjarnason I. (2008) Leukocytapheresis for the treatment of IBD. *Nat Clin Pract Gastroenterol Hepatol* **5:** 509–16.

Suzuki Y, Yoshimura N, Fukuda K, *et al.* (2006) A retrospective search for predictors of clinical response to selective granulocyte and monocyte apheresis in patients with ulcerative colitis. *Dig Dis Sci* **51:** 2031–8.

Suzuki Y, Yoshimura N, Saniabadi AR, Saito Y. (2004) Selective granulocyte and monocyte adsorptive apheresis as a first-line treatment for steroid naïve patients with active ulcerative colitis: A prospective uncontrolled study. *Dig Dis Sci* **49:** 565–71.

Takemoto K, Kuriyama M, Kato J, *et al.* (2009) Ratio of platelet reduction is an early predictive factor for the effectiveness of leukocytapheresis for ulcerative colitis patients. *Ther Apher Dial* **13:** 6–13.

Tanaka T, Okanobu H, Kuga Y, *et al.* (2010) Clinical and endoscopic features of responders and non-responders to adsorptive leucocytapheresis: A report based on 120 patients with active ulcerative colitis. *Gastroenterol Clin Biol* **34:** 687–95.

Tanaka T, Okanobu H, Yoshimi S, *et al.* (2008) In patients with ulcerative colitis, adsorptive depletion of granulocytes and monocytes impacts mucosal level of neutrophils and clinically is most effective in steroid naïve patients. *Dig Liver Dis* **40:** 731–6.

Voudoukis E, Karmiris K, Koutroubakis IE. (2014) Multipotent role of platelets in inflammatory bowel diseases: A clinical approach. *World J Gastroenterol* **20:** 3180–90.

Waitz G, Petermann S, Liebe S, *et al.* (2008) Reduction of dendritic cells by granulocyte and monocyte adsorption apheresis in patients with ulcerative colitis. *Dig Dis Sci* **53:** 2507–15.

Xavier RJ, Podolsky DK. (2007) Unravelling the pathogenesis of inflammatory bowel disease. *Nature* **448:** 427–34.

Yamaji K, Yang K, Tsuda H, Hashimoto H. (2002) Fluctuations in the peripheral blood leukocyte and platelet counts in leukocytapheresis in healthy volunteers. *Ther Apher* **6:** 402–12.

Yokoyama Y, Fukunaga K, Fukuda Y, *et al.* (2007) Demonstration of low-regulatory CD25High+CD4+ and high-pro-inflammatory CD28-CD4+ T-Cell subsets in patients with ulcerative colitis: Modified by selective granulocyte and monocyte adsorption apheresis. *Dig Dis Sci* **52:** 2725–31.

Yokoyama Y, Kawai M, Fukunaga K, *et al.* (2013) Looking for predictive factors of clinical response to adsorptive granulocyte and monocyte apheresis in patients with ulcerative colitis: Markers of response to GMA. *BMC Gastroenterol* **13:** 27.

Yoshida H, Granger DN. (2009) Inflammatory bowel disease: A paradigm for the link between coagulation and inflammation. *Inflamm Bowel Dis* **15:** 1245–55.

Youssefian T, Drouin A, Massé JM, *et al.* (2002) Host defense role of platelets: Engulfment of HIV and Staphylococcus aureus occurs in a specific subcellular compartment and is enhanced by platelet activation. *Blood* **99:** 4021–9.

Index